Elementary
Differential Equations

TEXTBOOKS in MATHEMATICS

Series Editors: Al Boggess and Ken Rosen

PUBLISHED TITLES

ABSTRACT ALGEBRA: A GENTLE INTRODUCTION
Gary L. Mullen and James A. Sellers

ABSTRACT ALGEBRA: AN INTERACTIVE APPROACH, SECOND EDITION
William Paulsen

ABSTRACT ALGEBRA: AN INQUIRY-BASED APPROACH
Jonathan K. Hodge, Steven Schlicker, and Ted Sundstrom

ADVANCED LINEAR ALGEBRA
Hugo Woerdeman

ADVANCED LINEAR ALGEBRA
Nicholas Loehr

ADVANCED LINEAR ALGEBRA, SECOND EDITION
Bruce Cooperstein

APPLIED ABSTRACT ALGEBRA WITH MAPLE™ AND MATLAB®, THIRD EDITION
Richard Klima, Neil Sigmon, and Ernest Stitzinger

APPLIED DIFFERENTIAL EQUATIONS: THE PRIMARY COURSE
Vladimir Dobrushkin

APPLIED DIFFERENTIAL EQUATIONS WITH BOUNDARY VALUE PROBLEMS
Vladimir Dobrushkin

APPLIED FUNCTIONAL ANALYSIS, THIRD EDITION
J. Tinsley Oden and Leszek Demkowicz

A BRIDGE TO HIGHER MATHEMATICS
Valentin Deaconu and Donald C. Pfaff

COMPUTATIONAL MATHEMATICS: MODELS, METHODS, AND ANALYSIS WITH MATLAB® AND MPI,
SECOND EDITION
Robert E. White

A COURSE IN DIFFERENTIAL EQUATIONS WITH BOUNDARY VALUE PROBLEMS, SECOND EDITION
Stephen A. Wirkus, Randall J. Swift, and Ryan Szypowski

A COURSE IN ORDINARY DIFFERENTIAL EQUATIONS, SECOND EDITION
Stephen A. Wirkus and Randall J. Swift

DIFFERENTIAL EQUATIONS: THEORY, TECHNIQUE, AND PRACTICE, SECOND EDITION
Steven G. Krantz

PUBLISHED TITLES CONTINUED

DIFFERENTIAL EQUATIONS: THEORY, TECHNIQUE, AND PRACTICE WITH BOUNDARY VALUE PROBLEMS
Steven G. Krantz

DIFFERENTIAL EQUATIONS WITH APPLICATIONS AND HISTORICAL NOTES, THIRD EDITION
George F. Simmons

DIFFERENTIAL EQUATIONS WITH MATLAB®: EXPLORATION, APPLICATIONS, AND THEORY
Mark A. McKibben and Micah D. Webster

DISCOVERING GROUP THEORY: A TRANSITION TO ADVANCED MATHEMATICS
Tony Barnard and Hugh Neill

DISCRETE MATHEMATICS, SECOND EDITION
Kevin Ferland

ELEMENTARY NUMBER THEORY
James S. Kraft and Lawrence C. Washington

THE ELEMENTS OF ADVANCED MATHEMATICS: FOURTH EDITION
Steven G. Krantz

ESSENTIALS OF MATHEMATICAL THINKING
Steven G. Krantz

EXPLORING CALCULUS: LABS AND PROJECTS WITH MATHEMATICA®
Crista Arangala and Karen A. Yokley

EXPLORING GEOMETRY, SECOND EDITION
Michael Hvidsten

EXPLORING LINEAR ALGEBRA: LABS AND PROJECTS WITH MATHEMATICA®
Crista Arangala

EXPLORING THE INFINITE: AN INTRODUCTION TO PROOF AND ANALYSIS
Jennifer Brooks

GRAPHS & DIGRAPHS, SIXTH EDITION
Gary Chartrand, Linda Lesniak, and Ping Zhang

INTRODUCTION TO ABSTRACT ALGEBRA, SECOND EDITION
Jonathan D. H. Smith

INTRODUCTION TO ANALYSIS
Corey M. Dunn

INTRODUCTION TO MATHEMATICAL PROOFS: A TRANSITION TO ADVANCED MATHEMATICS, SECOND EDITION
Charles E. Roberts, Jr.

INTRODUCTION TO NUMBER THEORY, SECOND EDITION
Marty Erickson, Anthony Vazzana, and David Garth

PUBLISHED TITLES CONTINUED

INVITATION TO LINEAR ALGEBRA
David C. Mello

LINEAR ALGEBRA, GEOMETRY AND TRANSFORMATION
Bruce Solomon

MATHEMATICAL MODELLING WITH CASE STUDIES: USING MAPLE™ AND MATLAB®, THIRD EDITION
B. Barnes and G. R. Fulford

MATHEMATICS IN GAMES, SPORTS, AND GAMBLING—THE GAMES PEOPLE PLAY, SECOND EDITION
Ronald J. Gould

THE MATHEMATICS OF GAMES: AN INTRODUCTION TO PROBABILITY
David G. Taylor

A MATLAB® COMPANION TO COMPLEX VARIABLES
A. David Wunsch

MEASURE AND INTEGRAL: AN INTRODUCTION TO REAL ANALYSIS, SECOND EDITION
Richard L. Wheeden

MEASURE THEORY AND FINE PROPERTIES OF FUNCTIONS, REVISED EDITION
Lawrence C. Evans and Ronald F. Gariepy

NUMERICAL ANALYSIS FOR ENGINEERS: METHODS AND APPLICATIONS, SECOND EDITION
Bilal Ayyub and Richard H. McCuen

ORDINARY DIFFERENTIAL EQUATIONS: AN INTRODUCTION TO THE FUNDAMENTALS
Kenneth B. Howell

PRINCIPLES OF FOURIER ANALYSIS, SECOND EDITION
Kenneth B. Howell

REAL ANALYSIS AND FOUNDATIONS, FOURTH EDITION
Steven G. Krantz

RISK ANALYSIS IN ENGINEERING AND ECONOMICS, SECOND EDITION
Bilal M. Ayyub

SPORTS MATH: AN INTRODUCTORY COURSE IN THE MATHEMATICS OF SPORTS SCIENCE AND SPORTS ANALYTICS
Roland B. Minton

A TOUR THROUGH GRAPH THEORY
Karin R. Saoub

TRANSFORMATIONAL PLANE GEOMETRY
Ronald N. Umble and Zhigang Han

TEXTBOOKS in MATHEMATICS

Elementary Differential Equations

Kenneth Kuttler

CRC Press
Taylor & Francis Group
Boca Raton London New York

CRC Press is an imprint of the
Taylor & Francis Group, an **informa** business

A CHAPMAN & HALL BOOK

CRC Press
Taylor & Francis Group
6000 Broken Sound Parkway NW, Suite 300
Boca Raton, FL 33487-2742

First issued in paperback 2022

© 2018 by Taylor & Francis Group, LLC
CRC Press is an imprint of Taylor & Francis Group, an Informa business

No claim to original U.S. Government works

Version Date: 20171016

ISBN 13: 978-1-03-247648-3 (pbk)
ISBN 13: 978-1-138-74091-4 (hbk)

DOI: 10.1201/9781315183138

Visit the Taylor & Francis Web site at
http://www.taylorandfrancis.com

and the CRC Press Web site at
http://www.crcpress.com

Contents

Preface   xiii

1 **Some Prerequisite Topics**   1
   1.1   Sets and Set Notation 1
   1.2   Well Ordering and Induction 3
   1.3   The Complex Numbers 5
   1.4   Polar Form of Complex Numbers 7
   1.5   Roots of Complex Numbers 8
   1.6   The Quadratic Formula 9
   1.7   The Complex Exponential 10
   1.8   The Fundamental Theorem of Algebra 11
   1.9   The Cauchy-Schwarz Inequality 13
   1.10   Exercises 14

I   **First Order Scalar Equations**   17

2   **The Idea of a Differential Equation**   19
   2.1   Examples and Slope Fields 19
   2.2   Classifying First Order Differential Equations 23
   2.3   Some Historical Observations 24
   2.4   Exercises 25

3   **Methods**   29
   3.1   First Order Linear Equations 29
   3.2   Bernouli Equations 37
   3.3   Separable Differential Equations, Stability 39
   3.4   Homogeneous Equations 47
   3.5   Exact Equations 48
   3.6   The Integrating Factor 50
   3.7   The Case Where M, N Are Affine Linear 54
   3.8   Linear and Nonlinear Differential Equations 57
   3.9   Computer Algebra Methods 60
     3.9.1   Maple 60
     3.9.2   Mathematica 61
     3.9.3   MATLAB ® 62

        3.9.4   Scientific Notebook 65
    3.10 Exercises   66

II   Scalar Linear Differential Equations, Methods   77

4   Homogeneous Linear Equations   79
    4.1 Factoring Polynomials 79
    4.2 Linear Equations 84
        4.2.1   Real Solutions to the Characteristic Equation 84
        4.2.2   Superposition and General Solutions 86
        4.2.3   Case of Complex Zeros 93
    4.3 Finding Another Solution 95
        4.3.1   Using Wronskians 95
        4.3.2   Reduction of Order 98
    4.4 Exercises   99

5   Nonhomogeneous Equations   105
    5.1 The General Solution to a Nonhomogeneous Equation 105
    5.2 Method of Undetermined Coefficients 106
    5.3 Method of Variation of Parameters 111
    5.4 The Equations of Undamped and Damped Oscillation 115
    5.5 Numerical Solutions   124
        5.5.1   MATLAB 124
        5.5.2   Scientific Notebook 127
    5.6 Exercises   127

6   Laplace Transform Methods   133
    6.1 Solving Initial Value Problems 135
    6.2 The Impulse Function 141
    6.3 The Convolution 142
    6.4 Why Does It Work? 143
    6.5 Automation with Computer Algebra 146
    6.6 Exercises   148

III   Series Methods   155

7   A Review of Power Series   157
    7.1 Operations on Power Series 159
    7.2 Some Other Theorems 166
    7.3 Exercises   169

8   Power Series Methods   175
    8.1 Second Order Linear Equations 175
    8.2   Differential Equations Near an Ordinary Point 177
    8.3 The Legendre Equation 183
    8.4 The Case Where $n(n+1)$ Is Replaced with λ 189
    8.5 The Euler Equations 191

8.6	Some Simple Observations on Power Series	194
8.7	Regular Singular Points	195
8.8	Finding the Solution	198
8.9	Method of Frobenius	204
8.10	The Bessel Equations	207
	8.10.1 The Case where $\nu = 0$	207
	8.10.2 The Case of ν Not an Integer	208
	8.10.3 Case Where ν is an Integer	210
8.11	Other Properties of Bessel Functions	212
8.12	Hermite Polynomials	215
8.13	Exercises	219

IV First Order Systems   223

9	**Methods for First Order Linear Systems**	**225**
9.1	Finding Solutions Using Eigenvalues	225
	9.1.1 Homogeneous Linear Equations	226
	9.1.2 The Case where A is Defective	233
9.2	Nonhomogeneous Problems	237
9.3	Laplace Transforms and First Order Systems	239
9.4	Fundamental Matrices and Laplace Transforms	242
9.5	Using MATLAB to Find Numerical Solutions	243
9.6	Scientific Notebook and Numerical Solutions	243
9.7	Exercises	244

10	**First Order Linear Systems, Theory**	**249**
10.1	Gronwall's Inequality	249
10.2	Fundamental Matrix	251
10.3	Exercises	259

11	**Theory of Ordinary Differential Equations**	**263**
11.1	Problems from Mechanics	263
11.2	Picard Iteration	267
11.3	Convergence of Picard Iterates	268
11.4	Linear Systems	273
11.5	Local Solutions	274
11.6	Continuous Dependence	276
11.7	Autonomous Equations	278
11.8	Numerical Methods, an Introduction	279
11.9	The Classical Runge-Kutta Algorithm	282
11.10	Computing Using Runge-Kutta Method	282
11.11	Exercises	283

12	**Equilibrium Points and Limit Cycles**	**289**
12.1	Stability with Graphing	289
12.2	Minimal Polynomial	292
12.3	Putzer's Method and Linear Systems	294

12.4 Stability and Eigenvalues 296
12.5 Lyapunov Functions 305
12.6 Periodic Orbits, Poincare Bendixon Theorem 309
12.7 Van der Pol Equation 310
12.8 Stable Manifold 314
12.9 Exercises 321

V Partial Differential Equations   331

13 Boundary Value Problems, Fourier Series   333

13.1 Boundary Value Problems 333
13.2 Eigenvalue Problems 334
13.3 Fourier Series 337
13.4 Mean Square Approximation 340
13.5 Pointwise Convergence of Fourier Series 343
        13.5.1 Explanation of Pointwise Convergence Theorem 344
        13.5.2 Mean Square Convergence 348
13.6 Integrating and Differentiating Fourier Series 351
13.7 Odd and Even Extensions 355
13.8 Exercises 356

14 Some Partial Differential Equations   365

14.1 Laplacian in Orthogonal Curvilinear
        Coordinates 365
14.2 Heat and Wave Equations 369
        14.2.1 Heat Equation 369
        14.2.2 The Wave Equation 375
14.3 Nonhomogeneous Problems 378
14.4 Laplace Equation 383
        14.4.1 Rectangles 383
        14.4.2 Circular Disks 386
14.5 Exercises 390

Appendix A MATLAB Syntax Summarized   395

A.1 Matrices in MATLAB 395
A.2 Summation in MATLAB 396
A.3 Graphing Slope Fields and Vector Fields 397
A.4 Using MATLAB to Find a Numerical Solution of ODE 397
A.5 Laplace Transforms and MATLAB 398
A.6 Plotting in MATLAB 399

Appendix B Calculus Review   401

B.1 The Limit of A Sequence 401
B.2 Cauchy Sequences 405
B.3 Continuity and the Limit of a Sequence 405
B.4 Integrals 407
B.5 Gamma Function 412

Appendix C Series   **415**
  C.1 Infinite Series of Numbers 415
      C.1.1 Basic Considerations 415
  C.2 More Tests for Convergence 421
      C.2.1 Convergence Because of Cancellation 421
      C.2.2 Ratio and Root Tests 423
  C.3 Double Series 424
  C.4 Taylor's Formula 429

Appendix D Review of Linear Algebra   **431**
  D.1 Systems of Equations 431
  D.2 Matrices 433
  D.3 Subspaces and Spans 436
  D.4 Application to Matrices 439
  D.5 Mathematical Theory of Determinants 440
      D.5.1 The Function sgn 440
      D.5.2 Determinants 443
      D.5.3 Definition of Determinants 443
      D.5.4 Permuting Rows or Columns 443
      D.5.5 A Symmetric Definition 445
      D.5.6 Alternating Property of the Determinant 445
      D.5.7 Linear Combinations and Determinants 446
      D.5.8 Determinant of a Product 447
      D.5.9 Cofactor Expansions 447
      D.5.10 Formula for the Inverse 449
      D.5.11 Cramer's Rule 451
      D.5.12 Upper Triangular Matrices 451
  D.6 Cayley-Hamilton Theorem 452
  D.7 Eigenvalues and Eigenvectors of a Matrix 454
      D.7.1 Definition of Eigenvectors and Eigenvalues 454
      D.7.2 Triangular Matrices 455
      D.7.3 Defective and Nondefective Matrices 456
      D.7.4 Diagonalization 459
  D.8 Schur's Theorem 461
  D.9 Direct Sums 464
  D.10 Block Diagonal Matrices 469

Appendix E Theory of Functions of Many Variables   **471**
  E.1 Closed and Open Sets 471
  E.2 Compactness 474
      E.2.1 Continuous Functions 477
      E.2.2 Convergent Sequences 480
      E.2.3 Continuity and the Limit of a Sequence 482
      E.2.4 The Extreme Value Theorem and Uniform Continuity 483
      E.2.5 Connected Sets 484

Appendix F Implicit Function Theorem   **489**

Appendix G The Jordan Curve Theorem   **493**

Appendix H Poincare Bendixon Theorem   **505**

Appendix I   Selected Solutions   **513**
   I.1   Prerequisite Topics 513
   I.2   The Idea of a Differential Equation 515
   I.3   Finding Solutions to First Order Scalar Equations 516
   I.4   Homogeneous Linear Equations 521
   I.5   Scalar Linear Nonhomogeneous Equations 525
   I.6   Laplace Transform Methods 529
   I.7   Power Series Theory 534
   I.8   Series Methods for Scalar O.D.E. 537
   I.9   First Order Linear Systems, Theory 540
   I.10   Methods for First Order Linear Systems 541
   I.11   Theory of Ordinary Differential Equations 544
   I.12   Equilibrium Points and Limit Cycles 548
   I.13   Boundary Value Problems and Fourier Series 556
   I.14   Some Partial Differential Equations 561

Bibliography   **565**

Index   **567**

Preface

This book is an introductory treatment of differential equations. It presents the standard material in an undergraduate course on differential equations, including all of the standard methods which have been a part of the subject since the time of Newton and the Bernoulli brothers. The emphasis in this book is on theory and methods and differential equations as a part of analysis.

I have included essentially all proofs of the theorems used for the sake of those who wonder why things are so and to make the book more useful as a reference. It seems to me that providing explanations which are likely not given in class is what a book should do, and is the main reason for having a book rather than a list of exercises. It also is an attempt to present ordinary differential equations as a part of mathematics, worth studying, rather than what it often is, merely some recipes to be used by people in physical science. Nevertheless, those who are only interested in methods can skip the sections in which the explanations are given, and this book is designed to allow for this option. The book gives substantial emphasis to methods which are generally presented first with theoretical considerations following.

The topics chosen are similar to those found in most other frequently used differential equations texts and presented in roughly the same order. One can switch the chapter on Laplace transforms with the material on power series if desired without noticing any discomfort. It is this way in several other books, but I thought that, since power series methods allow for regular singular points, and the Laplace transform methods do not, the latter topic fits better with the preceding two chapters at least in the organization of the book, but maybe this is not to everyone's taste, in which case simply consider the topics in whichever order desired.

I have introduced the use of computer algebra systems as a tool for finding solutions. It seems to me that the use of technology is the way people will solve differential equations these days. The book mentions all of the main computer algebra systems, but the emphasis is placed on MATLAB® and numerical methods which include graphing the solutions and obtaining tables of values. I have tried to give clear and explicit instructions on syntax in order to get these systems to give results. I have tested each one to be sure it works, and I have tried to limit the number of new constructions in the syntax, emphasizing brevity, repetition, and what seems to me to be the most basic syntax. I have not included the more specialized packages available but have limited the presentation to MATLAB and the symbolic math toolbox.

However, this book is about differential equations, not numerical analysis or MATLAB. Neither is this book on engineering or other physical science. I have mostly featured applications which are easily understood by a general audience in order to keep the book to a reasonable length while including complete explanations of the mathematics and emphasis on methods for finding solutions. This book makes no attempt to divert the reader with numerous examples taken from engineering and physical science, including extensive excursions into modeling. I assume the students will get these things in the appropriate courses and that their interests will be better served by simple and direct emphasis on methods and theory. I believe that differential equations is a subject worth studying for its own sake. This said, there are the usual examples like hanging chains, mixture problems, damped oscillations, and so forth.

There are several levels available in this book, which include nothing more than a catalog of methods which happen to work and on the other extreme, complete explanations of nearly all theorems mentioned in the book. This is why there is an extensive collection of appendices which even include things like theorems on the existence of the integral. It is absurd to present existence theorems for O.D.E. to those who do not even know why the integral of a continuous function exists, and unfortunately this is descriptive of most students who take differential equations, including those majoring in math. Similar considerations apply to the other topics presented in the appendices, especially determinants which are typically not given a complete treatment in linear algebra texts. These things have at least as much interest as their applications to differential equations. I hope that these appendices will be a useful resource for their own sake. I doubt that they appear together in any other book, but they all have an important link to O.D.E. These topics are chosen because they will not have been seen (in my experience) by undergraduate students, although they have a direct bearing on understanding differential equations.

The book uses standard notation. However, the symbol $\equiv$ is used to indicate that a statement is part of a definition.

MATLAB $^{\circledR}$ Is a registered trademark of The Math Works, Inc.
For product information please contact:
The MathWorks, Inc.
3 Apple Hill Drive
Natick, MA, 01760-2098 USA
Tel: 508-647-7000
Fax: 508-647-7001
E-mail: info@mathworks.com
Web: www.mathworks.com

Chapter 1

Some Prerequisite Topics

The reader should be familiar with most of the topics in this chapter. However, it is often the case that set notation is not familiar and so a short discussion of this is included first. Complex numbers are then considered in somewhat more detail. You cannot escape the use of complex numbers in differential equations and since this topic tends to be neglected, I have included an introduction here.

1.1 Sets and Set Notation

A set is just a collection of things called elements. Often these are also referred to as points in calculus. For example, $\{1, 2, 3, 8\}$ would be a set consisting of the elements 1,2,3, and 8. To indicate that 3 is an element of $\{1, 2, 3, 8\}$, it is customary to write $3 \in \{1, 2, 3, 8\}$. $9 \notin \{1, 2, 3, 8\}$ means 9 is not an element of $\{1, 2, 3, 8\}$. Sometimes a rule specifies a set. For example, you could specify a set as all integers larger than 2. This would be written as $S = \{x \in \mathbb{Z} : x > 2\}$. This notation says: the set of all integers, x, such that $x > 2$.

If A and B are sets with the property that every element of A is an element of B, then A is a subset of B. For example, $\{1, 2, 3, 8\}$ is a subset of $\{1, 2, 3, 4, 5, 8\}$, in symbols, $\{1, 2, 3, 8\} \subseteq \{1, 2, 3, 4, 5, 8\}$. It is sometimes said that "A is contained in B" or even "B contains A". The same statement about the two sets may also be written as $\{1, 2, 3, 4, 5, 8\} \supseteq \{1, 2, 3, 8\}$.

The union of two sets is the set consisting of everything which is an element of at least one of the sets, A or B. As an example of the union of two sets $\{1, 2, 3, 8\} \cup \{3, 4, 7, 8\} = \{1, 2, 3, 4, 7, 8\}$ because these numbers are those which are in at least one of the two sets. In general

$$A \cup B \equiv \{x : x \in A \text{ or } x \in B\}.$$

Be sure you understand that something that is in both A and B is in the union. It is not an exclusive or.

The intersection of two sets, A and B, consists of everything that is in both of the sets. Thus, $\{1, 2, 3, 8\} \cap \{3, 4, 7, 8\} = \{3, 8\}$ because 3 and 8 are those elements the two sets have in common. In general,

$$A \cap B \equiv \{x : x \in A \text{ and } x \in B\}.$$

1

The symbol $[a, b]$, where a and b are real numbers, denotes the set of real numbers x, such that $a \leq x \leq b$ and $[a, b)$ denotes the set of real numbers such that $a \leq x < b$. (a, b) consists of the set of real numbers x such that $a < x < b$ and $(a, b]$ indicates the set of numbers x such that $a < x \leq b$. $[a, \infty)$ means the set of all numbers x such that $x \geq a$ and $(-\infty, a]$ means the set of all real numbers which are less than or equal to a. These sorts of sets of real numbers are called intervals. The two points a and b are called endpoints of the interval. Other intervals such as $(-\infty, b)$ are defined by analogy to what was just explained. In general, the curved parenthesis indicates the end point it sits next to is not included while the square parenthesis indicates this end point is included. The reason that there will always be a curved parenthesis next to ∞ or $-\infty$ is that these are not real numbers. Therefore, they cannot be included in any set of real numbers.

A special set which needs to be given a name is the empty set also called the null set, denoted by $\emptyset$. Thus, $\emptyset$ is defined as the set which has no elements in it. Mathematicians like to say the empty set is a subset of every set. The reason they say this is that if it were not so, there would have to exist a set A, such that $\emptyset$ has something in it which is not in A. However, $\emptyset$ has nothing in it and so the least intellectual discomfort is achieved by saying $\emptyset \subseteq A$.

If A and B are two sets, $A \setminus B$ denotes the set of things which are in A but not in B. Thus,

$$A \setminus B \equiv \{x \in A : x \notin B\}.$$

When one is considering sets A which are subsets of a given set S one denotes as A^C the set $S \setminus A$. It means nothing more than the set of elements of S which are not in A.

To illustrate the use of this notation relative to intervals consider three examples of inequalities. Their solutions will be written in the notation just described.

Example 1.1.1 *Solve the inequality* $2x + 4 \leq x - 8$

$x \leq -12$ is the answer. This is written in terms of an interval as $(-\infty, -12]$.

Example 1.1.2 *Solve the inequality* $(x + 1)(2x - 3) \geq 0$.

The solution is $x \leq -1$ or $x \geq \dfrac{3}{2}$. In terms of set notation this is denoted by $(-\infty, -1] \cup [\dfrac{3}{2}, \infty)$.

Example 1.1.3 *Solve the inequality* $x(x + 2) \geq -4$.

This is true for any value of x. It is written as $\mathbb{R}$ or $(-\infty, \infty)$.

To take the union of $A_1, \cdots, A_n$ one often writes $\cup_{i=1}^{n} A_i$. If $\mathcal{C}$ is a set whose elements are subsets of some given set S, one often writes $\cup \mathcal{C}$ or more precisely, $\cup \{A : A \in \mathcal{C}\}$ to denote the set of all $s \in S$ which are in at least one of the sets of $\mathcal{C}$. Similarly, one writes $\cap \mathcal{C}$ or more precisely, $\cap \{A : A \in \mathcal{C}\}$ to denote those $s \in S$ which are in all of the sets of $\mathcal{C}$.

An important relation is known as DeMorgan's laws. It comes directly from the definition of union and intersection of sets. You should verify that these relations hold.

$$(\cup\{A : A \in \mathcal{C}\})^C = \cap\{A^C : A \in \mathcal{C}\}$$
$$(\cap\{A : A \in \mathcal{C}\})^C = \cup\{A^C : A \in \mathcal{C}\}$$

For example, consider the second. If S is the set which contains all of the sets of $\mathcal{C}$, then if s is in the left, it means that s must fail to be in one of the sets $A \in \mathcal{C}$ and so s must be in the right. If s is in the right, then it means that $s \in A^C$ for some $A \in \mathcal{C}$. But this means s fails to be in some set of $\mathcal{C}$ which is to say that it is not in all of the sets of $\mathcal{C}$ which is to say that it is in the left. Thus the second of the two relations holds. The first is similar.

1.2 Well Ordering and Induction

Mathematical induction and well ordering are two extremely important principles in math. They are often used to prove significant things which would be hard to prove otherwise.

Definition 1.2.1 *A set is well ordered if every nonempty subset S contains a smallest element z having the property that $z \leq x$ for all $x \in S$.*

Axiom 1.2.2 *Any set of integers larger than a given number is well ordered.*

In particular, the natural numbers defined as

$$\mathbb{N} \equiv \{1, 2, \cdots\}$$

is well ordered.

The above axiom implies the principle of mathematical induction. The symbol $\mathbb{Z}$ denotes the set of all integers. Note that if a is an integer, then there are no integers between a and $a + 1$.

Theorem 1.2.3 *(Mathematical induction) A set $S \subseteq \mathbb{Z}$, having the property that $a \in S$ and $n + 1 \in S$ whenever $n \in S$ contains all integers $x \in \mathbb{Z}$ such that $x \geq a$.*

Proof: Let T consist of all integers larger than or equal to a which are not in S. The theorem will be proved if $T = \emptyset$. If $T \neq \emptyset$ then by the well ordering principle, there would have to exist a smallest element of T, denoted as b. It must be the case that $b > a$ since by definition, $a \notin T$. Thus $b \geq a + 1$, and so $b - 1 \geq a$ and $b - 1 \notin S$ because if $b - 1 \in S$, then $b - 1 + 1 = b \in S$ by the assumed property of S. Therefore, $b - 1 \in T$ which contradicts the choice of b as the smallest element of T. ($b - 1$ is smaller.) Since a contradiction is obtained by assuming $T \neq \emptyset$, it must be the case that $T = \emptyset$ and this says that every integer at least as large as a is also in S. ∎

Mathematical induction is a very useful device for proving theorems about the integers.

Example 1.2.4 *Prove by induction that $\sum_{k=1}^{n} k^2 = \dfrac{n(n+1)(2n+1)}{6}$.*

By inspection, if $n = 1$ then the formula is true. The sum yields 1 and so does the formula on the right. Suppose this formula is valid for some $n \geq 1$ where n is an integer. Then

$$\sum_{k=1}^{n+1} k^2 = \sum_{k=1}^{n} k^2 + (n+1)^2 = \frac{n(n+1)(2n+1)}{6} + (n+1)^2.$$

The step going from the first to the second line is based on the assumption that the formula is true for n. This is called the induction hypothesis. Now simplify the expression in the second line,

$$\frac{n(n+1)(2n+1)}{6} + (n+1)^2.$$

This equals

$$(n+1)\left(\frac{n(2n+1)}{6} + (n+1)\right)$$

and

$$\frac{n(2n+1)}{6} + (n+1) = \frac{6(n+1) + 2n^2 + n}{6} = \frac{(n+2)(2n+3)}{6}$$

Therefore,

$$\sum_{k=1}^{n+1} k^2 = \frac{(n+1)(n+2)(2n+3)}{6} = \frac{(n+1)((n+1)+1)(2(n+1)+1)}{6},$$

showing the formula holds for $n+1$ whenever it holds for n. This proves the formula by mathematical induction.

Example 1.2.5 *Show that for all $n \in \mathbb{N}$,* $\dfrac{1}{2} \cdot \dfrac{3}{4} \cdots \dfrac{2n-1}{2n} < \dfrac{1}{\sqrt{2n+1}}.$

If $n = 1$ this reduces to the statement that $\dfrac{1}{2} < \dfrac{1}{\sqrt{3}}$ which is obviously true. Suppose then that the inequality holds for n. Then

$$\frac{1}{2} \cdot \frac{3}{4} \cdots \frac{2n-1}{2n} \cdot \frac{2n+1}{2n+2} < \frac{1}{\sqrt{2n+1}} \frac{2n+1}{2n+2} = \frac{\sqrt{2n+1}}{2n+2}.$$

The theorem will be proved if this last expression is less than $\dfrac{1}{\sqrt{2n+3}}$. This happens if and only if

$$\left(\frac{1}{\sqrt{2n+3}}\right)^2 = \frac{1}{2n+3} > \frac{2n+1}{(2n+2)^2}$$

which occurs if and only if $(2n+2)^2 > (2n+3)(2n+1)$ and this is clearly true which may be seen from expanding both sides. This proves the inequality.

Let's review the process just used. If S is the set of integers at least as large as 1 for which the formula holds, the first step was to show $1 \in S$ and then that whenever $n \in S$, it follows $n+1 \in S$. Therefore, by the principle of mathematical induction, S contains $[1, \infty) \cap \mathbb{Z}$, all positive integers. In doing an inductive proof of this sort, the set S is normally not mentioned. One just verifies the steps above. First show the thing is true for some $a \in \mathbb{Z}$ and then verify that whenever it is true for m it follows it is also true for $m+1$. When this has been done, the theorem has been proved for all $m \geq a$.

1.3 The Complex Numbers

Recall that a real number is a point on the real number line. Just as a real number should be considered as a point on the line, a complex number is considered a point in the plane which can be identified in the usual way using the Cartesian coordinates of the point. Thus (a, b) identifies a point whose x coordinate is a and whose y coordinate is b. In dealing with complex numbers, such a point is written as $a + ib$. For example, in the following picture, I have graphed the point $3 + 2i$. You see it corresponds to the point in the plane whose coordinates are $(3, 2)$.

Multiplication and addition are defined in the most obvious way subject to the convention that $i^2 = -1$. Thus,

$\bullet 3 + 2i$

$$(a + ib) + (c + id) = (a + c) + i(b + d)$$

and

$$
\begin{aligned}
(a + ib)(c + id) &= ac + iad + ibc + i^2 bd \\
&= (ac - bd) + i(bc + ad).
\end{aligned}
$$

Every non zero complex number $a + ib$, with $a^2 + b^2 \neq 0$, has a unique multiplicative inverse.

$$\frac{1}{a + ib} = \frac{a - ib}{a^2 + b^2} = \frac{a}{a^2 + b^2} - i\frac{b}{a^2 + b^2}.$$

You should prove the following theorem.

Theorem 1.3.1 *The complex numbers with multiplication and addition defined as above form a field satisfying all the field axioms. These are the following list of properties.*

1. $x + y = y + x$ *(commutative law for addition)*

2. $x + 0 = x$ *(additive identity).*

3. *For each* $x \in \mathbb{R}$, *there exists* $-x \in \mathbb{R}$ *such that* $x + (-x) = 0$ *(existence of additive inverse).*

4. $(x + y) + z = x + (y + z)$ *(associative law for addition).*

5. $xy = yx$ *(commutative law for multiplication). You could write this as* $x \times y = y \times x$.

6. $(xy)z = x(yz)$ *(associative law for multiplication).*

7. $1x = x$ *(multiplicative identity).*

8. *For each* $x \neq 0$, *there exists* x^{-1} *such that* $xx^{-1} = 1$ *(existence of multiplicative inverse).*

9. $x(y + z) = xy + xz$ *(distributive law).*

Something which satisfies these axioms is called a field. The only fields of interest in this book are the field of complex numbers or the field of real numbers. You have seen in earlier courses that the real numbers also satisfies the above axioms. The field of complex numbers is denoted as $\mathbb{C}$ and the field of real numbers is denoted as $\mathbb{R}$. An important construction regarding complex numbers is the complex conjugate denoted by a horizontal line above the number. It is defined as follows.

$$\overline{a + ib} \equiv a - ib.$$

What it does is reflect a given complex number across the x axis. Algebraically, the following formula is easy to obtain.

$$
\begin{aligned}
\left(\overline{a + ib}\right)(a + ib) &= (a - ib)(a + ib) \\
&= a^2 + b^2 - i(ab - ab) = a^2 + b^2.
\end{aligned}
$$

Definition 1.3.2 *Define the absolute value of a complex number as follows.*

$$|a + ib| \equiv \sqrt{a^2 + b^2}.$$

Thus, denoting by z the complex number $z = a + ib$,

$$|z| = (z\bar{z})^{1/2}.$$

Also from the definition, if $z = x + iy$ and $w = u + iv$ are two complex numbers, then $|zw| = |z|\,|w|$. You should verify this.

Notation 1.3.3 *Recall the following notation.*

$$\sum_{j=1}^{n} a_j \equiv a_1 + \cdots + a_n$$

There is also a notation which is used to denote a product.

$$\prod_{j=1}^{n} a_j \equiv a_1 a_2 \cdots a_n$$

The triangle inequality holds for the absolute value for complex numbers just as it does for the ordinary absolute value.

Proposition 1.3.4 *Let z, w be complex numbers. Then the triangle inequality holds.*

$$|z + w| \le |z| + |w|,\ \ ||z| - |w|| \le |z - w|.$$

Proof: Let $z = x + iy$ and $w = u + iv$. First note that

$$z\bar{w} = (x + iy)(u - iv) = xu + yv + i(yu - xv)$$

and so $|xu + yv| \le |z\bar{w}| = |z|\,|w|$.

$$|z + w|^2 = (x + u + i(y + v))(x + u - i(y + v))$$

$$= (x + u)^2 + (y + v)^2 = x^2 + u^2 + 2xu + 2yv + y^2 + v^2$$
$$\leq |z|^2 + |w|^2 + 2 |z| |w| = (|z| + |w|)^2,$$

so this shows the first version of the triangle inequality. To get the second,

$$z = z - w + w, \; w = w - z + z$$

and so by the first form of the inequality

$$|z| \leq |z - w| + |w|, \; |w| \leq |z - w| + |z|$$

and so both $|z| - |w|$ and $|w| - |z|$ are no larger than $|z - w|$ and this proves the second version because $||z| - |w||$ is one of $|z| - |w|$ or $|w| - |z|$. ∎

With this definition, it is important to note the following. Be sure to verify this. It is not too hard but you need to do it.

Remark 1.3.5 : *Let $z = a + ib$ and $w = c + id$. Then*

$$|z - w| = \sqrt{(a - c)^2 + (b - d)^2}.$$

Thus the distance between the point in the plane determined by the ordered pair (a, b) and the ordered pair (c, d) equals $|z - w|$ where z and w are as just described.

For example, consider the distance between $(2, 5)$ and $(1, 8)$. From the distance formula this distance equals $\sqrt{(2 - 1)^2 + (5 - 8)^2} = \sqrt{10}$. On the other hand, letting $z = 2 + i5$ and $w = 1 + i8$, $z - w = 1 - i3$ and so $(z - w)\overline{(z - w)} = (1 - i3)(1 + i3) = 10$ so $|z - w| = \sqrt{10}$, the same thing obtained with the distance formula.

Also note that for $z = a + ib$ with a, b real, $a = \frac{1}{2}(z + \bar{z})$, $b = \frac{1}{2i}(z - \bar{z})$.

1.4 Polar Form of Complex Numbers

Complex numbers are often written in the so-called polar form which is described next. Suppose $z = x + iy$ is a complex number. Then

$$x + iy = \sqrt{x^2 + y^2}\left(\frac{x}{\sqrt{x^2 + y^2}} + i\frac{y}{\sqrt{x^2 + y^2}}\right).$$

Now note that

$$\left(\frac{x}{\sqrt{x^2 + y^2}}\right)^2 + \left(\frac{y}{\sqrt{x^2 + y^2}}\right)^2 = 1$$

and so

$$\left(\frac{x}{\sqrt{x^2 + y^2}}, \frac{y}{\sqrt{x^2 + y^2}}\right)$$

is a point on the unit circle. Therefore, there exists a unique angle $\theta \in [0, 2\pi)$ such that

$$\cos\theta = \frac{x}{\sqrt{x^2 + y^2}}, \; \sin\theta = \frac{y}{\sqrt{x^2 + y^2}}.$$

The polar form of the complex number is then $r(\cos\theta + i\sin\theta)$ where θ is this angle just described and $r = \sqrt{x^2 + y^2} \equiv |z|$.

$$r = \sqrt{x^2 + y^2} \qquad \qquad x + iy = r(\cos(\theta) + i\sin(\theta))$$

1.5   Roots of Complex Numbers

A fundamental identity is the formula of De Moivre which follows.

Theorem 1.5.1 *Let $r > 0$ be given. Then if n is a positive integer,*

$$[r(\cos t + i\sin t)]^n = r^n(\cos nt + i\sin nt).$$

Proof: It is clear the formula holds if $n = 1$. Suppose it is true for n.

$$[r(\cos t + i\sin t)]^{n+1} = [r(\cos t + i\sin t)]^n [r(\cos t + i\sin t)]$$

which by induction equals

$$= r^{n+1}(\cos nt + i\sin nt)(\cos t + i\sin t)$$

$$= r^{n+1}((\cos nt \cos t - \sin nt \sin t) + i(\sin nt \cos t + \cos nt \sin t))$$

$$= r^{n+1}(\cos(n+1)t + i\sin(n+1)t)$$

by the formulas for the cosine and sine of the sum of two angles. ∎

Corollary 1.5.2 *Let z be a non zero complex number. Then there are always exactly k k^{th} roots of z in $\mathbb{C}$.*

Proof: Let $z = x + iy$ and let $z = |z|(\cos t + i\sin t)$ be the polar form of the complex number. By De Moivre's theorem, a complex number

$$r(\cos\alpha + i\sin\alpha),$$

is a kth root of z if and only if

$$r^k(\cos k\alpha + i\sin k\alpha) = |z|(\cos t + i\sin t).$$

This requires $r^k = |z|$ and so $r = |z|^{1/k}$ and also both $\cos(k\alpha) = \cos t$ and $\sin(k\alpha) = \sin t$. This can only happen if

$$k\alpha = t + 2l\pi$$

for l an integer. Thus

$$\alpha = \frac{t + 2l\pi}{k}, l \in \mathbb{Z}$$

and so the kth roots of z are of the form

$$|z|^{1/k}\left(\cos\left(\frac{t + 2l\pi}{k}\right) + i\sin\left(\frac{t + 2l\pi}{k}\right)\right), l \in \mathbb{Z}.$$

Since the cosine and sine are periodic of period 2π, there are exactly k distinct numbers which result from this formula. ∎

Example 1.5.3 *Find the three cube roots of i.*

First note that $i = 1 \left(\cos \left(\frac{\pi}{2} \right) + i \sin \left(\frac{\pi}{2} \right) \right)$. Using the formula in the proof of the above corollary, the cube roots of i are

$$1 \left(\cos \left(\frac{(\pi/2) + 2l\pi}{3} \right) + i \sin \left(\frac{(\pi/2) + 2l\pi}{3} \right) \right)$$

where $l = 0, 1, 2$. Therefore, the roots are

$$\cos \left(\frac{\pi}{6} \right) + i \sin \left(\frac{\pi}{6} \right), \cos \left(\frac{5}{6} \pi \right) + i \sin \left(\frac{5}{6} \pi \right), \cos \left(\frac{3}{2} \pi \right) + i \sin \left(\frac{3}{2} \pi \right).$$

Thus the cube roots of i are $\frac{\sqrt{3}}{2} + i \left(\frac{1}{2} \right), \frac{-\sqrt{3}}{2} + i \left(\frac{1}{2} \right)$, and $-i$.

The ability to find k^{th} roots can also be used to factor some polynomials.

Example 1.5.4 *Factor the polynomial $x^3 - 27$.*

First find the cube roots of 27. By the above procedure using De Moivre's theorem, these cube roots are $3, 3 \left(\frac{-1}{2} + i \frac{\sqrt{3}}{2} \right)$, and $3 \left(\frac{-1}{2} - i \frac{\sqrt{3}}{2} \right)$. Therefore, $x^3 - 27 =$

$$(x - 3) \left(x - 3 \left(\frac{-1}{2} + i \frac{\sqrt{3}}{2} \right) \right) \left(x - 3 \left(\frac{-1}{2} - i \frac{\sqrt{3}}{2} \right) \right).$$

Note also $\left(x - 3 \left(\frac{-1}{2} + i \frac{\sqrt{3}}{2} \right) \right) \left(x - 3 \left(\frac{-1}{2} - i \frac{\sqrt{3}}{2} \right) \right) = x^2 + 3x + 9$ and so

$$x^3 - 27 = (x - 3) \left(x^2 + 3x + 9 \right)$$

where the quadratic polynomial $x^2 + 3x + 9$ cannot be factored without using complex numbers.

Note that even though the polynomial $x^3 - 27$ has all real coefficients, it has some complex zeros, $\frac{-1}{2} + i \frac{\sqrt{3}}{2}$ and $\frac{-1}{2} - i \frac{\sqrt{3}}{2}$. These zeros are complex conjugates of each other. It is **always** this way. You should show this is the case. To see how to do this, see Problems 18 and 19 below.

1.6 The Quadratic Formula

The quadratic formula

$$x = \frac{-b \pm \sqrt{b^2 - 4ac}}{2a}$$

gives the solutions x to

$$ax^2 + bx + c = 0$$

where a, b, c are real numbers. It holds even if $b^2 - 4ac < 0$. This is easy to show from the above. There are exactly two square roots to this number $b^2 - 4ac$ from the above methods using De Moivre's theorem. These roots are of the form

$$\sqrt{4ac - b^2}\left(\cos\left(\frac{\pi}{2}\right) + i\sin\left(\frac{\pi}{2}\right)\right) = i\sqrt{4ac - b^2}$$

and

$$\sqrt{4ac - b^2}\left(\cos\left(\frac{3\pi}{2}\right) + i\sin\left(\frac{3\pi}{2}\right)\right) = -i\sqrt{4ac - b^2}$$

Thus the solutions, according to the quadratic formula are still given correctly by the above formula.

Do these solutions predicted by the quadratic formula continue to solve the quadratic equation? Yes, they do. You only need to observe that when you square a square root of a complex number z, you recover z. Thus

$$a\left(\frac{-b + \sqrt{b^2 - 4ac}}{2a}\right)^2 + b\left(\frac{-b + \sqrt{b^2 - 4ac}}{2a}\right) + c$$

$$= a\left(\frac{1}{2a^2}b^2 - \frac{1}{a}c - \frac{1}{2a^2}b\sqrt{b^2 - 4ac}\right) + b\left(\frac{-b + \sqrt{b^2 - 4ac}}{2a}\right) + c$$

$$= \left(-\frac{1}{2a}\left(b\sqrt{b^2 - 4ac} + 2ac - b^2\right)\right) + \frac{1}{2a}\left(b\sqrt{b^2 - 4ac} - b^2\right) + c = 0$$

Similar reasoning shows directly that $\frac{-b - \sqrt{b^2 - 4ac}}{2a}$ also solves the quadratic equation.

What if the coefficients of the quadratic equation are actually complex numbers? Does the formula hold even in this case? The answer is yes. This is a hint on how to do Problem 28 below, a special case of the fundamental theorem of algebra, and an ingredient in the proof of some versions of this theorem.

Example 1.6.1 *Find the solutions to $x^2 - 2ix - 5 = 0$.*

Formally, from the quadratic formula, these solutions are

$$x = \frac{2i \pm \sqrt{-4 + 20}}{2} = \frac{2i \pm 4}{2} = i \pm 2.$$

Now you can check that these really do solve the equation. In general, this will be the case. See Problem 28 below.

1.7   The Complex Exponential

It was shown above that every complex number can be written in the form

$$r\left(\cos\theta + i\sin\theta\right)$$

where $r \geq 0$. Laying aside the zero complex number, this shows that every non zero complex number is of the form $e^\alpha\left(\cos\beta + i\sin\beta\right)$. We write this in the form $e^{\alpha + i\beta}$.

Having done so, does it follow that the expression preserves the most important property of the function $t \rightarrow e^{(\alpha+i\beta)t}$ for t real, that

$$\left(e^{(\alpha+i\beta)t}\right)' = (\alpha + i\beta) e^{(\alpha+i\beta)t}?$$

By the definition just given which does not contradict the usual definition in case $\beta = 0$ and the usual rules of differentiation in calculus,

$$\left(e^{(\alpha+i\beta)t}\right)' = \left(e^{\alpha t}\left(\cos\left(\beta t\right) + i\sin\left(\beta t\right)\right)\right)'$$
$$= e^{\alpha t}\left[\alpha\left(\cos\left(\beta t\right) + i\sin\left(\beta t\right)\right) + \left(-\beta\sin\left(\beta t\right) + i\beta\cos\left(\beta t\right)\right)\right]$$

Now consider the other side. From the definition it equals

$$(\alpha + i\beta)\left(e^{\alpha t}\left(\cos\left(\beta t\right) + i\sin\left(\beta t\right)\right)\right) = e^{\alpha t}\left[(\alpha + i\beta)\left(\cos\left(\beta t\right) + i\sin\left(\beta t\right)\right)\right]$$

$$= e^{\alpha t}\left[\alpha\left(\cos\left(\beta t\right) + i\sin\left(\beta t\right)\right) + \left(-\beta\sin\left(\beta t\right) + i\beta\cos\left(\beta t\right)\right)\right]$$

which is the same thing. This is of fundamental importance in differential equations. It shows that there is no change in going from real to complex numbers for ω in the consideration of the problem $y' = \omega y$, $y(0) = 1$. The solution is always $e^{\omega t}$. The formula just discussed, that

$$e^{\alpha}\left(\cos\beta + i\sin\beta\right) = e^{\alpha+i\beta}$$

is Euler's formula.

1.8   The Fundamental Theorem of Algebra

The fundamental theorem of algebra states that every non-constant polynomial having coefficients in $\mathbb{C}$ has a zero in $\mathbb{C}$. If $\mathbb{C}$ is replaced by $\mathbb{R}$, this is not true because of the example, $x^2 + 1 = 0$. This theorem is a very remarkable result and notwithstanding its title, all the most straightforward proofs depend on either analysis or topology. It was first mostly proved by Gauss in 1797. The first complete proof was given by Argand in 1806. The proof given here follows Rudin [20]. See also Hardy [13] for a similar proof, more discussion and references. The shortest proof is found in the theory of complex analysis. First I will give an informal explanation of this theorem which shows why it is reasonable to believe in the fundamental theorem of algebra even without a rigorous proof.

Theorem 1.8.1 *Let $p(z) = a_n z^n + a_{n-1}z^{n-1} + \cdots + a_1 z + a_0$ where each a_k is a complex number and $a_n \neq 0, n \geq 1$. Then there exists $w \in \mathbb{C}$ such that $p(w) = 0$.*

To begin with, here is the informal explanation. Dividing by the leading coefficient a_n, there is no loss of generality in assuming that the polynomial is of the form

$$p(z) = z^n + a_{n-1}z^{n-1} + \cdots + a_1 z + a_0$$

If $a_0 = 0$, there is nothing to prove because $p(0) = 0$. Therefore, assume $a_0 \neq 0$. From the polar form of a complex number z, it can be written as $|z|(\cos\theta + i\sin\theta)$. Thus, by DeMoivre's theorem,

$$z^n = |z|^n (\cos(n\theta) + i\sin(n\theta))$$

It follows that z^n is some point on the circle of radius $|z|^n$

Denote by C_r the circle of radius r in the complex plane which is centered at 0. Then if r is sufficiently large and $|z| = r$, the term z^n is far larger than the rest of the polynomial. It is on the circle of radius $|z|^n$ while the other terms are on circles of fixed multiples of $|z|^k$ for $k \leq n - 1$. Thus, for r large enough, $A_r = \{p(z) : z \in C_r\}$ describes a closed curve which misses the inside of some circle having 0 as its center. It won't be as simple as suggested in the following picture, but it will be a closed curve thanks to De Moivre's theorem and the observation that the cosine and sine are periodic. Now shrink r. Eventually, for r small enough, the non constant terms are negligible and so A_r is a curve which is contained in some circle centered at a_0 which has 0 on the outside.

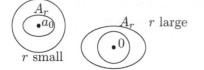

Thus it is reasonable to believe that for some r during this shrinking process, the set A_r must hit 0. It follows that $p(z) = 0$ for some z.

For example, consider the polynomial $x^3 + x + 1 + i$. It has no real zeros. However, you could let $z = r(\cos t + i\sin t)$ and insert this into the polynomial. Thus you would want to find a point where

$$(r(\cos t + i\sin t))^3 + r(\cos t + i\sin t) + 1 + i = 0 + 0i$$

Expanding this expression on the left to write it in terms of real and imaginary parts, you get on the left

$$r^3\cos^3 t - 3r^3\cos t \sin^2 t + r\cos t + 1 + i\left(3r^3\cos^2 t \sin t - r^3\sin^3 t + r\sin t + 1\right)$$

Thus you need to have both the real and imaginary parts equal to 0. In other words, you need to have $(0,0) =$

$$\left(r^3\cos^3 t - 3r^3\cos t \sin^2 t + r\cos t + 1, \, 3r^3\cos^2 t \sin t - r^3\sin^3 t + r\sin t + 1\right)$$

for some value of r and t. First here is a graph of this parametric function of t for $t \in [0, 2\pi]$ on the left, when $r = 4$. Note how the graph misses the origin $0 + i0$. In fact, the closed curve is in the exterior of a circle which has the point $0 + i0$ on its inside.

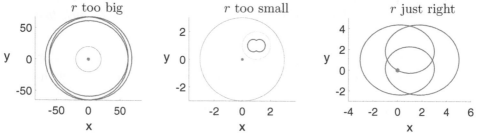

Next is the graph when $r = .5$. Note how the closed curve is included in a circle which has $0 + i0$ on its outside. As you shrink r you get closed curves. At first, these closed curves enclose $0 + i0$ and later, they exclude $0 + i0$. Thus one of them should pass through this point. In fact, consider the curve which results when $r = 1.386$ which is the graph on the right. Note how for this value of r the curve passes through the point $0 + i0$. Thus for some t, $1.386 (\cos t + i \sin t)$ is a solution of the equation $p(z) = 0$ or very close to one.

Now here is a short rigorous proof for those who have studied analysis.

Proof: Suppose the nonconstant polynomial $p(z) = a_0 + a_1 z + \cdots + a_n z^n, a_n \neq 0$, has no zero in $\mathbb{C}$. Since $\lim_{|z| \to \infty} |p(z)| = \infty$, there is a z_0 with

$$|p(z_0)| = \min_{z \in \mathbb{C}} |p(z)| > 0$$

Then let $q(z) = \frac{p(z+z_0)}{p(z_0)}$. This is also a polynomial which has no zeros and the minimum of $|q(z)|$ is 1 and occurs at $z = 0$. Since $q(0) = 1$, it follows $q(z) = 1 + a_k z^k + r(z)$ where $r(z)$ consists of higher order terms. Here a_k is the first coefficient which is nonzero. Choose a sequence, $z_n \to 0$, such that $a_k z_n^k < 0$. For example, let $-a_k z_n^k = (1/n)$. Then $|q(z_n)| \leq 1 - 1/n + |r(z_n)| < 1$ for all n large enough because the higher order terms in $r(z_n)$ converge to 0 faster than z_n^k. This is a contradiction. ■

1.9 The Cauchy-Schwarz Inequality

This is a fundamental result which says that if you have real numbers $a_1, \cdots, a_n$ and $b_1, \cdots, b_n$ then

$$\sum_{i=1}^{n} a_i b_i \leq \left(\sum_{i=1}^{n} a_i^2 \right)^{1/2} \left(\sum_{i=1}^{n} b_i^2 \right)^{1/2} \tag{1.1}$$

To prove this, consider

$$p(t) = \sum_{i=1}^{n} (a_i + t b_i)^2 \geq 0$$

The inequality is obvious if each $b_i = 0$ so suppose this is not the case. It is a polynomial in t of degree 2 which is always nonnegative. It equals

$$\sum_i a_i^2 + 2 \sum_i a_i b_i t + t^2 \sum_i b_i^2 \geq 0$$

Thus $p(t) = 0$ has either one real root or none. Hence, by the quadratic formula,

$$4 \left(\sum_i a_i b_i \right)^2 - 4 \left(\sum_i a_i^2 \right) \left(\sum_i b_i^2 \right) \leq 0$$

which is the same as the Cauchy* Schwarz[†] inequality.

1.10   Exercises

1. Verify the first of DeMorgan's laws, $(\cup \{A : A \in \mathcal{C}\})^C = \cap \{A^C : A \in \mathcal{C}\}$ where $\mathcal{C}$ consists of a set whose elements are subsets of a given set S.

2. Prove by induction that $\sum_{k=1}^{n} k^3 = \frac{1}{4}n^4 + \frac{1}{2}n^3 + \frac{1}{4}n^2$.

3. Prove by induction that whenever $n \geq 2, \sum_{k=1}^{n} \frac{1}{\sqrt{k}} > \sqrt{n}$.

4. Prove by induction that $1 + \sum_{i=1}^{n} i \, (i!) = (n+1)!$.

5. The binomial theorem states $(x + y)^n = \sum_{k=0}^{n} \binom{n}{k} x^{n-k} y^k$ where

$$\binom{n+1}{k} = \binom{n}{k} + \binom{n}{k-1} \text{ if } k \in [1, n], \quad \binom{n}{0} \equiv 1 \equiv \binom{n}{n}$$

Prove the binomial theorem by induction. Next show that

$$\binom{n}{k} = \frac{n!}{(n-k)!k!}, \quad 0! \equiv 1$$

6. Let $z = 5 + i9$. Find z^{-1}.

7. Let $z = 2 + i7$ and let $w = 3 - i8$. Find $zw, z + w, z^2$, and w/z.

8. Give the complete solution to $x^4 + 16 = 0$.

9. Graph the complex cube roots of 8 in the complex plane. Do the same for the four fourth roots of 16.

10. If z is a complex number, show there exists ω a complex number with $|\omega| = 1$ and $\omega z = |z|$.

---

*Augustin-Louis Cauchy (1789-1857), was one of the most prolific of all mathematicians. His name is associated with topics from pure mathematics to mechanics. In continuum mechanics, you see Cauchy's theorem which is about the existence of a stress tensor. In calculus, you find the Cauchy mean value theorem, the Cauchy condensation test, Cauchy sequences, Cauchy product, etc. He did work on finite groups, determinants, mechanics, differential equations, celestial mechanics, complex analysis, possibly his most famous contribution, and too many other areas to name here in this short footnote. He wrote some 800 papers filling 24 volumes and 5 textbooks. Cauchy was also able to speak and read many languages. He is known for beginning the process of making analysis rigorous. His initial training was as a civil engineer, and he became primarily interested in mathematics later.

Cauchy was a devout Catholic and royalist. He acted according to principles rather than expediency. This caused him a lot of trouble because of his refusal to do the convenient thing and take a loyalty oath when he felt he could not do so. As a result, he spent many years in exile.

†Karl Hermann Amandus Schwarz (1843-1921) was known for his work in complex analysis. He also worked on differential geometry and calculus of variations. The version of this inequality done by him involved integrals and appeared in about 1888 although Bunyakovsky (1804-1889) did an earlier version of it in 1859. Cauchy did the version for sums much earlier.

11. De Moivre's theorem says $[r\left(\cos t + i\sin t\right)]^{n} = r^{n}\left(\cos nt + i\sin nt\right)$ for n a positive integer. Does this formula continue to hold for all integers n, even negative integers? Explain.

12. You already know formulas for $\cos\left(x + y\right)$ and $\sin\left(x + y\right)$ and these were used to prove De Moivre's theorem. Now using De Moivre's theorem, derive a formula for $\sin\left(5x\right)$ and one for $\cos\left(5x\right)$.

13. If z and w are two complex numbers and the polar form of z involves the angle θ while the polar form of w involves the angle ϕ, show that in the polar form for zw the angle involved is $\theta + \phi$. Also, show that in the polar form of a complex number z, $r = |z|$.

14. Factor $x^{3} + 8$ as a product of linear factors.

15. Write $x^{3} + 27$ in the form $\left(x + 3\right)\left(x^{2} + ax + b\right)$ where $x^{2} + ax + b$ cannot be factored any more using only real numbers.

16. Completely factor $x^{4} + 16$ as a product of linear factors.

17. Factor $x^{4} + 16$ as the product of two quadratic polynomials each of which cannot be factored further without using complex numbers.

18. If z, w are complex numbers prove $\overline{zw} = \overline{z}\,\overline{w}$ and then show by induction that $\overline{\prod_{j=1}^{n} z_{j}} = \prod_{j=1}^{n} \overline{z_{j}}$. Also verify that $\overline{\sum_{k=1}^{m} z_{k}} = \sum_{k=1}^{m} \overline{z_{k}}$. In words this says the conjugate of a product equals the product of the conjugates and the conjugate of a sum equals the sum of the conjugates.

19. Suppose $p\left(x\right) = a_{n}x^{n} + a_{n-1}x^{n-1} + \cdots + a_{1}x + a_{0}$ where all the a_{k} are real numbers. Suppose also that $p\left(z\right) = 0$ for some $z \in \mathbb{C}$. Show it follows that $p\left(\overline{z}\right) = 0$ also.

20. Show that $1 + i, 2 + i$ are the only two zeros to $p\left(x\right) = x^{2} - \left(3 + 2i\right)x + \left(1 + 3i\right)$ so the zeros do not necessarily come in conjugate pairs if the coefficients are not real.

21. I claim that $1 = -1$. Here is why.

$$-1 = i^{2} = \sqrt{-1}\sqrt{-1} = \sqrt{\left(-1\right)^{2}} = \sqrt{1} = 1.$$

This is clearly a remarkable result but is there something wrong with it? If so, what is wrong?

22. De Moivre's theorem is really a grand thing. I plan to use it now for rational exponents, not just integers.

$$1 = 1^{\left(1/4\right)} = \left(\cos 2\pi + i\sin 2\pi\right)^{1/4} = \cos\left(\pi/2\right) + i\sin\left(\pi/2\right) = i.$$

Therefore, squaring both sides it follows $1 = -1$ as in the previous problem. What does this tell you about De Moivre's theorem? Is there a profound difference between raising numbers to integer powers and raising numbers to non integer powers?

23. Review Problem 11 at this point. Now here is another question: If n is an integer, is it always true that $(\cos\theta - i\sin\theta)^n = \cos(n\theta) - i\sin(n\theta)$? Explain.

24. Suppose you have any polynomial in $\cos\theta$ and $\sin\theta$. By this I mean an expression of the form $\sum_{\alpha=0}^{m}\sum_{\beta=0}^{n} a_{\alpha\beta}\cos^\alpha\theta\sin^\beta\theta$ where $a_{\alpha\beta} \in \mathbb{C}$. Can this always be written in the form $\sum_{\gamma=-(n+m)}^{m+n} b_\gamma\cos\gamma\theta + \sum_{\tau=-(n+m)}^{n+m} c_\tau\sin\tau\theta$? Explain.

25. Suppose $p(x) = a_n x^n + a_{n-1}x^{n-1} + \cdots + a_1 x + a_0$ is a polynomial and it has n zeros,

$$z_1, z_2, \cdots, z_n$$

listed according to multiplicity. (z is a root of multiplicity m if the polynomial $f(x) = (x-z)^m$ divides $p(x)$ but $(x-z)f(x)$ does not.) Show that

$$p(x) = a_n(x - z_1)(x - z_2)\cdots(x - z_n).$$

26. Give the solutions to the following quadratic equations having real coefficients.

   (a) $x^2 - 2x + 2 = 0$ (d) $x^2 + 4x + 9 = 0$
   (b) $3x^2 + x + 3 = 0$
   (c) $x^2 - 6x + 13 = 0$ (e) $4x^2 + 4x + 5 = 0$

27. Give the solutions to the following quadratic equations having complex coefficients. Note how the solutions do not come in conjugate pairs as they do when the equation has real coefficients.

   (a) $x^2 + 2x + 1 + i = 0$ (d) $x^2 - 4ix - 5 = 0$
   (b) $4x^2 + 4ix - 5 = 0$
   (c) $4x^2 + (4 + 4i)x + 1 + 2i = 0$ (e) $3x^2 + (1 - i)x + 3i = 0$

28. Prove the fundamental theorem of algebra for quadratic polynomials having coefficients in $\mathbb{C}$. That is, show that an equation of the form $ax^2 + bx + c = 0$ where a, b, c are complex numbers, $a \neq 0$ has a complex solution. **Hint:** Consider the fact, noted earlier that the expressions given from the quadratic formula do in fact serve as solutions.

Part I

First Order Scalar Equations

Chapter 2

The Idea of a Differential Equation

2.1   Examples and Slope Fields

A differential equation is just an equation which involves an unknown function and some of its derivatives. Often the idea is to find the unknown function. Many models are expressed in terms of differential equations or systems of differential equations.

Example 2.1.1 *Newton's second law states that* $\mathbf{F} = m\mathbf{a}$ *where* $\mathbf{F}$ *is the force and* m *is the mass. The other letter is the acceleration. It is defined to be the derivative of the velocity* $\mathbf{v}$. *Thus this law is really a differential equation.* $\mathbf{F} = m\frac{d\mathbf{v}}{dt}$. *You might want to find the velocity for example.*

Example 2.1.2 *Suppose you have a falling body which experiences air resistance which is proportional to the velocity. Then if the positive direction is up, the equation for the motion of this body is*

$$\frac{dv}{dt} = -9.8 + kv$$

where here k *is the constant of proportionality and 9.8 is just the acceleration of gravity in meters per second squared. This is a differential equation.*

When you have such a differential equation of the form $y' = f(t, y)$ it is sometimes a useful idea to graph a slope field associated with the equation. This is because at some point (t, y), $f(t, y)$ gives the slope of the function $t \to y(t)$ at that point.

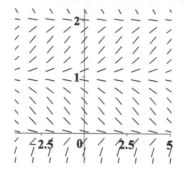

For example, consider $y' = (y - 1)(y)(2 - y)$. The diagram is a graph of the slope field for this differential equation. So if you pick a point (t, y) you could look at the graph and observe the slope of the solution to the differential equation which goes through that point. Note how the slopes of the little lines are close to 0 when y is close to 0, 2, or 1. Later we will learn how to find solutions to equations like the one whose slope field is given by this differential equation. The following picture illustrates how solutions to the equation $y' = (y - 1)(y)(2 - y)$ follow the slope fields. The curves are the graphs of functions $t \to y(t)$ where y is a solution to the differential equation and the little lines illustrate the slope of a solution at a point (t, y). The idea is that the solutions to the equation follow the slope lines.

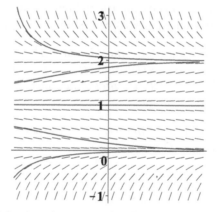

In coming up with the above graph, the computer actually used a more so-phisticated version of what was just said. That is, the solution follows the slopes. However, much of what is done in beginning courses on differential equations con-sists of finding formulas for the solutions in terms of known functions.

If you want to graph such a slope field, you should have a computer do it for you. This is because to get one which is usable, you will need to consider many points and also you want all of the little lines which indicate slope to be distinct. Thus, they should be scaled appropriately. In MATLAB, you can accomplish this with the following sequence of commands.

```
[a,b]=meshgrid(-1:.2:3,-1:.2:3);
c=(b-1).*b.*(2-b); d=(c.*c+1).^(1/2);
u=1./d; v=c./d;
quiver(a,b,u,v,'ShowArrowHead','off')
```

Be sure to type .* and not just *. This is because a,b are vectors and you want to do the operation on the individual entries of the vector. This will graph the slope field for $(y-1)(y)(2-y)$ shown above. That is at a point (x,y), it produces a little line which has slope equal to $(y-1)(y)(2-y)$. If you don't type in ,'ShowArrow-Head','off', then you will get little arrows which point in the direction of increasing t. This actually gives you a little more information so you might want to do this. In this example, you can see that if you start off at a point less than 1, the solution to $y' = (y-1)y(2-y)$ which follows the slope field will approach 0 in the limit. If you start off larger than 1, the solution to this differential equation will have 2 as a limit as $t \to \infty$.

If you have Scientific Notebook, the same problem of graphing this slope field is even easier. You simply type in math mode

$$\left[\frac{1}{\sqrt{((y-1)(y)(2-y))^2 + 1}}, \frac{(y-1)(y)(2-y)}{\sqrt{((y-1)(y)(2-y))^2 + 1}} \right]$$

and then click on compute, plot 2D, and vector field. Then in the dialog box attached to the graph, you can modify the number of vectors drawn. I don't know how to get rid of the arrows in this software, and the quality of the graph is not as good as what you can get in MATLAB, but it is much easier to use.

Instead of graphing slope fields, one can often see what is happening by graphing $f(y)$ in $y' = f(y)$ as illustrated in the following picture.

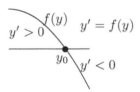

The point with a dot is where $f(y) = 0$. In the picture, you can see that y gets smaller if it is larger than y_0 and larger if it is smaller than y_0. Recall from calculus, the function is increasing if its derivative is positive and decreasing if its derivative is negative. Thus solutions to the differential equation $y' = f(y)$ will be drawn towards y_0 provided they start off sufficiently close to y_0. Consider what would happen if the slope of the graph of $f(y)$ were positive. This idea will be discussed much more later, but it is a good idea to begin thinking about this now.

The following are some easy examples of problems which involve differential equations.

Example 2.1.3 *The rate of growth of a sum of money in the bank is proportional to the amount of money in the bank, the constant of proportionality being the interest rate. What is a differential equation which expresses this?*

It is clear that this says nothing more than

$$\frac{dA}{dt} = kA$$

where A is the amount of money in the bank.

Example 2.1.4 *A pollutant enters a pond having volume V cubic meters at the rate of r kg per month where it is thoroughly mixed by swimming fish. Clean water flows into the pond at the rate of U cubic meters per month and a stream flows out* *of the pond at the rate of .8U cubic meters per month while .2U cubic meters of water evaporates each month. If initially there is no pollutant in the pond, write a differential equation with initial condition for the amount A in kg. of pollutant in the lake. Eventually, how much pollutant will be in the lake?*

You have

$$\frac{dA}{dt} = r + (\text{rate at which pollutant leaves})$$

So what is the rate at which it leaves? You have A/V kg. per cubic meter thanks to the fish. Therefore, you loose pollutant at the rate of $\frac{.8U}{V}A$ per month. Evaporation does not help because only water evaporates. Thus

$$\frac{dA}{dt} = r - \frac{.8U}{V}A$$

Since the lake started off clean, you would have the initial condition $A(0) = 0$. Thus the problem you would need to solve would involve

$$\frac{dA}{dt} = r - \frac{.8U}{V}A, \; A(0) = 0$$

which is called an initial value problem. Eventually, see Problem 2 on Page 25, you will have

$$A = \frac{rV}{.8U}$$

kg of pollutant in the lake.

Example 2.1.5 *The population that can be supported on earth will be denoted by M. Write differential equations which express the idea that the growth of population is proportional to both the number A and the difference between A and the maximum population.*

In this case, the words merely express the following differential equation.

$$\frac{dA}{dt} = kA(M - A)$$

This is a case of the logistic equation discussed later. This model has actually been used to predict the population of the earth, but it can be shown to be incorrect, although it seems reasonable.

Sometimes there is more than one differential equation involved.

Example 2.1.6 *On Isle Royal in Lake Superior, there is a population of wolves and moose. The wolves eat the moose and neither wolves nor moose can get off the island. Letting M be the number of moose and W the number of wolves, what is a reasonable model for population growth of wolves and moose?*

If you have more moose, then you would have more to reproduce and so you would think that

$$\frac{dM}{dt} = kM$$

However, you really should modify k to account for wolves because if a moose is eaten, then he/she will not be able to have baby moose. Therefore, k should involve some expression of this simple fact. One way to get this would be to let $k = \alpha \left(1 - \beta W\right)$. Then you would have

$$\frac{dM}{dt} = \alpha \left(1 - \beta W\right) M = \alpha M - \alpha \beta M W = aM - bMW$$

Now consider the growth of wolves. Without moose, they will starve and so will not be able to have baby wolves but if there is plentiful food, they will not starve and so the population will grow. Thus you would think that

$$\frac{dW}{dt} = kM$$

Of course, if there are no wolves, then they won't appear by magic. Hence you would adjust this equation as follows.

$$\frac{dW}{dt} = cWM$$

This looks fairly plausible until you begin to realize that if you have too many wolves, they will be competing for the food and there won't be enough to go around. Thus you will have starving wolves. To include this, you would write instead

$$\frac{dW}{dt} = cWM - dW$$

Thus we have the two equations, called the Lotka-Volterra equations.

$$\frac{dM}{dt} = aM - bMW, \ \frac{dW}{dt} = cWM - dW$$

These equations will be discussed more later. Hopefully, these examples show a way in which differential equations are encountered. Often, there may be more than one differential equation which seems reasonable, and to determine which is best, you would need to check to see which appears to give the best results. These issues require measurements and depend on considerations which are outside of mathematics. Just because something is plausible, does not mean it is either right or the only way to think of it.

2.2 Classifying First Order Differential Equations

To begin with here are some definitions which explain terminology related to differential equations. The differential equation

$$F'\left(x\right) = f\left(x\right)$$

for the unknown function F and the given function f is the simplest example of a differential equation and this is well solved by consideration of the integral. A solution is

$$F\left(x\right) = \int_{a}^{x} f\left(t\right) dt$$

for suitable a. However, there are many more interesting differential equations which cannot be solved so easily.

Definition 2.2.1 *Differential equations are equations involving an unknown function and some of its derivatives. A differential equation is first order if the highest order derivative of the unknown function found in the equation is 1. Thus a first order equation is one which is of the form*

$$f(t, y, y') = 0.$$

A second order differential equation is of the form

$$f(t, y, y', y'') = 0$$

with a similar definition holding for higher order equations.

First order differential equations are classified as being either **linear** or **nonlinear**.

Definition 2.2.2 *A first order linear differential equation is one which can be written in the form*

$$y' + p(t) y = q(t)$$

If it can't be written in this form, it is called a nonlinear equation. A second order equation is called linear if it can be written in the form

$$y'' + a(t) y' + b(t) y = c(t).$$

Example 2.2.3 $y' + t^2 y = \sin(t)$ *is first order linear while* $y' + y^2 = \sin(xy)$ *is nonlinear.*

Example 2.2.4 *Verify* $y = \tan(t)$ *is a solution of* $y' = 1 + y^2$.

This is easy. $y'(t) = \sec^2(t) = 1 + \tan^2(t) = 1 + y^2(t)$.

Of course it is trivial to verify something solves a differential equation. You just differentiate and plug in. A more interesting problem is in coming up with the solution to a differential equation in the first place. The next chapter contains techniques for doing this for various kinds of first order differential equations.

2.3   Some Historical Observations

In this short chapter, several problems were expressed in terms of differential equations. It was realized quite early (late 1600s) that differential equations and their solutions would help to solve many real world problems. For example, one of the interesting questions which was of interest at that time was in constructing clocks which would keep accurate time even when they wound down. This was because if it could be done, then they would have a way to determine longitude, not just latitude. This led to the tautochrone problem in which a curve was desired which had the property that a frictionless bead would slide to the bottom in the same time regardless of its initial position on the curve. This problem was solved in 1673 by Huygens and also by Newton in 1687. There was also the Brachistochrone problem which sought the shape of the curve joining two points along which such a frictionless bead would slide in minimum time. It was solved by the Bernouli brothers, Newton, Liebniz, and L'Hospital [11], but had been considered much earlier by

Gallileo. Also during this time, fundamental physical questions were current. One of the most important of these was the motion of the planets. Kepler had given laws concerning their motion and then Newton showed that Kepler's laws followed from his theory of gravity, an idea which had been considered earlier by Hooke. In every case, physics and astronomy were of fundamental interest. Later in the eighteenth century Lagrange wrote a book on mechanics which gives a system for producing appropriate differential equations. Laplace wrote a book on celestial mechanics. Daniel Bernouli used differential equations to study the spread of diseases. The subject of calculus of variations was also invented around this time by Euler and Lagrange. It leads to differential equations in a systematic manner. Many of these people did fundamental work on pure mathematics, but none of them lost their interest in engineering and physics, and it was differential equations which was the main tool. From these beginnings, differential equations have continued to have a fundamental place in science, engineering and math. Any time you know something about an unknown function in terms of how fast it changes, you tend to be involving differential equations.

The topics in the next chapter are mostly due to Euler, the Bernoulis, Newton, and Liebniz. They had remarkable insight and we use their ideas even now.

2.4 Exercises

1. Find $\lim_{t \to \infty} y(t)$ for $y(t)$ a solution to the differential equation, whenever it exists. That is, you need to determine initial conditions which lead to the limit.

   (a) $y' = -y$

   (b) $y' = y - \frac{1}{4}y^3$

   (c) $y' = \frac{1}{4}y^3 - y$

   (d) $y' = (1-y)(y-2)(y^2)$

   (e) $y' = -y + 3$

   (f) $y' = \frac{y-1}{y^2+1}$

   (g) $y' = \frac{1-y}{y^2+1}$

   (h) $y' = -\tan(y)$. On this one, only consider initial conditions where y is in $[-3/2, 3/2]$ because $\tan(\pi/2)$ is not defined.

2. Consider $y' = ay + b$. Explain why if $a < 0$, then every solution y to the equation appears to have the property that $\lim_{t \to \infty} y(t) = -b/a$.

3. Consider $y' = ay + b$. Explain why if $a > 0$, then the only solution to this equation which has a limit as $t \to \infty$ is the constant solution $y = -b/a$.

4. Below are graphs of slope fields which are labeled **A, B, C, D, E, F**. What follows are some differential equations. Label each with the appropriate slope field.

   (a) $y' = 1 - 2y$

   (b) $y' = (1-y)(y-2)$

   (c) $y' = 1 + y$

   (d) $y' = y(y-1)(y-2)$

   (e) $y' = y(1-y)(y-2)$

   (f) $y' = \sin(y)$

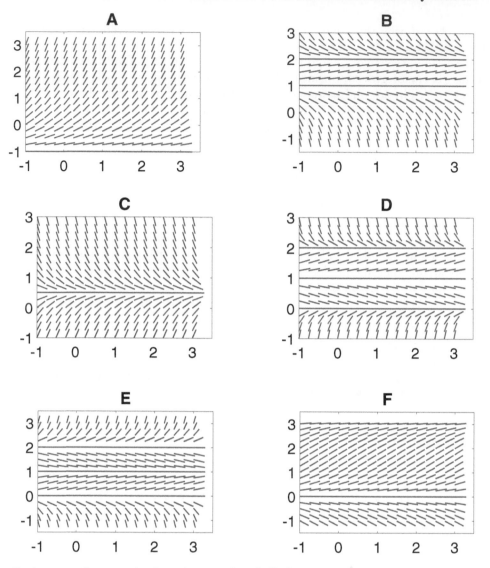

5. Assume that a rain drop is a perfect ball. It evaporates at a rate which is proportional to its surface area. Find a differential equation $dV/dt = f(V)$.

6. Newton's law of cooling states that the temperature T of an object changes in proportion to the difference between the temperature of the object and its surrounding temperature T_0. Express this law as a differential equation.

7. A small spherical object of mass m falling through the atmosphere has air resistance of the form kv^2 where v is the speed of the falling object and the force from gravity is mg. Express as a differential equation. The object will reach a terminal speed. Find it in terms of m, g, k.

8. Chemicals such as DDT degrade at a rate proportional to the amount present. Suppose the constant of proportionality is .5 and you begin with an amount A_0. Also suppose that the chemical is added to a field at the rate of 100

pounds per year. Write a differential equation satisfied by A. After a long time, how much of the chemical will be on the field? Does this answer depend on A_0?

9. The amount of money in a bank grows at the rate of $.05A$ where time is measured in years. At the same time, a person is spending this money at the rate of r dollars per year in a continuous manner. Let the person start with A_0. What must A_0 be if the person is to never run out of money? Show that if he has enough, he will not only not run out, he will become increasingly wealthy but if he has too little, then he will lose it all. You might want to consider graphing a slope field.

10. Consider the equation $y' = y$ with the initial condition $y(0) = 1$. From the equation,
$$y(0) = 1, y'(0) = y(0), y'' = y'$$
so $y''(0) = y'(0) = 1$. Continue this way. Explain why $y^{(n)}(0) = 1$. Now obtain a power series for $y(t)$. Explain why this process shows that there is only one solution to this equation and it is the power series you just obtained. Do you recognize this series from calculus? If not, do the following. Explain each step.
$$y' - y = 0, e^{-t}(y' - y) = 0, \frac{d}{dt}(e^{-t}y) = 0, \ e^{-t}y = C$$
$$y(t) = Ce^t, \ y(t) = e^t.$$

11. Consider $y'' + y = 0, y(0) = 0, y'(0) = 1$. Find a formula for $y^{(k)}(0)$ directly from the differential equation and obtain a power series for the solution to this equation. Do you recognize this series from calculus?

12. Do the same thing for $y'' + y = 0, y(0) = 1, y'(0) = 0$. Do you recognize this series from calculus?

13. You have a cement lined hole so water can only evaporate. If y is the depth of water in the hole, the area is $A(y)$ where $A'(y) > 0, \lim_{y \to \infty} A(y) = \infty, \lim_{y \to 0} A(y) = 0$. Water evaporates at a rate proportional to the surface area exposed. Suppose water is added slowly to this hole at a small but constant rate. Show that eventually, the level of the water must stabilize.

14. The equation $\frac{dy}{dt} = k(p - y)(q - y), \ k > 0, y(0) < p$ occurs in chemical reactions. To be specific, let $p = 1, q = 2$. Have MATLAB produce a graph of the slope field and determine the behavior of solutions to this differential equation. Explain why in the general case, if $p < q$, the reaction will continue until $y = p$.

15. A lake holding V cubic meters, is being polluted at the rate of r kg. per month, the pollutant being dissolved in run off from a factory which flows into the lake at the rate of U cubic meters per month. The pollutant is thoroughly mixed by fish swimming in the lake and a stream flows out of the lake at the rate of U cubic meters per month. Assuming the rate of evaporation is about the same as the rate water enters because of rain, write a differential equation for A the amount of pollutant in the lake.

16. Let T_1 be a closed tank of water containing V_1 cubic meters. Let T_2 also be a closed tank of water containing V_2 cubic meters. Polluted water containing k kg. of pollutant per cubic meter enters T_1 at the rate of U cubic meters per minute where it is stirred thoroughly. Then the water flows at the rate of U cubic meters per minute from T_1 to T_2 where it is also stirred and then out of tank T_2 at the same rate. If x is the amount of pollutant in T_1 and y the amount in T_2, write a pair of differential equations for x and y.

17. The next few problems involve a discrete version of differential equations. If you have an amount A in the bank for a payment period and r is the interest rate per payment period, then you will receive Ar in interest. Show that at the end of one payment period you have $A(1+r)$. Explain why after n payment periods, you have $A(1+r)^n$.

18. For an ordinary annuity, you have a sequence of payments which occur at the end of each payment period. Thus you begin with nothing and at the end of the first payment period, a payment of P is made. Then at the end of the next, another payment is made and so forth till n payment periods. The interest rate per payment period will be r. Determine a formula which will give the amount after n payment periods. **Hint:** Letting A_k be the amount after k periods, explain why

$$A_{k+1} = A_k(1+r) + P, \text{ so } A_{k+2} = A_{k+1}(1+r) + P$$

Subtracting these, you get

$$A_{k+2} = (2+r)A_{k+1} - (1+r)A_k \qquad (*)$$

Look for ρ such that ρ^k solves this. There should be two values. Explain why if ρ^k solves it, then so does $C\rho^k$ for C a constant. Also explain why if $\rho^k, \hat{\rho}^k$ both are solutions, then so is $C\rho^k + D\hat{\rho}^k$. This is called superposition. You have $A_0 = 0, A_1 = P$. These are initial conditions. Now find $\rho, \hat{\rho}, C, D$ to satisfy the difference equation $*$ and these initial conditions.

19. You want to pay off a loan of amount Q with a sequence of equal payments, the first occurring at the end of the first payment period, usually month. The rule is that you pay interest on the unpaid balance which is the amount still owed. If you owe Q_k at the end of the kth payment period, then in that payment period, you pay rQ_k where r is the interest rate per period. This is money thrown away. That which remains after removing rQ_k is what goes toward paying off the loan. Thus

$$Q_{k+1} = Q_k - (P - rQ_k) = Q_k(1+r) - P$$

So what is P if $Q_n = 0$ meaning that you pay it off in n payment periods? Determine this by using the technique of the above problem. First find Q_k and then write $Q_n = 0 = Q_{n-1}(1+r) - P$.

Chapter 3

Methods

3.1 First Order Linear Equations

The homogeneous first order constant coefficient linear differential equation is a differential equation of the form

$$y' + ay = 0. \tag{3.1}$$

It is arguably the most important differential equation in existence. Generalizations of it include the entire subject of linear differential equations and even many of the most important partial differential equations occurring in applications.

Here is how to find the solutions to this equation. Multiply both sides of the equation by e^{at}. Then use the product and chain rules to verify that

$$e^{at} (y' + ay) = \frac{d}{dt} \left(e^{at} y \right) = 0.$$

Therefore, since the derivative of the function $t \to e^{at} y(t)$ equals zero, it follows this function must equal some constant C. Consequently, $y e^{at} = C$ and so $y(t) = Ce^{-at}$. This shows that if there is a solution of the equation, $y' + ay = 0$, then it must be of the form Ce^{-at} for some constant, C. You should verify that every function of the form, $y(t) = Ce^{-at}$ is a solution of the above differential equation, showing that this yields all solutions. This proves the following theorem.

Theorem 3.1.1 *The solutions to the equation, $y' + ay = 0$ for a a real number consist of all functions of the form, Ce^{-at} where C is some constant.*

Example 3.1.2 *Radioactive substances decay in the following way. The rate of decay is proportional to the amount present. In other words, letting $A(t)$ denote the amount of the radioactive substance at time t, $A(t)$ satisfies the following initial value problem.*

$$A'(t) = -k^2 A(t), \ A(0) = A_0$$

where A_0 is the initial amount of the substance. What is the solution to the initial value problem?

Write the differential equation as $A'(t) + k^2 A(t) = 0$. From Theorem 3.1.1 the solution is

$$A(t) = Ce^{-k^2 t}$$

and it only remains to find C. Letting $t = 0$, it follows $A_0 = A(0) = C$. Thus $A(t) = A_0 \exp(-k^2 t)$.

Now here is another problem which is a little harder because it has something extra added in at the end.

Example 3.1.3 *Find solutions to $y' = 2y + 1$.*

Here is how you do it:

1. Write as $y' - 2y = 1$

2. Find an "**Integrating Factor**" $\int (-2) \, dt = -2t$. Note that I didn't bother to add in the arbitrary constant. This is because it does not matter. You don't care about finding all integrating factors. You just need one. Then an integrating factor is e^{-2t}.

3. Multiply both sides of the equation by the integrating factor.

$$e^{-2t}(y' - 2y) = \frac{d}{dt}\left(e^{-2t}y(t)\right) = e^{-2t}(1)$$

Note that the first equal sign follows from the product rule and the chain rule. **This is why we multiply by the integrating factor, to get the derivative of something equal to something known.**

4. Take antiderivatives of both sides.

$$e^{-2t}y(t) = \int e^{-2t}dt = -\frac{1}{2}e^{-2t} + C$$

Thus

$$y(t) = -\frac{1}{2} + Ce^{2t}$$

This time you need to be sure to keep the constant of integration because it **does** matter.

Note that by varying C you get different solutions to the differential equation. Now here are graphs of a few of these solutions along with the slope field.

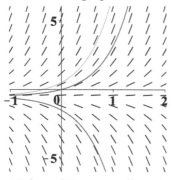

Note how the solutions follow the slope field. How do you determine the "right value" of C? This involves an

$$\boxed{\textbf{INITIAL CONDITION}}.$$

An initial condition involves specifying a particular point which is to lie on the graph of the solution to the differential equation. Then you can see from the picture that, having made this specification, the rest of the graph should be determined by the need to follow the slope field. When you have specified the initial condition as well as the differential equation, the problem is called an **initial value problem.**

Example 3.1.4 *Find the solution to the initial value problem*

$$y' = 2y + 1, \ y(1) = 2$$

From the above example, all solutions are of the form $y = -\frac{1}{2} + Ce^{2t}$. It is now just a matter of finding the value of C which will cause the given point $(1, 2)$, expressed by saying that $y(1) = 2$, to lie on the graph of y. Thus you need to have $2 = -\frac{1}{2} + Ce^2$. Then you just need to solve this equation for C. This yields $C = \frac{5}{2e^2}$. Therefore, $\boxed{\text{the}}$ solution to the initial value problem is

$$y = -\frac{1}{2} + \frac{5}{2e^2}e^{2t}$$

Note the use of the definite article. There is **only one** solution to this initial value problem although there are infinitely many solutions to the differential equation, three of which were graphed above. This uniqueness property will be discussed more later, but for now, you can see roughly why this is. It comes from the need for the solution to follow the slope field, so if you specify a point on the curve, you have essentially determined it.

Example 3.1.5 *Find the solution to the initial value problem*

$$y' + 2ty = \sin(t)e^{-t^2}, \ y(0) = 3.$$

1. Find the integrating factor. $\int 2t \, dt = t^2$. Integrating factor: $\exp(t^2) = e^{t^2}$.

2. Multiply both sides by the integrating factor.

$$\exp(t^2)(y' + 2ty) = \frac{d}{dt}(\exp(t^2)y) = \sin(t)$$

3. Take $\int$ of both sides.

$$\exp(t^2)y(t) = -\cos(t) + C$$

4. Solve for y

$$y = \exp(-t^2)(C - \cos(t))$$

5. Find C to satisfy the initial condition.

$$3 = C - 1, \ C = 4.$$

6. Place value of C you just found in the formula for y

$$y = \exp(-t^2)(4 - \cos(t))$$

Now at this point, you should check and see if it works. It needs to solve both the initial condition and the differential equation.

Example 3.1.6 *Find the solutions to*

$$y' + a(t)y = b(t).$$

1. Find integrating factor. $A(t) + C \equiv \int a(t)$. Integrating factor: $\exp(A(t))$

2. Multiply by $\exp(A(t))$

$$\overbrace{\exp(A(t))(y' + a(t)y) = \frac{d}{dt}(\exp(A(t))y)}^{\text{chain rule and product rule}} = \exp(A(t))b(t)$$

3. Do $\int$ to both sides. Pick $F(t) \in \int \exp(A(t))b(t)\,dt$.

$$\exp(A(t))y(t) = F(t) + C$$

$$y(t) = \exp(-A(t))F(t) + C\exp(-A(t))$$

This proves the following theorem.

Theorem 3.1.7 *The solutions to the equation, $y' + a(t)y = b(t)$ consist of all functions of the form $y(t) = e^{-A(t)}F(t) + e^{-A(t)}C$ where $F(t) \in \int e^{A(t)}b(t)\,dt$ and C is a constant, $A'(t) = a(t)$.*

Finally, here is a uniqueness theorem.

Theorem 3.1.8 *If $a(t)$ is a continuous function, there is at most one solution to the initial value problem, $y' + a(t)y = b(t)$, $y(r) = y_0$.*

Proof: If there were two solutions y_1 and y_2, then letting $w = y_1 - y_2$, it follows $w' + a(t)w = 0$ and $w(r) = 0$. Then multiplying both sides of the differential equation by $e^{A(t)}$ where $A'(t) = a(t)$, it follows $\left(e^{A(t)}w\right)' = 0$ and so $e^{A(t)}w(t) = C$ for some constant, C. However, $w(r) = 0$ and so this constant can only be 0. Hence $w = 0$ and so $y_1 = y_2$. ∎

Finally, consider the general linear initial value problem.

Definition 3.1.9 *A linear differential equation is one which is of the form*

$$y' + a(t)y = b(t)$$

where a, b are continuous. The corresponding initial value problem is

$$y' + a(t)y = b(t), \quad y(t_0) = y_0.$$

Now here are the steps for solving the initial value problem.

1. Find the integrating factor $\int a(t)\,dt \equiv A(t) + C$. The integrating factor is $\exp(A(t)) = e^{A(t)}$.

2. Multiply both sides by the integrating factor.

$$\exp(A(t))(y'(t) + a(t)y(t)) = \frac{d}{dt}(\exp(A(t))y(t)) = \exp(A(t))b(t)$$

Why is this so? It involves the chain rule and the product rule.

$$\begin{aligned}
\frac{d}{dt}(\exp(A(t))y(t)) &= \exp(A(t))A'(t)y(t) + \exp(A(t))y'(t) \\
&= \exp(A(t))a(t)y(t) + \exp(A(t))y'(t) \\
&= \exp(A(t))(y'(t) + a(t)y(t))
\end{aligned}$$

3. Next do $\int_{t_0}^t$ to both sides.

$$\int_{t_0}^t \frac{d}{ds} \left(\exp\left(A\left(s\right)\right) y\left(s\right) \right) ds = \int_{t_0}^t \exp\left(A\left(s\right)\right) b\left(s\right) ds$$

Then by the fundamental theorem of calculus,

$$\exp\left(A\left(t\right)\right) y\left(t\right) - \exp\left(A\left(t_0\right)\right) y\left(t_0\right) = \int_{t_0}^t \exp\left(A\left(s\right)\right) b\left(s\right) ds$$

and so, you can solve for $y\left(t\right)$ and get

$$
\begin{aligned}
y\left(t\right) &= \exp\left(-A\left(t\right)\right) \exp\left(A\left(t_0\right)\right) y\left(t_0\right) + \exp\left(-A\left(t\right)\right) \int_{t_0}^t \exp\left(A\left(s\right)\right) b\left(s\right) ds \\
&= \exp\left(A\left(t_0\right) - A\left(t\right)\right) y_0 + \int_{t_0}^t \exp\left(A\left(s\right) - A\left(t\right)\right) b\left(s\right) ds
\end{aligned}
$$

This shows that if the linear initial value problem has a solution, then it must be of the above form. Hence there is at most one solution to the initial value problem. Does the above formula actually give a solution to the initial value problem? Let $y\left(t\right)$ be given by that formula. Then

$$y\left(t_0\right) = \exp\left(0\right) y_0 + \int_{t_0}^{t_0} \exp\left(A\left(s\right) - A\left(t\right)\right) b\left(s\right) ds = y_0$$

so the initial condition holds. Does it solve the differential equation? By the chain rule and the fundamental theorem of calculus,

$$
\begin{aligned}
y'\left(t\right) &= \left(-A'\left(t\right)\right) \exp\left(A\left(t_0\right) - A\left(t\right)\right) y_0 + \exp\left(-A\left(t\right)\right) \exp\left(A\left(t\right)\right) b\left(t\right) \\
&\quad + \left(-A'\left(t\right)\right) \exp\left(-A\left(t\right)\right) \int_{t_0}^t \exp\left(A\left(s\right)\right) b\left(s\right) ds \\[2mm]
&= \left(-a\left(t\right)\right) \exp\left(A\left(t_0\right) - A\left(t\right)\right) y_0 + \exp\left(-A\left(t\right)\right) \exp\left(A\left(t\right)\right) b\left(t\right) \\
&\quad + \left(-a\left(t\right)\right) \exp\left(-A\left(t\right)\right) \int_{t_0}^t \exp\left(A\left(s\right)\right) b\left(s\right) ds = -a\left(t\right) y\left(t\right) + b\left(t\right)
\end{aligned}
$$

so it also is a solution of the linear initial value problem.

Example 3.1.10 *This example illustrates a different notation for differential equations. Find the solutions to*

$$x dy + \left(2xy - x \sin x\right) dx = 0$$

The idea is you divide by dx and so the exact meaning is

$$xy' + 2xy = x \sin\left(x\right)$$

Then

$$y' + 2y = \sin x, \quad \left(e^{2x} y\right)' = e^{2x} \sin x$$

$$e^{2x} y = \int e^{2x} \sin(x) \, dx = \frac{1}{5} e^{2x} (2 \sin x - \cos x) + C$$

$$y = \frac{1}{5} (2 \sin x - \cos x) + C e^{-2x}$$

The reason for writing it this way is that sometimes you want to find x as a function of y and this notation is neutral in terms of which variable is the independent variable.

Example 3.1.11 *A radioactive substance decays in such a way that the rate of change of the amount of the substance is a constant multiple of the amount present, the constant being negative. Thus $\frac{dA}{dt} = -kA$. There is a certain sample of decaying material. Measurements are taken after 5 years and it is found that there is about $9/10$ of the original amount present. Find the half life of this material. The half life is the amount of time it takes for half of it to have decayed.*

From the equation, $A = A_0 e^{-kt}$. Then $\frac{9}{10} A_0 = A_0 e^{-k(5)}$. Solving this for k yields $\frac{-\ln(.9)}{5} = k$ and so the amount of time to have half of what was started with is T given as a solution to the following equation.

$$e^{-\left(\frac{-\ln(.9)}{5}\right)(T)} = \frac{1}{2}, \text{ so } T = \frac{\ln(.5)}{\ln(.9)/5} = 32.894$$

This kind of thing is associated not just with radioactive material but with other chemicals as well. They degrade over time according to such an equation.

Example 3.1.12 *The ancient Babylonians were fascinated with the idea of compound interest. They were interested in how long it would take an initial amount to double. One can understand compound interest compounded continuously using the same kind of differential equation as the above only this time the constant is positive and is the interest rate. Thus*

$$\frac{dA}{dt} = kA$$

If the interest rate is 20% per year compounded continuously, how long will it take for an initial amount to double in size?

From the equation, $A = A_0 e^{.2t}$ where A_0 is the initial amount. Then you want to find T such that $2A_0 = A_0 e^{.2T}$ and so

$$T = \frac{\ln 2}{.2} = 5.0 \ln 2 = 3.4657$$

If the rate is r per year and you have n years and the interest is compounded at the end of each year rather than continuously, then the amount is given by the formula $(1 + r)^n = A(n)$. Anciently, they used this kind of thing because they did not have differential equations. If the interest rate is 20% compounded monthly, then the amount after n years is $A_0 \left(1 + \frac{.2}{12}\right)^{12n}$ where A_0 is the initial amount. If $n = 3.5$, a use of a calculator shows that

$$\left(1 + \frac{.2}{12}\right)^{12(3.5)} = 2.0022$$

which is very similar to compounding the interest continuously. The rational for this formula is that if it is compounded monthly, then the interest rate per month is .2/12. Each successive month is called a payment period.

Example 3.1.13 *A lake contains one million gallons of water. A gas tank starts to leak upstream and contaminated water mixed with gasoline starts flowing into the lake at the rate of 1000 gallons per month. This is mixed well due to large numbers of fish in the lake and water flows out at the same rate. The amount of gasoline in the contaminated water varies due to the demand for gas at the gas station and the concentration of gasoline in the contaminated water is $(1 + \sin(t))$ grams per gallon. Find a formula for the concentration of gasoline in the lake in grams per gallon as a function of time in months after a long time.*

Let A be the amount of gas in the lake. Then

$$\frac{dA}{dt} = (1 + \sin(t)) \times 1000 - \frac{A}{10^6} 1000 = 1000 (1 + \sin(t)) - \frac{1}{10^3} A$$

Rather than worry with the stupid numbers, write this as

$$A' + aA = b(1 + \sin(t)), \quad A(0) = 0$$

Following the procedure for finding solutions to a linear equation,

$$\left(e^{at} A\right)' = b(1 + \sin(t)) e^{at}$$

Now it follows that, taking antiderivatives of both sides,

$$e^{at} A = b \left(\frac{e^{at}}{a^3 + a} \left(a^2 \sin t - a \cos t + a^2 + 1\right) \right) + C$$

Since $A(0) = 0$, it follows that

$$0 = b \left(\frac{1}{a^3 + a} \left(-a + a^2 + 1\right) \right) + C$$

and so $C = -\frac{b}{a^3 + a} \left(a^2 - a + 1\right)$. Therefore,

$$A = b e^{-at} \left(\frac{e^{at}}{a^3 + a} \left(a^2 \sin t - a \cos t + a^2 + 1\right) \right) + \frac{\left(a - a^2 - 1\right) b e^{-at}}{a^3 + a}$$

Now placing in the formula the values of a and b and then simplifying the result it follows that A equals

$$10^6 e^{-0.001 t} \left(1.0 e^{0.001 t} - 0.001 e^{0.001 t} \cos t + 1.0 \times 10^{-6} e^{0.001 t} \sin t - 0.999\right)$$

Then, dividing by the number of gallons in the lake, this yields for the number of grams per gallon

$$e^{-0.001 t} \left(1.0 e^{0.001 t} - 0.001 e^{0.001 t} \cos t + 1.0 \times 10^{-6} e^{0.001 t} \sin t - 0.999\right)$$

After a long time, the terms having the negative exponential will disappear in the limit and this yields for the number of grams per gallon the formula

$$1 - 0.001 \cos t + 1.0 \times 10^{-6} \sin t$$

Note that this yields approximately 1 gram per gallon. Compare to the concentration of the incoming water. The concentration of the incoming water oscillates about 1 and so does the concentration of gas in the lake, although the oscillations are much much smaller. This is due to the large number of gallons in the lake. You might have expected this but you could not have predicted exact values without the differential equation.

Example 3.1.14 *A pumpkin is launched* $30°$ *from the horizontal at a speed of 60 feet per second. It is acted on by the force of gravity which delivers an acceleration which is 32 feet per second squared and an acceleration due to air resistance which we assume is .2 times the speed which acts in the opposite direction to the direction of motion. Describe the position of the pumpkin as a function of time.*

Let the initial position be at $(0,0)$ and let the coordinates of the point be $(x(t), y(t))$. What is the initial velocity? It is $\left(30\sqrt{3}, 30\right)$. Then the acceleration is given by

$$(x''(t), y''(t)) = -32(0,1) - .2(x'(t), y'(t))$$

Thus $y'' + .2y' = -32$. Let's solve for y'.

$$\left(e^{.2t}y'\right)' = (-32)e^{.2t}$$

$$e^{.2t}y'(t) = \frac{-32}{.2}e^{.2t} + C, \text{ so } y'(t) = -160 + Ce^{-.2t}$$

So what is C? When $t = 0$, we get $C - 160 = 30$ and so $C = 190$. Hence

$$y'(t) = 190e^{-.2t} - 160$$

$$y(t) = -160t - 950e^{-0.2t} + D$$

What is D? When $t = 0$ we want $y(0) = 0$ and so $D = 950$. Thus

$$y(t) = -160t - 950e^{-0.2t} + 950$$

As to x,

$$x'' + .2x' = 0 \text{ so } \left(x'e^{.2t}\right)' = 0$$

and so $x'(t) = Ce^{-.2t}$. To satisfy the initial condition, $x'(t) = 30\sqrt{3}e^{-.2t}$. Then

$$x(t) = \frac{30\sqrt{3}}{-(1/5)}e^{-.2t} + D$$

What is D? to satisfy the initial condition for the position, $D = 150\sqrt{3}$ and so

$$x(t) = -150\sqrt{3}e^{-.2t} + 150\sqrt{3}$$

The position of the pumpkin is

$$(x(t), y(t)) = \left(-150\sqrt{3}e^{-.2t} + 150\sqrt{3}, -160t - 950e^{-0.2t} + 950\right)$$

The following is a summary of the above discussion.

Procedure 3.1.15 *To solve the first order linear differential equation*

$$y' + a(t) y = f(t),$$

do the following:

1. *Find $A(t) \in \int a(t) \, dt$. That is, find $A(t)$ such that $A'(t) = a(t)$.*

2. *Multiply both sides by the integrating factor $e^{A(t)}$.*

3. *The above step yields*

$$\left(e^{A(t)} y(t)\right)' = e^{A(t)} f(t)$$

4. *Do $\int dt$ to both sides. Then choose the arbitrary constant to satisfy a given initial condition.*

3.2 Bernouli Equations

Some kinds of nonlinear equations can be changed to get a linear equation. An equation of the form

$$y' + a(t) y = b(t) y^{\alpha}$$

is called a Bernouli equation*. The trick is to define a new variable, $z = y^{1-\alpha}$. Then $y^{\alpha} z = y$ and so $z' = (1 - \alpha) y^{-\alpha} y'$ which implies $\frac{1}{(1-\alpha)} y^{\alpha} z' = y'$. Then

$$\frac{1}{(1-\alpha)} y^{\alpha} z' + a(t) y^{\alpha} z = b(t) y^{\alpha}$$

and so

$$z' + (1 - \alpha) a(t) z = (1 - \alpha) b(t).$$

Now this is a linear equation for z. Solve it and then use the transformation to find y.

Example 3.2.1 *Solve $y' + y = ty^3$.*

You let $z = y^{-2}$ and make the above substitution. Thus $zy^3 = y$ and

$$z' = (-2) y^{-3} y', \quad y' = -\frac{1}{2} y^3 z'$$

and so $-\frac{1}{2} y^3 z' + y^3 z = ty^3$. Hence, cancelling the y^3, $z' - 2z = (-2) t$. Then

$$\frac{d}{dt} \left(e^{-2t} z\right) = -2te^{-2t}$$

and so

$$e^{-2t} z = te^{-2t} + \frac{1}{2} e^{-2t} + C$$

---

*This is named after Jacob Bernoulli (1654-1705), one of a whole family of Swiss mathematicians. Others were Johann I and II Daniel, and Nicolaus.

and so
$$y^{-2} = z = t + \frac{1}{2} + Ce^{2t}$$

and so
$$y^2 = \frac{1}{t + \frac{1}{2} + Ce^{2t}}.$$

When you get this far, it is a good idea to check and see if it works. After all, this is the point of the manipulations, to get the answer. If you get the answer, then if there is a mistake, it is no longer terribly relevant.

$$2yy' = \frac{d}{dt}\left(\frac{1}{t + \frac{1}{2} + Ce^{2t}}\right) = -\frac{8Ce^{2t} + 4}{(2t + 2Ce^{2t} + 1)^2}$$

$$y' = -\frac{8Ce^{2t} + 4}{2y\left(2t + 2Ce^{2t} + 1\right)^2}$$

Then

$$
\begin{aligned}
y' + y &= -\frac{8Ce^{2t} + 4}{2y\left(2t + 2Ce^{2t} + 1\right)^2} + y \\
&= -\frac{8Ce^{2t} + 4}{2y\left(2t + 2Ce^{2t} + 1\right)^2} + \frac{2y^2\left(2t + 2Ce^{2t} + 1\right)^2}{2y\left(2t + 2Ce^{2t} + 1\right)^2} \\
&= -\frac{8Ce^{2t} + 4}{2y\left(2t + 2Ce^{2t} + 1\right)^2} + \frac{2\left(\frac{1}{t + \frac{1}{2} + Ce^{2t}}\right)\left(2t + 2Ce^{2t} + 1\right)^2}{2y\left(2t + 2Ce^{2t} + 1\right)^2} \\
&= 4\frac{t}{y\left(2t + 2Ce^{2t} + 1\right)^2} = \frac{t}{y\left(t + Ce^{2t} + \frac{1}{2}\right)^2} = \frac{t}{y}y^4 = ty^3
\end{aligned}
$$

so it appears to work.

The following procedure gives a summary of the above.

Procedure 3.2.2 *To solve the Bernouli equation*

$$y' + a(t)\,y = b(t)\,y^\alpha, \ \alpha \neq 1$$

do the following:

1. *Change the variable. Let $z = y^{1-\alpha}$. Then $z' = (1-\alpha)\,y^{-\alpha}y', y^\alpha z = y$.*

2. *Place in the equation.*

$$\frac{1}{1-\alpha}y^\alpha z' + a(t)\,y^\alpha z = b(t)\,y^\alpha$$

3. *Cancel the y^α and solve the linear equation for z.*

3.3 Separable Differential Equations, Stability

Separable differential equations also occur quite often in applications and they are fairly easy to deal with. This section gives a discussion of these equations.

Definition 3.3.1 *Separable differential equations are those which can be written in the form*

$$\frac{dy}{dx} = \frac{f(x)}{g(y)}.$$

The reason these are called separable is that if you formally cross multiply,

$$g(y)\, dy = f(x)\, dx$$

and the variables are "separated". The x variables are on one side and the y variables are on the other.

Proposition 3.3.2 *If $G'(y) = g(y)$ and $F'(x) = f(x)$, then if the equation, $F(x) - G(y) = c$ specifies y as a differentiable function of x, then $x \to y(x)$ solves the separable differential equation*

$$\frac{dy}{dx} = \frac{f(x)}{g(y)}. \tag{3.2}$$

Proof: Differentiate both sides of $F(x) - G(y) = c$ with respect to x. Using the chain rule,

$$F'(x) - G'(y)\frac{dy}{dx} = 0.$$

Therefore, since $F'(x) = f(x)$ and $G'(y) = g(y)$, $f(x) = g(y)\frac{dy}{dx}$ which is equivalent to (3.2). ∎

Definition 3.3.3 *The curves $F(x) - G(y) = c$ for various values of c are called **integral curves or solution curves**. It makes sense to think of these as giving a solution if, near a point on the level curve, one variable is a function of the other.*

Example 3.3.4 *Find the solution to the initial value problem,*

$$y^2 y' = x, \ y(0) = 1.$$

This is a separable equation and in fact, $y^2 dy = x dx$ so the solution to the differential equation is of the form $\frac{y^3}{3} - \frac{x^2}{2} = C$ and it only remains to find the constant C. To do this, you use the initial condition. Letting $x = 0$, it follows $\frac{1}{3} = C$ and so

$$\frac{y^3}{3} - \frac{x^2}{2} = \frac{1}{3}$$

The following picture shows how the integral curves follow the tangent field.

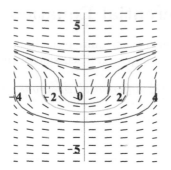

Sometimes, you can't expect to solve for one of the variables in terms of the other. In other words, the integral curve might not be a function of one variable. Here is a nice example from [5].

Example 3.3.5 *Find integral curves for the equation*

$$y' = \frac{x^2}{(1 - y^2)}$$

Separating variables, you get $(1 - y^2)\, dy = x^2 dx$ and so the integral curves are of the form

$$\left(y - \frac{y^3}{3}\right) - \frac{x^3}{3} = C$$

Here is a picture of a few of these integral curves along with the slope field.

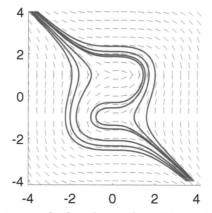

I used MATLAB to graph the above. One thing might be helpful to mention about MATLAB. It is very good at manipulating matrices and vectors and there is distinctive notation used to accomplish this. For example say you type

$$x=[1,2,3]; \; y=[2,3,4]; \; x.*y$$

and then press "enter". You will get 2,6,12. Of course you would get an error if you wrote x*y. Similarly, type

$$[2,4,6,8]./[1,2,3,4]$$

and press "enter". This yields $2, 2, 2, 2$. Of course [2,4,6,8]/[1,2,3,4] doesn't make any sense.

You can get graphs of some integral curves in MATLAB by typing the following and then "enter". You don't have to type it on two lines, but if you want to do so, to get to a new line, you press "shift" and "enter".

```
>> [x,y]=meshgrid(-4:.1:4,-4:.1:4);
z=y-(y.^3/3+x.^3/3);contour(x,y,z,[-.5,-1,-.3,1,2])
```

Example 3.3.6 *What is the equation of a hanging chain?*

Consider the following picture of a portion of this chain.

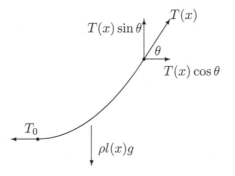

In this picture, ρ denotes the density of the chain which is assumed to be constant and g is the acceleration due to gravity. $T(x)$ and T_0 represent the magnitude of the tension in the chain at t and at 0 respectively, as shown. Let the bottom of the chain be at the origin as shown. If this chain does not move, then all these forces acting on it must balance. In particular,

$$T(x)\sin\theta = l(x)\rho g, \ T(x)\cos\theta = T_0.$$

Therefore, dividing these yields

$$\frac{\sin\theta}{\cos\theta} = l(x)\overbrace{\rho g/T_0}^{\equiv c}.$$

Now letting $y(x)$ denote the y coordinate of the hanging chain corresponding to x,

$$\frac{\sin\theta}{\cos\theta} = \tan\theta = y'(x).$$

Therefore, this yields

$$y'(x) = cl(x).$$

Now differentiating both sides of the differential equation,

$$y''(x) = cl'(x) = c\sqrt{1+y'(x)^2}$$

and so

$$\frac{y''(x)}{\sqrt{1+y'(x)^2}} = c.$$

Let $z(x) = y'(x)$ so the above differential equation becomes

$$\frac{z'(x)}{\sqrt{1+z^2}} = c.$$

Therefore, $\int \frac{z'(x)}{\sqrt{1+z^2}}\,dx = cx + d$. Change the variable in the antiderivative letting $u = z(x)$ and this yields

$$\int \frac{z'(x)}{\sqrt{1+z^2}}\,dx = \int \frac{du}{\sqrt{1+u^2}} = \sinh^{-1}(u) + C = \sinh^{-1}(z(x)) + C.$$

Therefore, combining the constants of integration,

$$\sinh^{-1}(y'(x)) = cx + d$$

and so

$$y'(x) = \sinh(cx + d).$$

Therefore,

$$y(x) = \frac{1}{c}\cosh(cx + d) + k$$

where d and k are some constants and $c = \rho g/T_0$. Curves of this sort are called catenaries. Note these curves result from an assumption that the only forces acting on the chain are as shown.

The next example has to do with population models. It was mentioned earlier. The idea is that if there were infinite resources, population growth would satisfy the differential equation

$$\frac{dy}{dt} = ky$$

where k is a constant. However, resources are not infinite and so k should be modified to be consistent with this. Instead of k, one writes $r\left(1 - \frac{y}{K}\right)$ which will cause the population growth to decrease as soon as y exceeds K. Of course the problem with this is that we are not sure whether K itself is dependent on other factors not included in the model.

Example 3.3.7 *The equation*

$$\frac{dy}{dt} = r\left(1 - \frac{y}{K}\right)y, \ r, K > 0$$

is called the logistic equation. It models population growth. You see that the right side is equal to 0 at the two values $y = K$ and $y = 0$.

This is a separable equation. Thus

$$\frac{dy}{\left(1 - \frac{y}{K}\right)y} = rdt$$

Now you do $\int$ to both sides. This requires partial fractions on the left.

$$\frac{1}{\left(1 - \frac{y}{K}\right)y} = \frac{1}{K - y} + \frac{1}{y}$$

Therefore,

$$\ln(y) - \ln(K - y) = rt + C$$

if $0 < y < K$. If $y > K$, you get

$$\ln(y) - \ln(y - K) = rt + C$$

Therefore, the integral curves are of the form

$$\ln\left(\frac{y}{K - y}\right) = rt + C$$

so changing the name of the constant C, it follows that for $y < K$, the integral curves are described by the following function.

$$y = K\frac{Ce^{rt}}{Ce^{rt} + 1}, \quad C > 0$$

In case $y > K$, these curves are described by

$$y = K\frac{Ce^{rt}}{Ce^{rt} - 1}, \quad C > 0$$

What follows is a picture of the slope field along with some of these integral curves in case $r = 1$ and $K = 10$.

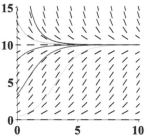

The bottom axis is the t axis. Note how all the integral curves in the picture approach K as t increases. This is why K is called a **stable equilibrium point.**

Definition 3.3.8 *Consider the equation $\frac{dy}{dt} = f(y)$. Then y_0 is called an equilibrium point if $f(y_0) = 0$. Note that the solution to the initial value problem $y' = f(y), y(t_0) = y_0$ is just $y = y_0$. An equilibrium point is stable if whenever y_1 is close enough to y_0, it follows that the solution to the initial value problem $y' = f(y)$, $y(0) = y_1$ stays close to y_0 for all $t > 0$. It is asymptotically stable if whenever y_1 is close enough to y_0, it follows that for y the solution to the initial value problem, $y' = f(y)$, $y(0) = y_1$ satisfies $\lim_{t\to\infty} y(t) = y_0$. The equilibrium point y_0 is unstable if there are initial conditions close to y_0 but the solution does not stay close to y_0. That is, there exists $\varepsilon > 0$ such that for any $\delta > 0$ there is y_1 with $|y_1 - y_0| < \delta$ but the solution to $y' = f(y), y(0) = y_1$ has the property that for some $t > 0, |y(t) - y_0| \geq \varepsilon$. An equilibrium point y_0 is semi-stable if it is stable from one side and unstable from the other.*

Now observe that $y = 0$ is the solution which results if you begin with the initial condition $y(0) = 0$. If there is nothing to start with, it can't grow. However, if you have any other positive number for $y(0)$, then you see that the solution curve approaches the stable point K. You can see this, not just by looking at the picture but also by taking the limit as $t \to \infty$ in the above formulae.

One of the interesting things about this equation is that it is possible to determine K the maximum capacity, by taking measurements at three equally spaced times. Suppose you do so at times $t, 2t, 3t$ and obtain y_1, y_2, y_3 respectively. Assume you are in the region where $y < K$. In an actual experiment, this is where you would be. Let $\lambda \equiv e^{rt}$. Then from the above formula for y, you have the equations

$$KC\lambda = y_1 \left(C\lambda + 1\right), KC\lambda^2 = y_2 \left(C\lambda^2 + 1\right), KC\lambda^3 = y_3 \left(C\lambda^3 + 1\right)$$

Then divide the second equation by λ and compare with the first. This shows that $\lambda = y_2/y_1$. Next divide the top equation by $C\lambda$ and the last by $C\lambda^3$. This yields

$$K = y_1 \left(1 + \frac{1}{C\lambda}\right) = y_3 \left(1 + \frac{1}{C\lambda^3}\right)$$

Now it becomes possible to solve for C. This yields

$$C = \frac{\left(y_1^3 y_3 - y_1^2 y_2^2\right)}{y_1 y_2^3 - y_2^3 y_3}$$

Then substitute this in to the first equation. This obtains

$$K \left(\frac{\left(y_1^3 y_3 - y_1^2 y_2^2\right)}{y_1 y_2^3 - y_2^3 y_3}\right) \frac{y_2}{y_1} = y_1 \left(\frac{\left(y_1^3 y_3 - y_1^2 y_2^2\right)}{y_1 y_2^3 - y_2^3 y_3} \left(\frac{y_2}{y_1}\right) + 1\right)$$

Then you can solve this for K. After some simplification, it yields

$$\frac{y_2^2 y_3 - y_1^2 y_3}{y_2^2 - y_1 y_3} = K$$

Note how the equilibrium point K was stable in the above example. There were only two equilibrium points, K and 0. The equilibrium point 0 was unstable because if the integral curve started near 0 but slightly positive, it tended to increase to K. Here is another harder example. In this example, there are three equilibrium points.

Example 3.3.9 $\frac{dy}{dt} = -r \left(1 - \frac{y}{T}\right) \left(1 - \frac{y}{K}\right) y, \ r > 0, 0 < T < K.$

This is a separable equation.

$$\frac{dy}{\left(1 - \frac{y}{T}\right) \left(1 - \frac{y}{K}\right) y} = -rdt$$

The partial fractions expansion is

$$\frac{1}{\left(1 - \frac{y}{T}\right) \left(1 - \frac{y}{K}\right) y} = \frac{1}{K - T} \left(\frac{K}{T - y} - \frac{T}{K - y}\right) + \frac{1}{y}$$

Therefore,

$$\frac{-1}{K - T} K \ln|T - y| + \frac{1}{K - T} T \ln|K - y| + \ln|y| = -rt + C$$

Consider the case where $r = 1, T = 5, K = 10$. Then you get

$$\ln \left(\left|\frac{(10 - y) y}{(5 - y)^2}\right|\right) = -t + C$$

There are cases, depending on where y is. Suppose first that $y \in (0, 10)$. Then you get for a different C

$$\frac{(10 - y)\, y}{(5 - y)^2} = Ce^{-t}, \ C > 0$$

You could solve this for y if you like and get

$$y = \frac{1}{Ce^{-t} + 1}\left(-5\sqrt{Ce^{-t} + 1} + 5Ce^{-t} + 5\right)$$

or

$$y = \frac{1}{Ce^{-t} + 1}\left(5\sqrt{Ce^{-t} + 1} + 5Ce^{-t} + 5\right)$$

On the other hand, if $y > 10$, you get

$$\frac{(y - 10)\, y}{(5 - y)^2} = Ce^{-t}$$

The following is a picture of some integral curves and the slope field. You see how the equilibrium points 10 and 0 are both stable but the equilibrium point 5 is not.

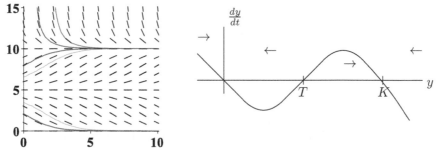

Is there a systematic way to figure this out without doing lots of computer generated pictures? The answer is yes! Furthermore, it is very easy to do. Consider the right side of the equation. If you graph the function $z = f(y)$, you get something which looks like the right side of the above.

Look at the graph. When $y \in (0, T)$, you have the slope of the graph is negative and so, from the equation, $\frac{dy}{dt}$ is negative and so $t \to y(t)$ is decreasing. (Remember calculus.) If $y \in (T, K)$, then the graph is positive and so $\frac{dy}{dt}$ is positive which requires that $t \to y(t)$ is increasing. When $y \in (K, \infty)$, the graph is negative and so $t \to y(t)$ is decreasing. Thus T is unstable, K is stable while 0 is also stable. I have not considered the case where $y < 0$ because this is not too interesting in the example which typically describes y as a population of something. However, you can see from the graph that if $y < 0$, then $t \to y(t)$ is increasing.

In general, you can consider $y' = f(y)$ and the equilibrium points. The following picture is descriptive of the situation. Such an equation is called **autonomous** because the function on the right depends only on y and not on t.

Pieces of the graph of f

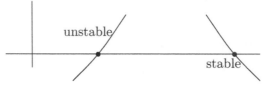

Proposition 3.3.10 *Suppose f is continuous with continuous derivative and that $f(y_0) = 0, f'(y_0) < 0$. Then y_0 is asymptotically stable.*

Proof: By continuity of f', there is $\delta > 0$ such that for $y \in (y_0 - \delta, y_0 + \delta) = I, f'(y) \leq -2\eta, \eta > 0$. Thus

$$
\begin{aligned}
f(y) &= f(y_0) + f'(y_0)(y - y_0) + o(y - y_0) \\
&= f'(y_0)(y - y_0) + o(y - y_0)
\end{aligned}
$$

Then if $y_1 \in I$, and if $y(t)$ is the solution to the equation $y' = f(y)$ having this initial condition, then

$$
\begin{aligned}
y(t) - y_0 &= y_1 - y_0 + \int_0^t f(y(s))\, ds \\
&= y_1 - y_0 + \int_0^t f'(y_0)(y(s) - y_0)\, ds + \int_0^t o(y(s) - y_0)\, ds
\end{aligned}
$$

We can also assume δ is small enough that $|o(y - y_0)| < \eta |y - y_0|$ for $y \in I$. Say $y_1 > y_0$. Then by assumption, $t \to y(t)$ is decreasing since $y' = f(y) < 0$ and so if $y(t)$ fails to converge to y_0, there would exist $\varepsilon > 0$ which is the limit of $y(t) - y_0$. Then

$$
\varepsilon + \int_0^t -f'(y_0)(y(s) - y_0)\, ds \leq y_1 - y_0 + \eta \int_0^t (y(s) - y_0)
$$

Thus

$$
\varepsilon + \int_0^t \eta\varepsilon \leq y_1 - y_0
$$

which is impossible because as $t \to \infty$, the left side is unbounded. Similar reasoning shows asymptotic stability if $y_1 < y_0$. ∎

Procedure 3.3.11 *To solve a separable equation which can be placed in the form*

$$
f(y)\, dy + g(x)\, dx = 0
$$

do the following:

1. *Place an $\int$ in front each term on the left and do what it seems to say.*

2. *You get $F(y) + G(x) = C$. This is the general solution. Now choose C to satisfy any initial condition which may be given.*

3. *An equilibrium point for $y' = f(y)$ is a point y_0 where $f(y_0) = 0$. It is an asymptotically stable equilibrium if $f'(y_0) < 0$ and unstable if $f'(y_0) > 0$.*

3.4 Homogeneous Equations

Sometimes equations can be made separable by changing the variables appropriately. This occurs in the case of the so called homogeneous equations, those of the form

$$y' = f\left(\frac{y}{x}\right).$$

When this sort of equation occurs, there is an easy trick which will allow you to consider a separable equation.

You define a new variable,

$$u \equiv \frac{y}{x}.$$

Thus $y = ux$ and so

$$y' = u'x + u = f(u).$$

Thus

$$\frac{du}{dx}x = f(u) - u$$

and so

$$\frac{du}{f(u) - u} = \frac{dx}{x}.$$

The variables have now been separated and you go to work on it in the usual way. This method is due to Leibniz[†] and dates from around 1691.

Example 3.4.1 *Find the solutions of the equation*

$$y' = \frac{y^2 + xy}{x^2}.$$

First note this is of the form

$$y' = \left(\frac{y}{x}\right)^2 + \left(\frac{y}{x}\right).$$

Let $u = \frac{y}{x}$ so $y = xu$. Then

$$u'x + u = u^2 + u$$

and so, separating the variables yields

$$\frac{du}{u^2} = \frac{dx}{x}$$

Hence

$$-\frac{1}{u} = \ln|x| + C$$

and so

$$\frac{y}{x} = u = \frac{1}{K - \ln|x|}$$

where $K = -C$. Hence

$$y(x) = \frac{x}{K - \ln|x|}$$

[†]Gottfried Wilhelm (von) Leibniz, (1646-1716) is credited with Newton as being one of the inventors of calculus. There was much controversy over who did it first. It is likely that Newton did it first, but Leibniz had superior notation. The notation $\frac{dy}{dx}$ for the derivative and the notation for integrals is due to him. Like many of these men, he was interested in many other subjects besides mathematics, such as philosophy, theology, geology, and medicine.

Procedure 3.4.2 *To solve a homogeneous equation, one which can be placed in the form*

$$y' = f\left(\frac{y}{x}\right),$$

do the following:

1. *Define a new variable $v = y/x$. Then $y = xv$ and so $y' = v + xv'$.*

2. *Plug in to the equation.*

$$v + xv' = f(v), \quad x\frac{dv}{dx} = f(v) - v$$

$$\frac{dv}{f(v) - v} = \frac{dx}{x}$$

This is separable. Place $\int$ before each side and do what it says. Then choose the constant of integration to satisfy any initial condition which may be present.

3.5   Exact Equations

Sometimes you have a differential equation of the form

$$M(x, y)\, dx + N(x, y)\, dy = 0$$

where $N_x = M_y$. In this happy situation, one can find a function of two variables $f(x, y)$ such that

$$f_x(x, y) = M(x, y), \quad f_y(x, y) = N(x, y) \tag{3.3}$$

and the solution to the equation is of the form

$$f(x, y) = C \tag{3.4}$$

where C is a constant. This function f is called a scalar potential or potential for short.

These equations are called **exact.** Why does $*$ yield a solution? Say the above relation defines y as a function of x. Then using the chain rule,

$$f_x(x, y) + f_y(x, y)\frac{dy}{dx} = 0$$

and so

$$f_x(x, y)\, dx + f_y(x, y)\, dy \;=\; 0$$
$$M(x, y)\, dx + N(x, y)\, dy \;=\; 0$$

It is easy to see that if there exists a C^2 function f with the property that $f_x = M, f_y = N$, then $N_x = M_y$. This follows because $M_y = f_{xy}$ and $N_y = f_{yx}$. By equality of mixed partial derivatives, you need to have $M_y = N_x$. In fact, if this last condition holds, then there will generally be such a potential function $f(x, y)$.

Why is it that if $N_x = M_y$ then there exists f with the properties described? Let

$$f(x,y) \equiv \int_0^x M(t,y)\,dt + N(0,y).$$

Then $f_x(x,y) = M(x,y)$, and formally differentiating across the integral,

$$
\begin{aligned}
f_y(x,y) &= \int_0^x M_y(t,y)\,dt + N(0,y) = \int_0^x N_x(t,y)\,dt \\
&= N(x,y) - N(0,y) + N(0,y) = N(x,y)
\end{aligned}
$$

In general, this process of $\left(\frac{\partial}{\partial y} \int_0^x M(t,y)\,dt = \int_0^x M_y(t,y)\,dt \right)$ has not been proved, but in examples, it will be obviously true. Also, it is formally true when you think of the integral as a sort of sum and use the fact that the derivative of a sum is the sum of the derivatives.

Example 3.5.1 *Find the solutions to*

$$(\cos(x) + 2xy)\,dx + x^2 dy = 0$$

You see that this is exact ($2x = 2x$). Then the $f(x,y)$ satisfies $f_x(x,y) = \cos(x) + 2xy$ and so $f(x,y) = \sin(x) + x^2 y + g(y)$. Then taking the partial derivative with respect to y, it follows that $x^2 + g'(y) = x^2$ and so is suffices to let $g(y) = 0$. Then the solutions to this differential equation are

$$\sin(x) + x^2 y = C$$

where C is a constant which would be determined by some sort of an initial condition.

Example 3.5.2 *In the above example, determine C if $(x,y) = \left(\frac{\pi}{2}, 0\right)$ is to be on the curve which yields a solution to the differential equation.*

You need to have $1 = C$ because

$$\sin\left(\frac{\pi}{2}\right) + \left(\frac{\pi}{2}\right)^2 \cdot 0 = C$$

and so the solution in this case is $\sin(x) + x^2 y = 1$.

All of the examples of this sort of thing are similar. Exact equations are easy to solve. Physically the solution to these equations is really a statement about the energy being constant.

Procedure 3.5.3 *To solve an exact equation*

$$M(x,y)\,dx + N(x,y)\,dy = 0$$

do the following:

1. *Check to see if it really is exact by seeing if $N_x = M_y$. If it is, find a scalar potential $f(x,y)$ such that $f_x = M, f_y = N$.*

2. *The general solution is $f(x,y) = C$. Choose C to satisfy initial conditions.*

3.6   The Integrating Factor

It turns out that theoretically, this is the most general method for solving equations

$$m\left(x,y\right)dx + n\left(x,y\right)dy = 0$$

I want to stress the word "theoretically" however. If the above equation is not exact, the idea is to multiply by a function μ which will make it exact. Thus it would be sufficient to have

$$\left(\mu m\right)_y = \left(\mu n\right)_x$$

The function μ is called an integrating factor. In other words, it is required that

$$\mu_y m + \mu m_y = \mu_x n + \mu n_x \tag{3.5}$$

This is called a first order linear partial differential equation and we don't know how to solve them. However, we don't need to find all solutions, just one which works. The idea is to look for $\mu = \mu\left(x\right)$ or $\mu = \mu\left(y\right)$. For us, if there is no such easy solution, the method has failed. So what would happen if there is a solution $\mu = \mu\left(x\right)$? then you would have

$$\mu'\left(x\right) = \mu\left(x\right)\frac{m_y\left(x,y\right) - n_x\left(x,y\right)}{n\left(x,y\right)}$$

and so there will be such an integrating factor if

$$\frac{m_y\left(x,y\right) - n_x\left(x,y\right)}{n\left(x,y\right)}$$

depends only on x. Similarly, there will be an integrating factor $\mu = \mu\left(y\right)$ if

$$\frac{n_x - m_y}{m}$$

depends only on y.

Example 3.6.1 *Find the solutions to*

$$\left(2y^3 + 2y\right)dx + \left(3xy^2 + x\right)dy = 0$$

The equation is clearly not exact so we look for an integrating factor.

$$\mu_y\left(2y^3 + 2y\right) + \mu\left(6y^2 + 2\right) = \mu_x\left(3xy^2 + x\right) + \mu\left(3y^2 + 1\right)$$

We look for one which depends on only one variable. Let's try to find $\mu = \mu\left(y\right)$ first. If there is such a solution, then

$$\mu'\left(y\right)\left(2y^3 + 2y\right) + \mu\left(6y^2 + 2\right) = \mu\left(3y^2 + 1\right)$$

so it looks like there is such a solution.

$$\mu'\left(y\right) = \frac{-3y^2 - 1}{2y^3 + 2y}\mu$$

Thus

$$\frac{d\mu}{\mu} = \frac{-3y^2 - 1}{2y^3 + 2y} dy \tag{3.6}$$

Then

$$\ln(\mu) = \int \frac{-3y^2 - 1}{2y^3 + 2y} dy = -\frac{1}{2} \ln(y^3 + y)$$

An integrating factor would be $1/\sqrt{y^3 + y}$. This looks really ugly. Let's try and find one which depends only on x. Then

$$\mu\left(6y^2 + 2 - \left(3y^2 + 1\right)\right) = \mu'(x)\left(3xy^2 + x\right)$$

$$\mu\left(3y^2 + 1\right) = \mu'(x)\left(3xy^2 + x\right)$$

$$\mu'(x) = \mu(x)\frac{1}{x}$$

Thus $\mu = x$ is also an integrating factor. Which would you rather use? Multiply by x. The equation is now

$$\left(2xy^3 + 2yx\right)dx + \left(3x^2y^2 + x^2\right)dy = 0$$

and it is an exact equation so you are in the situation of the preceding section. You find a scalar potential. The manipulations explained in the last section yield $x^2y^3 + x^2y$ as a scalar potential. Then the solutions are

$$x^2y^3 + x^2y = C$$

All of these are the same. You begin with (3.5) and look for solutions. In particular you look for solutions that depend on only one variable. If you can find one, then the problem has been reduced to that of the preceding section. If you can't find such a solution, then you give up. Under general conditions, it can be proved that solutions exist but as usual in mathematics, there is a big gap between knowing something exists and finding it. However, here is something nice which was discovered by Euler back in the 1700s. It is called Euler's identity along with the more famous one involving complex numbers.[‡]

Lemma 3.6.2 *A function $M(x, y)$ is homogeneous of degree α if $M(tx, ty) = t^\alpha M(x, y)$. For such a function,*

$$\alpha M(x, y) = x\frac{\partial M}{\partial x}(x, y) + y\frac{\partial M}{\partial y}(x, y)$$

[‡]Euler (1707-1783) (pronounced "oiler") was a Swiss mathematician, a student of Johann Bernoulli. He is one of the most important mathematicians to ever live. He wrote more mathematics than anyone else, some 530 books and papers in all areas of the subject. His very unusual memory allowed him to continue doing mathematical research even after he went blind in 1766. Many of the ideas in this book are due to him. Like many of the other great mathematicians of his time Euler's interests were not limited to mathematics. His work is also very important in engineering and physics. A remarkable amount of notation is due to him or popularized by him. Included in this list is the summation symbol Σ, e, π, i, and $f(x)$. Like many of his time, he was a very religious man who believed the Bible was inspired. He had incredible insight but like most of us, he made mistakes because he sometimes neglected issues related to convergence. However, the need for this sort of thing was not well understood in his time. Euler died in St. Petersburg.

Proof: You use the chain rule to differentiate both sides of the equation

$$M\left(tx, ty\right) = t^{\alpha} M\left(x, y\right)$$

with respect to t. Thus

$$\alpha t^{\alpha-1} M\left(x, y\right) = x \frac{\partial M}{\partial x}\left(tx, ty\right) + y \frac{\partial M}{\partial y}\left(tx, ty\right)$$

Now let $t = 1$. ∎

The reason this is pretty nice is that if you have the equation

$$M\left(x, y\right) dx + N\left(x, y\right) dy = 0$$

and both M and N are homogeneous of degree α, then

$$\frac{1}{xM + yN}$$

is an integrating factor. Here $M_x = \frac{\partial M}{\partial x}$. We verify this next. It is so if

$$\left(\frac{N}{xM + yN}\right)_x = \left(\frac{M}{xM + yN}\right)_y$$

By the quotient rule, this will be so if and only if

$$N_x\left(xM + yN\right) - N\left(M + xM_x + yN_x\right) =$$

$$M_y\left(xM + yN\right) - M\left(xM_y + N + yN_y\right)$$

In both sides of the above equation, some terms cancel and it follows that the desired result follows if and only if

$$xMN_x - \left(NM + xM_xN\right) = yNM_y - \left(MN + yMN_y\right)$$

and this happens if and only if

$$xMN_x - xM_xN = yNM_y - yMN_y$$

which happens if and only if

$$MxN_x + MyN_y = NyM_y + NxM_x$$

if and only if

$$M\left(xN_x + yN_y\right) = N\left(yM_y + xM_x\right)$$

But this is true because by Euler's identity, $xN_x + yN_y = \alpha N$ and $yM_y + xM_x = \alpha M$ so the above is just $\alpha NM = \alpha NM$. Of course it is assumed that $xM + yN \neq 0$ in the above.

Example 3.6.3 *Find the integral curves for*

$$\left(x^2 + xy\right) dx + \left(y^2 + x^2\right) dy = 0$$

Of course this can be written as a homogeneous equation and the technique for solving these can be used. However, let's use this new technique which says that an integrating factor is

$$\frac{1}{x\left(x^2 + xy\right) + y\left(y^2 + x^2\right)} = \frac{1}{x^3 + 2x^2 y + y^3}$$

Then multiplying by this yields an exact equation.

$$\frac{x^2 + xy}{x^3 + 2x^2 y + y^3}dx + \frac{y^2 + x^2}{x^3 + 2x^2 y + y^3}dy = 0$$

Unfortunately, it is too complicated for me to solve this conveniently. However, knowing that it is exact allows the use of the formula derived in showing that if $M_y = N_x$ then the equation was exact. Thus the integral curves are of the form

$$\int_0^x M\left(t, y\right) dt + N\left(0, y\right)$$

$$= \int_0^x \frac{t^2 + ty}{t^3 + 2t^2 y + y^3}dt + \frac{1}{y} = C$$

Now we consider an easier one.

Example 3.6.4 *Find the integral curves for*

$$\left(xy + y^2\right) dx + x^2 dy = 0$$

The integrating factor is

$$\frac{1}{xy\left(2x + y\right)}$$

and so the equation to solve is

$$\frac{1}{x\left(2x + y\right)}\left(x + y\right) dx + \frac{x}{y\left(2x + y\right)}dy = 0$$

Then integrating the first term with respect to x, the scalar potential is of the form

$$f\left(x, y\right) = \ln|x| - \frac{1}{2}\ln\left|x + \frac{1}{2}y\right| + g\left(y\right)$$

Then differentiating with respect to y,

$$-\frac{1}{2\left(2x + y\right)} + g'\left(y\right) = \frac{x}{y\left(2x + y\right)}$$

$$g'\left(y\right) = \frac{1}{2y}$$

and so $g\left(y\right) = \frac{1}{2}\ln|y|$ will work. Thus the integral curves are of the form

$$\ln|x| - \frac{1}{2}\ln\left|x + \frac{1}{2}y\right| + \frac{1}{2}\ln|y| = C$$

You could simplify this if desired.

Procedure 3.6.5 *To solve*

$$M(x, y)\, dx + N(x, y)\, dy = 0$$

using an integrating factor, do the following:

1. *Look for an integrating factor μ which is a function of x alone. You do this if*

$$\frac{M_y - N_x}{N}$$

   does not depend on y. In this case, you solve

$$\mu'(x) = \mu(x) \left(\frac{M_y - N_x}{N} \right)$$

   which is a separable equation. Solve and choose constant to satisfy initial condition. If this doesn't work,

2. *Look for an integrating factor μ which is a function of y alone. You do this if*

$$\frac{N_x - M_y}{M}$$

   does not depend on x. In this case, you solve

$$\mu'(y) = \frac{N_x - M_y}{M} \mu(y)$$

   which is a separable equation. Solve and choose constant to satisfy initial condition.

3. *If neither of these work, check to see if M, N are both homogeneous of the same degree. If they are, you could use either the methods of homogeneous equations or Euler's formula for the integrating factor*

$$\frac{1}{xM + yN}.$$

4. *If none of the above works, give up. You don't know how to do it. The integrating factor exists, but you don't know how to find it.*

3.7   The Case Where M, N Are Affine Linear

Something which often occurs is an equation of the form

$$(px + qy + r)\, dx + (\alpha x + \beta y + \gamma)\, dy = 0$$

It doesn't quite fit anything in the earlier discussion. It won't be exact, homogeneous, or separable or linear. However, one can massage it to get something which is homogeneous. This is illustrated in some examples.

Example 3.7.1 *Find the integral curves for*

$$(x + 2y + 3)\, dx + (2x - y + 1)\, dy = 0$$

Of course the problem is those constants $3, 1$ so it is reasonable to change variables. Let $u = x - a, v = y - b$ where we choose a, b in an auspicious manner to get the constants to disappear. First, $dx = du, dy = dv$. Then in terms of the new variables,

$$(u + a + 2\,(v + b) + 3)\, dx + (2\,(u + a) - (v + b) + 1)\, dy \;\; = \;\; 0$$
$$(u + 2v + (a + 2b + 3))\, dx + (2u - v + (2a - b + 1))\, dy \;\; = \;\; 0$$

and we want

$$a + 2b + 3 = 0$$
$$2a - b + 1 = 0$$

Hence we should let $a = -1$ and $b = -1$. Then with this, the equations reduce to

$$(u + 2v)\, du + (2u - v)\, dv = 0$$

This is now a homogeneous equation, or we could use the integrating factor described earlier, but, in this case, it is also an exact equation. A scalar potential is

$$\frac{u^2}{2} + 2uv - \frac{v^2}{2},$$

and so the integral curves for the original equation would be

$$\frac{1}{2}\,(x + 1)^2 + 2\,(x + 1)\,(y + 1) - \frac{1}{2}\,(y + 1)^2 = C$$

The example illustrates what to do in general. You just change the variables to remove those constant terms and then obtain a homogeneous equation which can be solved by a variety of methods.

Example 3.7.2 *Find the integral curves for*

$$(x + y + 2)\, dx + (2x - y + 4)\, dy = 0$$

As before, let $u = x - a, v = y - b$ and write in terms of these new variables. Thus

$$(u + v + (a + b + 2))\, du + (2u - v + (2a - b + 4))\, dv = 0$$

Then you need

$$a + b + 2 = 0$$
$$2a - b + 4 = 0$$

Thus $a = -2, b = 0$. The new equation then is

$$(u + v)\, du + (2u - v)\, dv = 0$$

This can be considered as a homogeneous equation.

$$\frac{dv}{du} = \frac{(u + v)}{v - 2u} = \frac{1 + \frac{v}{u}}{\frac{v}{u} - 2}$$

Then let $z = \frac{v}{u}$ and do the usual substitution. This yields

$$u\frac{dz}{du} = \frac{1}{z-2}\left(-z^2 + 3z + 1\right) \tag{3.7}$$

and so, separating the variables,

$$\frac{2-z}{z^2 - 3z - 1}dz = \frac{du}{u}$$

Then after much work one obtains integral curves of the form

$$\ln\left(\frac{v}{u} - \frac{1}{2}\sqrt{13} - \frac{3}{2}\right)^{\frac{1}{26}\sqrt{13}} + \ln\frac{1}{\left(\frac{v}{u} + \frac{1}{2}\sqrt{13} - \frac{3}{2}\right)^{\frac{1}{26}\sqrt{13}}}$$

$$+ \ln\frac{1}{\sqrt{\frac{v}{u} - \frac{1}{2}\sqrt{13} - \frac{3}{2}}} + \ln\frac{1}{\sqrt{\frac{v}{u} + \frac{1}{2}\sqrt{13} - \frac{3}{2}}} - \ln|u| = C$$

Then you plug in what u, v are in terms of x, y.

Actually, it was real easy to do this. The computer algebra system did it for me. Here is one which is not so ugly.

Example 3.7.3 *Find the integral curve which contains the given ordered pair.*

$$(6x - y - 4)\,dx = (y - 2x)\,dy,\ (2, 2)$$

The equation is

$$\frac{dy}{dx} = \frac{6x - y - 4}{y - 2x}$$

Now let $x = u + a, y = v + b$. Then we choose a, b such that in terms of the new variables the equation becomes homogeneous. Thus we need

$$6a - b - 4 = 0$$
$$b - 2a = 0$$

Thus we let $a = 1, b = 2$. Then the equation is

$$\frac{dv}{du} = \frac{6u - v}{v - 2u} \tag{3.8}$$

This is a homogeneous equation. Let $z = \frac{v}{u}$. Then

$$uz' = \frac{6 - z}{z - 2} - z = \frac{1}{z - 2}\left(-z^2 + z + 6\right)$$

Separating the variables,

$$\frac{(2 - z)\,dz}{z^2 - z - 6} = \frac{du}{u}$$

This is easily solved,

$$C - \left(\frac{4}{5}\ln|z + 2| + \frac{1}{5}\ln|z - 3|\right) = \ln|u|$$

The in terms of the original variables,

$$C = \left(\frac{4}{5} \ln \left| \frac{y-2}{x-1} + 2 \right| + \frac{1}{5} \ln \left| \frac{y-2}{x-1} - 3 \right| \right) + \ln |x-1|$$

Then to contain the ordered pair, you need

$$C = \frac{4}{5} \ln 2 + \frac{1}{5} \ln 3 = \frac{1}{5} \ln (48)$$

Procedure 3.7.4 *To solve affine linear equations of the form*

$$(px + qy + r) \, dx + (\alpha x + \beta y + \gamma) \, dy = 0,$$

do the following:

1. *Change the variables $u = x - a$, $v = y - b$, plug in and choose a, b to make the resulting equation homogeneous.*

2. *Solve the resulting homogeneous equation. Then substitute back in $x - a$ for u and $y - b$ for v. Pick the constant to satisfy initial conditions.*

3.8 Linear and Nonlinear Differential Equations

Recall initial value problems for linear differential equations are those of the form

$$y' + p(t) y = q(t), \; y(t_0) = y_0 \tag{3.9}$$

where $p(t)$ and $q(t)$ are continuous functions of t. Then if $t_0 \in [a, b]$, an interval, there exists a unique solution to the initial value problem given above which is defined for all $t \in [a, b]$. The following theorem which is really something of a review gives a proof.

Theorem 3.8.1 *Let $[a, b]$ be an interval containing t_0 and let $p(t)$ and $q(t)$ be continuous functions defined on $[a, b]$. Then there exists a unique solution to (3.9) valid for all $t \in [a, b]$.*

Proof: Let $P'(t) = p(t)$, $P(t_0) = 0$. For example, let $P(t) \equiv \int_{t_0}^{t} p(s) \, ds$. Then multiply both sides of the differential equation by $\exp(P(t))$. This yields

$$(y(t) \exp(P(t)))' = q(t) \exp(P(t))$$

and so, integrating both sides from t_0 to t,

$$y(t) \exp(P(t)) - y_0 = \int_{t_0}^{t} q(s) \exp(P(s)) \, ds$$

and so

$$y(t) = \exp(-P(t)) y_0 + \exp(-P(t)) \int_{t_0}^{t} q(s) \exp(P(s)) \, ds$$

which shows that if there is a solution to (3.9), then the above formula gives that solution. Thus there is at most one solution. Also, you see the above formula makes perfect sense on the whole interval. Since the steps are reversible, this shows $y(t)$ given in the above formula is a solution. You should provide the details. Use the fundamental theorem of calculus. ∎

It is not so simple for a nonlinear initial value problem of the form

$$y' = f(t, y), \; y(t_0) = y_0.$$

Theorem 3.8.2 *Let f and $\frac{\partial f}{\partial y}$ be continuous in some rectangle, $a < t < b, c < y < d$ containing the point (t_0, y_0). Then there exists a unique local solution to the initial value problem*

$$y' = f(t, y), \; y(t_0) = y_0.$$

This means there exists an interval, I such that $t_0 \in I \subseteq (a, b)$ and a unique function, y defined on this interval which solves the above initial value problem on that interval.

A much more general theorem will be proved later. Also, in the above, it suffices to say that f is continuous on the given rectangle and that for $y, z \in [c, d]$, $t \in [a, b]$,

$$|f(t, y) - f(t, z)| \leq K|y - z|$$

for some $K > 0$. This is called a Lipschitz condition. For now, note that it is reasonable to believe the conclusion of this theorem. Start with the point (t_0, y_0) and follow the slope field as illustrated in many of the above examples. The problem is, sometimes you can't extend the solution as far as you might like.

Example 3.8.3 *Solve $y' = 1 + y^2, y(0) = 0$.*

This satisfies the conditions of Theorem 3.8.2. Therefore, there is a unique solution to the above initial value problem defined on some interval containing 0. However, in this case, we can solve the initial value problem and determine exactly what happens. The equation is separable.

$$\frac{dy}{1 + y^2} = dt$$

and so $\arctan(y) = t + C$. Then from the initial condition, $C = 0$. Therefore, the solution to the equation is $y = \tan(t)$. Of course this function is defined on the interval $\left(-\frac{\pi}{2}, \frac{\pi}{2}\right)$. It is impossible to extend it further because it has an asymptote at the two ends of this interval.

Theorem 3.8.2 does not say that the local solution can never be extended beyond some small interval. Sometimes it can. It depends very much on the nonlinear equation. For example, the initial value problem

$$y' = 1 + y^2 - \varepsilon y^3, \; y(0) = y_0$$

turns out to have a solution on $\mathbb{R}$. Here ε is a small positive number. You might think about why this is so. It is related to the fact that in this new equation, the extra term prevents y' from becoming unbounded.

Also, you don't know whether the interval of existence is symmetric about the point at which the initial condition is given.

Example 3.8.4 *Solve* $y' = (y - 1)^2$, $y(0) = 0$.

The equation is separable

$$\frac{dy}{(y-1)^2} = dt$$

Then integrating and using the initial condition,

$$\frac{1}{1-y} = t + 1$$

Thus

$$y = \frac{t}{t+1}$$

which makes sense on $(-1, \infty)$.

The next one looks a lot like the above, but has a solution on the whole real line.

Example 3.8.5 *Consider* $y' = 1 - y^2$, $y(0) = 0$.

You can verify that

$$y = \frac{e^{2t} - 1}{e^{2t} + 1}$$

is the solution and it makes sense for all t.

Hopefully, this has demonstrated that all sorts of things can happen when you are considering nonlinear equations. However, it gets even worse.

If you assume less on f in the above theorem, you sometimes can get existence but not uniqueness for the initial value problem. In the next example $\frac{\partial f}{\partial y}$ is not continuous near $(0, 0)$.

Example 3.8.6 *Find the solutions to the initial value problem*

$$y' = y^{1/3}, \ y(0) = 0.$$

The equation is separable so $\frac{dy}{y^{1/3}} = dt$ and so the solutions are of the form

$$\frac{3}{2} y^{2/3} = t + C.$$

Letting $C = 0$ from the initial condition, one solution is $y = \left(\frac{2}{3}t\right)^{3/2}$ for $t > 0$. However, you can see that $y = 0$ is also a solution. Thus uniqueness is violated. Note there are two solutions to the initial value problem and both exist and solve the initial value problem on all of $[0, \infty)$.

Observation 3.8.7 *What are the main differences between linear and nonlinear equations? Linear initial value problems have an interval of existence which is the same as the interval on which the functions in the equation are continuous. Nonlinear initial value problems sometimes don't. Solutions to linear initial value problems are unique. This is not always true for nonlinear equations although if in the nonlinear equation, f and $\partial f / \partial y$ are both continuous, then you at least get uniqueness as well as existence on some possibly small interval of undetermined length.*

3.9   Computer Algebra Methods

One can use computer algebra systems to solve such equations. In this section, the use of various systems will be discussed. The intent here is to give a reasonably simple way to obtain these solutions, not to give all possible ways to use these systems. In this book, I will be emphasizing MATLAB. However, other systems will be discussed in this section. One very easy to use system which behaves a lot like MATLAB is Scientific Notebook, which is actually based on mupad. I will mention its use also.

3.9.1   Maple

Maple is a well-known computer algebra system and it has something called dsolve which produces solutions to initial value problems. The following illustrates what you type in to have it do this for you. Before typing in the syntax below, you need to go to file and new and then workshop mode. If you click on "text" in the upper left corner, what you type appears in red. If you click on math, it will be in italics and will appear in standard math notation. **If you want the commands to appear on new lines, you do shift enter to get this to happen.** However, there is no harm in using a single line. Since you have selected workshop mode, you will see a red > on the left. After this, type the following:

de:=diff(y(x),x)=4*y(x) - y(x)^2;
Y:=rhs(dsolve({de,y(0)=1},y(x)));
plot(Y,x=0..2, thickness=2);

Now you press "enter". The letters "rhs" are used to assign the value of the right side to the left. Then it will solve the equation and also give a graph of the solution. If you look at what you typed, you will see how to adjust the length of the interval on which the graph will be drawn. Here is what results from the above.

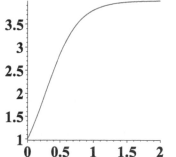

It also gives the solution is $y = \frac{4}{1+3e^{-4x}}$ and the graph of the solution is shown for $x \in [0,2]$. In what you typed above, $diff(y(x),x)$ signifies $y'(x)$. The rest should be pretty self explanatory. You can change the differential equation if you want. If you want to write in y'', this is signified by $diff(y(x),x\$2)$, and I think you can see what it would be for $y''', y^{(4)}$ and so forth. In principle, you can do various kinds of differential equations this way. Of course the problem is that neither you nor the computer algebra system may be able to find a simple analytic solution. In this case, there may still be a solution, but you won't have an answer for it in terms of known elementary functions. Maple can do this case also.

Before showing the syntax for obtaining a numerical solution with Maple, consider the above claim that even though you might not have a solution in terms of a simple formula, you might still have a solution. Say you have $y' = f(t,y), y(0) = y_0$ and you want the solution on some interval $[0,a]$. One way to obtain an approximate solution would be to replace the derivative with a difference quotient as follows:

$$\frac{y_i - y_{i-1}}{h} = f(t_i, y_i), i \geq 1$$

where here you have a uniform partition of $[0, a], t_0 < t_1 < \cdots < t_n = a$ where $h = t_i - t_{i-1}$. I will leave it to you to see that there is a solution to the above discrete problem. Then if you want, you could define a function $y_n(t)$ to be a piecewise linear function which equals y_i at t_i. This is an example of a numerical solution. The above method is called the Euler method. It isn't as good as what Maple will use.

Suppose you wanted to use Maple to solve the initial value problem.

$$y' = -\left(y + .01y^5\right), \; y(0) = 1$$

Then you would type in the following after the red >. Remember, if you want the commands on new lines, you press shift enter.

```
with(plots):
de:=diff(y(x),x)=-(y(x)+.01*y(x)^5);
Y:=dsolve({de,y(0)=1},y(x), numeric);
odeplot(Y,[x,y(x)],0..5);
```

Note the top line which says with(plots): You have to include this so it knows how to plot the solution you are going to get. Then after typing in the above stuff, you press return. What you get is a graph of this solution.

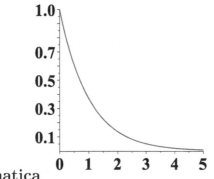

3.9.2 Mathematica

In using Mathematica, you need to open a new notebook to begin.

The syntax is different in Mathematica than in Maple. This software is obsessed with capital letters and square brackets. Thus you must write Sin[x] and not sin(x) or sin[x]. Otherwise Mathematica will not recognize this as a function. A nice thing about it is that you press enter to get on a new line and shift enter or click on evaluate cells on the top to get it to do something. This is different than Maple which requires "shift enter" to get to a new line and does something when you press "enter". Like Maple, it has a dsolve command but here you must write it as DSolve. I am going to consider a simple Bernoulli equation.

$$y' = y - y^3, y(0) = 2$$

Now here is the syntax

de=DSolve[{y'[x]==y[x]-y[x]^3, y[0]==2},y[x],x]
Plot[y[x]/.de,{x,-.5,5}]

After entering this, you press "shift enter" or click on the top where it says evaluate cells. This causes it to do something. What it does is to give you a formula for the solution

$$y[x] \rightarrow \frac{2e^x}{\sqrt{-3+4e^{2x}}}$$

and it then produces the following graph.

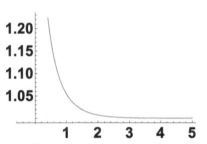

It refuses to graph the initial condition at $x = 0$. This is odd because a few computations will show that the above is indeed the solution to the differential equation and makes perfect sense at $x = 0$. There is no asymptote at 0 which the graph seems to imply. Maple gives a correct graph however as well as a correct solution. If you graph on a smaller interval, Mathematica will also give a good graph.

What of numerical methods? Of course these are available for Mathematica also. To consider the same differential equation, you would type the following.

NDSolve[{y'[x]==y[x]-y[x]^3, y[0]==2},y,{x,0,5}]
Plot[Evaluate[y[x]/. %],{x,0,5}]

Then you press "shift enter" and it does something. The thing it does is to give the same graph that the DSolve command gave. However, if you graph on a smaller interval, it gives a good graph which indicates the initial condition.

MATLAB, discussed next, gives a good graph for the numerical solution as does Scientific Notebook, also discussed later. These produce a graph of the function which shows the initial condition on the graph.

3.9.3   MATLAB ®

Another frequently used computer algebra system is MATLAB. You can also use this to find solutions to the initial value problem. With MATLAB, you don't have to be sure to select Worksheet mode as in Maple or a Notebook as in Mathematica. You just open it and type in what you want after >>. This is very nice. As with Maple, if you want commands to appear on separate lines, you use "shift enter". Mathematica is the only one which uses "shift enter" to cause the software to do something.

The basic version of MATLAB is sufficient to do the numerical procedures discussed. In order to do procedures which involve commands like "syms" you will need to have the symbolic math toolbox also. In particular, you need this toolbox for the first example given here in which "dsolve" is used, but not for the numerical procedures mentioned next.

Here is what you type to get MATLAB to compute the solution to

$$y' = y - .01y^2, y\left(0\right) = 2.$$

After the $>>$ you type the following:

$$\text{syms y(t); y(t)=dsolve(diff(y,t)==y - .01*y\hat{\ }2, y(0)==2)}$$

After typing in the above, you press enter and here is what results.

>>syms y(t); y(t)=dsolve(diff(y,t)==y-.01*y^2,y(0)==2)
y(t) =
100/(exp(log(49) - t)+1)

If you want a graph of this solution, this is also easy to get. After doing the above, type in the following to the right of $>>$

$$\text{ezplot(y(t),[0,3])}$$

and then press "enter" to obtain the graph of the solution on the interval $[0, 3]$.

Similarly, you can ask for numerical solutions in case you can't find an analytical solution. MATLAB can find these also. For example, if you wanted to solve on the interval $[0, 2]$ the initial value problem

$$y' = y - .01y^5, \ y(0) = 1,$$

You would do the following: After $>>$ you type

$$\text{f=@(x,y) y-.01*y\hat{\ }5;}$$

Next, type the following on a new line:

$$\text{[x,y]=ode45(f,[0,2],1)} \hspace{3cm} (*)$$

and on the next new line,

$$\text{plot(x,y)}$$

and press "enter". This will give a large table of values of x followed by values of y which comes from using a suitable numerical method named ode45 and it will also plot the solution.

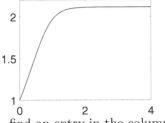

If you don't want to see this large table of values, simply place a ; at the end of $*$. This will cause MATLAB to defer displaying the table even though it knows about it.

If you placed ; at the end of $*$, and decide you would like to see $y(.5)$ for example, you ask for the table of values. This is done by typing [x,y] after $>>$ and then "enter" to see the whole table and simply scroll down to find an entry in the column for x which is close to .5. There is also another way to find the values using the deval function.

Another thing which is pretty easy to do in MATLAB is to change the initial conditions and graph the two solutions on the same set of axes. The above gives you a graph of $y(x)$ for $x \in [0, 2]$. It has defined the function y at least at many points. Now you can simply define another solution with a different initial condition as follows.

$$\text{[x1,y1]=ode45(f,[0,2],2)}$$

and press return. This will define the function x1 $\to$ y1(x1). Then to graph both on the same axes, you would type

$$\text{plot(x,y,x1,y1)}$$

and both will appear. You can do as many of these as you want of course. If you wanted to do a lot of graphs all at once, you can also have this done. You would do the following:

>> f=@(t,x) [x-x^3];
hold on
for z=-2:.5:2
[t,x]=ode45(f,[0,4],z);
plot(t,x)
end

Then press "enter" and you will get graphs of solutions for initial conditions

$$-2, -1.5, -1, -.5, \cdots, 2$$

With the above, which is solving

$$x' = x - x^3, \ x(0) = z$$

for various values of z, you get the following graph.

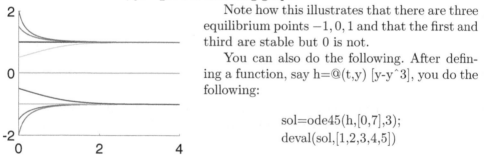

Note how this illustrates that there are three equilibrium points $-1, 0, 1$ and that the first and third are stable but 0 is not.

You can also do the following. After defining a function, say h=@(t,y) [y-y^3], you do the following:

sol=ode45(h,[0,7],3);
deval(sol,[1,2,3,4,5])

then press "enter". You should get the values of y at the points $1, 2, 3, 4, 5$. Remember that to place on a new line, you use "shift enter". You could also use any other symbol for "sol".

Another thing I have noticed when using MATLAB is that it sometimes puts the graph behind the command window so you don't see it till you shrink the command window.

To adjust the appearance of the graph which results, you go to the graph and click on file and then export setup. You can make changes in a dialog box and do things like change the thickness of the lines and the size of the font very easily. Then you can save it as an eps file or several other kinds of files.

Also, when you are done, type >> clear all or close all and then "enter". Then type >> clf and "enter" to get rid of any graphs it may have done and press "enter". To clear the screen, type >>clc and then press "enter". This is a very good idea because if you want to do something else, you don't want MATLAB to be confused about what you mean and it will be confused if it can.

3.9.4 Scientific Notebook

This is an easy to use version of another computer algebra system called mupad which is somewhat like MATLAB. (In fact, mupad is part of the symbolic math toolbox which is also used by MATLAB.) The difference is that you use standard notation and there is virtually no complicated syntax to fuss with. For example, suppose you want to find the solution to

$$y' = y - y^2, \ y(0) = 2.$$

Then you go to the tool bar and select a matrix which is 2×1 and fill in the differential equation on the top and the initial condition on the bottom. The symbols should all be in red indicating it is a math expression. To adjust this, click on the T which is on the top tool bar if it is not a red M and this will change to red M. However, this will be automatic if you are just filling in a matrix.

$$y' = y - y^2$$
$$y(0) = 2$$

Then you click on one of the cells of the matrix to place the cursor there and then click on "compute" solve ODE and exact. It will ask you for the name of the independent variable. I named it t. The result is as follows.

$$\begin{matrix} y' = y - y^2 \\ y(0) = 2 \end{matrix} \text{, Exact solution is : } \left\{ -\frac{1}{\frac{1}{2}e^{-t} - 1} \right\}$$

If you want to graph it, you select it and paste it on another line, and click on the expression. Now you go to compute and then plot2D and select rectangular. This will give you a graph of the function. This software often works and is very easy to use when it does.

If you want to do a numerical solution, the steps are similar. Enter the equation and initial condition as before. Then click on the matrix and go to compute, solveODE and numeric. This will return a function called y as shown below.

$$\begin{matrix} y' = y - y^2 \\ y(0) = 2 \end{matrix} \text{, Functions defined: } y$$

Of course, you would want to graph y and so you do the same thing. Place y on a new line, click on it, and then select compute, plot2D, and ODE. You can also simply click on compute, plot2D, and then ODE and this will produce the graph right after the y. Either way, the graph looks like the following for $t \in [0, 4]$. Actually, I used MATLAB to produce this graph. The ones produced in Scientific Notebook don't look quite as nice, and you can't export 2 dimensional graphs as eps files.

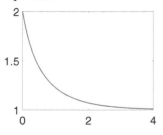

To clear the definition of the function, you click compute, definitions, clear definitions.

It is very fun to use, but sometimes it stalls and nothing happens. You end up having to close the program and open it again to put a stop to this. This can also happen with MATLAB if there is something amiss in the syntax, for example but with MATLAB, you press control c to put an end to it. The advantage Scientific

Notebook has over all the other computer algebra software is that there is no syntax to worry about. You just use standard math notation and it gives the solution in standard math notation also. This is particularly helpful when you are trying to find explicit formulas for the solutions rather than numerical solutions. However, it sometimes has trouble with surprisingly easy problems.

3.10   Exercises

Linear Equations

1. Find all solutions to the following linear equations. You may need to leave answers in terms of integrals on some of them.

   (a) $y' + 2ty = e^{-t^2}$

   (b) $y' - ty = e^t$

   (c) $y' + \cos(t) y = \cos(t)$

   (d) $y' + ty = \sin(t)$

   (e) $y' + \frac{1}{t-1}y = \frac{1}{(t-1)^2}$

   (f) $y' + \tan(t) y = \cos(t)$

   (g) $y' - \tan(t) y = \sec(t)$

   (h) $y' - \tan(t) y = \sec^2(t)$

2. In the above linear equations find the solution to the initial value problems when $y(0)$ equals the following numbers.

   (a) 1

   (b) 2

   (c) 3

   (d) 4

   (e) -2

   (f) 12

   (g) -3

   (h) -2

3. Solve the following initial value problem. $ty' - y = \frac{1}{t^2}, y(1) = 2$. Would it make any sense to give the initial condition at $t = 0$?

4. Solve the following initial value problem. $ty' + y = \frac{1}{t}, y(-1) = 2$. Would it make any sense to give the initial condition at $t = 0$? **Hint:** You need to remember that $\int \frac{1}{t} dt = \ln|t| + C$.

5. Solve the following initial value problems.

   (a) $\ln(t) y' + \frac{1}{t}y = \ln(t), \ y(2) = 3$.

   (b) $\ln(t) y' - \frac{1}{t}y = \ln^2(t), \ y(2) = 3$.

   (c) $y' + \tan(t) y = \cos^3(t), \ y(0) = 4$.

   (d) $\cosh(t) y' + \sinh(t) y = \sinh(t), \ y(0) = -4$

6. You have the equation $y' + p(t) y = q(t)$ where $P'(t) = p(t)$. Give a formula for all solutions to this differential equation.

7. The height of an object at time t is $y(t)$. It falls from an airplane at 30,000 feet which is traveling East at 500 miles per hour and is acted on by gravity which we will assume has acceleration equal to 32 feet per second squared and air resistance which we will suppose yields an acceleration equal to .1 times the speed of the falling object opposite to the direction of motion. If its initial velocity is in the direction of motion of the airplane, find a formula for the position of the object as a function of t in feet.

8. Solve the following differential equations. Give the general solution.

(a) $(x^3 + y)\, dx - x\, dy = 0$

(b) $y\, dx + (x - y)\, dy = 0$ **Hint:** You might look for x as a function of y.

(c) $y(y^2 - x)\, dy = dx$

(d) $2y\, dx = (x^2 - 1)(dx - dy)$

(e) $L\frac{di}{dt} + Ri = E\sin(\omega t)$. Here L, R, E are positive constants. L symbolizes inductance and R resistance while i is the current.

9. For compounding interest n times in one year which has interest rate r per year, the amount after t years is given by $A_0\left(1 + \frac{r}{n}\right)^{tn}$. Show that

$$\lim_{n \to \infty} \left(1 + \frac{r}{n}\right)^{tn} = e^{rt},$$

thus giving the same conclusion as mentioned in the chapter.

10. Consider the equation $y' + 2ty = t, y(0) = 32.76$. Find $\lim_{t \to \infty} y(t)$.

11. Although the gas supply was shut off, the air in the building continued to circulate. When the gas was shut off, the temperature in the building was 70 and after five hours, the temperature had fallen to a chilly 50 degrees. If the outside temperature was at 10 degrees, what is the constant in Newton's law of cooling?

12. A radioactive substance decays according to how much is present. Thus the equation is $A' = -kA$. If after 40 years, there is 5/6 of the amount initially there still present, what is the half life of this substance?

13. You have the following initial value problem $y' + y = \sin t$, $y(0) = y_0$. Letting y be the solution to this initial value problem, find a function $u(t)$ which does not depend on y_0 and $\lim_{t \to \infty} |y(t) - u(t)| = 0$.

14. A pond which holds V cubic meters is being polluted at the rate of $10 + \sin(2\pi t)$ kg per year. The periodic source represents seasonal variability. The total volume of the lake is constant because it loses $\frac{1}{4}V$ cubic meters per year and gains the same. After a long time, what is the average amount of pollutant in this lake in a year?

Bernouli Equations

15. Solve the following initial value problems involving Bernouli equations.

(a) $y' + 2xy = xy^3,\quad y(1) = 2$

(b) $y' + \sin(t)\, y = \sin(t)\, y^2,\quad y(1) = 1$

(c) $y' + 2y = x^2 y^3,\quad y(1) = -1$

(d) $y' - 2x^3 y = x^3 y^{-1},\quad y(1) = 1$

(e) $y' + y = x^2 y^{-2},\quad y(1) = 1$

(f) $y' + x^3 y = x^3 y^{-2},\quad y(1) = -1$

16. Consider $y' = py - qy^2, y(0) = \frac{p}{mq}$ where p, q are positive and $m > 1$. Solve this Bernouli equation and also find $\lim_{t \to \infty} y(t)$.

17. Consider $y' = 3y - y^3,\ y(0) = 1$. Solve this Bernouli equation and find $\lim_{t \to \infty} y(t)$.

18. Find the solution to the Bernouli equation $y' = (\cos t + 1) y - y^3,\ y(0) = 1$. **Hint:** You may have to leave the solution in terms of an integral.

19. Actually the drag force of a small object moving through the air is proportional not to the speed but to the square of the speed. Thus a falling object would satisfy the following equation for downward velocity. $v' = g - kv^2$. Here g is acceleration of gravity in whatever units are desired. Find $\lim_{t \to \infty} v(t)$ in terms of g, k. **Hint:** Look at the equation.

20. A Riccati equation is like a Bernouli equation except you have an extra function added in. These are of the form $y' = a(t) + b(t) y + c(t) y^2$. If you have a solution, y_1, show that $y(t) = y_1(t) + \frac{1}{v(t)}$ will be another solution provided v satisfies a suitable first order linear equation. Thus the set of all such y will involve a constant of integration and so can be regarded as a general solution to the Riccati equation. These equations result in a very natural way when you consider $y' = f(t, y)$ and approximate $f(t, y)$ by fixing t and approximating the resulting function of y with a second order Taylor polynomial.

Separable Equations

21. Solve the following initial value problems involving separable equations. The ordered pair given is to be included in the solution curve.

(a) $x^2 dx + (y^2 + 1) dy = 0,\ (1, 1)$

(b) $xy dx + (y^2 + 1) dy = 0,\ (1, 1)$

(c) $xy dx + (y^2 + 1) dy = 0,\ (1, -1)$

(d) $y dx + (y^2 + 1) x dy = 0,\ (1, 2)$

(e) $0 = \cos(y) dx + \tan(x) dy,\ \left(\frac{\pi}{2}, \frac{\pi}{4}\right)$

(f) $xy dx = (y^2 + 1) dy,\ (1, 1)$

(g) $xy dx = (y^2 - 1) dy,\ (2, 1)$

22. Find all integral curves of the equation $yx dx + e^{-x^2} dy = 0$. Graph several.

23. Find all integral curves of the equation $yx dx + \frac{1}{\ln(1+x^2)} y^3 dy = 0$. Graph several.

24. Give the integral curves to the equation $v' = g - kv^2$ mentioned above where g is acceleration of gravity and k a positive constant.

25. You have a collection of hyperbolas $x^2 - y^2 = C$ where each choice of C leads to a different hyperbola. Find another collection of curves which intersect these at a right angle. **Hint:** Say you have $f(x, y) = C$ is one of these. If

you are at a point where the relation defines y as a function of x, and (x, y) is a point on one of these hyperbolas just mentioned, then $\frac{dy}{dx}$ should have a relation to the tangent line to $x^2 - y^2 = C$. Since the two curves are to be perpendicular, you should have the product of their slopes equal to -1. Thus $\left(\frac{dy}{dx}\right)\left(\frac{x}{y}\right) = -1$.

26. Generalize the above problem. Suppose you have a family of level curves $f(x, y) = C$ and you want another family of curves which is perpendicular to this family of curves at every point of intersection. Find a differential equation which will express this condition. Recall that two curves are perpendicular if the products of the slopes of the tangent lines to the two curves equals -1.

27. Find and determine the stability of the equilibrium points for the following separable equations.

    (a) $y' = y^2 (y - 1)$

    (b) $y' = (y + 1)(y - 1)(y + 2)$

    (c) $y' = \sin(y)$

    (d) $y' = \cos(y)$

    (e) $y' = \ln(1 + y^2)$

    (f) $y' = e^{2y} - 1$

    (g) $y' = 1 - e^{y^2}$

28. The force on an object of mass m acted on by the earth having mass M is given by Newton's formula kmM/r^2 where k is the gravitation constant first calculated by Cavendish[§] in 1798. Letting R be the radius of the earth and letting g denote the acceleration of gravity on the earth's surface, show that $kM = R^2 g$. Now suppose a large gun having its muzzle at the surface of the earth is fired away from the center of the earth such that the projectile has velocity v_0. Explain why

$$\frac{dv}{dt} = -\frac{R^2 g}{(R + r)^2}$$

where r is the distance to the surface of the earth and here $v = v(t)$ the speed of the projectile at time t when it is at a distance of r from the surface of the earth. Next explain why

$$v\frac{dv}{dr} = -\frac{R^2 g}{(R + r)^2}$$

The two variables are v and r. Separate the variables and find the solution to this differential equation given that the initial speed is v_0 as stated above. Show that the maximum distance from the surface of the earth is given by

$$R\left(\frac{Rg}{Rg - \frac{1}{2}v_0^2} - 1\right)$$

provided that $Rg > \frac{1}{2}v_0^2$. What is the smallest value of v_0 such that the projectile will leave the earth and never return?

[§] For about 100 years, since the time Newton claimed the existence of this gravitation constant, no one knew what it was. Henry Cavendish did an extremely sensitive experiment in 1797-1798 to determine it. It involved lead balls mirrors telescopes and a torsion balance. He was a chemist who also found ways to make hydrogen. He did many other very precise experiments in physics and chemistry.

29. The Grompertz equation is $\frac{dy}{dt} = ry \ln\left(\frac{K}{y}\right)$. Find the solutions to this equation with initial condition $y(0) = y_0$. Also identify all equilibrium solutions and their stability. Also verify the inequality $ry \ln\left(\frac{K}{y}\right) \geq ry\left(1 - \frac{y}{K}\right)$ for $y \in [0, K]$. Explain why for a given initial condition $y_0 \in (0, K)$, the solution to the Grompertz equation should be at least as large as the solution to the logistic equation.

30. You have a population which satisfies the logistic equation $y' = ry\left(1 - \frac{y}{K}\right)$ and the initial condition is $y(0) = \alpha K$ where $0 < \alpha < 1/2$. How long will it take for the population to double?

31. An equilibrium point is called semi-stable if it is stable from one side and not stable from the other. Sketch the appearance of $f(y)$ near y_0 if y_0 is a semi-stable equilibrium point. Here $f(y_0) = 0$ and the differential equation is $y' = f(y)$.

32. Consider the differential equation $y' = a - y^2$ where a is a real number. Show that there are no equilibrium solutions if $a < 0$ but there are two of them if $a > 0$ and only one if $a = 0$. Discuss the stability of the two equilibrium points when $a > 0$. What about stability of equilibrium when $a = 0$?

33. Do exactly the same problem when $y' = ay - y^3$. This time show there are three equilibrium points when $a > 0$ and only one if $a < 0$. Discuss the stability of these points.

34. Do the same problem if $y' = ay - y^2$. These three problems illustrate something called bifurcation which is when the nature of the solutions changes dramatically when some parameter changes.

Homogeneous Equations

35. Find the solution curve to the following differential equations which contains the given point.

   (a) $y' = \frac{1}{x(2x+y)}(x+y)^2$, $(1, 1)$

   (b) $y' = -\frac{1}{x(x-2y)}(x^2 - xy + 2y^2)$, $(2, 0)$

   (c) $y' = \frac{1}{4x^2+yx}(x^2 + 4xy + y^2)$, $(-1, 1)$

   (d) $y' = -\frac{1}{3x^2-xy}(x^2 - 3xy + y^2)$, $(1, 1)$

   (e) $y' = \frac{1}{x(y+5x)}(x^2 + 5xy + y^2)$, $(-1, -1)$

   (f) $y' = \frac{1}{x(3y+2x)}(x^2 + 2xy + 3y^2)$, $(-2, 3)$

   (g) $y' = \frac{1}{x(4y-x)}(x^2 - xy + 4y^2)$, $(3, -2)$

36. Find the solution curve to the following ODEs which contains the given point.

   (a) $y' = \frac{1}{x^2}(x^2 + y^2 + xy)$, $(1, 1)$

   (b) $y' = \frac{1}{x^2}(4x^2 + y^2 + xy)$, $(2, 0)$

   (c) $y' = \frac{1}{x^2}(x^2 + 9y^2 + xy)$, $(3, 1)$

(d) $y' = \frac{1}{x^2}\left(4x^2 + 2y^2 + xy\right), (-1, 1)$

37. Find the solution curve to the following ODEs which contains the given point.

(a) $-(x + y)\, dx + (x + 2y)\, dy = 0, (1, 1)$

(b) $(x - y)\, dx + (x + 3y)\, dy = 0, (2, 1)$

(c) $(4x + y)\, dx + (x + 2y)\, dy = 0, (-1, 2)$

(d) $-(3x + y)\, dx + (x - y)\, dy = 0, (3, 2)$

(e) $(3x - 4y)\, dx + \left(4x - \frac{4}{3}y\right) dy = 0, (3, 1)$

(f) $(-y)\, dx + (4y - x)\, dy = 0, (0, 2)$

(g) $\left(-2x - \frac{31}{4}y\right) dx + \left(x - \frac{9}{4}y\right) dy = 0, (-1, 2)$.

38. Find all solutions to $y' + \sin\left(\frac{y}{x}\right) = 1$. **Hint:** You might need to leave the answer in terms of integrals.

39. Solve: $x^2 dy + \left(4x^2 - xy + 5y^2\right) dx = 0, y(3) = -1$.

40. Solve: $x^2 dy + \left(7x^2 - xy + 4y^2\right) dx = 0, y(2) = -1$.

41. Solve: $x^2 dy + \left(6x^2 - xy + 3y^2\right) dx = 0, y(-1) = 1$.

42. Solve: $\left(x^3 - 7x^2 y - 5y^3\right) dx + \left(7x^3 + 5xy^2\right) dy = 0, y(3) = -2$.

Exact Equations and Integrating Factor

43. Find the solution curve to the following ODEs which contain the given point. First verify that the equation is exact.

(a) $(2xy + 1)\, dx + x^2 dy = 0, (1, 1)$

(b) $(2x \sin y + 1)\, dx + \left(x^2 \cos y\right) dy = 0, \left(1, \frac{\pi}{2}\right)$

(c) $(2x \sin y - \sin x)\, dx + \left((\cos y) x^2 + 1\right) dy = 0, (0, 0)$

(d) $\left(\frac{y}{xy+1}\right) dx + \frac{1}{xy+1}(x + xy + 1)\, dy = 0, (1, 1)$

(e) $\left(y^2 \cos xy^2 + 1\right) dx + \left(2xy \cos xy^2 + 1\right) dy = 0, (1, 0)$

(f) $\left(y\left(\tan^2 xy + 1\right) + y \cos xy\right) dx + \left(x\left(\tan^2 xy + 1\right) + x \cos xy + 1\right) dy = 0,$ $(0, 1)$

44. Find the solution curve to the following ODEs which contains the given point.

(a) $\left(2y^3 + 2\right) dx + \left(3xy^2\right) dy = 0, (1, 1)$

(b) $\left(2y^3 + 2y + 2 \cos\left(x^2\right)\right) dx + \left(3xy^2 + x\right) dy = 0, (1, 1)$

(c) $\left(2xy^2 + y + 2xy \cos x^2\right) dx + \left(2 \sin x^2 + 3x^2 y + 2x\right) dy = 0, (2, 1)$

(d) $3y^4 dx + \left(4xy^3 + \frac{5y^4}{x^2}\right) dy = 0, (1, 2)$

(e) $\left(5x^4 y + 4x^3 y^3\right) dx + \left(3x^5 + 5x^4 y^2\right) dy = 0, (1, 1)$

(f) $\left(8x^4 y^6 + 3x^3\right) dx + \left(12x^5 y^5 + 3xy^2\right) dy = 0, (-1, 2)$

45. Explain why every separable ODE can be considered as an exact ODE.

46. Suppose you have a family of level curves $f(x, y) = C$ where C is a constant. Also suppose that f is a harmonic function. That is $f_{xx} + f_{yy} = 0$. Consider the problem of finding another family of level curves such that each of these is perpendicular to the original level curves $f(x, y) = C$ at any point on both of them. Show that the appropriate equation to solve is $0 = f_y dx - f_x dy$. Verify that this is an exact equation. Thus there exists $g(x, y)$ such that the solutions are $g(x, y) = C$.

M, N Both Affine Linear

47. Find the integral curve for the following differential equation which contains the given point. These are also exact so you could use either method.

  (a) $(2x + y - 3) dx + (x + y - 3) dy = 0, (1, 6)$

  (b) $(y - x + 2) dx + ((x - y) - 2) dy = 0, (3, 2)$

  (c) $(x + y - 3) dx + (x + 3y - 7) dy = 0, (2, 2)$

  (d) $(2x + y - 8) dx + (x + y - 7) dy = 0, (-2, 1)$

  (e) $(x + y - 2) dx + (x + 3y - 4) dy = 0 = 0, (4, 1)$

  (f) $(y - 2x + 5) dx + (x + y + 2) dy = 0, (1, 1)$

  (g) $(y - 4x + 3) dx + (x - 5y + 4) dy = 0, (2, 1)$

48. Find the integral curves for the following differential equation.

  (a) $(2y - x) dx = (4x + y - 9) dy$

  (b) $(5x + 4y - 13) dx = (8x + y - 10) dy$

  (c) $(3x - 2y + 1) dx = (y - 4x - 3) dy$

  (d) $(4y - 4x + 4) dx = (8x + y + 11) dy$

  (e) $(2y - x - 3) dx = (4x + y + 21) dy$

  (f) $(5y - 6x + 23) dx = (10x + y - 29) dy$

An Assortment of Exercises

49. Solve: $y' + 3\cos(t) y = 4 (\cos t) e^{-3\sin t}$, $y(0) = 1$.

50. Solve: $y' + \tan(t) y = \cos(t), y(0) = -2$.

51. Solve: $x^2 dy + (4x^2 - xy + 3y^2) dx = 0$, $y(2) = -2$.

52. Solve: $\left(\frac{7}{2}y - 2x\right) dx + \left(x - \frac{9}{4}y\right) dy = 0$ which contains the point $(x, y) = (1, 2)$.

53. Solve: $x^2 dy + (3x^2 - xy + 2y^2) dx = 0$, $y(2) = -3$.

54. Solve: $(x^3 - 6x^2 y - y^3) dx + (6x^3 + xy^2) dy = 0$, $y(2) = -3$. Graph the integral curve.

55. Solve: $(2y - 3x) dx + \left(2x - \frac{4}{3}y\right) dy = 0$ which contains the point $(x, y) = (1, 2)$. Graph the integral curve.

56. Solve: $y' + 5\cos(3t)\, y = 2e^{-(5/3)\sin 3t}\cos 3t$, $y(0) = 2$.

57. Solve: $x^2 dy + \left(5x^2 - xy + 5y^2\right) dx = 0$, $y(-2) = -2$.

58. Solve: $\left(3x + \frac{19}{4}y\right) dx + \left(-4x - \frac{9}{4}y\right) dy = 0$ which contains the point $(x, y) = (1, 2)$.

59. Solve: $\left(x^3 - 3x^2 y - y^3\right) dx + \left(3x^3 + xy^2\right) dy = 0$, $y(3) = -1$.

60. Solve: $(y)\, dx + (x + 4y)\, dy = 0$ which contains the point $(x, y) = (1, 2)$.

61. Solve: $5\left(t^6\right) y + y' = -5t^6 e^{t^7}$, $y(1) = 1$.

62. Solve: $x^2 dy + \left(6x^2 - xy + 5y^2\right) dx = 0$, $y(3) = 3$.

63. Find the solutions to the equation $y' + y\left(3\cos t\right) = 3\left(\cos t\right) e^{-3\sin t}$.

64. Solve: $(y - 2x)\, dx + \left(\frac{9}{2}y - x\right) dy = 0$ which contains the point $(x, y) = (1, 2)$.

65. Solve: $x^2 dy + \left(2x^2 - xy + y^2\right) dx = 0$, $y(2) = -1$.

66. Find the solutions to the equation $y' + 2ty = te^{t^2}$.

67. Solve: $\left(\frac{7}{3}y - 2x\right) dx + \left(x - \frac{4}{3}y\right) dy = 0$ which contains the point $(x, y) = (1, 2)$.

68. Solve: $y' + \tan(2t)\, y = \cos 2t$, $y(0) = 2$.

69. Find the general solution to the equation

$$y' + \left(4x^3 + x^2 + 3x\right) y = \exp\left(-x^4 - \frac{1}{3}x^3 - \frac{3}{2}x^2\right)\ln(x + 1)$$

70. Show that the following initial value problem fails to have a unique solution.

$$y' = y^{1/(2n+1)}, y(0) = 0, n \text{ a positive integer.}$$

71. Sometimes you have an equation of the form

$$y'' = f(y, y')$$

and you are looking for a function $t \to y(t)$ so the independent variable is missing. These can be massaged into a first order equation as follows. Let $v = y'$ and then you have

$$v' = f(y, v)$$

Now $\frac{dv}{dt} = \frac{dv}{dy}\frac{dy}{dt} = \frac{dv}{dy}v$. Thus we have

$$v\frac{dv}{dy} = f(y, v)$$

which is now a first order differential equation. Use this technique to solve the following problems. This won't always work. It is a gimmick which sometimes works.

(a) $y'' + 2y' = 0, y(0) = 1, y'(0) = 0$

(b) $y'' = y'(2y + 1), y(0) = 0, y'(0) = 1$

(c) $y'' = 2yy', y(0) = 0, y'(0) = 1$

(d) $y'' = y'(1 - 3y^2), y(0) = 1, y'(0) = 0$

(e) $y'y'' = 2, y(0) = 1, y'(0) = 2$

(f) $y'' = 2y, y(0) = 1, y'(0) = 2$

(g) $y'y'' + 3y = 0, y(0) = y'(0) = 1$

(h) $(1 + 3t^2) y'' + 6ty' - \frac{3}{t^2} = 0, y'(1) = 1, y(1) = 2$. **Hint:** This is not like the above but $\frac{d}{dt}((1 + 3t^2) y')$ gives the first two terms.

(i) $yy'' + (y')^2 = 0$. Give a general solution involving two constants of integration.

(j) $y'' + y(y')^2 = 0$. Give a general solution involving two constants of integration.

(k) $y''y^2 - 2y(y')^2 = 0$, Give a general solution involving two constants of integration.

(l) $y''y^3 - 3y'y^2 = 0$, Give a general solution involving two constants of integration.

(m) $3(y')^2 y''y^2 + 2y(y')^4 = 0$, Give a general solution involving two constants of integration.

72. Explain how you would proceed to solve an equation of the form $y'' = f(t, y')$ where the function you are looking for is $t \to y(t)$. How many independent constants would you have in a general solution?

Computer Algebra Problems

73. Give a graph of the solution to the following initial value problem on the interval $[0, 5]$. $y' = -y^3 + 3y^2 + 2$, $y(0) = 0$.

74. Give a graph of the solution to the following initial value problem on the interval $[0, 5]$.$y' = -y^3 + xy^2 + 1$, $y(0) = 1$.

75. Solve the following initial value problems and give a graph of each on $[0, 3]$ on the same axes. $y' = \frac{1}{10}y(5 - y), y(0) = .3, y' = \frac{1}{10}y(5 - y), y(0) = .5, y' = \frac{1}{10}y(5 - y), y(0) = -.3$.

76. Give a graph of the solutions to the differential equation $y' = ty^2 - (.1)y^3$ on the interval $[0, 5]$ which result from the initial conditions $y(0) = 1, 0, 2, -3$.

77. Give a graph of the solution to $y' = x(y^2)^{3/4} - xy^3 + 1$, $y(0) = 0$.

78. Use a computer algebra system to obtain a solution to the initial value problem

$$y' = \frac{y^3}{x^3 + 8y^3}, \ y(0) = 1$$

You may have to obtain a numerical solution in terms of a graph. It is true that the equation is homogeneous, but it might be too hard to carry out the computations. Scientific notebook has trouble with this one.

79. Use a computer algebra system to obtain the graph of the solution to the initial value problem $x^2 y' = 4x^2 + xy + y^2, y(4) = 1$.

80. Find the solution to the following initial value problem, either a graph or a formula. Then graph it

$$y' = xy + \sin(x) - \frac{1}{10} y^2, \ y(0) = 1$$

81. When you use MATLAB or other computer algebra system to find a numerical solution to a differential equation, you are using a fairly sophisticated numerical method. The most primitive method for obtaining numerical solutions to $y' = f(t, y)$ is called Euler's method. In this method, one has a step size h and partitions the time interval into $t_0 < t_1 < \cdots < t_n = T, t_{j+1} = t_j + h$. Then letting y_0 be the initial condition, Euler's method goes like this. You iterate the following process.

$$k = f(t_i, y_i), \ y_{i+1} = y_i + hk, \ t_{i+1} = t_i + h$$

When you get to t_n, you stop. Your solution consists of a function y which interpolates the points (t_i, y_i) meaning $y_i = y(t_i)$. You can easily get MATLAB to do this for you. Here is the case of $y' = y, y(0) = 1$.

```
f=@(t,y) y; h=.01; y(1)=1; t(1)=0;
hold on; for j=1:500;
k=f(t(j),y(j)); y(j+1)=y(j)+h*k; t(j+1)=t(j)+h;
end; plot(t,y); [t(501),y(501)]
```

The first line is defining the function $f(t, y) = y$. Thus the real solution is e^t. The number $y(501)$ is the Euler solution at 5. Compare with e^5.

82. Suppose you have the initial value problem $y' = y, y(0) = y_0$. You know the solution is $e^t y_0$. Consider the interval $[0, t]$. Consider for $k \leq n$

$$y_{k+1} = y_k + \frac{t}{n} y_k$$

Show that this is the same as finding $y_1, \cdots, y_n$ where

$$\frac{y_{k+1} - y_k}{t/n} = y_k$$

In place of $y'(s) = y(s)$, you have $\frac{y_{k+1} - y_k}{t/n} = y_k$. Now show that $y_n = \left(1 + \frac{t}{n}\right)^n y_0$. What is the limit as $n \to \infty$?

83. Suppose on an interval $[a, a+h]$, you have $y'(t) = f(t, y(t))$ and $z(t) = y(a) + (t-a) f(a, y(a))$. Suppose also the solution y has bounded continuous second derivatives. Show that $|y(h) - z(h)| < Ch^2$ for some constant C. You will need to use Taylor's theorem. This is the local error for the Euler method.

The Linear Part of the Book in Condensed Form

84. Recall Schur's theorem from linear algebra. If you have not seen it, it is Theorem D.8.5 in the appendix. It says that for A an $n \times n$ matrix, there is a unitary matrix $U, U^*U = I$ such that $U^*AU = T$ where T is upper triangular. Using this, and what you know about solving linear equations, show that if $t \to \mathbf{f}(t)$ is a continuous vector valued function, there exists a unique solution to the initial value problem

$$\mathbf{x}' = A\mathbf{x} + \mathbf{f}, \ \mathbf{x}(0) = \mathbf{x}_0 \tag{*}$$

85. Now show that if A is an $n \times n$ matrix, there exists a unique matrix $\Phi(t)$ whose entries are differentiable such that

$$\Phi'(t) = A\Phi(t), \ \Phi(0) = I$$

This is called the fundamental matrix. Other ways to obtain the fundamental matrix are discussed later. **Hint:** Let $\mathbf{x}_0 = \mathbf{e}_i$, $\mathbf{f} = \mathbf{0}$ for i^{th} column of Φ. $\Phi'(t)$ means the matrix obtained by differentiating all entries of $\Phi(t)$.

86. Consider the equation

$$y^{(n)} + a_{n-1}y^{(n-1)} + \cdots + a_1 y' + a_0 y = f$$

with initial condition given on $y^{(k)}, k < n$. Show it can be written as

$$\begin{pmatrix} x(1) \\ x(2) \\ \vdots \\ x(n) \end{pmatrix}' = \begin{pmatrix} 0 & 1 & \cdots & & 0 \\ 0 & 0 & \ddots & & \vdots \\ \vdots & \ddots & \ddots & & 1 \\ -a_0 & \cdots & & -a_{n-2} & -a_{n-1} \end{pmatrix} \begin{pmatrix} x(1) \\ x(2) \\ \vdots \\ x(n) \end{pmatrix} + \begin{pmatrix} 0 \\ 0 \\ \vdots \\ f \end{pmatrix}$$

with initial condition on the vector $\begin{pmatrix} x(1) & x(2) & \cdots & x(n-1) & x(n) \end{pmatrix}^T$.

Part II

Scalar Linear Differential Equations, Methods

Chapter 4

Homogeneous Linear Equations

4.1 Factoring Polynomials

We really can't factor polynomials in general. However, it is traditional to base the theory of differential equations on being able to factor polynomials. The idea is to take a very easy analysis problem and make it into a very hard, possibly impossible algebra problem. The advantage of doing this is that it can give us, when it works, formulas for the solutions and this is a very good thing. This short section is a review of how to factor polynomials. Sometimes you can do it. First of all, here is a fundamental theorem from algebra called the rational root theorem. For a proof, see any good algebra book.

Theorem 4.1.1 *(rational root theorem) Let $a_n x^n + \cdots + a_1 x + a_0 = 0$ where each a_i is an integer and $a_n, a_0 \neq 0$. Then* $\boxed{IF}$ *the equation has any rational solutions, then they are of the form*

$$\pm \frac{factor\ of\ a_0}{factor\ of\ a_n}.$$

The way it works is this. If it has rational roots, then you can factor the polynomial using the techniques about to be presented. If it does not, then you really have no way to factor the polynomial in the general case and therefore, the techniques used in this part of the book to find the solutions will not apply.

In dealing with polynomials, a fundamental result is explained in the next lemma which is called the Euclidean algorithm for polynomials. We say that two polynomials are the same if they have the same coefficients.

Lemma 4.1.2 *Let $f(\lambda)$ and $g(\lambda) \neq 0$ be polynomials. Then there exists a polynomial, $q(\lambda)$ such that*

$$f(\lambda) = q(\lambda) g(\lambda) + r(\lambda)$$

where the degree of $r(\lambda)$ is less than the degree of $g(\lambda)$ or $r(\lambda) = 0$. These polynomials $q(\lambda)$ and $r(\lambda)$ are unique.

Proof: Suppose that $f(\lambda) - q(\lambda) g(\lambda)$ is never equal to 0 for any $q(\lambda)$. If it is, then the conclusion follows. Now suppose $r(\lambda) = f(\lambda) - q(\lambda) g(\lambda)$ and the degree of $r(\lambda)$ is $m \geq n$ where n is the degree of $g(\lambda)$. Say the leading term of $r(\lambda)$ is $b\lambda^m$ while the leading term of $g(\lambda)$ is $\hat{b}\lambda^n$. Then letting $a = b/\hat{b}$, $a\lambda^{m-n} g(\lambda)$ has the same leading term as $r(\lambda)$. Thus the degree of $r_1(\lambda) \equiv r(\lambda) - a\lambda^{m-n} g(\lambda)$ is no more than $m - 1$. Then

$$r_1(\lambda) = f(\lambda) - \left(q(\lambda) g(\lambda) + a\lambda^{m-n} g(\lambda) \right) = f(\lambda) - \left(\overbrace{q(\lambda) + a\lambda^{m-n}}^{q_1(\lambda)} \right) g(\lambda)$$

Denote by S the set of polynomials $f(\lambda) - g(\lambda) l(\lambda)$. Out of all these polynomials, there exists one which has smallest degree $r(\lambda)$. Let this take place when $l(\lambda) = q(\lambda)$. Then by the above argument, the degree of $r(\lambda)$ is less than the degree of $g(\lambda)$. Otherwise, there is one which has smaller degree. Thus $f(\lambda) = g(\lambda) q(\lambda) + r(\lambda)$.

As to uniqueness, if you have $r(\lambda), \hat{r}(\lambda), q(\lambda), \hat{q}(\lambda)$ which work, then you would have

$$(\hat{q}(\lambda) - q(\lambda)) g(\lambda) = r(\lambda) - \hat{r}(\lambda)$$

Now if the polynomial on the right is not zero, then neither is the one on the left. Hence this would involve two polynomials which are equal although their degrees are different. This is impossible. Hence $r(\lambda) = \hat{r}(\lambda)$ and so, matching coefficients implies that $\hat{q}(\lambda) = q(\lambda)$. ∎

Let a be a number and $p(x)$ a polynomial. Then by the Euclidean algorithm, there exist polynomials $q(x)$ and $r(x)$ with the degree of $r(x) = 0$ or else $r(x) = 0$ such that $p(x) = (x - a) q(x) + r(x)$.

Definition 4.1.3 $(x - a)$ *is a factor of a polynomial* $p(x)$ *if there is a polynomial* $q(x)$ *such that* $p(x) = (x - a) q(x)$.

This implies the following theorem.

Theorem 4.1.4 $(x - a)$ *is a factor of* $p(x)$ *if and only if* $p(a) = 0$.

Proof: By the division algorithm, there exists a unique polynomial $q(x)$ such that

$$p(x) = (x - a) q(x) + r(x)$$

where $r(x) = 0$ or else $r(x)$ is a nonzero constant. If $(x - a)$ is a factor of $p(x)$, then $r(x) = 0$ and so $p(a) = 0$. If $p(a) = 0$, then it follows that $r(a) = 0$. However, $r(x)$ is a constant, so it must equal 0. Hence $(x - a)$ is a factor of $p(x)$. ∎

Consider the problem of finding $r(x)$ as well as the polynomial $q(x)$. Say

$$p(x) = a_n x^n + a_{n-1} x^{n-1} + \cdots + a_1 x + a_0$$

and

$$q(x) = b_{n-1} x^{n-1} + b_{n-1} x^{n-2} + \cdots + b_1 x + b_0$$

You are given what the a_k are and you want to find the b_j. You have

$$a_n x^n + a_{n-1} x^{n-1} + \cdots + a_1 x + a_0$$
$$= (x - a) \left(b_{n-1} x^{n-1} + b_{n-2} x^{n-2} + \cdots + b_1 x + b_0 \right) + r \tag{4.1}$$

Comparing the coefficients of x^n on both sides, you must have $b_{n-1} = a_n$. Now consider the coefficient of x^{n-1} on both sides. This must be the same and so $a_{n-1} = -ab_{n-1} + b_{n-2}$ and so you must have $b_{n-2} = a_{n-1} + ab_{n-1}$. You could have done this with the coefficient of x^k for any $k \leq n$ obtaining $a_k = -ab_k + b_{k-1}$, $b_{k-1} = a_k + ab_k$. Summarizing this, for each $1 \leq k < n, b_{n-1} = a_n$, $b_{k-1} = a_k + ab_k$. Thus this will give all the b_k. What about the r? The constant term in both sides of (4.1) must be the same and so $r - ab_0 = a_0$ and so $r = ab_0 + a_0$.

A simple algorithm which will compute these b_k and r using the above considerations is the method of synthetic division. I will outline it in general terms and then show how to implement it for specific examples. First write the coefficients of the polynomial across the top line, the coefficients of the highest powers of x on the left and falling toward the lowest powers on the right. Be sure to include a 0 in the appropriate position if some power of x does not appear. For example, if you have the polynomial $x^4 + 3x^2 + 2$, then on the top line you would write

$$1 \quad 0 \quad 3 \quad 0 \quad 2 \, .$$

Then follow the procedure indicated in the following, starting at the left and working toward the right. In the above example, if you wanted to evaluate when $x = 2$, you would write

	1	0	3	0	2
2					

In the general case, you would write

	a_n	a_{n-1}	a_{n-2}	$\cdots$	a_0
a				$\cdots$	
				$\cdots$	

Place a_n in the bottom of the second column.

	a_n	a_{n-1}	a_{n-2}	$\cdots$	a_0
a		aa_n		$\cdots$	
	a_n			$\cdots$	

Then place $aa_n = ab_{n-1}$ right below the a_{n-1} and add a_{n-1} to this which you place in the bottom of the third column.

	a_n	a_{n-1}	a_{n-2}	$\cdots$	a_0
a		aa_n		$\cdots$	
	a_n	$aa_n + a_{n-1}$		$\cdots$	

You take this, which by the above discussion equals b_{n-2}, multiply it by a and place in the fourth column right under the a_{n-2} and then add to a_{n-2}. This gives b_{n-3}.

Continue doing this process till you get to the end and the entry in the bottom right position is the value of the polynomial evaluated at a.

	a_n	a_{n-1}	a_{n-2}	$\cdots$	a_0
a		ab_{n-1}	ab_{n-2}	$\cdots$	ab_0
	$a_n = b_{n-1}$	$\overbrace{a_{n-1} + ab_{n-1}}^{b_{n-2}}$	$\overbrace{a_{n-2} + ab_{n-2}}^{b_{n-3}}$	$\cdots$	$\overbrace{a_0 + ab_0}^{r}$

Example 4.1.5 *Let* $p(x) = x^4 - 9x^3 + 12x^2 - 3x + 7$. *Find* $p(8)$.

Of course you could just plug in 8 and see what happens.

$$(8)^4 - 9(8)^3 + 12(8)^2 - 3(8) + 7 = 239$$

Now Let's do it the other way.

	1	-9	12	-3	7
8		8	-8	32	8×29
	1	-1	4	29	239

Note that the computations are much easier than doing things like 8^4. This is an easier way to evaluate a polynomial at various values than simply plugging the value in to the polynomial. Also I have found additional information from the bottom row. $x^4 - 9x^3 + 12x^2 - 3x + 7 = (x^3 - x^2 + 4x + 29)(x - 8) + 239$.

The technique of synthetic division and the rational root theorem can be used to factor polynomials sometimes. The rational root theorem can be used to identify possible zeros which are rational numbers. Thus if a is one of those zeros, the r in (4.1) must equal 0. Therefore, the above algorithm will identify a polynomial $q(x)$ such that $p(x) = (x - a)q(x)$ and you will have made at least a first step in factoring the polynomial.

Example 4.1.6 *Let* $p(x) = 3x^4 + 8x^3 - 18x^2 + 60$. *Find* $p(-5)$.

This time I will just use synthetic division.

	3	8	-18	0	60
-5		-15	35	-85	425
	3	-7	17	-85	485

and so $p(-5) = 485$. Thus it is also true that

$$p(x) = (x + 5)(3x^3 - 7x^2 + 17x - 85) + 485$$

Example 4.1.7 *Factor the polynomial* $p(x) = x^3 - 4x^2 + 5x - 2$.

First identify the possible rational roots. These are $\pm 1, \pm 2$. I don't know which of these are really roots and if I did, I would also need to find the polynomial $q(x)$ such that

$$x^3 - 4x^2 + 5x - 2 = (x - a)q(x)$$

A possible rational root is 1. Try it:

	1	-4	5	-2
1		1	-3	2
	1	-3	2	0

This worked. Therefore, the above algorithm also gives

$$p(x) = (x - 1)\left(x^2 - 3x + 2\right)$$

At this point, you could either use the quadratic formula to factor the quadratic polynomial or you could use another application of the rational root theorem. This is what I will do. The possible rational roots are the same so I will try 1 again.

	1	-3	2
1		1	-2
	1	-2	0

and so it worked. Thus $x^2 - 3x + 2 = (x - 1)(x - 2)$ and so $p(x) = (x - 1)^2 (x - 2)$.

Example 4.1.8 *Factor the polynomial* $p(x) = 2x^4 + 7x^3 + x^2 - 7x - 3$ *if possible.*

The possible rational roots are $\pm 1, \pm \dfrac{3}{2}, \pm 3, \pm \dfrac{1}{2}$. I think I have found all possibilities. Now it is necessary to try them. Let's try $-1/2$ first.

	2	7	1	-7	-3
$-1/2$		-1	-3	1	3
	2	6	-2	-6	0

and so $-1/2$ is a root. Furthermore, from the above

$$2x^4 + 7x^3 + x^2 - 7x - 3$$

$$= \left(x + \frac{1}{2}\right)\left(2x^3 + 6x^2 - 2x - 6\right)$$

Now I can use the rational root theorem to identify possible rational roots for the second of the above terms in the product. It looks like the possible ones include the original possibilities and those multiplied by 2. I shall try 1.

	2	6	-2	-6
1		2	8	6
	2	8	6	0

It was a good choice. Thus

$$p(x) = \left(x + \frac{1}{2}\right)(x - 1)\left(2x^2 + 8x + 6\right)$$

At this point, I will factor out the 2 and write it as follows.

$$p(x) = (2x + 1)(x - 1)\left(x^2 + 4x + 3\right)$$

That last one is easy to factor but if not, I could use the quadratic formula to factor it. Thus

$$p(x) = (2x + 1)(x - 1)(x + 3)(x + 1).$$

Example 4.1.9 *Factor $x^4 - 1$ as far as possible.*

The possible rational roots are ± 1. Let's try 1

1	0	0	0	-1
1		1	1	1	1
	1	1	1	1	0

and so $x^4 - 1 = (x - 1)\left(x^3 + x^2 + x + 1\right)$. At this point, you could observe that

$$x^3 + x^2 + x + 1 = x^2(x + 1) + (x + 1) = (x + 1)\left(x^2 + 1\right)$$

and so $x^4 - 1 = (x - 1)(x + 1)\left(x^2 + 1\right)$ and this is as far as you can go if you want the polynomials to have real coefficients because $x^2 + 1$ has no real roots.

It is important to understand that factoring polynomials is a useful skill, but there is no guarantee that you will be able to factor a given polynomial. Sometimes you can't do it. What then? If you are limited to the methods discussed here which are the traditional methods, you simply give up.

4.2   Linear Equations

4.2.1   Real Solutions to the Characteristic Equation

These linear differential equations are of the form

$$y^{(n)} + a_{n-1}(t) y^{(n-1)} + \cdots + a_1(t) y' + a_0(t) y = f$$

where f is some function. In case each a_k is a **constant** we will give a technique for finding the solutions.

When $f = 0$ the equation is called **homogeneous.** To find solutions to this equation having **constant coefficients,** you look for a solution in the form $y = e^{rt}$ and then try to choose r in such a way that it works. Here is a simple example.

Example 4.2.1 *Find solutions to the homogeneous equation*

$$y'' - 3y' + 2y = 0.$$

Following the above suggestion, you look for $y = e^{rt}$. Then plugging this in to the equation yields

$$r^2 e^{rt} - 3r e^{rt} + 2e^{rt} = e^{rt}\left(r^2 - 3r + 2\right) = e^{rt}(r - 2)(r - 1) = 0$$

Now it is clear this happens exactly when $r = 2, 1$. Therefore, both $y = e^{2t}$ and $y = e^t$ solve the equation.

How would this work in general? You have

$$y^{(n)} + a_{n-1} y^{(n-1)} + \cdots + a_1 y' + a_0 y = 0$$

and you look for a solution in the form $y = e^{rt}$. Plugging this in to the equation yields

$$e^{rt}\left(r^n + a_{n-1} r^{n-1} + \cdots + a_1 r + a_0\right) = 0$$

so you need to choose r such that the following **characteristic equation** is satisfied

$$r^n + a_{n-1}r^{n-1} + \cdots + a_1r + a_0 = 0.$$

Then when this is done, $y = e^{rt}$ will be a solution to the equation. The question of what to do when r is not real will be dealt with a little later. However, from Section 1.7, it is still the case that e^{rt} is a solution even if r is complex. The difficulty is that it is desired to write the solution in terms of real functions. This is considered more later.

Example 4.2.2 *Find solutions to the equation*

$$y''' - 6y'' + 11y' - 6y = 0$$

First you write the characteristic equation

$$r^3 - 6r^2 + 11r - 6 = 0$$

and find the solutions to the equation, $r = 1, 2, 3$ in this case. Then some solutions to the differential equation are $y = e^t, y = e^{2t}$, and $y = e^{3t}$.

What happens in the case of a repeated root of the characteristic equation? Here is an example.

Example 4.2.3 *Find solutions to the equation*

$$y''' - y'' - y' + y = 0$$

In this case the characteristic equation is $r^3 - r^2 - r + 1 = 0$ and when the polynomial is factored this yields $(r + 1)(r - 1)^2 = 0$. Therefore, $y = e^{-t}$ and $y = e^t$ are both solutions to the equation. Now in this case $y = te^t$ is also a solution. This is because the solutions to the characteristic equation are $1, 1, -1$ where 1 is listed twice because of the $(r - 1)^2$ in the factored characteristic polynomial. Corresponding to the first occurrence of 1 you get e^t and corresponding to the second occurrence you get te^t.

If the factored characteristic polynomial were of the form $(r + 1)^2 (r - 1)^3$, you would write $e^{-t}, te^{-t}, e^t, te^t, t^2e^t$. This is described in the following procedure

Procedure 4.2.4 *To find solutions to the homogeneous equation*

$$y^{(n)} + a_{n-1}y^{(n-1)} + \cdots + a_1y' + a_0y = 0,$$

You find solutions r to the characteristic equation

$$r^n + a_{n-1}r^{n-1} + \cdots + a_1r + a_0 = 0$$

and then $y = e^{rt}$ will be a solution to the differential equation. For every λ a repeated zero of order k, meaning $(r - \lambda)^k$ occurs in the factored characteristic polynomial, you also obtain as solutions the additional functions.

$$te^{\lambda t}, t^2 e^{\lambda t}, \cdots, t^{k-1}e^{rt}$$

Why do we care about these other functions? This involves the notion of general solution or a fundamental set of solutions. You notice the above procedure always delivers exactly n solutions to the differential equation. It turns out that to obtain all possible solutions you need all n.

Letting $Dy = y'$, we write $(D - a) y$ defined as $y' - ay$ and for k a positive integer, $(D - a)^k y$ is defined as $(D - a) (D - a)^{k-1} y$. If the characteristic equation is of the form

$$(r - a_1)^{r_1} \cdots (r - a_m)^{r_m} = 0$$

This corresponds to the differential equation being of the form

$$(D - a_1)^{r_1} \cdots (D - a_m)^{r_m} y = 0$$

Then the following lemma gives some idea why the above procedure holds.

Lemma 4.2.5 *Let* $0 \le m < k$. *Then* $(D - a)^k t^m e^{at} = 0$.

Proof: It is clearly true if $k = 1$. Suppose it is true for some k and $m < k + 1$ so $m - 1 < k$

$$
\begin{aligned}
(D - a)^{k+1} t^m e^{at} &= (D - a)^k \left(m t^{m-1} e^{at} + a t^m e^{at} - a t^m e^{at} \right) \\
&= (D - a)^k \left(m t^{m-1} e^{at} \right) = 0
\end{aligned}
$$

by induction. ∎

Thus $e^{at}, t e^{at}, \cdots, t^{k-1} e^{at}$ are each solutions to the equation $(D - a)^k y = 0$ and consequently each are solutions to the differential equation

$$(D - b_1)^{l_1} \cdots (D - b_r)^{l_r} (D - a)^k y = 0$$

where the characteristic equation is of the form $(r - b_1)^{l_1} \cdots (r - b_r)^{l_r} (r - a)^k = 0$. Then for each $(D - b_j)^{l_j}$ you do something similar. This gives you a total of $n = l_1 + l_2 + \cdots + k$ solutions.

This is a short explanation why the above procedure works.

4.2.2   Superposition and General Solutions

This is concerned with differential equations of the form

$$Ly (t) \equiv y^{(n)} (t) + a_{n-1} (t) y^{(n-1)} (t) + \cdots + a_1 (t) y' (t) + a_0 (t) y (t) = 0 \quad (4.2)$$

in which the functions $a_k (t)$ are continuous. The fundamental thing to observe about L is that it is linear.

Definition 4.2.6 *Suppose* L *satisfies the following condition in which* a *and* b *are numbers and* y_1, y_2 *are functions.*

$$L (ay_1 + by_2) = aLy_1 + bLy_2$$

Then L *is called a linear operator.*

Proposition 4.2.7 *Let* L *be given in (4.2). Then* L *is linear.*

Proof: To save space, note that L can be written in summation notation as

$$Ly(t) = y^{(n)}(t) + \sum_{k=0}^{n-1} a_k(t) y^{(k)}(t)$$

Then letting a, b be numbers and y_1, y_2 functions,

$$L(ay_1 + by_2)(t) \equiv (ay_1 + by_2)^{(n)}(t) + \sum_{k=0}^{n-1} a_k(t)(ay_1 + by_2)^{(k)}(t)$$

Now remember from calculus that the derivative of a sum is the sum of the derivatives and also the derivative of a constant times a function is the constant times the derivative of the function. Therefore,

$$L(ay_1 + by_2)(t) = ay_1^{(n)}(t) + by_2^{(n)}(t) + \sum_{k=0}^{n-1} a_k(t)\left(ay_1^{(k)}(t) + by_2^{(k)}(t)\right)$$

$$= ay_1^{(n)}(t) + a\sum_{k=0}^{n-1} a_k(t) y_1^{(k)}(t) + by_2^{(n)}(t) + b\sum_{k=0}^{n-1} a_k(t) y_2^{(k)}(t)$$

which equals $aLy_1(t) + bLy_2(t)$, so L is linear. ∎

Corollary 4.2.8 *If L is linear, then for a_k scalars and y_k functions,*

$$L\left(\sum_{k=1}^{m} a_k y_k\right) = \sum_{k=1}^{m} a_k L y_k.$$

Proof: The statement that L is linear applies to $m = 2$. Suppose you have three functions.

$$L(ay_1 + by_2 + cy_3) = L((ay_1 + by_2) + cy_3)$$

$$= L(ay_1 + by_2) + cLy_3 = aLy_1 + bLy_2 + cLy_3$$

Thus the conclusion holds for $m = 3$. Following the same pattern just illustrated, you see it holds for $m = 4, 5, \cdots$ also.

More precisely, assuming the conclusion holds for m,

$$L\left(\sum_{k=1}^{m+1} a_k y_k\right) = L\left(\sum_{k=1}^{m} a_k y_k + a_{m+1} y_{m+1}\right)$$

$$= L\left(\sum_{k=1}^{m} a_k y_k\right) + a_{m+1} L y_{m+1}$$

and assuming the conclusion holds for m, the above equals

$$\sum_{k=1}^{m} a_k L y_k + a_{m+1} L y_{m+1} = \sum_{k=1}^{m+1} a_k L y_k \quad \blacksquare$$

The principle of superposition applies to any linear operator in any context. Here the operator is the one defined above but the same result applies to any other example of a linear operator. The following is called the principle of superposition. It says that if you have some solutions to $Ly = 0$ you can multiply them by constants and add up the products and you will still have a solution.

Theorem 4.2.9 *Let L be a linear operator and suppose $Ly_k = 0$ for $k = 1, 2, \cdots, m$. Then if $a_1, \cdots, a_m$ are scalars,*

$$L\left(\sum_{k=1}^{m} a_k y_k\right) = 0$$

Proof: This follows because L is linear.

$$L\left(\sum_{k=1}^{m} a_k y_k\right) = \sum_{k=1}^{m} a_k L y_k = \sum_{k=1}^{m} a_k 0 = 0. \ \blacksquare$$

Example 4.2.10 *Find many solutions to the equation $y'' - 2y' + y = 0$.*

Recall how you do this. You write down the characteristic equation $r^2 - 2r + 1 = 0$ finding the solutions are $r = 1, 1$, there being a repeated zero. Then you know both e^t and te^t are solutions. It follows from the principle of superposition that any function of the form $C_1 e^t + C_2 te^t$ is a solution.

Consider the above example. You can pick the C_1 and C_2 any way you want so you have indeed found many solutions. What is the obvious question to ask at this point? In case you are not sure, here it is:

You have lots of solutions but do you have them all?

The answer to this question comes from linear algebra and a fundamental existence and uniqueness theorem. First, here is the fundamental existence and uniqueness theorem.

Theorem 4.2.11 *Let L be given in (4.2) and let $y_0, y_1, \cdots, y_{n-1}$ be given numbers and $(a, b), \infty \le a < b \le \infty$, be an interval on which each $a_k(t)$ in the definition of L is continuous. Also let $f(t)$ be a function which is continuous on (a, b). Then if $c \in (a, b)$, there exists a unique solution y to the initial value problem*

$$Ly = f, \ y(c) = y_0, y'(c) = y_1, \cdots, y^{(n-1)}(c) = y_{n-1}.$$

I will present a proof of a generalization of this important result later. It also follows from Problem 10 on Page 260 and Problem 86 on Page 76.

Example 4.2.12 *Two solutions of $L(y) = y'' - 3y' + 2y = 0$ are e^{2t} and e^t. Why are all solutions of the form $y_C = C_1 e^{2t} + C_2 e^t$?*

Let z be a solution to the equation $L(y) = 0$. Can you solve the equations

$$y_C(0) = z(0), \ y_C'(0) = z'(0)?$$

In other words, can you solve $C_1 + C_2 = z(0)$, $2C_1 + C_2 = z'(0)$? Yes because this involves the system

$$\begin{pmatrix} y_1(0) & y_2(0) \\ y_1'(0) & y_2'(0) \end{pmatrix} \begin{pmatrix} C_1 \\ C_2 \end{pmatrix} = \begin{pmatrix} 1 & 1 \\ 2 & 1 \end{pmatrix} \begin{pmatrix} C_1 \\ C_2 \end{pmatrix} = \begin{pmatrix} z(0) \\ z'(0) \end{pmatrix}$$

which has a solution because the matrix is invertible. Therefore, $y_{\mathbf{C}}(t) = z(t)$ when $t = 0$ and both $y_{\mathbf{C}}$ and z satisfy $Ly = 0$. By the uniqueness part of the above theorem, $y_{\mathbf{C}} = z$ and so this is indeed the general solution because all solutions are of the form $y_{\mathbf{C}}$ for suitable choice of $\mathbf{C}$. More generally, we have the following definition. *

Definition 4.2.13 *Let $y_1, \cdots, y_n$ be functions which have $n-1$ derivatives. Then*

$$W(y_1(t), \cdots, y_n(t))$$

is defined by the following determinant.

$$\det \begin{pmatrix} y_1(t) & y_2(t) & \cdots & y_n(t) \\ y_1'(t) & y_2'(t) & \cdots & y_n'(t) \\ \vdots & \vdots & & \vdots \\ y_1^{(n-1)}(t) & y_2^{(n-1)}(t) & \cdots & y_n^{(n-1)}(t) \end{pmatrix}$$

This determinant is called the Wronskian.

Example 4.2.14 *Find the Wronskian of the functions t^2, t^3, t^4.*

By the above definition

$$W(t^2, t^3, t^4) \equiv \det \begin{pmatrix} t^2 & t^3 & t^4 \\ 2t & 3t^2 & 4t^3 \\ 2 & 6t & 12t^2 \end{pmatrix} = 2t^6$$

Now the way to tell whether you have all possible solutions is contained in the following fundamental theorem. Sometimes this theorem is referred to as the Wronskian alternative.

Theorem 4.2.15 *Let L be given in (4.2) and suppose each $a_k(t)$ in the definition of L is continuous on (a, b), some interval such that $\infty \le a < b \le \infty$. Suppose for $i = 1, 2, \cdots, n, Ly_i = 0$. Then for any choice of scalars $C_1, \cdots, C_n$, $\sum_{i=1}^n C_i y_i$ is a solution of the equation $Ly = 0$. All possible solutions of this equation are obtained in this form if and only if for some $c \in (a, b)$,*

$$W(y_1(c), y_2(c), \cdots, y_n(c)) \ne 0.$$

*Furthermore, $W(y_1(t), y_2(t), \cdots, y_n(t))$ is either **always** equal to 0 for all $t \in (a, b)$ or **never** equal to 0 for any $t \in (a, b)$.*

---

*Jozef Maria Hoene Wronski (1776-1853) was a Polish philosopher, and mathematician. He was known for ridiculously grandiose ideas and speculation about a very wide variety of topics. The Wronskian may be his most significant contribution to mathematics. It was named this in 1882 and it appears to have been a part of Wronski's work on series.

In the case that all possible solutions are obtained as the above sum, we say $\sum_{i=1}^n C_i y_i$ is the **general solution** and that the functions y_i form a **fundamental set of solutions.**

Proof: Suppose for some $c \in (a,b)$, $W(y_1(c), y_2(c), \cdots, y_n(c)) \neq 0$. Suppose $Lz = 0$. Then consider the numbers $z(c), z'(c), \cdots, z^{(n-1)}(c)$. Since

$$W(y_1(c), y_2(c), \cdots, y_n(c)) \neq 0,$$

it follows that the matrix

$$M(c) \equiv \begin{pmatrix} y_1(c) & y_2(c) & \cdots & y_n(c) \\ y_1'(c) & y_2'(c) & \cdots & y_n'(c) \\ \vdots & \vdots & & \vdots \\ y_1^{(n-1)}(c) & y_2^{(n-1)}(c) & \cdots & y_n^{(n-1)}(c) \end{pmatrix}$$

has an inverse. Therefore, there exists a unique $\mathbf{C} \equiv \begin{pmatrix} C_1 & \cdots & C_n \end{pmatrix}^T$ such that

$$M(c)\mathbf{C} = \begin{pmatrix} z(c) & z'(c) & \cdots & z^{(n-1)}(c) \end{pmatrix}^T \qquad (*)$$

Now consider the function $y(t) = \sum_{k=1}^n C_k y_k(t)$. By the principle of superposition, $Ly = 0$ and in addition it follows from $*$ that $y^{(k)}(c) = z^{(k)}(c)$ for $k = 0, 1, \cdots, n-1$. By the uniqueness part of Theorem 4.2.15, $y(t) = z(t)$. Therefore, since $z(t)$ was arbitrary, this has shown that all solutions are obtained by varying the constants in the sum $\sum_{k=1}^n C_k y_k(t)$. This shows that if $W(y_1(c), y_2(c), \cdots, y_n(c)) \neq 0$ for some $c \in (a,b)$ then the general solution is obtained.

Suppose now that for some $c \in (a,b)$, $W(y_1(c), y_2(c), \cdots, y_n(c)) = 0$. I will show that in this case the general solution is **not obtained.** Since this Wronskian is equal to 0, the matrix $M(c)$ defined above is not invertible. Therefore, there exists $\mathbf{z} = \begin{pmatrix} z_0 & z_1 & \cdots & z_{n-1} \end{pmatrix}^T$ with no solution $\mathbf{C} = \begin{pmatrix} C_1 & C_2 & \cdots & C_n \end{pmatrix}^T$ to the system of equations

$$M(c)\mathbf{C} = \mathbf{z} \qquad (**)$$

From the existence part of Theorem 4.2.15, there exists z such that $Lz = 0$ and it satisfies the initial conditions

$$z(c) = z_0, z'(c) = z_1, \cdots, z^{(n-1)}(c) = z_{n-1}.$$

Therefore, there is no way to write $z(t)$ in the form

$$\sum_{k=1}^n C_k y_k(t) \qquad (4.3)$$

because if you could do so, you could differentiate $n-1$ times and plug in $t = c$ and get a solution to $**$ which has no solution. Therefore, when

$$W(y_1(c), y_2(c), \cdots, y_n(c)) = 0,$$

for any c, you do not get all the solutions to $Ly = 0$ by looking at sums of the form in (4.3) for various choices of $C_1, C_2, \cdots, C_n$.

Why does the Wronskian either vanish for all $t \in (a, b)$ or for no $t \in (a, b)$? This is because (4.3) either yields all possible solutions or it does not. If it does, then the Wronskian cannot equal zero at any $c \in (a, b)$. If it does not yield all possible solutions, then at any point $c \in (a, b)$ the Wronskian cannot be nonzero there. Hence it must be zero there. The following picture illustrates this alternative. The curved line represents the graph of the Wronskian and the picture is what **can't happen** for the reason labeled on the picture. Either it is or it isn't the general solution. You cannot have it both ways.

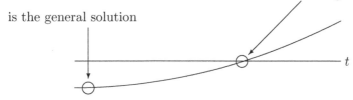

The above theorem shows that the span of finitely many solutions to $Ly = 0$ yields all of the solutions to this equation. It also turns out that these solutions, called a fundamental set of solutions are a basis for the vector space of all solutions. This is discussed in Problem 19 on Page 102. Thus the set of solutions has dimension n when L is an nth order differential operator.

Example 4.2.16 *Find all solutions of the equation*

$$y^{(4)} - 2y^{(3)} + 2y' - y = 0.$$

In this case the characteristic equation is $r^4 - 2r^3 + 2r - 1 = 0$. The polynomial factors.

$$r^4 - 2r^3 + 2r - 1 = (r - 1)^3 (r + 1)$$

Therefore, you can find 4 solutions $e^t, te^t, t^2 e^t, e^{-t}$. The general solution will be

$$C_1 e^t + C_2 t e^t + C_3 t^2 e^t + C_4 e^{-t} \tag{4.4}$$

if and only if the Wronskian of these functions is non zero at some point. First obtain the Wronskian. This is

$$\det \begin{pmatrix} e^t & te^t & t^2 e^t & e^{-t} \\ e^t & e^t + te^t & 2te^t + t^2 e^t & -e^{-t} \\ e^t & 2e^t + te^t & 2e^t + 4te^t + t^2 e^t & e^{-t} \\ e^t & 3e^t + te^t & 6e^t + 6te^t + t^2 e^t & -e^{-t} \end{pmatrix}$$

Now you only have to check at one point. I pick $t = 0$. Then it reduces to

$$\det \begin{pmatrix} 1 & 0 & 0 & 1 \\ 1 & 1 & 0 & -1 \\ 1 & 2 & 2 & 1 \\ 1 & 3 & 6 & -1 \end{pmatrix} = -16 \neq 0$$

Therefore, (4.4) is the general solution.

Example 4.2.17 *Does there exist any interval (a, b) containing 0 and continuous functions $a_2(t), a_1(t), a_0(t)$ such that the functions t^2, t^3, t^4 are solutions of the equation*

$$y''' + a_2(t)y'' + a_1(t)y' + a_0y = 0?$$

The answer is **NO**. This follows from Example 4.2.14 in which it is shown these functions have Wronskian equal to $2t^6$ which equals zero at 0 but is nonzero elsewhere. Therefore, if such functions and an interval containing 0 did exist, it would violate the conclusion of Theorem 4.2.15.

Is there an easy way to tell if the Wronskian is nonzero? Yes if you are looking at two solutions of a second order equation.

Proposition 4.2.18 *Let $a_1(t), a_0(t)$ be continuous on an interval (a, b) and suppose $y_1(t), y_2(t)$ are two functions which solve*

$$y'' + a_1(t)y' + a_0(t)y = 0.$$

Then the general solution is

$$C_1 y_1(t) + C_2 y_2(t)$$

if and only if the ratio $y_2(t)/y_1(t)$ is not constant.

Proof: First suppose the ratio is not constant. Then the derivative of the quotient is nonzero at some point t. Hence by the quotient rule,

$$0 \neq \frac{y_2'(t)y_1(t) - y_1'(t)y_2(t)}{y_1(t)^2} = \frac{W(y_1(t), y_2(t))}{y_1(t)^2}$$

and this shows that at some point $W(y_1(t), y_2(t)) \neq 0$. Therefore, the general solution is obtained.

Conversely, if the ratio is constant, then by the quotient rule as above, the Wronskian equals 0. ■

Example 4.2.19 *Show the general solution of the equation*

$$y'' + 2y' + 2y = 0$$

is

$$C_1 e^{-t} \cos(t) + C_2 e^{-t} \sin(t).$$

You can check and see both of these functions are solutions of the differential equation. Furthermore, their ratio is not a constant. Therefore, by Proposition 4.2.18 the above is the general solution. You know the general solution is also of the form

$$C_1 e^{(-1+i)t} + C_2 e^{(-1-i)t}$$

because both of these functions are solutions and if you check, you will find their Wronskian is nonzero. Indeed, this is obvious because their ratio is not constant. Of course this is less desirable because the functions used have complex values. Recall from Section 1.7

$$e^{(-1+i)t} = e^{-t}(\cos t + i\sin t), \ e^{(-1-i)t} = e^{-t}(\cos t - i\sin t)$$

4.2.3   Case of Complex Zeros

Consider the following equation

$$y'' - 2y' + 2y = 0$$

The characteristic equation for this ordinary differential equation is $r^2 - 2r + 2 = 0$ and the solutions to this equation are $r = 1 \pm i$ where i is the imaginary number which when squared gives -1. You would want the solutions to be $e^{\alpha t}$ where α is one of the complex numbers $1 + i$ or $1 - i$.

Why is it a good idea to look for solutions to a linear constant coefficient equation in the form $e^{\alpha t}$? It is because when you differentiate $e^{\alpha t}$ you get $\alpha e^{\alpha t}$ and so every time you differentiate it, it brings down another factor of α. Finally you obtain a polynomial in α times $e^{\alpha t}$ and then you simply cancel the $e^{\alpha t}$ and find α. This is the case if α is real. Thus we want to define $e^{(a+ib)t}$ in such a way that its derivative is $(a + ib) e^{(a+ib)t}$. Also, to conform to the case where α is real, we require $e^{(a+ib)0} = 1$. Thus it is desired to find a function $y(t)$ which satisfies the following two properties.

$$y(0) = 1, \ y'(t) = (a + ib) y(t). \tag{4.5}$$

In Section 1.7 it was shown that one way to do this is to define

$$e^{(a+ib)t} = e^{at} (\cos bt + i \sin bt).$$

Next it is shown that this is the **only** way to do this.

Proposition 4.2.20 *Let $y(t) = e^{at} (\cos(bt) + i \sin(bt))$. Then $y(t)$ is a solution to (4.5) and furthermore, this is the only function which satisfies the conditions of (4.5).*

Proof: It is easy to see that if $y(t)$ is as given above, then it satisfies the desired conditions. First

$$y(0) = e^0 (\cos(0) + i \sin(0)) = 1.$$

Next

$$\begin{aligned} y'(t) &= ae^{at} (\cos(bt) + i \sin(bt)) + e^{at} (-b \sin(bt) + ib \cos(bt)) \\ &= ae^{at} \cos bt - e^{at} b \sin bt + i \left(ae^{at} \sin bt + e^{at} b \cos bt \right) \end{aligned}$$

On the other hand,

$$(a + ib) \left(e^{at} (\cos(bt) + i \sin(bt)) \right)$$
$$= ae^{at} \cos bt - e^{at} b \sin bt + i \left(ae^{at} \sin bt + e^{at} b \cos bt \right)$$

which is the same thing. Remember $i^2 = -1$.

It remains to verify that this is the only function which satisfies (4.5). Suppose $y_1(t)$ is another function which works. Then letting $z(t) \equiv y(t) - y_1(t)$, it follows

$$z'(t) = (a + ib) z(t), \ z(0) = 0.$$

Now $z(t)$ has a real part and an imaginary part, $z(t) = u(t) + iv(t)$. Then $\overline{z}(t) \equiv u(t) - iv(t)$ and

$$\overline{z}'(t) = (a - ib) \overline{z}(t), \ \overline{z}(0) = 0$$

Then $|z(t)|^2 = z(t)\overline{z}(t)$ and by the product rule,

$$
\begin{aligned}
\frac{d}{dt}|z(t)|^2 &= z'(t)\overline{z}(t) + z(t)\overline{z}'(t) = (a+ib)z(t)\overline{z}(t) + (a-ib)z(t)\overline{z}(t) \\
&= (a+ib)|z(t)|^2 + (a-ib)|z(t)|^2 = 2a|z(t)|^2,\ |z(0)|^2 = 0.
\end{aligned}
$$

Therefore, since this is a first order linear equation for $|z(t)|^2$, it follows the solution is $|z(t)|^2 = 0e^{2at} = 0$. Thus $z(t) = 0$ and so $y(t) = y_1(t)$. ■

Definition 4.2.21 *The formula $e^{a+ib} = e^a(\cos(b) + i\sin(b))$ is called **Euler's formula**. You can get it by letting $t = 1$ in the above proposition.*

Note that the function $e^{(a+ib)t}$ is never equal to 0. This is because its absolute value is e^{at}(why?).

Now everything involving real solutions to the characteristic equation extends to the case where the solutions are complex with no change.

Example 4.2.22 *Find the general solution to*

$$
y^{(4)} - 4y^{(3)} + 8y'' - 8y' + 4y = 0
$$

It turns out that the characteristic equation $r^4 - 4r^3 + 8r^2 - 8r + 4$ can be factored. It equals

$$
(r - (1+i))^2 (r - (1-i))^2
$$

For those who might think there is a special system for finding such a factorization, let me disabuse your minds right now. I know it factors because I cooked it up. We can't factor general polynomials and it is disingenuous to pretend we can. Anyway, the general solution is then

$$
C_1 e^{(1+i)t} + C_2 e^{(1-i)t} + C_3 t e^{(1-i)t} + C_4 t e^{(1+i)t}
$$

Then in terms of familiar functions this is

$$
\begin{aligned}
&C_1 e^t(\cos(t) + i\sin(t)) + C_2 t e^t(\cos(t) + i\sin(t)) \\
&+ C_3 e^t(\cos(t) - i\sin(t)) + C_4 t e^t(\cos(t) - i\sin(t))
\end{aligned}
$$

Of course we don't want to include i. There is nothing mathematically wrong with doing so of course, but the main interest is in real functions. Therefore, it is desirable to write this in terms of real functions. Notice that linear combinations of $e^t(\cos(t) + i\sin(t))$ and $e^t(\cos(t) - i\sin(t))$ are the same set of functions as linear combinations of the functions $e^t\cos t$ and $e^t\sin t$. It follows that the general solution is

$$
D_1 e^t \cos(t) + D_2 t e^t \sin(t) + D_3 t e^t \cos(t) + D_4 e^t \sin(t)
$$

Other examples are similar. When you have a differential operator with real coefficients and $a + ib$ occurs as a root of the characteristic polynomial, this will lead to the two terms

$$
Ce^{at}\cos(t) + De^{at}\sin(t)
$$

in the description of the general solution to $Ly = 0$. If $a + ib$ has multiplicity m, then you will also need to include terms of the form

$$Ct^k e^{at} \cos(t) + Dt^k e^{at} \sin(t)$$

for all $k \leq m - 1$.

Also note that if the differential operator has all real coefficients, the roots to the characteristic equation come in conjugate pairs. Recall why this is so. The complex conjugate, $\overline{a + ib} \equiv a - ib$ has the properties

$$\overline{z + w} = \bar{z} + \bar{w}, \quad \overline{z\, w} = \bar{z}\, \bar{w}$$

Therefore, if $z^n + a_{n-1} z^{n-1} + \cdots + a_0 = 0$ where all the $a_i = 0$, then it follows that

$$
\begin{aligned}
0 &= \bar{0} = \overline{z^n + a_{n-1} z^{n-1} + \cdots + a_0} = \bar{z}^n + \bar{a}_{n-1} \bar{z}^n + \cdots + \bar{a}_0 \\
&= \bar{z}^n + a_{n-1} \bar{z}^n + \cdots + a_0
\end{aligned}
$$

showing that $\bar{z}$ is also a root whenever z is.

Procedure 4.2.23 *When a complex eigenvalue $\lambda = a + ib$ occurs as a root to the characteristic equation of multiplicity m for a homogeneous linear constant real coefficient differential equation, then so does its complex conjugate $\bar{\lambda} = a - ib$ and corresponding to every pair*

$$C_{1k} t^k e^{\lambda t} + C_{2k} t^k e^{\bar{\lambda} t}, \ 0 \leq k \leq m - 1$$

in the general solution, you write

$$C_{1k} t^k e^{at} \cos(bt) + C_{2k} t^k \sin(bt), \ 0 \leq k \leq m - 1$$

The entire reason for doing this is to express the solution in terms of real valued functions rather than complex valued functions. Then the sum of all such functions for $k \leq m - 1$ involving arbitrary C_{ik} is the general solution.

4.3 Finding Another Solution

4.3.1 Using Wronskians

Suppose you have found a solution y to the differential equation

$$y'' + p(t) y' + q(t) y = 0 \tag{4.6}$$

and you would like to find a solution z to the equation

$$z'' + p(t) z' + q(t) z = f(t). \tag{4.7}$$

How could you go about doing it? In case $f(t) = 0$, can you arrange it so that if z is the new solution then $Cz + Dy$ for C, D constants yields the general solution of (4.6)? The answer is yes, and there are easy ways to do it. Suppose z is this new solution. Let $W(y, z)(t)$ be the Wronskian which has the first column corresponding to the known solution y. Then a short computation yields

$$W' = (z'y - y'z)' = z''y - y''z$$

If you know $W(y, z)$, then you know $z'y - y'z$ and this yields a first order linear differential equation for z which you know how to solve. You also know from Theorem 4.2.11 that there is a solution to (4.7). The following lemma shows how to find $W(y, z)$ for y a solution to (4.6) and z a solution to (4.7).

Lemma 4.3.1 *Let y be a solution to (4.6). Then if $z'' + p(t)z' + q(t)z = f(t)$, it follows that $W(y, z)$ is a solution to the following first order linear differential equation.*

$$W(y, z)' + p(t)W(y, z) = y(t)f(t)$$

Proof: If $z'' + p(t)z(t) + q(t)z(t) = f(t)$ and y is a solution to (4.6), then multiplying the equation for z by y and the equation for y by z yields

$$z''y + pz'y + qzy = fy, \quad y''z + py'z + qzy = 0$$

From $*$, $W' = z''y - y''z$. Then subtracting the second from the first,

$$W'(t) + p(t)W(t) = f(t)y(t). \blacksquare$$

This motivates the following procedure.

Procedure 4.3.2 *Suppose y is a known solution to*

$$y'' + p(t)y' + q(t)y = 0 \tag{4.8}$$

To find a solution z which solves

$$z'' + p(t)z' + q(t)z = f$$

do the following: For

$$W = \begin{vmatrix} y & z \\ y' & z' \end{vmatrix}$$

Find a nonzero solution $W(t)$ of

$$W'(t) + p(t)W(t) = f(t)y(t)$$

Then solve for z in the equation

$$z'y - zy' = W$$

To find the general solution to (4.8), just let $f = 0$ in the above and the general solution is $Cz + Dy$ for C, D constants. It will be the general solution because W is not zero.

BE SURE THAT THE EQUATION IS IN THE FORM DESCRIBED IN THE ABOVE PROCEDURE. THIS MEANS THE COEFFICIENT OF y'' IS 1!

Here is an example.

Example 4.3.3 *You can check and see that $y = t^{-1}$ is a solution of the equation*

$$t^2 y'' + 3ty' + y = 0$$

Find a solution to

$$t^2 y'' + 3ty' + y = t^2 \tag{*}$$

I will use the above procedure. First write the equation in the correct form.

$$y'' + \frac{3}{t}y' + \frac{1}{t^2}y = 0, \ z'' + \frac{3}{t}z' + \frac{1}{t^2}z = 1$$

Now solve

$$W' + \frac{3}{t}W = \frac{1}{t}$$

The integrating factor is t^3.

$$(t^3W)' = t^2, \ t^3W = \frac{1}{3}t^3 + C$$

To find a non-constant solution W, let $C = 0$ for simplicity and then $W = \frac{1}{3}$. Now solve

$$z'y - y'z = z'\frac{1}{t} + \frac{1}{t^2}z = \frac{1}{3}$$

$$z' + \frac{1}{t}z = \frac{t}{3}, \ (tz)' = \frac{t^2}{3}, \ tz(t) = \frac{t^3}{9}, \ z(t) = \frac{t^2}{9}$$

This is a solution to $*$.

Example 4.3.4 *In the above example, find all solutions to the equation*

$$t^2y'' + 3ty' + y = 0$$

Writing the equation in the right form,

$$y'' + \frac{3}{t}y' + \frac{1}{t^2}y = 0, \ p(t) = \frac{3}{t}.$$

One solution is $1/t$. Then, following the procedure,

$$W' + \frac{3}{t}W = 0, \ (Wt^3)' = 0, Wt^3 = C, \ W(t) = \frac{C}{t^3}$$

$$z'\frac{1}{t} + \frac{1}{t^2}z = \frac{C}{t^3}, \ z' + \frac{1}{t}z = \frac{C}{t^2}, \ (tz)' = \frac{C}{t}, \ z(t)t = C\ln(t) + D$$

so $z(t) = C\frac{\ln t}{t} + D\frac{1}{t}$.

Example 4.3.5 e^t *is a solution to the equation* $y'' - (1+t)y' + ty = 0$. *Find the general solution.*

From the above procedure,

$$W' - (1+t)W = 0, \ \left(We^{-(t+t^2/2)}\right)' = 0$$

$$W(t) = Ce^{t+t^2/2} = z'y - y'z = e^t z' - e^t z$$

Thus

$$z' - z = Ce^{t^2/2}, \ (e^{-t}z)' = Ce^{t^2/2}e^{-t}, \ z(t)e^{-t} = C\int_0^t e^{-s+s^2/2}ds + D$$

$$z(t) = Ce^t\int_0^t e^{-s+s^2/2}ds + De^t$$

4.3.2   Reduction of Order

Another approach to these problems is to use the method of reduction of order. Recall that the problem is to find a solution z to

$$z'' + p(t)z' + q(t)z = f \tag{4.9}$$

given that y is a solution to

$$y'' + p(t)y' + q(t)y = 0 \tag{4.10}$$

In the method or reduction of order, you look for $z = vy$ where v is an unknown function which is to be found, not a constant. Then

$$z = vy, \ \ z' = v'y + vy', \ \ z'' = v''y + 2v'y' + vy''$$

Then substituting into Equation (4.9),

$$v''y + 2v'y' + vy'' + p(t)(v'y + vy') + q(t)vy = f$$

Now collect the terms which are multiplied by v. These add to 0 because y solves (4.10). This yields

$$v''y + v'(2y' + p(t)y) = f$$

Divide this by y to get

$$v'' + v'\left(\frac{2y'}{y} + p(t)\right) = \frac{1}{y}f \tag{4.11}$$

Letting $w = v'$, this is a first order equation in w,

$$w' + \left(\frac{2y'}{y} + p(t)\right)w = \frac{1}{y}f$$

You solve this linear equation for w and then do another integration to get v. The new solution is then vy. In the special case that $f = 0$ so you are just trying to find another solution to (4.10), you just let $f = 0$ and use

$$w' + \left(\frac{2y'}{y} + p(t)\right)w = 0$$

Procedure 4.3.6 *To use the method of reduction of order to find a solution to*

$$z'' + p(t)z' + q(t)z = f(t)$$

given that

$$y'' + p(t)y' + q(t)y = 0, \tag{*}$$

Find a solution to the first order equation

$$w' + \left(\frac{2y'}{y} + p(t)\right)w = \frac{1}{y}f$$

and then let $v \in \int w(t)\,dt$, v nonconstant. The solution is then $z = vy$. The general solution to $*$ is then $Cz + Dy$ for C, D constants in case $f = 0$.

BE SURE THAT THE EQUATION IS IN THE FORM DESCRIBED IN THE ABOVE PROCEDURE. THIS MEANS THE COEFFICIENT OF y'' IS 1!

Example 4.3.7 *Consider the differential equation $y'' + 2y' + y = 0$. The characteristic equation is then $r^2 + 2r + 1 = 0$ and so it has a repeated root. $r = -1$. Recall that the general solution is then*

$$C_1 e^{-t} + C_2 t e^{-t}$$

Prove that this is the case using the methods above.

Here you need to solve

$$w' + \left(\frac{-2e^{-t}}{e^{-t}} + 2\right) w = 0, \ w' = 0, \ w = C,$$

$$v \in \int w, \ v = Ct + D$$

Then the general solution is

$$Cte^{-t} + De^{-t}$$

which was argued earlier. The ratio of the two functions is not constant and both are solutions so this is the general solution.

Example 4.3.8 *t is a solution of $y'' + ty' - y = 0$. Give the general solution.*

Following the procedure,

$$w' + \left(\frac{2}{t} + t\right) w = 0, \ \left(e^{t^2/2} t^2 w\right)' = 0, \ e^{t^2/2} t^2 w = C$$

$$v = C \int_1^t e^{-s^2/2} s^{-2} ds + D, \ z = t \left(C \int_1^t e^{-s^2/2} s^{-2} ds + D\right)$$

$$z = Ct \int_1^t e^{-s^2/2} s^{-2} ds + Dt$$

Note that $\lim_{t\to 0} t \int_1^t e^{-s^2/2} s^{-2} ds = \lim_{t\to 0} \frac{e^{-t^2/2} t^{-2}}{-1/t^2} = 1$. The ratio of the two functions is not constant so this is the general solution.

Definition 4.3.9 *An equation of the form $y'' + p(t) y' + q(t) y = 0$ is called a homogeneous equation and one of the form $y'' + p(t) y' + q(t) y = f$ where $f \neq 0$ is called a nonhomogeneous equation. Also, an expression of the form*

$$L(y) = a_n y^{(n)} + \cdots + a_1 y' + a_0 y$$

will be called a differential operator. It will always be assumed that the a_i are continuous functions of t. More generally, an equation of the form $Ly = f$ is called a nonhomogeneous equation.

The next chapter is on nonhomogeneous equations.

4.4 Exercises

1. Find the general solution to the equations with the given initial conditions.

(a) $14y + 9y' + y'' = 0$,
$y(0) = -2, y'(0) = 1$

(b) $-12y - 22y' - 6y''$,
$y(0) = -1, y'(0) = 4$

(c) $23y' - 20y - 6y''$,
$y(0) = 2, y'(0) = 2$

(d) $8y'' - 10y' - 12y$,
$y(0) = 1, y'(0) = 6$

(e) $10y' - 3y + 8y''$,
$y(0) = 2, y'(0) = 5$

(f) $y'' - 3y' - 10y = 0$,
$y(-1) = 2, y'(-1) = 3$

(g) $12y + 8y' + y'' = 0$,
$y(2) = -1, y'(2) = 2$

(h) $4y' - 5y + y'' = 0$,
$y(-2) = -1, y'(-2) = 2$

(i) $3y' - 18y + y'' = 0$,
$y(2) = -1, y'(2) = 7$

(j) $y'' - 4y = 0$,
$y(3) = -1, y'(3) = 7$

(k) $y'' - 4y = 0$,
$y(0) = 1, y'(0) = 6$

(l) $y'' - y' - 6y = 0$,
$y(1) = -1, y'(1) = 1$

(m) $y'' - 4y = 0$,
$y(3) = 1, y'(3) = 6$

(n) $7y + 8y' + y'' = 0$,
$y(-2) = 0, y'(-2) = 2$

(o) $21y + 10y' + y'' = 0$,
$y(1) = -1, y'(1) = 4$

2. Find the general solution to the equations with the given initial conditions.

(a) $2y' - 8y + y'' = 0$,
$y(2) = 1, y'(2) = 4$

(b) $9y + 6y' + y'' = 0$,
$y(0) = 1, y'(0) = 5$

(c) $4y - 4y' + y'' = 0$,
$y(0) = -2, y'(0) = 2$

(d) $y + 2y' + y'' = 0$,
$y(0) = -2, y'(0) = 7$

(e) $16y + 8y' + y'' = 0$,
$y(-1) = -2, y'(-1) = 1$

(f) $16y + 8y' + y'' = 0$,
$y(-2) = -1, y'(-2) = 7$

(g) $61y - 12y' + y'' = 0$,
$y(0) = 2, y'(0) = 4$

(h) $13y - 4y' + y'' = 0$,
$y(0) = 2, y'(0) = 3$

(i) $29y - 10y' + y'' = 0$,
$y(0) = 0, y'(0) = 6$

(j) $41y + 10y' + y'' = 0$,
$y(0) = 0, y'(0) = 5$

(k) $25y + 8y' + y'' = 0$,
$y(0) = -2, y'(0) = 3$

(l) $29y + 10y' + y'' = 0$,
$y(0) = -1, y'(0) = 3$

(m) $13y + 6y' + y'' = 0$,
$y(3) = 1, y'(3) = 7$

(n) $26y - 2y' + y'' = 0$,
$y(-2) = 2, y'(-2) = 1$

(o) $45y + 6y' + y'' = 0$,
$y(3) = -1, y'(3) = 6$

Finding Another Solution

3. A differential equation is given along with a solution. Find the general solution.

(a) $y'(4t - 3) - 12y + ty'' = 0$, e^{-4t}.

(b) $12y + 8y' + y'' = 0$, e^{-2t}.

(c) $ty'' - 2y + y'(t - 2) = 0$, e^{-t}.

(d) $y'(2t - 1) - 2y + ty'' = 0$, $1 - 2t$.

   (e) $8y - y'(4t + 2) + ty'' = 0$, $16t^2 + 8t + 2$.

   (f) $y'' \sin t - y'(\cos t + 2\sin t) + 2y \cos t = 0$, e^{2t}.

   (g) $y'' \sin t - y'(\cos t + 5\sin t) + 5y \cos t = 0$, e^{5t}.

   (h) $x^2 y'' + xy' + \left(x^2 - \frac{1}{4}\right) y = 0$, $y(x) = x^{-(1/2)} \sin x$.

   (i) $\left(x^3 + x^2\right) y'' - 2x^2 y' - 6y = 0$, $y(x) = x^3$.

4. One solution to $\left(x^3 + x^2\right) y'' - 2x^2 y' - 6y = 0$ is $y(x) = x^3$. Find a solution to
$$\left(x^3 + x^2\right) y'' - 2x^2 y' - 6y = -\left(2x^2 + 6x\right)$$

5. One solution to $\left(x^3 + 4x^2 + 5x + 2\right) y'' + (x + 1) y' - (6x + 15) y = 0$ is $(x + 1)^3$. Find a solution to
$$\left(x^3 + 4x^2 + 5x + 2\right) y'' + (x + 1) y' - (6x + 15) y = -6x^2 - 2x + 31$$

6. One solution to $\left(x^2 + 1\right) y'' + xy' - \left(x + 1 + x^2\right) y = 0$ is e^x. Find a solution to $\left(x^2 + 1\right) y'' + xy' - \left(x + 1 - x^2\right) y = -x^4 - x^3 + 3x^2 + 2$

7. A solution to $\left(x^4 - x^2\right) y'' + \left(x^5 + x^2\right) y' - \left(2x^4 + 2x^2 + 2x - 2\right) y = 0$ is x^2. Find a solution to
$$\left(x^4 - x^2\right) y'' + \left(x^5 + x^2\right) y' - \left(2x^4 + 2x^2 + 2x - 2\right) y = x^7 + 4x^5 + x^4 - 4x^3$$

8. A solution to $\left(x^4 - x^2\right) y'' + \left(x^5 + x^2\right) y' - \left(2x^4 + 2x^2 + 2x - 2\right) y = 0$ is x^2. Find the general solution. **Hint:** You might need to leave in terms of an integral.

Some Higher Order Problems

9. Find the general solution to the equation $15y - 52y' + 19y'' + 6y''' = 0$. Then determine the solution which also satisfies the initial condition $y(0) = 2, y'(0) = 7, y''(0) = 3$

10. Find the general solution to the equation $6y + 11y' - 4y'' - 4y''' = 0$. Then determine the solution which also satisfies the initial condition $y(0) = 0, y'(0) = 4, y''(0) = 1$

11. Find the general solution to the equation $6y - 8y' - 10y'' + 4y''' = 0$. Then determine the solution which also satisfies the initial condition $y(-2) = -1, y'(-2) = 2, y''(-2) = 0$

12. Find the general solution to the equation $3y^{(4)} - 2y - 9y' - 9y'' + y''' = 0$. Then determine the solution which also satisfies the initial condition $y(0) = -3, y'(0) = 0, y''(0) = 0, y'''(0) = 0$.

13. Find the general solution to the equation $-2y^{(4)} + 54y - 27y' - 18y'' + 13y''' = 0$. Then determine the solution which also satisfies the initial condition $y(0) = 0, y'(0) = -3, y''(0) = -3, y'''(0) = 0$.

14. Find the general solution to the equation $y^{(4)} + 16y - 32y' + 24y'' - 8y''' = 0$. Then determine the solution which also satisfies the initial condition $y(0) = 1, y'(0) = -3, y''(0) = 1, y'''(0) = 7$.

15. Find the general solution to the equation $y^{(4)} + 81y + 108y' + 54y'' + 12y''' = 0$. Then determine the solution which also satisfies the initial condition $y(0) = 2, y'(0) = 3, y''(0) = 3, y'''(0) = 1$.

16. Find the general solution to the equation $y^{(4)} + y + 4y' + 6y'' + 4y''' = 0$. Then determine the solution which also satisfies the initial condition $y(0) = 1, y'(0) = -2, y''(0) = 2, y'''(0) = 7$.

17. Find the general solution to the equation $y^{(4)} - 6y - 5y' + 5y'' + 5y''' = 0$. Then determine the solution which also satisfies the initial condition $y(0) = 0, y'(0) = 0, y''(0) = 0, y'''(0) = 0$.

18. e^x is a solution to the equation

$$xy'' - (x + N) y' + Ny = 0, \ N \text{ a positive integer.}$$

Use the method of reduction of order or direct verification so show that another solution is of the form $e^x \int_0^x t^N e^{-t} dt$. Hence $\frac{-1}{6} e^x \int_0^x t^N e^{-t} dt - e^x$ is also a solution. Now show that another solution is of the form $\sum_{k=0}^N \frac{x^k}{k!}$. This interesting problem is in [5].

Some Theoretical Problems

19. You have the linear operator defined in the chapter

$$Ly(t) \equiv y^{(n)}(t) + a_{n-1}(t) y^{(n-1)}(t) + \cdots + a_1(t) y'(t) + a_0(t) y(t)$$

Suppose you have found the general solution to the equation $Ly = 0$ and it is of the form

$$\sum_{i=1}^n C_i y_i \text{ on an interval } (a, b)$$

where $Ly_i = 0$ for each y_i. Show that $\{y_1, \cdots, y_n\}$ is a linearly independent set of functions on this interval (a, b). Recall what this means. It means that if $\sum_{i=1}^n C_i y_i(t) = 0$ for all t, then each $C_i = 0$. That is, if $\sum_{i=1}^n C_i y_i$ is the zero function, then each $C_i = 0$.

20. Consider the two functions $y_1(t) = t^2$ and $y_2(t) = |t| t$. Show that

$$W(y_1, y_2)(t) = 0$$

but that these two functions are actually linearly independent on the interval $(-2, 2)$. Does this contradict the theorems in the chapter about the Wronskian?

21. Suppose you have some smooth functions having all necessary derivatives $\{y_1, \cdots, y_n\}$ on an interval (a, b). Also suppose that at some point $\hat{t} \in (a, b), W(y_1, \cdots, y_n)(\hat{t}) \neq 0$. Show that then it follows that these functions are indeed linearly independent.

22. Here are two functions. $y_1(x) = x^3$ and $y_2(x) = x^4$. Their Wronskian equals

$$\begin{vmatrix} x^3 & x^4 \\ 3x^2 & 4x^3 \end{vmatrix} = x^6$$

This equals 0 at $x = 0$ but is nonzero everywhere else. Does this contradict the theorem about the Wronskian in the chapter? Didn't this theorem say that the Wronskian either vanished identically or never? Yet here we have two perfectly smooth functions defined on the entire real line and this condition does not hold. What is the problem?

23. Verify that

$$\frac{d}{dt} \begin{vmatrix} y_1 & y_2 \\ y_1' & y_2' \end{vmatrix} = \begin{vmatrix} y_1 & y_2 \\ y_1'' & y_2'' \end{vmatrix}$$

Describe the general result for an $n \times n$ matrix. If you use the description of the determinant in Appendix D, you should be able to prove your result.

24. Now suppose you have solutions $y_1, \cdots, y_n$ to the equation $Ly = 0$ where

$$Ly = y^{(n)} + a_{n-1}(t) y^{(n-1)} + \cdots + a_1(t) y' + a_0(t) y$$

Verify using the above problem that

$$W(y_1, \cdots, y_n)'(t) = -a_{n-1}(t) W(y_1, \cdots, y_n)(t)$$

You will need to use the differential equation $Ly = 0$ and properties of determinants. Prove the following fundamental formula known as Abel's formula.

$$W(y_1, \cdots, y_n)(t) = Ce^{A(t)}$$

where $A'(t) = -a_{n-1}(t)$. Show how this implies the Wronskian alternative of the chapter directly and verifies that n solutions to the differential equation $Ly = 0$ yield a general solution if and only if $W(y_1, \cdots, y_n)(t) \neq 0$ for some t.

25. An Euler equation is one which is of the form

$$t^2 y'' + aty' + by = 0$$

where we are interested in the solution for $t > 0$. Let $t = e^s$. Then show that for $y(t) = y(s)$,

$$\frac{d^2 y}{dt^2} = -\frac{1}{t^2} \frac{dy}{ds} + \frac{1}{t^2} \frac{d^2 y}{ds^2}, \quad \frac{dy}{dt} = \frac{1}{t} \frac{dy}{ds}$$

Show that the Euler equation can be studied in the form

$$y''(s) + (a - 1) y'(s) + by(s) = 0$$

Use this transformation to find the general solution to the following Euler equations.

(a) $t^2 y'' - 2y = 0$

(b) $t^2 y'' + 3ty' + 2y = 0$

(c) $t^2 y'' + 3ty' + y = 0$

(d) $y'' t^2 - y't + 3y = 0$

(e) $y'' t^2 - 3y't + 4y = 0$

(f) $y'' t^2 + 3y't + 10y = 0$

(g) $y'' t^2 + 5y't = 0$

(h) $y'' t^2 - y't + 2y = 0$

(i) $y'' t^2 - y't + y = 0$

(j) $y'' t^2 + 3y't + y = 0$

(k) $y'' t^2 + 5y't + 4y = 0$

(l) $y'' t^2 + y't - 9y = 0$

(m) $y'' t^2 - 2y't + 2y = 0$

(n) $y'' t^2 - y't + y = 0$

(o) $y'' t^2 - 3y't + 4y = 0$

(p) $y'' t^2 + 5y't + 4y = 0$

(q) $y'' t^2 + 3y't + y = 0$

(r) $y'' t^2 - 5y't + 9y = 0$

26. Show that one can find a solution to $z'' + p(t) z' + q(t) z = f(t)$ by the following procedure. Find $W(t) \in e^{-P(t)} \int e^{P(s)} y(s) f(s) \, ds$ where y is a solution to $y'' + p(t) y + q(t) y = 0$ and $P'(t) = p(t)$. Then a solution to the equation of interest is obtained by

$$z \in y(t) \int \frac{W(s)}{y^2(s)} \, ds$$

27. In the method of Wronskians for finding another solution, show that if y is a known nonzero solution to the homogeneous equation as described there and if z is a function such that $W(y, z)' + p(t) W(y, z) = f(t) y(t)$, then z will be a solution to $z'' + p(t) z' + q(t) z = f(t)$.

Chapter 5

Nonhomogeneous Equations

5.1 The General Solution to a Nonhomogeneous Equation

Let $Ly = f$ be a nonhomogeneous equation. Here L is a linear differential operator as in Definition 4.3.9. A **particular solution** is a function y_p such that $Ly_p = f$. Then we have the following fundamental theorem.

Theorem 5.1.1 *Suppose the general solution to $Ly = 0$ is*

$$\sum_{i=1}^{n} C_i y_i$$

where $L(y_i) = 0$. Also suppose $Ly_p = f$ so that y_p is a particular solution. Then the general solution to $Ly = f$ is

$$\sum_{i=1}^{n} C_i y_i + y_p \tag{5.1}$$

Proof: First note that $L\left(\sum_{i=1}^{n} C_i y_i + y_p\right) = \sum_{i=1}^{n} C_i L(y_i) + Ly_p = f$ so the above formula gives a lot of solutions. Does it give them all? Suppose $Lz = f$. Then $L(z - y_p) = Lz - Ly_p = f - f = 0$ and so $z - y_p$ is a solution to $Ly = 0$. Therefore, there exist constants C_i such that

$$z - y_p = \sum_{i=1}^{n} C_i y_i$$

so $z = y_p + \sum_{i=1}^{n} C_i y_i$ and therefore, (5.1) gives all the solutions to $Ly = f$. ∎

This means that if you have the general solution to $Ly = 0$ and a single particular solution to $Ly = f$, then you can find the general solution to $Ly = f$. Thus it is desired to get methods for finding a particular solution.

5.2 Method of Undetermined Coefficients

This is a very good method for solving $Ly = f$ when it works. However, you have to be careful to only try to apply it when it works.

You be sure L is a constant coefficient linear differential operator or else the method does not apply. You also must be sure that f is of the right form. It must be of the form e^{at}, $\sin(bt)$, $\cos(bt)$, $e^{at}\cos(bt)$, $e^{at}\sin(bt)$, or a polynomial times any of these.

Warning 5.2.1 *If the differential operator does not have constant co-efficients, **you must not even imagine using this method. Failure to head this warning will result in calamity. It is hard to overstate the dire consequences of trying to use this method on such a differential operator.***

Example 5.2.2 *Find the solution to $y''' - 2y'' + y' - 2y = \cos(t)$.*

This is the sort of thing you can apply the method to. First solve the homogeneous equation. The solution to the homogeneous equation is

$$C_1 e^{2t} + C_2 \cos(t) + C_3 \sin(t)$$

Now you look for a particular solution in the form $y_p = A\cos(t) + B\sin(t)$. This is because when you differentiate this expression and multiply by constants and add such together, you always get something of this form and the function $\cos(t)$ is also of this form. However, this is

$$\boxed{\textbf{doomed to failure}}$$

because one of the terms in the expression, $\cos(t)$, solves the homogeneous equation and so you must modify y_p as follows. You multiply by t.

$$y_p = At\cos(t) + Bt\sin(t)$$

This one will work because none of the terms solve the homogeneous equation. Then

$$y_p' = A\cos(t) - At\sin(t) + B\sin(t) + Bt\cos(t)$$
$$y_p'' = -2A\sin(t) - At\cos(t) + 2B\cos(t) - Bt\sin(t)$$

and

$$y_p''' = -3A\cos(t) + At\sin(t) - 3B\sin(t) - Bt\cos(t)$$

Now plug in to the equation and require that things work. This yields the following for the left side of the differential equation.

$-3A\cos(t) + At\sin(t) - 3B\sin(t) - Bt\cos(t)$
$-2(-2A\sin(t) - At\cos(t) + 2B\cos(t) - Bt\sin(t))$
$+ (A\cos(t) - At\sin(t) + B\sin(t) + Bt\cos(t))$
$-2(At\cos(t) + Bt\sin(t)) =$
$-2A\cos t - 2B\sin t + 4A\sin t - 4B\cos t$

Thus you must have

$$-2A\cos t - 2B\sin t + 4A\sin t - 4B\cos t = \cos(t)$$

This requires

$$-2A - 4B = 1, \quad -2B + 4A = 0$$

and the solution is $A = -\frac{1}{10}, B = -\frac{1}{5}$. Thus the general solution to this nonhomogeneous problem is the sum of the general solution to the homogeneous problem with a particular solution just found. It is

$$y = C_1 e^{2t} + C_2 \cos(t) + C_3 \sin(t) - \frac{1}{10} t \cos(t) - \frac{1}{5} t \sin(t).$$

There is another way to do this problem which sometimes seems a little easier. It involves using Euler's formula. I will solve instead

$$z''' - 2z'' + z' - 2z = e^{it} \tag{5.2}$$

because the real part of e^{it} equals $\cos(t)$. Thus I will solve for z and then take its real part. In this case the particular solution I would try is

$$z_p = A e^{it}$$

which will not work because $e^{it} = \cos(t) + i \sin(t)$ and both $\cos(t)$ and $\sin(t)$ solve the homogeneous equation. Therefore,

$$z_p = A t e^{it}$$

and now I only have to find A.

$$z_p' = A e^{it} + i A t e^{it}$$

$$z_p'' = 2Aie^{it} - A t e^{it}$$
$$z_p''' = -3A e^{it} - A t i e^{it}.$$

Now plug in to the equation. This yields for the left side of the equation

$$-3A e^{it} - A t i e^{it} - 2\left(2Aie^{it} - A t e^{it}\right)$$

$$+ \left(A e^{it} + i A t e^{it}\right) - 2 A t e^{it} = (-2 - 4i) A e^{it}$$

and this needs to equal the right side of (5.2). Thus

$$A(-2 - 4i) = 1$$

and so, solving for A yields

$$A = -\frac{1}{10} + \frac{1}{5} i$$

It follows

$$z_p = \left(-\frac{1}{10} + \frac{1}{5} i\right) t e^{it}$$

$$= -\frac{1}{10} t \cos t - \frac{1}{5} t \sin t + i \left(\frac{1}{5} t \cos t - \frac{1}{10} t \sin t\right)$$

The real part is

$$y_p = -\frac{1}{10}t\cos(t) - \frac{1}{5}t\sin(t)$$

which is the same answer for the particular solution as before. Do these problems either way you want. However, this way will be a little easier in some cases. I think it was less trouble in the above example. In the next example, it will certainly be much less trouble.

Example 5.2.3 *Solve*

$$y'' + 2y' + 5y = 2e^{-t}\sin(2t)$$

The solution to the homogeneous equation is

$$C_1 e^{-t}\cos(2t) + C_2 e^{-t}\sin(2t)$$

This time it will certainly be much easier to consider the particular solution as the imaginary part of a solution to the equation

$$z'' + 2z' + 5z = 2e^{(-1+2i)t}$$

Note by Euler's identity

$$2e^{(-1+2i)t} = 2e^{-t}\cos(2t) + i2e^{-t}\sin(2t)$$

The particular solution should be of the form

$$z_p = Ae^{(-1+2i)t}$$

but this is not usable because its real and imaginary parts solve the homogeneous equation and so by superposition it also solves the homogeneous equation. Thus

$$z_p = Ate^{(-1+2i)t}$$

Now

$$
\begin{aligned}
z_p' &= At(-1+2i)e^{(-1+2i)t} + Ae^{(-1+2i)t} \\
z_p'' &= At(-1+2i)^2 e^{(-1+2i)t} + 2A(-1+2i)e^{(-1+2i)t}
\end{aligned}
$$

Plugging this in to the left side of the differential equation yields

$$At(-1+2i)^2 e^{(-1+2i)t} + 2A(-1+2i)e^{(-1+2i)t} +$$

$$2\left(At(-1+2i)e^{(-1+2i)t} + Ae^{(-1+2i)t}\right) + 5\left(Ate^{(-1+2i)t}\right)$$

$$= 4iAe^{(-1+2i)t}$$

and this needs to equal the right side. Thus

$$4iAe^{(-1+2i)t} = 2e^{(-1+2i)t}$$

and so $A = \frac{1}{2i} = -\frac{1}{2}i$. Therefore,

$$z_p = \left(-\frac{1}{2}i\right) t \left(e^{(-1+2i)t}\right) = \frac{1}{2} t e^{-t} \sin(2t) - \frac{1}{2} i t e^{-t} \cos(2t)$$

and we want the imaginary part. Thus

$$y_p = -\frac{1}{2} t e^{-t} \cos(2t)$$

and so the general solution is

$$-\frac{1}{2} t e^{-t} \cos(2t) + C_1 e^{-t} \cos(2t) + C_2 e^{-t} \sin(2t)$$

Note that if we had picked the real part of z_p, this would have given a particular solution to

$$y'' + 2y' + 5y = 2e^{-t} \cos(2t)$$

Thus the general solution to this equation is

$$C_1 e^{-t} \cos(2t) + C_2 e^{-t} \sin(2t) + \frac{1}{2} t e^{-t} \sin(2t).$$

Example 5.2.4 *Find a particular solution to the equation*

$$y'''' - y'' = e^t + t. \tag{5.3}$$

In this case the right side is not of the right form to apply the method. Remember, the right side must be a polynomial times a cosine or a sine or a polynomial times an exponential as described above. If you misuse this method, you will have bad results. However, if you split the right side then you can use the method on each piece. First, the solution to the homogeneous equation is

$$C_1 e^t + C_2 e^{-t} + C_3 + C_4 t$$

Now consider

$$y'''' - y'' = e^t.$$

The form for a particular solution should be $y_p = Ae^t$ but this cannot work because it is a solution to the homogeneous equation so the real particular solution is

$$y_p = Ate^t.$$

Then

$$\begin{aligned} y_p' &= Ae^t + Ate^t \\ y_p'' &= 2Ae^t + Ate^t \\ y_p''' &= 3Ae^t + Ate^t \\ y_p'''' &= 4Ae^t + Ate^t \end{aligned}$$

and so plugging in to the equation yields

$$4Ae^t + Ate^t - \left(2Ae^t + Ate^t\right) = 2Ae^t = e^t$$

and so $A = \frac{1}{2}$. Thus the particular solution to this piece is

$$y_{p_1} = \frac{1}{2}te^t$$

Now consider the other piece

$$y'''' - y'' = t$$

In this case you would think

$$y_p = A + Bt$$

but it is wrong because it has a term which solves the homogeneous equation. Multiply by t again. This yields

$$y_p = At + Bt^2$$

and this might very well also be the wrong thing because it has a term which solves the homogeneous equation. Multiply by t again. This yields

$$y_p = At^2 + Bt^3$$

and none of the terms solve the homogeneous equation so this will work out.

$$y'_p = 2At + 3Bt^2, \ y''_p = 2A + 6Bt, y'''_p = 6B, y''''_p = 0$$

and so plugging in to the equation yields

$$2A + 6Bt = t$$

Therefore, $A = 0$ and $B = \frac{1}{6}$. Thus $y_{p_2} = \frac{1}{6}t^3$ and so the original equation, (5.3) has a particular solution,

$$y_p = \frac{1}{6}t^3 + \frac{1}{2}te^t$$

In summary, how do we use this method to find particular solutions?

Procedure 5.2.5 *To find a particular solution by the method of undetermined coefficients, do the following.*

1. *Check to see that the differential operator has all constant coefficients.* **If it doesn't, DON'T USE THE METHOD.**

2. *Check to see if the right side is a polynomial times e^{kt} for $k \in \mathbb{C}$ or a polynomial times either $\sin(at)$ or $\cos(at)$ or a polynomial times $e^{kt}\cos(at)$ or a polynomial times $e^{kt}\sin(at)$.* **If this is not so, DON'T USE THE METHOD.** *If the right side is the sum of things like this,* **solve separate problems** *corresponding to each term which* **is** *like this. Use the following steps to deal with each summand. Then add your particular solutions together to get a particular solution for the original problem.*

3. *Write the proposed particular solutions in terms of arbitrary constants which are to be determined. This needs to be something that will remain of the desired form when you do a constant coefficient differential operator to it. If any of the terms solves the homogeneous equation, then it won't work. Multiply the **whole expression** by t and check the terms again. Continue doing this till you find one where none of the terms solves the homogeneous equation. The result is the form of the desired particular solution.*

4. *Substitute the expression you just got into the nonhomogeneous differential equation and require that it works. Solve for the constants.*

5.3   Method of Variation of Parameters

In this case the equation is of the form

$$Ly \equiv y^{(n)} + a_{n-1}(t) y^{(n-1)} + \cdots + a_1(t) y' + a_0(t) y = f(t)$$

and you assume a general solution is known to $Ly = 0$ which is of the form

$$\sum_{k=1}^{n} C_k y_k.$$

$$\boxed{\textbf{NOTE THAT THE COEFFICIENT OF } y^{(n)} \textbf{ IS } 1.}$$

If this is not so, divide to make it so. Failure to make this happen will result in a disaster. If you observe carefully the following explanation, you will see why this is so.

Then you look for a solution to the nonhomogeneous equation which is of the form

$$y_p = \sum_{k=1}^{n} A_k y_k.$$

where instead of constants, the A_k are functions. The idea is to differentiate and plug in and make the functions A_k such that the above is a particular solution.

$$y_p' = \overbrace{\sum_{k=1}^{n} A_k' y_k}^{=0} + \sum_{k=1}^{n} A_k y_k'$$

As indicated, I will set the first sum equal to 0. Therefore,

$$y_p'' = \sum_{k=1}^{n} A_k' y_k' + \sum_{k=1}^{n} A_k y_k''$$

and again set the first sum to equal 0. Each time set equal to 0 the expression which involves A_k'. Then continuing this way, finally results in

$$y_p^{(n-1)} = \sum_{k=1}^{n} A_k' y_k^{(n-2)} + \sum_{k=1}^{n} A_k y_k^{(n-1)}$$

and as before, set the first term equal to 0. Then

$$y_p^{(n)} = \sum_{k=1}^{n} A_k' y_k^{(n-1)} + \sum_{k=1}^{n} A_k y_k^{(n)}$$

Plugging in to the equation and using the fact that each y_k is a solution of the homogeneous equation, the terms involving A_k all cancel and you are left with

$$\sum_{k=1}^{n} A_k' y_k^{(n-1)}$$

which must then equal $f(t)$. Thus this is a particular solution if A_k' solves the following system of equations.

$$\sum_{k=1}^{n} A_k' y_k = 0, \ \sum_{k=1}^{n} A_k' y_k' = 0, \ \cdots, \ \sum_{k=1}^{n} A_k' y_k^{(n-2)} = 0, \ \sum_{k=1}^{n} A_k' y_k^{(n-1)} = f(t)$$

Let $M(t)$ denote the matrix you take the determinant of to get the Wronskian and let $M_i(f)(t)$ denote the matrix which replaces the ith column with the column

$$\begin{pmatrix} 0 & 0 & \cdots & 0 & f(t) \end{pmatrix}^T.$$

Thus

$$M(t) = \begin{pmatrix} y_1 & y_2 & \cdots & y_n \\ y_1' & y_2' & \cdots & y_n' \\ \vdots & \vdots & & \vdots \\ y_1^{(n-1)} & y_2^{(n-1)} & \cdots & y_n^{(n-1)} \end{pmatrix}$$

and the above system is of the form

$$M(t) \begin{pmatrix} A_1' \\ A_2' \\ \vdots \\ A_n' \end{pmatrix} = \begin{pmatrix} 0 \\ 0 \\ \vdots \\ f \end{pmatrix}$$

and so by Cramer's rule,

$$A_k'(t) = \frac{\det(M_k(f)(t))}{\det(M(t))} = \frac{\det(M_k(f)(t))}{W(y_1, \cdots, y_n)(t)}$$

and so

$$A_k(t) = \int_a^t \frac{\det(M_k(f)(s))}{W(y_1, \cdots, y_n)(s)} ds + C_k.$$

Therefore, a particular solution is

$$y_p(t) = \sum_{k=1}^{n} y_k(t) \int_a^t \frac{\det(M_k(f)(s))}{W(y_1, \cdots, y_n)(s)} ds$$

or in another form,

$$\int_a^t \left(\sum_{k=1}^{n} \frac{\det(M_k(f)(s)) y_k(t)}{W(y_1, \cdots, y_n)(s)} \right) ds$$

This explains the following procedure.

Procedure 5.3.1 *To find a particular solution using the method of variation of parameters, do the following:*

1. *First be sure that the coefficient of $y^{(n)}$ is a 1.* **If it is not, divide to make it so.**

2. *Find the general solution to the homogeneous equation*

$$y^{(n)} + a_{n-1}(t) y^{(n-1)} + \cdots + a_1(t) y' + a_0(t) y = 0$$

   as $\sum_{i=1}^{n} C_i y_i$. It is the parameters C_i which are replaced with $A_i(t)$ in order to obtain a particular solution to

$$y^{(n)} + a_{n-1}(t) y^{(n-1)} + \cdots + a_1(t) y' + a_0(t) y = f(t)$$

3. *Assuming the y_i are defined near the point a where the initial data is given, a particular solution is of the form*

$$y_p(t) = \sum_{i=1}^{n} A_i(t) y_i(t)$$

   where

$$A_k'(t) = \frac{\det (M_k(f)(t))}{\det (M(t))}$$

   Here $\det (M(t))$ is the Wronskian of the functions $\{y_1, \cdots, y_n\}$

$$M(t) = \begin{pmatrix} y_1(t) & \cdots & y_n(t) \\ \vdots & & \vdots \\ y_1^{(n-1)}(t) & \cdots & y_n^{(n-1)}(t) \end{pmatrix}$$

   The matrix $M_k(t)$ is obtained from $M(t)$ by replacing the kth column with the column

$$\begin{pmatrix} 0 & 0 & \cdots & 0 & f(t) \end{pmatrix}^T$$

Here is an example.

Example 5.3.2 *Find the general solution to*

$$t^2 y'' + 3ty' + y = t^4$$

given one solution to the homogeneous equation is $y_1(t) = 1/t$.

We could use the method of reduction of order to solve the whole thing but I will start with finding another solution to the homogeneous equation using reduction of order or Abel's formula and then will use the method of variation of parameters. You can't use the method of undetermined coefficients because the differential operator does not have constant coefficients. Thus, dividing by t^2 to make the leading coefficient equal to 1,

$$y'' + \frac{3}{t} y' + \frac{1}{t^2} y = t^2.$$

By the result of Example 4.3.4, the general solution to the homogeneous equation is

$$C_1 t^{-1} + C_2 \frac{\ln t}{t}$$

Now it follows the Wronskian of these two functions is

$$\begin{vmatrix} \frac{1}{t} & \frac{\ln(t)}{t} \\ -\frac{1}{t^2} & \frac{1-\ln(t)}{t^2} \end{vmatrix} = \frac{1}{t^3}$$

The particular solution is

$$y_p(t) = A\frac{1}{t} + B\left(\frac{\ln t}{t}\right)$$

where

$$A' = \frac{\begin{vmatrix} 0 & \frac{\ln(t)}{t} \\ t^2 & \frac{1-\ln(t)}{t^2} \end{vmatrix}}{1/t^3} = -(\ln t) t^4$$

and

$$B' = \frac{\begin{vmatrix} \frac{1}{t} & 0 \\ -\frac{1}{t^2} & t^2 \end{vmatrix}}{1/t^3} = t^4$$

Therefore, neglecting the constants of integration which only will contribute to terms which satisfy the homogeneous equation,

$$A = -\frac{1}{5}(\ln t) t^5 + \frac{1}{25}t^5, \ B = \frac{t^5}{5}$$

It follows

$$y_p(t) = -\frac{1}{5}(\ln t) t^4 + \frac{1}{25}t^4 + \frac{1}{5}(\ln t) t^4 = \frac{1}{25}t^4.$$

Example 5.3.3 *Find the general solution to* $y'' + y = \ln\left(1 + t^2\right)$.

The solution to the homogeneous equation is $C_1 \cos t + C_2 \sin t$ and the Wronskian of these functions is 1. Then a particular solution is of the form

$$y_p = A(t)\cos(t) + B(t)\sin(t)$$

where

$$A'(t) = \frac{\begin{vmatrix} 0 & \sin t \\ \ln(1+t^2) & \cos t \end{vmatrix}}{1} = -\sin(t)\ln(1+t^2)$$

$$B'(t) = \frac{\begin{vmatrix} \cos t & 0 \\ -\sin t & \ln(1+t^2) \end{vmatrix}}{1} = \cos(t)\ln(1+t^2)$$

Hence a particular solution is

$$y_p(t) = -\int_0^t \sin(s)\ln\left(1+s^2\right)ds\cos(t) + \int_0^t \cos(s)\ln\left(1+s^2\right)ds\sin(t)$$

You could also combine these to write

$$
\begin{aligned}
y_p(t) &= \int_0^t \ln\left(1+s^2\right)\left[\sin t \cos s - \cos t \sin s\right] ds \\
&= \int_0^t \ln\left(1+s^2\right)\sin\left(t-s\right) ds
\end{aligned}
$$

What if $\ln\left(1+t^2\right)$ had been replaced with $f(t)$? Then everything would be the same and you could conclude that

$$
y_p(t) = \int_0^t f(s)\sin\left(t-s\right) ds
$$

Thus, in the case of the example, the general solution is

$$
C_1 \cos t + C_2 \sin t + \int_0^t \ln\left(1+s^2\right)\sin\left(t-s\right) ds
$$

5.4 The Equations of Undamped and Damped Oscillation

In this section, a simple mechanical example is discussed which is modeled in terms of a differential equation. First here is a short discussion of Newton's laws.

Definition 5.4.1 *Let* $\mathbf{r}(t)$ *denote the position of an object. Then the acceleration of the object is defined to be* $\mathbf{r}''(t)$.

Newton's* first law is: "Every body persists in its state of rest or of uniform motion in a straight line unless it is compelled to change that state by forces impressed on it."

Newton's second law is:
$$
\mathbf{F} = m\mathbf{a} = m\mathbf{r}''(t) \tag{5.4}
$$
where $\mathbf{a}$ is the acceleration and m is the mass of the object.

Newton's third law states: "To every action there is always opposed an equal reaction; or, the mutual actions of two bodies upon each other are always equal, and directed to contrary parts."

---

*Isaac Newton (1642-1727), Lucasian Professor of Mathematics at Cambridge, is often credited with inventing calculus although most of the ideas were in existence earlier. However, he made major contributions to the subject partly in order to study physics and astronomy. He formulated the laws of gravity, made major contributions to optics, and stated the fundamental laws of mechanics listed here. He invented a version of the binomial theorem when he was only 23 years old and built a reflecting telescope to solve a difficult problem present in optics. He showed that Kepler's laws for the motion of the planets came from calculus and his laws of gravitation. In 1686 he published an important book, Principia, in which many of his ideas are found. One of the things which is most interesting about Newton and his contemporaries is their extremely wide collection of interests which, in the case of Newton, included theology and alchemy as well as traditional topics in science and math. Newton was a heretic (Snobelen, British Journal for the history of science, 1999) who did not believe in the Christian creeds. At that time, you could be hanged for promoting some of the ideas Newton held. He ended up as a government official, master of the mint but continued his mathematical research [11].

Of these laws, only the second two are independent of each other, the first law being implied by the second. The third law says roughly that if you apply a force to something, the thing applies the same force back.

The second law is the one of most interest. Note that the statement of this law depends on the concept of the derivative because the acceleration is defined as a derivative. Newton used calculus and these laws to solve profound problems involving the motion of the planets and other problems in mechanics.

Consider a garage door spring. These springs exert a force which resists extension. Attach such a spring to the ceiling and attach a mass m, to the bottom end of the spring as shown in the following picture. Any mass will do. It does not have to be a small elephant.

The weight of this mass mg, is a downward force which extends the spring, moving the bottom end of the spring downward a distance l where the upward force exerted by the spring exactly balances the downward force exerted on the mass by gravity. It has been experimentally observed that as long as the extension z, of such a spring is not too great, the restoring force exerted by the spring is of the form kz where k is some constant which depends on the spring. (It would be different for a slinky than for a garage door spring.) This is known as Hooke's[†] law which is the simplest model for elastic materials. Therefore, $mg = kl$. Now let y be the displacement from this equilibrium position of the bottom of the spring with the positive direction being down. Thus the acceleration of the spring is y''. The extension of the spring in terms of y is $(l + y)$. Then Newton's second law along with Hooke's law imply

$$my'' = k(l + y) - mg$$

and since $kl - mg = 0$, this yields

$$my'' + ky = 0.$$

Dividing by m and letting $\omega^2 = k/m$ yields the equation for undamped oscillation,

$$y'' + \omega^2 y = 0.$$

Based on physical reasoning just presented, there should be a solution to this equation. It is the displacement of the bottom end of a spring from the equilibrium position. However, it is not enough to base questions of existence in mathematics

---

[†]Robert Hooke (1635-1703) is best remembered for this law. Like Newton, he studied gravity and light and built telescopes. He also built vacuum pumps, microscopes, and watches. Like many of his time he was interested in how to determine longitude.

on physical intuition, although it is sometimes done. The following theorem gives the necessary existence and uniqueness results. The equation is the equation of undamped oscillations. It occurs in modeling a weight on a spring but it also occurs in many other physical settings.

In seeking to understand oscillations, it is good to write things in terms of amplitude and phase shift. In general, when you have

$$k \cos (\omega t) + l \sin (\omega t),$$

this equals

$$\sqrt{k^2 + l^2} \left(\frac{k}{\sqrt{k^2 + l^2}} \cos (\omega t) + \frac{l}{\sqrt{k^2 + l^2}} \sin (\omega t) \right) = \sqrt{k^2 + l^2} \cos (\omega t - \phi)$$

for a suitable phase shift ϕ having

$$\cos \phi = \frac{k}{\sqrt{k^2 + l^2}}, \quad \sin \phi = \frac{l}{\sqrt{k^2 + l^2}}$$

which is possible because $\left(\frac{k}{\sqrt{k^2+l^2}}, \frac{l}{\sqrt{k^2+l^2}} \right)$ is a point on the unit circle. This formula simplifies the above formula and makes it easier to understand what is happening in terms of things which are physically meaningful.

Theorem 5.4.2 *The initial value problem*

$$y'' + \omega^2 y = 0, \ y(0) = y_0, y'(0) = y_1 \tag{5.5}$$

has a unique solution and this solution is

$$y(t) = y_0 \cos (\omega t) + \frac{y_1}{\omega} \sin (\omega t). \tag{5.6}$$

Proof: You should verify that (5.6) does indeed provide a solution to the initial value problem. It only remains to verify uniqueness. Suppose then that y_1 and y_2 both solve the initial value problem (5.5). Let $w = y_1 - y_2$. Then you should verify that $w'' + \omega^2 w = 0$, $w(0) = 0 = w'(0)$. Then multiplying both sides of the differential equation by w' it follows

$$w''w' + \omega^2 w w' = 0.$$

However, $w''w' = \frac{1}{2} \frac{d}{dt} (w')^2$ and $w'w = \frac{1}{2} \frac{d}{dt} (w)^2$ so the above equation reduces to

$$\frac{1}{2} \frac{d}{dt} \left((w')^2 + w^2 \right) = 0.$$

Therefore, $(w')^2 + w^2$ is equal to some constant. However, when $t = 0$, this shows the constant must be zero. Therefore, $y_1 - y_2 = w = 0$. ∎

Example 5.4.3 *Find the general solution to*

$$y'' + \omega_0^2 y = A \cos (\omega t), \ \omega \neq \omega_0$$

From the above theorem, the solution to the homogeneous equation is

$$C_1 \cos(\omega_0 t) + C_2 \sin(\omega_0 t)$$

Then it remains to find a particular solution. Consider

$$z'' + \omega_0^2 z = A e^{i\omega t}$$

A particular solution is of the form $B e^{i\omega t}$. You don't need to multiply by t because this is not a solution to the homogeneous equation. Now substitute this in to the equation.

$$-\omega^2 B e^{i\omega t} + \omega_0^2 B e^{i\omega t} = A e^{i\omega t}$$

and so

$$B = \frac{A}{\omega_0^2 - \omega^2}$$

Thus

$$z_p = \frac{A}{\omega_0^2 - \omega^2}(\cos \omega t + i \sin \omega t)$$

The real part is the desired particular solution. Therefore, the general solution is

$$C_1 \cos(\omega_0 t) + C_2 \sin(\omega_0 t) + \frac{A}{\omega_0^2 - \omega^2}\cos(\omega t)$$

Consider what happens when ω_0 is close to ω and $y(0) = y'(0) = 0$. Then

$$C_1 = -\frac{A}{\omega_0^2 - \omega^2}, C_2 = 0.$$

From the above theorem

$$y = \frac{A}{\omega_0^2 - \omega^2}(\cos(\omega t) - \cos(\omega_0 t))$$

From trigonometry,

$$y = \frac{2A}{\omega_0^2 - \omega^2}\sin\left(\frac{\omega_0 - \omega}{2}t\right)\sin\left(\frac{\omega + \omega_0}{2}t\right)$$

This is very interesting. Here is a graph of the sort of thing which is happening. There is a slowly varying amplitude times a rapidly varying expression. I will let $\omega = 3$ and $\omega_0 = 2.8$ and $A = 1$ for convenience.

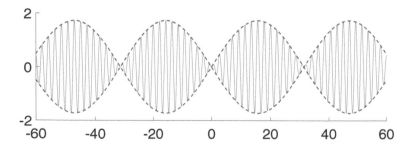

The dashed line is the graph of $\frac{2A}{\omega_0^2-\omega^2}\sin\left(\frac{\omega_0-\omega}{2}t\right)$ and $-\frac{2A}{\omega_0^2-\omega^2}\sin\left(\frac{\omega_0-\omega}{2}t\right)$ for these values of ω_0, ω, and A. It represents the slowly varying amplitude. When you have two frequencies close together, they can interact in such a way as to produce a loud low frequency sound called a beat. The number ω is called the angular velocity.

Now consider the differential equation of damped oscillation. In the example of the object bobbing on the end of a spring,

$$my'' = -ky$$

where k was the spring constant and m was the mass of the object. Suppose the object is also attached to a dash pot, a device which resists motion like a shock absorber on a car, or the vibrations are taking place in some sort of fluid which resists motion. You know how shock absorbers work. If the car is just sitting still the shock absorber applies no force to the car. It only gives a force in response to up and down motion of the car and you assume this force is proportional to the velocity and opposite the velocity. Thus in the spring example, you would have

$$my'' = -ky - \delta^2 y'$$

where δ^2 is the constant of proportionality of the resisting force. Dividing by m and adjusting the coefficients, such damped oscillation satisfies an equation of the form

$$y'' + by' + a^2 y = 0. \tag{5.7}$$

Actually this is a general homogeneous second order equation, more general than what results from damped oscillation. Concerning the solutions to this equation, the following theorem is given. In this theorem the first case is referred to as the under-damped case. The second case is called the critically damped case and the third is called the over-damped case.

Theorem 5.4.4 *Suppose $\frac{b^2}{4} - a^2 < 0$. Then all solutions of (5.7) are of the form*

$$e^{-(b/2)t}\left(C_1\cos\left(\omega t\right) + C_2\sin\left(\omega t\right)\right). \tag{5.8}$$

where $\omega = \frac{1}{2}\sqrt{4a^2 - b^2}$ and C_1 and C_2 are constants. In the case that $\frac{b^2}{4} - a^2 = 0$, the solutions of (5.7) are of the form

$$e^{-(b/2)t}\left(C_1 + C_2 t\right). \tag{5.9}$$

In the case that $\frac{b^2}{4} - a^2 > 0$ the solutions are of the form

$$e^{-(b/2)t}\left(C_1 e^{-rt} + C_2 e^{rt}\right), \tag{5.10}$$

where $r = \frac{1}{2}\sqrt{b^2 - 4a^2}$.

Proof: Let $z = e^{(b/2)t}y$ and write (5.7) in terms of z. Thus,

$$\frac{b^2}{4}e^{-(b/2)t}z + z''e^{-(b/2)t} - be^{-(b/2)t}z' + be^{-(b/2)t}z' - \frac{b^2}{2}e^{-(b/2)t}z + a^2 e^{-(b/2)t}z = 0$$

Then dividing by the exponential,

$$\frac{b^2}{4}z + z'' - bz' + bz' - \frac{b^2}{2}z + a^2 z = 0$$

and so z is a solution to the equation

$$z'' + \left(a^2 - \frac{b^2}{4} \right) z = 0. \tag{5.11}$$

If $\frac{b^2}{4} - a^2 < 0$, then by Theorem 5.4.2, $z(t) = C_1 \cos(\omega t) + C_2 \sin(\omega t)$ where $\omega = \sqrt{a^2 - \frac{b^2}{4}} = \frac{1}{2}\sqrt{4a^2 - b^2}$. Therefore,

$$y = e^{-(b/2)t}\left(C_1 \cos(\omega t) + C_2 \sin(\omega t) \right)$$

as claimed. Next suppose $\frac{b^2}{4} - a^2 > 0$. This is the over damped case. In this case, you have

$$z'' - \left(\frac{b^2}{4} - a^2 \right) z = 0$$

and so the solutions z are

$$C_1 e^{\frac{1}{2}\sqrt{(b^2 - 4a^2)}\,t} + C_2 e^{-\frac{1}{2}\sqrt{(b^2 - 4a^2)}\,t}$$

Hence, in this case, you get

$$y = C_1 e^{-(b/2)t} e^{\frac{1}{2}\sqrt{(b^2 - 4a^2)}\,t} + C_2 e^{-(b/2)t} e^{-\frac{1}{2}\sqrt{(b^2 - 4a^2)}\,t}$$

You see that in this case, there are no oscillations and as $t \to \infty$ you have $y \to 0$.

Finally consider the critically damped case where $\frac{b^2}{4} - a^2 = 0$. In this case, you have no oscillations either and

$$z(t) = C_1 + C_2 t$$

and so

$$y(t) = C_1 e^{-(b/2)t} + C_2 t e^{-(b/2)t} \quad \blacksquare$$

Note that if you had $y'' + a^2 y = 0$, the angular velocity is a but when there is a small damping term, the angular velocity changes slightly from a to $\sqrt{a^2 - \frac{b^2}{4}}$.

Here is a picture of the graph of an under-damped oscillation.

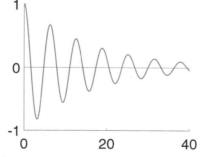

It is a graph of the solution to

$$y'' + \frac{1}{8}y' + y = 0, \ y(0) = 1, y'(0) = 0$$

Example 5.4.5 *An important example of these equations occurs in an electrical circuit having a capacitor, a resistor, and an inductor in series as shown in the following picture.*

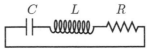

The voltage drop across the inductor is $L\frac{di}{dt}$ where i is the current and L is the inductance. The voltage drop across the resistor is Ri where R is the resistance. This is according to Ohm's law. The voltage drop across the capacitor is $v = \frac{Q}{C}$ where Q is the charge on the capacitor and C is a constant called the capacitance. The current equals the rate of change of the charge on the capacitor. Thus $i = Q' = Cv'$. When these voltages are summed, you must get zero because there is no voltage source in the circuit. Thus $L\frac{di}{dt} + Ri + \frac{Q}{C} = 0$ and written in terms of the voltage drop across the capacitor, this becomes $LCv'' + CRv' + v = 0$, a second order linear differential equation of the sort discussed above.

Next consider the situation of forced and damped oscillation.

Example 5.4.6 *Find the solution to $y'' + \frac{1}{8}y' + y = \cos(2t)$ where $y(0) = 1, y'(0) = 0$.*

Let's find a particular solution first. It is easiest to find the real part of a particular solution for

$$z'' + \frac{1}{8}z' + z = e^{2it}$$

Look for $z_p = Ae^{2it}$. This is not a solution to the homogeneous equation so you don't have to multiply by t. Substitute this in to the equation.

$$-\left(3 - \frac{1}{4}i\right)Ae^{2it} = e^{2it}$$

Then

$$A = \frac{1}{\frac{1}{4}i - 3} = -\frac{48}{145} - \frac{4}{145}i$$

Thus

$$z = \left(-\frac{48}{145} - \frac{4}{145}i\right)(\cos(2t) + i\sin(2t))$$

$$= \left(-\frac{48}{145}\right)\cos(2t) + \left(\frac{4}{145}\right)\sin(2t) + i\,(\text{something})$$

Therefore, the desired particular solution is

$$\left(-\frac{48}{145}\right)\cos(2t) + \left(\frac{4}{145}\right)\sin(2t)$$

Next it is required to find the general solution to the homogeneous equation. This is

$$C_1\left(\cos\frac{1}{16}\sqrt{255}t\right)e^{-\frac{1}{16}t} - C_2\left(\sin\frac{1}{16}\sqrt{255}t\right)e^{-\frac{1}{16}t}$$

Next you need to find the constants. The general solution is

$$C_1 \left(\cos \frac{1}{16} \sqrt{255} t \right) e^{-\frac{1}{16} t} - C_2 \left(\sin \frac{1}{16} \sqrt{255} t \right) e^{-\frac{1}{16} t} +$$

$$+ \left(-\frac{48}{145} \right) \cos (2t) + \left(\frac{4}{145} \right) \sin (2t)$$

From the initial conditions,

$$1 = C_1 - \frac{48}{145}$$

so $C_1 = \frac{193}{145} = 1.331$ Next you have to take the derivative and get 0 when you plug in $t = 0$. This gives $C_2 = -\frac{13}{7395} \sqrt{255} = -2.8072 \times 10^{-2}$. It follows that the solution is

$$y(t) = \quad 1.331 \left(\cos (0.99804 t) \right) e^{-\frac{1}{16} t} + \left(2.8072 \times 10^{-2} \right) \left(\sin (0.99804) t \right) e^{-\frac{1}{16} t}$$

$$+ \left(-\frac{48}{145} \right) \cos (2t) + \left(\frac{4}{145} \right) \sin (2t)$$

Suppose you did not have the damping term. Then the solution is

$$\frac{4}{3} \cos t - \frac{1}{3} \cos 2t = 1.3333 \cos t - 0.33333 \cos (2t)$$

Notice how the two solutions are close, at least for small t. One is being damped and the other is not.

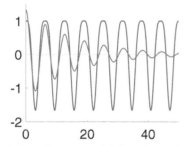

In the solution, the terms which come from the homogeneous equation disappear as $t \to \infty$ and this is why these terms are called "transients". Thus, after a substantial time, the solution to the forced damped equation is essentially given by

$$\left(-\frac{48}{145} \right) \cos (2t) + \left(\frac{4}{145} \right) \sin (2t)$$

Then letting $y(t)$ be the solution to the damped equation given above, here is a graph of $y(t) - \left(\left(-\frac{48}{145} \right) \cos (2t) + \left(\frac{4}{145} \right) \sin (2t) \right)$.

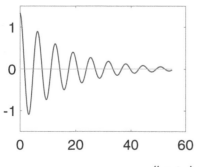

You see how the difference between the solution and this steady state solution becomes very small as t increases.

In general, consider the interesting case where the forcing function is periodic of the form $F_0 \cos (\omega t)$ and the equation is of the form

$$my'' + \gamma y' + ky = F_0 \cos (\omega t)$$

and the equation is under damped. Dividing by m we will consider the equation of the form

$$y'' + by' + a^2 y = G \cos (\omega t), \; b > 0$$

The characteristic equation for finding solutions to the homogeneous equation is

$$r^2 + br + a^2 = 0$$

and the general solution is of the form $e^{-(b/2)t}$ times terms which involve cosines and sines. Thus these terms all converge to 0 as $t \to \infty$. They are transient terms. However, the particular solution obtained from the use of variation of constants will not converge to 0 as $t \to \infty$ and it is this which is of interest. Consider

$$z'' + bz' + a^2 z = Ge^{i\omega t}$$

Look for $z_p = Ae^{i\omega t}$. It is not necessary to multiply by t because the expression is not a solution to the homogeneous equation. Substitute this into the equation. This yields

$$Aa^2 e^{it\omega} - A\omega^2 e^{it\omega} + iAb\omega e^{it\omega} = Ge^{i\omega t}$$

Therefore, you need

$$A\left(a^2 - \omega^2 + ib\omega\right) = G$$

and so

$$A = \frac{G}{a^2 - \omega^2 + ib\omega} = \frac{G\left(a^2 - \omega^2 - ib\omega\right)}{\left(a^2 - \omega^2\right)^2 + b^2\omega^2}$$

Thus

$$z_p = G\left[\frac{a^2 - \omega^2}{\left(a^2 - \omega^2\right)^2 + b^2\omega^2} - i\frac{b\omega}{\left(a^2 - \omega^2\right)^2 + b^2\omega^2}\right]\left(\cos\left(\omega t\right) + i\sin\left(\omega t\right)\right)$$

and its real part is

$$y_p = G\left(\frac{a^2 - \omega^2}{\left(a^2 - \omega^2\right)^2 + b^2\omega^2}\cos\left(\omega t\right) + \frac{b\omega}{\left(a^2 - \omega^2\right)^2 + b^2\omega^2}\sin\left(\omega t\right)\right) \qquad (5.12)$$

To gain more insight, it is a good idea to write this differently. Equation (5.12) can be written in terms of an amplitude and phase shift. The amplitude equals

$$\sqrt{\left(\frac{a^2 - \omega^2}{\left(a^2 - \omega^2\right)^2 + b^2\omega^2}\right)^2 + \left(\frac{b\omega}{\left(a^2 - \omega^2\right)^2 + b^2\omega^2}\right)^2} = \frac{1}{\sqrt{\left(a^2 - \omega^2\right)^2 + b^2\omega^2}}$$

Then the expression reduces to

$$\frac{1}{\sqrt{\left(a^2 - \omega^2\right)^2 + b^2\omega^2}}\cos\left(\omega t - \phi\right) \qquad (5.13)$$

for a suitable ϕ. The amplitude of this expression is

$$\frac{1}{\sqrt{\left(a^2 - \omega^2\right)^2 + b^2\omega^2}}$$

and you see that this is as large as possible when the denominator is as small as possible. You can control this by changing ω. Note also that for ω large, the

amplitude of this will be small. To make this as large as possible, you would take
the derivative of the expression in the denominator and set it equal to 0. This yields

$$\omega = \sqrt{a^2 - \frac{1}{2}b^2}$$

If you substitute this in to (5.13), you get the following as the solution corresponding to the value of ω which will maximize the amplitude.

$$\frac{1}{b\sqrt{a^2 - \frac{1}{4}b^2}} \cos\left(\left(\sqrt{a^2 - \frac{1}{2}b^2}\right)t - \phi\right)$$

Note that as $b \to 0+$, the amplitude is unbounded. As to the phase shift ϕ, from
(5.12),

$$\tan\phi = \frac{b\omega}{a^2 - \omega^2}$$

so as $b \to 0+$, this phase shift converges to 0.

In the following picture of the graphs of long time solutions, each graph has
$a = 1$ and b equals respectively $.1, .4, .7$, and 1. Note how as the damping controlled
by b gets smaller, the resulting graph of the long time solution gets larger, and also
how the phase shift translates the graph horizontally as b changes.

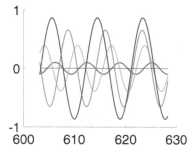

Note that the undamped equation is of the
form

$$y'' + a^2 y = G\cos(\omega t)$$

and the value of ω which will maximize the amplitude will be $\omega = a$. Letting $a = 1$ as above,
the above methods show that the general solution will be of the form

$$C_1 \cos(t) + C_2 \sin(t) + \frac{G}{2} t \sin(t)$$

so the damped equation for small damping constant is trying to approximate this
undamped solution but there is a difference. This undamped solution is unbounded
while the slightly damped solution might be large, but it won't be unbounded.
However, the frequency of the vibrations will be close to 1. In the above graphs,
the frequencies maximizing the amplitude were $.997, .935$, and $.707$ for $b = .1, .5$,
and 1, respectively.

5.5   Numerical Solutions

5.5.1   MATLAB

One can use MATLAB to find numerical solutions to these linear equations even
when it is not possible to factor the polynomials.

Example 5.5.1 *Find the solution to the initial value problem*

$$y''' + y'' - 2y' + y = \sin(t)$$
$$y(0) = 0, \ y'(0) = 1, \ y''(0) = 0$$

In this case, the characteristic equation is $r^3 + r^2 - 2r + 1 = 0$. It has no rational roots. However, it is perfectly possible to find the solution numerically.

In MATLAB, one writes it as a first order system as follows. You let $y(1) = y, y(2) = y(1)', y(3) = y(2)'$

$$y(1)' = y(2), \ y(2)' = y(3)$$
$$y(3)' + y(3) - 2y(2) + y(1) = \sin(t)$$

$$y(1)(0) = 0, \ y(2)(0) = 1, \ y(3)(0) = 0$$

Then by analogy with the order one case, one types in the following after the $>>$.

```
>> f=@(t,y)[y(2);y(3);-y(3)+2*y(2)-y(1)+sin(t)];
[t,y]=ode45(f,[0,10],[0;1;0]);
plot(t,y)
```

Remember that to type on a new line, you press "shift enter". Then you press "enter" and see the graphs. It will plot $t, y(1), y(2), y(3)$ on the same axes. Note how the variable y is used in two totally different ways in the above syntax.

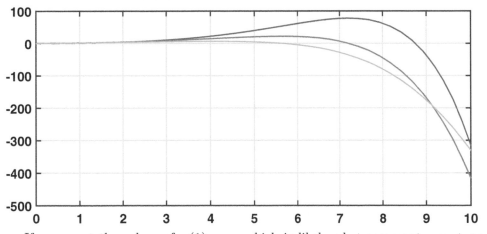

If you want the values of $y(1) = y$, which is likely what you want, you type [t,y] and press "enter". This will give you four columns, the first will be values of t between 0 and 10. The second will be the corresponding values of $y = y(1)$ and the others will be the values of $y(2), y(3)$. Thus you are finding the values of y' in $y(2)$ column and y'' in the $y(3)$ column. These are each graphed in the above picture. The one you are most interested in is the graph for y and is on the top in this case. You can identify which variable is being graphed by looking at the table. As just mentioned, the others are graphs for the derivative and second derivative.

If you want a graph of $y = y(1)$, you do the following.

```
f=@(t,y)[y(2);y(3);-y(3)+2*y(2)-y(1)+sin(t)];
[t,y]=ode45(f,[0,10],[0;1;0]);
plot(t,y(:,1))
```

then press "enter". The plot will be of $y(1)$. To see the values of this function at the points, 1,2,3,4,5, on a new line, type

$$\text{interp1}(t,[y(:,1)],[1,2,3,4,5])$$

That symbol after the p in the above is a one, not an l.

When you use functions y(i) with i an index, the i needs to be a positive integer. Otherwise, MATLAB will have a fit. You can also do the following to get values.

$$\text{f=@(t,y)[y(2);y(3);-y(3)+2*y(2)-y(1)+sin(t)];}$$
$$\text{sol=ode45(f,[0,7],[1;0;0]);}$$
$$\text{deval(sol,[1,2,3,4,5])}$$

then press "enter". This will give you five columns of numbers which give the values of $y(1), y(2)$, and $y(3)$ at these five numbers, $1,2,3,4,5$. You probably are most interested in the values of $y(1)$ at the five values. You can use any letter in place of sol.

So you see that you can completely understand the solutions to such differential equations by using this numerical procedure. You may not have a neat formula for the answer, but you can obtain its graph and its values at any points you desire.

The following is a procedure for changing one of these higher order linear equations into a first order system which is the right way to study differential equations anyway.

Procedure 5.5.2 *Consider the equation*

$$y^{(n)} + a_{n-1}y^{(n-1)} + \cdots + a_1y' + a_0y = f$$

To write as a first order system, do the following.

$$\begin{pmatrix} x(1) \\ x(2) \\ \vdots \\ x(n) \end{pmatrix} = \begin{pmatrix} y \\ y' \\ \vdots \\ y^{(n-1)} \end{pmatrix}$$

Then

$$\begin{pmatrix} x(1) \\ x(2) \\ \vdots \\ x(n-1) \\ x(n) \end{pmatrix}' = \begin{pmatrix} x(2) \\ x(3) \\ \vdots \\ x(n) \\ y^{(n)} \end{pmatrix} = \begin{pmatrix} x(2) \\ x(3) \\ \vdots \\ x(n) \\ -(a_{n-1}x(n) + \cdots + a_1x(2) + a_0x(1)) + f \end{pmatrix}$$

In terms of matrices,

$$\begin{pmatrix} x(1) \\ x(2) \\ \vdots \\ x(n-1) \\ x(n) \end{pmatrix}' = \begin{pmatrix} 0 & 1 & 0 & \cdots & 0 \\ 0 & 0 & 1 & \ddots & \vdots \\ \vdots & \ddots & \ddots & \ddots & 0 \\ 0 & \cdots & 0 & 0 & 1 \\ -a_0 & -a_1 & \cdots & -a_{n-2} & -a_{n-1} \end{pmatrix} \begin{pmatrix} x(1) \\ x(2) \\ \vdots \\ x(n-1) \\ x(n) \end{pmatrix} + \begin{pmatrix} 0 \\ 0 \\ \vdots \\ 0 \\ f \end{pmatrix}$$

5.5.2 Scientific Notebook

This will work very well for such linear equations in which you are not able to find roots to the characteristic equation. It is much easier to use than MATLAB. You simply enter the following column vector and then click on it and then click compute, ode, numeric and it gives a function called y

$$y''' + y'' - 2y' + y = \sin(t)$$
$$\begin{array}{l} y(0) = 0 \\ y'(0) = 1 \\ y''(0) = 0 \end{array} \quad , \text{ Functions defined: } y$$

Then to graph y, you type y and then with the cursor next to and right of y you click on compute, plot2D, and ODE. It will produce a graph of the numerical solution. Also, if you want y evaluated at 4, you simply type in $y(4)$ and then "evaluate numerical" on the display, the button which looks like $\stackrel{\#}{=}?$. This gives $y(4) = 22.138$.

The selection of the time interval $[0, 5]$ for the graph is automatic. If you want the graph on a larger interval, say $[0, 8]$, you only need to change the interval in the dialog box for the graph which you bring up by double clicking on the graph. If you want a table of values, this is also easy. Say you want the values of y for t equal to some integers. You do the following. Type y and use the parentheses from the tool bar and inside these, you place the column vector shown. This comes from entering a 10×1 matrix and filling in the entries as shown. The parentheses will expand to fit the vector. Then you click on $\stackrel{\#}{=}?$ and it will evaluate at each of the integer values to give the following.

$$y \begin{pmatrix} 1 \\ 3 \\ 5 \\ 7 \\ 9 \end{pmatrix} = \begin{pmatrix} 1.2861 \\ 10.568 \\ 39.781 \\ 77.073 \\ -36.243 \end{pmatrix}$$

This software is very convenient and will work well for linear equations. It sometimes has problems with nonlinear equations. The results are identical with what is determined by MATLAB in the above.

When you have your results, you should clear the definitions. Click compute, definitions, clear definitions. That way you can use y again in another problem.

Obviously, this kind of thing is a black box approach to getting a solution and begs much more consideration of the mathematics involved. This is considered briefly in a later chapter. In general, these kinds of things are dealt with in numerical analysis courses and deserve a far more detailed treatment than what is in this book.

5.6 Exercises

1. Find the solution to the equations and given initial condition.

   (a) $y'' - 2y' - 3y = -t - 3e^{-2t}$, $y(0) = -1, y'(0) = 4$.
   (b) $y'' - y' - 6y = 2e^{-t} - t$, $y(0) = -1, y'(0) = 2$.

    (c) $y'' - 2y' - 3y = t + e^{-2t}, y(0) = 1, y'(0) = 3.$

    (d) $3y - 4y' + y'' = 12\cos t - 6\sin t, y(0) = 1, y'(0) = 2.$

    (e) $y'' - y' - 6y = 9\cos 3t + 45\sin 3t, y(0) = 5, y'(0) = -9.$

    (f) $6y + 5y' + y'' = 30\cos 3t - 6\sin 3t, \ y(0) = 8, y'(0) = -17.$

    (g) $y'' - y = -6\sin t, y(0) = 5, y'(0) = -6.$

2. Find the general solution to the equation $y'' - 9y = -26\sin 2t.$

3. Find the general solution to the equation $3y + 4y' + y'' = 8\sin t - 4\cos t.$

4. Find the general solution to the equation $3y - 4y' + y'' = 8\cos t + 4\sin t - 4t\cos t - 8t\sin t.$

5. Find the general solution to the equation $y'' - y' - 6y = 15e^{-2t}.$

6. Find the solution to the equations and initial conditions

    (a) $2y - 3y' + y'' = 2e^{2t}, \ y(0) = 0, y'(0) = 6.$

    (b) $y'' - y = 4e^{-t} - 2\cos t, y(0) = 0 = y'(0).$

    (c) $y'' - 2y' - 3y = -12e^{3t}, y(0) = 5, y'(0) = 4.$

    (d) $y'' - 4y = 12e^{-2t}, \ y(0) = 3, y'(0) = 7.$

    (e) $y'' - 9y = 18e^{-3t}, \ y(0) = -4, y'(0) = 21.$

    (f) $2y - 3y' + y'' = 4e^t(t - 1), y(0) = 1, y'(0) = 0.$

    (g) $y'' - y' - 6y = 4e^{3t}(5t + 1), \ y(0) = -9, y'(0) = 3.$

    (h) $y'' - 4y = -6e^{-2t}(4t - 1), \ y(0) = -3, y'(0) = 22.$

    (i) $y' - 6y + y'' = 10e^{-3t}, \ y(-1) = 5e^{-2} + 4e^3, y'(-1) = 10e^{-2} - 14e^3.$

    (j) $2y' - 3y + y'' = -12e^{-3t}, \ y(-2) = 2e^{-2} - 9e^6, y'(-2) = 2e^{-2} + 30e^6.$

    (k) $y'' - y' - 6y = 5e^{3t}, \ y(3) = 4e^{-6} + e^9, y'(3) = 4e^9 - 8e^{-6}.$

    (l) $y' - 2y + y'' = -6e^{-2t}, \ y(3) = 5, y'(3) = 2.$

    (m) $6y - 5y' + y'' = -2e^{3t}, \ y(2) = 1, y'(2) = -2.$

7. Find the solution to the following initial value problems.

    (a) $y'' - y' - 6y = -15e^{3t}, \ y(-1) = -3, y'(-1) = -3.$

    (b) $y'' - y' - 6y = 10e^{-2t}, \ y(1) = 2, y'(1) = -2.$

    (c) $5y - 4y' + y'' = -2e^{2t}, \ y(0) = 5, y'(0) = 2.$

    (d) $58y - 6y' + y'' = 100e^{2t}, \ y(0) = 3, y'(0) = 2.$

    (e) $10y - 6y' + y'' = -10e^t, \ y(0) = 1, y'(0) = 1.$

    (f) $45y - 6y' + y'' = -144e^{-3t}, \ y(0) = -5, y'(0) = -3.$

    (g) $4y + y'' = -12\sin 2t, \ y(0) = -1, y'(0) = 1.$

    (h) $49y + y'' = -28\sin 7t, \ y(0) = 0, y'(0) = 0.$

    (i) $9y + y'' = -18\sin 3t, \ y(0) = -2, y'(0) = 2.$

(j) $49y + y'' = 14 \sin 7t$, $y(0) = -1, y'(0) = 3$.

(k) $4y + y'' = -4 \sin 2t$, $y(0) = 5, y'(0) = 1$.

(l) $16y + y'' = 16 \sin 4t$, $y(0) = 1, y'(0) = 2$.

(m) $9y + y'' = 8 \cos t - 12 \sin 3t$, $y(0) = 2, y'(0) = -3$.

(n) $y + y'' = 4 \sin t - 3 \cos 2t$, $y(0) = 1, y'(0) = -2$.

(o) $4y + y'' = 8 \sin 2t - 5 \cos 3t$, $y(0) = -2, y'(0) = -3$.

(p) $9y + y'' = -12 \sin 3t$, $y(0) = -3, y'(0) = -3$.

(q) $y + y'' = 4 \sin t - 8 \cos 3t$, $y(0) = 2, y'(0) = -2$.

(r) $9y + y'' = 5 \cos 2t + 12 \sin 3t$, $y(0) = -2, y'(0) = -3$.

(s) $9y + y'' = -12 \sin 3t$, $y(0) = -2, y'(0) = 1$.

(t) $4y + y'' = 3 \cos t - 8 \sin 2t$, $y(0) = 5, y'(0) = -3$.

(u) $4y + y'' = 3 \cos t + 8 \sin 2t$, $y(0) = -3, y'(0) = -3$.

(v) $9y + y'' = -12 \sin 3t$, $y(0) = 5, y'(0) = -3$.

8. Find the long time solution for the equation $13y - 4y' + y'' = \sin 3t - 2 \cos 3t$. Recall that this is the solution which results when the transients have become so small that they are not relevant.

9. Find the long time solution for the equation $34y - 10y' + y'' = 3 \cos 3t - 3 \sin 3t$. Recall that this is the solution which results when the transients have become so small that they are not relevant.

10. Find the long time solution for the equation $13y + 4y' + y'' = \cos 3t + 2 \sin 3t$. Recall that this is the solution which results when the transients have become so small that they are not relevant.

11. Find the long time solution for the equation $20y - 8y' + y'' = 3 \cos 2t + 2 \sin 2t$. Recall that this is the solution which results when the transients have become so small that they are not relevant.

12. Find the general solution to the equation $3y - y' - 3y'' + y''' = 8e^{3t} - 4e^t$ Then find the solution to this equation which satisfies the initial conditions $y(0) = 0, y'(0) = 1, y''(0) = 0$.

13. Find the general solution to the equation $11y' - 6y - 6y'' + y''' = 2e^t + 2e^{3t}$ Then find the solution to this equation which satisfies the initial conditions $y(0) = 0, y'(0) = 1, y''(0) = 0$.

14. Find the general solution to the equation $y'' - 2y' + y''' = 3e^t + 6e^{-2t}$ Then find the solution to this equation which satisfies the initial conditions $y(0) = 0, y'(0) = 1, y''(0) = 0$.

15. Find the general solution to the equation $y'' - 4y' - 4y + y''' = 4e^{-2t}$ Then find the solution to this equation which satisfies the initial conditions $y(0) = 1, y'(0) = 0, y''(0) = 0$.

16. Find the general solution to the equation $12y - 4y' - 3y'' + y''' = -4e^{2t}$ Then find the solution to this equation which satisfies the initial conditions $y(0) = 1, y'(0) = 0, y''(0) = 0$.

17. Find the general solution to the equation $y'''' + 81y = \sin(3t)$.

18. Find the general solution to the equation $y'''' - 16y = -2\cos 2t - \sin 2t$.

19. Find the general solution to the equation $y'''' - 256y = 2\cos 4t - 3\sin 4t$.

20. Find the general solution to the equation $y'''' - 256y = 3\cos 4t + \sin 4t$.

21. Find a particular solution to the equation $25y + y'' = -\frac{1}{\cos 5t}$.

22. Find a particular solution to the equation $16y + y'' = \frac{1}{\cos 4t}$.

23. Find a particular solution to the equation $9y + 6y' + y'' = 2e^{-3t}\ln 3t$, $t > 0$.

24. Find a particular solution to the equation $9y + 6y' + y'' = -2e^{-3t}\ln 3t$, $t > 0$.

25. Find a particular solution to the equation $25y + 10y' + y'' = e^{-5t}\ln 3t$, $t > 0$.

26. Find a particular solution to the equation $12y + 7y' + y'' = \tan(t^4 + 1)$, $t > 0$.
    Leave your answer in terms of integrals of the form $\int_0^t f(s)\,ds$.

27. Find a particular solution to the equation $y'' - 2y' - 3y = \cosh(t^2)$, $t > 0$.
    Leave your answer in terms of integrals of the form $\int_0^t f(s)\,ds$.

28. The equation $y''t^2 + 4y't + 2y = 0$, $t > 0$ has a solution of the form $y = \frac{1}{t}$.
    Find another solution. Then determine the general solution to the equation
    $y''t^2 + 4y't + 2y = t^5\ln t^2$.

29. The equation $y''t^2 + 3y't - 3y = 0$, $t > 0$ has a solution of the form $y = \frac{1}{t^3}$.
    Find another solution. Then determine the general solution to the equation
    $y''t^2 + 3y't - 3y = t^4\ln\frac{1}{t}$.

30. The equation $y''t^2 + 6y't + 6y = 0$, $t > 0$ has a solution of the form $y = \frac{1}{t^2}$.
    Find another solution. Then determine the general solution to the equation
    $y''t^2 + 6y't + 6y = t^7\ln\frac{1}{t}$.

31. The equation $y''t^2 + y't - y = 0$, $t > 0$ has a solution of the form $y = t$.
    Find another solution. Then determine the general solution to the equation
    $y''t^2 + y't - y = t^2\ln t^3$.

32. The equation $y''t^2 - 5y't + 9y = 0$, $t > 0$ has a solution of the form $y = t^3$.
    Find another solution. Then determine the general solution to the equation
    $y''t^2 - 5y't + 9y = t^4$.

33. The equation $y''t^2 + 5y't + 4y = 0$, $t > 0$ has a solution of the form $y = \frac{1}{t^2}$.
    Find another solution. Then determine the general solution to the equation
    $y''t^2 + 5y't + 4y = t^6$.

34. Solve $4y + y'' = t$, $y(0) = 0, y(\pi) = 0$. This is an example of a boundary value
    problem. Determine all possible solutions.

35. Solve $y + y'' = 2\cos t$, $y(0) = 0, y(\pi) = 0$. This is an example of a boundary
    value problem. Determine all possible solutions.

36. Solve $y + y'' = 2\cos t$, $y(0) = 0, y(\pi/2) = 0$. This is an example of a boundary value problem. Determine all possible solutions.

37. Solve $4y + y'' = \sin^2 2t$, $y(0) = 0, y'(0) = 1$. You can use $\sin^2(2t) = \frac{1 - \cos(4t)}{2}$.

38. We wrote the higher order linear equation $y^{(n)} + a_{n-1}y^{(n-1)} + \cdots + a_1 y' + a_0 y = f$ as a first order system of the form $\mathbf{x}' = A\mathbf{x} + \mathbf{f}$ where $A =$

$$
\begin{pmatrix}
0 & 1 & 0 & \cdots & & 0 \\
0 & 0 & 1 & \ddots & & \vdots \\
\vdots & \ddots & \ddots & \ddots & & 0 \\
0 & \cdots & 0 & 0 & & 1 \\
-a_0 & -a_1 & \cdots & -a_{n-2} & & -a_{n-1}
\end{pmatrix}
$$

The characteristic polynomial for A is $\det(\lambda I - A)$. Remember that this is what you use to find eigenvalues in linear algebra. Show that this characteristic polynomial is $\lambda^n + a_{n-1}\lambda^{n-1} + \cdots + a_1\lambda + a_0$. The above matrix is called a companion matrix of this polynomial. Thus, when you are finding the r such that e^{rt} is a solution to the equation, you are really finding eigenvalues of the above matrix.

Using Computer Algebra Systems

39. Use a computer algebra system to obtain a graph on $[0, 5]$ of the solution to the following initial value problems.

   (a) $y'' + 3y' + 2y = \ln(1 + t^2)$, $y(0) = 1, y'(0) = 2$.
   (b) $y''' + 2y'' - 5y' + 6y = t\sin(t) + e^t$, $y(0) = 1, y'(0) = 0, y''(0) = -3$.
   (c) $y'' + (1 + .1t^2)y' - y = \sin(\ln(1 + y^2))$, $y(0) = 1, y'(0) = 0$. On this one, graph on $[0, 2]$.
   (d) $y'' + (1 + t^2)y' - ty = \sin(\ln(1 + y^2))$, $y(0) = 0, y'(0) = 1$.
   (e) $y'''' + 2y''' - y' + 7y = \ln(2 + \sin(t))$, $y(0) = 1, y'(0) = y''(0) = y'''(0) = 0$.
   (f) $y'''' + 2y''' - y' + 7y = \ln(2 + \cos(t))$, $y(0) = 0, y'(0) = 1, y''(0) = y'''(0) = 0$.

40. Enter the following in MATLAB.

```
hold on
for a=0:.1:1
fa=@(t,y)[y(2);-(a*y(2)+4*y(1)+sin(2*t))];
[t,y]=ode45(fa,[0:.05:16],[1;0]);
plot(t,y(:,1))
end
```

Then press "enter" and observe the graph which results. What did you just obtain? Why does this gorgeous graph look the way it does?

41. This is like the above but uses larger damping constants. Enter the following in MATLAB.

```
hold on
for a=0:2:8
ka=@(t,x)[x(2);-(a*x(2)+4*x(1))+sin(2*t)];
[t,x]=ode45(ka,[0:.05:12],[1;0]);
plot(t,x(:,1))
end
```

Then press "enter" and observe the graph which results. What did you just obtain? Why does this gorgeous graph look the way it does?

42. The numerical methods don't require the equation to be linear. Duffing's equation is of the form $y'' + \alpha y' + \beta y + \varepsilon y^3 = f(t)$ where usually ε is small. It models a vibrating pendulum in which the string is replaced with a spring. You can imagine that the motion could be pretty complicated. For simplicity, let $\alpha, \beta, \varepsilon$ all be 1 and let $f(t) = 0$. Solve numerically and graph on $[0, 5]$ the solution to the initial value problems coming from this case of Duffing's equation and the initial conditions $y(0) = 0, y'(0) = z$ for $z = .3, .6, .9, 1.2, 1.5$.

43. Now let the initial data for the above equation be $y(0) = 0, y'(0) = 1$ and keep $\alpha = \beta = 1$ but consider solutions for

$$\varepsilon = -1.5, -1.2, -.9, -.6, -.3, 0$$

The solution for $\varepsilon = 0$ is the one which corresponds to the linear equation of damped oscillation. This shows how the solutions change with respect to the addition of this nonlinear term. Graph on $[0, 12]$. Next consider the same problem for the same values of ε but this time, let $\alpha = .1$. Only graph on a much smaller interval, say $[0, 2.5]$. Why are the graphs so different?

Chapter 6

Laplace Transform Methods

There is a popular algebraic technique which may be the fastest way to find closed form solutions to the initial value problem. This method of Laplace* transforms succeeds so well because of the algebraic technique of partial fractions and the fact that the Laplace transform is a linear mapping.

This presentation will emphasize the algebraic procedures. The analytical questions are not trivial and are given a discussion in optional sections.

Definition 6.0.1 *Let f be a function defined on $[0, \infty)$ which has **exponential growth**, meaning that*

$$|f(t)| \leq Ce^{\lambda t}$$

*for some real λ. Then the **Laplace transform** of f, denoted by $\mathcal{L}(f)$ is defined as*

$$\mathcal{L}f(s) = \int_0^\infty e^{-ts} f(t)\, dt$$

for all s sufficiently large. It is customary to write this transform as $F(s)$ or $\mathcal{L}f(s)$ and the function as $f(t)$ instead of f.

Lemma 6.0.2 *$\mathcal{L}$ is a linear mapping in the sense that if f, g have exponential growth, then for all s large enough and a, b scalars,*

$$\mathcal{L}(af(t) + bg(t))(s) = a\mathcal{L}f(s) + b\mathcal{L}g(s)$$

Proof: Let f, g be two functions having exponential growth. Then for s large enough,

$$
\begin{aligned}
\mathcal{L}(af(t) + bg(t)) &\equiv \int_0^\infty e^{-ts}(af(t) + bg(t))\, dt \\
&= a\int_0^\infty e^{-ts} f(t)\, dt + b\int_0^\infty e^{-ts} g(t)\, dt = a\mathcal{L}f(s) + b\mathcal{L}g(s) \quad \blacksquare
\end{aligned}
$$

---

*Pierre-Simon, marquis de Laplace (1749-1827) had interests in mathematics, physics, probability, and astronomy. He wrote a major book called celestial mechanics. There is also the Laplacian named after him, and Laplace's equation in potential theory. The expansion of a determinant along a row or column is called Laplace expansion. He was also involved in the development of the metric system. It is hard to overstate the importance of his contributions to mathematics and the other subjects which interested him. [11]

The usefulness of this method in solving differential equations, comes from the following observation.

$$\mathcal{L}\left(x'\left(t\right)\right) = \int_0^\infty x'\left(t\right)e^{-ts}dt = x\left(t\right)e^{-st}\big|_0^\infty + \int_0^\infty se^{-st}x\left(t\right)dt = -x\left(0\right) + s\mathcal{L}x\left(s\right).$$

In the table, $\Gamma\left(p+1\right)$ denotes the gamma function

$$\Gamma\left(p+1\right) = \int_0^\infty e^{-t}t^p dt$$

The function $u_c\left(t\right)$ denotes the step function which equals 1 for $t > c$ and 0 for $t < c$.

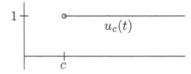

The expression in Formula 20.) is defined as follows

$$\int \delta\left(t-c\right)f\left(t\right)dt = f\left(c\right)$$

It models an impulse and is sometimes called the Dirac delta function. There is no such function but it is called this anyway. In the following, n will be a positive integer and $f * g\left(t\right) \equiv \int_0^t f\left(t-u\right)g\left(u\right)du$. Also, $F\left(s\right)$ will denote $\mathcal{L}\left\{f\left(t\right)\right\}$ the Laplace transform of the function $t \to f\left(t\right)$.

Table of Laplace Transforms

$f\left(t\right)$	$F\left(s\right)$	$f\left(t\right)$	$F\left(s\right)$
1.) 1	$1/s$	12.) $e^{at}\sinh bt$	$\frac{b}{\left(s-a\right)^2-b^2}$
2.) e^{at}	$1/\left(s-a\right)$	13.) $e^{at}\cosh bt$	$\frac{s-a}{\left(s-a\right)^2-b^2}$
3.) t^n	$\frac{n!}{s^{n+1}}$	14) $t^n e^{at}$	$\frac{n!}{\left(s-a\right)^{n+1}}$
4.) $t^p,\ p > -1$	$\frac{\Gamma\left(p+1\right)}{s^{p+1}}$	15.) $u_c\left(t\right)$	$\frac{e^{-cs}}{s}$
5.) $\sin at$	$\frac{a}{s^2+a^2}$	16.) $u_c\left(t\right)f\left(t-c\right)$	$e^{-cs}F\left(s\right)$
6.) $\cos at$	$\frac{s}{s^2+a^2}$	17.) $e^{ct}f\left(t\right)$	$F\left(s-c\right)$
7.) e^{ibt}	$\frac{s+ib}{s^2+b^2}$	18.) $f\left(ct\right)$	$\frac{1}{c}F\left(\frac{s}{c}\right)$
8.) $\sinh at$	$\frac{a}{s^2-a^2}$	19.) $f * g\left(t\right)$	$F\left(s\right)G\left(s\right)$
9.) $\cosh at$	$\frac{s}{s^2-a^2}$	20.) $\delta\left(t-c\right)$	e^{-cs}
10.) $e^{at}\sin bt$	$\frac{b}{\left(s-a\right)^2+b^2}$	21.) $f'\left(t\right)$	$sF\left(s\right)-f\left(0\right)$
11.) $e^{at}\cos bt$	$\frac{s-a}{\left(s-a\right)^2+b^2}$	22.) $\left(-t\right)^n f\left(t\right)$	$\frac{d^n F}{ds^n}\left(s\right)$

6.1 Solving Initial Value Problems

The Laplace transform and the above table can be used to solve initial value problems very easily. To do this, here is a lemma. In dealing with these transforms, it is customary to abuse notation slightly and write $f(t)$ as the name of a function rather than the value of the function f at t.

Lemma 6.1.1 *The following holds. For $f^{(k)}$ the kth derivative of f,*

$$\mathcal{L}\left(f^{(k)}(t)\right) = -\left(\begin{array}{c} f^{(k-1)}(0) + sf^{(k-2)}(0) + s^2 f^{(k-3)}(0) + \\ \cdots + s^{k-1} f(0) \end{array}\right) + s^k \mathcal{L}(f(t))$$

Proof: This is true for $k = 1$ which was shown above. Suppose true for k. Then by induction

$$\mathcal{L}\left(f^{(k+1)}(t)\right) = -\left(\begin{array}{c} f'^{(k-1)}(0) + sf'^{(k-2)}(0) + s^2 f'^{(k-3)}(0) + \\ \cdots + s^{k-1} f'(0) \end{array}\right) + s^k \mathcal{L}(f'(t))$$

Now

$$s^k \mathcal{L}(f'(t)) = s^k \left(-f(0) + s\mathcal{L}(f(t))\right)$$

Therefore, $\mathcal{L}\left(f^{(k+1)}(t)\right) =$

$$-\left(f^k(0) + sf^{(k-1)}(0) + s^2 f^{(k-2)}(0) + \cdots + s^{k-1} f'(0)\right) - s^k f(0) + s^{k+1}\mathcal{L}(f(t))$$

which was to be shown. ∎

This makes it possible to solve initial value problems for differential equations which have constant coefficients.

Example 6.1.2 *Find the solution to the initial value problem*

$$y'' + 3y' + 2y = e^{-t} + \sin(t), \ \ y(0) = 1, y'(0) = 2.$$

Take the Laplace transform of both sides. You can look up these things in the table and use the above lemma.

$$-(y'(0) + sy(0)) + s^2 Y(s) + 3(-y(0)) + 3sY(s) + 2Y(s) = \frac{1}{s+1} + \frac{1}{s^2+1}$$

Now you put in the initial conditions.

$$-(2+s) + s^2 Y(s) + 3(-1) + 3sY(s) + 2Y(s) = \frac{1}{s+1} + \frac{1}{s^2+1}$$

Thus

$$Y(s)\left(s^2 + 3s + 2\right) = \frac{1}{s+1} + \frac{1}{s^2+1} + 3 + (2+s)$$

Hence, solving for $Y(s)$,

$$Y(s) = \frac{s^4 + 6s^3 + 7s^2 + 7s + 7}{\left(s^3 + s^2 + s + 1\right)\left(s^2 + 3s + 2\right)}$$

Now all you have to do is to find the function $y(t)$ whose Laplace transform is the above. This is based on the idea that if two functions have the same Laplace transforms, then they are the same function. This is false of course, but if they are continuous functions, then you might expect it to be true and this kind of thing can be shown either with methods of complex variables or using the Weierstrass theorem from real analysis. We do not concern ourselves with significant mathematical questions here. Only manipulations are of significance at this time. How can we go backwards in the table? That expression on the right does not occur in the table. This is why we take the partial fractions decomposition of the rational function and hope that this will yield terms which occur in the table. The above rational function equals

$$\frac{7}{2(s+1)} - \frac{\frac{3}{10}s - \frac{1}{10}}{s^2+1} + \frac{1}{(s+1)^2} - \frac{11}{5(s+2)}$$

Now each of these is close enough to something in the table that we can go backwards and find $y(t)$. This yields

$$\frac{7}{2}e^{-t} - \frac{3}{10}\cos(t) + \frac{1}{10}\sin(t) + te^{-t} - \frac{11}{5}e^{-2t} = y(t)$$

That was pretty easy. This is why people like this method. You save a lot of trouble by simply looking things up in a table.

So where do the things in the table come from? You simply do the necessary integrals once and obtain the results in the table. I will consider a few of these.

1. Formula 2:

$$\int_0^\infty e^{-st}e^{at}dt = \frac{e^{at-st}}{a-s}\Big|_0^\infty = \frac{1}{s-a}, \; s > a$$

2. Formula 3: For $p > -1$, change variables as shown.

$$\int_0^\infty t^p e^{-st}dt = \frac{1}{s^{p+1}}\int_0^\infty e^{-u}u^p du \equiv \frac{\Gamma(p+1)}{s^{p+1}}$$

   Note that if p is a positive integer, $\Gamma(p+1) = p!$. You should verify this by induction and integration by parts.

3. Formulas 5, 6, 10, and 11 can all be done at once by looking at real and imaginary parts of the following integral which is worked out here. For $s > a$,

$$\int_0^\infty e^{-st}e^{(a+ib)t}dt = -e^{at+ibt-st}\frac{ib-a+s}{a^2-2as+b^2+s^2}\Big|_0^\infty = \frac{ib-a+s}{a^2-2as+b^2+s^2}$$

   Now take the real and imaginary parts

$$\frac{s-a}{(s-a)^2+b^2} + i\frac{b}{(s-a)^2+b^2}$$

   From Euler's formula, you get Formulas 10 and 11 by taking imaginary and real parts respectively. You get Formulas 5 and 6 by letting $a = 0$. Formula 7 results from letting $a = 0$ and using the top line of the above argument. Thus you get

$$\int_0^\infty e^{-st}e^{ibt}dt = \frac{ib+s}{b^2+s^2}$$

4. Formulas 8 and 9 can be obtained by noting that $\cosh(at)$ is the even part of e^{at} and $\sinh(at)$ is the odd part. Then from the above

$$\int_0^\infty e^{-st} e^{at} dt = \frac{1}{s-a}$$

Taking the even part of a function is linear and similar for the odd part. Thus the even part of what was just obtained will yield 9 and the odd part will yield 8. Thus 9 is

$$\frac{\frac{1}{s-a} + \frac{1}{-s-a}}{2} = \frac{a}{s^2 - a^2}$$

Taking the odd part yields 8

$$\frac{\frac{1}{s-a} - \frac{1}{-s-a}}{2} = \frac{s}{s^2 - a^2}$$

5. Formulas 12 and 13 are routine. You just need to recall that

$$\sinh(bt) = \frac{1}{2}\left(e^{bt} - e^{-bt}\right), \cosh(bt) = \frac{1}{2}\left(e^{bt} + e^{-bt}\right)$$

Then you simply compute the improper integrals. However, you can also see that these formulas should hold by replacing b with ib in Formulas 10 and 11. This goes as follows.

$$\sin(bt) = \frac{e^{ibt} - e^{-ibt}}{2i} \text{ so } \sin(ibt) = \frac{e^{-bt} - e^{bt}}{2i} = -\frac{1}{i}\sinh(bt)$$

Thus, formally, the Laplace transform of $e^{at}\sinh(bt)$ is the Laplace transform of

$$-ie^{at}\sin(ibt)$$

which formally equals

$$\frac{-i(ib)}{(s-a)^2 - b^2} = \frac{b}{(s-a)^2 - b^2}$$

and one can obtain the other involving the $\cosh(bt)$ the same way.

6. Formula 14 holds if $n = 0$. Recall $0! \equiv 1$. It is also clearly true for $n = 1$. A short integration by parts yields

$$\int_0^\infty e^{-st} t e^{at} dt = \frac{1}{(s-a)^2}$$

Now suppose you have the case where n is a positive integer and you have shown it true for $n-1$. Then for $s > a$, it follows from induction that

$$\int_0^\infty e^{-st} t^n e^{at} dt = \int_0^\infty e^{(a-s)t} t^n dt = \frac{e^{(a-s)t}}{a-s} t^n \Big|_0^\infty + n \int_0^t \frac{e^{(a-s)t}}{s-a} t^{n-1} dt$$

$$= \frac{n}{s-a} \int_0^t e^{(a-s)t} t^{n-1} dt = \frac{n}{s-a} \frac{(n-1)!}{(s-a)^n} = \frac{n!}{(s-a)^{n+1}}$$

7. Formula 15:

$$\int_0^\infty e^{-st} u_c(t)\, dt = \int_c^\infty e^{-st} dt = \frac{e^{-cs}}{s}$$

8. Formula 16:

$$\int_0^\infty u_c(t) f(t-c) e^{-st} dt = e^{-sc} \int_1^\infty f(t-c) e^{-s(t-c)} dt$$

$$= e^{-sc} \int_0^\infty f(u) e^{-su} du = e^{-sc} F(s)$$

9. Formula 19 will involve interchanging the order of integration. This the most profound of the formulas in this table. $f * g$ is called the convolution of f and g.

$$\int_0^\infty f * g(t) e^{-st} dt = \int_0^\infty \int_0^t f(t-u) g(u)\, du e^{-st} dt =$$

$$\int_0^\infty \int_u^\infty f(t-u) g(u) e^{-st} dt du = \int_0^\infty e^{-su} g(u) \int_u^\infty f(t-u) e^{-s(t-u)} dt du$$

$$= \int_0^\infty e^{-su} g(u) \int_0^\infty f(x) e^{-sx} dx du = G(s) F(s)$$

10. Formula 20 is obvious if you make the definition given above for the symbol $\delta(t-c)$. This notion that it is some sort of function which picks out a single value of the function is complete nonsense. It should be thought of as a measure. However, we will use the traditional notation here.

Now let's use the above theory to consider what happens when you have $Ly = f$ and f comes in pieces, maybe having jumps. You can still use the table to find the solutions. First here is a contrived example.

Example 6.1.3 *Solve the initial value problem*

$$2y'' - y' - y = f(t), \ y(0) = 1, y'(0) = 0$$

where

$$f(t) = \begin{cases} \sin(t-1) \ \ if \ t \in [0,1] \\ \quad t-1 \ if \ t > 1 \end{cases}$$

Then

$$f(t) = (1 - u_1(t)) \sin(t-1) + u_1(t)(t-1)$$

This is because $(1 - u_1(t))$ equals 1 on $(0,1)$ and 0 elsewhere and $u_1(t)$ equals 1 for $t > 1$ and zero elsewhere. Take the Laplace transform of both sides. On the right, use Formula 16. Also $\sin(t-1) = \sin t \cos 1 - \cos t \sin 1$. Thus,

$$2\left(-y'(0) - sy(0) + s^2 Y(s)\right) - \left(-y(0) + sY(s)\right) - Y(s)$$

$$= \cos 1 \frac{s}{s^2+1} - \sin 1 \frac{1}{s^2+1} - e^{-s} \frac{s}{s^2+1} + e^{-s} \frac{1}{s^2}$$

Now plug in for the initial conditions.

$$\left(2s^2 - s - 1\right) Y\left(s\right) = \cos 1 \frac{s}{s^2 + 1} - \sin 1 \frac{1}{s^2 + 1}$$

$$-e^{-s} \frac{s}{s^2 + 1} + e^{-s} \frac{1}{s^2} + 2s - 1$$

Then you have

$$Y\left(s\right) = \cos 1 \frac{s}{\left(s^2 + 1\right)\left(2s^2 - s - 1\right)} - \sin 1 \frac{1}{\left(s^2 + 1\right)\left(2s^2 - s - 1\right)}$$

$$-e^{-s} \frac{s}{\left(s^2 + 1\right)\left(2s^2 - s - 1\right)} + e^{-s} \frac{1}{s^2\left(2s^2 - s - 1\right)} + \frac{2s - 1}{\left(2s^2 - s - 1\right)}$$

Here are the partial fractions expansions.

$$\frac{s}{\left(s^2 + 1\right)\left(2s^2 - s - 1\right)} = \frac{1}{6\left(s - 1\right)} - \frac{\frac{3}{10}s + \frac{1}{10}}{s^2 + 1} + \frac{4}{15\left(2s + 1\right)}$$

$$\frac{1}{\left(s^2 + 1\right)\left(2s^2 - s - 1\right)} = \frac{\frac{1}{10}s - \frac{3}{10}}{s^2 + 1} + \frac{1}{6\left(s - 1\right)} - \frac{8}{15\left(2s + 1\right)}$$

$$\frac{1}{s^2\left(2s^2 - s - 1\right)} = \frac{1}{3\left(s - 1\right)} - \frac{8}{3\left(2s + 1\right)} + \frac{1}{s} - \frac{1}{s^2}$$

$$\frac{2s - 1}{\left(2s^2 - s - 1\right)} = \frac{1}{3\left(s - 1\right)} + \frac{4}{3\left(2s + 1\right)}$$

Thus, massaging the partial fractions terms a little bit so that they yield things which can be seen in the table,

$$Y\left(s\right) = \cos 1 \left(\frac{1}{6\left(s - 1\right)} - \frac{\frac{3}{10}s + \frac{1}{10}}{s^2 + 1} + \frac{4}{30\left(s + 1/2\right)}\right) -$$

$$\sin 1 \left(\frac{\frac{1}{10}s - \frac{3}{10}}{s^2 + 1} + \frac{1}{6\left(s - 1\right)} - \frac{8}{30\left(s + 1/2\right)}\right) -$$

$$e^{-s} \left(\frac{1}{6\left(s - 1\right)} - \frac{\frac{3}{10}s + \frac{1}{10}}{s^2 + 1} + \frac{4}{30\left(s + 1/2\right)}\right) +$$

$$e^{-s} \left(\frac{1}{3\left(s - 1\right)} - \frac{8}{6\left(s + 1/2\right)} + \frac{1}{s} - \frac{1}{s^2}\right) + \frac{1}{3\left(s - 1\right)} + \frac{4}{6\left(s + 1/2\right)}$$

Then consulting the table, it follows that

$$y\left(t\right) = \cos 1 \left[\frac{1}{6}e^t - \frac{3}{10}\cos\left(t\right) - \frac{1}{10}\sin\left(t\right) + \frac{4}{30}e^{-(1/2)t}\right] -$$

$$\sin 1 \left[\frac{1}{10}\cos\left(t\right) - \frac{3}{10}\sin\left(t\right) + \frac{1}{6}e^t - \frac{8}{30}e^{-(1/2)t}\right] -$$

$$u_1\left(t\right) \left[\frac{1}{6}e^{t-1} - \frac{3}{10}\cos\left(t - 1\right) - \frac{1}{10}\sin\left(t - 1\right) + \frac{4}{30}e^{-(1/2)(t-1)}\right] +$$

$$u_1\left(t\right)\left[\frac{1}{3}e^{t-1} - \frac{8}{6}e^{-(1/2)(t-1)} + 1 - (t-1)\right] + \frac{1}{3}e^t + \frac{4}{6}e^{-(1/2)t}$$

What follows are harder examples in which you need to massage more in order to use the table.

Example 6.1.4 *Solve the initial value problem*

$$2y'' - y' - y = f\left(t\right), \; y\left(0\right) = y'\left(0\right) = 0$$

where

$$f\left(t\right) = \begin{cases} \sin\left(t\right) \; if \, t \in [0,1] \\ t \; if \, t > 1 \end{cases}$$

First write the right side in terms of things which can be looked up in the table to avoid heroics. A function which is 1 on $(0,1)$ and zero elsewhere is $1 - u_1\left(t\right)$. A function which is 1 for $t > 1$ and zero elsewhere is $u_1\left(t\right)$. Therefore, the right side equals $\sin\left(t\right)\left(1 - u_1\left(t\right)\right) + tu_1\left(t\right)$. You don't worry about what happens at a single point. The integral does not see this. Therefore, it does not matter. Now this equals

$$\sin\left(t\right) + \left(t - 1\right)u_1\left(t\right) + u_1\left(t\right) - u_1\left(t\right)\left[\sin\left(t - 1\right)\cos\left(1\right) + \cos\left(t - 1\right)\sin\left(1\right)\right]$$

These are each terms which can be looked up in the table. Thus, since the initial conditions are both 0,

$$\left(2s^2 - s - 1\right)Y\left(s\right) = \frac{1}{s^2 + 1} + e^{-s}\frac{1}{s^2} + e^{-s}\frac{1}{s} - e^{-s}\cos\left(1\right)\frac{1}{s^2 + 1} - e^{-s}\sin\left(1\right)\frac{s}{s^2 + 1}$$

$$Y\left(s\right) = \frac{1}{\left(s^2 + 1\right)\left(2s^2 - s - 1\right)} + e^{-s}\frac{1}{s^2\left(2s^2 - s - 1\right)} + e^{-s}\frac{1}{s\left(2s^2 - s - 1\right)}$$

$$- \cos\left(1\right)e^{-s}\frac{1}{\left(s^2 + 1\right)\left(2s^2 - s - 1\right)} - \sin\left(1\right)e^{-s}\frac{s}{\left(s^2 + 1\right)\left(2s^2 - s - 1\right)}$$

Here are the necessary partial fractions expansions.

$$\frac{1}{\left(s^2 + 1\right)\left(2s^2 - s - 1\right)} = \frac{\frac{1}{10}s - \frac{3}{10}}{s^2 + 1} + \frac{1}{6\left(s - 1\right)} - \frac{8}{30\left(s + 1/2\right)}$$

$$\frac{1}{s^2\left(2s^2 - s - 1\right)} = \frac{1}{3\left(s - 1\right)} - \frac{8}{6\left(s + 1/2\right)} + \frac{1}{s} - \frac{1}{s^2}$$

$$\frac{1}{s\left(2s^2 - s - 1\right)} = \frac{1}{3\left(s - 1\right)} + \frac{4}{6\left(s + 1/2\right)} - \frac{1}{s}$$

$$\frac{s}{\left(s^2 + 1\right)\left(2s^2 - s - 1\right)} = \frac{1}{6\left(s - 1\right)} - \frac{\frac{3}{10}s + \frac{1}{10}}{s^2 + 1} + \frac{4}{30\left(s + 1/2\right)}$$

Now it is just a matter of using the table.

$$y\left(t\right) = \frac{1}{10}\cos\left(t\right) - \frac{3}{10}\sin\left(t\right) + \frac{1}{6}e^t - \frac{8}{30}e^{-(1/2)t} +$$

$$u_1(t)\left[\frac{1}{3}e^{t-1} - \frac{8}{6}e^{-(1/2)(t-1)} + 1 - (t-1)\right] +$$

$$u_1(t)\left[\frac{1}{3}e^{t-1} + \frac{4}{6}e^{-(1/2)(t-1)} - 1\right] -$$

$$\cos(1)u_1(t)\left[\frac{1}{10}\cos(t-1) - \frac{3}{10}\sin(t-1) + \frac{1}{6}e^{t-1} - \frac{8}{30}e^{-(1/2)(t-1)}\right] -$$

$$\sin(1)u_1(t)\left[\frac{1}{6}e^{t-1} - \frac{3}{10}\cos(t-1) - \frac{1}{10}\sin(t-1) + \frac{4}{30}e^{-(1/2)(t-1)}\right]$$

6.2 The Impulse Function

There is no such thing as a function with the properties attributed to $\delta(t-c)$. This is just a useful convention. The physically meaningful idea is this. There exists a function f_δ which is positive, $\int_0^\infty f_\delta(t)\,dt = 1$ and f_δ equals zero off the interval $[0, \delta]$. This is true for each or some sequence of positive δ converging to 0. Then let g be an arbitrary continuous function which has an improper Riemann integral on $[0, \infty)$. Then consider for $c > 0$

$$\int_0^\infty g(t)f_\delta(t-c)\,dt$$

Since $\int_0^\infty f_\delta(t-c)\,dt = 1$,

$$\left| g(c) - \int_0^\infty g(t)f_\delta(t-c)\,dt \right| = \left| \int_0^\infty g(c)f_\delta(t-c)\,dt - \int_0^\infty g(t)f_\delta(t-c)\,dt \right|$$

$$\leq \int_c^{c+\delta} |g(c) - g(t)|\,f_\delta(t-c)\,dt$$

Now let $\varepsilon > 0$ be given. Then there exists a $\delta > 0$ such that if $|c - t| < \delta$, then $|g(c) - g(t)| < \varepsilon$. This is be continuity of g. Therefore, if δ is this small,

$$\left| g(c) - \int_0^\infty g(t)f_\delta(t-c)\,dt \right| \leq \int_c^{c+\delta} \varepsilon f_\delta(t-c)\,dt = \varepsilon$$

It follows that

$$\lim_{\delta \to 0+} \int_0^\infty g(t)f_\delta(t-c)\,dt = g(c)$$

Therefore, if you have a function which acts on a very small interval of time but is large enough that its integral equals 1, then you can approximate its effect by using the formal definition of $\delta(t-c)$,

$$\int_0^\infty g(t)f_\delta(t-c)\,dt \approx \int_0^\infty \delta(t-c)g(t)\,dt \equiv g(c).$$

What would $2\delta(t-c)$ mean? It would mean

$$\int_0^\infty 2\delta(t-c)f(t)\,dt = 2f(c)$$

Note that you have something here which is defined more in terms of what it does than the standard sort of formula. This takes a little getting used to, but it is really the way all functions are. They are defined in terms of what they do to something. The difference here is that the thing it does something to is a function and we use the standard notation for convenience.

Example 6.2.1 *Find the solution to the initial value problem*

$$y'' + 2y' + y = 2\delta (t - 3), \ y(0) = y'(0) = 0$$

The interpretation, is that there is a forcing function which acts on a very small interval of time containing 3 and this function has integral equal to 2. It is being approximated by $2\delta (t - 3)$. This is because it will have essentially the same effect on e^{-st} as the actual function, so we use the above impulse function for simplicity. Thus, since the initial data is 0, $(s^2 + 2s + 1) Y(s) = 2e^{-3s}$. Then

$$Y(s) = 2\frac{e^{-3s}}{(s + 1)^2}$$

Hence

$$y(t) = 2u_3(t)(t - 3) e^{-(t-3)}$$

Note that until $t = 3$, the solution equals 0 which is should do. After all, till this time, the right side equals 0 and then, since the initial conditions are 0, the solution must equal 0 thanks to uniqueness.

6.3   The Convolution

This features Formula 17. It is best to consider it in terms of an example.

Example 6.3.1 *Solve the following initial value problem*

$$3y'' - 5y' - 2y = f(t), \ y(0) = y_0, y'(0) = y_1$$

Here $f(t)$ is some function and the idea is to find a description of the solution to this initial value problem. Note that this is completely general in the sense that y_0, y_1 and $f(t)$ are arbitrary.

Take Laplace transform of both sides.

$$3\left(-y'(0) - sy(0) + s^2 Y(s)\right) - 5\left(-y(0) + sY(s)\right) - 2Y(s) = F(s)$$

Place the values of the initial data in to the equation.

$$3\left(-y_1 - sy_0 + s^2 Y(s)\right) - 5\left(-y_0 + sY(s)\right) - 2Y(s) = F(s)$$

Now solve for $Y(s)$.

$$Y(s)\left(3s^2 - 5s - 2\right) = 3y_1 + (3s - 5)y_0 + F(s)$$

Hence

$$Y(s) = \frac{1}{3s^2 - 5s - 2}F(s) + \frac{3}{3s^2 - 5s - 2}y_1 + \frac{3s - 5}{3s^2 - 5s - 2}y_0$$

First find the partial fractions expansions

$$\frac{1}{3s^2 - 5s - 2} = \frac{1}{7(s-2)} - \frac{3}{21(s+1/3)}, \ \frac{3s-5}{3s^2 - 5s - 2} = \frac{1}{7(s-2)} + \frac{18}{7(3s+1)}$$

Then the function whose Laplace transform is $\frac{1}{3s^2 - 5s - 2}$ is $\frac{1}{7}e^{2t} - \frac{1}{7}e^{-(1/3)t}$. Using the formula for the convolution integral, it follows that

$$
\begin{aligned}
y(t) &= \int_0^t \left(\frac{1}{7}e^{2(t-s)} - \frac{1}{7}e^{-(1/3)(t-s)}\right) f(s)\, ds \\
&\quad + 3y_1 \left(\frac{1}{7}e^{2t} - \frac{1}{7}e^{-(1/3)t}\right) + \left(\frac{1}{7}e^{2t} + \frac{6}{7}e^{-(1/3)t}\right) y_0
\end{aligned}
$$

Of course it is assumed that the function $f(t)$ has exponential growth to justify the above formula. However, it will end up holding even if this condition is not known. That is because, assuming f is continuous, you could confine your attention to an interval $[0, T]$ on which f is bounded and then set f to equal 0 off this interval. The result would have exponential growth. Then the above formula would be valid for $t \in [0, T]$. But T is arbitrary. Therefore, this formula always works.

6.4 Why Does It Work?

The whole approach for Laplace transforms is based on the assertion that if $\mathcal{L}(f) = \mathcal{L}(g)$, then $f = g$. In the exercises, you will show that this is not even true! However, I will show here that if f, g are continuous, then it will be true. Actually, it is shown here that if $\mathcal{L}(f) = 0$, and f is continuous, then $f = 0$. This can be done a variety of ways. Here we will use a fundamental theorem called the Weierstrass approximation theorem. Warning: This uses somewhat more analysis than beginning calculus. You will need to use Theorem E.2.23 which says that on a compact set, a continuous function is uniformly continuous. This should have been studied in calculus since it is essential to understanding the Riemann integral, but is normally left out which is why it is in the appendix.

Lemma 6.4.1 *The following estimate holds for $x \in [0, 1]$ and $m \geq 2$.*

$$\sum_{k=0}^{m} \binom{m}{k} (k - mx)^2 x^k (1-x)^{m-k} \leq \frac{1}{4}m$$

Proof: First of all, from the binomial theorem

$$\sum_{k=0}^{m} \binom{m}{k} \left(e^{t(k-mx)}\right) x^k (1-x)^{m-k} = e^{-tmx} \sum_{k=0}^{m} \binom{m}{k} \left(e^{tk}\right) x^k (1-x)^{m-k}$$

$$= e^{-tmx} \left(1 - x + xe^t\right)^m = e^{-tmx} g(t)^m, \ g(0) = 1, g'(0) = g''(0) = x$$

Take a partial derivative with respect to t twice.

$$\sum_{k=0}^{m} \binom{m}{k} (k - mx)^2 e^{t(k-mx)} x^k (1-x)^{m-k}$$

$$= (mx)^2 e^{-tmx} g(t)^m + 2(-mx) e^{-tmx} mg(t)^{m-1} g'(t)$$
$$+e^{-tmx} \left[m(m-1) g(t)^{m-2} g'(t)^2 + mg(t)^{m-1} g''(t) \right]$$

Now let $t = 0$ and note that the right side is $m(x - x^2) \leq m/4$ for $x \in [0,1]$. Thus

$$\sum_{k=0}^{m} \binom{m}{k} (k - mx)^2 x^k (1-x)^{m-k} = mx - mx^2 \leq m/4 \quad \blacksquare$$

With this preparation, here is the first version of the Weierstrass approximation theorem. You will need to know about uniform continuity. Let $\|f\|_\infty = \max_{x \in [0,1]} |f(x)|$.

Theorem 6.4.2 *Let f be continuous on $[0,1]$ and let*

$$p_m(x) \equiv \sum_{k=0}^{m} \binom{m}{k} x^k (1-x)^{m-k} f\left(\frac{k}{m}\right).$$

Then these polynomials converge uniformly to f on $[0,1]$. That is, for every $\varepsilon > 0$, there exists N such that if $m \geq N$, then

$$\max\{|p_m(x) - f(x)|, x \in [0,1]\} < \varepsilon$$

Proof: Let $\|f\|_\infty$ denote the largest value of $|f|$. By uniform continuity of f, (Theorem E.2.23) there exists $\delta > 0$ such that if $|x - x'| < \delta$, then $|f(x) - f(x')| < \varepsilon/2$. By the binomial theorem,

$$|p_m(x) - f(x)| = \left| \sum_{k=0}^{m} \binom{m}{k} x^k (1-x)^{m-k} \left(f\left(\frac{k}{m}\right) - f(x) \right) \right|$$
$$\leq \sum_{k=0}^{m} \binom{m}{k} x^k (1-x)^{m-k} \left| f\left(\frac{k}{m}\right) - f(x) \right|$$

$$\leq \sum_{\left|\frac{k}{m} - x\right| < \delta} \binom{m}{k} x^k (1-x)^{m-k} \left| f\left(\frac{k}{m}\right) - f(x) \right|$$
$$+ 2\|f\|_\infty \sum_{\left|\frac{k}{m} - x\right| \geq \delta} \binom{m}{k} x^k (1-x)^{m-k}$$

Therefore,

$$\leq \sum_{k=0}^{m} \binom{m}{k} x^k (1-x)^{m-k} \frac{\varepsilon}{2} + 2\|f\|_\infty \sum_{(k-mx)^2 \geq m^2 \delta^2} \binom{m}{k} x^k (1-x)^{m-k} \leq$$

$$\frac{\varepsilon}{2} + 2\|f\|_\infty \frac{1}{m^2 \delta^2} \sum_{k=0}^{m} \binom{m}{k} (k-mx)^2 x^k (1-x)^{m-k} \leq \frac{\varepsilon}{2} + 2\|f\|_\infty \frac{1}{4} m \frac{1}{\delta^2 m^2} < \varepsilon$$

provided m is large enough. $\blacksquare$

Lemma 6.4.3 *Suppose q is a continuous function defined on* $[0, 1]$. *Also suppose that for all* $n = 0, 1, 2, \cdots$,

$$\int_0^1 q(x) x^n dx = 0$$

Then it follows that $q = 0$.

Proof: By assumption, for $p(x)$ any polynomial, $\int_0^1 q(x) p(x) dx = 0$. Now let $\{p_n(x)\}$ be a sequence of polynomials which converge uniformly to $q(x)$ by Theorem 6.4.2. Say

$$\max_{x \in [0,1]} |q(x) - p_n(x)| < \frac{1}{n}$$

Then

$$
\int_0^1 q^2(x)\, dx = \int_0^1 q(x)(q(x) - p_n(x))\, dx + \overbrace{\int_0^1 q(x) p_n(x)\, dx}^{=0}
$$

$$
\leq \int_0^1 |q(x)(q(x) - p_n(x))|\, dx \leq \int_0^1 |q(x)|\, dx \frac{1}{n}
$$

Since n is arbitrary, it follows that $\int_0^1 q^2(x)\, dx = 0$ and by continuity, it must be the case that $q(x) = 0$ for all x since otherwise, there would be a small interval on which $q^2(x)$ is positive and so the integral could not have been 0 after all. ∎

Lemma 6.4.4 *Suppose* $|f(t)| \leq Ce^{-\delta t}$ *for some* $\delta > 0$ *and all* $t > 0$ *and also that f is continuous. Suppose that* $\int_0^\infty e^{-st} f(t)\, dt = 0$ *for all* $s > 0$. *Then* $f = 0$.

Proof: First note that $\lim_{t \to \infty} |f(t)| = 0$. Next change the variable letting $x = e^{-t}$ and so $x \in [0, 1]$. Then this reduces to $\int_0^1 x^{s-1} f(-\ln(x))\, dx$. Now if you let $q(x) = f(-\ln(x))$, it is not defined when $x = 0$, but $x = 0$ corresponds to $t \to \infty$. Thus $\lim_{x \to 0+} q(x) = 0$. Defining $q(0) \equiv 0$, it follows that it is continuous and for all $n = 0, 1, 2, \cdots$, $\int_0^1 x^n q(x)\, dx = 0$ and so $q(x) = 0$ for all x from Lemma 6.4.3. Thus $f(-\ln(x)) = 0$ for all $x \in (0, 1]$ and so $f(t) = 0$ for all $t \geq 0$. ∎

Now suppose only that $|f(t)| \leq Ce^{rt}$ so f has exponential growth and that for all s sufficiently large, $\mathcal{L}(f) = 0$. Does it follow that $f = 0$? Say this holds for all $s \geq s_0$ where also $s_0 > r$. Then consider $\hat{f}(t) \equiv e^{-s_0 t} f(t)$. Then if $s > 0$,

$$
\int_0^\infty e^{-st} \hat{f}(t)\, dt = \int_0^\infty e^{-st} e^{-s_0 t} f(t)\, dt = \int_0^\infty e^{-(s+s_0)t} f(t)\, dt = 0
$$

because $s + s_0$ is large enough for this to happen. It follows from Lemma 6.4.4 that $\hat{f} = 0$. But this implies that $f = 0$ also. This proves the following fundamental theorem. The rest of what is needed is in the exercises including a different approach in Problem 29.

Theorem 6.4.5 *Suppose f has exponential growth and is continuous on* $[0, \infty)$. *Suppose also that for all s large enough,* $\mathcal{L}(f)(s) = 0$. *Then* $f = 0$.

For more on Laplace transforms, see the old book of Widder [21].

6.5   Automation with Computer Algebra

This procedure will also be introduced later in the context of first order systems.

You can automate the whole process using MATLAB. It is tedious to look things up in a table when the computer algebra system will do it for you right away. However, you will need the symbolic math toolbox for MATLAB.

$$2y'' - y' - y = f(t), \; y(0) = 1, y'(0) = 0$$

First make the coefficient of y'' a 1.

$$y'' - \frac{1}{2}y' - \frac{1}{2}y = \frac{1}{2}f(t), \; y(0) = 1, y'(0) = 0$$

Now use the scheme presented earlier in Procedure 5.5.2 for writing as a first order system. It is less hassle if you do this.

$$\left(\begin{array}{c} x(1) \\ x(2) \end{array} \right)' = \left(\begin{array}{cc} 0 & 1 \\ 1/2 & 1/2 \end{array} \right) \left(\begin{array}{c} x(1) \\ x(2) \end{array} \right) + \left(\begin{array}{c} 0 \\ \frac{1}{2}f(t) \end{array} \right)$$

Now note that, just as for a scalar function,

$$\mathcal{L}(\mathbf{x}'(t)) = sX(s) - \mathbf{x}(0)$$

where $X(s)$ is the vector whose entries are the Laplace transforms of the corresponding entries of $\mathbf{x}(t)$. Also, for A an $n \times n$ matrix,

$$\mathcal{L}(A\mathbf{x}(t)) = AX(s)$$

This is left as an exercise. Thus, if you have

$$\mathbf{x}' = A\mathbf{x} + \mathbf{f}(t),$$

then, taking Laplace transforms,

$$sX(s) - \mathbf{x}(0) = AX(s) + F(s), \; (sI - A)X(s) = \mathbf{x}(0) + F(s)$$

and so

$$X(s) = (sI - A)^{-1}(\mathbf{x}(0) + F(s))$$

You could get MATLAB to do it all at once as follows. Suppose you have a differential equation of order 3 and you found A as described above to consider as a first order system. Say

$$A = \left(\begin{array}{ccc} a11 & a12 & a13 \\ a21 & a22 & a33 \\ a31 & a32 & a33 \end{array} \right)$$

Suppose your initial condition is $(1, 2, 3)^T$ and $\mathbf{f}(t) = (f1(t), f2(t), f3(t))^T$. In the present context, only $f3(t)$ will be nonzero. Then you would enter the following.

```
syms s t ; a=[1;2;3];
b=s*[1 0 0;0 1 0;0 0 1]-[a11 a12 a13;a21 a22 a23;a31 a32 a33];
c=[f1(t);f2(t);f3(t)]; simplify(ilaplace(inv(b)*(a+laplace(c))))
```

For the identity matrix on the second line, you can type eye(3) where 3 is the size of the identity.

Example 6.5.1 *Solve the initial value problem*

$$2y'' - y' - y = \sin t, \; y(0) = 1, y'(0) = 2$$

In this case, you divide by the 2 to make the coefficient of y'' a one. Then as a system, you have

$$\begin{pmatrix} X(1) \\ X(2) \end{pmatrix}' = \begin{pmatrix} 0 & 1 \\ 1/2 & 1/2 \end{pmatrix} \begin{pmatrix} X(1) \\ X(2) \end{pmatrix} + \begin{pmatrix} 0 \\ (1/2)\sin t \end{pmatrix}$$

Then following the above syntax, adjusting for the size of the matrices, one gets the following:

>>syms s t ; a=[1;2]; b=s*eye(2)-[0 1;1/2 1/2];
c=[0;(1/2)*sin(t)]; simplify(ilaplace(inv(b)*(a+laplace(c))))
ans =
cos(t)/10 - (14*exp(-t/2))/15 + (11*exp(t))/6 - (3*sin(t))/10
(7*exp(-t/2))/15 - (3*cos(t))/10 + (11*exp(t))/6 - sin(t)/10

It is the top line you want. Remember the $X(1) = y$.

When you have the right side given in terms of step functions, it will still work but you have to write things differently so that MATLAB will know what you mean. Suppose

$$f(t) = \begin{cases} \sin(t-1) \text{ if } t \in [0,1] \\ t - 1 \text{ if } t > 1 \end{cases}$$

Then you need to write this as follows:

$$f(t) = (1 - \text{heaviside}(t-1))\sin(t-1) + (t-1)\,\text{heaviside}(t-1)$$

or

$$f(t) = \text{heaviside}(1-t)\sin(t-1) + (t-1)\,\text{heaviside}(t-1) \qquad (*)$$

Here the Heaviside function is defined as follows.

$$\text{heaviside}(t-a) = \begin{cases} 1 \text{ if } t > a \\ 0 \text{ if } t < a \end{cases}$$

You can remember it because, although there was a person called Heaviside, the function has one side of the jump at a heavier than the other. If you use this $f(t)$ of $*$ as the right side of the above example, you get the following from MATLAB for $X(1)$ which is the desired solution:

cos(t - 1)/5 - (2*exp(-t/2))/3 + 4*heaviside(t - 1) - (3*sin(t - 1))/5
+ (5*exp(t))/3 - (4*exp(-t/2)*sin(1))/15 - 2*t*heaviside(t - 1)
- (32*heaviside(t - 1)*exp(1/2 - t/2))/15 + (cos(1)*exp(t))/3
- (exp(t)*sin(1))/3 - (cos(t - 1)*heaviside(t - 1))/5
+ (heaviside(t - 1)*exp(t - 1))/3 + (3*heaviside(t - 1)*sin(t - 1))/5
- (8*cos(1)*exp(-t/2))/15

You can also consider this scheme with impulse functions.

Example 6.5.2 *Find the solution to*

$$2y'' - y' - y = \delta(t-1), \; y(0) = 1, y'(0) = 2$$

To include an impulse function at 1 as above, you write dirac(t-1). Then the syntax for having MATLAB solve this for you is as follows.

syms s t ; a=[1;2]; b=s*[1 0;0 1]-[0 1;1/2 1/2];
c=[0;(1/2)*dirac(t-1)]; simplify(ilaplace(inv(b)*(a+laplace(c))))

Then what results is

ans =

(5*exp(t))/3 - (2*exp(-t/2))/3 + heaviside(t - 1)*(exp(t - 1)/3 - exp(1/2 - t/2)/3)

exp(-t/2)/3 + (5*exp(t))/3 + heaviside(t - 1)*(exp(t - 1)/3 + exp(1/2 - t/2)/6)

You want the line on the top. Thus, in usual math notation, this is

$$\frac{1}{3}\left(5e^t - 2e^{-t/2}\right) + \text{heaviside}\,(t-1)\left(e^{\frac{1}{3}t-\frac{1}{3}} - e^{\frac{3}{2}-\frac{3}{2}t}\right)$$

Thus, after the impulse, you get $\frac{1}{3}\left(5e^t - 2e^{-t/2}\right) + e^{\frac{1}{3}t-\frac{1}{3}} - e^{\frac{3}{2}-\frac{3}{2}t}$ and before the impulse, you get $\frac{1}{3}\left(5e^t - 2e^{-t/2}\right)$

In Scientific Notebook, it is even easier if you don't have impulse functions and step functions. I will illustrate with an example.

Exercise 6.5.3 *Find the solution to* $2y'' - 3y' + y = e^t + \sin(t), y(0) = 1, y'(0) = 0.$

You do the following: Enter a 3×1 matrix and fill in the slots as shown. Then click on the matrix and go to compute, ODE, and then laplace. This gives the following result.

$$\begin{aligned} 2y'' - 3y' + y &= e^t + \sin(t) \\ y(0) &= 1 \\ y'(0) &= 0 \end{aligned} \quad ,$$

Laplace solution is: $\left\{ \frac{3}{10}\cos t - \frac{5}{2}e^t - \frac{1}{10}\sin t + \frac{16}{5}e^{\frac{1}{2}t} + te^t \right\}$

They have set it up to do everything for you and all you have to do is click and let it give the answer. It is of course a good idea to check to see if it really did give the right answer.

6.6   Exercises

1. The Laplace transform is given. Compute the inverse Laplace transform of the given functions. You can use the table or a computer algebra system.

   (a) $-\frac{2}{(s-3)^2-3}$

   (b) $\frac{5s+12}{s^2+5s+6}$

   (c) $3\frac{s}{s^2+9}$

   (d) $-\frac{2s+1}{s^2+9}$

   (e) $\frac{2s-2}{s^2+6s+5}$

   (f) $\frac{\left(7s^3+18s^2+26s+12\right)}{s(2s^3+11s^2+13s+4)}$

2. The Laplace transform is given. Compute the inverse Laplace transform of the given functions. You can use the table or a computer algebra system.

(a) $\frac{2}{(s+4)^2-1}$

(b) $3\frac{s-3}{-s^2+s+2}$

(c) $-4\frac{s}{s^2+1}$

(d) $\frac{1}{s^2+16}$

(e) $1/s$

(f) $-\frac{1}{s(s^2-1)}\left(7s^2+7s-4\right)$

3. Recall that for y a function, the Laplace transform $Y(s)$ is given by $\int_0^\infty y(t)e^{-st}dt$. Consider taking the derivative of $Y(s)$. One would write

$$\frac{Y(s+h)-Y(s)}{h} = \int_0^\infty y(t)\frac{e^{-(s+h)t}-e^{-st}}{h}dt$$

Then one would take $\lim_{h\to 0}$ and if you can interchange the lim and the integral, this would yield a formula for $Y'(s)$. In fact, this can be justified. (Theorem B.5.5) Show that $Y^{(n)}(s) = \mathcal{L}(t^n(-1)^n y(t))$.

4. One can use the above problem to find the Laplace transform of some awful functions which are such that finding the Laplace transform directly from the definition would be a lot of trouble. Find the Laplace transform of the following. **Hint:** For example, from the table, $\mathcal{L}(e^{2t}) = \frac{1}{s-2}$. Now use the above problem. If $Y(s) = \mathcal{L}(e^{2t})$, then $Y'(s) = -\mathcal{L}(te^{2t})$.

(a) te^{at}

(b) $t^2\sin(at)$

(c) $t^2\cos(at)$

(d) $te^{at}\cos(bt)$

(e) $te^{at}\sin(bt)$

5. Using Laplace transforms, give the solution to the following initial value problems.

(a) $4y - 4y' + y'' = 5t^3e^{2t}$, $y(0) = 1, y'(0) = 5$

(b) $4y - 4y' + y'' = 4t^4e^{2t}$, $y(0) = 1, y'(0) = 4$

(c) $9y - 6y' + y'' = 2t^2e^{3t}$, $y(0) = 1, y'(0) = 2$

(d) $4y - 4y' + y'' = 3t^3e^{2t}$, $y(0) = -1, y'(0) = 3$

(e) $4y - 4y' + y'' = te^{2t}$, $y(0) = -1, y'(0) = 1$

(f) $y - 2y' + y'' = 5t^5e^t$, $y(0) = 2, y'(0) = 5$

(g) $3y' - 4y + y'' = e^{3t}\sin 7t$, $y(0) = 1, y'(0) = 1$

(h) $y'' - 2y' - 3y = e^t\sin 2t$, $y(0) = 1, y'(0) = 2$

(i) $3y + 4y' + y'' = e^{-3t}\sin 4t$, $y(0) = -1, y'(0) = -3$

(j) $2y' - 3y + y'' = e^{-t}\sin 3t$, $y(0) = -1, y'(0) = -2$

In many of the following problems, you can leave your answer in terms of integrals if things seem too long.

6. Using Laplace transforms, give a solution to the following initial value problems.

(a) $58y + 6y' + y'' = -e^{-4t}\sin 3t, y(0) = -1, y'(0) = 1$

(b) $13y - 4y' + y'' = e^{-5t} \sin 2t$, $y(0) = -1$, $y'(0) = -3$

(c) $10y + 6y' + y'' = -e^{-2t} \sin 3t$, $y(0) = 1$, $y'(0) = 1$.

(d) $13y - 4y' + y'' = e^{-2t} \sin 2t$, $y(0) = 1$, $y'(0) = -2$

(e) $58y - 6y' + y'' = e^{-4t} \sin 3t$, $y(0) = 1$, $y'(0) = 3$

(f) $29y + 4y' + y'' = (\cos 2t) e^{-3t}$, $y(0) = -1$, $y'(0) = 3$

(g) $25y + 6y' + y'' = -4 \cosh 3t$, $y(0) = -1$, $y'(0) = 2$

(h) $29y - 4y' + y'' = -5 \sinh 2t$, $y(0) = 1$, $y'(0) = 1$

(i) $40y + 4y' + y'' = -\cosh 2t$, $y(0) = 1$, $y'(0) = -3$

(j) $34y - 6y' + y'' = -2 \sinh 3t$, $y(0) = 1$, $y'(0) = 2$

(k) $26y - 2y' + y'' = 3te^t$, $y(0) = 1$, $y'(0) = 1$

(l) $50y + 2y' + y'' = 2te^{-t}$, $y(0) = 1$, $y'(0) = 1$

7. Using Laplace transforms, give a solution to the following initial value problem.

$$y'' - 4y' + 3y = f(t), \ y(0) = 1, y'(0) = 2$$

where

$$f(t) = \begin{cases} 1 & if \ \ 0 < t < 2 \\ \sin(t-2) & if \ \ \ t \geq 2 \end{cases}$$

8. Using Laplace transforms, give a solution to the following initial value problem. $y'' - 4y' + 3y = f(t)$, $y(0) = 1, y'(0) = 2$ where

$$f(t) = \begin{cases} 1 & if \ \ 0 < t < 2 \\ \sin(t) & if \ \ \ t \geq 2 \end{cases}$$

9. Using Laplace transforms, give a solution to the following initial value problem. $2y' - 3y + y'' = f(t)$, $y(0) = 3, y'(0) = 1$ where

$$f(t) = \begin{cases} -1 & if \ \ 0 < t < 4 \\ t & if \ \ \ t \geq 4 \end{cases}$$

10. Using Laplace transforms, give a solution to the following initial value problem. $2y' - 3y + y'' = f(t)$, $y(0) = -2, y'(0) = 2$ where

$$f(t) = \begin{cases} -2 & if \ \ 0 < t < 2 \\ t & if \ \ \ t \geq 2 \end{cases}$$

11. Using Laplace transforms, give a solution to the following initial value problem. $y'' - y = f(t)$, $y(0) = 4, y'(0) = 1$ where

$$f(t) = \begin{cases} -3 & if \ \ 0 < t < 2 \\ t & if \ \ \ t \geq 2 \end{cases}$$

12. Using Laplace transforms, give the general solution to the following equations.

(a) $y'' - y = \delta(t - 4)$

(b) $y' - 20y + y'' = \delta(t - 2)$

(c) $3y + 4y' + y'' = \delta(t - 5)$

(d) $58y + 6y' + y'' = \delta(t - 1)$

(e) $40y - 4y' + y'' = 2\delta(t - 5)$

(f) $13y - 6y' + y'' = 3\delta(t - 4)$

(g) $10y - 2y' + y'' = \delta(t - 2)$

13. Using Laplace transforms, give a particular solution to the following equations.

(a) $y^{(4)} - 3y - 4y' + 2y'' + 4y''' = 3e^{-t}$

(b) $y^{(4)} - 32y + 22y'' + 9y''' = 3e^{-4t}$

(c) $y^{(4)} + 8y - 4y' - 6y'' + y''' = e^{-2t}$

(d) $y^{(4)} - 3y + 2y' - 2y'' + 2y''' = 2\cos t$

(e) $y^{(4)} + 108y - 63y' + 21y'' - 7y''' = 3\cos 3t$

(f) $y^{(4)} - 27y - 18y' + 6y'' - 2y''' = 2\cos 3t$

(g) $y^4 + 27y + 36y' + 12y'' + 4y''' = 3\cos 3t$

14. Using the convolution integral write the solution to the following initial value problems in terms of an integral involving the unknown function $f(t)$.

(a) $12y - 7y' + y'' = f(t), y(0) = -2, y'(0) = -4.$

(b) $15y - 8y' + y'' = f(t), y(0) = -1, y'(0) = -2.$

(c) $8y - 6y' + y'' = f(t), y(0) = 3, y'(0) = -2.$

(d) $13y + 4y' + y'' = f(t), y(0) = -1, y'(0) = -4.$

(e) $13y + 6y' + y'' = f(t), y(0) = 1, y'(0) = -1.$

(f) $13y + 6y' + y'' = f(t), y(0) = 1, y'(0) = -4.$

(g) $29y + 10y' + y'' = f(t), y(0) = -2, y'(0) = -2.$

15. Find $y(t)$ where $y(t) =$

(a) $-4 + 4\int_0^t \sin(u - t) y(u) \, du$

(b) $4 + 2\int_0^t \sin(u - t) y(u) \, du$

(c) $-3 + 2\int_0^t \sin(4u - 4t) y(u) \, du$

(d) $4 + \int_0^t \sin(5u - 5t) y(u) \, du$

(e) $-2 + 5\int_0^t \exp(2u - 2t) y(u) \, du$

(f) $-2 + 4\int_0^t \exp(3u - 3t) y(u) \, du$

(g) $-2 + 3\int_0^t \exp(5u - 5t) y(u) \, du$

16. Find $\int_0^\infty e^{-t^2} dt$ exactly. It equals $\frac{1}{2}\sqrt{\pi}$. Explain why. **Hint:** First explain why it exists. Then let it equal I. Explain why

$$I^2 = \int_0^\infty \int_0^\infty e^{-(t^2 + s^2)} dt ds$$

Now use polar coordinates to find this.

17. Recall that the Laplace transform of $f(t) = t^p, p > -1$ is $\frac{\Gamma(p+1)}{s^{p+1}}$. This was discussed in the chapter. Now find $\Gamma(1/2)$. Recall the definition of the Γ function

$$\Gamma(\alpha) \equiv \int_0^\infty e^{-t} t^{\alpha-1} dt$$

Hint: This will involve changing the variable. Let $t = u^2$ and use the result of the previous problem.

18. This problem is going to be a little elaborate. It is from [5]. It gives an interesting way to use the convolution integral to determine a differential equation. Suppose you have an increasing function defined on $[0, 1]$ which goes from $(0, 0)$ to (a, b). Using conservation of energy, explain why the speed v of a bead sliding down the curve with no friction from (a, b) to a point on the curve at a height y is $v(y) = \sqrt{2g(b-y)}$ where g is the acceleration of gravity. Letting s be the length of the graph of this function measured from (a, b), $ds = \sqrt{1 + \left(\frac{dx}{dy}\right)^2} dy$. Thus $v = \frac{ds}{dt} = \sqrt{2g(b-y)}$ and so $dt = \frac{ds}{\sqrt{2g(b-y)}}$ Thus

$$dt = \frac{\sqrt{1 + \left(\frac{dx}{dy}\right)^2} dy}{\sqrt{2g(b-y)}}$$

Letting $f(y) = \sqrt{1 + \left(\frac{dx}{dy}\right)^2}$, it follows that the total time to slide to $(0, 0)$ equals

$$T = \int_0^b \frac{f(y)}{\sqrt{2g(b-y)}} dy$$

The idea is to have an equation for the curve $x = g(y)$ such that no matter where you start the bead sliding on the graph of $x = g(y)$, the time is always the same T_0 to arrive at $(0, 0)$. This is called the tautochrone problem. It was of great interest because it was desired to construct a way to build a clock which would not lose its accuracy when it ran down. Use the convolution integral and the above problems to determine a differential equation for $g(y)$.

19. In the above problem, the differential equation you should have gotten is

$$\frac{dx}{dy} = \sqrt{\frac{2}{\pi} \frac{g}{y} T_0^2 - 1}$$

Letting $\alpha^2 = \frac{2}{\pi} g T_0^2$, find the solutions to the separable differential equation

$$\frac{dx}{dy} = \sqrt{\frac{\alpha^2}{y} - 1} = \sqrt{\frac{1}{y}(\alpha^2 - y)}$$

Hint: Variables separate as $dx = \sqrt{\frac{1}{y}(\alpha^2 - y)} dy$. Now let $u = \sqrt{\alpha^2 - y}$. Now replace α and obtain solutions for the tautochrone problem in terms of T_0. To be specific, let $(0, 0)$ be on the graph of $x = g(y)$.

20. The entire method of Laplace transforms is based on the observation that if $\mathcal{L}(f) = \mathcal{L}(g)$ for all s sufficiently large, then $f = g$. Show that this is equivalent to saying that if $\mathcal{L}(f) = 0$ for all s sufficiently large, then $f = 0$. Now show that this last is not even true! However, show that it might be true if it is known that f is continuous. In fact, it can be shown that if $\mathcal{L}(f) = 0$, then $f(t) = 0$ for almost all t. There is a precise way of making sense of this last term which is considered in analysis. Thus there is nothing wrong with the methods presented in this chapter. For more on this, see the next few problems, in particular Problem 29.

21. Suppose $|f(t)|$ has exponential growth and is continuous except for jumps. Let $g(t) \equiv \int_0^t f(u)\,du$. Show that $\mathcal{L}(g) = \frac{1}{s}\mathcal{L}(f)$.

22. Suppose $|f(t)|$ has exponential growth and is continuous except for jumps. Also suppose that $\mathcal{L}(f)(s) = 0$ for all s large enough. Show that then $f = 0$ except for jumps. **Hint:** Consider $g(t) \equiv \int_0^t f(u)\,du$. Then g is continuous. Apply Theorem 6.4.5 to g. Then consider the fundamental theorem of calculus.

23. Theorem 6.4.2 said that every function f continuous on $[0,1]$ is the uniform limit of a sequence of polynomials. That is,

$$\lim_{n\to\infty}\left(\max_{x\in[0,1]}|f(x) - p_n(x)|\right) = 0$$

Show that this is also true with $[0,1]$ replaced with $[a,b]$. This is actually the Weierstrass approximation theorem. The one in the text is a special case.

24. Find a formula for $\mathcal{L}((D-a)^m y)$. In particular, show that it is of the form

$$(s-a)^m \mathcal{L}(y) - \sum_{k=0}^{m-1} c_k (s-a)^k$$

where c_k are constants depending on the initial data. Use this to explain why the general solution to $(D-a)^m y = 0$ is of the form $\sum_{k=0}^{m-1} C_k t^k e^{at}$ for some constants C_k. This gives another explanation why the procedure for dealing with repeated roots works.

25. It was shown that $\mathcal{L}(\sin(t)) = \frac{1}{1+s^2}$. Show that it makes sense to take $\mathcal{L}\left(\frac{\sin t}{t}\right)$. Show that

$$\int_0^\infty \frac{\sin(t)}{t} e^{-st}\,dt = \int_s^\infty \frac{1}{1+u^2}\,du \qquad (*)$$

Now show that the derivative with respect to s is the same on both sides. Use Theorem B.5.5 to take the derivative inside the integral. Therefore, the two functions of s differ by a constant. Show the constant is 0 by taking $\lim_{s\to\infty}$ on both sides. Now letting $s = 0$, conclude that you should have

$$\int_0^\infty \frac{\sin(t)}{t}\,dt = \int_0^\infty \frac{1}{1+u^2} = \frac{1}{2}\pi$$

In doing this, you should also verify that in fact the left side of the above inequality makes sense.

26. Assuming that everything makes sense, show that for $\mathcal{L}(f)(s) = F(s), \mathcal{L}(g)(s) = G(s)$,

$$\int_0^\infty F(t) g(t) ds = \int_0^\infty f(s) G(s) ds$$

27. It is reasonable to expect that $\mathcal{L}\left(\sum_{k=0}^\infty a_k t^k\right) = \sum_{k=0}^\infty a_k \mathcal{L}(t^k)$. This can be used to find Laplace transforms of functions which have a power series. Find Laplace transforms of the following functions using their power series expansions.

   (a) $\sin(t^2)$

   (b) $\cos(t^2)$

   (c) $t^p \sin t$

   (d) $t^p \cos t$

28. Suppose $F(s)$ is the Laplace transform of $f(t)$. Show that for c a positive number,

   (a) $\mathcal{L}(f(ct)) = \frac{1}{c} F\left(\frac{s}{c}\right)$

   (b) $F(cs) = \mathcal{L}\left(\frac{1}{c} f\left(\frac{t}{c}\right)\right)$

   (c) If $a > 0$ then $F(as + b) = \mathcal{L}\left(\left(\frac{1}{a} e^{-bt/a}\right) f\left(\frac{t}{a}\right)\right)$

29. Here is another approach for showing the method of Laplace transforms works in Widder [21]. Do the following in which k will be a positive integer:

   (a) Verify that $\frac{k^{k+1}}{k!} \int_0^\infty \left(e^{-u} u\right)^k du = 1$. **Hint:** Recall $\Gamma(k+1) = k!$

   (b) Show using a ratio test argument that $\lim_{k\to\infty} \frac{k^{k+1}}{k!} \left(e^{-a} a\right)^k = 0$ for $0 < a < 1$ and that $u \to u e^{-u}$ is increasing on $(0, 1)$.

   (c) Next show that if $|\phi(u)| \le C e^{\lambda u}$, $\lim_{k\to\infty} \frac{k^{k+1}}{k!} \int_0^\infty \left(e^{-u} u\right)^k \phi(u) du = \phi(1)$ provided ϕ is continuous at 1. Then show that for each $t > 0$ where ϕ is continuous, $\lim_{k\to\infty} \frac{k^{k+1}}{k!} \int_0^\infty \left(e^{-u} u\right)^k \phi(tu) du = \phi(t)$. **Hint:** Use part a. and split the integral into $[0, 1 - \delta], [1 - \delta, 1 + \delta], [1 + \delta, \infty]$ where on the middle interval, $|\phi(u) - \phi(1)| < \varepsilon$.

   (d) Now let $f(s) = \mathcal{L}(\phi(t))$ where ϕ has exponential growth and is piecewise continuous. Assuming you can differentiate under the integral, (Theorem B.5.5) show that

$$\frac{(-1)^k}{k!} \left[f^{(k)}\left(\frac{k}{t}\right)\right] \left(\frac{k}{t}\right)^{k+1} = \frac{k^{k+1}}{k!} \frac{1}{t^{k+1}} \int_0^\infty u^k e^{-ku/t} \phi(u) du$$

$$= \frac{k^{k+1}}{k!} \int_0^\infty \left(e^{-u} u\right)^k \phi(ut) du$$

   Now use the above to see that if ϕ is continuous at t, $\phi(t)$ is the limit of the left side. How does this justify the Laplace transform procedure?

Part III

Series Methods

Part III

Series Methods

Chapter 7

A Review of Power Series

Power series methods provide a way to solve ordinary differential equations which have coefficients which are not constant. In addition to this, these methods can include situations when the equation is degenerate in the sense that one cannot solve for the highest order derivative of the unknown function because its coefficient vanishes, thus making the earlier methods inapplicable. This chapter gives a review of power series.

In calculus you likely used Taylor polynomials to approximate known functions such as $\sin x$ and $\ln(1 + x)$. A much more exciting idea is to use infinite series as definitions of possibly new functions.

Definition 7.0.1 *Let $\{a_k\}_{k=0}^{\infty}$ be a sequence of numbers. The expression*

$$\sum_{k=0}^{\infty} a_k (x - a)^k \tag{7.1}$$

is called a Taylor series centered at a. This is also called a power series centered at a.

In the above definition, x is a variable. Thus you can put in various values of x and ask whether the resulting series of numbers converges. Defining D to be the set of all values of x such that the resulting series does converge, define a new function f defined on D as

$$f(x) \equiv \sum_{k=0}^{\infty} a_k (x - a)^k.$$

This might be a totally new function, one which has no name. Nevertheless, much can be said about such functions. The following lemma is fundamental in considering the form of D which always turns out to be an interval centered at a which may or may not contain either end point.

Lemma 7.0.2 *Suppose $z \in D$. Then if $|x - a| < |z - a|$, then $x \in D$ also and furthermore, the series $\sum_{k=0}^{\infty} |a_k| |x - a|^k$ converges.*

Proof: Let $1 > r = |x - a| / |z - a|$. The nth term test implies

$$\lim_{n \to \infty} |a_n| |z - a|^n = 0$$

and so for all n large enough,

$$|a_n| |z - a|^n < 1$$

so for such n,

$$|a_n| |x - a|^n = |a_n| |z - a|^n \frac{|x - a|^n}{|z - a|^n} \leq \frac{|x - a|^n}{|z - a|^n} < r^n$$

Therefore, $\sum_{k=0}^{\infty} |a_k| |x - a|^k$ converges by comparison with the geometric series $\sum r^n$. ∎

With this lemma, the following fundamental theorem is obtained.

Theorem 7.0.3 *Let $\sum_{k=0}^{\infty} a_k (x - a)^k$ be a Taylor series. Then there exists $r \leq \infty$ such that the Taylor series converges absolutely if $|x - a| < r$. Furthermore, if $|x - a| > r$, the Taylor series diverges.*

Proof: Let

$$r \equiv \sup \{|y - a| : y \in D\}.$$

Then if $|x - a| < r$, it follows there exists $z \in D$ such that $|z - a| > |x - a|$ since otherwise, r wouldn't be as defined. In fact $|x - a|$ would then be an upper bound to $\{|y - a| : y \in D\}$. Therefore, by the above lemma $\sum_{k=0}^{\infty} |a_k| |x - a|^k$ converges and this proves the first part of this theorem.

Now suppose $|x - a| > r$. If $\sum_{k=0}^{\infty} a_k (x - a)^k$ converges then by the above lemma, r fails to be an upper bound to $\{|y - a| : y \in D\}$ and so the Taylor series must diverge as claimed. ∎

From now on D will be referred to as the interval of convergence and r of the above theorem as the radius of convergence. Determining which points of $\{x : |x - a| = r\}$ are in D requires the use of specific convergence tests and can be quite hard. However, the determination of r tends to be pretty easy.

Example 7.0.4 *Find the interval of convergence of the Taylor series $\sum_{n=1}^{\infty} \frac{x^n}{n}$.*

Use Corollary C.2.8.

$$\lim_{n \to \infty} \left(\frac{|x|^n}{n} \right)^{1/n} = \lim_{n \to \infty} \frac{|x|}{\sqrt[n]{n}} = |x|$$

because $\lim_{n \to \infty} \sqrt[n]{n} = 1$ and so if $|x| < 1$ the series converges. The endpoints require special attention. When $x = 1$ the series diverges because it reduces to $\sum_{n=1}^{\infty} \frac{1}{n}$. At the other endpoint however, the series converges, because it reduces to $\sum_{n=1}^{\infty} \frac{(-1)^n}{n}$ and the alternating series test applies and gives convergence.

Example 7.0.5 *Find the radius of convergence of $\sum_{n=1}^{\infty} \frac{n^n}{n!} x^n$.*

Apply the ratio test. Taking the ratio of the absolute values of the $(n+1)^{th}$ and the nth terms

$$\frac{\frac{(n+1)^{(n+1)}}{(n+1)n!}|x|^{n+1}}{\frac{n^n}{n!}|x|^n} = (n+1)^n |x| \, n^{-n} = |x| \left(1 + \frac{1}{n}\right)^n \to |x| \, e$$

Therefore the series converges absolutely if $|x| \, e < 1$ and diverges if $|x| \, e > 1$. Consequently, $r = 1/e$.

7.1 Operations on Power Series

It is desirable to be able to differentiate, integrate, and multiply power series. The following theorem says one can differentiate power series in the most natural way on the interval of convergence, just as you would differentiate a polynomial. This theorem may seem obvious, but it is a serious mistake to think this. You usually cannot differentiate an infinite series whose terms are functions even if the functions are themselves polynomials. The following is special and pertains to power series. It is an example of the interchange of two limits, in this case, the limit involved in taking the derivative and the limit of the sequence of finite sums.

Theorem 7.1.1 *Let $\sum_{n=0}^{\infty} a_n (x-a)^n$ be a Taylor series having radius of convergence $r > 0$ and let*

$$f(x) \equiv \sum_{n=0}^{\infty} a_n (x-a)^n \tag{7.2}$$

for $|x-a| < r$. Then

$$f'(x) = \sum_{n=0}^{\infty} a_n n (x-a)^{n-1} = \sum_{n=1}^{\infty} a_n n (x-a)^{n-1} \tag{7.3}$$

and this new differentiated power series, the derived series, has radius of convergence equal to r.

Proof: First it will be shown that the series on the right in (7.3) has the same radius of convergence as the original series. Thus let $|x-a| < r$ and pick y such that

$$|x-a| < |y-a| < r.$$

Then

$$\lim_{n\to\infty} |a_n| \, |y-a|^{n-1} = \lim_{n\to\infty} |a_n| \, |y-a|^n = 0$$

because

$$\sum_{n=0}^{\infty} |a_n| \, |y-a|^n < \infty$$

and so, for n large enough,

$$|a_n| \, |y-a|^{n-1} < 1.$$

Therefore, for large enough n,

$$|a_n|\, n\, |x - a|^{n-1} = |a_n|\, |y - a|^{n-1}\, n \left|\frac{x - a}{y - a}\right|^{n-1} \leq n \left|\frac{x - a}{y - a}\right|^{n-1}$$

and so $\sum_{n=1}^{\infty} n \left|\frac{x-a}{y-a}\right|^{n-1}$ converges by the ratio test. By the comparison test, it follows $\sum_{n=1}^{\infty} a_n n\, (x - a)^{n-1}$ converges absolutely for any x satisfying $|x - a| < r$. Therefore, the radius of convergence of the derived series is at least as large as that of the original series. On the other hand, if $\sum_{n=1}^{\infty} |a_n|\, n\, |x - a|^{n-1}$ converges then by the comparison test, $\sum_{n=1}^{\infty} |a_n|\, |x - a|^{n-1}$ and therefore $\sum_{n=1}^{\infty} |a_n|\, |x - a|^{n}$ also converges which shows the radius of convergence of the derived series is no larger than that of the original series. It remains to verify the assertion about the derivative.

Let $|x - a| < r$ and let $r_1 < r$ be close enough to r that

$$x \in (a - r_1, a + r_1) \subseteq [a - r_1, a + r_1] \subseteq (a - r, a + r).$$

Thus, letting $r_2 \in (r_1, r)$,

$$\sum_{n=0}^{\infty} |a_n|\, r_1^n,\ \sum_{n=0}^{\infty} |a_n|\, r_2^n < \infty \tag{7.4}$$

Letting y be close enough to x, it follows both x and y are in $[a - r_1, a + r_1]$. Then considering the difference quotient,

$$\frac{f(y) - f(x)}{y - x} = \sum_{n=0}^{\infty} a_n\, (y - x)^{-1}\, [(y - a)^n - (x - a)^n] = \sum_{n=1}^{\infty} a_n n z_n^{n-1} \tag{7.5}$$

where the last equation follows from the mean value theorem and z_n is some point between $x - a$ and $y - a$. Therefore,

$$\frac{f(y) - f(x)}{y - x} = \sum_{n=1}^{\infty} a_n n z_n^{n-1}$$

$$= \sum_{n=1}^{\infty} a_n n \left(z_n^{n-1} - (x - a)^{n-1}\right) + \sum_{n=1}^{\infty} a_n n\, (x - a)^{n-1}$$

$$= \sum_{n=2}^{\infty} a_n n\, (n - 1)\, w_n^{n-2}\, (z_n - (x - a)) + \sum_{n=1}^{\infty} a_n n\, (x - a)^{n-1} \tag{7.6}$$

where w_n is between z_n and $x - a$. Thus w_n is between $x - a$ and $y - a$ and so

$$w_n + a \in [a - r_1, a + r_1]$$

which implies $|w_n| \leq r_1$. The first sum on the right in (7.6) therefore satisfies

$$\left| \sum_{n=2}^{\infty} a_n n\, (n - 1)\, w_n^{n-2}\, (z_n - (x - a)) \right| \leq |y - x| \sum_{n=2}^{\infty} |a_n|\, n\, (n - 1)\, |w_n|^{n-2}$$

$$\leq |y - x| \sum_{n=2}^{\infty} |a_n|\, n\, (n - 1)\, r_1^{n-2}$$

$$= |y - x| \sum_{n=2}^{\infty} |a_n| \, r_2^{n-2} n \, (n-1) \left(\frac{r_1}{r_2} \right)^{n-2}$$

Now from (7.4), $|a_n| \, r_2^{n-2} < 1$ for all n large enough. Therefore, for such n,

$$|a_n| \, r_2^{n-2} n \, (n-1) \left(\frac{r_1}{r_2} \right)^{n-2} \leq n \, (n-1) \left(\frac{r_1}{r_2} \right)^{n-2}$$

and the series $\sum n \, (n-1) \left(\frac{r_1}{r_2} \right)^{n-2}$ converges by the ratio test. Therefore, there exists a constant C independent of y such that

$$\sum_{n=2}^{\infty} |a_n| \, n \, (n-1) \, r_1^{n-2} = C < \infty$$

Consequently, from (7.6)

$$\left| \frac{f(y) - f(x)}{y - x} - \sum_{n=1}^{\infty} a_n n \, (x-a)^{n-1} \right| \leq C \, |y - x| \, .$$

Taking the limit as $y \to x$ (7.3) follows. ∎

As an immediate corollary, it is possible to characterize the coefficients of a Taylor series.

Corollary 7.1.2 *Let $\sum_{n=0}^{\infty} a_n \, (x-a)^n$ be a Taylor series with radius of convergence $r > 0$ and let*

$$f(x) \equiv \sum_{n=0}^{\infty} a_n \, (x-a)^n \, . \tag{7.7}$$

Then

$$a_n = \frac{f^{(n)}(a)}{n!} \, . \tag{7.8}$$

Proof: From (7.7), $f(a) = a_0 \equiv f^{(0)}(a) / 0!$. From Theorem 7.1.1,

$$f'(x) = \sum_{n=1}^{\infty} a_n n \, (x-a)^{n-1} = a_1 + \sum_{n=2}^{\infty} a_n n \, (x-a)^{n-1} \, .$$

Now let $x = a$ and obtain that $f'(a) = a_1 = f'(a) / 1!$. Next use Theorem 7.1.1 again to take the second derivative and obtain

$$f''(x) = 2a_2 + \sum_{n=3}^{\infty} a_n n \, (n-1) \, (x-a)^{n-2}$$

let $x = a$ in this equation and obtain $a_2 = f''(a) / 2 = f''(a) / 2!$. Continuing this way proves the corollary. ∎

This also shows the coefficients of a Taylor series are unique. That is, if

$$\sum_{k=0}^{\infty} a_k \, (x-a)^k = \sum_{k=0}^{\infty} b_k \, (x-a)^k$$

for all x in some interval, then $a_k = b_k$ for all k.

Example 7.1.3 *Find the power series for* $\sin(x)$*, and* $\cos(x)$ *centered at 0 and give the interval of convergence.*

First consider $f(x) = \sin(x)$. Then $f'(x) = \cos(x)$, $f''(x) = -\sin(x)$, $f'''(x) = -\cos(x)$ etc. Therefore, from Taylor's formula, Theorem C.4.1 on Page 429,

$$f(x) = 0 + x + 0 - \frac{x^3}{3!} + 0 + \frac{x^5}{5!} + \cdots + \frac{x^{2n+1}}{(2n+1)!} + \frac{f^{(2n+2)}(\xi_n)}{(2n+2)!}$$

where ξ_n is some number between 0 and x. Furthermore, this equals either $\pm \sin(\xi_n)$ or $\pm \cos(\xi_n)$ and so its absolute value is no larger than 1. Thus

$$\left| \frac{f^{(2n+2)}(\xi_n)}{(2n+2)!} \right| \leq \frac{1}{(2n+2)!}.$$

By the ratio test, it follows that

$$\sum_{n=0}^{\infty} \frac{1}{(2n+2)!} < \infty$$

and so by the comparison test,

$$\sum_{n=0}^{\infty} \left| \frac{f^{(2n+2)}(\xi_n)}{(2n+2)!} \right| < \infty$$

also. Therefore, by the nth term test $\lim_{n\to\infty} \frac{f^{(2n+2)}(\xi_n)}{(2n+2)!} = 0$. This implies

$$\sin(x) = \sum_{k=0}^{n} (-1)^k \frac{x^{2k+1}}{(2k+1)!} + \frac{f^{(2n+2)}(\xi_n)}{(2n+2)!}$$

and the last term converges to zero as $n \to \infty$ for any value of x and therefore,

$$\sin(x) = \sum_{k=0}^{\infty} (-1)^k \frac{x^{2k+1}}{(2k+1)!}$$

for all $x \in \mathbb{R}$. By Theorem 7.1.1, you can differentiate both sides, doing the series term by term and obtain

$$\cos(x) = \sum_{k=0}^{\infty} (-1)^k \frac{x^{2k}}{(2k)!}$$

for all $x \in \mathbb{R}$.

Example 7.1.4 *Find the sum* $\sum_{k=1}^{\infty} k2^{-k}$.

It may not be obvious what this sum equals but with the above theorem it is easy to find. From the formula for the sum of a geometric series, $\frac{1}{1-t} = \sum_{k=0}^{\infty} t^k$ if $|t| < 1$. Differentiate both sides to obtain

$$(1-t)^{-2} = \sum_{k=1}^{\infty} kt^{k-1}$$

whenever $|t| < 1$. Let $t = 1/2$. Then

$$4 = \frac{1}{(1 - (1/2))^2} = \sum_{k=1}^{\infty} k 2^{-(k-1)}$$

and so if you multiply both sides by 2^{-1},

$$2 = \sum_{k=1}^{\infty} k 2^{-k}.$$

The following is a very important example known as the binomial series.

Example 7.1.5 *Find a Taylor series for the function $(1 + x)^\alpha$ centered at 0 valid for $|x| < 1$.*

Use Theorem 7.1.1 to do this. First note that if $y(x) \equiv (1 + x)^\alpha$, then y is the unique solution of the following initial value problem.

$$y' - \frac{\alpha}{(1 + x)} y = 0, \ y(0) = 1. \tag{7.9}$$

The strategy for finding the Taylor series of this function consists of finding a series which solves the initial value problem above. Let

$$y(x) \equiv \sum_{n=0}^{\infty} a_n x^n \tag{7.10}$$

be a solution to (7.9). Of course it is not known at this time whether such a series exists. However, the process of finding it will demonstrate its existence. From Theorem 7.1.1 and the initial value problem

$$(1 + x) \sum_{n=0}^{\infty} a_n n x^{n-1} - \sum_{n=0}^{\infty} \alpha a_n x^n = 0$$

and so

$$\sum_{n=1}^{\infty} a_n n x^{n-1} + \sum_{n=0}^{\infty} a_n (n - \alpha) x^n = 0$$

Changing the order variable of summation in the first sum

$$\sum_{n=0}^{\infty} a_{n+1} (n + 1) x^n + \sum_{n=0}^{\infty} a_n (n - \alpha) x^n = 0$$

and from Corollary 7.1.2 and the initial condition for (7.9) this requires

$$a_{n+1} = \frac{a_n (\alpha - n)}{n + 1}, a_0 = 1. \tag{7.11}$$

Therefore, from (7.11) and letting $n = 0$, $a_1 = \alpha$. Then using (7.11) again along with this information, $a_2 = \frac{\alpha(\alpha-1)}{2}$. Using the same process, $a_3 = \frac{\left(\frac{\alpha(\alpha-1)}{2}\right)(\alpha-2)}{3} = \frac{\alpha(\alpha-1)(\alpha-2)}{3!}$. By now you can spot the pattern. In general,

$$a_n = \frac{\overbrace{\alpha(\alpha - 1) \cdots (\alpha - n + 1)}^{n \text{ of these factors}}}{n!}.$$

Therefore, the candidate for the Taylor series is

$$y\left(x\right) = \sum_{n=0}^{\infty} \frac{\alpha\left(\alpha - 1\right) \cdots \left(\alpha - n + 1\right)}{n!} x^n.$$

Furthermore, the above discussion shows this series solves the initial value problem on its interval of convergence. It only remains to show the radius of convergence of this series equals 1. It will then follow that this series equals $\left(1 + x\right)^\alpha$ because of uniqueness of the initial value problem. To find the radius of convergence, use the ratio test. Thus the ratio of the absolute values of $\left(n + 1\right)^{st}$ term to the absolute value of the nth term is

$$\frac{\left|\frac{\alpha(\alpha-1)\cdots(\alpha-n+1)(\alpha-n)}{(n+1)n!}\right| |x|^{n+1}}{\left|\frac{\alpha(\alpha-1)\cdots(\alpha-n+1)}{n!}\right| |x|^n} = |x| \frac{|\alpha - n|}{n + 1} \to |x|$$

showing that the radius of convergence is 1 since the series converges if $|x| < 1$ and diverges if $|x| > 1$.

The expression $\frac{\alpha(\alpha-1)\cdots(\alpha-n+1)}{n!}$ is often denoted as $\binom{\alpha}{n}$. With this notation, the following theorem has been established.

Theorem 7.1.6 *Let α be a real number and let $|x| < 1$. Then*

$$\left(1 + x\right)^\alpha = \sum_{n=0}^{\infty} \binom{\alpha}{n} x^n.$$

There is a very interesting issue related to the above theorem which illustrates the limitation of power series. The function $f\left(x\right) = \left(1 + x\right)^\alpha$ makes sense for all $x > -1$ but one is only able to describe it with a power series on the interval $\left(-1, 1\right)$. Think about this. The above technique is a standard one for obtaining solutions of differential equations and this example illustrates a deficiency in the method.

To completely understand power series, it is necessary to take a course in complex analysis. You may have noticed the prominent role played by geometric series. This is no accident. It turns out that the right way to consider Taylor series is through the use of geometric series and something called the Cauchy integral formula of complex analysis. However, these are topics for another course.

You can also integrate power series on their interval of convergence.

Theorem 7.1.7 *Let $f\left(x\right) = \sum_{n=0}^{\infty} a_n \left(x - a\right)^n$ and suppose the radius of convergence is $r > 0$. Then if $|y - a| < r$,*

$$\int_a^y f\left(x\right) dx = \sum_{n=0}^{\infty} \int_a^y a_n \left(x - a\right)^n dx = \sum_{n=0}^{\infty} \frac{a_n \left(y - a\right)^{n+1}}{n + 1}.$$

Proof: Define $F\left(y\right) \equiv \int_a^y f\left(x\right) dx$ and $G\left(y\right) \equiv \sum_{n=0}^{\infty} \frac{a_n (y-a)^{n+1}}{n+1}$. By Theorem 7.1.1 and the fundamental theorem of calculus,

$$G'\left(y\right) = \sum_{n=0}^{\infty} a_n \left(y - a\right)^n = f\left(y\right) = F'\left(y\right).$$

Therefore, $G(y) - F(y) = C$ for some constant. But $C = 0$ because $F(a) - G(a) = 0$. ■

Next consider the problem of multiplying two power series.

Theorem 7.1.8 *Let $\sum_{n=0}^{\infty} a_n (x-a)^n$ and $\sum_{n=0}^{\infty} b_n (x-a)^n$ be two power series having radii of convergence r_1 and r_2, both positive. Then*

$$\left(\sum_{n=0}^{\infty} a_n (x-a)^n \right) \left(\sum_{n=0}^{\infty} b_n (x-a)^n \right) = \sum_{n=0}^{\infty} \left(\sum_{k=0}^{n} a_k b_{n-k} \right) (x-a)^n$$

whenever $|x-a| < r \equiv \min(r_1, r_2)$.

Proof: By Theorem 7.0.3 both series converge absolutely if $|x-a| < r$. Therefore, by Theorem C.3.8

$$\left(\sum_{n=0}^{\infty} a_n (x-a)^n \right) \left(\sum_{n=0}^{\infty} b_n (x-a)^n \right) =$$

$$\sum_{n=0}^{\infty} \sum_{k=0}^{n} a_k (x-a)^k b_{n-k} (x-a)^{n-k} = \sum_{n=0}^{\infty} \left(\sum_{k=0}^{n} a_k b_{n-k} \right) (x-a)^n . ■$$

The significance of this theorem in terms of applications is that it states you can multiply power series just as you would multiply polynomials and everything will be all right on the common interval of convergence.

This theorem can be used to find Taylor series which would perhaps be hard to find without it. Here is an example.

Example 7.1.9 *Find the Taylor series for $e^x \sin x$ centered at $x = 0$.*

Using Example 7.1.3 on Page 162 and the power series from calculus, all that is required is to multiply

$$\left(\overbrace{1 + x + \frac{x^2}{2!} + \frac{x^3}{3!} \cdots}^{e^x} \right) \left(\overbrace{x - \frac{x^3}{3!} + \frac{x^5}{5!} + \cdots}^{\sin x} \right)$$

From the above theorem the result should be

$$x + x^2 + \left(-\frac{1}{3!} + \frac{1}{2!} \right) x^3 + \cdots = x + x^2 + \frac{1}{3} x^3 + \cdots$$

You can continue this way and get the following to a few more terms.

$$x + x^2 + \frac{1}{3} x^3 - \frac{1}{30} x^5 - \frac{1}{90} x^6 - \frac{1}{630} x^7 + \cdots$$

I do not see a pattern in these coefficients but I can go on generating them as long as I want. (In practice this tends to not be very long.) I also know the resulting power series will converge for all x because both the series for e^x and the one for $\sin x$ converge for all x.

Example 7.1.10 *Find the Taylor series for* $\tan x$ *centered at* $x = 0$.

Let's suppose it has a Taylor series $a_0 + a_1 x + a_2 x^2 + \cdots$. Then

$$\left(a_0 + a_1 x + a_2 x^2 + \cdots\right) \overbrace{\left(1 - \frac{x^2}{2} + \frac{x^4}{4!} + \cdots\right)}^{\cos x} = \left(x - \frac{x^3}{3!} + \frac{x^5}{5!} + \cdots\right).$$

Using the above, $a_0 = 0$, $a_1 x = x$ so $a_1 = 1$, $\left(0\left(\frac{-1}{2}\right) + a_2\right) x^2 = 0$ so $a_2 = 0$.

$$\left(a_3 - \frac{a_1}{2}\right) x^3 = \frac{-1}{3!} x^3$$

so $a_3 - \frac{1}{2} = -\frac{1}{6}$ so $a_3 = \frac{1}{3}$. Clearly one can continue in this manner. Thus the first several terms of the power series for tan are

$$\tan x = x + \frac{1}{3} x^3 + \cdots.$$

You can go on calculating these terms and find the next two yielding

$$\tan x = x + \frac{1}{3} x^3 + \frac{2}{15} x^5 + \frac{17}{315} x^7 + \cdots$$

This is a very significant technique because, as you see, there does not appear to be a very simple pattern for the coefficients of the power series for $\tan x$. Of course there are some issues here about whether $\tan x$ even has a power series, but if it does, the above must be it. In fact, $\tan(x)$ will have a power series valid on some interval centered at 0 and this becomes completely obvious when one uses methods from complex analysis but it is not too obvious at this point. Note also that what has been accomplished is to divide the power series for $\sin x$ by the power series for $\cos x$ just like they were polynomials.

7.2   Some Other Theorems

First recall Theorem C.3.8 on Page 428. For convenience, the version of this theorem which is of interest here is listed below.

Theorem 7.2.1 *Suppose* $\sum_{i=0}^{\infty} a_i$ *and* $\sum_{j=0}^{\infty} b_j$ *both converge absolutely. Then*

$$\left(\sum_{i=0}^{\infty} a_i\right)\left(\sum_{j=0}^{\infty} b_j\right) = \sum_{n=0}^{\infty} c_n \text{ where } c_n = \sum_{k=0}^{n} a_k b_{n-k}.$$

Furthermore, $\sum_{n=0}^{\infty} c_n$ *converges absolutely.*

Proof: It only remains to verify the last series converges absolutely. By Theorem C.3.5 on Page 426 and letting p_{nk} be as defined there,

$$\sum_{n=0}^{\infty} |c_n| = \sum_{n=0}^{\infty} \left|\sum_{k=0}^{n} a_k b_{n-k}\right| \leq \sum_{n=0}^{\infty} \sum_{k=0}^{n} |a_k| |b_{n-k}| = \sum_{n=0}^{\infty} \sum_{k=0}^{\infty} p_{nk} |a_k| |b_{n-k}|$$

$$= \sum_{k=0}^{\infty} \sum_{n=0}^{\infty} p_{nk} |a_k| |b_{n-k}| = \sum_{k=0}^{\infty} \sum_{n=k}^{\infty} |a_k| |b_{n-k}| = \sum_{k=0}^{\infty} |a_k| \sum_{n=0}^{\infty} |b_n| < \infty. \blacksquare$$

The theorem is about multiplying two series. What if you wanted to consider $\left(\sum_{n=0}^{\infty} a_n\right)^p$ where p is a positive integer maybe larger than 2? Is there a similar theorem to the above?

Definition 7.2.2 *Define*

$$\sum_{k_1+\cdots+k_p=m} a_{k_1} a_{k_2} \cdots a_{k_p}$$

as follows. Consider all ordered lists of nonnegative integers $k_1, \cdots, k_p$ which have the property that $\sum_{i=1}^{p} k_i = m$. For each such list of integers, form the product $a_{k_1} a_{k_2} \cdots a_{k_p}$ and then add all these products.

Note that

$$\sum_{k=0}^{n} a_k a_{n-k} = \sum_{k_1+k_2=n} a_{k_1} a_{k_2}$$

Therefore, from the above theorem, if $\sum a_i$ converges absolutely, it follows

$$\left(\sum_{i=0}^{\infty} a_i\right)^2 = \sum_{n=0}^{\infty} \left(\sum_{k_1+k_2=n} a_{k_1} a_{k_2}\right).$$

It turns out that a similar theorem holds replacing 2 with p.

Theorem 7.2.3 *Suppose $\sum_{n=0}^{\infty} a_n$ converges absolutely. Then*

$$\left(\sum_{n=0}^{\infty} a_n\right)^p = \sum_{m=0}^{\infty} c_{mp} \text{ where } c_{mp} \equiv \sum_{k_1+\cdots+k_p=m} a_{k_1} \cdots a_{k_p}.$$

Proof: First note this is obviously true if $p = 1$ and is also true if $p = 2$ from the above theorem. Now suppose this is true for p and consider $\left(\sum_{n=0}^{\infty} a_n\right)^{p+1}$. By induction and the above theorem on the Cauchy product, $\left(\sum_{n=0}^{\infty} a_n\right)^{p+1} =$

$$\left(\sum_{n=0}^{\infty} a_n\right)^p \left(\sum_{n=0}^{\infty} a_n\right) = \left(\sum_{m=0}^{\infty} c_{mp}\right) \left(\sum_{n=0}^{\infty} a_n\right) = \sum_{n=0}^{\infty} \left(\sum_{k=0}^{n} c_{kp} a_{n-k}\right)$$

$$= \sum_{n=0}^{\infty} \sum_{k=0}^{n} \sum_{k_1+\cdots+k_p=k} a_{k_1} \cdots a_{k_p} a_{n-k} = \sum_{n=0}^{\infty} \sum_{k_1+\cdots+k_{p+1}=n} a_{k_1} \cdots a_{k_{p+1}} \ \blacksquare$$

This theorem implies the following corollary for power series.

Corollary 7.2.4 *Let*

$$\sum_{n=0}^{\infty} a_n (x - a)^n$$

be a power series having radius of convergence $r > 0$. Then if $|x - a| < r$,

$$\left(\sum_{n=0}^{\infty} a_n (x - a)^n\right)^p = \sum_{n=0}^{\infty} b_{np} (x - a)^n \text{ where } b_{np} \equiv \sum_{k_1+\cdots+k_p=n} a_{k_1} \cdots a_{k_p}.$$

Proof: Since $|x - a| < r$, the series $\sum_{n=0}^{\infty} a_n (x - a)^n$, converges absolutely. Therefore, the above theorem applies and

$$\left(\sum_{n=0}^{\infty} a_n (x - a)^n \right)^p = \sum_{n=0}^{\infty} \left(\sum_{k_1 + \cdots + k_p = n} a_{k_1} (x - a)^{k_1} \cdots a_{k_p} (x - a)^{k_p} \right)$$

$$= \sum_{n=0}^{\infty} \left(\sum_{k_1 + \cdots + k_p = n} a_{k_1} \cdots a_{k_p} \right) (x - a)^n . \blacksquare$$

With this theorem it is possible to consider the question raised in Example 7.1.10 on Page 166 about the existence of the power series for $\tan x$. This question is clearly included in the more general question of when $\left(\sum_{n=0}^{\infty} a_n (x - a)^n \right)^{-1}$ has a power series.

Lemma 7.2.5 *Let* $f(x) = \sum_{n=0}^{\infty} a_n (x - a)^n$, *a power series having radius of convergence* $r > 0$. *Suppose also that* $f(a) = 1$. *Then there exists* $r_1 > 0$ *and* $\{b_n\}$ *such that for all* $|x - a| < r_1$,

$$\frac{1}{f(x)} = \sum_{n=0}^{\infty} b_n (x - a)^n .$$

Proof: By continuity, there exists $r_1 > 0$ such that if $|x - a| < r_1$, then

$$\sum_{n=1}^{\infty} |a_n| \, |x - a|^n < 1.$$

Now pick such an x. Then

$$\frac{1}{f(x)} = \frac{1}{1 + \sum_{n=1}^{\infty} a_n (x - a)^n} = \frac{1}{1 + \sum_{n=0}^{\infty} c_n (x - a)^n}$$

where $c_n = a_n$ if $n > 0$ and $c_0 = 0$. Then

$$\left| \sum_{n=1}^{\infty} a_n (x - a)^n \right| \leq \sum_{n=1}^{\infty} |a_n| \, |x - a|^n < 1 \tag{7.12}$$

and so from the formula for the sum of a geometric series,

$$\frac{1}{f(x)} = \sum_{p=0}^{\infty} (-1)^p \left(\sum_{n=0}^{\infty} c_n (x - a)^n \right)^p .$$

By Corollary 7.2.4, this equals

$$\sum_{p=0}^{\infty} \sum_{n=0}^{\infty} b_{np} (x - a)^n \text{ where } b_{np} = \sum_{k_1 + \cdots + k_p = n} c_{k_1} \cdots c_{k_p} . \tag{7.13}$$

Thus $|b_{np}| \le \sum_{k_1 + \cdots + k_p = n} |c_{k_1}| \cdots |c_{k_p}| \equiv B_{np}$ and so by Theorem 7.2.3,

$$\sum_{p=0}^{\infty} \sum_{n=0}^{\infty} |b_{np}| |x-a|^n \le \sum_{p=0}^{\infty} \sum_{n=0}^{\infty} B_{np} |x-a|^n = \sum_{p=0}^{\infty} \left(\sum_{n=0}^{\infty} |c_n| |x-a|^n \right)^p < \infty$$

by (7.12) and the formula for the sum of a geometric series. Since the series of (7.13) converges absolutely, Theorem C.3.5 on Page 426 implies the series in (7.13) equals

$$\sum_{n=0}^{\infty} \left(\sum_{p=0}^{\infty} b_{np} \right) (x-a)^n$$

and so, let $\sum_{p=0}^{\infty} b_{np} \equiv b_n$. ∎

With this lemma, the following theorem is easy to obtain.

Theorem 7.2.6 *Let $f(x) = \sum_{n=0}^{\infty} a_n (x-a)^n$, a power series having radius of convergence $r > 0$. Suppose also that $f(a) \ne 0$. Then there exists $r_1 > 0$ and $\{b_n\}$ such that for all $|x-a| < r_1$,*

$$\frac{1}{f(x)} = \sum_{n=0}^{\infty} b_n (x-a)^n.$$

Thus also $1/f(a) = b_0$ so $f(a) = 1/b_0$.

Proof: Let $g(x) \equiv f(x)/f(a)$ so that $g(x)$ satisfies the conditions of the above lemma. Then by that lemma, there exists $r_1 > 0$ and a sequence $\{b_n\}$ such that

$$\frac{f(a)}{f(x)} = \sum_{n=0}^{\infty} b_n (x-a)^n$$

for all $|x-a| < r_1$. Then

$$\frac{1}{f(x)} = \sum_{n=0}^{\infty} \tilde{b}_n (x-a)^n$$

where $\tilde{b}_n = b_n/f(a)$. ∎

There is a very interesting question related to r_1 in this theorem. One might think that if $|x-a| < r$, the radius of convergence of $f(x)$ and if $f(x) \ne 0$ it should be possible to write $1/f(x)$ as a power series centered at a. Unfortunately this is not true. Consider $f(x) = 1 + x^2$. In this case $r = \infty$ but the power series for $1/f(x)$ converges only if $|x| < 1$. What happens is this, $1/f(x)$ will have a power series that will converge for $|x-a| < r_1$ where r_1 is the distance between a and the nearest singularity or zero of $f(x)$ in the complex plane. In the case of $f(x) = 1 + x^2$ this function has a zero at $x = \pm i$. This is just another instance of why the natural setting for the study of power series is the complex plane. To read more on power series, you should see the book by Apostol [2] or any text on complex variable.

7.3 Exercises

1. Find the radius of convergence of the following.

(a) $\sum_{k=1}^{\infty} \left(\frac{x}{2}\right)^n$

(d) $\sum_{n=0}^{\infty} \frac{(3n)^n}{(3n)!} x^n$

(b) $\sum_{k=1}^{\infty} \sin\left(\frac{1}{n}\right) 3^n x^n$

(e) $\sum_{n=0}^{\infty} \frac{(2n)^n}{(2n)!} x^n$

(c) $\sum_{k=0}^{\infty} k! x^k$

(f) $\sum_{n=0}^{\infty} \frac{(n)^n}{(n)!} x^n$

2. Find $\sum_{k=1}^{\infty} k 2^{-k}$

4. Find $\sum_{k=1}^{\infty} \frac{2^{-k}}{k}$

3. Find $\sum_{k=1}^{\infty} k^2 3^{-k}$

5. Find $\sum_{k=1}^{\infty} \frac{3^{-k}}{k}$

6. Find where the series $\sum_{k=1}^{\infty} \frac{1-e^{-kx}}{k}$ converges.

7. Find the power series centered at 0 for the function $1/\left(1 + x^2\right)$ and give the radius of convergence. Where does the function make sense? Where does the power series equal the function?

8. Find a power series for the function $f(x) \equiv \frac{\sin(\sqrt{x})}{\sqrt{x}}$ for $x > 0$. Where does $f(x)$ make sense? Where does the power series you found converge?

9. Use the power series technique which was applied in Example 7.1.5 to consider the initial value problem $y' = y, y(0) = 1$. This yields another way to obtain the power series for e^x.

10. Use the power series technique on the initial value problem $y' + y = 0$, $y(0) = 1$. What is the solution to this initial value problem?

11. Use the power series technique to find the first several nonzero terms in the power series solution to the initial value problem $y'' + xy = 0$, $y(0) = 0, y'(0) = 1$. Tell where your solution gives a valid description of a solution for the initial value problem. **Hint:** This is a little different but you proceed the same way as in Example 7.1.5.

12. Suppose the function e^x is defined in terms of a power series, $e^x \equiv \sum_{k=0}^{\infty} \frac{x^k}{k!}$. Use Theorem C.3.8 on Page 428 to show directly the usual law of exponents,

$$e^{x+y} = e^x e^y.$$

Be sure to check all the hypotheses.

13. Define the following function:*

$$f(x) \equiv \begin{cases} e^{-\left(1/x^2\right)} & \text{if } x \neq 0 \\ 0 & \text{if } x = 0 \end{cases}.$$

---

*Surprisingly, this function is very important to those who use modern techniques to study differential equations. One needs to consider test functions which have the property of having infinitely many derivatives but vanish outside of some interval. The theory of complex variables can be used to show that there are no examples of such functions if they have a valid power series expansion. It even becomes a little questionable whether such strange functions exist at all. Nevertheless, they do, there are enough of them, and it is this very example which is used to show this.

Show that $f^{(k)}(x)$ exists for all k and for all x. Show also that $f^{(k)}(0) = 0$ for all $k \in \mathbb{N}$. Therefore, the power series for $f(x)$ is of the form $\sum_{k=0}^{\infty} 0x^k$ and it converges for all values of x. However, it fails to converge to $f(x)$ except at the single point $x = 0$.

14. Let $f_n(x) \equiv \left(\frac{1}{n} + x^2\right)^{1/2}$. Show that for all x, $||x| - f_n(x)| \leq \frac{1}{\sqrt{n}}$. Now show $f'_n(0) = 0$ for all n and so $f'_n(0) \to 0$. However, the function $f(x) \equiv |x|$ has no derivative at $x = 0$. Thus even though $f_n(x) \to f(x)$ for all x, you cannot say that $f'_n(0) \to f'(0)$.

15. The above problem gives an example of uniform convergence of a sequence of functions $\{f_n(x)\}$ to a given function. Here is the general definition.

Definition 7.3.1 *Let S be a subset of $\mathbb{R}$. Let $\{f_n\}$ be a sequence of functions defined on S. Then f_n converges uniformly to a function f means: For every $\varepsilon > 0$ there exists N such that if $n \geq N$, then*

$$\sup_{x \in S} |f_n(x) - f(x)| < \varepsilon.$$

The symbol $\sup_{x \in S} g(x)$ means the least upper bound if it exists and ∞ if there is no upper bound.

show that if f_n converges uniformly to f on S and if f_n is continuous at x, then so is f.

16. In Section B.4 of the appendix, it is shown that if f is continuous on $[a, b]$ a bounded interval, then $\int_a^b f(x)\,dx$ exists. Show that if you have a sequence $\{f_n\}$ of continuous functions on $[a, b]$ which converges uniformly to f on $[a, b]$, then

$$\lim_{n \to \infty} \int_a^b f_n(x)\,dx = \int_a^b f(x)\,dx$$

17. If $f(x) \geq 0$ for all $x \in [a, b]$ and $f \in R([a, b])$, why is $\int_a^b f(x)\,dx \geq 0$? Verify that

$$\int_a^b |f(x)|\,dx \geq \left| \int_a^b f(x)\,dx \right|.$$

18. Let $a < b < c$ and suppose f is continuous. Show that

$$\int_a^b f(x)\,dx + \int_b^c f(x)\,dx = \int_a^c f(x)\,dx$$

Hint: Reduce to sums.

19. Let the functions $f_n(x)$ be given in Problem 14 and consider

$$g_1(x) = f_1(x), \quad g_n(x) = f_n(x) - f_{n-1}(x) \text{ if } n > 1.$$

Show that for all x, $\sum_{k=1}^{\infty} g_k(x) = |x|$ and that $g'_k(0) = 0$ for all k. Therefore, you cannot differentiate the series term by term and get the right answer.[†]

20. Use the theorem about the binomial series to give a proof of the binomial theorem

$$(a+b)^n = \sum_{k=0}^{n} \binom{n}{k} a^{n-k} b^k$$

whenever n is a positive integer.

21. You know $\int_0^x \frac{1}{t+1} dt = \ln|1+x|$. Use this and Theorem 7.1.7 to find the power series for $\ln|1+x|$ centered at 0. Where does this power series converge? Where does it converge to the function $\ln|1+x|$?

22. You know $\int_0^x \frac{1}{t^2+1} dt = \arctan x$. Use this and Theorem 7.1.7 to find the power series for $\arctan x$ centered at 0. Where does this power series converge? Where does it converge to the function $\arctan x$?

23. Find the power series for $\sin(x^2)$ by plugging in x^2 where ever there is an x in the power series for $\sin x$. How do you know this is the power series for $\sin(x^2)$?

24. Find the first several terms of the power series for $\sin^2(x)$ by multiplying the power series for $\sin(x)$. Next use the trig. identity $\sin^2(x) = \frac{1-\cos(2x)}{2}$ and the power series for $\cos(2x)$ to find the power series.

25. Find the power series for $f(x) = \frac{1}{\sqrt{1-x^2}}$.

26. It is hard to find $\int_0^1 e^{x^2} dx$ because you don't have a convenient antiderivative for the integrand. Replace e^{x^2} with an appropriate power series and estimate this integral.

27. Do the same as the previous problem for $\int_0^1 \sin(x^2) dx$.

28. Find $\lim_{x \to 0} \frac{\tan(\sin x) - \sin(\tan x)}{x^7}$.[‡]

29. Consider the function $S(x) \equiv \sum_{n=1}^{\infty} (-1)^{n+1} \frac{x^{2n-1}}{(2n-1)!}$. This is the power series for $\sin(x)$ but pretend you don't know this. Show that the series for $S(x)$ converges for all $x \in \mathbb{R}$. Also show that S satisfies the initial value problem $y'' + y = 0$, $y(0) = 0$, $y'(0) = 1$.

30. Consider the function $C(x) \equiv \sum_{n=0}^{\infty} (-1)^n \frac{x^{2n}}{(2n)!}$. This is the power series for $\cos(x)$ but pretend you do not know this. Show that the series for $C(x)$ converges for all $x \in \mathbb{R}$. Also show that S satisfies the initial value problem $y'' + y = 0$, $y(0) = 1$, $y'(0) = 0$.

[†]How bad can this get? It can be much worse than this. In fact, there are functions which are continuous everywhere and differentiable nowhere. We typically don't have names for them but they are there just the same. Every such function can be written as an infinite sum of polynomials which of course have derivatives at every point. Thus it is nonsense to differentiate an infinite sum term by term without a theorem of some sort.

[‡]This is a wonderful example. You should plug in small values of x using a calculator and see what you get using modern technology.

31. Show there is at most one solution to the initial value problem $y'' + y = 0$, $y(0) = a$, $y'(0) = b$ and find the solution to this problem in terms of $C(x)$ and $S(x)$. Also show directly from the series descriptions for $C(x)$ and $S(x)$ that $S'(x) = C(x)$ and $C'(x) = -S(x)$.

32. Using uniqueness of the initial value problem, show that

$$C(x+y) = C(x)C(y) - S(x)S(y)$$

and

$$S(x+y) = S(x)C(y) + S(y)C(x).$$

Do this in the following way: Fix y and consider the function $f(x) \equiv C(x+y)$ and

$$g(x) = C(x)C(y) - S(x)S(y).$$

Then show both f and g satisfy the same initial value problem and so they must be equal. Do the other identity the same way. Also show $S(-x) = -S(x)$ and $C(-x) = C(x)$ and $S(x)^2 + C(x)^2 = 1$. This last claim is really easy. Just take the derivative and see $S^2 + C^2$ must be constant.

33. You know $S(0) = 0$ and $C(0) = 1$. Show there exists $T > 0$ such that on $(0, T)$ both $S(x)$ and $C(x)$ are positive but $C(T) = 0$ while $S(T) = 1$. (We usually refer to T as $\frac{\pi}{2}$.) To do this, note that $S'(0) > 0$ and so S is an increasing function on some interval. Therefore, C is a decreasing function on that interval because of $S^2 + C^2 = 1$. If C is bounded below by some positive number, then S must be unbounded because $S' = C$. However this would contradict $S^2 + C^2 = 1$. Therefore, $C(T) = 0$ for some T. Let T be the first time this occurs. You fill in the mathematical details of this argument. Next show that on $(T, 2T)$, $S(x) > 0$ and $C(x) < 0$ and on $(2T, 3T)$, both $C(x)$ and $S(x)$ are negative. Finally, show that on $(3T, 4T)$, $C(x) > 0$ and $S(x) < 0$. Also show $C(x + 2T) = C(x)$ and $S(x + 2T) = S(x)$. Do all this without resorting to identifying $S(x)$ with $\sin x$ and $C(x)$ with $\cos x$. Finally explain why $\sin x = S(x)$ for all x and $C(x) = \cos x$ for all x.

34. Bessel's equation of order n is the differential equation

$$x^2 y'' + xy' + (x^2 - n^2) y = 0.$$

Show that a solution of the form $\sum_{k=0}^{\infty} a_k x^k$ exists in the case where $n = 0$. Show that this function $J_0(x)$ is defined as

$$J_0(x) = \sum_{k=0}^{\infty} (-1)^k \frac{x^{2k}}{(k!)^2 \, 2^{2k}}$$

and verify that the series converges for all real x.

35. Explain why the function $y(x) = \tan(x)$ solves the initial value problem

$$y' = 1 + y^2, \ y(0) = 0.$$

Thus $y(0) = 0$. Then from the equation, $y'(0) = 1$, $y'' = 2yy'$ and so $y''(0) = 0$, etc. Explain these assertions and then tell how to use this differential equation to find a power series for $\tan(x)$.

Note: Problems 29-33 outline a way to define the circular functions with no reference to plane geometry.

36. In calculus, you should have seen Taylor's theorem. If you have not seen this theorem see Theorem C.4.1 where an explanation is also given. Now suppose that $\mathbf{x} = (x_1, \cdots, x_n)$ is a point in $\mathbb{R}^n$ and that f is a smooth function which is defined near $\mathbf{x}$. By "smooth" I mean that it has as many continuous partial derivatives as desired. Taylor's formula is concerned with approximating $f(\mathbf{x} + \mathbf{h})$ where $\sum_{i=1}^n |h_i|^2$ is sufficiently small. In the following, $f_{,i}$ will denote the partial derivative with respect to x_i. Go over and fill in the loose ends to the following argument: For $t \in [0, 1]$,

$$g(t) \equiv f(\mathbf{x} + t\mathbf{h}) - f(\mathbf{x})$$

$$g(1) = \sum_{k=1}^m \frac{1}{k!} g^{(k)}(0) + \frac{1}{(m+1)!} g^{(m+1)}(t), \ t \in (0,1)$$

Thus we only need to get a formula for $g^{(k)}(0)$. By the chain rule from calculus,

$$g'(t) = \sum_{i_1=1}^n f_{,i_1}(\mathbf{x} + t\mathbf{h}) h_{i_1}, \ g'(t) = \sum_{i_1=1}^n f_{,i_1}(\mathbf{x}) h_{i_1}$$

$$g''(t) = \sum_{i_1,i_2} f_{,i_1 i_2}(\mathbf{x} + t\mathbf{h}) h_{i_1} h_{i_2}, \ \sum_{i_1,i_2} f_{,i_1 i_2}(\mathbf{x}) h_{i_1} h_{i_2}$$

$$\vdots$$

$$g^{(k)}(t) = \sum_{i_1,i_2,\cdots i_k} f_{,i_1 i_2 \cdots i_k}(\mathbf{x} + t\mathbf{h}) h_{i_1} h_{i_2} \cdots h_{i_k}$$

Thus for some $t \in (0, 1)$,

$$f(\mathbf{x} + \mathbf{h}) - f(\mathbf{x}) = \sum_{k=1}^m \frac{1}{k!} \sum_{i_1,i_2,\cdots i_k} f_{,i_1 i_2 \cdots i_k}(\mathbf{x}) h_{i_1} h_{i_2} \cdots h_{i_k}$$

$$+ \frac{1}{m!} \sum_{i_1,i_2,\cdots i_m} f_{,i_1 i_2 \cdots i_k}(\mathbf{x} + t\mathbf{h}) h_{i_1} h_{i_2} \cdots h_{i_k}$$

If $\mathbf{f}$ were vector valued, the result would be the same except the $t \in (0, 1)$ would depend on the component of $\mathbf{f}$. In other words, there would be such a formula for each component of $\mathbf{f}$.

Chapter 8

Power Series Methods

8.1   Second Order Linear Equations

Second order linear equations are those which can be written in the following form.

$$y'' + p(x)y' + q(x)y = 0 \tag{8.1}$$

By Theorem 4.2.11, the initial value problem for one of these in which it is assumed that p, q are continuous has a unique solution. Recall this is the problem consisting of the differential equation along with initial conditions.

$$y'' + p(x)y' + q(x)y = 0 \tag{8.2}$$

$$y(a) = y_0, y'(a) = y_1.$$

Suppose y_1 and y_2 are two solutions to (8.2). Then from the Wronskian alternative presented earlier,

$$\Phi(x) = \begin{pmatrix} y_1(x) & y_2(x) \\ y_1'(x) & y_2'(x) \end{pmatrix}$$

either is invertible for all x or is not invertible for any x. As explained earlier, the general solution to the differential equation is of the form

$$C_1 y_1 + C_2 y_2$$

if and only if $\det \Phi(x) = W(y_1, y_2)(x) \neq 0$ at some point x.

Also recall the convenient way to check whether this Wronskian is nonzero which involves observing whether the ratio of y_1 and y_2 is a constant.

$$W(y_1, y_2) = (y_1)^2 \left(\frac{y_2}{y_1} \right)' = y_2' y_1 - y_2 y_1' \tag{8.3}$$

and so, the Wronskian is nonzero if the ratio is not constant. Thus when this happens, the general solution of (8.2) can be obtained in the form

$$C_1 y_1 + C_2 y_2 \tag{8.4}$$

175

for suitable constants, C_1, C_2.

Also recall the method for finding another solution which was presented earlier in Procedure 4.3.2. This is reviewed here.

Suppose y_2 and y_1 are two solutions to (8.2). Then both of the following equations must hold.

$$y_2 y_1'' + p(x) y_2 y_1' + q(x) y_2 y_1 = 0,$$

and

$$y_1 y_2'' + p(x) y_1 y_2' + q(x) y_1 y_2 = 0.$$

Subtracting the first from the second yields

$$y_1 y_2'' - y_2 y_1'' + p(x)(y_1 y_2' - y_2 y_1') = 0.$$

Now the term, $y_1 y_2'' - y_2 y_1''$ equals $(y_1 y_2' - y_2 y_1')' = W(y_1, y_2)'$ and so this reduces to

$$W' + p(x) W = 0,$$

a first order linear equation for the Wronskian W. Letting $P'(x) = p(x)$, the solution to this equation is

$$W(y_1, y_2)(x) = Ce^{-P(x)}.$$

Note this shows the Wronskian either vanishes identically $(C = 0)$ or not at all $(C \neq 0)$, as claimed earlier. By dividing by C in case it is nonzero, we can assume y_2 is such that $C = 1$. This formula, called Abel's formula, can be used to find the general solution. This was the method of Wronskians presented earlier. See Procedure 4.3.2. We review this here.

Theorem 8.1.1 *If y_1 solves the equation (8.2), then the general solution is given by (8.4) where y_2 is a solution to the differential equation,*

$$y_2' y_1 - y_2 y_1' = e^{-P(x)}. \tag{8.5}$$

Proof: Let $\tilde{y}_2$ be the solution to an initial value problem for (8.2) such that

$$\det \begin{pmatrix} y_1(a) & \tilde{y}_2(a) \\ y_1'(a) & \tilde{y}_2'(a) \end{pmatrix} \neq 0$$

Then from Abel's formula, there is a nonzero constant C such that

$$\tilde{y}_2' y_1 - \tilde{y}_2 y_1' = Ce^{-P(x)}$$

Now let $y_2 = \tilde{y}_2 / C$. ∎

This theorem says that in order to find the general solution given one solution y_1, it suffices to find a solution y_2 to the differential equation (8.5) and then the general solution will be given by

$$C_1 y_1 + C_2 y_2.$$

See also Problem 27 on Page 104.

8.2 Differential Equations Near an Ordinary Point

The problem is to find the solution to something like this.

$$y'' + p(x)y' + q(x)y = 0 \tag{8.6}$$

$$y(a) = y_0, y'(a) = y_1 \tag{8.7}$$

given $p(x)$ and $q(x)$ are **analytic** near the point a. The term analytic near a means that both functions have a valid power series

$$p(x) = \sum_{k=0}^{\infty} p_k (x-a)^k , \quad q(x) = \sum_{k=0}^{\infty} q_k (x-a)^k$$

which are valid for $|x-a| < r$ for some $r > 0$. This and other terminology is in the following definition. Such a point a is called an ordinary point.

Definition 8.2.1 *A function f is analytic near a if for all $|x-a|$ small enough,*

$$f(x) = \sum_{k=0}^{\infty} a_k (x-a)^k$$

That is, f is correctly given by a power series for x near a. The radius of convergence of a power series $\sum_{k=0}^{\infty} a_k (x-a)^k$ is a nonnegative number r such that if $|x-a| < r$, then the series converges.

The right place to study analytic functions is in the context of the theory of complex analysis. When this is done, it becomes obvious that, for example, the composition of analytic functions is analytic and many other things of a similar nature. Then there is a nice theorem which follows.

Theorem 8.2.2 *In (8.6) suppose p and q are analytic near a and that r is the minimum of the radii of convergence of these functions. Then the initial value problem (8.6) - (8.7) has a unique solution given by a power series*

$$y(x) = \sum_{k=0}^{\infty} a_k (x-a)^k$$

which is valid at least for all $|x-a| < r$.

I am not proving this at this time, but when you look for a power series using the procedures to be presented, it will often be clear that what you obtain does indeed converge and will give a solution to the equation.

Example 8.2.3 *Where does there exist a power series solution to the initial value problem*

$$y'' + \frac{1}{x^2+1}y = 0, y(0) = 1, y'(0) = 0?$$

The answer is there exists a power series solution for all x satisfying

$$|x| < 1$$

This is because the function $\frac{1}{1+x^2}$ has a power series,

$$\sum_{k=0}^{\infty} (-1)^k x^{2k}$$

which converges for all $|x| < 1$. An easy way to see this is to note that the function is undefined when $x = i$ and that i is the closest point to 0 such that the function is undefined. The distance between 0 and i is 1 and so the radius of convergence will be 1. You can also use the root test or ratio test to verify this.

Now here is a variation.

Example 8.2.4 *On what interval does there exist a power series solution to the initial value problem*

$$y'' + \frac{1}{x^2+1}y = 0, y(3) = 1, y'(3) = 0?$$

In this case, you observe that $1/(1+x^2)$ is zero at i or $-i$ and the distance between either of these points and 3 is $\sqrt{10}$. Therefore, the initial value problem will have a solution on the interval

$$|x - 3| < \sqrt{10}$$

The radius of convergence is $\sqrt{10}$.

These examples illustrate why power series methods are unsatisfactory. The initial value problem has a solution for all $x \in \mathbb{R}$. However, if you insist on using power series methods, you can only get the solution on a small interval because of something happening in the complex plane which you couldn't care less about.

The following example shows how you can find the first several terms of a power series solution.

Example 8.2.5 *Find several terms of the power series solution to*

$$y'' + \frac{1}{x^2+1}y = 0, y(3) = 1, y'(3) = 0?$$

The way this is usually done is to plug in a power series and compute the coefficients. I will illustrate another simple way to do the same thing which is adequate for finding several terms of the power series solution.

$$y(x) = \sum_{k=0}^{\infty} a_k (x-3)^k$$

From calculus you know

$$a_k = \frac{y^{(k)}(3)}{k!}$$

so all we have to do is find the derivatives of y. We can do this from the equation itself along with the initial conditions. Thus

$$a_0 = y(3) = 1, \ a_1 = y'(3) = 0$$

How do you find $y''(3)$? Use the equation.

$$y''(3) = -\frac{1}{1+3^2}y(3) = -\frac{1}{10}$$

and so

$$a_2 = \left(-\frac{1}{10}\right)/2 = \frac{-1}{20}$$

Now let's find $y'''(3)$. From the equation,

$$y'''(x) + y'(x)\left(\frac{1}{1+x^2}\right) + y(x)\left(-\frac{2}{(1+x^2)^2}x\right) = 0$$

Plug in $x = 3$. This gives

$$y'''(3) + y'(3)\left(\frac{1}{10}\right) + y(3)\left(-\frac{2}{100}3\right) = 0$$

$$y'''(3) + 0\left(\frac{1}{10}\right) + 1\left(-\frac{2}{100}3\right) = 0$$

and so

$$y'''(3) = \frac{6}{100} = \frac{3}{50}$$

Therefore since $3! = 6$,

$$a_3 = \left(\frac{3}{50}\right)/6 = \frac{1}{100}$$

The first four terms of the power series solution for this equation are

$$y(x) = 1 - \frac{1}{20}(x-3)^2 + \frac{1}{100}(x-3)^3 + \cdots$$

You could keep on finding terms for this series but you know by the above theorem that the convergence takes place if $|x - 3| < \sqrt{10}$. Note I didn't find a general formula for the power series. It was too much trouble. There is one and it is easy to find as many terms as needed. Furthermore, in applications this is often all that is needed anyway because after all, the series only represents the solution to the differential equation for x pretty close to 3.

Example 8.2.6 *Do the above example another way by plugging in the power series to the equation.*

Now let's do it another way by plugging in a power series. If you do this, you need to use the power series for the function $1/(1+x^2)$ expanded about 3. After doing some fussy work you find

$$1/\left(1+x^2\right) = \frac{1}{10} - \frac{3}{50}\left(x-3\right) + \frac{13}{500}\left(x-3\right)^2$$
$$-\frac{6}{625}\left(x-3\right)^3 + \frac{79}{25\,000}\left(x-3\right)^4 + O\left(\left(x-3\right)^5\right)$$

I will just find the first several terms by matching coefficients.

$$y = \sum_{k=0}^{\infty} a_k\left(x-3\right)^k, y' = \sum_{k=1}^{\infty} a_k k\left(x-3\right)^{k-1},$$
$$y'' = \sum_{k=2}^{\infty} a_k k\left(k-1\right)\left(x-3\right)^{k-2}$$

Plugging in the power series, we get

$$\sum_{k=2}^{\infty} a_k k\left(k-1\right)\left(x-3\right)^{k-2} + \left(\frac{1}{10} - \frac{3}{50}\left(x-3\right) + \frac{13}{500}\left(x-3\right)^2 - \frac{6}{625}\left(x-3\right)^3\right.$$

$$\left. + \frac{79}{25\,000}\left(x-3\right)^4\right) \sum_{k=0}^{\infty} a_k\left(x-3\right)^k = 0$$

Now we match the terms. First, the constant terms. Recall $a_1 = 0$ and $a_0 = 1$.

$$2a_2 + \frac{1}{10}a_0 = 0$$

so $a_2 = -1/20$. Next consider the first order terms.

$$a_3 6 + \frac{1}{10}a_1 + \left(\frac{-3}{50}\right)a_0 = 0$$

and so $a_3 = \frac{3}{6\times50} = \frac{1}{100}$. Next you could consider the second order terms but I am sick of doing this. Therefore, we get up to first order terms the following expression for the power series.

$$y = 1 - \frac{1}{20}\left(x-3\right)^2 + \frac{1}{100}\left(x-3\right)^3$$

Note that in this simple problem it would be better to consider the equation in the form

$$\left(1+x^2\right)y'' + y = 0, \ y\left(0\right) = 1, y'\left(0\right) = 0.$$

To solve it in this form you write $1 + x^2$ as a power series

$$1 + x^2 = 10 + 6\left(x-3\right) + \left(x-3\right)^2$$

Then you plug in the series for y and y'' as before.

$$\left(10 + 6\left(x-3\right) + \left(x-3\right)^2\right) \sum_{k=2}^{\infty} a_k k\left(k-1\right)\left(x-3\right)^{k-2} + \sum_{k=0}^{\infty} a_k\left(x-3\right)^k = 0$$

and then match powers of $(x - 3)$. This problem is actually easy enough that you could get a nice recurrence relation.

The reason the power series method works for solving ordinary differential equations is dependent on the fact that a power series can be differentiated term by term on its radius of convergence. This is a theorem in calculus which is seldom proved these days. See Theorem 7.1.1. The following procedure tells how to get solutions to these differential equations with a ordinary point at 0.

Procedure 8.2.7 *To find the general solution to an ordinary differential equation having an ordinary point at 0, find a recurrence relation if possible and then let one solution correspond to $a_0 = 1, a_1 = 0$ and the other by letting $a_0 = 0, a_1 = 1$. If you can't find a convenient recurrence relation, just list the first several terms of the two power series for the solutions, one corresponding to $a_0 = 1, a_1 = 0$ and the other by letting $a_0 = 0, a_1 = 1$. These will be approximations to the desired solutions. If you have initial conditions given, you just use these to find the constants.*

Example 8.2.8 *The Legendre equation is*

$$\left(1 - x^2\right) y'' - 2xy' + \alpha \left(\alpha + 1\right) y = 0$$

Describe the radius of convergence of solutions for which the initial data is given at $x = 0$. Repeat the problem if the initial data is given at $x = 5$.

You divide by $\left(1 - x^2\right)$ and note that the nearest "singular" point of

$$2x / \left(1 - x^2\right)$$

occurs when $x = 1$ or -1 and so the power series solutions to this equation with initial data given at $x = 0$ converge on $(-1, 1)$. In the case where the initial data is given at 5, the series solution will converge for $|x - 5| < 4$. In other words, on the interval $(1, 9)$. Actually, in this example, you sometimes get polynomials for the solution and so the power series converges on all of $\mathbb{R}$.

Example 8.2.9 *Determine a lower bound for the radius of convergence of series solutions of the differential equation*

$$\left(1 + x^2\right) y'' + 2xy' + 4x^2 y = 0$$

about the point 0 and -7 and give the intervals on which the solutions will be sure to converge.

Divide by $1 + x^2$. Then you have two functions

$$\frac{2x}{1 + x^2}, \frac{4x^2}{1 + x^2}$$

and the power series centered at 0 for these functions converge if $|x| < 1$ so when the initial data is given at 0, the power series converge on $(-1, 1)$. Now consider the case where the initial condition is given at -7. In this case you need

$$|x - (-7)| < \sqrt{7^2 + 1} = \sqrt{50}$$

and so the series will converge on the interval

$$\left(-7 - \sqrt{50}, -7 + \sqrt{50}\right)$$

Perhaps this is not the first thing you would think of.

Example 8.2.10 *Where will power series solutions to the equation*

$$y'' + \sin(x)\, y' + \left(1 + x^2\right) y = 0$$

converge?

These will converge for all values of x because the power series for $\sin(x)$ converges for all x and so does the power series for $\left(1 + x^2\right)$ and this is true no matter where the initial data is given.

Could you find the first several terms of a power series solution for the initial value problem

$$y'' + \sin(x)\, y' + \left(1 + x^2\right) y = 0$$
$$y(0) = 1,\ y'(0) = 1 \tag{8.8}$$

First find the general solution as described in the above procedure. Using the above techniques, the solution to

$$y'' + \sin(x)\, y' + \left(1 + x^2\right) y = 0$$
$$y(0) = 1,\ y'(0) = 0$$

is $1 - \frac{1}{2}x^2 + \frac{1}{24}x^4 + O\left(x^5\right)$ and the solution to

$$y'' + \sin(x)\, y' + \left(1 + x^2\right) y = 0$$
$$y(0) = 0,\ y'(0) = 1$$

is $x - \frac{1}{3}x^3 + O\left(x^6\right)$ and so the general solution is

$$C_1 \left(1 - \frac{1}{2}x^2 + \frac{1}{24}x^4 + O\left(x^5\right)\right) + C_2 \left(x - \frac{1}{3}x^3 + O\left(x^6\right)\right)$$

Now taking the derivative, you have that the derivative of this is of the form

$$C_1 \left(-x + \frac{4}{24}x^3 + O\left(x^4\right)\right) + C_2 \left(1 - x^2 + O\left(x^5\right)\right)$$

and so, you need to have $C_1 = 1$ and $C_2 = 1$ if you want to have the solution to (8.8). Thus,

$$1 + x - \frac{1}{2}x^2 - \frac{1}{3}x^3 + \frac{1}{24}x^4 + O\left(x^5\right)$$

Of course it is much easier to not find the general solution but to simply work with the given initial conditions. You know that the solution is of the form

$$1 + x + \sum_{k=2}^{\infty} a_k x^k$$

where $a_2 = y''(0)/2$. This is determined by the equation. $y''(0) + 0 + 1 = 0$ and so $y''(0) = -1$. Thus this yields for the first three terms

$$1 + x - \frac{1}{2}x^2 + \cdots$$

Then to find the next term, you differentiate the equation

$$y''' + \cos(x) y' + \sin(x) y'' + 2xy + \left(1 + x^2\right) y' = 0$$

Then putting in $x = 0$,

$$y'''(0) + 1 + 1 = 0, \; y'''(0) = -2$$

and so the first four terms are

$$1 + x - \frac{1}{2}x^2 + \left(\frac{-2}{3!}\right) x^3 + O\left(x^4\right)$$

$$= 1 + x - \frac{1}{2}x^2 - \frac{1}{3}x^3 + O\left(x^4\right)$$

You continue in this way to obtain the desired solution.

8.3   The Legendre Equation

In this short section, the polynomial solutions to the important Legendre equation will be considered in more detail at the regular point $x = 0$. As mentioned above, this equation is

$$\left(1 - x^2\right) y'' - 2xy' + n(n+1) y = 0$$

where n is a positive integer. For some more on the applications and significance of this equation, see [4].

First is the observation that this equation has polynomial solutions.

Proposition 8.3.1 *For n a nonnegative integer, the Legendre equation has a polynomial solution.*

Proof: First note that $y = 1$ is a solution if $n = 0$. From the above discussion, it has a power series solution $\sum_{k=0}^{\infty} a_k x^k$. The idea is to show that it has one which terminates. Substituting this power series in to the equation,

$$\left(\sum_{k=0}^{\infty} a_k k(k-1) x^{k-2} - \sum_{k=0}^{\infty} a_k k(k-1) x^k\right) - \sum_{k=0}^{\infty} 2a_k k x^k + \sum_{k=0}^{\infty} n(n+1) a_k x^k = 0$$

Then changing the variable of summation in the first term,

$$\sum_{k=0}^{\infty} a_{k+2}(k+2)(k+1) x^k - \left(\sum_{k=0}^{\infty} a_k k(k-1) x^k + \sum_{k=0}^{\infty} 2a_k k x^k\right)$$

$$+ \sum_{k=0}^{\infty} n(n+1) a_k x^k = 0$$

It follows there is the following recurrence relation.

$$a_{k+2}(k+2)(k+1) = a_k((k(k-1)+2k)-n(n+1))$$
$$= a_k(k(k+1)-n(n+1))$$

Thus

$$a_{k+2} = \frac{k(k+1)-n(n+1)}{(k+2)(k+1)}a_k$$

Clearly this gives polynomial solutions. Suppose n is even. Then let $a_0 \neq 0$ and $a_1 = 0$. The above recurrence relation will imply that all odd terms equal 0. As to the even terms, they would be zero if $k \geq n$. Similar considerations will apply if n is odd. Thus there is a polynomial solution to the above equation. ■

You see how it was obtained. Depending on whether n is odd or even, you will get a single polynomial multiplied by a_0 if n is even or a single polynomial multiplied by a_1 if n is odd. Thus we can specify the polynomial solution by giving a desired value for the polynomial at $x = 1$ or at some other point. It is customary to pick $x = 1$ and to let the polynomial have the value of 1 at this point, thus harmonizing with letting the polynomial equal the constant 1 when $n = 0$.

The next proposition is an interesting formula for producing these polynomials. This is called Rodrigues' formula.

Proposition 8.3.2 *The polynomial $\frac{d^n}{dx^n}(x^2-1)^n$ is a solution to the Legendre equation for n. If one multiplies by $1/(n!2^n)$, the resulting polynomial has value equal to 1 when $x = 1$. Thus we define $p_n(x) \equiv \frac{1}{n!2^n}\frac{d^n}{dx^n}(x^2-1)^n$.*

Proof: Consider $\frac{d^n}{dx^n}(x^2-1)^n$. It is certainly a polynomial. Recall the Leibniz formula for taking derivatives.

$$(fg)^{(n)} = \sum_{k=0}^{n}\binom{n}{k}f^{(n-k)}g^{(k)}$$

Let $v = (x^2-1)^n$ then differentiating both sides, it follows that

$$v' = n(x^2-1)^{n-1}2x, \text{ and so } (x^2-1)v' = 2xnv$$

Then use the Leibniz formula to differentiate this $n+1$ times. Thus

$$\sum_{k=0}^{n}\binom{n}{k}v^{(n+1-k)}(x^2-1)^{(k)} = 2n\sum_{k=0}^{n}\binom{n}{k}v^{(n-k)}x^{(k)}$$

Now only a few terms survive. From the usual notation used above $f^{(0)}(x) = f(x)$. Thus you get

$$v^{(n+1)}(x^2-1)+2nxv^{(n)}+2\frac{n(n-1)}{2}v^{(n-1)} = 2xnv^{(n)}+2n^2v^{(n-1)}$$

Then combining terms yields

$$(x^2-1)v^{(n+1)}+(-n(n+1))v^{(n-1)} = 0$$

Differentiate this again.

$$2xv^{(n+1)} + \left(x^2 - 1\right)v^{n+2} + \left(-n\left(n+1\right)\right)v^{(n)} = 0$$

Now multiply by -1 to get

$$\left(1 - x^2\right)v^{(n+2)} - 2xv^{(n+1)} + n\left(n+1\right)v^{(n)} = 0$$

Let

$$
\begin{aligned}
f_n\left(x\right) &= \frac{d^n}{dx^n}\left(\left(x-1\right)^n\left(x+1\right)^n\right) \\
&= \sum_{k=0}^{n}\binom{n}{k}\frac{d^{n-k}}{dx^{n-k}}\left(\left(x-1\right)^n\right)\frac{d^k}{dx^k}\left(\left(x+1\right)^n\right) \\
&= \frac{d^n}{dx^n}\left(\left(x-1\right)^n\right)\left(x+1\right)^n + \sum_{k=1}^{n-1}\binom{n}{k}\frac{d^{n-k}}{dx^{n-k}}\left(\left(x-1\right)^n\right)\frac{d^k}{dx^k}\left(\left(x+1\right)^n\right) \\
&\quad + \left(\left(x-1\right)^n\right)\frac{d^n}{dx^n}\left(\left(x+1\right)^n\right)
\end{aligned}
$$

The last term is 0 when $x = 1$ and the middle terms are also. Thus the only term which survives is the first and when $x = 1$ this gives $n!2^n$. Therefore, the constant is $1/\left(n!2^n\right)$ and $\frac{1}{n!2^n}f_n\left(x\right) = p_n\left(x\right)$. ∎

Next to be considered is a generating function for these polynomials.

$$\Phi\left(x, h\right) \equiv \left(1 - 2xh + h^2\right)^{-1/2}.$$

Then one can show that $\left(1 - x^2\right)\Phi_{xx} - 2x\Phi_x + h\left(h\Phi\right)_{hh} = 0$. Here are the steps. Letting D_x or D_h denote the partial derivative with respect to the indicated variable,

$$D_x\left(\left(1 - 2xh + h^2\right)^{-1/2}\right) = \frac{h}{\left(h^2 - 2xh + 1\right)^{\frac{3}{2}}}$$

$$D_x\left(\frac{h}{\left(h^2 - 2xh + 1\right)^{\frac{3}{2}}}\right) = 3\frac{h^2}{\left(h^2 - 2xh + 1\right)^{\frac{5}{2}}}$$

$$
\begin{aligned}
&hD_h\left(D_h\left(\left(h\left(1 - 2xh + h^2\right)^{-1/2}\right)\right)\right) \\
&= -\frac{h}{\left(h^2 - 2xh + 1\right)^{\frac{5}{2}}}\left(-2h^2x + hx^2 + 3h - 2x\right)
\end{aligned}
$$

And so the expression of interest is

$$
\begin{aligned}
&\left(1 - x^2\right)\frac{3h^2}{\left(h^2 - 2xh + 1\right)^{\frac{5}{2}}} - 2x\left(\frac{h}{\left(h^2 - 2xh + 1\right)^{\frac{3}{2}}}\right) \\
&+ \left(-\frac{h}{\left(h^2 - 2xh + 1\right)^{\frac{5}{2}}}\left(-2h^2x + hx^2 + 3h - 2x\right)\right)
\end{aligned}
$$

This equals

$$\left(1 - x^2\right) 3 \frac{h^2}{\left(h^2 - 2xh + 1\right)^{\frac{5}{2}}} - 2x \left(\frac{h\left(h^2 - 2xh + 1\right)}{\left(h^2 - 2xh + 1\right)^{\frac{5}{2}}}\right)$$

$$+ \left(-\frac{h}{\left(h^2 - 2xh + 1\right)^{\frac{5}{2}}} \left(-2h^2 x + hx^2 + 3h - 2x\right)\right)$$

$$\frac{h}{\left(h^2 - 2xh + 1\right)^{\frac{5}{2}}} \left(-2h^2 x + hx^2 + 3h - 2x\right)$$

$$+ \left(-\frac{h}{\left(h^2 - 2xh + 1\right)^{\frac{5}{2}}} \left(-2h^2 x + hx^2 + 3h - 2x\right)\right) = 0$$

Now for small h, it follows from the binomial theorem from calculus that there is a power series of the following form.

$$\Phi\left(x, h\right) \equiv \left(1 - 2xh + h^2\right)^{-1/2} = \sum_{k=0}^{\infty} a_k\left(x\right) h^k$$

In fact, the $a_k\left(x\right)$ will be polynomials. This follows from the binomial theorem. You will get $\left(h^2 - 2xh\right)$ raised to various powers and this will always be a polynomial in x.

Now use the above identity to write

$$0 = \sum_{k=0}^{\infty} \left(1 - x^2\right) a_k''\left(x\right) h^k - \sum_{k=0}^{\infty} 2x a_k'\left(x\right) h^k + h \left(h \sum_{k=0}^{\infty} a_k\left(x\right) h^k\right)_{hh}$$

$$= \sum_{k=0}^{\infty} \left(1 - x^2\right) a_k''\left(x\right) h^k - \sum_{k=0}^{\infty} 2x a_k'\left(x\right) h^k + h \left(\sum_{k=0}^{\infty} a_k\left(x\right) \left(k + 1\right) k h^{k-1}\right)$$

$$= \sum_{k=0}^{\infty} \left(1 - x^2\right) a_k''\left(x\right) h^k - \sum_{k=0}^{\infty} 2x a_k'\left(x\right) h^k + \left(\sum_{k=0}^{\infty} a_k\left(x\right) \left(k + 1\right) k h^k\right)$$

Hence $a_k\left(x\right)$ is a polynomial and satisfies the Legendre equation for k and also. Letting $x = 1$,

$$\sum_{k=0}^{\infty} a_k\left(1\right) h^k = \left(1 - 2h + h^2\right)^{-1/2} = \frac{1}{1 - h} = \sum_{k=0}^{\infty} h^k$$

showing that each $a_k\left(1\right) = 1$. Hence these are the Legendre polynomials. This is called a generating function for the Legendre polynomials. This is the content of the above theorem.

Theorem 8.3.3 *Let $p_k\left(x\right)$ be the Legendre polynomial for k. That is, it is a polynomial, equals 1 when $x = 1$, and satisfies the Legendre differential equation. Then for small h,*

$$\left(1 - 2xh + h^2\right)^{-1/2} = \sum_{k=0}^{\infty} p_k\left(x\right) h^k$$

Now with this result, one can obtain an interesting and significant relation between the different Legendre polynomials. Taking the partial derivative with respect to h of each side,

$$\frac{x-h}{(h^2 - 2xh + 1)^{\frac{3}{2}}} = \sum_{k=1}^{\infty} k p_k(x) h^{k-1}$$

Then

$$\left(1 - 2xh + h^2\right) \frac{x-h}{(h^2 - 2xh + 1)^{\frac{3}{2}}} = (x-h)\left(1 - 2xh + h^2\right)^{-1/2}$$

$$\left(1 - 2xh + h^2\right) \frac{x-h}{(h^2 - 2xh + 1)^{\frac{3}{2}}} = \left(1 - 2xh + h^2\right) \sum_{k=1}^{\infty} k p_k(x) h^{k-1}$$

Now you can collect terms and compare. Using the result of the above theorem,

$$(x-h)\left(1 - 2xh + h^2\right)^{-1/2} = (x-h)\sum_{k=0}^{\infty} p_k(x) h^k$$

$$= \left(1 - 2xh + h^2\right) \sum_{k=1}^{\infty} k p_k(x) h^{k-1}$$

Now one collects terms to obtain

$$\sum_{k=0}^{\infty} x p_k(x) h^k - \sum_{k=0}^{\infty} p_k(x) h^{k+1}$$

$$= \sum_{k=1}^{\infty} k p_k(x) h^{k-1} - 2\sum_{k=1}^{\infty} k x p_k(x) h^k + \sum_{k=1}^{\infty} k p_k(x) h^{k+1}$$

Thus

$$\sum_{k=0}^{\infty} (x + 2kx) p_k(x) h^k = \sum_{k=1}^{\infty} k p_k(x) h^{k-1} + \sum_{k=1}^{\infty} k p_k(x) h^{k+1}$$

$$+ \sum_{k=0}^{\infty} p_k(x) h^{k+1}$$

$$\sum_{k=0}^{\infty} (x + 2kx) p_k(x) h^k = \sum_{k=0}^{\infty} (k+1) p_{k+1}(x) h^k + \sum_{k=0}^{\infty} (k+1) p_k(x) h^{k+1}$$

$$= \sum_{k=0}^{\infty} (k+1) p_{k+1}(x) h^k + \sum_{k=1}^{\infty} k p_{k-1}(x) h^k$$

$$= \sum_{k=0}^{\infty} (k+1) p_{k+1}(x) h^k + \sum_{k=0}^{\infty} k p_{k-1}(x) h^k$$

Then for $k \geq 1$,

$$(k+1) p_{k+1}(x) = (2k+1) x p_k(x) - k p_{k-1}(x) \tag{8.9}$$

As noted earlier $p_0(x) = 1$. Also, from the generating function, the polynomial $p_1(x)$ is just the partial derivative of the generating function with respect to h evaluated when $h = 0$. Thus $p_1(x) = x$. Now one can use this recursion relation to find as many Legendre polynomials as desired.

$$p_0(x) = 1, p_1(x) = x, p_2(x) = \frac{1}{2}\left(3x^2 - 1\right), \cdots$$

Also from this relation, one can determine a formula for

$$\int_{-1}^{1} p_k(x)^2\, dx$$

From Problem 4, $\int_{-1}^{1} p_m(x)\, p_n(x)\, dx = 0$ if $m \neq n$. Multiply (8.9) by $p_{k+1}(x)$ and then integrate. Thus

$$
\begin{aligned}
(k+1)\int_{-1}^{1} p_{k+1}(x)^2\, dx &= (2k+1)\int_{-1}^{1} xp_k(x)\, p_{k+1}(x)\, dx \\
&\quad -k\int_{-1}^{1} p_{k+1}(x)\, p_{k-1}(x)\, dx \qquad (8.10) \\
&= (2k+1)\int_{-1}^{1} xp_k(x)\, p_{k+1}(x)\, dx \qquad (8.11)
\end{aligned}
$$

Replace k with $k-1$.

$$k\int_{-1}^{1} p_k(x)^2\, dx = (2k-1)\int_{-1}^{1} xp_{k-1}(x)\, p_k(x)\, dx \qquad (8.12)$$

Now multiply (8.9) by $p_{k-1}(x)$ and integrate. Then this yields

$$
\begin{aligned}
(k+1)\int_{-1}^{1} p_{k+1}(x)\, p_{k-1}(x)\, dx &= (2k+1)\int_{-1}^{1} xp_k(x)\, p_{k-1}(x)\, dx \\
&\quad -k\int_{-1}^{1} p_{k-1}(x)^2\, dx
\end{aligned}
$$

Hence

$$k\int_{-1}^{1} p_{k-1}(x)^2\, dx = (2k+1)\int_{-1}^{1} xp_k(x)\, p_{k-1}(x)\, dx$$

Thus from (8.12)

$$\frac{k}{2k+1}\int_{-1}^{1} p_{k-1}(x)^2\, dx = \int_{-1}^{1} xp_{k-1}(x)\, p_k(x)\, dx = \frac{k}{2k-1}\int_{-1}^{1} p_k(x)^2\, dx$$

$$\int_{-1}^{1} p_k(x)^2\, dx = \frac{2k-1}{2k+1}\int_{-1}^{1} p_{k-1}(x)^2\, dx$$

Iterating this, we get

$$\int_{-1}^{1} p_k(x)^2\, dx = \frac{2k-1}{2k+1}\frac{2k-3}{2k-1}\cdots\frac{1}{3}\int_{-1}^{1} p_0(x)^2\, dx = \frac{2}{2k+1}.$$

This has proved the following theorem which gives a collection of orthogonal polynomials.

Theorem 8.3.4 *Let $p_n(x)$ be the polynomial solution to the Legendre equation*

$$\left(1 - x^2\right) y'' - 2xy' + n(n+1) = 0.$$

such that $p_n(1) = 1$. Then if $q_n(x) = \frac{\sqrt{2n+1}}{\sqrt{2}} p_n(x)$

$$\int_{-1}^{1} q_n(x) \, q_m(x) \, dx = \begin{cases} 1 \ \text{if } n = m \\ 0 \ \text{if } n \neq m \end{cases}$$

Also, we have the following recurrence relation.

$$(k+1) p_{k+1}(x) = (2k+1) x p_k(x) - k p_{k-1}(x)$$

8.4   The Case Where $n(n+1)$ Is Replaced with λ

Here we consider what happens in Legendre's equation if you replace $n(n+1)$ with positive λ which is not necessarily of this form.

You still have the following recurrence relation

$$a_{k+2} = \frac{k(k+1) - \lambda}{(k+2)(k+1)} a_k, \ k \geq 0$$

only now maybe λ is not of the desired form. However, the power series

$$y = \sum_{k=0}^{\infty} a_k x^k$$

at least converges to a solution which is valid on $(-1, 1)$. The idea here is to show that the solutions blow up badly at an endpoint unless λ has the form $n(n+1)$ for n a nonnegative integer.

Following Hochstadt [14] one compares with a power series which comes from a recurrence relation

$$b_{k+2} = \frac{k(k+1)}{(k+2)(k+1)} b_k = \frac{k}{k+2} b_k \text{ for } k \geq 1.$$

Then let b_0, b_1 be given and define $b_2 = \frac{b_0}{2}$ and the remaining terms be given by the above recurrence relation. Thus

$$b_4 = \frac{2}{4} b_2 = \frac{1}{2} \frac{b_0}{2} = \frac{1}{4} b_0$$

$$b_3 = \frac{1}{3} b_1, \ b_5 = \frac{3}{5} \frac{1}{3} b_1 = \frac{1}{5} b_1$$

$$b_6 = \frac{4}{6} b_4 = \frac{4}{6} \frac{1}{4} b_0 = \frac{1}{6} b_0$$

continuing this way, you see that

$$b_{2k} = \frac{b_0}{2k}, k \geq 1, \quad b_{2k+1} = \frac{b_1}{2k+1}, k \geq 0$$

Then

$$\left| \frac{a_{k+2}}{b_{k+2}} \right| = \frac{\left| \frac{k(k+1)-\lambda}{(k+2)(k+1)} \right|}{\left| \frac{k(k+1)}{(k+2)(k+1)} \right|} \left| \frac{a_k}{b_k} \right| = \frac{1}{k^2+k} \left| k^2 + k - \lambda \right| \left| \frac{a_k}{b_k} \right|$$

$$= \left| 1 - \frac{\lambda}{k(1+k)} \right| \left| \frac{a_k}{b_k} \right|$$

Then

$$\ln\left(\left| \frac{a_{k+2}}{b_{k+2}} \right| \right) = \ln\left(\left| \frac{a_k}{b_k} \right| \right) + \ln\left| 1 - \frac{\lambda}{k(1+k)} \right|$$

You have such a thing for even k and for odd k. Therefore, you find that

$$\ln\left| \frac{a_m}{b_m} \right| = \sum_{i=1}^{m-1} \ln\left| 1 - \frac{\lambda}{k(1+k)} \right| \tag{*}$$

Note that if λ is not of the form $n(n+1)$ or if the condition for obtaining a polynomial solution is violated because, for example, if n is even and $a_1 \neq 0$, then there will be infinitely many nonzero terms in the above sum, all negative. Now for $\lambda > 0$,

$$\lim_{k \to \infty} \frac{\left| \ln\left(1 - \frac{\lambda}{k(k+1)} \right) \right|}{\lambda/(k(k+1))} = 1$$

and so the sum

$$\sum_{i=1}^{\infty} \ln\left| 1 - \frac{\lambda}{k(1+k)} \right|$$

converges with

$$\sum_{i=1}^{\infty} \frac{\lambda}{k(k+1)}$$

the sum involving the ln converging to a negative number $-\sigma$. Then from $*$,

$$\left| \frac{a_m}{b_m} \right| = \exp\left(\sum_{i=1}^{m-1} \ln\left| 1 - \frac{\lambda}{k(1+k)} \right| \right)$$

Letting $m \to \infty$,

$$\lim_{m \to \infty} \left| \frac{a_m}{b_m} \right| = e^{-\sigma} > 0$$

It follows that

$$\sum_k a_k(-1)^k, \sum_k a_k, \sum_k b_k, \sum_k b_k(-1)^k$$

all fail to converge absolutely at $1, -1$. This is obvious for the b_k series because it is just a p series for $p = 1$ and the comparison test applies to give what was just claimed about absolute convergence. However, if you look at 1, and consider the factor in the recurrence relation,

$$\frac{k(k+1)-\lambda}{(k+2)(k+1)} > 0$$

for all large k. This is eventually positive. Therefore, for large k, either the $a_k 1^k$ or the $a_k (-1)^k$ must have constant sign. Thus the above result about absolute divergence applies to show that the solution is not defined either at 1 or at -1. This explanation has shown the following.

Proposition 8.4.1 *For the equation*

$$\left(\left(1 - x^2\right) y'\right)' + \lambda y = 0$$

If $\lambda \neq n(n+1)$ for n a nonnegative integer, then every solution fails to be defined at either 1 or -1. If $\lambda = n(n+1)$ and n is even but $a_1 \neq 0$, then the resulting solution also fails to be defined at either 1 or -1. If $\lambda = n(n+1)$ and n is odd, but $a_0 \neq 0$, then the resulting solution also fails to be defined either at 1 or at -1.

In short, the only solutions to the Legendre equation $\left(\left(1 - x^2\right) y'\right)' + \lambda y = 0$ which are defined on the whole of $[-1, 1]$, are polynomials and λ must be of the form $n(n+1)$ for n a nonnegative integer. The initial condition must also be chosen judiciously in order to obtain the polynomial solution. For a totally different approach to this, see Problem 30 on Page 362.

8.5 The Euler Equations

The next step is to progress from ordinary points to regular singular points. The simplest equation to illustrate the concept of a regular singular point is the so called Euler equation, sometimes called a Cauchy Euler equation. These kinds of equations are fundamentally different than what has been studied earlier because near 0, there is no way to write them in the form discussed earlier,

$$y'' + p(x) y' + q(x) y = 0$$

where p, q are continuous functions. Therefore, other procedures for studying them must be developed.

Definition 8.5.1 *A differential equation is called an Euler equation if it can be written in the form*

$$x^2 y'' + axy' + by = 0.$$

Solving a Cauchy Euler equation is really easy. You look for a solution in terms $y = x^r$ and try to choose r in such a way that it solves the equation. Plugging this in to the above equation,

$$x^2 r(r-1) x^{r-2} + xar x^{r-1} + bx^r = 0$$

This reduces to

$$x^r (r(r-1) + ar + b) = 0$$

and so you have to solve the equation

$$r(r-1) + ar + b = 0$$

to find the values of r. If these values of r are different, say $r_1 \neq r_2$ then the general solution must be

$$C_1 x^{r_1} + C_2 x^{r_2}$$

because the Wronskian of the two functions will be nonzero. I know this because the ratio of the two functions is not a constant so the derivative of their ratio is not 0. However, by the quotient rule, the numerator is ± 1 times the Wronskian.

Example 8.5.2 *Find the general solution to $x^2 y'' - 2xy' + 2y = 0$.*

You plug in x^r and look for r. Then as above this yields

$$r(r-1) - 2r + 2 = r^2 - 3r + 2 = 0$$

and so the two values of r are $1, 2$. Therefore, the general solution to this equation is

$$C_1 x + C_2 x^2.$$

Of course there are three cases for solutions to the so called indicial equation

$$r(r-1) + ar + b = 0$$

Either the zeros are distinct and real, distinct and complex or repeated. Consider the case where they are distinct and complex next.

Example 8.5.3 *Find the general solution to $x^2 y'' + 3xy' + 2y = 0$.*

This time you have

$$r^2 + 2r + 2 = 0$$

and the solutions are $r = -1 \pm i$. How do we interpret

$$x^{-1+i}, x^{-1-i}?$$

It is real easy. You assume always that $x > 0$ since otherwise the leading coefficient could vanish. Then

$$x^{-1+i} = e^{\ln(x)(-1+i)} = e^{-\ln(x) + i\ln(x)}$$

and by Euler's formula this equals

$$
\begin{aligned}
x^{-1+i} &= e^{\ln(x^{-1})} \left(\cos\left(\ln\left(x\right)\right) + i\sin\left(\ln\left(x\right)\right) \right) \\
&= \frac{1}{x} \left(\cos\left(\ln\left(x\right)\right) + i\sin\left(\ln\left(x\right)\right) \right)
\end{aligned}
$$

Corresponding to x^{-1-i} we get something similar.

$$x^{-1-i} = \frac{1}{x} \left(\left(\cos\left(\ln\left(x\right)\right) - i\sin\left(\ln\left(x\right)\right) \right) \right)$$

Adding these together and dividing by 2 to get the real part, the principle of superposition implies

$$\frac{1}{x} \cos\left(\ln\left(x\right)\right)$$

is a solution. Then subtracting them and dividing by $2i$ you get

$$\frac{1}{x}\sin\left(\ln\left(x\right)\right)$$

is a solution. Hence anything of the form

$$C_1\frac{1}{x}\cos\left(\ln\left(x\right)\right) + C_2\frac{1}{x}\sin\left(\ln\left(x\right)\right)$$

is a solution. Is this the general solution? Of course. This follows because the ratio of the two functions is not constant and this implies their Wronskian is nonzero.

In the general case, suppose the solutions of the indicial equation

$$r\left(r-1\right) + ar + b = 0$$

are $\alpha \pm i\beta$. Then the general solution for $x > 0$ is

$$C_1 x^\alpha \cos\left(\beta\ln\left(x\right)\right) + C_2 x^\alpha \sin\left(\beta\ln\left(x\right)\right)$$

Finally consider the case where the zeros of the indicial equation are real and repeated. Note I have included all cases because since the coefficients of this equation are real, the zeros come in conjugate pairs if they are not real. Suppose then that x^r is a solution of

$$x^2 y'' + axy' + by = 0$$

Then if $z\left(x\right)$ is another solution which is not a multiple of x^r, you would have the following by Theorem 8.1.1.

$$\begin{vmatrix} x^r & z \\ rx^{r-1} & z' \end{vmatrix} = e^{-a\ln(x)} = x^{-a}$$

and so

$$z'x^r - zrx^{r-1} = x^{-a}$$

and so

$$z' - \frac{1}{x}rz = x^{-a-r}$$

Then doing the usual thing for first order linear equations,

$$\frac{d}{dx}\left(x^{-r}z\right) = x^{-a-r}x^{-r} = x^{-a-2r}$$

and so if $-a - 2r + 1 \neq 0$,

$$\left(x^{-r}z\right) = \frac{x^{-a-2r+1}}{-a-2r+1}$$

Therefore, z is a multiple of x^{r_2} for some $r_2 \neq r$, which is assumed not to happen. Therefore, $a + 2r = 1$. In this case, you get

$$x^{-r}z = \ln\left(x\right) + C$$

and so another solution is

$$z = x^r \ln\left(x\right)$$

Example 8.5.4 *Find the general solution of the equation*

$$x^2 y'' + 3xy' + y = 0.$$

In this case the indicial equation is

$$r(r-1) + 3r + 1 = r^2 + 2r + 1 = 0$$

and there is a repeated zero, $r = -1$. Therefore, the general solution is

$$y = C_1 x^{-1} + C_2 \ln(x) x^{-1}.$$

This is pretty easy isn't it?

How would things be different if the equation was of the form

$$(x-a)^2 y'' + a(x-a) y' + by = 0?$$

The answer is that is wouldn't be any different. You could just define a new independent variable $t \equiv (x-a)$ and then the equation in terms of t becomes

$$t^2 z'' + atz + bz = 0$$

where $z(t) \equiv y(x) = y(t+a)$. You can always reduce these sorts of equations to the case where the singular point is at 0. However, you might not want to do this. If not, you look for a solution in the form $y = (x-a)^r$, plug in and determine the correct value of r. In the case of real and distinct zeros you get

$$y = C_1 (x-a)^{r_1} + C_2 (x-a)^{r_2}$$

In the case where $r = \alpha \pm i\beta$ you get

$$y = C_1 (x-a)^\alpha \cos(\beta \ln(x-a)) + C_2 (x-a)^\alpha \sin(\beta \ln(x-a))$$

for the general solution for $x > a$

In the case where r is a repeated zero, you get

$$y = C_1 (x-a)^r + C_2 \ln(x-a)(x-a)^r.$$

8.6 Some Simple Observations on Power Series

This section is a review of a few facts about power series which should have been learned in calculus.

Theorem 8.6.1 *Suppose $f(x) = \sum_{n=0}^\infty a_n (x-a)^n$ for x near a and suppose $a_0 \neq 0$. Then*

$$f(x)^{-1} = \frac{1}{a_0} + h(x)$$

where $h(x) = \sum_{n=1}^\infty b_n (x-a)^n$ so $h(a) = 0$.

Proof: It follows from Theorem 7.2.6 that $f(x)^{-1}$ has a power series representation near a and $f(a)^{-1} = 1/a_0$. ∎

Theorem 8.6.2 *Suppose $f(x) = \sum_{n=0}^{\infty} a_n x^n$ and $g(x) = \sum_{n=0}^{\infty} b_n x^n$ for x near 0. Then $f(x) g(x)$ also has a power series near 0 and in fact,*

$$f(x) g(x) = \sum_{n=0}^{\infty} \left(\sum_{k=0}^{n} a_{n-k} b_k \right) x^n.$$

Proof: This follows from Theorem 7.2.1 applied to the two power series. Both converge absolutely on the intersection of their intervals of convergence by Theorem 7.0.3, and so this theorem gives the desired result. ∎

8.7 Regular Singular Points

First of all, here is the definition of what a regular singular point is.

Definition 8.7.1 *A differential equation has a regular singular point at 0 if the equation can be written in the form*

$$x^2 y'' + x b(x) y' + c(x) y = 0 \tag{8.13}$$

where

$$b(x) = \sum_{n=0}^{\infty} b_n x^n, \ \sum_{n=0}^{\infty} c_n x^n = c(x)$$

for all x near 0. More generally, a differential equation

$$P(x) y'' + Q(x) y' + R(x) y = 0 \tag{8.14}$$

where P, Q, R are analytic near a has a regular singular point at a if it can be written in the form

$$(x - a)^2 y'' + (x - a) b(x) y' + c(x) y = 0 \tag{8.15}$$

where

$$b(x) = \sum_{n=0}^{\infty} b_n (x - a)^n, \ \sum_{n=0}^{\infty} c_n (x - a)^n = c(x)$$

for all $|x - a|$ small enough. The equation (8.14) has a singular point at a if $P(a) = 0$.

The following table emphasizes the similarities between the Euler equations and the regular singular point equations. I have featured the point 0. If you are interested in another point a, you just replace x with $x - a$ everywhere it occurs.

	Euler Equation	Regular Singular Point
Form of equation	$x^2 y'' + x b_0 y' + c_0 y = 0$	$x^2 y'' + x (b_0 + b_1 x + \cdots) y'$ $+ (c_0 + c_1 x + \cdots) y = 0$
Indicial Equation	$r(r-1) + b_0 r + c_0 = 0$	$r(r-1) + b_0 r + c_0 = 0$
One solution	$y = x^r$	$y = x^r \sum_{k=0}^{\infty} a_k x^k, \ a_0 = 1.$

$$\tag{8.16}$$

Recognizing Regular Singular Points

How do you know a singular differential equation can be written a certain way? In particular, how can you recognize a regular singular point when you see one? Suppose

$$P(x) y'' + Q(x) y' + R(x) y = 0$$

where all of P, Q, R are analytic functions near a. How can you tell if it has a regular singular point at a? Here is how. It has a regular singular point at a if

$$\lim_{x \to a} (x - a) \frac{Q(x)}{P(x)} \text{ exists}$$

$$\lim_{x \to a} (x - a)^2 \frac{R(x)}{P(x)} \text{ exists}$$

If these conditions hold, then by theorems in complex analysis it will be the case that

$$(x - a) \frac{Q(x)}{P(x)} = \sum_{n=0}^{\infty} b_n (x - a)^n,$$

and

$$(x - a)^2 \frac{R(x)}{P(x)} = \sum_{n=0}^{\infty} c_n (x - a)^n$$

for x near a. Indeed, equations of this form reduce to the form in (8.15) upon dividing by $P(x)$ and multiplying by $(x - a)^2$.

Example 8.7.2 *Find the regular singular points of the equation and find the singular points.*

$$x^3 (x - 2)^2 (x - 1)^2 y'' + (x - 2) \sin(x) y' + (1 + x) y = 0$$

The singular points are $0, 2, 1$. Let's consider 0 first.

$$\lim_{x \to 0} x \frac{(x - 2) \sin(x)}{x^3 (x - 2)^2 (x - 1)^2}$$

does not exist. Therefore, 0 is not a regular singular point. I don't have to check any further. Now consider the singular point 2.

$$\lim_{x \to 2} (x - 2) \frac{(x - 2) \sin(x)}{x^3 (x - 2)^2 (x - 1)^2} = \frac{1}{8} \sin 2$$

and

$$\lim_{x \to 2} (x - 2)^2 \frac{1 + x}{x^3 (x - 2)^2 (x - 1)^2} = \frac{3}{8}$$

and so yes, 2 is a regular singular point. Now consider 1.

$$\lim_{x \to 1} (x - 1) \frac{(x - 2) \sin(x)}{x^3 (x - 2)^2 (x - 1)^2}$$

does not exist so 1 is not a regular singular point. Thus the above equation has only one regular singular point and this is where $x = 2$.

Example 8.7.3 *Find the regular singular points of*

$$x \sin(x) y'' + 3 \tan(x) y' + 2y = 0$$

The singular points are $0, n\pi$ where n is an integer. Let's consider a point at $n\pi$ where $n \neq 0$. To be specific, let's let $n = 3$.

$$\lim_{x \to 3\pi} (x - 3\pi) \frac{3 \tan(x)}{x \sin(x)} = 0$$

Similarly the limit exists for other values of n. Now consider

$$\lim_{x \to 3\pi} (x - 3\pi)^2 \frac{2}{x \sin(x)} = 0$$

Similarly the limit exists for other values of n. What about 0?

$$\lim_{x \to 0} x \frac{3 \tan(x)}{x \sin(x)} = 3$$

and

$$\lim_{x \to 0} x^2 \frac{2}{x \sin(x)} = 2$$

so it appears all these singular points are regular singular points.

Example 8.7.4 *Find the regular singular points of*

$$x^2 \sin(x) y'' + 3 \tan(x) y' + 2y = 0$$

Let's look at $x = 0$ first. The equation has the same singular points.

$$\lim_{x \to 0} x \frac{3 \tan(x)}{x^2 \sin(x)} = \text{undefined}$$

so 0 is not a regular singular point.

$$\lim_{x \to 3\pi} (x - 3\pi) \frac{3 \tan(x)}{x^2 \sin(x)} = 0$$

and the situation is similar for other singular points $n\pi$. Also

$$\lim_{x \to 3\pi} (x - 3\pi)^2 \frac{2}{x^2 \sin(x)} = 0$$

with similar result for arbitrary $n\pi$ where $n \neq 0$. Thus in this case 0 is not a regular singular point but $n\pi$ is a regular singular point for all integers $n \neq 0$.

In general, if you have an equation which has a regular singular point at a so that the equation can be massaged to give something of the form

$$(x - a)^2 y'' + (x - a) b(x) y' + c(x) y = 0$$

you could always define a new variable $t \equiv (x - a)$ and letting $z(t) = y(x)$, you could rewrite the equation in terms of t in the form

$$t^2 z'' + tb(a + t) z' + c(a + t) z = 0$$

and thereby reduce to the case where the regular singular point is at 0. Thus there is no loss of generality in concentrating on the case where the regular singular point is at 0. In addition, the most important examples are like this. Therefore, from now on, I will consider this case. This just means you have all the series in terms of powers of x rather than the more general powers of $x - a$.

8.8 Finding the Solution

Suppose you have reduced the equation to

$$x^2 y'' + xp(x) y' + q(x) y = 0 \tag{8.17}$$

where each of p, q is analytic near 0. Then letting

$$p(x) = b_0 + b_1 x + \cdots$$

$$q(x) = c_0 + c_1 x + \cdots$$

you see that for small x the equation should be approximately equal to

$$x^2 y'' + x b_0 y' + c_0 y = 0$$

which is an Euler equation. This would have a solution in the form x^r where

$$r(r-1) + b_0 r + c_0 = 0,$$

the indicial equation for the Euler equation, and so it is not unreasonable to look for a solution to the equation in (8.17) which is of the form The values of r are called the exponents of the singularity.

$$x^r \sum_{k=0}^{\infty} a_k x^k, \ a_0 \neq 0.$$

You perturb the coefficients of the Euler equation to get (8.17) and so it is not unreasonable to think you should look for a solution to (8.17) of the above form.

Example 8.8.1 *Find the general solution to the equation*

$$x^2 y'' + x\left(1 + x^2\right) y' - 2y = 0.$$

The associated Euler equation is of the form

$$x^2 y'' + xy' - 2y = 0$$

and so the indicial equation is

$$r(r-1) + r - 2 = 0 \tag{8.18}$$

so $r = \sqrt{2}, r = -\sqrt{2}$. Then you would look for a solution in the form

$$y = x^r \sum_{k=0}^{\infty} a_k x^k = \sum_{k=0}^{\infty} a_k x^{k+r}$$

where $r = \pm\sqrt{2}$. Plug in to the equation.

$$x^2 \sum_{k=0}^{\infty} a_k (k+r)(k+r-1) x^{k+r-2}$$

$$+x\left(1+x^2\right)\sum_{k=0}^{\infty}a_k\left(k+r\right)x^{k+r-1}-2\sum_{k=0}^{\infty}a_k x^{k+r}=0$$

This simplifies to

$$\sum_{k=0}^{\infty}a_k\left(k+r\right)\left(k+r-1\right)x^{k+r}+\sum_{k=0}^{\infty}a_k\left(k+r\right)x^{k+r} \qquad (8.19)$$

$$+\sum_{k=0}^{\infty}a_k\left(k+r\right)x^{k+r+2}-2\sum_{k=0}^{\infty}a_k x^{k+r}=0$$

The lowest order term is the x^r term and it yields

$$a_0\left(r\right)\left(r-1\right)+a_0\left(r\right)-2a_0=0$$

but this is just $a_0\left(r\left(r-1\right)+r-2\right)=0$. Since r is one of the zeros of (8.18), there is no restriction on the choice of a_0. In fact, as discussed below, this lack of a requirement on a_0 is equivalent to finding the right value of r. Next consider the x^{r+1} terms. There are no such terms in the third of the above sums just as there were no x^r terms in this sum. Then

$$a_1\left(\left(1+r\right)\left(r\right)+\left(1+r\right)-2\right)=0$$

Now if r solves (8.18) then $1+r$ does not do so because the two solutions to this equation do not differ by an integer. Therefore, the above equation requires $a_1=0$. At this point we can give a recurrence relation for the other a_k. To do this, change the variable of summation in the third sum of (8.19) to obtain

$$\sum_{k=0}^{\infty}a_k\left(k+r\right)\left(k+r-1\right)x^{k+r}+\sum_{k=0}^{\infty}a_k\left(k+r\right)x^{k+r}$$

$$+\sum_{k=2}^{\infty}a_{k-2}\left(k-2+r\right)x^{k+r}-2\sum_{k=0}^{\infty}a_k x^{k+r}=0$$

Thus for $k\geq 2$,

$$a_k\left[\left(k+r\right)\left(k+r-1\right)+\left(k+r\right)-2\right]+a_{k-2}\left(k-2+r\right)=0$$

Hence for $k\geq 2$,

$$a_k=\frac{-a_{k-2}\left(k-2+r\right)}{\left[\left(k+r\right)\left(k+r-1\right)+\left(k+r\right)-2\right]}=\frac{-a_{k-2}\left(k-2+r\right)}{\left[\left(k+r\right)\left(k+r-1\right)+\left(k+r\right)-2\right]}$$

and we take $a_0\neq 0$ while $a_1=0$. Now let's find the first several terms of two independent solutions, one for $r=\sqrt{2}$ and the other for $r=-\sqrt{2}$. Let $a_0=1$ for simplicity. Then the above recurrence relation shows that since $a_1=0$ all the odd terms equal 0. Also

$$a_2=\frac{-r}{\left[\left(2+r\right)\left(2+r-1\right)+\left(2+r\right)-2\right]}=-\frac{r}{\left[\left(2+r\right)\left(1+r\right)+r\right]}$$

while

$$a_4 = \frac{-\left(-\frac{r}{[(2+r)(1+r)+r]}\right)(4-2+r)}{[(4+r)(4+r-1)+(4+r)-2]} = \frac{r}{[2+4r+r^2]}\frac{2+r}{[14+8r+r^2]}$$

Continuing this way, you can get as many terms as you want. Now let's put in the two values of r to obtain the beginning of the two solutions. First let $r = \sqrt{2}$

$$y_1(x) = x^{\sqrt{2}}\left(1 + \left(-\frac{\sqrt{2}}{[(2+\sqrt{2})(\sqrt{2}+1)+\sqrt{2}]}\right)x^2 + \right.$$

$$\left. + \left(\frac{\sqrt{2}}{[4+4\sqrt{2}]}\frac{2+\sqrt{2}}{[16+8\sqrt{2}]}\right)x^4\cdots\right)$$

the solution which corresponds to $r = -\sqrt{2}$ is

$$y_2(x) = x^{-\sqrt{2}}\left(1 + \left(\frac{\sqrt{2}}{[(2-\sqrt{2})(1-\sqrt{2})-\sqrt{2}]}\right)x^2 + \right.$$

$$\left. \sqrt{2}\frac{-2+\sqrt{2}}{[4-4\sqrt{2}][16-8\sqrt{2}]}x^4 + \cdots\right)$$

Then the general solution is

$$C_1 y_1 + C_2 y_2$$

and this is valid for $x > 0$.

Generalities

For an equation

$$x^2 y'' + x p(x) y' + q(x) y = 0$$

having a regular singular point at 0, one looks for solutions in the form

$$y(x) = \sum_{n=0}^{\infty} a_n x^{r+n} \tag{8.20}$$

where r is a constant which is to be determined, in such a way that $a_0 \neq 0$. It turns out that such equations **always** have such solutions although solutions of this sort are not always enough to obtain the general solution to the equation. The constant r is called the exponent of the singularity because the solution is of the form

$$x^r a_0 + \text{ higher order terms.}$$

Thus the behavior of the solution to the equation given above is like x^r for x near the singularity, 0.

If you require that (8.20) solves (8.17) and plug in, you obtain using Theorem 8.6.2

$$\sum_{n=0}^{\infty}(r+n)(r+n-1)a_n x^{n+r} + + \sum_{n=0}^{\infty}\left(\sum_{k=0}^{n} a_k(k+r)b_{n-k}\right)x^{n+r}$$

$$+\sum_{n=0}^{\infty}\left(\sum_{k=0}^{n}c_{n-k}a_k\right)x^{n+r}=0. \tag{8.21}$$

Since $a_0 \neq 0$,

$$p(r) \equiv r(r-1) + b_0 r + c_0 = 0 \tag{8.22}$$

and this is called the indicial equation. (Note it is the indicial equation for the Euler equation which comes from deleting all the nonconstant terms in the power series for $p(x)$ and $q(x)$.) Also the following equation must hold for $n = 1, \cdots$.

$$p(n+r) a_n = -\sum_{k=0}^{n-1} a_k (k+r) b_{n-k} - \sum_{k=0}^{n-1} c_{n-k} a_k \equiv f_n(a_i, b_i, c_i) \tag{8.23}$$

These equations are all obtained by setting the coefficient of x^{n+r} equal to 0.

There are various cases depending on the nature of the solutions to this indicial equation. I will always assume the zeros are real, but will consider the case when the zeros are distinct and do not differ by an integer and the case when the zeros differ by a non negative integer.

It turns out that the nature of the problem changes according to which of these cases holds. You can see why this is the case by looking at the equations (8.22) and (8.23). If r_1, r_2 solve the indicial equation and $r_1 - r_2 \neq$ an integer, then with r in equation (8.23) replaced by either r_1 or r_2, for $n = 1, \cdots$, $p(n+r) \neq 0$ and so there is a unique solution to (8.23) for each $n \geq 1$ once $a_0 \neq 0$ has been chosen. Therefore, in this case that $r_1 - r_2 \neq$ an integer, equation (8.13) has a general solution in the form

$$C_1 \sum_{n=0}^{\infty} a_n x^{n+r_1} + C_2 \sum_{n=0}^{\infty} b_n x^{n+r_2}, a_0, b_0 \neq 0.$$

It is obvious this is the general solution because the ratio of the two solutions is non constant. As pointed out earlier, this requires their Wronskian to be nonzero.

On the other hand, if $r_1 - r_2 =$ an integer, then there exists a unique solution to (8.23) for each $n \geq 1$ if r is replaced by the larger of the two zeros, r_1. Therefore, in this case there is always a solution of the form

$$y_1(x) = \sum_{n=0}^{\infty} a_n x^{n+r_1}, \ a_0 = 1, \tag{8.24}$$

but you might very well hit a snag when you attempt to find a solution of this form with r_1 replaced with the smaller of the two zeros r_2 due to the possibility that for some $m \geq 1$, $p(m + r_2) = p(r_1) = 0$ without the right side of (8.23) vanishing. In the case when both zeros are equal, there is only one solution of the form in (8.24) since there is always a unique solution to (8.23) for $n \geq 1$. Therefore, in the case when $r_1 - r_2 =$ a non negative integer either 0 or some positive integer, you must consider other solutions. I will use Abel's formula to find the second solution. The equation solved by these two solutions is

$$x^2 y'' + x p(x) y' + q(x) y = 0$$

and dividing by x^2 to place in the right form for using Abel's formula,

$$y'' + \frac{1}{x}p(x)y' + \frac{1}{x^2}q(x)y = 0$$

Thus letting y_1 be the solution of the form in (8.24), and y_2 another solution, it follows from the Procedure 4.3.2 that

$$y_2 \in y_1 \int \frac{1}{y_1^2}e^{-P}dx$$

where $P(x) \in \int x^{-1}p(x)\,dx$. Thus

$$P(x) \in \int \left(\frac{b_0}{x} + b_1 + b_2x + \cdots\right)dx = b_0\ln x + b_1x + b_2x^2/2 + \cdots$$

and so

$$-P(x) = \ln x^{-b_0} + k(x)$$

for $k(x)$ some analytic function, $k(0) = 0$. Therefore,

$$e^{-P(x)} = e^{\ln\left(x^{-b_0}\right)+k(x)} = x^{-b_0}g(x)$$

for $g(x)$ some analytic function, $g(0) = 1$. Therefore,

$$y_2 \in y_1(x) \int \frac{1}{y_1^2}\left(x^{-b_0}g(x)\right)dx, \ g(0) = 1. \tag{8.25}$$

Next it is good to understand y_1 and r_1 in terms of b_0. Consider the zeros to the indicial equation,

$$r(r-1) + b_0r + c_0 = r^2 - r + b_0r + c_0 = 0.$$

It is given that $r_1 = r_2 + m$ where m is a non negative integer. Thus the left side of the above equals

$$(r - r_2)(r - r_2 - m) = r^2 - 2rr_2 - rm + r_2^2 + r_2m$$

and so

$$-2r_2 - m = b_0 - 1$$

which implies

$$r_2 = \frac{1 - b_0}{2} - \frac{m}{2}$$

and hence

$$r_1 = r_2 + m = \frac{1 - b_0}{2} + \frac{m}{2}$$

$$y_1(x) = x^{\frac{1-b_0+m}{2}}\sum_{n=0}^{\infty}a_nx^n, \ a_0 = 1 \tag{8.26}$$

Now from Theorem 8.6.1 and looking at (8.26) $y_1(x)^{-2}$ is of the form

$$\frac{1}{x^{1-b_0+m}\left(\sum_{n=0}^{\infty}a_nx^n\right)^2} = x^{b_0-1-m}(1 + h(x))$$

where $h(x)$ is analytic, $h(0) = 0$. Therefore, (8.25) is

$$y_2(x) \in y_1(x) \int x^{b_0 - 1 - m} (1 + h(x)) \left(x^{-b_0} g(x) \right) dx$$

$$y_2(x) \in y_1(x) \int x^{-1-m} (1 + l(x)) dx, \ l(0) = 0 \tag{8.27}$$

Now suppose that $m > 0$. Then,

$$\frac{y_2(x)}{y_1(x)} = \frac{-x^{-m}}{m} + \sum_{n=1}^{m-1} A_n \frac{x^{n-m}}{n-m} + A_m \ln(x) + \sum_{n=m+1}^{\infty} A_n \frac{x^{n-m}}{n-m}.$$

It follows

$$y_2 = A_m \ln(x) y_1 + x^{-m} \left(\frac{-1}{m} + \sum_{n=1}^{\infty} B_n x^n \right) \overbrace{x^{r_1} \sum_{n=0}^{\infty} a_n x^n}^{y_1}.$$

Where $B_n = \frac{A_n}{n-m}$ for $n \neq m$. Therefore, y_2 has the following form.

$$y_2 = A_m \ln(x) y_1 + x^{r_2} \sum_{n=0}^{\infty} C_n x^n.$$

If $m = 0$ so there is a repeated zero to the indicial equation then (8.27) implies

$$\frac{y_2}{y_1} = \ln x + \sum_{n=1}^{\infty} \frac{A_n}{n} x^n + A_0$$

where A_0 is a constant of integration. Thus, the second solution is of the form

$$y_2 = \ln(x) y_1 + x^{r_2} \sum_{n=0}^{\infty} C_n x^n.$$

The following theorem summarizes the above discussion.

Procedure 8.8.2 *Let (8.13) be an equation with a regular singular point and let r_1 and r_2 be real solutions of the indicial equation, (8.22) with $r_1 \geq r_2$. Then if $r_1 - r_2$ is not equal to an integer, the general solution (8.13) may be written in the form :*

$$C_1 \sum_{n=0}^{\infty} a_n x^{n+r_1} + C_2 \sum_{n=0}^{\infty} b_n x^{n+r_2}$$

where we can have $a_0 = 1$ and $b_0 = 1$. If $r_1 = r_2 = r$ then the general solution of (8.13) may be obtained in the form

$$C_1 \overbrace{\sum_{n=0}^{\infty} a_n x^{n+r}}^{y_1} + C_2 \left(\ln(x) \overbrace{\sum_{n=0}^{\infty} a_n x^{n+r}}^{y_1} + \sum_{n=0}^{\infty} C_n x^{n+r} \right)$$

where we may take $a_0 = 1$. If $r_1 - r_2 = m$, a positive integer, then the general solution to (8.13) may be written as

$$C_1 \left(\overbrace{\sum_{n=0}^{\infty} a_n x^{n+r_1}}^{y_1} \right) + C_2 \left(k \ln (x) \overbrace{\left(\sum_{n=0}^{\infty} a_n x^{n+r_1} \right)}^{y_1} + x^{r_2} \sum_{n=0}^{\infty} C_n x^n \right),$$

where k may or may not equal zero and we may take $a_0 = 1$.

This procedure indicates what one should look for in the various cases. In the next section, a different procedure will be described.

8.9   Method of Frobenius

There is a method of Frobenius* which can be used to consider the case where there is a repeated root to the indicial equation or there are two whose difference is an integer which might be easier. The equation of interest here is

$$Ly = x^2 y'' + xp(x) y' + q(x) y = 0$$

where p, q are analytic near 0. Consider the following function

$$y(x, s) = \sum_{k=0}^{\infty} a_k(s) x^{k+s} \tag{8.28}$$

When μ is a root of the indicial equation

$$s(s-1) + p(0) s + q(0) = 0$$

$x \to y(x, \mu)$ will be a solution to the equation $Ly = 0$. In this case $a_0(\mu) \neq 0$ and there are recurrence relations which hold:

$$(k+s)(k+s-1) a_k(s) + \sum_{i=0}^{k} \frac{1}{(k-i)!} p^{(k-i)}(0)(i+s) a_i(s)$$
$$+ \sum_{i=0}^{k} \frac{1}{(k-i)!} q^{(k-i)}(0) a_i(s) = 0, \ k \geq 1 \tag{8.29}$$

where s is replaced with μ to obtain a solution. These recurrence relations come from plugging in the power series to the equation and matching the terms of order $k + s$ for $k \geq 1$ as discussed earlier.

Thus a choice of $a_0(s)$ determines all other $a_k(s)$. Note that the order $k + s$ terms involve

$$((k+s)(k+s-1) + p(0)(k+s) + q(0)) a_k(s)$$
$$= \text{ expression for } a_i(s), i < k$$

---

*Ferdinand Georg Frobenius (1849-1917) is known for his contributions to differential equations and group theory. He is also the first to give a complete proof of the Cayley-Hamilton theorem which says that $q(A) = 0$ whenever $q(\lambda)$ is the characteristic polynomial. His name is also associated with some important topics in matrix theory. Along with Peirce, he proved a very important representation theorem for division algebras.

The expression multiplying $a_k(s)$ is nonzero provided $s + k$ is not a root to the indicial equation, which will be the case if s is close to but not equal to μ. Retain these recurrence relations for arbitrary s near μ, a root to the indicial equation for $k \geq 1$. Thus these recurrence relations imply that Ly is of the form

$$Ly = L\left(\sum_{k=0}^{\infty} a_k(s) x^{k+s}\right) = a_0(s)(s(s-1) + p(0)s + q(0))x^s$$

since the x^{k+s} terms vanish for $k \geq 1$. You would get 0 on the right if $s = \mu$.

Suppose first that

$$s(s-1) + p(0)s + q(0) = (s - \mu)^2$$

so there is a repeated root to the indicial equation. Then

$$Ly = a_0(s)(s - \mu)^2 x^s$$

The idea is as follows. Since L is linear, it is at least formally the case that

$$\frac{\partial}{\partial s}Ly = L\frac{\partial y}{\partial s}$$

Then

$$L\left(\frac{\partial y}{\partial s}\right) = \frac{\partial}{\partial s}(Ly)$$

$$= \left(2(s-\mu)x^s + (s-\mu)^2 \ln(x) x^s\right)a_0(s) + (s-\mu)^2 x^s a_0'(s)$$

Now if you let $s = \mu$, this says that $L\left(\frac{\partial y}{\partial s}\right) = 0$. That is, $\frac{\partial y}{\partial s}|_{s=\mu}$ is a solution to the equation $Ly = 0$.

Will the general solution be obtained from these two solutions, $y(x, \mu)$ and $\frac{\partial y}{\partial s}|_{s=\mu}$? This is clearly true because

$$\frac{\partial}{\partial s}\sum_{k=0}^{\infty} a_k(s) x^{k+s} = \sum_{k=0}^{\infty} (a_k'(s) + a_k(s)\ln(x)) x^{k+s}$$

Then inserting $s = \mu$ the new solution is

$$\sum_{k=0}^{\infty} (a_k'(\mu) + a_k(\mu)\ln(x)) x^{k+s}$$

and this is clearly not a multiple of the first solution since $a_0(\mu) \neq 0$, so the general solution will be a linear combination of these two, the recurrence relation being

$$(k+s)(k+s-1)a_k(s) + \sum_{i=0}^{k} \frac{1}{(k-i)!} p^{(k-i)}(0)(i+s)a_i(s)$$
$$+ \sum_{i=0}^{k} \frac{1}{(k-i)!} q^{(k-i)}(0) a_i(s) = 0, \; k \geq 1$$

Now suppose that there are two solutions to the indicial equation and they are $\mu, \mu - m$ where m is a positive integer. Then there is one solution which is of the form $\sum_{k=0}^{\infty} a_k(\mu) x^{k+\mu}, a_0(\mu) \neq 0$.

Then since the two roots are $\mu, \mu - m$, you have the indicial polynomial of the form

$$I(s) = (s - \mu)(s - (\mu - m)) = (s - \mu)^2 + m(s - \mu)$$

Then in (8.28), assign $a_0(s) \equiv s - \mu + m$ with all the other terms in the series equal zero because of a suitable recurrence relation. Then one would have

$$Ly = \left((s - \mu)^2 + m(s - \mu)\right) \left(\overbrace{s - \mu + m}^{a_0(s) \neq 0}\right) x^s = (s - \mu)(s - \mu + m)^2 x^s$$

Then in addition to the standard solution coming from the series of (8.28) in which $s = \mu$ and the necessary recurrence relations hold, one can consider the solution $\partial y / \partial s$. Then as above, since L is linear,

$$L \frac{\partial y}{\partial s} = \frac{\partial}{\partial s} Ly = \frac{\partial}{\partial s}\left((s - \mu)(s - \mu + m)^2 x^s\right)$$

$$= \frac{\partial}{\partial s}\left((s - \mu)(s - \mu + m)^2\right) x^s + \left((s - \mu)(s - \mu + m)^2\right) \ln(x) x^s$$

Then if s is assigned to equal $\mu - m$ this gives 0 because in the first term, you have $(s - \mu + m)^2$. Thus you have another solution. What of this other solution? Is it a multiple of the first, or can it be used with the first to give the general solution? In this case of two roots, $\mu, \mu - m$, you get the following using $a_0(s) = s - \mu + m$.

$$y(x, s) = \sum_{k=0}^{\infty} a_k(s) x^{k+s}$$

$$\frac{\partial}{\partial s} y(x, s) = \left(x^s + \left(\overbrace{s - \mu + m}^{a_0(s)}\right) \ln(x) x^s\right) + \sum_{k=1}^{\infty} (a_k'(s) + a_k(s) \ln(x)) x^{k+s}$$

Then when s is assigned to equal $\mu - m$, this yields the new solution

$$x^{\mu - m} + \sum_{k=1}^{\infty} (a_k'(\mu - m)(\mu - m) + a_k(\mu - m)(\mu - m) \ln(x)) x^{k+\mu - m}$$

Will this be a multiple of the original solution? No, it can't be because the lowest order term of the original solution is $a_0(\mu)x^\mu$ so one of these is not a multiple of the other. This justifies the following procedure.

Procedure 8.9.1 *To find the general solution to*

$$x^2 y'' + xp(x) y' + q(x) y = 0$$

let

$$y(x, s) \equiv \sum_{k=0}^{\infty} a_k(s) x^{k+s}$$

where $\sum_{k=0}^{\infty} a_k(\mu) x^{k+\mu}$ is a solution to the differential equation.

If the roots to the indicial equation,

$$s(s-1) + p(0)s + q(0) = 0$$

μ, λ *are different and do not differ by an integer, then there will be a general solution of the form*

$$C_1 y(x, \mu) + C_2 y(x, \lambda)$$

where the lowest order terms satisfy $a_0(\mu), a_0(\lambda) = 1$.

If the roots are the same, μ, *then there will be a general solution of the form*

$$C_1 y(x, \mu) + C_2 \frac{\partial y}{\partial s}(x, \mu)$$

where $a_0(s) = 1$.

If the roots are μ *and* $\mu - m$ *for* m *a positive integer, then in the series for* y, *let*

$$a_0(s) \equiv (s - \mu + m) = (s - (\mu - m))$$

There will be a general solution of the form

$$C_1 y(x, \mu) + C_2 \frac{\partial y}{\partial s}(x, \mu - m)$$

8.10 The Bessel Equations

The Bessel differential equations are

$$x^2 y'' + xy' + (x^2 - \nu^2) y = 0$$

Obviously this has a regular singular point at 0 and from (8.16), the indicial equation is

$$r(r-1) + r - \nu^2 = r^2 - \nu^2 = 0$$

Thus the two indices of singularity are $\pm \nu$. There are various cases according to whether ν is 0, not an integer, or an integer.

8.10.1 The Case where $\nu = 0$

First consider the case where $\nu = 0$. In this case, there exists a solution of the form $\sum_{n=0}^{\infty} a_n x^n$ and it is required to find the constants a_n. Plugging into the equation one gets

$$x^2 \sum_{n=0}^{\infty} a_n n(n-1) x^{n-2} + x \sum_{n=0}^{\infty} a_n n x^{n-1} + \sum_{n=0}^{\infty} a_n x^{n+2} = 0$$

Then change the variable of summation in the last sum. This yields

$$\sum_{n=0}^{\infty} a_n n(n-1) x^n + \sum_{n=0}^{\infty} a_n n x^n + \sum_{n=2}^{\infty} a_{n-2} x^n = 0$$

It follows that there is no restriction on a_0, a_1 but for $n \geq 2$,

$$a_n \left(n \left(n - 1 \right) + n \right) + a_{n-2} = a_n n^2 + a_{n-2} = 0$$

Thus $a_n = -\frac{a_{n-2}}{n^2}$.

Taking $a_0 = 1, a_1 = 0$, it follows that all odd terms equal 0 and

$$a_2 = \frac{-1}{4}, a_4 = \frac{1}{2^2} \frac{1}{4^2}, a_6 = -\frac{1}{2^2} \frac{1}{4^2} \frac{1}{6^2}, \cdots$$

The pattern is now fairly clear:

$$a_{2n} = (-1)^n \frac{1}{2^n \left(n! \right)^2}$$

Then this solution is

$$J_0 \left(x \right) = \sum_{k=0}^{\infty} (-1)^k \frac{1}{2^k \left(k! \right)^2} x^{2k} \tag{8.30}$$

Then by Theorem 8.8.2, the general solution is of the form

$$C_1 J_0 \left(x \right) + C_2 \left(\ln \left(x \right) J_0 \left(x \right) + \sum_{n=0}^{\infty} C_n x^n \right)$$

for suitable choice of the C_n. Thus one is bounded near $x = 0$ and the other is unbounded near $x = 0$. In fact, it is customary to let the second solution be a complicated linear combination of these two solutions. When this is done, the function which results is known as $Y_0 \left(x \right)$. Then $J_0 \left(x \right)$ is the Bessel function of the first kind and the $Y_0 \left(x \right)$ is called the Bessel function of the second kind. Here are graphs of these functions.

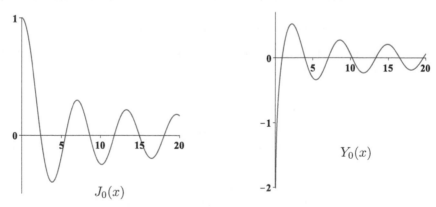

$J_0(x)$ $Y_0(x)$

8.10.2   The Case of ν Not an Integer

Next consider the case where ν is not an integer. This time, the series is of the form

$$\sum_{n=0}^{\infty} a_n x^{n+\nu}$$

Substituting into the equation,

$$x^2 \sum_{n=0}^{\infty} a_n (n + \nu) (n + \nu - 1) x^{n+\nu-2} + x \sum_{n=0}^{\infty} a_n (n + \nu) x^{n+\nu-1}$$

$$+ \sum_{n=0}^{\infty} a_n x^{n+\nu+2} - \sum_{n=0}^{\infty} \nu^2 a_n x^{n+\nu} = 0$$

Thus a little simplification yields

$$\sum_{n=0}^{\infty} a_n (n + \nu)^2 x^{n+\nu} + \sum_{n=2}^{\infty} a_{n-2} x^{n+\nu} - \sum_{n=0}^{\infty} \nu^2 a_n x^{n+\nu} = 0$$

Then we need to have $a_1 = 0$ but let $a_0 = 1$. Then for $n \geq 2$,

$$a_n \left((n + \nu)^2 - \nu^2 \right) = -a_{n-2} \text{ so } a_n = \frac{-a_{n-2}}{(n + \nu)^2 - \nu^2} = \frac{-a_{n-2}}{n (n + 2\nu)} \qquad (8.31)$$

Thus all the odd terms are 0 and the first several terms are as follows.

$$a_0 = 1, \ a_2 = -\frac{1}{2 (2 + 2\nu)}, \ a_4 = \frac{1}{2 (2 + 2\nu)} \frac{1}{4 (4 + 2\nu)}, \cdots$$

The pattern seems clear at this point. Thus

$$a_{2n} = \frac{(-1)^n \, 1}{(2 \cdot 4 \cdot \cdots \cdot 2n) (2 + 2\nu) (4 + 2\nu) \cdots (2n + 2\nu)}$$

$$= \frac{(-1)^n \, 1}{2^{2n} n! \, (1 + \nu) (2 + \nu) \cdots (n + \nu)}$$

That product $(1 + \nu) (2 + \nu) \cdots (n + \nu)$ in the bottom will be denoted as $(n + \nu)_n$. Then this reduces to

$$a_{2n} = \frac{(-1)^n}{2^{2n} n! \, (n + \nu)_n}$$

Thus a solution corresponding to ν is

$$x^\nu + \sum_{k=1}^{\infty} \frac{(-1)^k}{2^{2k} k! \, (k + \nu)_k} x^{2k+\nu}$$

Then this is massaged a little more. It is multiplied by the constant $\frac{1}{\Gamma(\nu+1)2^\nu}$. Recall that the Gamma function satisfies

$$\Gamma (\alpha) \alpha = \Gamma (\alpha + 1)$$

Applying this rule repeatedly in the above sum yields

$$J_\nu (x) = \sum_{k=0}^{\infty} \frac{(-1)^k}{k! \Gamma (k + \nu + 1)} \left(\frac{x}{2} \right)^{2k+\nu}$$

You can verify directly that

$$J_{-\nu}(x) = \sum_{k=0}^{\infty} \frac{(-1)^k}{k!\Gamma(k-\nu+1)} \left(\frac{x}{2}\right)^{2k-\nu}$$

is also a solution to the Bessel equation. The definition of $\Gamma(k-\nu+1)$ when the argument is negative is defined in terms of the property of the gamma function which was responsible for making $J_\nu(x)$ be a solution, $\Gamma(\alpha+1) = \alpha\Gamma(\alpha)$. Thus, for example, if $-\nu+1 < 0, \Gamma(-\nu+1)(-\nu+1)\cdots(-\nu+m) = \Gamma(-\nu+1+m)$ where m is large enough that $-\nu+1+m > 0$. Since ν is not an integer, $-\nu+k$ is never zero so there is never a difficulty in encountering something which does not make sense.

The Bessel function of the first kind J_ν converges to 0 as $x \to 0+$ while $J_{-\nu}$ is unbounded as $x \to 0+$. Consequently, their ratio cannot be a constant and so the general solution is obtained as linear combinations of these two solutions. Of course everything changes if ν is a positive integer. In this case, the second solution fails to even make sense because you could have $k - \nu = 0$ and $\Gamma(0)$ is not even defined.

In fact, what people tabulate is a linear combination of these two solutions

$$Y_\nu(x) \equiv \frac{\cos(\pi\nu) J_\nu(x) - J_{-\nu}(x)}{\sin(\pi\nu)}$$

It is called the Weber function or the Neumann function. The main thing to notice here is that it is unbounded as $x \to 0$.

8.10.3   Case Where ν is an Integer

Let $\nu = m$ a positive integer. Then you still get one solution which is of the form

$$x^m + \sum_{k=1}^{\infty} \frac{(-1)^k}{2^{2k} k! (k+m)_k} x^{2k+m}$$

and multiplying by a constant as above, you can obtain

$$J_m(x) = \sum_{k=0}^{\infty} \frac{(-1)^k}{k!\Gamma(k+m+1)} \left(\frac{x}{2}\right)^{2k+m} = \sum_{k=0}^{\infty} \frac{(-1)^k}{k!(k+m)!} \left(\frac{x}{2}\right)^{2k+m}$$

as one solution. In fact, you could consider simply replacing m with $-m$ in the above, but this will not work out. It won't work out roughly because $\Gamma(k-m+1) = \pm\infty$ for $k+1 \leq m$. Thus the sum will reduce to

$$\sum_{k=m}^{\infty} \frac{(-1)^k}{k!\Gamma(k-m+1)} \left(\frac{x}{2}\right)^{2k-m}$$

Changing the variable of summation to $k = j + m$, this becomes

$$(-1)^m \sum_{j=0}^{\infty} \frac{(-1)^j}{(j-m)!\Gamma(j+1)} \left(\frac{x}{2}\right)^{2j+m} = (-1)^m \sum_{j=0}^{\infty} \frac{(-1)^j}{\Gamma(j-m+1)j!} \left(\frac{x}{2}\right)^{2j+m}$$

$$= (-1)^m J_m(x)$$

so the new solution obtained by replacing m with $-m$ is nothing more than $(-1)^m$ times the old solution. It follows that there is no way to obtain the general solution as a linear combination of these two. The second solution must involve a logarithmic term and will therefore, be unbounded near 0. However, it is convenient to define

$$J_{-m}(x) \equiv (-1)^m J_m(x)$$

The way this is dealt with is to define the second solution as

$$Y_m(x) \equiv \lim_{\nu \to m} Y_\nu(x)$$

because the limit does exist for all $x > 0$.

One other important consideration is easy to get which is that the solutions to Bessel's equation must oscillate about 0 like sines and cosines.

Proposition 8.10.1 *Let y be a solution of the Bessel equation*

$$x^2 y'' + x y' + (x^2 - \nu^2) y = 0$$

Then y has infinitely many zeros.

Proof: Use the transformation of Problem 25 on Page 103. This changes the independent variable to s where $x = e^s$. Thus, letting $y(s) = y(x)$, it will suffice to show that $s \to y(s)$ has infinitely many zeros. Doing the transformation yields the following differential equation for $s \to y(s)$

$$y''(s) + (e^{2s} - \nu^2) y(s) = 0$$

Obviously for all s large enough, $e^{2s} - \nu^2 > 1$. Consider now the equation

$$z'' + z = 0$$

The idea is to show that if a, b are successive zeros of z for a large enough that for $s > a$, $e^{2s} - \nu^2 > 1$ it follows that y must have a zero in $[a, b]$. Since z has infinitely many zeros, it follows that so does y.

Without loss of generality, assume z is positive on (a, b). If it isn't, multiply by -1 to make this happen. The solution z is a linear combination of sines and cosines. It can be written in the form

$$z = A \cos(s - \phi)$$

and so $z'(a) > 0$ and $z'(b) < 0$.

If y has no zeros on $[a, b]$, then again, without loss of generality, let y be positive on $[a, b]$.

$$z'' y - y'' z + (1 - (e^{2s} - \nu^2)) yz = 0$$

Thus on the open interval (a, b),

$$W(y, z)' = ((e^{2s} - \nu^2) - 1) yz > 0$$

where $W(y, z)$ is the Wronskian. It follows from the mean value theorem that $W(y, z)(a) < W(y, z)(b)$. Then

$$\begin{vmatrix} y(a) & 0 \\ y'(a) & z'(a) \end{vmatrix} < \begin{vmatrix} y(b) & 0 \\ y'(b) & z'(b) \end{vmatrix}$$

$$\text{positive} = y(a) z'(a) < y(b) z'(b) = \text{ negative},$$

a contradiction. ∎

For the purposes of this book, this will suffice. The main message is that there are two independent solutions, one bounded near 0 and the other unbounded as described above. Both oscillate about 0 and have infinitely many zeros. In many applications, the unbounded one is of no interest based on physical considerations.

8.11   Other Properties of Bessel Functions

Recall that for m a nonnegative integer,

$$J_m(x) = \sum_{k=0}^{\infty} \frac{(-1)^k}{k!(k+m)!} \left(\frac{x}{2}\right)^{2k+m}$$

and that if $-m$ is negative,

$$J_{-m}(x) = (-1)^m J_m(x)$$

This $J_m(x)$ was the bounded solution for the Bessel equation. Note that an infinite sum of these functions is absolutely convergent. Indeed,

$$
\begin{aligned}
|J_m(x)| &\leq \sum_{k=0}^{\infty} \frac{1}{k!m!} \left(\left|\frac{x}{2}\right|\right)^{2k+m} \leq \frac{1}{m!} \left|\frac{x}{2}\right|^m \sum_{k=0}^{\infty} \frac{1}{k!} \left(\left(\frac{x}{2}\right)^2\right)^k \\
&= \frac{1}{m!} \left|\frac{x}{2}\right|^m \exp\left(x^2/4\right)
\end{aligned}
\tag{8.32}
$$

Therefore, it is permissible to sum the various series which result in what follows in any order desired. See Section C.3.

Now for $t \neq 0$,

$$e^{\frac{xt}{2}} = \sum_{l=0}^{\infty} \frac{1}{l!} \left(\frac{x}{2}\right)^l t^l, \quad e^{-\frac{x}{2t}} = \sum_{k=0}^{\infty} \frac{1}{k!} (-1)^k \left(\frac{x}{2}\right)^k t^{-k}$$

We multiply these two series. This will involve many terms which can be added in any order thanks to absolute convergence, Section C.3. To get t^m for $m \geq 0$, you need to multiply terms $l = m + k$ times the term for t^{-k} in the second sum. Thus you get for this term

$$t^m \sum_{k=0}^{\infty} \frac{1}{(m+k)!} \left(\frac{x}{2}\right)^{m+k} \frac{1}{k!} (-1)^k \left(\frac{x}{2}\right)^k$$

$$= t^m \sum_{k=0}^{\infty} (-1)^k \frac{1}{k!(m+k)!} \left(\frac{x}{2}\right)^{2k+m} = t^m J_m(x)$$

This gives the terms t^m for $m \geq 0$.

What of the terms involving $m < 0$? To get these terms, you need to have $l - k = m$ so you need $k = l - m$. Thus the sum which results for these terms is

$$t^m \sum_{l=0}^{\infty} \frac{1}{l!} \left(\frac{x}{2}\right)^l \frac{1}{(l-m)!} (-1)^{l-m} \left(\frac{x}{2}\right)^{l-m} = t^m \sum_{l=0}^{\infty} \frac{(-1)^{l-m}}{l!(l-m)!} \left(\frac{x}{2}\right)^{2l-m}$$

$$= (-1)^m t^m \sum_{l=0}^{\infty} \frac{(-1)^l}{l! (l-m)!} \left(\frac{x}{2}\right)^{2l-m} = (-1)^m t^m J_{-m}(x)$$

Therefore,

$$e^{\frac{xt}{2}} e^{-\frac{x}{2t}} = e^{(x/2)(t-1/t)}$$

must equal the sum of t^m terms for $m \geq 0$ and the sum of t^m terms for $m < 0$. It follows that

$$
\begin{aligned}
e^{(x/2)(t-1/t)} &= J_0(x) + \sum_{m=1}^{\infty} t^m J_m(x) + \sum_{m=1}^{\infty} (-1)^m t^{-m} J_m(x) \\
&= J_0(x) + \sum_{m=1}^{\infty} J_m(x) \left(t^m + (-1)^m t^{-m}\right)
\end{aligned}
$$

Now recall that $J_{-m}(x) = (-1)^m J_m(x)$ and so

$$e^{(x/2)(t-1/t)} = J_0(x) + \sum_{m=1}^{\infty} J_m(x) t^m + \sum_{m=1}^{\infty} t^{-m} J_{-m}(x) = \sum_{m=-\infty}^{\infty} t^m J_m(x)$$

That is, $J_m(x)$ is just the mth coefficient of the series for $e^{(x/2)(t-1/t)}$. This has proved the following interesting result on the generating function for Bessel equations.

Theorem 8.11.1 *For m an integer and $J_m(x) = (-1)^m J_{-m}(x)$, we have the following generating function for these Bessel functions.*

$$e^{(x/2)(t-1/t)} = \sum_{m=-\infty}^{\infty} t^m J_m(x) \tag{8.33}$$

In addition to this, there is an addition formula

$$J_m(x+y) = \sum_{k=-\infty}^{\infty} J_{m-k}(x) J_k(y) \tag{8.34}$$

Proof: It remains to obtain the above addition formula. This is remarkably easy to obtain.

$$e^{((x+y)/2)(t-1/t)} = \sum_{m=-\infty}^{\infty} t^m J_m(x+y)$$

$$e^{((x+y)/2)(t-1/t)} = e^{(x/2)(t-1/t)} e^{(y/2)(t-1/t)}$$

$$= \sum_{l=-\infty}^{\infty} t^l J_l(x) \sum_{k=-\infty}^{\infty} t^k J_k(y)$$

and in this product, the t^m term is the sum of products for which $l + k = m$. That is,

$$J_m(x+y) = \sum_{k=-\infty}^{\infty} J_k(y) J_{m-k}(x)$$

This shows the addition formula. ∎

Of course t was completely arbitrary as long as it is not zero. Thus let it equal $e^{i\theta}$ in (8.33). Then from Euler's identity, $e^{i\theta} = (\cos(\theta) + i\sin(\theta))$,

$$e^{(x/2)(2i\sin\theta)} = \sum_{m=-\infty}^{\infty} \left(e^{i\theta}\right)^m J_m(x)$$

Then using Euler's identity again,

$$\cos(x\sin(\theta)) + i\sin(x\sin(\theta)) = \sum_{m=-\infty}^{\infty} (\cos(m\theta) + i\sin(m\theta)) J_m(x)$$

Equating real and imaginary parts,

$$\cos(x\sin(\theta)) = \sum_{m=-\infty}^{\infty} \cos(m\theta) J_m(x)$$

$$\sin(x\sin(\theta)) = \sum_{m=-\infty}^{\infty} \sin(m\theta) J_m(x)$$

Now recall from trig. identities,

$$\cos(a)\cos(b) + \sin(a)\sin(b) = \cos(a - b)$$

multiply the top by $\cos(n\theta)$ and the bottom by $\sin(n\theta)$ and add. Thus

$$\cos(n\theta - x\sin(\theta)) = \sum_{m=-\infty}^{\infty} \cos(n\theta - m\theta) J_m(x)$$

Because of the uniform convergence of the partial sums of the above series which follows from computations like those in (8.32), one can interchange $\int_0^\pi$ with the infinite summation, Problem 16 on Page 171. This yields

$$\int_0^\pi \cos(n\theta - x\sin(\theta)) \, d\theta = \pi J_n(x)$$

because, unless $n = m$, $\int_0^\pi \cos(n\theta - m\theta) \, d\theta = 0$. Therefore, this yields the very important integral identity for $J_n(x)$,

$$J_n(x) = \frac{1}{\pi} \int_0^\pi \cos(n\theta - x\sin(\theta)) \, d\theta \qquad (8.35)$$

The interchange of the integral with the summation follows from noting that the sums of the form $\sum_{m=-k}^{k} \cos(n\theta - m\theta) J_m(x)$ converge uniformly on $[0, \pi]$ to the infinite sum thanks to the $1/m!$ in the estimates of (8.32). Thus, from the fact that the integral is linear,

$$\int_0^\pi \sum_{m=-\infty}^{\infty} \cos(n\theta - m\theta) J_m(x) \, d\theta = \int_0^\pi \lim_{k\to\infty} \sum_{m=-k}^{k} \cos(n\theta - m\theta) J_m(x) \, d\theta$$

$$= \lim_{k\to\infty} \sum_{m=-k}^{k} \int_0^\pi \cos(n\theta - m\theta) J_m(x) \, d\theta = \sum_{m=-\infty}^{\infty} \int_0^\pi \cos(n\theta - m\theta) J_m(x) \, d\theta$$

Theorem 8.11.2 *Let n be a positive integer. Then*

$$J_n(x) = \frac{1}{\pi} \int_0^\pi \cos(n\theta - x\sin(\theta)) \, d\theta$$

How do you compute $J_n(x)$? You can't get it the usual way very conveniently because the leading term vanishes at 0. This integral will give an easy way to do it. For example,

$$J_4(6) = \frac{1}{\pi} \int_0^\pi \cos(4\theta - 6\sin(\theta)) \, d\theta = 0.357\,64$$

I just did the integral numerically in Scientific Notebook and got the answer easily. One can also produce a graph of $x \to J_4(x)$ very easily in this software by graphing the function of x given by $\frac{1}{\pi} \int_0^\pi \cos(4\theta - x\sin(\theta)) \, d\theta$. To do this, you simply type the expression in math mode and then select plot 2d. It has to work at it a little but will produce the graph. It knows that the variable is x and acts accordingly. In the exercises is a problem on how to do this in MATLAB. It is more elaborate.

There are whole books written on Bessel functions, [12].

8.12 Hermite Polynomials

A related topic to the above is the Hermite polynomials. Here is a version in which they depend on a parameter. They will be presented and then some of their properties are developed. It turns out that they are solutions of an ordinary differential equation having variable coefficients. These polynomials have some remarkable properties including those presented here which make them of great interest in probability theory.

From calculus, the product rule implies the following Leibniz rule. It is left as an exercise.

Theorem 8.12.1 *Suppose f, g are functions which have n derivatives. Then*

$$\frac{d^n}{dx^n}(fg) = \sum_{k=0}^n \binom{n}{k} f^{(n-k)} g^{(k)}$$

Here

$$\binom{n}{k} = \frac{n!}{k!(n-k)!}$$

Definition 8.12.2 *You can also consider Hermite polynomials which depend on λ. These are defined as follows:*

$$H_n(x, \lambda) \equiv \frac{(-\lambda)^n}{n!} e^{\frac{1}{2\lambda}x^2} \frac{\partial^n}{\partial x^n}\left(e^{-\frac{1}{2\lambda}x^2}\right)$$

You can see clearly that these are polynomials in x. For example, let $n = 2$. Then you would have from the above definition.

$$H_0(x, \lambda) = 1, \quad H_1(x, \lambda) = \frac{(-\lambda)^1}{1!} e^{\frac{1}{2\lambda}x^2} \frac{\partial}{\partial x}\left(e^{-\frac{1}{2}\frac{x^2}{\lambda}}\right) = x$$

$$H_2\left(x,\lambda\right) \equiv \frac{\left(-\lambda\right)^2}{2!} e^{\frac{1}{2\lambda}x^2} \frac{\partial^2}{\partial x^2}\left(e^{-\frac{1}{2\lambda}x^2}\right) = \frac{1}{2}x^2 - \frac{1}{2}\lambda$$

The idea is you end up with polynomials of degree n times $e^{-x^2/2\lambda}$ in the derivative part and then this cancels with $e^{x^2/2\lambda}$ to leave you with a polynomial of degree n. Also the leading term will always be $\frac{x^n}{n!}$ which is easily seen from the above. Then there are some relationships satisfied by these.

Say $n > 1$ in what follows.

$$\frac{\partial}{\partial x}H_n\left(x,\lambda\right) = \frac{x\left(-\lambda\right)^n e^{\frac{1}{2}\frac{x^2}{\lambda}}}{\lambda} \frac{\partial^n}{\partial x^n}\left(e^{-\frac{1}{2\lambda}x^2}\right) + \frac{\left(-\lambda\right)^n}{n!} e^{\frac{1}{2\lambda}x^2} \frac{\partial^n}{\partial x^n}\left(\frac{\partial}{\partial x}e^{-\frac{1}{2\lambda}x^2}\right)$$

$$= \frac{x\left(-\lambda\right)^n e^{\frac{1}{2}\frac{x^2}{\lambda}}}{\lambda} \frac{\partial^n}{\partial x^n}\left(e^{-\frac{1}{2\lambda}x^2}\right) + \frac{\left(-\lambda\right)^n}{n!} e^{\frac{1}{2\lambda}x^2} \frac{\partial^n}{\partial x^n}\left(-\frac{x}{\lambda}e^{-\frac{1}{2}\frac{x^2}{\lambda}}\right)$$

Now since $n > 1$, that last term reduces to

$$\frac{\left(-\lambda\right)^n}{n!} e^{\frac{1}{2\lambda}x^2}\left[-\frac{x}{\lambda}\frac{\partial^n}{\partial x^n}\left(e^{-\frac{1}{2}\frac{x^2}{\lambda}}\right) + n\frac{\partial}{\partial x}\left(-\frac{x}{\lambda}\right)\frac{\partial^{n-1}}{\partial x^{n-1}}\left(e^{-\frac{1}{2}\frac{x^2}{\lambda}}\right)\right]$$

this by Leibniz formula. Thus this cancels with the first term to give

$$\frac{\partial}{\partial x}H_n\left(x,\lambda\right) = \frac{\left(-\lambda\right)^n n}{n!}\left(-\frac{1}{\lambda}\right) e^{\frac{1}{2\lambda}x^2} \frac{\partial^{n-1}}{\partial x^{n-1}}\left(e^{-\frac{1}{2}\frac{x^2}{\lambda}}\right)$$

$$= \frac{\left(-\lambda\right)^{n-1}}{\left(n-1\right)!} e^{\frac{1}{2\lambda}x^2} \frac{\partial^{n-1}}{\partial x^{n-1}}\left(e^{-\frac{1}{2}\frac{x^2}{\lambda}}\right) \equiv H_{n-1}\left(x,\lambda\right)$$

In case of $n = 1$, this appears to also work. $\frac{\partial}{\partial x}H_1\left(x,\lambda\right) = 1 = H_0\left(x,\lambda\right)$ from the above computations. This shows that

$$\frac{\partial}{\partial x}H_n\left(x,\lambda\right) = H_{n-1}\left(x,\lambda\right)$$

Next, is the claim that

$$\left(n+1\right)H_{n+1}\left(x,\lambda\right) = xH_n\left(x,\lambda\right) - \lambda H_{n-1}\left(x,\lambda\right)$$

If $n = 1$, this says that

$$2H_2\left(x,\lambda\right) = xH_1\left(x,\lambda\right) - \lambda H_0\left(x,\lambda\right) = x^2 - \lambda$$

and so the formula does indeed give the correct description of $H_2\left(x,\lambda\right)$ when $n = 1$. Thus assume $n > 1$ in what follows. The left side equals

$$\frac{\left(-\lambda\right)^{n+1}}{n!} e^{\frac{1}{2\lambda}x^2} \frac{\partial^{n+1}}{\partial x^{n+1}}\left(e^{-\frac{1}{2\lambda}x^2}\right)$$

This equals

$$\frac{\left(-\lambda\right)^{n+1}}{n!} e^{\frac{1}{2\lambda}x^2} \frac{\partial^n}{\partial x^n}\left(-\frac{x}{\lambda}e^{-\frac{1}{2}\frac{x^2}{\lambda}}\right)$$

Now by Liebniz formula,

$$
= \frac{(-\lambda)^{n+1}}{n!} e^{\frac{1}{2\lambda}x^2} \left[-\frac{x}{\lambda} \frac{\partial^n}{\partial x^n} e^{-\frac{1}{2}\frac{x^2}{\lambda}} + n \left(\frac{-1}{\lambda} \right) \frac{\partial^{n-1}}{\partial x^{n-1}} \left(e^{-\frac{1}{2}\frac{x^2}{\lambda}} \right) \right]
$$

$$
= \frac{(-\lambda)^{n+1}}{n!} e^{\frac{1}{2\lambda}x^2} \left(-\frac{x}{\lambda} \frac{\partial^n}{\partial x^n} e^{-\frac{1}{2}\frac{x^2}{\lambda}} \right) + \frac{(-\lambda)^{n+1}}{n!} e^{\frac{1}{2\lambda}x^2} n \left(\frac{-1}{\lambda} \right) \frac{\partial^{n-1}}{\partial x^{n-1}} \left(e^{-\frac{1}{2}\frac{x^2}{\lambda}} \right)
$$

$$
= x \frac{(-\lambda)^n}{n!} e^{\frac{1}{2\lambda}x^2} \frac{\partial^n}{\partial x^n} e^{-\frac{1}{2}\frac{x^2}{\lambda}} + \frac{(-\lambda)^n}{(n-1)!} e^{\frac{1}{2\lambda}x^2} \frac{\partial^{n-1}}{\partial x^{n-1}} \left(e^{-\frac{1}{2}\frac{x^2}{\lambda}} \right)
$$

$$
= x H_n(x, \lambda) - \lambda H_{n-1}(x, \lambda)
$$

which shows the formula is valid for all $n \geq 1$.

Next is the claim that

$$
H_n(-x, \lambda) = (-1)^n H_n(x, \lambda)
$$

This is easy to see from the observation that

$$
\frac{\partial}{\partial x} = \frac{\partial}{\partial(-x)} (-1)
$$

Thus if it involves n derivatives, you end up multiplying by $(-1)^n$.

Finally is the claim that

$$
\frac{\partial}{\partial \lambda} H_n(x, \lambda) = -\frac{1}{2} \frac{\partial^2}{\partial x^2} H_n(x, \lambda)
$$

It is certainly true for $n = 0, 1, 2$. So suppose it is true for all $k \leq n$. Then from earlier claims and induction,

$$
(n+1) H_{(n+1)\lambda}(x, \lambda) = x H_{n\lambda}(x, \lambda) - H_{(n-1)}(x, \lambda) - \lambda H_{(n-1)\lambda}(x, \lambda)
$$

$$
= x \left(\frac{-1}{2} \right) H_{nxx} - H_{n-1} + \lambda \frac{1}{2} H_{(n-1)xx} = x \left(\frac{-1}{2} \right) H_{n-2} - H_{n-1} + \lambda \frac{1}{2} H_{(n-3)}
$$

$$
= -\frac{1}{2} (x H_{n-2} - \lambda H_{n-3} + 2H_{n-1}) = -\frac{1}{2} ((n-1) H_{n-1} + 2H_{n-1})
$$

$$
= -\frac{1}{2} ((n+1) H_{n-1})
$$

comparing the ends,

$$
H_{(n+1)\lambda} = -\frac{1}{2} H_{n-1} = -\frac{1}{2} H_{(n+1)xx}
$$

This proves the following theorem.

Theorem 8.12.3 *Let $H_n(x, \lambda)$ be defined by*

$$
H_n(x, \lambda) \equiv \frac{(-\lambda)^n}{n!} e^{\frac{1}{2\lambda}x^2} \frac{\partial^n}{\partial x^n} \left(e^{-\frac{1}{2\lambda}x^2} \right)
$$

for $\lambda > 0$. Then the following properties are valid.

$$\frac{\partial}{\partial x} H_n\left(x, \lambda\right) = H_{n-1}\left(x, \lambda\right) \tag{8.36}$$

$$\left(n + 1\right) H_{n+1}\left(x, \lambda\right) = x H_n\left(x, \lambda\right) - \lambda H_{n-1}\left(x, \lambda\right) \tag{8.37}$$

$$H_n\left(-x, \lambda\right) = \left(-1\right)^n H_n\left(x, \lambda\right) \tag{8.38}$$

$$\frac{\partial}{\partial \lambda} H_n\left(x, \lambda\right) = -\frac{1}{2} \frac{\partial^2}{\partial x^2} H_n\left(x, \lambda\right) \tag{8.39}$$

With this theorem, one can also prove the following.

Theorem 8.12.4 *The Hermite polynomials are the coefficients of a certain power series. Specifically,*

$$\exp\left(tx - \frac{1}{2}t^2\lambda\right) = \sum_{n=0}^{\infty} H_n\left(x, \lambda\right) t^n$$

Proof: Replace H_n with K_n which really are the coefficients of the power series and then show $K_n = H_n$. Thus

$$\exp\left(tx - \frac{1}{2}t^2\lambda\right) = \sum_{n=0}^{\infty} K_n\left(x, \lambda\right) t^n$$

Then $K_0 = 1 = H_0\left(x\right)$. Also $K_1\left(x\right) = x = H_1\left(x\right)$.

$$\frac{\partial}{\partial t}\left(\exp\left(tx - \frac{1}{2}t^2\lambda\right)\right) = \exp\left(tx - \frac{1}{2}t^2\lambda\right)\left(x - t\lambda\right)$$

$$= \sum_{n=0}^{\infty} x K_n\left(x, \lambda\right) t^n - \sum_{n=0}^{\infty} \lambda K_n\left(x, \lambda\right) t^{n+1} = \sum_{n=0}^{\infty} x K_n\left(x, \lambda\right) t^n - \sum_{n=1}^{\infty} \lambda K_{n-1}\left(x, \lambda\right) t^n$$

Also,

$$\frac{\partial}{\partial t}\left(\exp\left(tx - \frac{1}{2}t^2\lambda\right)\right) = \sum_{n=1}^{\infty} n K_n\left(x, \lambda\right) t^{n-1} = \sum_{n=0}^{\infty} \left(n + 1\right) K_{n+1}\left(x, \lambda\right) t^n$$

It follows that for $n \geq 1$,

$$\left(n + 1\right) K_{n+1}\left(x, \lambda\right) = x K_n\left(x, \lambda\right) - \lambda K_{n-1}\left(x, \lambda\right)$$

Thus the first two K_0, K_1 coincide with H_0 and H_1, respectively. Then since both K_n and H_n satisfy the recursion relation (8.37), it follows that $K_n = H_n$ for all n. ∎

8.13 Exercises

1. The Hermite equation is
$$y'' - xy' + ny = 0$$

   Verify that if $n = 0$ or a positive integer, then this equation always has a polynomial solution. These are called Hermite polynomials.

2. If you have two polynomial solutions to the Hermite equation above, $p_m(x)$ corresponding to m in the equation and $p_n(x)$ corresponding to n in the equation, $n \neq m$, show that

$$\int_{-\infty}^{\infty} e^{-x^2} p_m(x) p_n(x) \, dx = 0$$

3. The equation
$$\left(1 - x^2\right) y'' - 2xy' + n(n+1) y = 0$$

   is Legendre's equation. Note that 0 is an ordinary point for this equation. Show that for n a non-negative integer, this equation has polynomial solutions. It is in the chapter, but go over it yourself. Also explain why this equation has a regular singular point at $1, -1$.

4. In the above problem, suppose $p_k(x)$ and $p_l(x)$ are solutions, to the equations corresponding to $n = k, l$ respectively. Show that

$$\int_{-1}^{1} p_k(x) p_l(x) \, dx = 0$$

   Thus this gives an example of a collection of orthogonal polynomials. It is in the chapter but go through the details yourself.

5. The Legendre polynomials are given in the above problem but one multiplies by a constant so that the result satisfies $p_n(1) = 1$. The purpose of this problem is to find the constant. **Hint:** Use the Leibniz formula on $\left(x^2 - 1\right)^n = (x-1)^n (x+1)^n$.

6. The equation $\left(1 - x^2\right) y'' - xy' + n^2 y = 0$ is called the Chebychev equation. Find solutions to this equation. That is, specify a recurrence relation and two solutions. Explain why there exist polynomial solutions to this equation.

7. The equation $\left(1 - x^2\right) y'' - 3xy' + n(n+2) y = 0$ is also called the Chebychev equation. Find solutions to this equation. That is, specify a recurrence relation and two solutions. Explain why there exist polynomial solutions to this equation.

8. Specify two solutions to the following differential equation by determining a recurrence relation and then describing how to obtain two solutions.

(a) $y'' (x^2 + 1) + 5xy' + 2y = 0.$

(b) $y'' (x^2 + 1) + xy' + 3y = 0.$

(c) $y'' (x^2 + 1) + 7xy' + 4y = 0.$

(d) $y'' (1 - 3x^2) + 6xy' + 4y = 0.$

(e) $y'' - 5x^2 y' - 4xy = 0.$

(f) $y'' - 2x^2 y' - xy = 0.$

(g) $y'' + x^2 y' + 2xy = 0.$

(h) $y'' - 3x^2 y' - xy = 0.$

(i) $y'' + 2x^2 y' - 4xy = 0.$

9. Find the solution to the initial value problem $y'' + \sin(x) y' + \cos(3x) y = 0$ along with the initial conditions $y(0) = 1, y'(0) = -1$. You just need to find the first terms of the power series solution up to x^4.

10. Find the solution to the initial value problem $y'' + \tan(2x) y' + \cos(3x) y = 0$ along with the initial conditions $y(0) = -1, y'(0) = 2$. You just need to find the first terms of the power series solution up to x^4.

11. Find the solution to the initial value problem $y'' + \tan(5x) y' + \sec(3x) y = 0$ along with the initial conditions $y(0) = -2, y'(0) = 3$. You just need to find the first terms of the power series solution up to x^4.

12. Find the general solution to the following Euler equations.

(a) $y'' x^2 - 3y' x + 3y = 0.$

(b) $y'' x^2 + 4y' x - 4y = 0.$

(c) $y'' x^2 + 2y' x - 6y = 0.$

(d) $y'' x^2 + 6y' x + 6y = 0.$

(e) $y'' x^2 + 4y' x - 4y = 0.$

(f) $y'' x^2 - 3y' x + 4y = 0.$

(g) $y'' x^2 + 5y' x + 4y = 0.$

(h) $y'' x^2 - 5y' x + 9y = 0.$

(i) $y'' x^2 - 3y' x + 4y = 0.$

(j) $y'' x^2 - 3y' x + 4y = 0.$

(k) $y'' x^2 - y' x + y = 0.$

(l) $y'' x^2 + 7y' x + 10y = 0.$

(m) $y'' x^2 + 7y' x + 10y = 0.$

(n) $y'' x^2 + 9y' x + 32y = 0.$

(o) $y'' x^2 + 11y' x + 26y = 0.$

(p) $y'' x^2 + 11y' x + 34y = 0.$

13. The hypergeometric equation is

$$x(1 - x) y'' + (\gamma - (1 + \alpha + \beta) x) y' - \alpha\beta y = 0$$

Show it has a regular singular point at 0 and that the roots of the indicial equation are 0 and $1 - \gamma$.

14. In the above example, change the independent variable as follows: $t = 1/x$. Determine the equation which results in terms of t and show that the resulting equation has a regular singular point at 0 and that the roots of the indicial equation are α, β. **Hint:** You need to show that $y''(x) = y''(t) t^4 + 2t^3 y'(t), \ y'(x) = -t^2 y'(t)$. When you let $t = 0$, you are looking at the "point at infinity". Thus you are showing that the "point at infinity" is a regular singular point.

15. Consider the Bessel equation in which $\nu = 1/2$. In this case, the roots of the indicial equation differ by an integer. Nevertheless, there are two solutions,

neither of which involves a logarithm. Verify that for ν not an integer,

$$x^{-\nu} + \sum_{k=1}^{\infty} \frac{(-1)^k}{2^{2k} k! \, (k-\nu)_k} x^{2k-\nu}$$

does indeed yield a solution to the Bessel equation.

16. Show that for $\nu = 1/2$, one solution to the Bessel equation is $x^{-1/2} \sin(x)$. What is the other solution? Verify your answer. Show that one of these solutions is bounded and in fact converges to 0 as $x \to 0+$ while the other is unbounded as $x \to 0+$.

17. Explain why in every case, if you have a general solution to the Bessel equation, one of the solutions will be unbounded as $x \to 0$ and the other must converge to 0 as $x \to 0+$.

18. The Laguerre differential equation is

$$xy'' + (1-x)\,y' + my = 0$$

Show that when m is a nonnegative integer, there always exists a polynomial which is a solution to this differential equation. Letting $p_k(x), p_l(x)$ be polynomial solutions corresponding to $m = k, l$ respectively, show that

$$\int_0^{\infty} e^{-x} p_k(x) \, p_l(x) \, dx = 0, \ \ k \neq l$$

19. Prove Theorem 8.12.1, Leibniz rule.

20. Show that for $H_n(x, \lambda)$ a Hermite polynomial and for H' denoting the derivative with respect to x,

$$\lambda H''_{n+1} - x H'_{n+1} + (n+1) H_{n+1} = 0$$

21. Let $\lambda = 1$ in the above problem and let $y_n = H_n$. Thus

$$y''_n - x y'_n + n y_n = 0.$$

Let a boundary condition consist of y_n, y'_n have polynomial growth as $x \to \pm\infty$. Show that if $n \neq m$, then

$$\int_{-\infty}^{\infty} e^{-x^2/2} y_n(x) \, y_m(x) \, dx = 0$$

22. Using Procedure 8.9.1 give a description of the general solution to the Bessel equation

$$Ly = x^2 y'' + xy' + (x^2 - m^2)\,y = 0$$

You have one solution

$$\sum_{k=0}^{\infty} \frac{(-1)^k}{k! \, \Gamma(k+m+1)} \left(\frac{x}{2}\right)^{2k+m}$$

and you need to find an unbounded solution. Show that such an unbounded solution exists. **Hint:** In this case, the two roots are $-m, m$ and they differ by the integer $2m$.

23. Suppose you have any linear second order differential equation $Ly = 0$ in which there is a general solution $C_1 y + C_2 z$ such that $W(y, x) \neq 0$ for $x \in [a, b]$. Show that if $y(x) = 0$, then $y'(x) \neq 0$. Why does this show that given a zero of a nonzero solution to the Bessel equation, or any other second order linear differential equation, there is a next zero?

24. Problem 23 on Page 153 is the Weierstrass approximation theorem. Now consider the Legendre polynomials $p_0(x), p_1(x), \cdots$. Recall that these are orthogonal in the sense that

$$\int_{-1}^{1} p_k(x) p_j(x) \, dx = 0$$

unless $j = k$. Explain why $\{p_0(x), p_1(x), \cdots, p_n(x)\}$ is linearly independent. Now if f is continuous on $[-1, 1]$, and if $\varepsilon > 0$, explain why there is n and scalars c_k such that

$$\max\left\{\left| f(x) - \sum_{k=1}^{n} c_k p_k(x) \right| : x \in [-1, 1]\right\} < \varepsilon$$

25. Consider the equation $x^3 y'' + 2xy' + y = 0$. Explain why it does not have a regular singular point at 0. Show that the only possible nonzero power series solution to this has radius of convergence equal to 0. In fact there really isn't any such series solution to this problem.

26. Consider the Bessel function $J_m(x)$ for m a positive integer. Recall the summation formula.

$$J_m(x + y) = \sum_{k=-\infty}^{\infty} J_{m-k}(x) J_k(y),$$

$$J_m(x) = \sum_{k=0}^{\infty} \frac{(-1)^k}{k! \, (k+m)!} \left(\frac{x}{2}\right)^{2k+m}, \quad J_m(x) = (-1)^m J_{-m}(x)$$

Explain why J_m is even if m is even and J_m is odd if m is odd. Next let $m = 0$ and see what comes out of the summation formula. Then let $y = -x$ to obtain an inequality which shows that all the J_n are bounded. Show in particular that each $J_n(x)$ has the property that $|J_n(x)| \leq 1/\sqrt{2}$ if $n > 0$.

27. Use the integral formula for the Bessel function to graph $J_4(x)$ for $x \in [0, 20]$. Here is the syntax which will work for this. You put in the new lines. hold on for k=1:201 f=@(t,k)cos(4*t-((k-1)*.1)*sin(t)); y(k)=pi^(-1)*integral(@(t)f(t,k),0,pi); x(k)=(k-1)*.1; plot(x,y,'linewidth',2) end.

Part IV

First Order Systems

Chapter 9

Methods for First Order Linear Systems

9.1 Finding Solutions Using Eigenvalues

Earlier, the techniques of solving higher order linear differential equations were discussed. These techniques were based on factoring polynomials thus reducing an easy analysis problem to an impossible algebra problem. This overall approach is continued here. First is an important definition.

Definition 9.1.1 *Let A be an $n \times n$ matrix. Then $\Phi(t)$ is called **the** fundamental matrix if*

$$\Phi'(t) = A\Phi(t), \ \Phi(0) = I$$

It was shown in Problem 85 on Page 76 that the fundamental matrix always exists and is unique. I will discuss the theoretical question on existence of the fundamental matrix another way later.

What is meant by the above symbols? The idea is that $\Phi(t)$ is a matrix whose entries are differentiable functions of t. The meaning of $\Phi'(t)$ is the matrix whose entries are the derivatives of the entries of $\Phi(t)$. For example, abusing notation slightly,

$$\begin{pmatrix} t & t^2 \\ \sin(t) & \tan(t) \end{pmatrix}' = \begin{pmatrix} 1 & 2t \\ \cos(t) & \sec^2(t) \end{pmatrix}.$$

What are some properties of this derivative? Does the product rule hold for example?

Lemma 9.1.2 *Suppose $\Phi(t)$ is $m \times n$ and $\Psi(t)$ is $n \times p$ and these are differentiable matrices. Then*

$$(\Phi(t)\Psi(t))' = \Phi'(t)\Psi(t) + \Phi(t)\Psi'(t)$$

Proof: By definition,

$$\left((\Phi(t)\Psi(t))'\right)_{ij} = \left((\Phi(t)\Psi(t))_{ij}\right)' = \left(\sum_k \Phi(t)_{ik}\Psi(t)_{kj}\right)'$$

$$= \sum_k \Phi'(t)_{ik} \Psi(t)_{kj} + \sum_k \Phi(t)_{ik} \Psi'(t)_{kj} = (\Phi'(t)\Psi(t))_{ij} + (\Phi(t)\Psi'(t))_{ij}$$

and so the conclusion follows. ■

An algebraic method involving eigenvalues and eigenvectors will be used to explicitly find the fundamental matrix when the eigenvalue problem can be solved. Later two other ways of obtaining the fundamental matrix will be presented.

First order systems of equations are equations which are of the form

$$\mathbf{x}' = A\mathbf{x} + \mathbf{f}, \ \mathbf{x}(a) = \mathbf{x}_a$$

As discussed earlier, all of the higher order linear equations can be written this way as first order systems. See Problem 86 on Page 76 or the procedure for reducing to a first order system in Procedure 5.5.2 on Page 126.

However, not every first order system comes to you this way. For example, consider

$$\begin{pmatrix} x'(t) \\ y'(t) \end{pmatrix} = \begin{pmatrix} 1 & -1 \\ 1 & 1 \end{pmatrix} \begin{pmatrix} x(t) \\ y(t) \end{pmatrix} + \begin{pmatrix} t^2 \\ \sin(t) \end{pmatrix}$$

9.1.1   Homogeneous Linear Equations

These are linear systems of equations which are of the form

$$\mathbf{x}' = A\mathbf{x}, \ \mathbf{x} \in \mathbb{R}^n, \ A \text{ an } n \times n \text{ matrix}$$

The idea is to look for a solution in the form

$$\mathbf{x}(t) = \mathbf{u}e^{\lambda t}$$

where $\mathbf{u}$ is a fixed nonzero vector. Substituting this into the desired equation yields

$$\lambda \mathbf{u}e^{\lambda t} = A\mathbf{u}e^{\lambda t}$$

Divide both sides by $e^{\lambda t}$ and this gives

$$A\mathbf{u} = \lambda \mathbf{u}$$

Thus you need to have $\mathbf{u}$ be an eigenvector of A and λ an eigenvalue. This is how you can find a solution.

Now if you have several solutions $\mathbf{x}_1(t), \cdots, \mathbf{x}_m(t)$ and scalars a_k,

$$\left(\sum_{k=1}^m a_k \mathbf{x}_k \right)' = \sum_{k=1}^m a_k \mathbf{x}_k' = \sum_{k=1}^m a_k A\mathbf{x}_k = A \left(\sum_{k=1}^m a_k \mathbf{x}_k \right)$$

Hence linear combinations of solutions are also solutions. This is the principle of superposition discussed earlier in the context of higher order scalar problems.

Theorem 9.1.3 *(Superposition) If* $\mathbf{x}_i' = A\mathbf{x}_i$ *then*

$$\left(\sum_{k=1}^m a_k \mathbf{x}_k \right)' = A \left(\sum_{k=1}^m a_k \mathbf{x}_k \right)$$

Suppose then that you have found n solutions $\{\mathbf{x}_1, \cdots, \mathbf{x}_n\}$. Let

$$\Phi(t) = \begin{pmatrix} \mathbf{x}_1(t) & \cdots & \mathbf{x}_n(t) \end{pmatrix}$$

be the $n \times n$ matrix which has these as columns. Then the above theorem on superposition implies that for any choice of scalars $a_1, \cdots, a_n$,

$$\sum_{k=1}^{n} a_k \mathbf{x}_k(t) = \Phi(t)\,\mathbf{a}$$

is a solution. Here $\mathbf{a} = \begin{pmatrix} a_1 & \cdots & a_n \end{pmatrix}^T$ and is the column vector having a_i in the ith position. Recall how to multiply by matrices. Now there is the obvious question. You have found lots of solutions but:

Have you found $\boxed{\text{all}}$ solutions?

When you have found **all** solutions in this form, $\Phi(t)$ is called a fundamental matrix. **This one is a little different than Definition 9.1.1 because maybe** $\Phi(0) \neq I$. As earlier, the resolution of this question depends on the following fundamental existence and uniqueness theorem. A more general version of this is proved in Chapter 11.

Theorem 9.1.4 *Let A be an $n \times n$ matrix and let $\mathbf{g}(t)$ be a continuous vector for $t \in [a, b]$ and let $c \in [a, b]$ and $\mathbf{x}_0 \in \mathbb{R}^n$. Then there exists a unique solution to the initial value problem*

$$\mathbf{x}'(t) = A\mathbf{x}(t) + \mathbf{g}(t), \ \mathbf{x}(c) = \mathbf{x}_0$$

valid for $t \in [a, b]$.

Proof: In case $c = 0$, this follows from Problem 84 on Page 76. In the general case, just let $\mathbf{y}(s) \equiv \mathbf{x}(s + c), s \in [a - c, b - c]$, so that $\mathbf{y}(0) = \mathbf{x}(c)$. Thus there is a unique solution to the above initial value problem is equivalent to the existence and uniqueness of the initial value problem on $[a - c, b - c]$,

$$\mathbf{y}'(s) = A\mathbf{y}(s) + \mathbf{g}(s + c), \ \mathbf{y}(0) = \mathbf{x}_0$$

which has a unique solution by that problem. ∎

Note that it makes no difference if $[a, b]$ is replaced with (a, b) since every open interval is the union of closed intervals. Thus, in particular, if $\mathbf{g}$ is continuous on $\mathbb{R}$ then the solution to the above initial value problem exists and is unique on $\mathbb{R}$.

Then the following major theorem gives the desired result.

Theorem 9.1.5 *Let $\mathbf{x}'_k = A\mathbf{x}_k$ and let $\Phi(t)$ be the matrix having $\mathbf{x}_k$ as the kth column. Then all possible solutions are obtained as*

$$\sum_{k=1}^{n} c_k \mathbf{x}_k = \Phi(t)\,\mathbf{c}, \ \mathbf{c} \in \mathbb{R}^n$$

if and only if $\det(\Phi(t)) \neq 0$ for some t. Furthermore, $\det(\Phi(t))$ either vanishes for all $t \in \mathbb{R}$ or for no $t \in \mathbb{R}$.

Proof: First suppose that $\det \Phi(t_0) \neq 0$ for some t_0. Let $\mathbf{z}' = A\mathbf{z}$. Then there exists $\mathbf{c}$ such that

$$\Phi(t_0)\,\mathbf{c} = \mathbf{z}(t_0)$$

Then by the principle of superposition, $\Phi(t)\,\mathbf{c}$ gives a solution of the differential equation which coincides with $\mathbf{z}$ at t_0. Therefore, $\mathbf{z}(t) = \Phi(t)\,\mathbf{c}$. This shows that $\Phi(t)\,\mathbf{c}$ for $\mathbf{c} \in \mathbb{R}^n$ gives all possible solutions.

Next suppose that $\det \Phi(t_0) = 0$. Then there exists $\mathbf{z}_0 \in \mathbb{R}^n$ such that there is no solution $\mathbf{c}$ to the equation

$$\Phi(t_0)\,\mathbf{c} = \mathbf{z}_0.$$

However, from the existence part of the above theorem, there exists $\mathbf{z}$ such that $\mathbf{z}' = A\mathbf{z}$ and $\mathbf{z}(t_0) = \mathbf{z}_0$. Hence $\mathbf{z}$ cannot be given in the form $\Phi(t)\,\mathbf{c}$ for any $\mathbf{c} \in \mathbb{R}^n$. Therefore, $\Phi(t)\,\mathbf{c}$ for $\mathbf{c} \in \mathbb{R}^n$ cannot give all possible solutions. Consider the following picture.

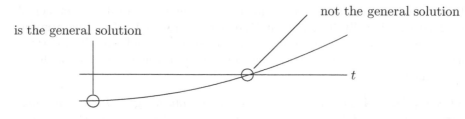

This picture illustrates what cannot occur for the graph of $t \rightarrow \det(\Phi(t))$. Either $\Phi(t)\,\mathbf{c}$ for $\mathbf{c} \in \mathbb{R}^n$ gives all solutions or it does not. Therefore, $\det(\Phi(t))$ either vanishes identically or not at all. ∎

This is the Wronskian alternative presented earlier and generalized to the situation of first order systems. Now what follows is a slight generalization of the notion of "fundamental matrix". Here we will not insist that $\Phi(0) = I$ and instead require the columns to be linearly independent for each t. The next definition gives a more general usage of the term "fundamental matrix".

Definition 9.1.6 *When $\Phi(t)\,\mathbf{c}$ for $\mathbf{c} \in \mathbb{R}^n$ yields all possible solutions to $\mathbf{x}' = A\mathbf{x}$, the matrix $\Phi(t)$ is called a fundamental matrix. Note that in this context, maybe $\Phi(0) \neq I$ but from what was just shown, $\det(\Phi(t)) \neq 0$.*

Note that $\Phi(t)$ is a fundamental matrix if and only if the set of all solutions to $\mathbf{x}' = A\mathbf{x}$ is obtained as linear combinations of the columns of $\Phi(t)$ which happens if and only if the columns of $\Phi(t)$ for some (all) t are linearly independent.

Proposition 9.1.7 *Let A be an $n \times n$ matrix. Then an $n \times n$ matrix $\Phi(t)$ is a fundamental matrix if and only if the columns of $\Phi(t)$ are solutions to the differential equation $\mathbf{x}' = A\mathbf{x}$ and for each (any) t, these columns are a linearly independent set. The latter case is equivalent to the general solution being of the form $\Phi(t)\,\mathbf{c}$ for $\mathbf{c}$ a constant vector.*

Let's go over why this is so and follows from the above discussion. If the columns of $\Phi(t)$ each solve the differential equation, then letting

$$\Phi(t) = \begin{pmatrix} \mathbf{v}_1(t) & \cdots & \mathbf{v}_n(t) \end{pmatrix}$$

then

$$\Phi'(t) = \left(\begin{array}{ccc} \mathbf{v}_1'(t) & \cdots & \mathbf{v}_n'(t) \end{array}\right) = \left(\begin{array}{ccc} A\mathbf{v}_1(t) & \cdots & A\mathbf{v}_n(t) \end{array}\right)$$

$$= A\left(\begin{array}{ccc} \mathbf{v}_1(t) & \cdots & \mathbf{v}_n(t) \end{array}\right) = A\Phi(t)$$

Therefore, it satisfies $\Phi' = A\Phi$. By Theorem 9.1.5, the columns are either always dependent for all t or always independent for all t. This last case being equivalent to the general solution being of the form $\Phi(t)\mathbf{c}$ for $\mathbf{c}$ a constant vector.

If $\Phi(t)$ is a fundamental matrix as in the above definition, then by Theorem 9.1.5, the general solution is of the form $\Phi(t)\mathbf{c}$ for $\mathbf{c}$ a constant vector. In this case, you could take $\mathbf{c} = \mathbf{e}_i$ to find that the ith column of $\Phi(t)$ is a solution to the differential equation.

Next the attempt is made to explicitly compute fundamental matrices. This is where we reduce an easy problem to a profoundly difficult one and declare we have made progress.

From a theoretical point of view there are exactly two cases of concern. Here they are.

1. There exists a basis for $\mathbb{C}^n$ of eigenvectors of A. ("Easy" case, A is nondefective.)

2. A is defective so there does not exist a basis for $\mathbb{C}^n$ of eigenvectors of A.

Even in the "easy" case, it isn't really easy because one must factor the characteristic polynomial which is often impossible. Therefore, at least for large systems, this method is a complete fraud. However, there is an easy to consider condition which will require that there exists a basis of eigenvectors although it won't help you to find it.

Theorem 9.1.8 *Suppose $A\mathbf{v}_i = \lambda_i \mathbf{v}_i, i = 1, \cdots, r$, $\mathbf{v}_i \neq 0$, and that if $i \neq j$, then $\lambda_i \neq \lambda_j$. Then the set of eigenvectors, $\{\mathbf{v}_1, \cdots, \mathbf{v}_r\}$ is linearly independent.*

Proof: Suppose the set of eigenvectors is not linearly independent. Then there exists $p \leq r$ such that

$$\sum_{i=1}^{p} c_i \mathbf{v}_i = \mathbf{0}, \text{ not all } c_i = 0. \tag{$*$}$$

We can assume that p is as small as possible for this to happen. Then $p > 1$ since otherwise, you would have $c_1 \mathbf{v}_1 = \mathbf{0}$ and dividing by $c_1, \mathbf{v}_1 = \mathbf{0}$ but eigenvectors are never $\mathbf{0}$. Also, $c_p \neq 0$ since otherwise, p is not the smallest such number. Then multiplying by λ_p and also doing A to both sides,

$$\sum_{i=1}^{p} c_i \lambda_p \mathbf{v}_i = \mathbf{0}, \ \sum_{i=1}^{p} c_i \lambda_i \mathbf{v}_i = \mathbf{0} \text{ so } \sum_{i=1}^{p-1} c_i (\lambda_p - \lambda_i) \mathbf{v}_i = \mathbf{0}$$

Since p was as small as possible, each $c_i (\lambda_p - \lambda_i) = 0$ and since the eigenvalues are distinct, each $c_i = 0$. Hence $*$ reduces to $c_p \mathbf{v}_p = \mathbf{0}, c_p \neq 0$ and so $\mathbf{v}_p = \mathbf{0}$ which is impossible because eigenvectors are never zero. ∎

Example 9.1.9 *The matrix*

$$A = \begin{pmatrix} 7 & 14 & 8 & -2 \\ -6 & -13 & -8 & 2 \\ 4 & 10 & 7 & -2 \\ -2 & -5 & -2 & 4 \end{pmatrix}$$

has distinct eigenvalues $l = 1, -1, 2, 3$ with corresponding eigenvectors,

$$\begin{pmatrix} -4 & 3 & -2 & 1 \end{pmatrix}^T, \begin{pmatrix} -3 & 3 & -2 & 1 \end{pmatrix}^T,$$
$$\begin{pmatrix} -2 & 2 & -2 & 1 \end{pmatrix}^T, \begin{pmatrix} -1 & 1 & -1 & 1 \end{pmatrix}^T$$

Therefore, a fundamental matrix is

$$\begin{pmatrix} -4e^t & -3e^{-t} & -2e^{2t} & -e^{3t} \\ 3e^t & 3e^{-t} & 2e^{2t} & e^{3t} \\ -2e^t & -2e^{-t} & -2e^{2t} & -e^{3t} \\ e^t & e^{-t} & e^{2t} & e^{3t} \end{pmatrix}$$

Thus if the $n \times n$ matrix A has n distinct eigenvalues, then the matrix is non-defective and there exists a basis of eigenvectors. However, there may be repeated eigenvalues and yet the matrix be nondefective. Also recall from Problem 85 on Page 76 and the methods used there involving Schur's theorem Theorem D.8.5 of the appendix, when A is a real matrix, the fundamental matrix $\Phi(t)$ is also real valued. This will be made more explicit later if you have not worked this problem.

So what happens when A is real but the eigenvalues are complex? Suppose that $A\mathbf{v} = (a + ib)\mathbf{v}$. Then recalling that the complex conjugate of a product is the product of the complex conjugates, it follows that

$$A\bar{\mathbf{v}} = \bar{A}\bar{\mathbf{v}} = (a - ib)\bar{\mathbf{v}}$$

It follows that for a real matrix, the complex eigenvalues come in complex conjugate pairs and so do the corresponding eigenvectors. Now letting $a + ib$ be a complex eigenvalue corresponding to the eigenvector $\mathbf{v}$, the set of linear combinations

$$C_1 \mathbf{v} e^{(a+ib)t} + C_2 \bar{\mathbf{v}} e^{(a-ib)t}, \quad C_1, C_2 \in \mathbb{C}, \text{ the complex numbers}$$

is the same as the set of linear combinations

$$K_1 \frac{1}{2} \left(\mathbf{v} e^{(a+ib)t} + \bar{\mathbf{v}} e^{(a-ib)t} \right) + K_2 \frac{1}{2i} \left(\mathbf{v} e^{(a+ib)t} - \bar{\mathbf{v}} e^{(a-ib)t} \right),$$

This is because

$$\mathbf{v} e^{(a+ib)t} = \frac{1}{2} \left(\mathbf{v} e^{(a+ib)t} + \bar{\mathbf{v}} e^{(a-ib)t} \right) + i \frac{1}{2i} \left(\mathbf{v} e^{(a+ib)t} - \bar{\mathbf{v}} e^{(a-ib)t} \right)$$
$$\bar{\mathbf{v}} e^{(a-ib)t} = \frac{1}{2} \left(\mathbf{v} e^{(a+ib)t} + \bar{\mathbf{v}} e^{(a-ib)t} \right) - i \frac{1}{2i} \left(\mathbf{v} e^{(a+ib)t} - \bar{\mathbf{v}} e^{(a-ib)t} \right)$$

Therefore, any linear combination involving $\mathbf{v} e^{(a+ib)t}$ and $\bar{\mathbf{v}} e^{(a-ib)t}$ is equal to one which involves instead the functions

$$\frac{1}{2} \left(\mathbf{v} e^{(a+ib)t} + \bar{\mathbf{v}} e^{(a-ib)t} \right), \frac{1}{2i} \left(\mathbf{v} e^{(a+ib)t} - \bar{\mathbf{v}} e^{(a-ib)t} \right)$$

The reason this is significant is that these functions have entries which are real. See Section 1.3. Therefore, you can replace the pairs of columns in $\Phi(t)$ corresponding to a complex eigenvalue with these real valued columns. The above vector valued functions are the real and imaginary parts of the solution $\mathbf{v}e^{(a+ib)t}$.

Here are some examples. The first example is of a matrix which is nondefective having real eigenvalues.

Example 9.1.10 *Find a fundamental matrix for*

$$A = \begin{pmatrix} -1 & -2 & -2 \\ -1 & 1 & -1 \\ 5 & 4 & 6 \end{pmatrix}$$

The eigenvalues are $1, 2, 3$ and eigenvectors are respectively

$$\begin{pmatrix} -1 & 0 & 1 \end{pmatrix}^T, \begin{pmatrix} 0 & -1 & 1 \end{pmatrix}^T, \begin{pmatrix} -1 & -1 & 3 \end{pmatrix}^T$$

Therefore, a fundamental matrix is

$$\Phi(t) = \begin{pmatrix} -e^t & 0 & -e^{3t} \\ 0 & -e^{2t} & -e^{3t} \\ e^t & e^{2t} & 3e^{3t} \end{pmatrix}$$

It follows that the general solution to the equation $\mathbf{x}' = A\mathbf{x}$ consists of vectors $\Phi(t)\mathbf{c}, \mathbf{c} \in \mathbb{R}^3$. In other words all linear combinations of the columns of $\Phi(t)$.

Now here is an example where the matrix is nondefective although it has a repeated eigenvalue.

Example 9.1.11 *Find a fundamental matrix for*

$$A = \begin{pmatrix} 2 & 1 & 1 \\ -2 & -1 & -2 \\ 2 & 2 & 3 \end{pmatrix}$$

The eigenvalues are $1,1,2$ listed according to multiplicity and eigenvectors are, respectively,

$$\begin{pmatrix} -1 & 1 & 0 \end{pmatrix}^T, \begin{pmatrix} -1 & 0 & 1 \end{pmatrix}^T, \begin{pmatrix} 1 & -2 & 2 \end{pmatrix}^T$$

Therefore, a fundamental matrix is

$$\Phi(t) = \begin{pmatrix} -e^t & -e^t & e^{2t} \\ e^t & 0 & -2e^{2t} \\ 0 & e^t & 2e^{2t} \end{pmatrix}$$

It follows that the general solution to the differential equation $\mathbf{x}' = A\mathbf{x}$ is $\Phi(t)\mathbf{c}, \mathbf{c} \in \mathbb{R}^3$. Thus the general solution to the equation $\mathbf{x}' = A\mathbf{x}$ consists of linear combinations of columns of $\Phi(t)$.

Next is an example where there are two complex eigenvalues.

Example 9.1.12 *Find a real fundamental matrix for*

$$
\begin{pmatrix}
1 & 0 & 0 \\
-3 & -2 & -5 \\
2 & 2 & 4
\end{pmatrix}
$$

The eigenvalues are 1, $1 - i$, and $1 + i$. The eigenvectors are, respectively,

$$
\begin{pmatrix} -1 & 1 & 0 \end{pmatrix}^T, \begin{pmatrix} 0 & -3 - i & 2 \end{pmatrix}^T, \begin{pmatrix} 0 & -3 + i & 2 \end{pmatrix}^T
$$

Consider the eigenvector which goes with $\lambda = 1 - i$. The solution which goes with this eigenvalue is

$$
\begin{pmatrix}
0 \\
-3 - i \\
2
\end{pmatrix} e^{(1-i)t}
$$

Now the trick is to replace the two occurrences of this and its conjugate with the real and imaginary parts. Thus we need to find the real and imaginary parts of the above expression. It equals

$$
e^t \left(\begin{pmatrix} 0 \\ -3 \\ 2 \end{pmatrix} + i \begin{pmatrix} 0 \\ -1 \\ 0 \end{pmatrix} \right) (\cos t - i \sin t)
$$

Thus the real part is

$$
e^t \left(\begin{pmatrix} 0 \\ -3 \\ 2 \end{pmatrix} \cos(t) + \begin{pmatrix} 0 \\ -1 \\ 0 \end{pmatrix} \sin(t) \right) = \begin{pmatrix} 0 \\ -e^t(3\cos t + \sin t) \\ 2(\cos t)e^t \end{pmatrix}
$$

and the imaginary part is

$$
e^t \left(\begin{pmatrix} 0 \\ -1 \\ 0 \end{pmatrix} \cos(t) - \begin{pmatrix} 0 \\ -3 \\ 2 \end{pmatrix} \sin(t) \right) = \begin{pmatrix} 0 \\ -e^t(\cos t - 3\sin t) \\ -2e^t \sin t \end{pmatrix}
$$

Therefore, a real fundamental matrix is

$$
\begin{pmatrix}
-e^t & 0 & 0 \\
e^t & -e^t(3\cos t + \sin t) & -e^t(\cos t - 3\sin t) \\
0 & 2(\cos t)e^t & -2e^t \sin t
\end{pmatrix}
$$

The real solutions to the differential equation $\mathbf{x}' = A\mathbf{x}$ are then

$$
\begin{pmatrix}
-e^t & 0 & 0 \\
e^t & -e^t(3\cos t + \sin t) & -e^t(\cos t - 3\sin t) \\
0 & 2(\cos t)e^t & -2e^t \sin t
\end{pmatrix}
\begin{pmatrix} c_1 \\ c_2 \\ c_3 \end{pmatrix}
$$

where the $c_i \in \mathbb{R}$.

9.1.2 The Case where A is Defective

Sometimes A is defective. This means it is not possible to find a basis of eigenvectors. This means you cannot find a fundamental matrix of the sort where the columns are $\mathbf{v}e^{\lambda t}$ for $A\mathbf{v} = \lambda\mathbf{v}$. You cannot find it because it is not there to be found. You must do something else. As pointed out above, this can only happen if you have a repeated eigenvalue. Suppose then that λ is repeated and the algebraic multiplicity of this eigenvalue is greater than the geometric multiplicity. A solution associated with this eigenvalue is $\mathbf{v}e^{\lambda t}$. Of course it would be tempting to say that another solution is $t\mathbf{v}e^{\lambda t}$ by analogy to the scalar case, but this will not even be a solution to $\mathbf{x}' = A\mathbf{x}$. Just try it and see. Something else is needed.

Let $\mathbf{v}$ be an eigenvector. Thus $A\mathbf{v} = \lambda\mathbf{v}$. A generalized eigenvector based on $\mathbf{v}$ is a vector $\mathbf{w}$ such that

$$(A - \lambda I)\,\mathbf{w} = \mathbf{v}$$

Then another solution to the differential equation is of the form

$$t\mathbf{v}e^{\lambda t} + \mathbf{w}e^{\lambda t} \tag{9.1}$$

That this is a solution to the differential equation is easily seen.

$$\left(t\mathbf{v}e^{\lambda t} + \mathbf{w}e^{\lambda t}\right)' = \mathbf{v}e^{\lambda t} + t\mathbf{v}\lambda e^{\lambda t} + \lambda\mathbf{w}e^{\lambda t}$$

$$A\left(t\mathbf{v}e^{\lambda t} + \mathbf{w}e^{\lambda t}\right) = t\lambda\mathbf{v}e^{\lambda t} + e^{\lambda t}\left(\lambda\mathbf{w} + \mathbf{v}\right)$$

which is the same thing. Therefore, you use this solution in (9.1) to fill one of the missing columns. Of course, you might have to look for even more such solutions. In general, suppose you have a chain of generalized eigenvectors $\mathbf{w}_1, \cdots, \mathbf{w}_k$ such that $(A - \lambda I)\,\mathbf{w}_j = \mathbf{w}_{j-1}$ and $(A - \lambda I)\,\mathbf{w}_1 = \mathbf{v}$ where $\mathbf{v}$ is an eigenvector for λ. Consider the following vector.

$$\frac{t^k}{k!}\mathbf{v}e^{\lambda t} + \sum_{j=1}^{k}\frac{t^{k-j}}{(k-j)!}\mathbf{w}_je^{\lambda t} = \mathbf{y}$$

Then

$$A\mathbf{y} = \lambda\frac{t^k}{k!}\mathbf{v}e^{\lambda t} + \sum_{j=1}^{k}\frac{t^{k-j}}{(k-j)!}\left(\lambda\mathbf{w}_j + \mathbf{w}_{j-1}\right)e^{\lambda t},\ \mathbf{w}_{j-1} \equiv \mathbf{v}$$

Then this equals

$$\lambda\frac{t^k}{k!}\mathbf{v}e^{\lambda t} + \sum_{j=2}^{k}\frac{t^{k-j}}{(k-j)!}\left(\lambda\mathbf{w}_j + \mathbf{w}_{j-1}\right)e^{\lambda t} + \frac{t^{k-1}}{(k-1)!}\left(\lambda\mathbf{w}_1 + \mathbf{v}\right)e^{\lambda t}$$

Also

$$\mathbf{y}' = \frac{t^k}{k!}\lambda\mathbf{v}e^{\lambda t} + \sum_{j=1}^{k}\frac{t^{k-j}}{(k-j)!}\lambda\mathbf{w}_je^{\lambda t} + \frac{t^{k-1}}{(k-1)!}\mathbf{v}e^{\lambda t} + \sum_{j=1}^{k-1}\frac{t^{k-(j+1)}}{((k-j)-1)!}\mathbf{w}_je^{\lambda t}$$

$$= \frac{t^k}{k!}\lambda\mathbf{v}e^{\lambda t} + \sum_{j=1}^{k}\frac{t^{k-j}}{(k-j)!}\lambda\mathbf{w}_je^{\lambda t} + \frac{t^{k-1}}{(k-1)!}\mathbf{v}e^{\lambda t} + \sum_{j=2}^{k}\frac{t^{k-j}}{(k-j)!}\mathbf{w}_{j-1}e^{\lambda t}$$

$$= \frac{t^k}{k!} \lambda \mathbf{v} e^{\lambda t} + \sum_{j=2}^{k} \frac{t^{k-j}}{(k-j)!} \left(\lambda \mathbf{w}_j + \mathbf{w}_{j-1} \right) e^{\lambda t} + \frac{t^{k-1}}{(k-1)!} \mathbf{v} e^{\lambda t} + \frac{t^{k-1}}{(k-1)!} \lambda \mathbf{w}_1 e^{\lambda t}$$

which is the same as $A\mathbf{y}$.

The rule is described by the following pattern.

$$\mathbf{v} e^{\lambda t}, \; t \mathbf{v} e^{\lambda t} + \mathbf{w}_1 e^{\lambda t}, \; \frac{t^2}{2} \mathbf{v} e^{\lambda t} + t \mathbf{w}_1 e^{\lambda t}, \; \frac{t^3}{3!} \mathbf{v} e^{\lambda t} + \frac{t^2}{2} \mathbf{w}_1 e^{\lambda t} + t \mathbf{w}_2 e^{\lambda t}, \cdots$$

You find as many of these as you need to equal the algebraic multiplicity of the given defective eigenvalue. Then you use them as columns for your fundamental matrix. Will there be enough of them to obtain a fundamental matrix? Will the resulting matrix have nonzero determinant? These are questions which are equivalent to the existence of the Jordan canonical form in linear algebra. Therefore, this issue will be left open.

To see a logically complete and algebraic description of this problem of finding fundamental matrices, see Chapter 12 or Problem 84 on Page 76. The approach in Chapter 12 is due to Putzer and has been around since the 1960s and it demonstrates conclusively, using only algebraic methods, the existence of a fundamental matrix using algebraic methods. Algebra is not the right way to study the theory of differential equations, but it is not as bad as this standard presentation given above suggests. So why do we use this awful method? It is because we crave closed form solutions and for fairly simple problems, it will be shorter than the other methods mentioned. Fortunately, there are now computer algebra systems which allow the determination of a fundamental matrix in the case where you can find it explicitly very easily. This method is presented later.

Here is an example of the awful situation when the matrix is defective.

Example 9.1.13 *Find the general solution for* $\mathbf{x}' = A\mathbf{x}$ *where* $A =$

$$\begin{pmatrix} -2 & 5 & 2 & -3 \\ -3 & 5 & 3 & -3 \\ -2 & 2 & 4 & -2 \\ -1 & 1 & 1 & 1 \end{pmatrix}$$

In this case, there is exactly one eigenvalue $\lambda = 2$. The eigenspace is of the form

$$s \begin{pmatrix} -2 & -1 & 0 & 1 \end{pmatrix}^T + t \begin{pmatrix} 3 & 2 & 1 & 0 \end{pmatrix}^T, t, s \in \mathbb{R}$$

Now we need to find generalized eigenvectors. This involves row reduction of

$$(A - 2I | \mathbf{v})$$

where $\mathbf{v}$ is some eigenvector. Let's just try to find one for $\mathbf{v}$ equal to the first in the above linear combination. Thus you need to row reduce the following.

$$\begin{pmatrix} -4 & 5 & 2 & -3 & -2 \\ -3 & 3 & 3 & -3 & -1 \\ -2 & 2 & 2 & -2 & 0 \\ -1 & 1 & 1 & -1 & 1 \end{pmatrix}$$

When this is done, you get

$$\begin{pmatrix} 1 & 0 & -3 & 2 & 0 \\ 0 & 1 & -2 & 1 & 0 \\ 0 & 0 & 0 & 0 & 1 \\ 0 & 0 & 0 & 0 & 0 \end{pmatrix}$$

and so there is no solution. **It didn't work.** Thus you need to worry about which eigenvector on which to base your generalized eigenvectors. You can't just pick one and expect it to work. It won't necessarily do so.

Do you see why it is not clear that you can always get solutions of this form involving eigenvectors and generalized eigenvectors? We need to base the generalized eigenvector on a general eigenvector. Thus we need to row reduce

$$\begin{pmatrix} -4 & 5 & 2 & -3 & -2s + 3t \\ -3 & 3 & 3 & -3 & -1s + 2t \\ -2 & 2 & 2 & -2 & t \\ -1 & 1 & 1 & -1 & s \end{pmatrix}$$

and choose s, t in such a way that there is a solution to the above system of equations symbolized by the above augmented matrix. After some row operations, you get

$$\begin{pmatrix} -1 & 1 & 1 & -1 & s \\ 0 & 1 & -2 & 1 & 3t - 6s \\ 0 & 0 & 0 & 0 & t - 2s \\ 0 & 0 & 0 & 0 & 0 \end{pmatrix} \quad \begin{pmatrix} -1 & 1 & 1 & -1 & 1 \\ 0 & 1 & -2 & 1 & 0 \\ 0 & 0 & 0 & 0 & 0 \\ 0 & 0 & 0 & 0 & 0 \end{pmatrix}$$

Thus, if there is to be a solution, you must have $t = 2s$. Let $s = 1$ and $t = 2$. Then to finish the job and find a generalized eigenvector you would continue row reducing

$$\begin{pmatrix} -1 & 1 & 1 & -1 & 1 \\ 0 & 1 & -2 & 1 & 0 \\ 0 & 0 & 0 & 0 & 0 \\ 0 & 0 & 0 & 0 & 0 \end{pmatrix}$$

this yields

$$\begin{pmatrix} 1 & 0 & -3 & 2 & -1 \\ 0 & 1 & -2 & 1 & 0 \\ 0 & 0 & 0 & 0 & 0 \\ 0 & 0 & 0 & 0 & 0 \end{pmatrix}$$

Now the generalized eigenvector is

$$\begin{pmatrix} 3z - 2w - 1 & 2z - w & z & w \end{pmatrix}^T$$

One of them is

$$\begin{pmatrix} -1 & 0 & 0 & 0 \end{pmatrix}^T$$

and so you get the following three solutions to the differential equation.

$$
C_1 \begin{pmatrix} -2 \\ -1 \\ 0 \\ 1 \end{pmatrix} e^{2t}, C_2 \begin{pmatrix} 3 \\ 2 \\ 1 \\ 0 \end{pmatrix} e^{2t}, C_3 \left(t \begin{pmatrix} -2 + 3\,(2) \\ -1 + 2\,(2) \\ 2 \\ 1 \end{pmatrix} e^{2t} + \begin{pmatrix} -1 \\ 0 \\ 0 \\ 0 \end{pmatrix} e^{2t} \right)
$$

$$
= C_1 \begin{pmatrix} -2 \\ -1 \\ 0 \\ 1 \end{pmatrix} e^{2t}, C_2 \begin{pmatrix} 3 \\ 2 \\ 1 \\ 0 \end{pmatrix} e^{2t}, C_3 \begin{pmatrix} 4te^{2t} - e^{2t} \\ 3te^{2t} \\ 2te^{2t} \\ te^{2t} \end{pmatrix}
$$

Alas, this is not enough. We need another solution. Another generalized eigenvector is required. We need to row reduce $(A - 2I\,|\,\mathbf{w})$ where $\mathbf{w}$ is a generalized eigenvector to get the next generalized eigenvector. To get a solution, we need to be sure that $\mathbf{w}$ is the most general generalized eigenvector. Thus the system to row reduce is

$$
\begin{pmatrix} -4 & 5 & 2 & -3 & -1 + 3z - 2w \\ -3 & 3 & 3 & -3 & 2z - w \\ -2 & 2 & 2 & -2 & z \\ -1 & 1 & 1 & -1 & w \end{pmatrix}
$$

and part of the problem is to find w, z such that there will be a solution. You can see right away that to have a solution $z = 2w$. Otherwise, the bottom two rows will lead to the system being inconsistent. Let's let $w = 1$ and $z = 2$ and seek a solution. Thus we would row reduce

$$
\begin{pmatrix} -4 & 5 & 2 & -3 & -1 + 3\,(2) - 2 \\ -3 & 3 & 3 & -3 & 2\,(2) - 1 \\ -2 & 2 & 2 & -2 & 2 \\ -1 & 1 & 1 & -1 & 1 \end{pmatrix} = \begin{pmatrix} -4 & 5 & 2 & -3 & 3 \\ -3 & 3 & 3 & -3 & 3 \\ -2 & 2 & 2 & -2 & 2 \\ -1 & 1 & 1 & -1 & 1 \end{pmatrix}
$$

The row reduced echelon form is

$$
\begin{pmatrix} 1 & 0 & -3 & 2 & -2 \\ 0 & 1 & -2 & 1 & -1 \\ 0 & 0 & 0 & 0 & 0 \\ 0 & 0 & 0 & 0 & 0 \end{pmatrix}
$$

and so a third generalized eigenvector is

$$
\begin{pmatrix} -2 & -1 & 0 & 0 \end{pmatrix}^T
$$

Then the third solution will be

$$
\left(\frac{t^2}{2} \begin{pmatrix} -2 + 3\,(2) \\ -1 + 2\,(2) \\ 2 \\ 1 \end{pmatrix} + t \begin{pmatrix} -1 + 3\,(2) - 2 \\ 2\,(2) - 1 \\ 2 \\ 1 \end{pmatrix} + \begin{pmatrix} -2 \\ -1 \\ 0 \\ 0 \end{pmatrix} \right) e^{2t}
$$

$$
= \begin{pmatrix} e^{2t} \left(2t^2 + 3t - 2 \right) \\ \frac{1}{2} e^{2t} \left(3t^2 + 6t - 2 \right) \\ te^{2t} \left(t + 2 \right) \\ \frac{1}{2} te^{2t} \left(t + 2 \right) \end{pmatrix}
$$

Thus the general solution is of the form

$$
C_1 \begin{pmatrix} -2 \\ -1 \\ 0 \\ 1 \end{pmatrix} e^{2t} + C_2 \begin{pmatrix} 3 \\ 2 \\ 1 \\ 0 \end{pmatrix} e^{2t} + C_3 \begin{pmatrix} 4te^{2t} - e^{2t} \\ 3te^{2t} \\ 2te^{2t} \\ te^{2t} \end{pmatrix} + C_4 \begin{pmatrix} e^{2t} \left(2t^2 + 3t - 2 \right) \\ \frac{1}{2} e^{2t} \left(3t^2 + 6t - 2 \right) \\ te^{2t} \left(t + 2 \right) \\ \frac{1}{2} te^{2t} \left(t + 2 \right) \end{pmatrix}
$$

This is the kind of thing which must result when you take the initial value problem which is one of the easiest problems in math and change it into the eigenvalue problem which is one of the hardest, all because of the love of closed form solutions. Differential equations is not part of algebra, but the above continues the tradition of pretending that it is.

9.2 Nonhomogeneous Problems

First order systems are of the form

$$
\mathbf{x}' = A\mathbf{x} + \mathbf{f}, \ \mathbf{x}(a) = \mathbf{x}_0
$$

Without the $\mathbf{f}$, this was discussed above. It involved finding a fundamental matrix $\Phi(t)$ such that the general solution to $\mathbf{x}' = A\mathbf{x}$ is given by $\Phi(t)\mathbf{c}$.

Theorem 9.2.1 *Suppose*

$$
\mathbf{y}'_p = A\mathbf{y}_p + \mathbf{f}
$$

Thus $\mathbf{y}_p$ is a particular solution to the nonhomogeneous problem. Then if $\mathbf{z}' = A\mathbf{z} + \mathbf{f}$ is any solution, it follows there exists $\mathbf{c} \in \mathbb{R}^n$ such that

$$
\mathbf{z} = \Phi(t)\mathbf{c} + \mathbf{y}_p
$$

Proof:

$$
\left(\mathbf{z} - \mathbf{y}_p \right)' = A\mathbf{z} + \mathbf{f} - \left(A\mathbf{y}_p + \mathbf{f} \right) = A\mathbf{z} - A\mathbf{y}_p = A\left(\mathbf{z} - \mathbf{y}_p \right)
$$

Therefore, $\mathbf{z} - \mathbf{y}_p$ is a solution to the homogeneous equation and so there exists $\mathbf{c}$ such that

$$
\mathbf{z} - \mathbf{y}_p = \Phi(t)\mathbf{c} \ \blacksquare
$$

From this theorem, it shows that to find the general solution to the nonhomogeneous problem, it suffices to find a particular solution to the nonhomogeneous problem. There is a very simple way to do this. You vary the constant $\mathbf{c}$. Recall that the usual product rule holds when multiplying matrices. This was shown above. Also note the following. Let $\Phi(t)$ be a fundamental matrix.

$$
\Phi(t) = \begin{pmatrix} \mathbf{v}_1(t) & \cdots & \mathbf{v}_n(t) \end{pmatrix}
$$

$$
\begin{aligned}
\Phi'(t) &= \begin{pmatrix} \mathbf{v}'_1(t) & \cdots & \mathbf{v}'_n(t) \end{pmatrix} = \begin{pmatrix} A\mathbf{v}_1(t) & \cdots & A\mathbf{v}_n(t) \end{pmatrix} \\
&= A\begin{pmatrix} \mathbf{v}_1(t) & \cdots & \mathbf{v}_n(t) \end{pmatrix} = A\Phi(t)
\end{aligned}
$$

Now look for a particular solution in the form $\Phi(t)\mathbf{v}(t)$. Thus from the product rule,

$$\Phi'(t)\mathbf{v}(t) + \Phi(t)\mathbf{v}'(t) = A\Phi(t)\mathbf{v}(t) + \mathbf{f}(t)$$

Thus,

$$A\Phi(t)\mathbf{v}(t) + \Phi(t)\mathbf{v}'(t) = A\Phi(t)\mathbf{v}(t) + \mathbf{f}(t)$$

Therefore,

$$\Phi(t)\mathbf{v}'(t) = \mathbf{f}(t)$$

and so $\Phi(t)\mathbf{v}(t)$ will be a particular solution if and only if

$$\mathbf{v}(t) \in \int \Phi(t)^{-1} f(t)\, dt$$

Example 9.2.2 *Here is a matrix.* $A = \begin{pmatrix} 3 & 4 \\ -2 & -3 \end{pmatrix}$. *Find the general solution to the system*

$$\mathbf{x}' = A\mathbf{x} + \mathbf{f}$$

where

$$\mathbf{f}(t) = \begin{pmatrix} \sin(t) \\ t \end{pmatrix}$$

First you need the general solution to the homogeneous equation. The eigenvalues are -1 and 1 and eigenvectors are, respectively,

$$\begin{pmatrix} -1 & 1 \end{pmatrix}^T, \begin{pmatrix} -2 & 1 \end{pmatrix}^T$$

Therefore, a fundamental matrix is

$$\Phi(t) = \begin{pmatrix} -e^{-t} & -2e^t \\ e^{-t} & e^t \end{pmatrix}$$

Next you need to find $\Phi(t)^{-1}$. This is the sort of thing where it may be convenient to use the formula for the inverse which comes in terms of the transpose of the cofactor matrix divided by the determinant.

$$\Phi(t)^{-1} = \begin{pmatrix} e^t & 2e^t \\ -e^{-t} & -e^{-t} \end{pmatrix}$$

Next

$$\Phi(t)^{-1}\mathbf{f}(t) = \begin{pmatrix} 2te^t + e^t \sin t \\ -te^{-t} - (\sin t)\,e^{-t} \end{pmatrix}$$

Now, you need to find an antiderivative of this.

$$\mathbf{v}(t) = \begin{pmatrix} -\frac{1}{2}e^t(\cos t - 4t - \sin t + 4) \\ \frac{1}{2}e^{-t}(2t + \cos t + \sin t + 2) \end{pmatrix} \in \int \begin{pmatrix} 2te^t + e^t \sin t \\ -te^{-t} - (\sin t)\,e^{-t} \end{pmatrix} dt$$

Then a particular solution is $\Phi(t)\mathbf{v}(t)$ which equals

$$\begin{pmatrix} -e^{-t} & -2e^t \\ e^{-t} & e^t \end{pmatrix}\begin{pmatrix} -\frac{1}{2}e^t(\cos t - 4t - \sin t + 4) \\ \frac{1}{2}e^{-t}(2t + \cos t + \sin t + 2) \end{pmatrix} = \begin{pmatrix} -\frac{1}{2}(8t + \cos t + 3\sin t) \\ 3t + \sin t - 1 \end{pmatrix}$$

If you check this, you will find that it works. Therefore, the general solution is

$$\left(\begin{array}{cc} -e^{-t} & -2e^{t} \\ e^{-t} & e^{t} \end{array} \right) \left(\begin{array}{c} c_1 \\ c_2 \end{array} \right) + \left(\begin{array}{c} -\frac{1}{2}\left(8t + \cos t + 3\sin t\right) \\ 3t + \sin t - 1 \end{array} \right)$$

where $c_i \in \mathbb{R}$.

Here is the explicit procedure.

Procedure 9.2.3 *To find the solutions to* $\mathbf{x}' = A\mathbf{x} + \mathbf{f}$*, find a fundamental matrix* $\Phi(t)$*, a matrix whose columns are solutions to* $\mathbf{x}' = A\mathbf{x}$ *which has an inverse for all t. Then the general solution is of the form*

$$\Phi(t)\mathbf{c} + \Phi(t)\mathbf{v}(t), \ \ \mathbf{c} \in \mathbb{C}^n \ is \ arbitrary$$

where

$$\mathbf{v}(t) \in \int \Phi(t)^{-1}\mathbf{f}(t)\,dt$$

9.3   Laplace Transforms and First Order Systems

If you are looking for a convenient way to get closed form solutions in the special cases when they can be found, probably the easiest way to do it is to use Laplace transforms. This is based on the following simple observation.

$$\int_0^\infty \mathbf{x}'(t)e^{-st}dt = -\mathbf{x}(0) + \int_0^\infty \mathbf{x}(t)e^{-st}dt$$

for all s large enough, which was noted earlier in the case of scalar valued functions under the assumption that the component functions of $\mathbf{x}(t)$ have exponential growth. You have

$$\mathbf{x}'(t) = A\mathbf{x}(t) + \mathbf{f}(t), \ \ \mathbf{x}(0) = \mathbf{x}_0$$

Then taking Laplace transforms, and letting $X(s)$ denote the Laplace transform of $\mathbf{x}(t)$,

$$sX(s) - \mathbf{x}_0 = AX(s) + F(s), \ \ F(s) = \mathcal{L}(\mathbf{f}(t))$$

and so

$$\begin{aligned} (sI - A)X(s) &= \mathbf{x}_0 + F(s) \\ X(s) &= (sI - A)^{-1}(\mathbf{x}_0 + F(s)) \end{aligned}$$

Then it is just a matter of taking inverse Laplace transforms of both sides to find $\mathbf{x}(t)$. Of course this can be done very easily in MATLAB. Here is an outline

>>syms s t; a=(enter initial vector here); b=(enter sI-A here); c=(enter $\mathbf{f}(t)$ here);

simplify(ilaplace(inv(b)*(a+laplace(c))))

Then press enter and it will give you $\mathbf{x}(t)$. You can enter the identity matrix as eye(n) where A is $n \times n$.

Example 9.3.1 *Here is a matrix.* $A = \begin{pmatrix} 3 & 4 \\ -2 & -3 \end{pmatrix}$. *Find the solution to the system*

$$\mathbf{x}' = A\mathbf{x} + \mathbf{f}, \ \mathbf{x}(0) = \begin{pmatrix} 1 \\ 2 \end{pmatrix}$$

where

$$\mathbf{f}(t) = \begin{pmatrix} \sin(t) \\ t \end{pmatrix}$$

In this case you just have to type in the following according to the above procedure.

syms s t; a=[1;2]; b=s*[1 0;0 1]-[3 4;-2 -3]; c=[sin(t);t];
simplify(ilaplace(inv(b)*(a+laplace(c))))
Then it gives
ans =
9*exp(t) - (15*exp(-t))/2 - cos(t)/2 - 4*t - (3*sin(t))/2
3*t + (15*exp(-t))/2 - (9*exp(t))/2 + sin(t) - 1
In standard notation, this is

$$\begin{pmatrix} 9e^t - \frac{1}{2}\cos t - 4t - \frac{3}{2}\sin t - \frac{15}{2}e^{-t} \\ 3t - \frac{9}{2}e^t + \sin t + \frac{15}{2}e^{-t} - 1 \end{pmatrix}$$

You can check that this is the solution to the initial value problem.

Example 9.3.2 Use the method of Laplace transforms to solve the following initial value problem.

$$\mathbf{x}' = \begin{pmatrix} 3 & 3 & 2 \\ -1 & 0 & -1 \\ 0 & -1 & 1 \end{pmatrix} \mathbf{x} + \begin{pmatrix} 1 \\ 1 \\ \sin(t) \end{pmatrix}, \ \mathbf{x}(0) = \begin{pmatrix} 0 \\ 1 \\ 0 \end{pmatrix}$$

The matrix is defective.

From the above procedure, you type in
syms s t; a=[0;1;0]; b=s*eye(3)-[3 3 2;-1 0 -1;0 -1 1]; c=[1;1;sin(t)];
simplify(ilaplace(inv(b)*(a+laplace(c))))
Then press "enter" and it gives the following.
ans =
(11*exp(2*t))/5 + (4*cos(t))/5 - 6*exp(t) + sin(t)/10 + (11*t*exp(t))/2 + 3
(11*exp(t))/2 - (3*cos(t))/10 - (11*exp(2*t))/5 - sin(t)/10 - 2
(11*exp(2*t))/5 - (7*cos(t))/10 + exp(t)/2 - (2*sin(t))/5 - (11*t*exp(t))/2 - 2
In usual notation, this is

$$\begin{pmatrix} \frac{4}{5}\cos t - 6e^t + \frac{1}{10}\sin t + \frac{11}{5}e^{2t} + \frac{11}{2}te^t + 3 \\ \frac{11}{2}e^t - \frac{3}{10}\cos t - \frac{1}{10}\sin t - \frac{11}{5}e^{2t} - 2 \\ \frac{1}{2}e^t - \frac{7}{10}\cos t - \frac{2}{5}\sin t + \frac{11}{5}e^{2t} - \frac{11}{2}te^t - 2 \end{pmatrix}$$

It was effortless to get this solution.

However, it is even easier in Scientific Notebook. You enter the following in math mode, click on the matrix and then: compute, solve ODE, laplace. It then produces the solution shown, although it might give y before x.

$$x' = 3x + 3y + 2z + 1$$

$$\begin{aligned} y' &= -x - z + 1 \\ z' &= z - y + \sin(t) \\ x(0) &= 0 \\ y(0) &= 1 \\ z(0) &= 0 \end{aligned} \quad , \quad \begin{aligned} x(t) &= \tfrac{4}{5}\cos t - 6e^t + \tfrac{1}{10}\sin t + \tfrac{11}{5}e^{2t} + \tfrac{11}{2}te^t + 3, \\ y(t) &= \tfrac{11}{2}e^t - \tfrac{3}{10}\cos t - \tfrac{1}{10}\sin t - \tfrac{11}{5}e^{2t} - 2, \\ z(t) &= \tfrac{1}{2}e^t - \tfrac{7}{10}\cos t - \tfrac{2}{5}\sin t + \tfrac{11}{5}e^{2t} - \tfrac{11}{2}te^t - 2 \end{aligned}$$

Another thing should be noted. There is a way to compute inverse Laplace transforms numerically which MATLAB knows how to do. This means that even if you don't have exact solutions to the characteristic equation, it can still give you an approximate solution.

Example 9.3.3 *Find the solution to the initial value problem*

$$\mathbf{x}' = A\mathbf{x} + \mathbf{f}(t)$$

$$A = \begin{pmatrix} 1 & 2 & 3 \\ 1 & 0 & 1 \\ 0 & -1 & 1 \end{pmatrix}, \ \mathbf{f}(t) = \begin{pmatrix} 1 \\ 1 \\ \sin t \end{pmatrix}$$

This is a problem because we can't factor the characteristic polynomial. However, one can make a small modification in what is entered in MATLAB and still get a useable solution. You type

syms s t; a=[0;1;0]; b=s*[1 0 0;0 1 0;0 0 1]-[1 2 3;1 0 1;0 -1 1]; c=[1;1;sin(t)];
vpa((ilaplace(inv(b)*(a+laplace(c)))),2)

Note how there is a vpa placed on the left instead of simplify. The result is as follows.

ans =

0.54*cos(t) - 0.53*exp(-1.1*t) + 0.24*sin(t) + exp(1.6*t)*cos(1.0*t)*(0.5 - 0.29i) + exp(1.6*t)*cos(1.0*t)*(0.5 + 0.29i)
+ exp(1.6*t)*sin(1.0*t)*(0.29 + 0.5i) + exp(1.6*t)*sin(1.0*t)*(0.29 - 0.5i) - 1.0

0.33*exp(-1.1*t) + 0.22*cos(t) + 0.3*sin(t) + exp(1.6*t)*cos(1.0*t)*(0.22 - 0.18i) + exp(1.6*t)*cos(1.0*t)*(0.22 + 0.18i) + exp(1.6*t)*sin(1.0*t)*(0.18 + 0.22i) + exp(1.6*t)*sin(1.0*t)*(0.18 - 0.22i)

0.16*exp(-1.1*t) - 0.24*cos(t) - 0.46*sin(t) + exp(1.6*t)*cos(1.0*t)*(0.043 + 0.24i) + exp(1.6*t)*cos(1.0*t)*(0.043 - 0.24i) - exp(1.6*t)*sin(1.0*t)*(0.24 - 0.043i) - exp(1.6*t)*sin(1.0*t)*(0.24 + 0.043i)

It is a column vector. Note the inclusion of i in some of the terms. This is inevitable because the method for producing the inverse Laplace transform involves the complex plane. This will not be pursued further in this book. You can check that these imaginary terms will cancel and this is a good thing, because the solution is real valued. In other words, one could simplify the above expression by getting rid of the imaginary parts.

9.4 Fundamental Matrices and Laplace Transforms

As a case of the above method for finding solutions to the initial value problem using Laplace transforms, consider the problem of finding the fundamental matrix. Recall that for A an $n \times n$ matrix, $\Phi(t)$ is a fundamental matrix if and only if

$$\Phi'(t) = A\Phi(t), \ \ \Phi(0) = I$$

Take Laplace transforms of both sides. This means you replace each entry with its Laplace transform. It is left as an exercise that if $\Psi(s)$ is the Laplace transform of $\Phi(t)$,

$$s\Psi(s) - I = A\Psi(s)$$

Therefore,

$$(sI - A)\Psi(s) = I, \ \ \Psi(s) = (sI - A)^{-1}$$

and so $\Phi(t)$ is the inverse Laplace transform of $\Psi(s)$. Recall that from the cofactor method for finding the inverse, you can find a formula for $(sI - A)^{-1}$. Thus you can find the fundamental matrix as follows.

syms s t; a=(enter sI-A here); simplify(ilaplace(inv(a)))

Then press "enter" and it will give you the fundamental matrix. Your installation of MATLAB must have the symbolic math toolbox to do this.

Example 9.4.1 *Find the fundamental matrix for*

$$\begin{pmatrix} 0 & 1 & 1 \\ -1 & 1 & -2 \\ 1 & 0 & 3 \end{pmatrix}$$

You type in the following and press "enter" to get the answer which is given.

syms s t; a=s*eye(3)-[0 1 1;-1 1 -2;1 0 3]; simplify(ilaplace(inv(a)))

ans =

[1, exp(t) - 1, exp(t) - 1]

[1/3 - exp(3*t)/3, (3*exp(t))/2 - exp(3*t)/6 - 1/3, (3*exp(t))/2 - (7*exp(3*t))/6 - 1/3]

[exp(3*t)/3 - 1/3, ((exp(t) - 1)^2*(exp(t) + 2))/6, (7*exp(3*t))/6 - exp(t)/2 + 1/3]

In standard notation, this says that the fundamental matrix is

$$\begin{pmatrix} 1 & e^t - 1 & e^t - 1 \\ \frac{1}{3} - \frac{1}{3}e^{3t} & \frac{3}{2}e^t - \frac{1}{6}e^{3t} - \frac{1}{3} & \frac{3}{2}e^t - \frac{7}{6}e^{3t} - \frac{1}{3} \\ \frac{1}{3}e^{3t} - \frac{1}{3} & \left(\frac{1}{6}e^t + \frac{1}{3}\right)\left(e^t - 1\right)^2 & \frac{7}{6}e^{3t} - \frac{1}{2}e^t + \frac{1}{3} \end{pmatrix}$$

Note how the identity matrix was entered by typing: eye(3). Also note how there is no fussing with generalized eigenvectors. However, you will not get a nice solution if the characteristic equation cannot be factored. In this case, you just replace simplify(ilaplace(inv(a))) with vpa(ilaplace(inv(a)),4) to get four digits in an approximate fundamental matrix.

This greatly simplifies the problem of finding a fundamental matrix for $\mathbf{x}' = A\mathbf{x}$. Just use MATLAB to make it simple and avoid all that nonsense with generalized eigenvectors.

In Scientific Notebook, it is even easier. Type in math mode $(sI - A)^{-1}$ and then evaluate to get the inverse. Then with the cursor in the end result, you go to compute transforms inverse Laplace and click on it to get the same thing. Both MATLAB and Scientific notebook easily beat the method of generalized eigenvectors. The output looks nicer in Scientific notebook.

9.5 Using MATLAB to Find Numerical Solutions

You can use MATLAB to find numerical solutions directly and it is typically the case that this is the only thing that can be done because, as noted earlier, we can't typically factor polynomials. Also, what of the case where the matrix has entries which are functions of t? Suppose you wish to solve

$$\begin{pmatrix} x \\ y \end{pmatrix}' = \begin{pmatrix} t^2 & \sin t \\ t+1 & 2 \end{pmatrix} \begin{pmatrix} x \\ y \end{pmatrix}, \begin{pmatrix} x(0) \\ y(0) \end{pmatrix} = \begin{pmatrix} 0 \\ 1 \end{pmatrix}$$

Then you would type in the following:

>>f=@(t,y)[t^2*y(1)+sin(t)*y(2);(t+1)*y(1)+2*y(2)];
[t,x]=ode45(f,[0,2],[0;1]); plot(t,x)

The vector $\begin{pmatrix} x & y \end{pmatrix}^T$ is denoted as y=$\begin{pmatrix} y(1) & y(2) \end{pmatrix}^T$. Then press "enter" and it will graph these functions on $[0, 2]$. If you want a table, you type in

>> s=ode45(f,[0,2],[0;1]);
deval(s,[0,.2,.4,.6,.8,1,1.2,1.4,1.6,1.8,2])

Then when you press "enter", you get a table of column vectors which give the values of the vector $\begin{pmatrix} y(1) \\ y(2) \end{pmatrix}$ at the specified values of t. You can force the differential equations solver to consider smaller step sizes by replacing ode45 material with something like "ode45(f,$\overbrace{[0:0.04:2]}^{\text{change here}}$,[0;1]);". This will make all the step sizes no larger than .04.

Then, when you have what you want, you ought to type "close all" and then "enter" and then type "clear vars" and "enter" and clf and then "enter" to get rid of any figures. This is so you can do something else without closing MATLAB and starting it over again. MATLAB remembers the functions which have been defined and so unless you do this, it may think you are referring to something other than what you want if you do another computation without closing it down.

9.6 Scientific Notebook and Numerical Solutions

Consider the same problem,

$$\begin{pmatrix} x \\ y \end{pmatrix}' = \begin{pmatrix} t^2 & \sin t \\ t+1 & 2 \end{pmatrix} \begin{pmatrix} x \\ y \end{pmatrix}, \begin{pmatrix} x(0) \\ y(0) \end{pmatrix} = \begin{pmatrix} 0 \\ 1 \end{pmatrix}$$

In this system, you just write the following in math (be sure the little T on the top is replaced with a red M), click on it and go to compute, solve ode, and numeric. This will return the functions x, y computed numerically. Here is what appears.

$$\begin{pmatrix} x' = t^2 x + y \sin t \\ y' = 2y + x(t+1) \\ x(0) = 0 \\ y(0) = 1 \end{pmatrix}, \text{ Functions defined: } x, y$$

If you want a table of values for y, you type a column vector, select it, and enclose with () and then "enter" to get the following. It is similar for x.

$$y \begin{pmatrix} 0 \\ .2 \\ .4 \\ .6 \\ .7 \end{pmatrix} = \begin{pmatrix} 1.0 \\ 1.4939 \\ 2.251 \\ 3.4521 \\ 4.3154 \end{pmatrix}, \quad x \begin{pmatrix} 0 \\ .2 \\ .4 \\ .6 \\ .7 \end{pmatrix} = \begin{pmatrix} 0 \\ 2.61 \times 10^{-2} \\ 0.13901 \\ 0.42513 \\ 0.68291 \end{pmatrix}$$

The software would not graph the functions. When you use this, you have defined some functions. You should then click: "compute, definitions, clear definitions" to clear out these definitions. While it may have some limitations, this software is easy to use and when it works, it does very well and seems to take a lot less time to use than a more conventional computer algebra system.

9.7   Exercises

In the following problems, you can do them by hand or you can use MATLAB and have it do all the work for you using Laplace transforms. If you use the latter technique, these will not be very difficult.

1. Find the fundamental matrix $\Phi(t)$ for $\Phi'(t) = A\Phi(t)$ and $\Phi(0) = I$. Here

$$A = \begin{pmatrix} -11 & -4 \\ 42 & 15 \end{pmatrix}$$

2. Find the fundamental matrix $\Phi(t)$ for $\Phi'(t) = A\Phi(t)$ and $\Phi(0) = I$. Here

$$A = \begin{pmatrix} -34 & -5 \\ 252 & 37 \end{pmatrix}$$

3. Find the fundamental matrix $\Phi(t)$ for $\Phi'(t) = A\Phi(t)$ and $\Phi(0) = I$. Here

$$A = \begin{pmatrix} -38 & -12 \\ 117 & 37 \end{pmatrix}$$

4. Find the solution of the initial value problem $\mathbf{x}' = A\mathbf{x} + \mathbf{f}, \mathbf{x}(0) = \mathbf{x}_0$ where

$$\mathbf{f}(t) = \begin{pmatrix} 5t \\ 1 \end{pmatrix}, \mathbf{x}_0 = \begin{pmatrix} 2 \\ -1 \end{pmatrix}, A = \begin{pmatrix} -8 & 30 \\ -2 & 8 \end{pmatrix}$$

5. Find the solution of the initial value problem $\mathbf{x}' = A\mathbf{x} + \mathbf{f}, \mathbf{x}(0) = \mathbf{x}_0$ where

$$\mathbf{f}(t) = \begin{pmatrix} 2t \\ 3 \end{pmatrix}, \mathbf{x}_0 = \begin{pmatrix} 1 \\ 2 \end{pmatrix}, A = \begin{pmatrix} -\frac{9}{2} & 5 \\ -\frac{15}{4} & \frac{11}{2} \end{pmatrix}$$

6. Find the solution of the initial value problem $\mathbf{x}' = A\mathbf{x} + \mathbf{f}, \mathbf{x}(0) = \mathbf{x}_0$ where

$$\mathbf{f}(t) = \begin{pmatrix} 2t \\ 5 \end{pmatrix}, \mathbf{x}_0 = \begin{pmatrix} -1 \\ 3 \end{pmatrix}, A = \begin{pmatrix} \frac{1}{2} & -1 \\ -\frac{5}{4} & -\frac{3}{2} \end{pmatrix}$$

7. Find the solution of the initial value problem $\mathbf{x}' = A\mathbf{x} + \mathbf{f}, \mathbf{x}(0) = \mathbf{x}_0$ where

$$\mathbf{f}(t) = \begin{pmatrix} 4t \\ 1 \end{pmatrix}, \mathbf{x}_0 = \begin{pmatrix} -2 \\ -2 \end{pmatrix}, A = \begin{pmatrix} 0 & 4 \\ -\frac{1}{2} & 3 \end{pmatrix}$$

8. Find the solution of the initial value problem $\mathbf{x}' = A\mathbf{x} + \mathbf{f}, \mathbf{x}(0) = \mathbf{x}_0$ where

$$\mathbf{f}(t) = \begin{pmatrix} 3 \\ 3t \end{pmatrix}, \mathbf{x}_0 = \begin{pmatrix} -2 \\ -2 \end{pmatrix}, A = \begin{pmatrix} 6 & -9 \\ 3 & -6 \end{pmatrix}$$

9. Find the solution of the initial value problem $\mathbf{x}' = A\mathbf{x} + \mathbf{f}, \mathbf{x}(0) = \mathbf{x}_0$ where

$$\mathbf{f}(t) = \begin{pmatrix} 1 \\ 2t \end{pmatrix}, \mathbf{x}_0 = \begin{pmatrix} 3 \\ -2 \end{pmatrix}, A = \begin{pmatrix} -\frac{16}{7} & \frac{15}{7} \\ \frac{10}{7} & \frac{9}{7} \end{pmatrix}$$

10. Find the solution of the initial value problem $\mathbf{x}' = A\mathbf{x} + \mathbf{f}, \mathbf{x}(0) = \mathbf{x}_0$ where

$$\mathbf{f}(t) = \begin{pmatrix} 3 \\ 7t \end{pmatrix}, \mathbf{x}_0 = \begin{pmatrix} 3 \\ 2 \end{pmatrix}, A = \begin{pmatrix} -\frac{5}{6} & -\frac{1}{2} \\ \frac{7}{18} & -\frac{13}{6} \end{pmatrix}$$

The eigenvalues are $-2, -1$.

11. Find the solution of the initial value problem $\mathbf{x}' = A\mathbf{x} + \mathbf{f}, \mathbf{x}(0) = \mathbf{x}_0$ where

$$\mathbf{f}(t) = \begin{pmatrix} e^{4t} \\ 4t \end{pmatrix}, \mathbf{x}_0 = \begin{pmatrix} 3 \\ -2 \end{pmatrix}, A = \begin{pmatrix} \frac{13}{7} & \frac{2}{7} \\ -\frac{4}{7} & \frac{22}{7} \end{pmatrix}$$

12. Find the solution of the initial value problem $\mathbf{x}' = A\mathbf{x} + \mathbf{f}, \mathbf{x}(0) = \mathbf{x}_0$ where

$$\mathbf{f}(t) = \begin{pmatrix} e^{5t} \\ 5t \end{pmatrix}, \mathbf{x}_0 = \begin{pmatrix} 3 \\ 3 \end{pmatrix}, A = \begin{pmatrix} 3 & 4 \\ -2 & -3 \end{pmatrix}$$

13. Find the solution of the initial value problem $\mathbf{x}' = A\mathbf{x} + \mathbf{f}, \mathbf{x}(0) = \mathbf{x}_0$ where

$$\mathbf{f}(t) = \begin{pmatrix} 3e^{4t} \\ 4t \end{pmatrix}, \mathbf{x}_0 = \begin{pmatrix} 3 \\ 2 \end{pmatrix}, A = \begin{pmatrix} 11 & -9 \\ 12 & -10 \end{pmatrix}$$

The eigenvalues are $-1, 2$.

14. Find the solution of the initial value problem $\mathbf{x}' = A\mathbf{x} + \mathbf{f}, \mathbf{x}(0) = \mathbf{x}_0$ where

$$\mathbf{f}(t) = \begin{pmatrix} 4e^{4t} \\ 4t \end{pmatrix}, \mathbf{x}_0 = \begin{pmatrix} 3 \\ -2 \end{pmatrix}, A = \begin{pmatrix} 0 & 3 \\ 3 & 0 \end{pmatrix}$$

15. Find the solution of the initial value problem $\mathbf{x}' = A\mathbf{x} + \mathbf{f}, \mathbf{x}(0) = \mathbf{x}_0$ where

$$\mathbf{f}(t) = \begin{pmatrix} 5e^{2t} \\ 2t \\ 1 \end{pmatrix}, \mathbf{x}_0 = \begin{pmatrix} 3 \\ -1 \\ 0 \end{pmatrix}, A = \begin{pmatrix} 7 & -6 & 36 \\ -4 & 5 & -24 \\ -2 & 2 & -11 \end{pmatrix}$$

The eigenvalues are $1, 1, -1$.

16. Find the solution of the initial value problem $\mathbf{x}' = A\mathbf{x} + \mathbf{f}, \mathbf{x}(0) = \mathbf{x}_0$ where

$$\mathbf{f}(t) = \begin{pmatrix} e^{2t} \\ 2t \\ 1 \end{pmatrix}, \mathbf{x}_0 = \begin{pmatrix} 3 \\ -1 \\ 0 \end{pmatrix}, A = \begin{pmatrix} -2 & 3 & -18 \\ 2 & -1 & 12 \\ 1 & -1 & 7 \end{pmatrix}$$

The eigenvalues are $1, 1, 2$.

17. Find the solution of the initial value problem $\mathbf{x}' = A\mathbf{x} + \mathbf{f}, \mathbf{x}(0) = \mathbf{x}_0$ where

$$\mathbf{f}(t) = \begin{pmatrix} e^t \\ t \\ 1 \end{pmatrix}, \mathbf{x}_0 = \begin{pmatrix} 1 \\ 0 \\ 1 \end{pmatrix}, A = \begin{pmatrix} 12 & 31 & -32 \\ -8 & -21 & 22 \\ -3 & -9 & 11 \end{pmatrix}$$

The eigenvalues are $1, -1, 2$.

18. Find the solution of the initial value problem $\mathbf{x}' = A\mathbf{x} + \mathbf{f}, \mathbf{x}(0) = \mathbf{x}_0$ where

$$\mathbf{f}(t) = \begin{pmatrix} 5e^{2t} \\ 2t \end{pmatrix}, \mathbf{x}_0 = \begin{pmatrix} 3 \\ -1 \end{pmatrix}, A = \begin{pmatrix} \frac{11}{9} & \frac{40}{9} \\ \frac{8}{9} & \frac{7}{9} \end{pmatrix}$$

19. Find the solution of the initial value problem $\mathbf{x}' = A\mathbf{x} + \mathbf{f}, \mathbf{x}(0) = \mathbf{x}_0$ where

$$\mathbf{f}(t) = \begin{pmatrix} \sin t \\ t \\ \cos t \end{pmatrix}, \mathbf{x}_0 = \begin{pmatrix} 1 \\ 0 \\ 0 \end{pmatrix}, A = \begin{pmatrix} 6 & 28 & 47 \\ -3 & -17 & -31 \\ 1 & 8 & 16 \end{pmatrix}$$

The eigenvalues are $1, 1, 3$.

20. Find the solution of the initial value problem $\mathbf{x}' = A\mathbf{x} + \mathbf{f}, \mathbf{x}(0) = \mathbf{x}_0$ where

$$\mathbf{f}(t) = \begin{pmatrix} \sin t \\ t \\ \cos t \end{pmatrix}, \mathbf{x}_0 = \begin{pmatrix} 0 \\ 1 \\ 0 \end{pmatrix}, A = \begin{pmatrix} 12 & 22 & 11 \\ -7 & -13 & -7 \\ 3 & 6 & 4 \end{pmatrix}$$

The eigenvalues are $1, 1, 1$.

21. Suppose A is a matrix and you want solutions to $t\mathbf{x}' = A\mathbf{x}$. Show that $\mathbf{z}t^\alpha$ is a solution if α is an eigenvalue and $\mathbf{z}$ is an eigenvector. Now use this to find general solutions to the following degenerate systems of equations in which $A =$

(a) $\begin{pmatrix} 3 & 4 \\ -2 & -3 \end{pmatrix}$

(b) $\begin{pmatrix} 4 & 6 \\ -3 & -5 \end{pmatrix}$

(c) $\begin{pmatrix} 3 & 2 \\ -1 & 0 \end{pmatrix}$

(d) $\begin{pmatrix} -5 & 6 \\ -3 & 4 \end{pmatrix}$

(e) $\begin{pmatrix} 6 & -13 & -22 \\ 8 & -21 & -38 \\ -3 & 9 & 17 \end{pmatrix}$

(f) $\begin{pmatrix} 7 & -12 & -18 \\ 10 & -19 & -30 \\ -4 & 8 & 13 \end{pmatrix}$

(g) $\begin{pmatrix} 14 & -15 & -15 \\ 21 & -22 & -21 \\ -9 & 9 & 8 \end{pmatrix}$

22. In the above problem, show these degenerate equations could have been solved by changing the independent variable to reduce to $t = e^s$. Then in terms of the new variable s show that the equation reduces to $\mathbf{x}' = A\mathbf{x}$.

23. There are three 10 liter tanks, T_1, T_2, and T_3. Polluted water having 1 kg. per liter of pollutant flows into T_1 at the rate of 1 liter per hour. From T_1 the water goes to T_2 at the rate of 5 liters per hour and from T_2 it goes to T_3 also at the rate of 5 liters per hour. Then from T_3 it flows out on the ground at the rate of 1 liters per hour and part goes to T_1 at the rate of 4 liters per hour. Let x_i be the number of kg. of pollutant in T_i. Each tank is well mixed. Find a differential equation which describes x_i and graph $(t, x_3(t))$ for $t \in [0, 20]$ with the initial condition consisting of each $x_i(0) = 0$. Find $\lim_{t \to \infty} x_3(t)$. Repeat with initial condition $x_i(0) = 12$.

24. Show that there is at most one solution to the problem $\Psi'(t) = A\Psi(t)$, $\Psi(0) = M$ where M is a given matrix. **Hint:** You might consider Problem 85 on Page 76..

25. Suppose $\Psi(t)$ is $p \times p$ and $\Psi'(t) = A\Psi(t)$. Express $\Psi(t)$ in terms of the fundamental matrix $\Phi(t)$ where $\Phi'(t) = A\Phi(t)$, $\Phi(0) = I$.

26. For Φ the fundamental matrix which satisfies $\Phi'(t) = A\Phi(t)$, $\Phi(0) = I$, recall that $\Phi(t)A = A\Phi(t)$. However, if you just know $\Psi' = A\Psi$, it does not follow that $\Psi A = A\Psi$. Consider

$$A = \begin{pmatrix} 1 & 0 \\ 0 & 2 \end{pmatrix}$$

and suppose $\Psi'(t) = A\Psi(t)$, $\Psi(0) = \begin{pmatrix} 0 & 1 \\ 1 & 0 \end{pmatrix}$. Find $\Psi(t)$ and then show that $A\Psi(t) \neq \Psi(t)A$. This shows that the special fundamental matrix which equals I when $t = 0$ is better than an ordinary fundamental matrix which just has the columns independent and satisfies $\Psi' = A\Psi$.

27. Let $\Psi(t)$ be a fundamental matrix for A. That is, the columns of $\Psi(t)$ consist of a set of linearly independent vectors for each t and each column is a solution to $\mathbf{x}' = A\mathbf{x}$. Let $\Phi(t)$ be **the**. That is, $\Phi'(t) = A\Phi(t)$, $\Phi(0) = I$. Show an easy way to obtain $\Phi(t)$ from $\Psi(t)$.

28. Let A be an $n \times n$ matrix having fundamental matrix $\Phi(t)$. That is $\Phi' = A\Phi$ and $\Phi(0) = I$. Also let $t \to \Psi(t)$ be a continuous function having values in the space of $n \times m$ matrices. Find a formula for the solution to the initial value problem $\Sigma' = A\Sigma + \Psi$, $\Sigma(0) = \Sigma_0$. **Hint:** You might consider the columns of Σ each one being the solution of an initial value problem.

29. Find fundamental matrices for the following matrices. You might want to use MATLAB or Scientific Notebook or some other computer algebra system.

(a) $\begin{pmatrix} -1 & 0 & -6 \\ 1 & 1 & 3 \\ 1 & 0 & 4 \end{pmatrix}$

(c) $\begin{pmatrix} 5 & 12 & -6 \\ -3 & -7 & 3 \\ -1 & -4 & 4 \end{pmatrix}$

(b) $\begin{pmatrix} -10 & -15 & -9 \\ 6 & 9 & 5 \\ 4 & 5 & 5 \end{pmatrix}$

(d) $\begin{pmatrix} 7 & 12 & 0 \\ -4 & -7 & 0 \\ -2 & -4 & 1 \end{pmatrix}$

30. Here are some first order systems. Find numerical solutions and determine their values on 4 equally spaced points in $[0, 2]$. Also give a graph of the two functions on $[0, 2]$.

(a) $\begin{pmatrix} x \\ y \end{pmatrix}' = \begin{pmatrix} t & \sin t \\ 1 & \cos t \end{pmatrix} \begin{pmatrix} x \\ y \end{pmatrix} + \begin{pmatrix} t \\ \sin t \end{pmatrix}$, $\begin{pmatrix} x(0) \\ y(0) \end{pmatrix} = \begin{pmatrix} 0 \\ 1 \end{pmatrix}$

(b) $\begin{pmatrix} x \\ y \end{pmatrix}' = \begin{pmatrix} t & t+1 \\ 1 & \cos t \end{pmatrix} \begin{pmatrix} x \\ y \end{pmatrix} + \begin{pmatrix} \cos t \\ t-1 \end{pmatrix}$, $\begin{pmatrix} x(0) \\ y(0) \end{pmatrix} = \begin{pmatrix} 1 \\ 1 \end{pmatrix}$

(c) $\begin{pmatrix} x \\ y \end{pmatrix}' = \begin{pmatrix} 0 & \sin t \\ 1 & t+1 \end{pmatrix} \begin{pmatrix} x \\ y \end{pmatrix} + \begin{pmatrix} t \\ (t-1)^2 \end{pmatrix}$, $\begin{pmatrix} x(0) \\ y(0) \end{pmatrix} = \begin{pmatrix} 1 \\ 0 \end{pmatrix}$

(d) $\begin{pmatrix} x \\ y \end{pmatrix}' = \begin{pmatrix} t-1 & t+1 \\ \cos t & (t-1)^2 \end{pmatrix} \begin{pmatrix} x \\ y \end{pmatrix} + \begin{pmatrix} \sin t \\ 2t \end{pmatrix}$, $\begin{pmatrix} x(0) \\ y(0) \end{pmatrix} = \begin{pmatrix} 1 \\ 0 \end{pmatrix}$

31. If $A(t)$ is a matrix with real valued continuous functions as entries, and $\mathbf{x}'(t) = A\mathbf{x}(t)$, show that $\bar{\mathbf{x}}'(t) = A(t)\bar{\mathbf{x}}(t)$ and that

$$\text{span}(\mathbf{x}, \bar{\mathbf{x}}) = \text{span}(\text{Re}(\mathbf{x}), \text{Im}(\mathbf{x})).$$

Chapter 10

First Order Linear Systems, Theory

The main goal in this short chapter is to understand the existence and use of the fundamental matrix. It always exists. This was done earlier in Problem 85 on Page 76 but this chapter gives a better way of understanding this, based more on analysis.

10.1 Gronwall's Inequality

To begin with, there is a fundamental and elementary result known as Gronwall's inequality which is useful in dealing with fundamental questions related to differential equations.

Lemma 10.1.1 *Let $\mu \geq 0$. Suppose for $t \leq \tau$*

$$a(t) \leq \mu \int_t^\tau a(s)ds + C \tag{10.1}$$

where $a(t) \geq 0$ for all $t \leq \tau$. Then

$$a(t) \leq C e^{\mu(\tau-t)} \tag{10.2}$$

If $t \geq \tau$, and for $a(t) \geq 0$,

$$a(t) \leq C + \mu \int_\tau^t a(s)\, ds, \tag{10.3}$$

then

$$a(t) \leq C e^{\mu(t-\tau)} \tag{10.4}$$

Proof: Consider the first version of this inequality. If $\mu = 0$ there is nothing to show. Say $\mu > 0$.

$$a(t) + \mu \int_\tau^t a(s)\, ds \leq C$$

Then let $A(t) \equiv \int_\tau^t a(s)\,ds$. The above is of the form

$$A'(t) + \mu A(t) \leq C$$

As in the earlier material on first order equations, multiply by the integrating factor $e^{\mu(t-\tau)}$. Then

$$\frac{d}{dt}\left(A(t)\,e^{\mu(t-\tau)}\right) \leq Ce^{\mu(t-\tau)}$$

Now do $\int_t^\tau$ to both sides. The inequality is preserved because $t < \tau$. By the fundamental theorem of calculus,

$$A(\tau) - A(t)\,e^{\mu(t-\tau)} \leq C\int_t^\tau e^{\mu(s-\tau)}ds = \frac{1}{\mu}C\left(-e^{\mu(t-\mu)} + 1\right)$$

Now $A(\tau) = 0$ so

$$-A(t) \leq \frac{1}{\mu}C\left(-1 + e^{\mu(\tau-t)}\right), \quad -\mu A(t) \leq C\left(-1 + e^{\mu(\tau-t)}\right)$$

It follows that

$$\mu \int_t^\tau a(s)\,ds \leq C\left(-1 + e^{\mu(\tau-t)}\right)$$

and so from the assumption,

$$a(t) \leq \mu \int_t^\tau a(s)ds + C \leq C\left(-1 + e^{\mu(\tau-t)}\right) + C = Ce^{\mu(\tau-t)}$$

Now consider the other case. In this case you let $A(t)$ be defined the same way and

$$A'(t) - \mu A(t) \leq C$$

Then multiply by $e^{-\mu(t-\tau)}$ and obtain

$$\frac{d}{dt}\left(e^{-\mu(t-\tau)}A(t)\right) \leq Ce^{-\mu(t-\tau)}$$

Then do $\int_\tau^t$ to both sides. In this case $t > \tau$ and so the inequality is preserved. Hence

$$A(t)\,e^{-\mu(t-\tau)} - 0 \leq C\int_\tau^t e^{-\mu(s-\tau)}ds = Ce^{\mu\tau}\frac{1}{\mu}\left(e^{-\mu\tau} - e^{-\mu t}\right) = C\frac{1}{\mu}\left(1 - e^{-\mu(t-\tau)}\right)$$

It follows

$$A(t) \equiv \int_\tau^t a(s)\,ds \leq C\frac{1}{\mu}\left(e^{\mu(t-\tau)} - 1\right)$$

Then by hypothesis,

$$a(t) \leq C + \mu A(t) \leq C + C\left(e^{\mu(t-\tau)} - 1\right) = Ce^{\mu(t-\tau)} \blacksquare$$

10.2   Fundamental Matrix

You want to find a matrix valued function $\Phi(t)$ such that

$$\Phi'(t) = A\Phi(t), \ \Phi(0) = I, \ A \text{ is } p \times p \tag{10.5}$$

Such a matrix is called a fundamental matrix. The main purpose for this chapter is to show that there **always exists** a unique fundamental matrix for any $p \times p$ matrix A. For a different method based on linear algebra, see Problem 85 on Page 76. You need Schur's theorem for this approach. That done here is more easily generalized.

What do we mean when we say that for $\{B_n\}$ a sequence of matrices

$$\lim_{n\to\infty} B_n = B?$$

We mean the obvious thing. The ij^{th} entry of B_n converges to the ij^{th} entry of B. One convenient way to ensure that this happens is to give a measure of distance between matrices which will ensure that it happens.

Definition 10.2.1 *For A, B matrices of the same size, define $\|A - B\|_\infty$ to be*

$$\max\{|A_{ij} - B_{ij}|, \ all \ ij\}$$

Thus

$$\|A\|_\infty = \max\{|A_{ij}|, \ all \ ij\}$$

Then to say that $\lim_{n\to\infty} B_n = B$ is the same as saying that

$$\lim_{n\to\infty} \|B_n - B\|_\infty = 0$$

Similarly, a vector valued function $t \to \mathbf{f}(t)$ is said to be continuous if each of its component functions is continuous. It is differentiable if each of its entries is differentiable. To differentiate, one just differentiates the components and the differentiated vector is what results. Thus, for example, for $\mathbf{f}(t) = \begin{pmatrix} t & \sin(t^2) & e^t \end{pmatrix}^T$,

$$\mathbf{f}'(t) = \begin{pmatrix} t \\ \sin(t^2) \\ e^t \end{pmatrix}' = \begin{pmatrix} 1 \\ 2t\cos(t^2) \\ e^t \end{pmatrix} = \lim_{h\to 0} \frac{\mathbf{f}(t+h) - \mathbf{f}(t)}{h}$$

and this is a continuous vector valued function because each entry is a continuous function of t. More generally,

$$\Phi'(t) = \lim_{h\to 0} \frac{\Phi(t+h) - \Phi(t)}{h}$$

where the limit exists in the sense that $\lim_{h\to 0} \left\| \Phi'(t) - \frac{\Phi(t+h) - \Phi(t)}{h} \right\|_\infty = 0$. This is the same as taking the derivative of each entry of $\Phi(t)$.

Here is a useful lemma.

Lemma 10.2.2 *If A, B_n, B are $p \times p$ matrices and $\lim_{n \to \infty} B_n = B$, then*

$$
\begin{aligned}
\lim_{n \to \infty} A B_n &= AB, \\
\lim_{n \to \infty} B_n A &= BA,
\end{aligned}
\tag{10.6}
$$

Also

$$
\|AB\|_\infty \le p \|A\|_\infty \|B\|_\infty
\tag{10.7}
$$

$$
\|A^k\|_\infty \le p^{k-1} \|A\|_\infty^k
\tag{10.8}
$$

for any positive integer k and

$$
\left\| \sum_{k=1}^m A_k \right\|_\infty \le \sum_{k=1}^m \|A_k\|_\infty
$$

For t a scalar,

$$
\|tA\|_\infty = |t| \|A\|_\infty
\tag{10.9}
$$

Also

$$
|A\mathbf{x}| \le \sqrt{p} \|A\|_\infty |\mathbf{x}|
\tag{10.10}
$$

and

$$
\|A + B\|_\infty \le \|A\|_\infty + \|B\|_\infty
$$

Proof: First consider the claim (10.7).

$$
\begin{aligned}
\|AB\|_\infty &\equiv \sup_{i,j} \left| \sum_k A_{ik} B_{kj} \right| \le \sup_{i,j} \sum_k \|A\|_\infty \|B\|_\infty \\
&= \sup_{i,j} p \|A\|_\infty \|B\|_\infty \le p \|A\|_\infty \|B\|_\infty
\end{aligned}
$$

Now consider (10.6). From what was just shown,

$$
\|AB_n - AB\|_\infty = \|A(B_n - B)\|_\infty \le p \|A\|_\infty \|B_n - B\|_\infty
$$

which is assumed to converge to 0. Similarly $B_n A \to BA$. This establishes the first part of the lemma. Now (10.8) follows by induction. Indeed, the result holds for $k = 1$. Suppose true for $n - 1$ for $n \ge 2$. Then

$$
\|A A^{n-1}\|_\infty \le p \|A\|_\infty \|A^{n-1}\|_\infty \le p \|A\|_\infty p^{n-2} \|A\|_\infty^{n-1} = p^{n-1} \|A\|_\infty^n .
$$

Consider the claim about the sum.

$$
\left| \left(\sum_{k=1}^m A_k \right)_{ij} \right| = \left| \sum_{k=1}^m (A_k)_{ij} \right| \le \sum_{k=1}^m \|A_k\|_\infty
$$

Since this holds for arbitrary ij, it follows that

$$
\left\| \sum_{k=1}^m A_k \right\|_\infty \le \sum_{k=1}^m \|A_k\|_\infty
$$

as claimed. The assertion (10.9) is obvious. Consider (10.10). Using the Cauchy-Schwarz inequality as needed,

$$
|A\mathbf{x}| \equiv \left| \sum_{j=1}^{p} A_{ij} x_j \right| \leq \|A\|_\infty \sum_{j=1}^{p} |x_j|
$$

$$
\leq \|A\|_\infty \left(\sum_{j=1}^{p} 1^2 \right)^{1/2} \left(\sum_{j=1}^{p} |x_j|^2 \right)^{1/2} \leq \sqrt{p}\, \|A\|_\infty\, |\mathbf{x}|
$$

Now consider the last claim.

$$
|A_{ij} + B_{ij}| \leq \|A\|_\infty + \|B\|_\infty
$$

and so,

$$
\|A + B\|_\infty \leq \|A\|_\infty + \|B\|_\infty \quad \blacksquare
$$

The last claim leads to the following important proposition.

Proposition 10.2.3 *Let A, B be matrices. Then*

$$
\|A - B\|_\infty \geq |\,\|A\|_\infty - \|B\|_\infty\,|
$$

Proof: To save notation, I will not include the subscript ∞. This is also to emphasize that the same argument would work with other definitions of the norm. Say $\|A\| \geq \|B\|$ without loss of generality. If it is the other way, just reverse the argument. Then

$$
\begin{aligned}
|\,\|A\| - \|B\|\,| &= \|A\| - \|B\| = \|A - B + B\| - \|B\| \\
&\leq \|A - B\| + \|B\| - \|B\| = \|A - B\| \quad \blacksquare
\end{aligned}
$$

You should observe from this that if $s \to \Psi(s)$ is continuous, then so is $s \to \|\Psi(s)\|_\infty$.

By analogy with the situation in calculus, consider the infinite sum

$$
\sum_{k=0}^{\infty} \frac{A^k t^k}{k!} \equiv \lim_{n \to \infty} \sum_{k=0}^{n} \frac{A^k t^k}{k!}
$$

where here A is a $p \times p$ matrix having real or complex entries. Then letting $m < n$, it follows from the above lemma that

$$
\left\| \sum_{k=0}^{n} \frac{A^k t^k}{k!} - \sum_{k=0}^{m} \frac{A^k t^k}{k!} \right\|_\infty = \left\| \sum_{k=m+1}^{n} \frac{A^k t^k}{k!} \right\|_\infty \leq \sum_{k=m+1}^{n} \frac{|t|^k}{k!} \|A^k\|_\infty
$$

$$
\leq \sum_{k=m}^{\infty} \frac{|t|^k}{k!} \|A^k\|_\infty \leq \sum_{k=m}^{\infty} \frac{|t|^k p^k \|A\|_\infty^k}{k!}
$$

Now the series $\sum_{k=0}^{\infty} \frac{|t|^k p^k \|A\|_\infty^k}{k!}$ converges and in fact equals $\exp(|t|\, p\, \|A\|_\infty)$. It follows from calculus that

$$
\lim_{m \to \infty} \sum_{k=m}^{\infty} \frac{|t|^k p^k \|A\|_\infty^k}{k!} = 0.
$$

This has shown that the ij^{th} entry of the partial sum $\sum_{k=0}^{n} \frac{A^k t^k}{k!}$ is a Cauchy sequence and hence by completeness of $\mathbb{C}$ or $\mathbb{R}$ it converges. See Theorem B.4.6 for the case of most interest here where the entries of A are all real. Therefore, the above limit exists. This is stated as the essential part of the following theorem.

Theorem 10.2.4 *Let $t \in [a, b] \subseteq \mathbb{R}$ where $b - a < \infty$. Then for each $t \in [a, b]$,*

$$\lim_{n \to \infty} \sum_{k=0}^{n} \frac{A^k t^k}{k!} \equiv \sum_{k=0}^{\infty} \frac{A^k t^k}{k!} \equiv \Phi(t), \ A^0 \equiv I,$$

exists. Furthermore, there exists a single constant C such that for $t_k \in [a, b]$, the infinite sum

$$\sum_{k=0}^{\infty} \frac{A^k t_k^k}{k!}$$

converges and in fact

$$\left\| \sum_{k=0}^{\infty} \frac{A^k t_k^k}{k!} \right\|_{\infty} \leq C$$

Proof: The convergence for $\sum_{k=0}^{\infty} \frac{A^k t^k}{k!}$ was just established.
Consider the estimate. From the above lemma,

$$\left\| \sum_{k=0}^{n} \frac{A^k t_k^k}{k!} \right\|_{\infty} \leq \sum_{k=0}^{n} \frac{p^k (|a| + |b|)^k \|A\|_{\infty}^k}{k!} \leq \sum_{k=0}^{\infty} \frac{p^k (|a| + |b|)^k \|A\|_{\infty}^k}{k!}$$
$$= \exp\left(p(|a| + |b|) \|A\|_{\infty}\right)$$

It follows that the ij^{th} entry of $\sum_{k=0}^{n} \frac{A^k t_k^k}{k!}$ has magnitude no larger than the right side of the above inequality. Also, a repeat of the above argument after Lemma 10.2.2 shows that the partial sums of the ij^{th} entry of $\sum_{k=0}^{\infty} \frac{A^k t_k^k}{k!}$ form a Cauchy sequence. Hence passing to the limit, it follows from calculus that

$$\left| \left(\sum_{k=0}^{\infty} \frac{A^k t^k}{k!} \right)_{ij} \right| \leq \exp\left(p(|a| + |b|) \|A\|_{\infty}\right)$$

Since ij is arbitrary, this establishes the inequality. ∎

Next consider the derivative of $\Phi(t)$. Why is $\Phi(t)$ a solution to the above (10.5)? By the mean value theorem,

$$\frac{\Phi(t + h) - \Phi(t)}{h} = \frac{1}{h} \sum_{k=0}^{\infty} \frac{(t + h)^k - t^k}{k!} A^k = \frac{1}{h} \sum_{k=0}^{\infty} \frac{k(t + \theta_k h)^{k-1} h}{k!} A^k$$
$$= A \sum_{k=1}^{\infty} \frac{(t + \theta_k h)^{k-1}}{(k-1)!} A^{k-1} = A \sum_{k=0}^{\infty} \frac{(t + \theta_k h)^k}{k!} A^k, \theta_k \in (0, 1)$$

Does this sum converge to $\sum_{k=0}^{\infty} \frac{t^k}{k!} A^k \equiv \Phi(t)$? If so, it will have been shown that $\Phi'(t) = A\Phi(t)$. By the mean value theorem again, for , $\eta_k \in (0,1)$

$$\left\| \sum_{k=0}^{\infty} \frac{(t+\theta_k h)^k}{k!} A^k - \sum_{k=0}^{\infty} \frac{t^k}{k!} A^k \right\|_{\infty} = \left\| h \sum_{k=0}^{\infty} \frac{k(t+\eta_k h)^{k-1} \theta_k}{k!} A^k \right\|_{\infty}$$

$$\leq \left\| h \sum_{k=0}^{\infty} \frac{k(t+\eta_k h)^{k-1}}{k!} A^k \right\|_{\infty}$$

Now for $|h| \leq 1$, the expression $t_k \equiv t + \eta_k h \in [t-1, t+1]$ and so by Theorem 10.2.4,

$$\left\| \sum_{k=0}^{\infty} \frac{(t+\theta_k h)^k}{k!} A^k - \sum_{k=0}^{\infty} \frac{t^k}{k!} A^k \right\|_{\infty} = \left\| h \sum_{k=0}^{\infty} \frac{k(t+\eta_k h)^{k-1}}{k!} A^k \right\|_{\infty} \leq C|h| \quad (10.11)$$

for some C. Then

$$\left\| \frac{\Phi(t+h) - \Phi(t)}{h} - A\Phi(t) \right\|_{\infty} = \left\| A \sum_{k=0}^{\infty} \frac{(t+\theta_k h)^k}{k!} A^k - A \sum_{k=0}^{\infty} \frac{t^k}{k!} A^k \right\|_{\infty}$$

This converges to 0 as $h \to 0$ by Lemma 10.2.2 and (10.11). Hence the ij^{th} entry of the difference quotient

$$\frac{\Phi(t+h) - \Phi(t)}{h}$$

converges to the ij^{th} entry of the matrix $A \sum_{k=0}^{\infty} \frac{t^k}{k!} A^k = A\Phi(t)$. In other words,

$$\Phi'(t) = A\Phi(t)$$

Now also it follows right away from the formula for the infinite sum that $\Phi(0) = I$. This proves most of the following theorem.

Theorem 10.2.5 *Let A be a real or complex $p \times p$ matrix. Then there exists a differentiable $p \times p$ matrix $\Phi(t)$ satisfying the following initial value problem.*

$$\Phi'(t) = A\Phi(t), \quad \Phi(0) = I. \tag{10.12}$$

This matrix is given by the infinite sum

$$\Phi(t) = \sum_{k=0}^{\infty} \frac{t^k}{k!} A^k$$

with the usual convention that $t^0 = 1, A^0 = I, 0! = 1$. In addition to this, $A\Phi(t) = \Phi(t)A$. If A is real, then so is $\Phi(t)$. Furthermore, there is only one such fundamental matrix $\Phi(t)$.

Proof: Why is $A\Phi(t) = \Phi(t)A$? This follows from the observation that A obviously commutes with the partial sums,

$$A \sum_{k=0}^{n} \frac{t^k}{k!} A^k = \sum_{k=0}^{n} \frac{t^k}{k!} A^k A = \sum_{k=0}^{n} \frac{t^k}{k!} A^{k+1},$$

and Lemma 10.2.2,

$$A\Phi(t) = \lim_{n\to\infty} A \sum_{k=0}^{n} \frac{t^k}{k!} A^k = \lim_{n\to\infty} \left(\sum_{k=0}^{n} \frac{t^k}{k!} A^k \right) A = \Phi(t) A.$$

The next claim follows from noting that A^n will have all real entries if A is real.

Consider the claim that there is only one solution to the initial value problem (10.12). In general, if $\Psi(s)$ is a continuous function of s, for $t > a$

$$\left\| \int_a^t \Psi(s) \, ds \right\|_\infty \leq \int_a^t \|\Psi(s)\|_\infty \, ds$$

This follows from the definition. When you integrate the matrix, you simply replace each entry with its integral. Then the result follows from the scalar version of the triangle inequality for integrals seen in calculus. Now the following lemma is the assertion about uniqueness.

Lemma 10.2.6 *There is at most one matrix $\Phi(t)$ satisfying*

$$\Phi'(t) = A\Phi(t), \ \ \Phi(0) = I$$

Proof: If there were two of them, say $\Phi(t), \Psi(t)$, then consider $\Phi(t) - \Psi(t) \equiv \Sigma(t)$. Then

$$\Sigma'(t) = A\Sigma(t), \ \ \Sigma(0) = 0$$

Thus $\Sigma(t) = \int_0^t A\Sigma(s) \, ds$. Then for $t > 0$, and the above properties of $\|\cdot\|_\infty$

$$\|\Sigma(t)\|_\infty \leq \int_0^t \|A\Sigma(s)\|_\infty \, ds \leq p \int_0^t \|A\|_\infty \|\Sigma(s)\|_\infty$$

It follows from Gronwall's inequality Lemma 10.1.1

$$\|\Sigma(t)\|_\infty \leq 0 e^{p\|A\|_\infty t} = 0$$

For $t < 0$

$$\|\Sigma(t)\|_\infty \leq \int_t^0 \|A\Sigma(s)\|_\infty \leq p \int_t^0 \|A\|_\infty \|\Sigma(s)\|_\infty \, ds$$

and so from Gronwall's inequality again,

$$\|\Sigma(t)\|_\infty \leq 0 e^{p\|A\|_\infty (0-t)} = 0$$

Therefore, $\Sigma(t) = 0$ for all t and so $\Phi(t) = \Psi(t)$ for all t. Thus there is at most one solution. ∎

Next is considered the case where A is replaced with $-A$. The above $\Phi(t)$ is like e^{At} so of course the inverse of this ought to be e^{-At} but is this really so? Here A is a $p \times p$ matrix. In fact everything works out just as it should.

$$\Psi(t) \equiv \sum_{k=0}^{\infty} \frac{(-A)^k t^k}{k!}$$

In the same way as above $\Psi'(t) = (-A)\Psi(t), \Psi(0) = I$, and $A\Psi(t) = \Psi(t) A$.

Lemma 10.2.7 $\Phi(t)^{-1} = \Psi(t)$

Proof:

$$\begin{aligned}(\Phi(t)\Psi(t))' &= \Phi'(t)\Psi(t) + \Phi(t)\Psi'(t) = A\Phi(t)\Psi(t) + \Phi(t)(-A)\Psi(t) \\ &= A\Phi(t)\Psi(t) - A\Phi(t)\Psi(t) = 0\end{aligned}$$

Therefore, $\Phi(t)\Psi(t)$ is a constant matrix. Just use the usual calculus facts on the entries of the matrix. This matrix can only be I because $\Phi(0) = \Psi(0) = I$. ∎

The formula in the following theorem is called the variation of constants formula. It contains the mathematical substance of much of what follows.

Theorem 10.2.8 *Let $t \to \mathbf{f}(t)$ be continuous and let $\mathbf{x}_0 \in \mathbb{R}^p$. Then there exists a unique solution to the equation*

$$\mathbf{x}' = A\mathbf{x} + \mathbf{f}, \ \mathbf{x}(0) = \mathbf{x}_0$$

This solution is given by the formula

$$\mathbf{x}(t) = \Phi(t)\mathbf{x}_0 + \Phi(t)\int_0^t \Psi(s)\mathbf{f}(s)\,ds$$

Where $\Phi(t)$ is the fundamental matrix described in Theorem 10.2.5 and $\Psi(t)$ is the fundamental matrix defined the same way from $-A$. The fundamental matrix satisfying

$$\Phi'(t) = A\Phi(t), \quad \Phi(0) = I$$

is obtained as follows. It's k^{th} column is the solution to

$$\mathbf{x}'_k = A\mathbf{x}_k, \ \mathbf{x}_k(0) = \mathbf{e}_k$$

where $\mathbf{e}_k$ is the vector with 1 in the kth position and 0 in every other position.

Proof: Suppose $\mathbf{x}' = A\mathbf{x} + \mathbf{f}$. Then $\mathbf{x}' - A\mathbf{x} = \mathbf{f}$. Multiply both sides by $\Psi(t)$. Then

$$\begin{aligned}(\Psi(t)\mathbf{x})' &= \Psi(t)\mathbf{x}'(t) + \Psi'(t)\mathbf{x}(t) = \Psi(t)\mathbf{x}'(t) - A\Psi(t)\mathbf{x}(t) \\ &= \Psi(t)\mathbf{x}'(t) - \Psi(t)A\mathbf{x}(t) = \Psi(t)(\mathbf{x}'(t) - A\mathbf{x}(t))\end{aligned}$$

Therefore,

$$(\Psi(t)\mathbf{x})' = \Psi(t)\mathbf{f}(t)$$

Hence

$$\Psi(t)\mathbf{x}(t) - \mathbf{x}_0 = \int_0^t \Psi(s)\mathbf{f}(s)\,ds$$

Therefore, multiplying on the left by $\Phi(t)$,

$$\mathbf{x}(t) = \Phi(t)\mathbf{x}_0 + \Phi(t)\int_0^t \Psi(s)\mathbf{f}(s)\,ds$$

Thus, if there is a solution, this is it. Thus there is at most one solution to the initial value problem.

Now if $\mathbf{x}(t)$ is given by the above formula, then

$$\mathbf{x}(0) = \Phi(0)\mathbf{x}_0 = \mathbf{x}_0$$

and also

$$\mathbf{x}'(t) = \Phi'(t)\mathbf{x}_0 + \Phi'(t)\int_0^t \Psi(s)\mathbf{f}(s)\,ds + \Phi(t)\Psi(t)\mathbf{f}(t)$$

$$= A\Phi(t)\mathbf{x}_0 + A\Phi(t)\int_0^t \Psi(s)\mathbf{f}(s)\,ds + \mathbf{f}(t) = A\mathbf{x}(t) + \mathbf{f}(t)$$

As to the last claim of columns of $\Phi(t)$, just apply what was just done to $\mathbf{x}_0 = \mathbf{e}_k$ and $\mathbf{f}(t) = \mathbf{0}$. Thus the solution to $\mathbf{x}'_k = A\mathbf{x}_k$, $\mathbf{x}_k(0) = \mathbf{e}_k$ is $\mathbf{x}_k(t) = \Phi(t)\mathbf{e}_k$ but on the right, this is just the kth column of $\Phi(t)$. Thus the kth column of $\Phi(t)$ is this $\mathbf{x}_k(t)$. ∎

It is also of interest to observe that $\Psi(t) = \Phi(-t)$. This is in the next proposition.

Proposition 10.2.9 *In the context of the above theorem,* $\Psi(t) = \Phi(-t) = \Phi(t)^{-1}$ *and for all* $s, t \in \mathbb{R}$,

$$\Phi(t+s) = \Phi(t)\Phi(s)$$

The above variation of constants formula has the form

$$\mathbf{x}(t) = \Phi(t)\mathbf{x}_0 + \Phi(t)\int_0^t \Psi(s)\mathbf{f}(s)\,ds = \Phi(t)\mathbf{x}_0 + \int_0^t \Phi(t-s)\mathbf{f}(s)\,ds$$

Proof: Fix s. Then letting $\mathbf{x}(t) \equiv \Phi(t+s)\mathbf{v}$,

$$\mathbf{x}'(t) = \Phi'(t+s)\mathbf{v} = A\Phi(t+s)\mathbf{v} = A\mathbf{x}(t),\ \mathbf{x}(0) = \Phi(s)\mathbf{v}$$

Let $\mathbf{y}(t) \equiv \Phi(t)\Phi(s)\mathbf{v}$. Then

$$\mathbf{y}'(t) = A\Phi(t)\Phi(s)\mathbf{v} = A\mathbf{y}(t),\ \mathbf{y}(0) = \Phi(s)\mathbf{v}$$

By uniqueness of solutions to the initial value problem, Theorem 10.2.8, $\mathbf{y}(t) = \mathbf{x}(t)$ for all t. Hence for any $\mathbf{v}$,

$$\Phi(t+s)\mathbf{v} = \Phi(t)\Phi(s)\mathbf{v}$$

Thus $\Phi(t+s) = \Phi(t)\Phi(s)$. Now $I = \Phi(0) = \Phi(t-t) = \Phi(t)\Phi(-t)$ and $I = \Phi(t)\Psi(t)$ so $\Psi(t) = \Phi(-t)$. The claim about the formula for the solution follows from this and what was just shown. ∎

Observation 10.2.10 *As a special case when A is a real or complex scalar, the above theorem shows that $\Phi(t)x_0 \equiv \sum_{k=0}^\infty \frac{t^k}{k!}A^k x_0$ is the one and only solution of the differential equation*

$$x'(t) = Ax(t),\ x(0) = x_0.$$

Another solution to this is e^{At} where this is defined in Section 1.7. It follows that in this special case, the uniqueness provision of the above theorem shows that

$$\sum_{k=0}^{\infty} \frac{t^k}{k!} A^k = e^{At}$$

in the special case that A is a real or complex number.

Proposition 10.2.11 *There is only one way to have $\left(e^{\lambda t}\right)' = \lambda e^{\lambda t}, e^0 = 1$ for all λ real or complex and this is to define*

$$e^{t(a+ib)} = e^{at}\left(\cos\left(bt\right) + i\sin\left(bt\right)\right)$$

Proof: It is desired to have $\left(e^{\lambda t}\right)' = \lambda e^{\lambda t}$ and so $t \to e^{\lambda t}$ is to be a solution to $x' = \lambda x, x\left(0\right) = 1$ and the solution is unique and is given above in terms of a series. It only remains to observe that $t \to e^{at}\left(\cos\left(bt\right) + i\sin\left(bt\right)\right)$ is a solution to the initial value problem which follows from a simple computation as was done earlier. ∎

10.3   Exercises

These exercises summarize the presentation of linear equations in the previous part of the book.

1. Suppose that $\mathbf{x}_i'\left(t\right) = A\mathbf{x}_i\left(t\right)$, $\mathbf{x}_i\left(t\right) = \mathbf{v}_i$ for $i = 1, \cdots, n$. Also suppose $\{\mathbf{v}_1, \cdots, \mathbf{v}_n\}$ is independent. Then let

$$\Theta\left(t\right) \equiv \left(\begin{array}{ccc} \mathbf{x}_1\left(t\right) & \cdots & \mathbf{x}_n\left(t\right) \end{array}\right).$$

   That is, $\Theta\left(t\right)$ is the matrix which has the $\mathbf{x}_i$ as columns. Explain why

$$\Theta'\left(t\right) = A\Theta\left(t\right), \ \Theta\left(0\right) = \left(\begin{array}{ccc} \mathbf{v}_1 & \cdots & \mathbf{v}_n \end{array}\right)$$

2. In the above problem, let $S = \left(\begin{array}{ccc} \mathbf{v}_1 & \cdots & \mathbf{v}_n \end{array}\right)$. Show that $\Theta\left(t\right) S^{-1} = \Phi\left(t\right)$, **the** fundamental matrix.

3. Consider the system of equations

$$\mathbf{x}' = \left(\begin{array}{ccc} 1 & 3 & 2 \\ 0 & 2 & 1 \\ 0 & -1 & 0 \end{array}\right) \mathbf{x}$$

   Show that

$$\left(\begin{array}{c} e^t + \frac{1}{2}t^2 e^t + 3te^t \\ e^t + te^t \\ -te^t \end{array}\right), \left(\begin{array}{c} \frac{1}{2}t^2 e^t + 3te^t \\ e^t + te^t \\ -te \end{array}\right), \left(\begin{array}{c} \frac{1}{2}t^2 e^t + 2te^t \\ te^t \\ e^t - te^t \end{array}\right)$$

   are each solutions of the differential equation. Use the above problem to determine a fundamental matrix. **Hint:** You better use the above problem. This matrix is **defective**.

4. Suppose you are looking for solutions to $\mathbf{x}' = A\mathbf{x}$ and you try $\mathbf{x}(t) = e^{\lambda t}\mathbf{v}$. Show that this is a solution to the equation if and only if $(\lambda, \mathbf{v})$ are an eigenvalue and eigenvector for A. If A is non defective, explain how this can be used to obtain a fundamental matrix using Problem 1.

5. Now suppose you have an eigenvalue λ and an eigenvector $\mathbf{v}$. Consider $te^{\lambda t}\mathbf{v} + \mathbf{u}e^{\lambda t}$ where $(A - \lambda I)\mathbf{u} = \mathbf{v}$. Show that this is also a solution to the differential equation. Show that $\{\mathbf{v}, \mathbf{u}\}$ is linearly independent. This gives another way to find solutions to the differential equation.

6. In the situation of the above problem, consider $\frac{1}{2}t^2 e^{\lambda t}\mathbf{v} + t\mathbf{u}e^{\lambda t} + \mathbf{w}e^{\lambda t}$ where

$$(A - \lambda I)\mathbf{w} = \mathbf{u}$$

and $(A - \lambda I)\mathbf{u} = \mathbf{v}$ while $(A - \lambda I)\mathbf{v} = \mathbf{0}$. Show that this longer expression is also a solution to the differential equation.

7. Suppose $\Theta'(t) = A\Theta(t)$. Show that $\det\Theta(t)$ either is nonzero for all t or $\det\Theta(t) = 0$ for all t.

8. An $n \times n$ matrix A is called normal if $AA^* = A^*A$. Show that every normal matrix is unitarily similar to a diagonal matrix D. That is, there is a unitary matrix U such that $U^*AU = D$. Thus all normal matrices are nondefective. **Hint:** Use Schur's theorem and first show that a normal upper triangular matrix is a diagonal matrix. Recall that A^* is the conjugate transpose of A.

9. Give a description of how to write an arbitrary n^{th} order differential equation

$$y^{(n)} + a_{n-1}y^{(n-1)} + \cdots + a_1 y' + a_0 y = f$$

in the form $\mathbf{x}' = A\mathbf{x} + \mathbf{F}$ where A is an $n \times n$ matrix. This is done earlier in Procedure 5.5.2, but try to come up with a way to do it on your own. See also Problem 86 on Page 76 which is the same as this problem.

10. Using the above problem, show that there exists a unique solution to the following initial value problem.

$$y^{(n)} + a_{n-1}y^{(n-1)} + \cdots + a_1 y' + a_0 y = f$$

$$y(0) = y_0, y'(0) = y_1, \cdots .y^{(n-1)}(0) = y_{n-1}$$

11. Prove that if $s \to \Psi(s)$ is continuous for $\Psi(s)$ a $p \times p$ matrix, then $s \to \|\Psi(s)\|_\infty$ is also continuous. **Hint:** Use the triangle inequality in the form given in Proposition 10.2.3.

12. It was shown directly that there is only one fundamental matrix $\Phi(t)$ satisfying

$$\Phi'(t) = A\Phi(t), \quad \Phi(0) = I \qquad\qquad (*)$$

This was done by a use of Gronwall's inequality. However, it was also shown that $\Phi(t)$ given by

$$\sum_{k=0}^{\infty} \frac{A^k t^k}{k!}$$

was a solution to the above and it was shown any solution to the initial value problem

$$\mathbf{x}' = A\mathbf{x} + \mathbf{f}, \ \mathbf{x}(0) = \mathbf{x}_0$$

is of a particular form involving the variation of constants formula and $\Phi(t)$.

$$\mathbf{x}(t) = \Phi(t)\mathbf{x}_0 + \Phi(t)\int_0^t \Psi(s)\mathbf{f}(s)\,ds$$

Explain why $\Phi(t)$ satisfies * if and only if the kth column $\mathbf{x}_k(t)$ is a solution of

$$\mathbf{x}' = A\mathbf{x}, \ \mathbf{x}(0) = \mathbf{e}_k$$

Thus $\mathbf{x}_k(t)$ equals $\Phi(t)\mathbf{e}_k$ where $\Phi(t)$ is defined in terms of that infinite series. Why does this imply that $\Phi(t)$ is uniquely determined? **Hint:** If $\Sigma(t)$ solves *, then its kth column solves the above initial value problem and so its kth column equals the kth column of $\Phi(t)$. Write the details.

13. It was shown above in the construction of the fundamental matrix as

$$\Phi(t) = \sum_{k=0}^{\infty} \frac{A^k t^k}{k!}$$

that if A is real, so is the fundamental matrix $\Phi(t)$. Here is a matrix which has eigenvalues $i, -i, 1$

$$\begin{pmatrix} 1 & -2 & -2 \\ 1 & -3 & -4 \\ 0 & 2 & 3 \end{pmatrix}$$

corresponding to these eigenvalues are the eigenvectors

$$\begin{pmatrix} 2 \\ 3-i \\ -2 \end{pmatrix}, \begin{pmatrix} 2 \\ 3+i \\ -2 \end{pmatrix}, \begin{pmatrix} 0 \\ -1 \\ 1 \end{pmatrix}$$

respectively. Then as explained above, the unique fundamental matrix is of the form

$$\begin{pmatrix} 2 & 2 & 0 \\ 3-i & 3+i & -1 \\ -2 & -2 & 1 \end{pmatrix} \begin{pmatrix} e^{it} & 0 & 0 \\ 0 & e^{-it} & 0 \\ 0 & 0 & e^t \end{pmatrix} \begin{pmatrix} 2 & 2 & 0 \\ 3-i & 3+i & -1 \\ -2 & -2 & 1 \end{pmatrix}^{-1}$$

$$= \begin{pmatrix} 2 & 2 & 0 \\ 3-i & 3+i & -1 \\ -2 & -2 & 1 \end{pmatrix} \begin{pmatrix} e^{it} & 0 & 0 \\ 0 & e^{-it} & 0 \\ 0 & 0 & e^t \end{pmatrix} \begin{pmatrix} \frac{1}{4}-\frac{1}{4}i & \frac{1}{2}i & \frac{1}{2}i \\ \frac{1}{4}+\frac{1}{4}i & -\frac{1}{2}i & -\frac{1}{2}i \\ 1 & 0 & 1 \end{pmatrix}$$

This is supposed to be real? Work it out and verify that it is indeed real. This must be the case because it was proved to be so. This is just verifying that indeed it does turn out to be real.

14. Suppose that $\Psi'(t) = A\Psi(t), \Phi'(t) = A\Phi(t)$ where $\Phi(t)$ is the standard fundamental matrix. Show that there is some constant c such that $\det(\Psi(t)) = c\det(\Phi(t))$ for all t. **Hint:** You might note that $\Sigma(t) \equiv \Psi(-t)$ solves $\Sigma'(t) = -A\Sigma(t)$. Then consider $\frac{d}{dt}(\Phi\Sigma)$.

15. It was shown in this chapter that the fundamental matrix for a $p \times p$ matrix A is $\Phi(t) = \sum_{k=0}^{\infty} \frac{A^k t^t}{k!}$. Thus $\lim_{n\to\infty} \Phi_n(t) = \Phi(t)$ where $\Phi_n(t) \equiv \sum_{k=0}^{n} \frac{A^k t^t}{k!}$. From Problem 85 on Page 76 there is only one $\Phi(t)$ and it makes perfect sense to take its Laplace transform since the construction there shows that each entry is of exponential growth. Show that for sufficiently large s, $\sum_{k=0}^{\infty} \frac{A^k}{s^{k+1}}$ exists and

$$\sum_{k=0}^{\infty} \mathcal{L}\left(\frac{A^k t^t}{k!}\right) = \sum_{k=0}^{\infty} \frac{A^k}{s^{k+1}} = (sI - A)^{-1} = \mathcal{L}(\Phi(t)) = \mathcal{L}\left(\sum_{k=0}^{\infty} \frac{A^k t^t}{k!}\right)$$

16. Suppose A is nondefective. Thus there is S such that

$$S^{-1}AS = D = \begin{pmatrix} \lambda_1 & & 0 \\ & \ddots & \\ 0 & & \lambda_p \end{pmatrix}$$

Show that the fundamental matrix is

$$\Phi(t) = S \begin{pmatrix} e^{\lambda_1 t} & & 0 \\ & \ddots & \\ 0 & & e^{\lambda_p t} \end{pmatrix} S^{-1}.$$

17. Explain in terms of partial fractions expansions of the entries of $(sI - A)^{-1}$ what it means for A to be defective. Assume all factorizations of the rational functions are accomplished over the complex field in order to avoid having to consider irreducible quadratic polynomials.

18. Recall Schur's theorem from linear algebra or consider Theorem D.8.5 in the appendix. Using this theorem, show that for A an $n \times n$ matrix, there exists a sequence $\{A_k\}_{k=1}^{\infty}$ of $n \times n$ matrices, each of which is nondefective with the property that the entries of A_k converge to the corresponding entries of A. It can be shown that the fundamental matrix of A_k will converge to the fundamental matrix of A. This will become more clear in the next chapter. It can also be proved easily from the explicit description of the fundamental matrix given in this chapter.

Chapter 11

Theory of Ordinary Differential Equations

11.1 Problems from Mechanics

In the preceding material, there has been considerable emphasis placed on finding analytic formulas for solutions to differential equations. In this chapter, fundamental considerations are presented which have to do with whether a solution even exists, and in any case; how to find such solutions. Lagrangian mechanics provides many examples of situations in which these are the two questions of most importance. This is not a book on modeling or mechanics so only a very brief introduction to this is given. The reason for doing so is to show that it is common to encounter equations which cannot be solved using standard techniques.

Let $\mathbf{y} = \mathbf{y}(\mathbf{x},t)$ where t signifies time and $\mathbf{x} \in U \subseteq \mathbb{R}^m$ for U an open set, while $\mathbf{y} \in \mathbb{R}^n$ and suppose $\mathbf{x}$ is a function of t. Physically, this corresponds to an object moving over a surface in $\mathbb{R}^n$, its position being $\mathbf{y}(\mathbf{x}, t)$. If we know about $\mathbf{x}(t)$ then we also know $\mathbf{y}$. More generally, we might have M masses, the position of mass α being $\mathbf{y}_\alpha$. For example, consider the pendulum in which there is only one mass.

in which $n = 2, l$ is fixed and $y^1 = l\sin\theta, y^2 = l - l\cos\theta$. Thus, in this simple example, $m = 1$ and $\mathbf{x} = \theta$. If l were changing in a known way with respect to t, then this would be of the form $\mathbf{y} = \mathbf{y}(\mathbf{x}, t)$. We seek differential equations for $\mathbf{x}$.

The kinetic energy is defined as

$$T \equiv \frac{1}{2}\sum_\alpha m_\alpha \dot{\mathbf{y}}_\alpha \cdot \dot{\mathbf{y}}_\alpha \qquad (*)$$

where the dot on the top signifies differentiation with respect to t. Thus, from the chain rule, T is a function of $\dot{\mathbf{x}}$. The following lemma is an important observation.

Lemma 11.1.1 *The following formula holds.*

$$\frac{\partial T}{\partial \dot{x}^k} = \sum_\alpha m_\alpha \dot{\mathbf{y}}_\alpha \cdot \frac{\partial \mathbf{y}_\alpha}{\partial x^k}.$$

263

Proof: From the chain rule,

$$\dot{\mathbf{y}}_\alpha = \sum_k \frac{\partial \mathbf{y}_\alpha}{\partial x^k} \dot{x}^k + \frac{\partial \mathbf{y}_\alpha}{\partial t} \qquad (**)$$

and so

$$\frac{\partial \dot{\mathbf{y}}_\alpha}{\partial \dot{x}^k} = \frac{\partial \mathbf{y}_\alpha}{\partial x^k}.$$

Therefore,

$$\frac{\partial T}{\partial \dot{x}^k} = \sum_\alpha m_\alpha \dot{\mathbf{y}}_\alpha \cdot \frac{\partial \dot{\mathbf{y}}_\alpha}{\partial \dot{x}^k} = \sum_\alpha m_\alpha \dot{\mathbf{y}}_\alpha \cdot \frac{\partial \mathbf{y}_\alpha}{\partial x^k} \quad \blacksquare$$

It follows from the above and the product and chain rule that

$$\frac{d}{dt}\left(\frac{\partial T}{\partial \dot{x}^k}\right) = \sum_\alpha m_\alpha \ddot{\mathbf{y}}_\alpha \cdot \frac{\partial \mathbf{y}_\alpha}{\partial x^k} +$$

$$\sum_\alpha m_\alpha \dot{\mathbf{y}}_\alpha \cdot \sum_r \frac{\partial^2 \mathbf{y}_\alpha}{\partial x^r \partial x^k} \dot{x}^r + \sum_\alpha m_\alpha \dot{\mathbf{y}}_\alpha \cdot \frac{\partial^2 \mathbf{y}_\alpha}{\partial t \partial x^k}. \qquad (***)$$

Also from the product rule,

$$\frac{\partial T}{\partial x^k} = m\dot{\mathbf{y}} \cdot \left(\frac{\partial \dot{\mathbf{y}}}{\partial x^k}\right)$$

But from **,

$$\frac{\partial \dot{\mathbf{y}}_\alpha}{\partial x^k} = \sum_r \frac{\partial^2 \mathbf{y}_\alpha}{\partial x^k \partial x^r} \dot{x}^r + \frac{\partial^2 \mathbf{y}_\alpha}{\partial x^k \partial t}$$

Thus

$$\frac{\partial T}{\partial x^k} = \sum_\alpha m_\alpha \dot{\mathbf{y}}_\alpha \cdot \left(\sum_r \frac{\partial^2 \mathbf{y}_\alpha}{\partial x^k \partial x^r} \dot{x}^r + \frac{\partial^2 \mathbf{y}_\alpha}{\partial x^k \partial t}\right)$$

$$= \sum_\alpha \sum_r m_\alpha \dot{\mathbf{y}}_\alpha \cdot \frac{\partial^2 \mathbf{y}_\alpha}{\partial x^k \partial x^r} \dot{x}^r + m_\alpha \dot{\mathbf{y}}_\alpha \cdot \frac{\partial^2 \mathbf{y}_\alpha}{\partial x^k \partial t}$$

From this and ***,

$$\frac{d}{dt}\left(\frac{\partial T}{\partial \dot{x}^k}\right) - \frac{\partial T}{\partial x^k} = \sum_\alpha m_\alpha \ddot{\mathbf{y}}_\alpha \cdot \frac{\partial \mathbf{y}_\alpha}{\partial x^k}$$

Now $\ddot{\mathbf{y}}_\alpha$ denotes the acceleration of the αth mass and so, by Newton's second law, if $\mathbf{F}$ is the force acting on the object,

$$\frac{d}{dt}\left(\frac{\partial T}{\partial \dot{x}^k}\right) - \frac{\partial T}{\partial x^k} = \sum_\alpha \mathbf{F}_\alpha \cdot \frac{\partial \mathbf{y}_\alpha}{\partial x^k} \qquad (11.1)$$

This is a particularly agreeable formula in case $\mathbf{F}_\alpha = \nabla \Phi_\alpha(\mathbf{y}) + \mathbf{g}_\alpha$ where $\mathbf{g}_\alpha$ is a force of constraint which causes motion to remain in the surface $\mathbf{x} \to \mathbf{y}_\alpha(\mathbf{x})$. Thus $\mathbf{g}_\alpha \cdot \frac{\partial \mathbf{y}_\alpha}{\partial x^k} = 0$. In this special case, you have

$$\frac{d}{dt}\left(\frac{\partial T}{\partial \dot{x}^k}\right) - \frac{\partial T}{\partial x^k} = \sum_\alpha \nabla \Phi_\alpha(\mathbf{y}) \cdot \frac{\partial \mathbf{y}_\alpha}{\partial x^k} = \sum_\alpha \frac{\partial}{\partial x^k}(\Phi_\alpha(\mathbf{y}))$$

Let Φ denote the total potential energy so $\Phi = \sum_\alpha \Phi_\alpha$. Now $\Phi_\alpha(\mathbf{y})$ does not depend on $\dot{\mathbf{x}}$, only on $\mathbf{x}$. Hence $\frac{\partial \Phi_\alpha(\mathbf{y})}{\partial \dot{x}_k} = 0$. It follows that in this special case,

$$\frac{d}{dt}\left(\frac{\partial(T - \Phi)}{\partial \dot{x}_k}\right) - \frac{\partial(T - \Phi)}{\partial x^k} = 0, \tag{11.2}$$

this for each k. This formula is due to Lagrange.*

Theorem 11.1.2 *Let $\mathbf{y}_\alpha(\mathbf{x}, t)$ denote the position of an object of mass m_α where $\mathbf{x}$ is a function of t. Let the kinetic energy be defined by*

$$T \equiv \frac{1}{2}\sum_\alpha m_\alpha \dot{\mathbf{y}}_\alpha \cdot \dot{\mathbf{y}}_\alpha.$$

Let the mass m_α be acted on by a force $\mathbf{F}_\alpha$. Then Newton's second law implies

$$\frac{d}{dt}\left(\frac{\partial T}{\partial \dot{x}^k}\right) - \frac{\partial T}{\partial x^k} = \sum_\alpha \mathbf{F}_\alpha \cdot \frac{\partial \mathbf{y}_\alpha}{\partial x^k} \tag{11.3}$$

In case $\mathbf{F}_\alpha = \nabla \Phi_\alpha + \mathbf{g}_\alpha$ where $\mathbf{g}_\alpha$ is a force of constraint so the total force comes from forces of constraint and the gradient of a potential function, then

$$\frac{d}{dt}\left(\frac{\partial(T - \Phi)}{\partial \dot{x}_k}\right) - \frac{\partial(T - \Phi)}{\partial x^k} = 0$$

Also, the above (11.3) implies Newton's second law.

Proof: The above derivation shows that Newton's law implies the above two formulas. On the other hand, if (11.3) holds, then in the case of one mass, the first part of the derivation which depended only on the chain rule and product rule shows

$$\frac{d}{dt}\left(\frac{\partial T}{\partial \dot{x}^k}\right) - \frac{\partial T}{\partial x^k} = m\ddot{\mathbf{y}} \cdot \frac{\partial \mathbf{y}}{\partial x^k}$$

Thus if (11.3) and there is no force of constraint, then $\mathbf{F} = m\ddot{\mathbf{y}}$ which is Newton's second law. ∎

Example 11.1.3 *In the case of the simple pendulum, $\mathbf{x} = \theta$ as shown in the picture and*

$$\begin{pmatrix} y^1 \\ y^2 \end{pmatrix} = \begin{pmatrix} l\sin\theta \\ l - l\cos\theta \end{pmatrix}$$

the force acting on weight being $mg(-\mathbf{j}) = \nabla(-mgy^2)$. Find the equation of motion of this pendulum.

---

*Joseph Louis Lagrange (1736-1813) was born in Italy but lived much of his life in France which is where he died. He made major contributions to analysis, number theory, and mechanics. His most famous work is likely Mécanique analytique. He invented the method of variation of parameters used earlier. With Euler, he invented the calculus of variations and also the method of Lagrange multipliers in order to include constraints. Lagrange was also involved in the development of the metric system.

$$T = \frac{1}{2}m \begin{pmatrix} l\cos(\theta)\,\theta' \\ l\sin(\theta)\,\theta' \end{pmatrix} \cdot \begin{pmatrix} l\cos(\theta)\,\theta' \\ l\sin(\theta)\,\theta' \end{pmatrix} = \frac{1}{2}ml^2\,(\theta')^2$$

Then $\Phi = -mg\,(l - l\cos\theta)$. $T - \Phi = \frac{1}{2}ml^2\,(\theta')^2 + mg\,(l - l\cos\theta)$. Thus the equation of motion of this pendulum is

$$\frac{d}{dt}\left(ml^2\theta'\right) - mgl\left(-\sin(\theta)\right) = 0$$

so

$$\theta'' + \frac{g}{l}\sin\theta = 0$$

This is an equation which doesn't have a simple analytic solution in terms of standard calculus type functions.

Example 11.1.4 *In the above simple pendulum, suppose there is a friction force* $-k(\mathbf{y})\,\dot{\mathbf{y}}$ *acting to impede the motion. What are equations of motion in this case?*

The following is from the chain rule.

$$\dot{\mathbf{y}} = \begin{pmatrix} l\cos\theta \\ l\sin\theta \end{pmatrix}\theta'$$

Denote $k(l\sin\theta, l - l\cos\theta)$ as $k(\theta)$ to save notation. Then it follows from (11.3) and the previous example that

$$\frac{d}{dt}\left(\frac{\partial}{\partial\theta'}\left(\frac{1}{2}ml^2\,(\theta')^2 + mg\,(l - l\cos\theta)\right)\right) -$$

$$\frac{\partial}{\partial\theta}\left(\frac{1}{2}ml^2\,(\theta')^2 + mg\,(l - l\cos\theta)\right) = -k(\theta)\,\theta'\begin{pmatrix} l\cos\theta \\ l\sin\theta \end{pmatrix} \cdot \begin{pmatrix} l\cos\theta \\ l\sin\theta \end{pmatrix}$$

and so

$$\frac{d}{dt}\left(ml^2\theta'\right) + mgl\sin(\theta) = -k(\theta)\,\theta'l^2$$

$$\theta'' + \frac{k(\theta)}{m}\theta' + \frac{g}{l}\sin(\theta) = 0$$

This is another equation for which we don't have a good way to obtain a simple analytic solution.

Example 11.1.5 *Given a force* $\mathbf{F}$ *what are the equations of motion in terms of* ρ, ϕ, θ, *the spherical coordinates?*

First, one needs the kinetic energy in terms of these spherical coordinates and their derivatives.

$$\begin{pmatrix} y^1 \\ y^2 \\ y^3 \end{pmatrix} = \begin{pmatrix} \rho\sin\phi\cos\theta \\ \rho\sin\phi\sin\theta \\ \rho\cos\phi \end{pmatrix}$$

$$\dot{\mathbf{y}} = \rho'\begin{pmatrix} \sin\phi\cos\theta \\ \sin\phi\sin\theta \\ \cos\phi \end{pmatrix} + \phi'\begin{pmatrix} \rho\cos\phi\cos\theta \\ \rho\cos\phi\sin\theta \\ -\rho\sin\phi \end{pmatrix} + \theta'\begin{pmatrix} -\rho\sin\phi\sin\theta \\ \rho\sin\phi\cos\theta \\ 0 \end{pmatrix}$$

Then $\dot{\mathbf{y}} \cdot \dot{\mathbf{y}}$ equals the following after much algebra.

$$(\rho')^2 + \frac{1}{2}\rho^2(\theta')^2 + \rho^2(\phi')^2 - \frac{1}{2}\rho^2(\theta')^2 \cos 2\phi$$

Thus the kinetic energy is of the form

$$T = \frac{1}{2}m\left((\rho')^2 + \frac{1}{2}\rho^2(\theta')^2 + \rho^2(\phi')^2 - \frac{1}{2}\rho^2(\theta')^2 \cos 2\phi\right)$$

Then if $\mathbf{F}\,(\rho\sin\phi\cos\theta, \rho\sin\phi\sin\theta, \rho\cos\phi)$ is the force acting on the object, one of the equations of motion is

$$\frac{1}{2}m\left[2\rho'' - \left(\rho\,(\theta')^2 + 2\rho\,(\phi')^2 - \rho\,(\theta')^2\cos(2\phi)\right)\right] = \mathbf{F}\cdot\left(\begin{array}{c} \sin\phi\cos\theta \\ \sin\phi\sin\theta \\ \cos\phi \end{array}\right)$$

the other two equations are similarly involved. You might want to write them. The point is, that if you want to do things in spherical coordinates, you end up with equations which do not fit well with what we have done. It is not clear how to get simple analytic solutions to these equations.

Thus one must consider the two questions mentioned above. Does there even exist a solution to the initial value problem and if so, how can you find it?

11.2 Picard Iteration

We suppose that $\mathbf{f} : [a,b] \times \mathbb{R}^n \to \mathbb{R}^n$ satisfies the following two conditions.

$$|\mathbf{f}\,(t,\mathbf{x}) - \mathbf{f}\,(t,\mathbf{x}_1)| \le K\,|\mathbf{x} - \mathbf{x}_1|, \tag{11.4}$$

$$\mathbf{f} \text{ is continuous.} \tag{11.5}$$

The first of these conditions is known as a Lipschitz condition.

Lemma 11.2.1 *Suppose* $\mathbf{x} : [a,b] \to \mathbb{R}^n$ *is a continuous function and* $c \in [a,b]$. *Then* $\mathbf{x}$ *is a solution to the initial value problem,*

$$\mathbf{x}' = \mathbf{f}\,(t,\mathbf{x})\,,\ \mathbf{x}\,(c) = \mathbf{x}_0 \tag{11.6}$$

if and only if $\mathbf{x}$ *is a solution to the integral equation,*

$$\mathbf{x}\,(t) = \mathbf{x}_0 + \int_c^t \mathbf{f}\,(s,\mathbf{x}\,(s))\,ds. \tag{11.7}$$

Proof: If $\mathbf{x}$ solves (11.7), then since $\mathbf{f}$ is continuous, we may apply the fundamental theorem of calculus to differentiate both sides and obtain $\mathbf{x}'\,(t) = \mathbf{f}\,(t,\mathbf{x}\,(t))$. Also, letting $t = c$ on both sides, gives $\mathbf{x}\,(c) = \mathbf{x}_0$. Conversely, if $\mathbf{x}$ is a solution of the initial value problem, we may integrate both sides from c to t to see that $\mathbf{x}$ solves (11.7). ∎

It follows from this lemma that we may study the initial value problem, (11.6) by considering the integral equation (11.7). The most famous technique for studying this integral equation is the method of Picard iteration. In this method, we start with an initial function, $\mathbf{x}_0(t) \equiv \mathbf{x}_0$ and then iterate as follows.

$$\mathbf{x}_1(t) \equiv \mathbf{x}_0 + \int_c^t \mathbf{f}(s, \mathbf{x}_0(s))\, ds = \mathbf{x}_0 + \int_c^t \mathbf{f}(s, \mathbf{x}_0)\, ds,$$

$$\mathbf{x}_2(t) \equiv \mathbf{x}_0 + \int_c^t \mathbf{f}(s, \mathbf{x}_1(s))\, ds,$$

and if $\mathbf{x}_{k-1}(s)$ has been determined,

$$\mathbf{x}_k(t) \equiv \mathbf{x}_0 + \int_c^t \mathbf{f}(s, \mathbf{x}_{k-1}(s))\, ds.$$

These are the Picard iterates. Under the conditions just described, these converge uniformly to a solution to the differential equation. This is discussed in the next section.

11.3 Convergence of Picard Iterates

It is no harder to allow all functions to have values in $\mathbb{C}$ or $\mathbb{C}^n$. The only difference is that now $|\cdot|$ refers to the distance in $\mathbb{C}^n$. However, this is not really more general because you can consider $\mathbb{C}$ as $\mathbb{R}^2$. The arguments used will work for general normed vector spaces.

Definition 11.3.1 *Let T be a subset of some $\mathbb{R}^m$ or $\mathbb{C}^m$, possibly all of $\mathbb{R}^m$ or $\mathbb{C}^m$. Let $BC(T; \mathbb{C}^n)$ denote the bounded continuous functions defined on T.[†] Then this is a vector space (linear space) with respect to the usual operations of addition and scalar multiplication of functions. Also, we can define a norm as follows:*

$$\|\mathbf{f}\| \equiv \sup_{t \in T} |\mathbf{f}(t)| < \infty.$$

This is a norm because it satisfies the axioms of a norm which are as follows:

$$\|\mathbf{f} + \mathbf{g}\| \leq \|\mathbf{f}\| + \|\mathbf{g}\|$$

$$\|\alpha \mathbf{f}\| = |\alpha|\, \|\mathbf{f}\|$$

$$\|\mathbf{f}\| \geq 0 \text{ and equals 0 if and only if } \mathbf{f} = \mathbf{0}$$

A sequence $\{\mathbf{f}_n\}$ in $BC(T; \mathbb{C}^n)$ is a Cauchy sequence if for every $\varepsilon > 0$ there exists M_ε such that if $m, n \geq M_\varepsilon$, then

$$\|\mathbf{f}_n - \mathbf{f}_m\| < \varepsilon$$

Such a normed linear space is called complete if every Cauchy sequence converges. Such a complete normed linear space is called a Banach space.

[†]In fact, they will be automatically bounded if the set T is a closed interval like [0,T], but the considerations presented here will work even when a compact set is not being considered.

Now there is another norm which works just as well in the case where $T \equiv [a, b]$, an interval. This is described in the following definition.

Definition 11.3.2 *For $\mathbf{f} \in BC([a, b]; \mathbb{C}^n)$, let $c \in [a, b]$, γ a real number. Then*

$$\|\mathbf{f}\|_\gamma \equiv \sup_{t \in [a,b]} \left| \mathbf{f}(t) e^{-|\gamma(t-c)|} \right|$$

Then this is a norm. The above definition corresponds to $\gamma = 0$.

Lemma 11.3.3 *$\|\cdot\|_\gamma$ is a norm for $BC([a, b]; \mathbb{C}^n)$ and $BC([a, b]; \mathbb{C}^n)$ is a complete normed linear space. Also, a sequence is Cauchy in $\|\cdot\|_\gamma$ if and only if it is Cauchy in $\|\cdot\|$.*

Proof: First consider the claim about $\|\cdot\|_\gamma$ being a norm. To simplify notation, let $T = [a, b]$. It is clear that $\|\mathbf{f}\|_\gamma = 0$ if and only if $\mathbf{f} = \mathbf{0}$ and $\|\mathbf{f}\|_\gamma \geq 0$. Also,

$$\|\alpha \mathbf{f}\|_\gamma \equiv \sup_{t \in T} \left| \alpha \mathbf{f}(t) e^{-|\gamma(t-c)|} \right| = |\alpha| \sup_{t \in T} \left| \mathbf{f}(t) e^{-|\gamma(t-c)|} \right| = |\alpha| \|\mathbf{f}\|_\gamma$$

so it does what is should for scalar multiplication. Next consider the triangle inequality.

$$\begin{aligned}
\|\mathbf{f} + \mathbf{g}\|_\gamma &= \sup_{t \in T} \left| (\mathbf{f}(t) + \mathbf{g}(t)) e^{-|\gamma(t-c)|} \right| \\
&\leq \sup_{t \in T} \left(\left| \mathbf{f}(t) e^{-|\gamma(t-c)|} \right| + \left| \mathbf{g}(t) e^{-|\gamma(t-c)|} \right| \right) \\
&\leq \sup_{t \in T} \left| \mathbf{f}(t) e^{-|\gamma(t-c)|} \right| + \sup_{t \in T} \left| \mathbf{g}(t) e^{-|\gamma(t-c)|} \right| \\
&= \|\mathbf{f}\|_\gamma + \|\mathbf{g}\|_\gamma
\end{aligned}$$

The rest follows from the next inequalities.

$$\begin{aligned}
\|\mathbf{f}\| &\equiv \sup_{t \in T} |\mathbf{f}(t)| = \sup_{t \in T} \left| \mathbf{f}(t) e^{-|\gamma(t-c)|} e^{|\gamma(t-c)|} \right| \leq e^{|\gamma(b-a)|} \|\mathbf{f}\|_\gamma \\
&\equiv e^{|\gamma(b-a)|} \sup_{t \in T} \left| \mathbf{f}(t) e^{-|\gamma(t-c)|} \right| \leq e^{|\gamma|(b-a)} \sup_{t \in T} |\mathbf{f}(t)| = e^{|\gamma|(b-a)} \|\mathbf{f}\| \quad \blacksquare
\end{aligned}$$

Now consider the general case where T is just some set.

Lemma 11.3.4 *The collection of functions $BC(T; \mathbb{C}^n)$ is a normed linear space (vector space) and it is also complete which means by definition that every Cauchy sequence converges.*

Proof: Showing that this is a normed linear space is entirely similar to the argument in the above for $\gamma = 0$ and $T = [a, b]$ is the case of interest here.

Let $\{\mathbf{f}_n\}$ be a Cauchy sequence. Then for each $\mathbf{t} \in T$, $\{\mathbf{f}_n(\mathbf{t})\}$ is a Cauchy sequence in $\mathbb{C}^n$. By completeness of $\mathbb{C}^n$ this converges to some $\mathbf{g}(\mathbf{t}) \in \mathbb{C}^n$. We need to verify that $\|\mathbf{g} - \mathbf{f}_n\| \to 0$ and that $\mathbf{g} \in BC(T; \mathbb{C}^n)$. Let $\varepsilon > 0$ be given. There exists M_ε such that if $m, n \geq M_\varepsilon$, then $\|\mathbf{f}_n - \mathbf{f}_m\| < \frac{\varepsilon}{4}$. Let $n > M_\varepsilon$. By Lemma C.3.3 which says you can switch supremums,

$$
\begin{aligned}
\sup_{t \in T} |\mathbf{g}(t) - \mathbf{f}_n(t)| &\leq \sup_{t \in T} \sup_{k \geq M_\varepsilon} |\mathbf{f}_k(t) - \mathbf{f}_n(t)| \\
&= \sup_{k \geq M_\varepsilon} \sup_{t \in T} |\mathbf{f}_k(t) - \mathbf{f}_n(t)| = \sup_{k \geq M_\varepsilon} \|\mathbf{f}_k - \mathbf{f}_n\| \leq \frac{\varepsilon}{4} \quad (*)
\end{aligned}
$$

Therefore,

$$
\sup_{t \in T} (|\mathbf{g}(t)| - |\mathbf{f}_n(t)|) \leq \sup_{t \in T} |\mathbf{g}(t) - \mathbf{f}_n(t)| \leq \frac{\varepsilon}{4}
$$

Hence

$$
\frac{\varepsilon}{4} \geq \sup_{t \in T} (|\mathbf{g}(t)| - |\mathbf{f}_n(t)|) = \sup_{t \in T} |\mathbf{g}(t)| - \inf_{t \in T} |\mathbf{f}_n(t)| \geq \sup_{t \in T} |\mathbf{g}(t)| - \|\mathbf{f}_n\|
$$

$$
\sup_{t \in T} |\mathbf{g}(t)| \leq \frac{\varepsilon}{4} + \|\mathbf{f}_n\| < \infty
$$

so in fact $\mathbf{g}$ is bounded. Now by the fact that $\mathbf{f}_n$ is continuous, there exists $\delta > 0$ such that if $|\mathbf{t} - \mathbf{s}| < \delta$, then

$$
|\mathbf{f}_n(\mathbf{t}) - \mathbf{f}_n(\mathbf{s})| < \frac{\varepsilon}{3}
$$

It follows that

$$
|\mathbf{g}(\mathbf{t}) - \mathbf{g}(\mathbf{s})| \leq |\mathbf{g}(\mathbf{t}) - \mathbf{f}_n(\mathbf{t})| + |\mathbf{f}_n(\mathbf{t}) - \mathbf{f}_n(\mathbf{s})| + |\mathbf{f}_n(\mathbf{s}) - \mathbf{g}(\mathbf{s})| \leq \frac{\varepsilon}{4} + \frac{\varepsilon}{3} + \frac{\varepsilon}{4} < \varepsilon
$$

Therefore, $\mathbf{g}$ is continuous at $\mathbf{t}$. Since $\mathbf{t}$ is arbitrary, this shows that $\mathbf{g}$ is continuous on T. Thus $\mathbf{g} \in BC(T; \mathbb{C}^n)$. By $*$, $\|\mathbf{f}_n - \mathbf{g}\| < \varepsilon$ when n is large enough so $\lim_{n \to \infty} \|\mathbf{f}_n - \mathbf{g}\| = 0$. $\blacksquare$

Definition 11.3.5 *When $\lim_{n \to \infty} \|\mathbf{f}_n - \mathbf{f}\| = 0$, we say that $\mathbf{f}_n$ converges uniformly to $\mathbf{f}$ and speak of uniform convergence. This norm is also called the uniform norm.*

Note that uniform convergence of continuous functions imparts continuity to the limit function. This is not true of pointwise convergence, that the sequence converges for each t, as can be seen by consideration of $f_n(t) = t^n$ for $t \in [0, 1]$. The limit function is discontinuous on this interval and is 0 on $[0, 1)$ and 1 at 1.

Now here is a major theorem called the Banach fixed point theorem.

Theorem 11.3.6 *Let $(X, \|\cdot\|)$ be a complete (Cauchy sequences converge.) normed linear space and let $F : X \to X$ be a contraction map. That is,*

$$
\|Fx - Fy\| \leq r|x - y|, \quad 0 \leq r < 0
$$

Then F has a unique fixed point, that is a point $x \in X$ such that $Fx = x$. In addition to this, if $\|Fx_0 - x_0\| < R(1 - r)$ and F is only defined on $\overline{B(x_0, R)}$ then F has a unique fixed point in this ball. Here $\overline{B(x_0, R)}$ signifies the set of all x such that $\|x - x_0\| \leq R$. Also, the sequence $\{F^n x_0\}$ converges.

Proof: Pick any $x_0 \in X$. Consider the sequence $\{F^n x_0\}$. I will argue that this is a Cauchy sequence. To see this, suppose $n, m \geq M$ with $n > m$ and consider the following which comes from the triangle inequality for the norm, $\|x + y\| \leq \|x\| + \|y\|$.

$$\|F^n x_0 - F^m x_0\| \leq \sum_{k=m}^{n-1} \|F^{k+1} x_0 - F^k x_0\|$$

Now $\|F^{k+1} x_0 - F^k x_0\| \leq$

$$r \|F^k x_0 - F^{k-1} x_0\| \leq r^2 \|F^{k-1} x_0 - F^{k-2} x_0\| \cdots \leq r^k \|F x_0 - x_0\|.$$

Using this in the above, $\|F^n x_0 - F^m x_0\| \leq$

$$\sum_{k=m}^{n-1} \|F^{k+1} x_0 - F^k x_0\| \leq \sum_{k=m}^{n-1} r^k \|F x_0 - x_0\| \leq \frac{r^m}{1-r} \|F x_0 - x_0\| \tag{11.8}$$

since $r < 1$, this is a Cauchy sequence. Hence it converges to some x. Therefore,

$$x = \lim_{n \to \infty} F^n x_0 = \lim_{n \to \infty} F^{n+1} x_0 = F \lim_{n \to \infty} F^n x_0 = Fx.$$

The third equality is a consequence of the following consideration. If $z_n \to z$, then

$$\|F z_n - F z\| \leq r \|z_n - z\|$$

so also $F z_n \to F z$. In the above, $F^n x_0$ plays the role of z_n and its limit plays the role of z.

The fixed point is unique because if you had two of them, $x, \hat{x}$, then

$$\|x - \hat{x}\| = \|F x - F \hat{x}\| \leq r \|x - \hat{x}\|$$

and so $x = \hat{x}$.

In the second case, let $m = 0$ in (11.8) and you get the estimate

$$\|F^n x_0 - x_0\| \leq \frac{1}{1-r} \|F x_0 - x_0\| < R$$

It is still the case that the sequence $\{F^n x_0\}$ is a Cauchy sequence and must therefore converge to some $x \in \overline{B(x_0, R)}$ which is a fixed point as before. The fixed point is unique because of the same argument as before. ∎

With this theorem is a fundamental existence and uniqueness result for ordinary differential equations.

Theorem 11.3.7 *Let* **f** *satisfy the Lipschitz condition (11.4) and the continuity condition (11.5). Then there exists a unique solution to the initial value problem, (11.6) on* $[a, b]$, $c \in [a, b]$. *Furthermore, the Picard iterates converge in* $BC([a, b], \mathbb{C}^n)$ *with respect to the norm* $\|\cdot\|$ *defined above.*

Proof: As explained above, this is equivalent to finding a solution to the integral equation

$$\mathbf{x}(t) = \mathbf{x}_0 + \int_c^t \mathbf{f}(s, \mathbf{x}(s)) \, ds, \ t \in [a, b]$$

Let a, b be finite but given and completely arbitrary, $c \in [a, b]$. Let $F\mathbf{x}(t) \equiv \mathbf{x}_0 + \int_c^t \mathbf{f}(s, \mathbf{x}(s))\, ds$ Thus

$$F : BC([a, b], \mathbb{C}^n) \to BC([a, b], \mathbb{C}^n)$$

Let $\|\cdot\|_\gamma$ be the new norm on $BC([a, b], \mathbb{C}^n)$.

$$\|\mathbf{f}\|_\gamma \equiv \sup_{t \in [a,b]} \left| \mathbf{f}(t) e^{-|\gamma(t-c)|} \right|$$

Note that

$$|\mathbf{x}(s) - \mathbf{y}(s)| = e^{|\gamma(s-c)|} e^{-|\gamma(s-c)|} |\mathbf{x}(s) - \mathbf{y}(s)| \leq e^{|\gamma(s-c)|} \|\mathbf{x} - \mathbf{y}\|_\gamma$$

Then

$$|F\mathbf{x}(t) - F\mathbf{y}(t)| \leq \left| \int_c^t |\mathbf{f}(s, \mathbf{x}(s)) - \mathbf{f}(s, \mathbf{y}(s))|\, ds \right| \leq \left| \int_c^t K|\mathbf{x}(s) - \mathbf{y}(s)|\, ds \right|$$

$$\leq K \left| \int_c^t e^{|\gamma(s-c)|} \|\mathbf{x} - \mathbf{y}\|_\gamma\, ds \right| = K\|\mathbf{x} - \mathbf{y}\|_\gamma \left| \int_c^t e^{|\gamma(s-c)|} ds \right| \qquad (*)$$

Now consider the case that $t \geq c$. The above reduces to

$$= K\|\mathbf{x} - \mathbf{y}\|_\gamma \int_c^t e^{|\gamma|(s-c)} ds = K\|\mathbf{x} - \mathbf{y}\|_\gamma \left(\frac{e^{|\gamma|(s-c)}}{|\gamma|} \Big|_c^t \right)$$

$$= K\|\mathbf{x} - \mathbf{y}\|_\gamma \left(\frac{e^{|\gamma|(t-c)}}{|\gamma|} - \frac{1}{|\gamma|} \right) = K\|\mathbf{x} - \mathbf{y}\|_\gamma \left(\frac{e^{|\gamma(t-c)|}}{|\gamma|} - \frac{1}{|\gamma|} \right)$$

Next consider $*$ when $t < c$. In this case, $*$ becomes

$$K\|\mathbf{x} - \mathbf{y}\|_\gamma \int_t^c e^{|\gamma|(c-s)} ds = K\|\mathbf{x} - \mathbf{y}\|_\gamma \left(\frac{e^{|\gamma|(c-s)}}{-|\gamma|} \Big|_t^c \right)$$

$$= K\|\mathbf{x} - \mathbf{y}\|_\gamma \left(\frac{1}{-|\gamma|} + \frac{e^{|\gamma|(c-t)}}{|\gamma|} \right)$$

$$\leq K\|\mathbf{x} - \mathbf{y}\|_\gamma \left(\frac{e^{|\gamma(t-c)|}}{|\gamma|} - \frac{1}{|\gamma|} \right)$$

It follows that for any $t \in [a, b]$,

$$e^{-|\gamma(t-c)|} |F\mathbf{x}(t) - F\mathbf{y}(t)| \leq K\|\mathbf{x} - \mathbf{y}\|_\gamma e^{-|\gamma(t-c)|} \left(\frac{e^{|\gamma(t-c)|}}{|\gamma|} - \frac{1}{|\gamma|} \right)$$

$$\leq K\|\mathbf{x} - \mathbf{y}\|_\gamma \frac{1}{|\gamma|}$$

and so

$$\|F\mathbf{x} - F\mathbf{y}\|_\gamma \leq K\|\mathbf{x} - \mathbf{y}\|_\gamma \frac{1}{|\gamma|}$$

so let $|\gamma| > 2K$ and this shows that F is a contraction map on $BC([a, b]; \mathbb{C}^n)$.

Thus there is a unique solution to the above integral equation and hence a unique solution to the initial value problem (11.6) on $[a, b]$.

As to the convergence of the Picard iterates, by Theorem 11.3.6, these converge in $BC\left([a, b], \mathbb{C}^n\right)$ with respect to $\|\cdot\|_\gamma$ and so they also converge in $BC\left([a, b], \mathbb{C}^n\right)$ with respect to the usual norm $\|\cdot\|$ by Lemma 11.3.3. ∎

For another approach to this problem, see Problem 17 in the exercises.

In the proof of completeness of $BC\left(T, \mathbb{C}^n\right)$ compactness of T was not needed.

Lemma 11.3.8 *For $\gamma > 0$, let*

$$E_\gamma \equiv \left\{\mathbf{x} \in BC\left([0, \infty), \mathbb{C}^n\right) : t \to e^{\gamma t}\mathbf{x}(t) \text{ is also in } BC\left([0, \infty), \mathbb{C}^n\right)\right\}$$

and let the norm be given by

$$\|\mathbf{x}\|_\gamma \equiv \sup\left\{\left|e^{\gamma t}\mathbf{x}(t)\right| : t \in [0, \infty)\right\}$$

Then E_γ is a Banach space.

Proof: Let $\{\mathbf{x}_k\}$ be a Cauchy sequence in E_γ. Then since $BC\left([0, \infty), \mathbb{C}^n\right)$ is a Banach space, there exists $\mathbf{y} \in BC\left([0, \infty), \mathbb{C}^n\right)$ such that $e^{\gamma t}\mathbf{x}_k(t)$ converges uniformly on $[0, \infty)$ to $\mathbf{y}(t)$. Therefore $e^{-\gamma t}e^{\gamma t}\mathbf{x}_k(t) = \mathbf{x}_k(t)$ converges uniformly to $e^{-\gamma t}\mathbf{y}(t)$ on $[0, \infty)$. Define $\mathbf{x}(t) \equiv e^{-\gamma t}\mathbf{y}(t)$. Then $\mathbf{y}(t) = e^{\gamma t}\mathbf{x}(t)$ and by definition, $\|\mathbf{x}_k - \mathbf{x}\|_\gamma \to 0$. ∎

11.4 Linear Systems

As an example of the above theorem, consider for $t \in [a, b]$ the system

$$\mathbf{x}' = A(t)\mathbf{x}(t) + \mathbf{g}(t), \ \mathbf{x}(c) = \mathbf{x}_0 \tag{11.9}$$

where $A(t)$ is an $n \times n$ matrix whose entries are continuous functions of t, $(a_{ij}(t))$ and $\mathbf{g}(t)$ is a vector whose components are continuous functions of t satisfies the conditions of Theorem 11.3.7 with $\mathbf{f}(t, \mathbf{x}) = A(t)\mathbf{x} + \mathbf{g}(t)$. To see this, let $\mathbf{x} = (x_1, \cdots, x_n)^T$ and $\mathbf{x}_1 = (x_{11}, \cdots, x_{1n})^T$. Then letting

$$M = \max\left\{\left|a_{ij}(t)\right| : t \in [a, b], i, j \le n\right\}, \ |\mathbf{f}(t, \mathbf{x}) - \mathbf{f}(t, \mathbf{x}_1)| = |A(t)(\mathbf{x} - \mathbf{x}_1)|$$

$$= \left|\left(\sum_{i=1}^n\left|\sum_{j=1}^n a_{ij}(t)(x_j - x_{1j})\right|^2\right)^{1/2}\right| \le M\left|\left(\sum_{i=1}^n\left(\sum_{j=1}^n|x_j - x_{1j}|\right)^2\right)^{1/2}\right|$$

$$\le M\left|\left(\sum_{i=1}^n n\sum_{j=1}^n|x_j - x_{1j}|^2\right)^{1/2}\right| = Mn\left(\sum_{j=1}^n|x_j - x_{1j}|^2\right)^{1/2} = Mn|\mathbf{x} - \mathbf{x}_1|.$$

Therefore, we can let $K = Mn$. This proves

Theorem 11.4.1 *Let $A(t)$ be a continuous $n \times n$ matrix and let $\mathbf{g}(t)$ be a continuous vector for $t \in [a, b]$ and let $c \in [a, b]$ and $\mathbf{x}_0 \in \mathbb{R}^n$. Then there exists a unique solution to (11.9) valid for $t \in [a, b]$.*

This includes more examples of linear equations than everything done earlier.

11.5   Local Solutions

The above gives global solutions based on an assumption that there is a Lipschitz constant. What if there isn't one? It turns out that in this case, one gets local solutions which exist on some interval. This situation is discussed next.

Lemma 11.5.1 *Let $D(\mathbf{x}_0, r) \equiv \{\mathbf{x} \in \mathbb{R}^n : |\mathbf{x} - \mathbf{x}_0| \leq r\}$ and suppose U is an open set containing $D(\mathbf{x}_0, r)$ such that $\mathbf{f} : U \to \mathbb{R}^n$ is $C^1(U)$. (Recall this means all partial derivatives of $\mathbf{f}$ exist and are continuous.) Then for $K = Mn$, where M denotes the maximum of $\left|\frac{\partial \mathbf{f}}{\partial x_i}(\mathbf{z})\right|$ for $\mathbf{z} \in D(\mathbf{x}_0, r)$, it follows that for all $\mathbf{x}, \mathbf{y} \in D(\mathbf{x}_0, r)$,*

$$|\mathbf{f}(\mathbf{x}) - \mathbf{f}(\mathbf{y})| \leq K|\mathbf{x} - \mathbf{y}|.$$

Proof: Let $\mathbf{x}, \mathbf{y} \in D(\mathbf{x}_0, r)$ and consider the line segment joining these two points, $\mathbf{x} + t(\mathbf{y} - \mathbf{x})$ for $t \in [0, 1]$. If we let $\mathbf{h}(t) = \mathbf{f}(\mathbf{x} + t(\mathbf{y} - \mathbf{x}))$ for $t \in [0, 1]$, then

$$\mathbf{f}(\mathbf{y}) - \mathbf{f}(\mathbf{x}) = \mathbf{h}(1) - \mathbf{h}(0) = \int_0^1 \mathbf{h}'(t)\, dt.$$

Also, by the chain rule,

$$\mathbf{h}'(t) = \sum_{i=1}^n \frac{\partial \mathbf{f}}{\partial x_i}(\mathbf{x} + t(\mathbf{y} - \mathbf{x}))(y_i - x_i).$$

Therefore, we must have

$$|\mathbf{f}(\mathbf{y}) - \mathbf{f}(\mathbf{x})| = \left| \int_0^1 \sum_{i=1}^n \frac{\partial \mathbf{f}}{\partial x_i}(\mathbf{x} + t(\mathbf{y} - \mathbf{x}))(y_i - x_i)\, dt \right|$$

$$\leq \int_0^1 \sum_{i=1}^n \left|\frac{\partial \mathbf{f}}{\partial x_i}(\mathbf{x} + t(\mathbf{y} - \mathbf{x}))\right| |y_i - x_i|\, dt \leq M \sum_{i=1}^n |y_i - x_i| \leq Mn|\mathbf{x} - \mathbf{y}|. \quad \blacksquare$$

Now consider the map, P which maps all of $\mathbb{R}^n$ to $D(\mathbf{x}_0, r)$ given as follows. For $\mathbf{x} \in D(\mathbf{x}_0, r)$, we have $P\mathbf{x} = \mathbf{x}$. For $\mathbf{x} \notin D(\mathbf{x}_0, r)$ we have $P\mathbf{x}$ will be the closest point in $D(\mathbf{x}_0, r)$ to $\mathbf{x}$. Such a closest point exists because $D(\mathbf{x}_0, r)$ is a closed and bounded set. Taking $f(\mathbf{y}) \equiv |\mathbf{y} - \mathbf{x}|$, it follows f is a continuous function defined on $D(\mathbf{x}_0, r)$ which must achieve its minimum value by the extreme value theorem from calculus.

Lemma 11.5.2 *For any pair of points, $\mathbf{x}, \mathbf{y} \in \mathbb{R}^n$ we have $|P\mathbf{x} - P\mathbf{y}| \leq |\mathbf{x} - \mathbf{y}|$.*

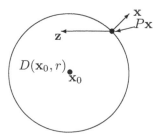

Proof: From the above picture it follows that for $\mathbf{z} \in D(\mathbf{x}_0, r)$ arbitrary, the angle between the vectors $\mathbf{x} - P\mathbf{x}$ and $\mathbf{z} - P\mathbf{x}$ is always greater than $\pi/2$ radians. Therefore, the cosine of this angle is always negative. Less geometrically, but more convincingly, you could consider $|\mathbf{x} - (P\mathbf{x} + t(\mathbf{z} - P\mathbf{x}))|^2, t \in [0, 1]$ and note that by definition, its minimum is achieved at $t = 0$ and so its derivative at $t = 0$ must be nonnegative. However, it equals

$$|\mathbf{x} - P\mathbf{x}|^2 - (\mathbf{x} - P\mathbf{x} \cdot \mathbf{z} - P\mathbf{x})t + t^2|\mathbf{z} - P\mathbf{x}|^2$$

and so on taking the derivative, and letting $t = 0$,

$$-(\mathbf{x} - P\mathbf{x} \cdot \mathbf{z} - P\mathbf{x}) \geq 0, \quad (\mathbf{x} - P\mathbf{x} \cdot \mathbf{z} - P\mathbf{x}) \leq 0.$$

It follows that

$$(\mathbf{y} - P\mathbf{y}) \cdot (P\mathbf{x} - P\mathbf{y}) \leq 0$$

and

$$(\mathbf{x} - P\mathbf{x}) \cdot (P\mathbf{y} - P\mathbf{x}) \leq 0.$$

Thus $(\mathbf{x} - P\mathbf{x}) \cdot (P\mathbf{x} - P\mathbf{y}) \geq 0$ and so if we subtract,

$$(\mathbf{x} - P\mathbf{x} - (\mathbf{y} - P\mathbf{y})) \cdot (P\mathbf{x} - P\mathbf{y}) \geq 0$$

which implies

$$(\mathbf{x} - \mathbf{y}) \cdot (P\mathbf{x} - P\mathbf{y}) \geq (P\mathbf{x} - P\mathbf{y}) \cdot (P\mathbf{x} - P\mathbf{y}) = |P\mathbf{x} - P\mathbf{y}|^2.$$

Now apply the Cauchy-Schwarz inequality to the left side of the above inequality to obtain

$$|\mathbf{x} - \mathbf{y}| \, |P\mathbf{x} - P\mathbf{y}| \geq |P\mathbf{x} - P\mathbf{y}|^2$$

which yields the claim of the lemma. ■

With this here is the local existence and uniqueness theorem.

Theorem 11.5.3 *Let $[a, b]$ be a closed interval and let U be an open subset of $\mathbb{R}^n$. Let $\mathbf{f} : [a, b] \times U \to \mathbb{R}^n$ be continuous and suppose that for each $t \in [a, b]$, the map $\mathbf{x} \to \frac{\partial \mathbf{f}}{\partial x_i}(t, \mathbf{x})$ is continuous. Also let $\mathbf{x}_0 \in U$ and $c \in [a, b]$. Then there exists an interval, $I \subseteq [a, b]$ such that $c \in I$ and there exists a unique solution to the initial value problem,*

$$\mathbf{x}' = \mathbf{f}(t, \mathbf{x}), \, \mathbf{x}(c) = \mathbf{x}_0 \tag{11.10}$$

valid for $t \in I$.

Proof: Consider the following picture.

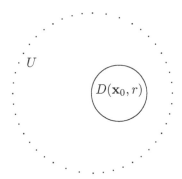

The large dotted circle represents U and the little solid circle represents $D(\mathbf{x}_0, r)$ as indicated. Here we have chosen r so small that $D(\mathbf{x}_0, r)$ is contained in U as shown. Now let P denote the projection map defined above. Consider the initial value problem

$$\mathbf{x}' = \mathbf{f}(t, P\mathbf{x}), \, \mathbf{x}(c) = \mathbf{x}_0. \tag{11.11}$$

From Lemma 11.5.1 and the continuity of $\mathbf{x} \to \frac{\partial \mathbf{f}}{\partial x_i}(t, \mathbf{x})$, we know there exists a constant, K such that if $\mathbf{x}, \mathbf{y} \in D(\mathbf{x}_0, r)$, then $|\mathbf{f}(t, \mathbf{x}) - \mathbf{f}(t, \mathbf{y})| \leq K |\mathbf{x} - \mathbf{y}|$ for all $t \in [a, b]$.

Therefore, by Lemma 11.5.2

$$|\mathbf{f}(t, P\mathbf{x}) - \mathbf{f}(t, P\mathbf{y})| \leq K |P\mathbf{x} - P\mathbf{y}| \leq K |\mathbf{x} - \mathbf{y}|.$$

It follows from Theorem 11.3.7 that (11.11) has a unique solution valid for $t \in [a, b]$. Since $\mathbf{x}$ is continuous, it follows that there exists an interval, I containing c such that for $t \in I$, we have $\mathbf{x}(t) \in D(\mathbf{x}_0, r)$. Therefore, for these values of t, we have $\mathbf{f}(t, P\mathbf{x}) = \mathbf{f}(t, \mathbf{x})$ and so there is a unique solution to (11.10) on I. ■

Now suppose $\mathbf{f}$ has the <u>property</u> that for every $R > 0$ there exists a constant, K_R such that for all $\mathbf{x}, \mathbf{x}_1 \in \overline{B(0, R)}$,

$$|\mathbf{f}(t, \mathbf{x}) - \mathbf{f}(t, \mathbf{x}_1)| \leq K_R |\mathbf{x} - \mathbf{x}_1|. \tag{11.12}$$

Corollary 11.5.4 *Let $\mathbf{f}$ satisfy (11.12) and suppose also that $(t, \mathbf{x}) \to \mathbf{f}(t, \mathbf{x})$ is continuous. Suppose now that $\mathbf{x}_0$ is given and there exists an estimate of the form $|\mathbf{x}(t)| < R$ for all $t \in [0, T)$ where $T \leq \infty$ on the local solution to*

$$\mathbf{x}' = \mathbf{f}(t, \mathbf{x}), \ \mathbf{x}(0) = \mathbf{x}_0. \tag{11.13}$$

Then there exists a unique solution to the initial value problem, (11.13) valid on $[0, T)$.

Proof: Replace $\mathbf{f}(t, \mathbf{x})$ with $\mathbf{f}(t, P_R\mathbf{x})$ where P_R is the projection onto $\overline{B(0, R)}$. Then by Theorem 11.3.7 there exists a unique solution to the system

$$\mathbf{x}' = \mathbf{f}(t, P_R\mathbf{x}), \ \mathbf{x}(0) = \mathbf{x}_0$$

valid on $[0, T_1]$ for every $T_1 < T$. Therefore, the above system has a unique solution on $[0, T)$ and from the estimate, $P_R\mathbf{x} = \mathbf{x}$. ■

11.6 Continuous Dependence

Recall Gronwall's inequality which was Lemma 10.1.1 on Page 249 presented earlier.

Theorem 11.6.1 *Suppose $u(t) \geq 0$ and for all $t \in [0, T]$,*

$$u(t) \leq u_0 + \int_0^t K u(s) \, ds. \tag{11.14}$$

where K is some constant. Then

$$u(t) \leq u_0 e^{Kt}. \tag{11.15}$$

One of the nice things about Gronwall's inequality is that it gives an important result about dependence on initial data.

Theorem 11.6.2 *Let $\mathbf{x} \to \mathbf{f}(t, \mathbf{x})$ satisfy a Lipschitz condition with constant K and suppose*

$$\lim_{n \to \infty} \mathbf{x}_{0n} = \mathbf{x}_0$$

Letting $\mathbf{x}(t, \mathbf{x}_0)$ be the solution to the initial value problem

$$\mathbf{x}' = \mathbf{f}(t, \mathbf{x}), \ \mathbf{x}(0) = \mathbf{x}_0$$

it follows that for each $t \geq 0$

$$\lim_{n \to \infty} |\mathbf{x}(t, \mathbf{x}_{0n}) - \mathbf{x}(t, \mathbf{x}_0)| = 0$$

If it is known that $\mathbf{f}(t, \mathbf{x})$ is bounded, $|\mathbf{f}(t, \mathbf{x})| \leq M$ then in fact $(t, \mathbf{x}_0) \to \mathbf{x}(t, \mathbf{x}_0)$ is continuous. Thus the solution is a continuous function of both t and initial condition $\mathbf{x}_0$.

Proof: From the integral equations solved,

$$\mathbf{x}(t, \mathbf{x}_{0n}) - \mathbf{x}(t, \mathbf{x}_0) = \mathbf{x}_{0n} - \mathbf{x}_0 + \int_0^t \mathbf{f}(s, \mathbf{x}(s, \mathbf{x}_{0n})) - \mathbf{f}(s, \mathbf{x}(s, \mathbf{x}_0)) \, ds$$

Therefore,

$$|\mathbf{x}(t, \mathbf{x}_{0n}) - \mathbf{x}(t, \mathbf{x}_0)| \leq |\mathbf{x}_{0n} - \mathbf{x}_0| + \int_0^t |\mathbf{f}(s, \mathbf{x}(s, \mathbf{x}_{0n})) - \mathbf{f}(s, \mathbf{x}(s, \mathbf{x}_0))| \, ds$$

$$\leq |\mathbf{x}_{0n} - \mathbf{x}_0| + \int_0^t K |\mathbf{x}(s, \mathbf{x}_{0n}) - \mathbf{x}(s, \mathbf{x}_0)| \, ds$$

By Gronwall's inequality, $|\mathbf{x}(t, \mathbf{x}_{0n}) - \mathbf{x}(t, \mathbf{x}_0)| \leq |\mathbf{x}_{0n} - \mathbf{x}_0| e^{Kt}$ which yields the desired result.

Consider the last claim. Say $\mathbf{f}(t, \mathbf{x}) \leq M$. Let $(t_n, \mathbf{x}_{0n}) \to (t, \mathbf{x}_0)$. Then

$$|\mathbf{x}(t_n, \mathbf{x}_{0n}) - \mathbf{x}(t, \mathbf{x}_0)| \leq |\mathbf{x}(t_n, \mathbf{x}_{0n}) - \mathbf{x}(t, \mathbf{x}_{0n})| + |\mathbf{x}(t, \mathbf{x}_{0n}) - \mathbf{x}(t, \mathbf{x}_0)|$$

Now

$$|\mathbf{x}(t_n, \mathbf{x}_{0n}) - \mathbf{x}(t, \mathbf{x}_{0n})| \leq \left| \int_t^{t_n} \mathbf{f}(s, \mathbf{x}(s, \mathbf{x}_{0n})) \, ds \right| \leq M |t - t_n|$$

Therefore,

$$|\mathbf{x}(t_n, \mathbf{x}_{0n}) - \mathbf{x}(t, \mathbf{x}_0)| \leq M |t - t_n| + |\mathbf{x}_{0n} - \mathbf{x}_0| e^{Kt} \ \blacksquare$$

In fact, one can say more. I will just sketch the details because these considerations depend on more analysis than is typically known in a beginning course. Assume $\mathbf{f}$ is a C^1 function, meaning that its partial derivatives exist and are continuous. Then you can conclude that $(t, \mathbf{x}_0) \to \mathbf{x}(t, \mathbf{x}_0)$ is a C^1 function if you assume that $\mathbf{x}(t, \mathbf{x}_0)$ stays in a bounded set. To see this, formally differentiate the differential equation.

$$\mathbf{x}'' = D_1\mathbf{f}(t, \mathbf{x}) + D_2\mathbf{f}(t, \mathbf{x})\mathbf{x}'(t), \ \mathbf{x}'(0) = \mathbf{f}(0, \mathbf{x}_0)$$

Note how $\mathbf{x}$ is now known and this is an equation for $\mathbf{x}'$. You consider

$$\mathbf{y}' = D_1\mathbf{f}(t, \mathbf{x}) + D_2\mathbf{f}(t, \mathbf{x})\mathbf{y}(t), \mathbf{y}(0) = \mathbf{f}(0, \mathbf{x}_0)$$

Now one just uses Lemma 11.5.1 and repeats the above argument with the dependent variable being $\mathbf{y}(t, \mathbf{x}_0)$. The function $(t, \mathbf{y}) \to D_1\mathbf{f}(t, \mathbf{x}) + D_2\mathbf{f}(t, \mathbf{x})\mathbf{y}$ will be Lipschitz continuous in $\mathbf{y}$ and so $\mathbf{y}$ is a continuous function of t and $\mathbf{x}_0$. Then you show that $\mathbf{y} = \mathbf{x}'$. This is done by noting that

$$\mathbf{y}(t) - \mathbf{x}'(t) = \int_0^t D_1\mathbf{f}(s, \mathbf{x}) + D_2\mathbf{f}(s, \mathbf{x})\mathbf{y}(s) \, ds - (\mathbf{f}(t, \mathbf{x}(t)) - \mathbf{f}(0, \mathbf{x}_0))$$

$$= \int_0^t D_1\mathbf{f}(s, \mathbf{x}) + D_2\mathbf{f}(s, \mathbf{x})\mathbf{y}(s) - (D_1\mathbf{f}(s, \mathbf{x}(s)) + D_2\mathbf{f}(s, \mathbf{x}(s))\mathbf{x}'(s)) \, ds$$

$$= \int_0^t D_2\mathbf{f}\left(s, \mathbf{x}\left(s\right)\right)\left(\mathbf{y}\left(s\right) - \mathbf{x}'\left(s\right)\right) ds$$

and then using Gronwall's inequality.

To find continuity of $\mathbf{x} \to D_2\mathbf{x}\left(t, \mathbf{x}\right)$, just do something similar.

$$\frac{1}{h}\left(\mathbf{x}\left(t, \mathbf{x}_0 + h\mathbf{e}_k\right) - \mathbf{x}\left(t, \mathbf{x}_0\right)\right) = \mathbf{e}_k + \int_0^t \frac{1}{h}\left(\mathbf{f}\left(s, \mathbf{x}\left(s, \mathbf{x}_0 + h\mathbf{e}_k\right)\right) - \mathbf{f}\left(s, \mathbf{x}\left(s, \mathbf{x}_0\right)\right)\right) ds$$

$$= \mathbf{e}_k + \int_0^t D\mathbf{f}\left(s, \mathbf{x}\left(s, \mathbf{x}_0\right)\right)\mathbf{e}_k ds + \int_0^t \mathbf{o}\left(h\right)/h ds$$

Then passing to a limit, one obtains $\frac{\partial \mathbf{x}}{\partial x_k}\left(t, \mathbf{x}_0\right) = \left(I + \int_0^t D\mathbf{f}\left(s, \mathbf{x}\left(s, \mathbf{x}_0\right)\right) ds\right)\mathbf{e}_k$ and this shows that these partial derivatives are also continuous. Thus $(t, \mathbf{x}_0) \to \mathbf{x}\left(t, \mathbf{x}_0\right)$ is indeed a C^1 function if you know that $(t, \mathbf{x})$ stays in a bounded set and $\mathbf{f}$ is C^1.

11.7 Autonomous Equations

The following definition tells what the term autonomous means. These equations have some very nice properties.

Definition 11.7.1 *The differential equation is autonomous if it is of the form*

$$\mathbf{x}' = \mathbf{f}\left(\mathbf{x}\right)$$

The special feature is that $\mathbf{f}$ does not depend on t.

The nice property of autonomous equations is next.

Proposition 11.7.2 *Let $\mathbf{f}$ satisfy a Lipschitz condition on every ball $B\left(\mathbf{x}_0, r\right)$. Let $\mathbf{x}\left(t, \mathbf{x}_0\right)$ denote the solution to*

$$\mathbf{x}' = \mathbf{f}\left(\mathbf{x}\right), \ \mathbf{x}\left(0\right) = \mathbf{x}_0$$

Then

$$\mathbf{x}\left(t, \mathbf{x}\left(s, \mathbf{x}_0\right)\right) = \mathbf{x}\left(t + s, \mathbf{x}_0\right) \tag{11.16}$$

Also if $\mathbf{x}_0 \neq \hat{\mathbf{x}}_0$, then $\mathbf{x}\left(t, \mathbf{x}_0\right) \neq \mathbf{x}\left(t, \hat{\mathbf{x}}_0\right)$.

Proof: There exist at least local solutions because the function is locally Lipschitz. As in Corollary 11.5.4 you simply consider $\mathbf{x}' = \mathbf{f}\left(P_R\mathbf{x}\right)$ where $\mathbf{x}\left(s, \mathbf{x}_0\right) \in B\left(\mathbf{0}, R\right)$ and note that both functions of t in (11.16) are local solutions to the initial value problem

$$\mathbf{x}' = \mathbf{f}\left(P_R\mathbf{x}\right), \ \mathbf{x}\left(0\right) = \mathbf{x}\left(s, \mathbf{x}_0\right)$$

By uniqueness, these are equal at least on $B\left(\mathbf{0}, R\right)$. Now simply enlarge R to conclude that these two are equal as long as the solution exists.

As to the second claim, suppose for some $t_1 > 0, \mathbf{x}\left(t_1, \mathbf{x}_0\right) = \mathbf{x}\left(t_1, \hat{\mathbf{x}}_0\right) = \mathbf{c}$ and let t_1 be the smallest positive number for which these are equal. Then consider the initial value problems

$$\mathbf{x}'\left(t\right) = \mathbf{f}\left(\mathbf{x}\right), \ \mathbf{x}\left(t_1\right) = \mathbf{c}, \ t \in \left[t_1 - \delta, t_1\right]$$

One can show that there is a unique solution to this initial value problem provided δ is small enough. This can be done in the same way as indicated above by considering the Lipschitz constant on a small ball and a projection map. But both $\mathbf{x}(t, \mathbf{x}_0)$ and $\mathbf{x}(t, \hat{\mathbf{x}}_0)$ are solutions to this and so these two must be equal on some small interval $[t_1 - \delta, t_1]$ contradicting the definition of t_1 as the smallest value where these are equal. More specifically, define for $t \in [t_1 - \delta, t_1]$ and $\mathbf{y} \in C([t_1 - \delta, t_1]; \mathbb{R}^n)$

$$F\mathbf{y}(t) = \mathbf{c} + \int_{t_1}^{t} \mathbf{f}(\mathbf{y}(s)) ds$$

Then

$$|F\mathbf{y}(t) - F\hat{\mathbf{y}}(t)| = \left| \int_{t_1}^{t} \mathbf{f}(\mathbf{y}(s)) - \mathbf{f}(\hat{\mathbf{y}}(s)) ds \right| \leq \int_{t}^{t_1} K |\mathbf{y}(s) - \hat{\mathbf{y}}(s)| ds$$

$$\leq \delta K \|y - \hat{y}\| < \frac{1}{2} \|\mathbf{y} - \hat{\mathbf{y}}\|$$

provided δ is small enough and so

$$\|F\mathbf{y} - F\hat{\mathbf{y}}\| \leq \frac{1}{2} \|\mathbf{y} - \hat{\mathbf{y}}\|$$

Thus the usual contraction mapping theorem discussed above holds and gives a unique solution to the initial value problem as a fixed point of F and contradicts the definition of t_1. ■

Observation 11.7.3 *The last assertion says that solution curves $t \to \mathbf{x}(t, \mathbf{x}_0)$ for different initial conditions do not cross for t in any closed interval. In other words, if two such solution curves start out different, they stay that way.*

11.8 Numerical Methods, an Introduction

Nothing very extensive will be considered here, just a few basic ideas, because this topic is best left to a numerical analysis course.

The above proofs of existence could in theory be used to produce numerical methods for finding the solution, but there are other ways which in cases of most interest yield much better results. It is like the definition of the integral versus Simpson's formula in calculus. You could theoretically compute approximations to the integral using Riemann sums, but you would typically be much better off using Simpson's formula. This is because Simpson's formula exploits the smoothness of the function while the Riemann sums do not. The same is true here. By the discussion after Theorem 11.6.2 one can conclude that the solutions to differential equations inherit smoothness from the function $\mathbf{f}(t, \mathbf{x})$ and if $\mathbf{f}$ is smooth then so is the solution $\mathbf{x}(t, \mathbf{x}_0)$.

The methods presented here will exploit this smoothness to produce error estimates between the real solution and the numerically determined solutions. Remember how this happened with the integral, Simpsons rule included an error estimate which was dependent on the size of the fourth derivative of the function and it was a very good estimate but it did depend on smoothness of the function being integrated. This is typical of error estimates. You need to have some sort of smoothness on the function approximated in order to get them. If you don't have error estimates, then how do you know whether what you have found is close to what you are looking for? Everything degenerates into speculation without these estimates.

Consider the following on the interval $[0, h]$. Let $y_1 \equiv y_0 + hf(0, y_0)$. How close will this be to the solution to the initial value problem $y' = f(t, y)$, $y(0) = y_0$? From Taylor's theorem, $y(h) = y(0) + y'(0)h + O(h^2)$. The notation $O(h^2)$ means $O(h^2)/h^2$ is bounded as $h \to 0$. Thus it converges to 0 like h^2.

Now from the differential equation and initial condition,

$$y(h) = y_0 + f(0, y_0)h + O(h^2), \ y_1 \equiv y_0 + f(0, y_0)h$$

Therefore, $y(h) - y_1 = O(h^2)$ which means that this converges to 0 like h^2 as $h \to 0$. This is called the local error.

Euler's method involves breaking the time interval into n subintervals each of length h and then doing the above on each of these subintervals. Thus it is

$$y_{k+1} = y_k + hf(t_k, y_k), \ k = 0, 1, \cdots, n-1, \ t_k \equiv kh$$

Then y_k is an approximation of $y(kh)$. At each time step, an additional error of $O(h^2)$ is incurred between $y(kh)$ and the numerically computed y_k by a repeat of the above argument using differentiability of f to account for the error between y_k and the desired value $y(kh)$, with y_k playing the role of y_0 and kh the role 0. Now $h = T/n$ and so after n applications of this formula, you would have a total error of at most $\frac{T}{h}O(h^2) = O(h)$. Thus, as $h \to 0$, the solution does at least converge to the solution to the differential equation and $|y(kh) - y_k| \leq Ch$. Of course this does not identify the constant C, but this could also be estimated with more effort.

There is no essential difference in using this method for vector valued functions. Thus $f(t, y)$ could be replaced with $\mathbf{f}(t, \mathbf{y})$ and the initial condition would then be a vector. There is no change in the above argument. The method is called Euler's method.

Procedure 11.8.1 *To find a numerical solution to*

$$\mathbf{y}' = \mathbf{f}(t, \mathbf{y}), \ \mathbf{y}(0) = \mathbf{y}_0$$

using Euler's method, do the following:

$$\mathbf{y}_{k+1} = \mathbf{y}_k + h\mathbf{f}(t_k, \mathbf{y}_k), \ k = 0, 1, \cdots, n-1, \ t_k \equiv kh$$

then there is some constant C such that $|\mathbf{y}(kh) - \mathbf{y}_k| < Ch$.

There is no essential change if the time interval is $[t_0, t_0 + nh]$ instead of $[0, nh]$. You just let $t_k \equiv t_0 + kh$. Alternatively, you could just change the independent variable to make the interval of interest $[0, nh]$, so there is no loss of generality in assuming the initial conditions are taking place at $t = 0$.

Now consider the following method.

$$y_{k+1} = y_k + \frac{h}{2}f(t_k, y_k) + \frac{h}{2}f(t_k + h, y_k + hf(t_k, y_k)), \ t_k \equiv kh$$

this for $k = 0, 1, \cdots, n-1$. Consider first the local error. For simplicity, let the time be 0.

$$y_1 = y_0 + \frac{h}{2} f(0, y_0) + \frac{h}{2} f(0 + h, y_0 + hf(0, y_0))$$

From the differential equation and the initial condition, and using Taylor's theorem for a function of two variables, (see Problem 36 on Page 174)

$$y_1 = y_0 + \frac{h}{2} y'(0) + \frac{h}{2} \left[f \overset{y'(0)}{(0, y_0)} + f_t(0, y_0) h + f_y(0, y_0) f \overset{y'(0)}{(0, y_0)} h + O\left(h^2\right) \right]$$

Now you notice that from the chain rule,

$$f_t(0, y_0) h + f_y(0, y_0) f(0, y_0) h = \frac{d}{dt}(f(\cdot, y)) |_{t=0} h = y''(0) h$$

thus the above reduces to

$$\begin{aligned} y_1 &= y(0) + \frac{h}{2} y'(0) + \frac{h}{2} y'(0) + \frac{h^2}{2} y''(0) + O\left(h^3\right) \\ &= y(0) + hy'(0) + \frac{h^2}{2} y''(0) + O\left(h^3\right) \end{aligned}$$

From the chain rule, it is also the case that

$$y(h) = y(0) + hy'(0) + \frac{h^2}{2} y''(0) + O\left(h^3\right)$$

Therefore, $|y(h) - y_1| \leq Ch^3$. Again, the errors will add and so you get

$$|y(kh) - y_k| \leq Ch^2.$$

What this does is to average an approximation to the slope at the right of the interval and at the beginning of the interval and use that instead of the slope taken only at the beginning of the interval. It is like using the trapezoid rule to approximate $\int_{t_0}^{t_0+h} y'(s) \, ds$ rather than a term in a Riemann sum. The error estimate is better, since for small h, h^2 is a lot smaller than h. One can write this as follows.

$$k_1 \equiv f(t_k, y_k), k_2 \equiv f(t_k + h, y_k + k_1 h), \ t_k = kh, \ y_{k+1} = y_k + \frac{h}{2}(k_1 + k_2)$$

This numerical method is called the improved Euler method. As with the Euler method, everything works with no change for systems of equations.

Procedure 11.8.2 *To find a numerical solution to*

$$\mathbf{y}' = \mathbf{f}(t, \mathbf{y}), \ \mathbf{y}(0) = \mathbf{y}_0$$

using using the improved Euler method, do the following:

$$\mathbf{k}_1 \equiv \mathbf{f}(t_k, \mathbf{y}_k), \mathbf{k}_2 \equiv \mathbf{f}(t_k + h, \mathbf{y}_k + \mathbf{k}_1 h), \ t_k = kh, \ \mathbf{y}_{k+1} = \mathbf{y}_k + \frac{h}{2}(\mathbf{k}_1 + \mathbf{k}_2)$$

then there is some constant C such that $|\mathbf{y}(kh) - \mathbf{y}_k| < Ch^2$.

Actually more is done. When using these kinds of algorithms, the length of the sub intervals is adjusted. For example, if $\mathbf{f}(t_k, \mathbf{y}_k)$ is very large, then it is a good idea to make the next interval smaller.

11.9 The Classical Runge-Kutta Algorithm

A more complicated version of the above which gives much better error estimates is the Runge-Kutta algorithm which dates from around 1901. Actually, there are many Runge-Kutta algorithms, but I will only discuss the most famous.

Procedure 11.9.1 *To find a numerical solution to*

$$\mathbf{y}' = \mathbf{f}(t, \mathbf{y}), \ \mathbf{y}(0) = \mathbf{y}_0$$

using using the Runge-Kutta algorithm, do the following: For $t_k = kh, k = 0, \cdots,$

$$\mathbf{k}_1 \equiv \mathbf{f}(t_k, \mathbf{y}_k), \mathbf{k}_2 \equiv \mathbf{f}\left(t_k + \frac{h}{2}, \mathbf{y}_k + \mathbf{k}_1 \frac{h}{2}\right),$$

$$\mathbf{k}_3 \equiv \mathbf{f}\left(t_k + \frac{h}{2}, \mathbf{y}_k + \mathbf{k}_2 \frac{h}{2}\right), \mathbf{k}_4 \equiv \mathbf{f}(t_k + h, \mathbf{y}_k + \mathbf{k}_3 h)$$

$$\mathbf{y}_{k+1} = \mathbf{y}_k + \frac{h}{6}(\mathbf{k}_1 + 2\mathbf{k}_2 + 2\mathbf{k}_3 + \mathbf{k}_4)$$

then there is some constant C *such that* $|\mathbf{y}(kh) - \mathbf{y}_k| < Ch^4$.

Notice that each $\mathbf{k}_i$ is an approximation of $\mathbf{y}'$ and the algorithm gives a weighted average of these values as an approximation to the desired integral $\int_{t_k}^{t_k+h} \mathbf{y}'(s)\, ds$. To show the difference between the various methods, the following picture considers the very simple example $y' = y, \ y(0) = 1$.

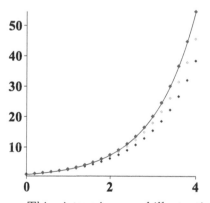

The exact solution is $y = e^t$. Now in the picture, the Euler method, improved Euler method, and Runge-Kutta methods are used to solve the differential equation numerically. The diamonds on the graph of $y = e^t$ are from the Runge-Kutta method and the other diamonds represent the result of using the other two methods. Note that the Runge-Kutta points appear to be exactly right while the other solutions stray from the true solution. When one is evaluating numerical algorithms, it is in general a good idea to experiment with them on problems with a known answer.

This picture is a good illustration of the superiority of the Runge-Kutta method. In all of these graphs, the step size was $h = .2$. It is known that the error between the Runge-Kutta solution and the true solution is $O(h^4)$ meaning that this error converges to 0 as fast as h^4. This is discussed in [19]. An outline is in Problem 16 below.

11.10 Computing Using Runge-Kutta Method

You can have MATLAB use the Runge-Kutta algorithm to numerically find a solution to a system of ordinary differential equations. I will illustrate with a first order system which comes from the Van der Pol equation. The exact equation studied is not too important at this point. My intent is to illustrate the syntax used. Here it is:

```
f=@(t,x)[x(2),-((x(1)^2-1)*x(2)+x(1))]; n=300; h=.05;
y(1,:)=[1,0]; t(1)=0;
hold on
for r=1:n
k1=f(t(r),y(r,:)); k2=f(t(r)+h/2,y(r,:)+k1*(h/2));
k3=f(t(r)+h/2,y(r,:)+k2*(h/2)); k4=f(t(r)+h,y(r,:)+k3*h);
y(r+1,:)=y(r,:)+(h/6)*(k1+2*k2+2*k3+k4); t(r+1)=t(r)+h;
end
plot(t,y(:,1),t,y(:,2))
interp1(t,[y(:,1),y(:,2)],[2,3,4,5,6])
```

Then press "enter". This will compute the solution to the differential equation

$$\left(\begin{array}{c} x \\ y \end{array} \right)' = \left(\begin{array}{c} y \\ -\left(\left(x^2 - 1\right) y + x\right) \end{array} \right), \left(\begin{array}{c} x\left(0\right) \\ y\left(0\right) \end{array} \right) = \left(\begin{array}{c} 1 \\ 0 \end{array} \right).$$

on $[0, 300 \times .05] = [0, 15]$. The x is denoted as $y\left(:,1\right)$, and y is denoted as $y\left(:,2\right)$. Actually, y is an $n \times 2$ matrix the first column being a list of values of x and the second column $y\left(:,2\right)$ being values of y. Note how the function f is defined as a row vector. In the bottom line, the approximate values of the row vector $\left(x\left(t\right), y\left(t\right)\right)$ is obtained at the values of t given by 2,3,4,5,6. This will deliver the following

-1.1961	-1.8675
-1.7280	0.4147
-0.9569	1.1587
0.9870	2.6183
1.9549	-0.3356

If you just want one of these functions, you could write interp1(t,y(:,1),[2,3,4,5,6,7]) instead. Note that after the p in "interp" it is the number 1, not the letter l.

You can see how to adapt this for any other function you would like. If you want to find the solution to a first order system for vectors in $\mathbb{R}^3$, you would simply have the initial condition of the form [1,1,0] for example, and the definition of the function would also involve three commas.

11.11 Exercises

1. In deriving the equation for the pendulum, it was assumed that the mass was attached to a massless rod. Keep the mass of mass m at the end but now let the rod be more realistic. Let it have length l and density equal to ρ where this density is per unit length of the rod. You imagine the rod is a very large number of masses all rigidly joined together. Then the kinetic energy will involve the moment of inertia of the rotating rod mass solid which will need to be computed using an integral. Similarly the potential energy

will need to be computed. However, the methodology is the same other than this. Lagrange did all of this back in the 1700s and he considered much more elaborate mechanical systems than this. By considering only one angle, I have limited things to "one degree of freedom".

2. To graph $y = e^t$ on $[-1, 3]$ you would use the following syntax in MATLAB.

$$t=-1:.1:3; \; x=\exp(t); \; \text{plot}(t,x)$$

If you want to make the graph thicker, you type plot(t,x,'LineWidth',3) where the 3 tells how thick you want it. It will be pretty thick if you use 3. Graphing other functions is done similarly. However, if you wanted to graph $x = t^3 + 2t$, you would use the following syntax at the second line: x=t.^3+2*t. This is because t is a vector and you want to do the operation on individual entries of the vector. If you want to graph two on the same axis and have different line widths, you could use the following syntax.

> t=-5:.1:2; s=-5:.2:2; x=exp(t); y=sin(s);
> hold on
> plot(t,x,'color','green','linewidth',2)
> plot(s,y,'color','red','linewidth',1)

Using MATLAB, graph the following pairs of functions on the same axes on the interval $[-1, 1]$. Make them have different line widths or colors.

(a) $\exp(t), \sin(t)$ (d) $\cos(3t), 1 - t^3$

(b) $2^t, \cos(t)$

(c) $\sin(2t), t^2 - 1$ (e) $t^2 - 2t + 3, \sin(t)$

3. Now consider the initial value problem $y' = y, y(0) = 1$. Use MATLAB and Euler's method to graph on the same axes for $t \in [0, 9]$ the real solution $y = \exp(t)$ and the numerical solutions from Euler's method with the number of steps equal to 30. If you like, you can make the two graphs have different line widths or colors to tell them apart.

4. Do the same problem with the improved Euler method.

5. Do the same problem with the Runge-Kutta method. For y the function determined by the Runge-Kutta method, compare $y(9)$ and $\exp(9)$. The step size wasn't even all that small.

6. In this problem, from [6] a known solution $y(t)$ is available and given. Letting $r(t)$ denote the Runge-Kutta solution, $\hat{e}(t)$ the improved Euler solution, and $e(t)$ the Euler solution, fill in the following table for each of the examples.

step size = .1	step size = .2
y(1)	y(1)
r(1)	r(1)
ê(1)	ê(1)
e(1)	e(1)

(a) $y' = 1 + t - y$, $y(0) = 0$, $(y(t) = t)$

(b) $y' = 2ty$, $y(0) = 2$, $\left(y(t) = 2e^{t^2} \right)$

(c) $y' = 1 + y^2 - t^2$, $y(0) = 0$, $(y(t) = t)$

(d) $y' = te^{-y} + \frac{t}{1+t^2}$, $y(0) = 0$, $\left(y(t) = \ln\left(1 + t^2\right) \right)$

(e) $y' = -1 + 2t + \frac{y^2}{(1+t^2)^2}$, $y(0) = 1$, $\left(y(t) = 1 + t^2 \right)$

7. Use the Runge-Kutta method with step size .2 to obtain graphs of the solutions to the following initial value problems.

(a) $y' = 5t - \sqrt{y}$, $y(0) = 1$,

(b) $y' = \sqrt{t + y}$, $y(0) = 2$,

(c) $y' = \frac{ty}{1+y^2}$, $y(0) = 1$,

(d) $y' = \frac{y^2 + ty}{1+y^2}$, $y(0) = 1$,

(e) $y' = \left(t^2 - y^2\right)\sin(y)$, $y(0) = 1$,

(f) $y' = e^{-ty}$, $y(0) = 0$,

8. In this chapter, are existence theorems for solutions to the initial value problem

$$\mathbf{x}' = \mathbf{f}(t, \mathbf{x}), \mathbf{x}(0) = \mathbf{x}_0$$

As pointed out earlier, sometimes all you can get is a local solution. For example, you might have $x' = x^2$, $x(0) = 1$. The solution is $x(t) = \frac{1}{1-t}$ so it only makes sense on $[0, 1)$. Use the Runge-Kutta method with step size .2 and do 300 iterations so you would be trying to get a solution on a larger interval than $[0, 1]$. What happens? Why? Now change the Runge-Kutta method as follows. Designate $h1 = .2$ at the beginning but now after the loop begins, adjust the step size as follows. $h = h1/((f(t(r), y(r))^2 + 1)$. Now see what happens. Be sure to type in clf and enter to clear the figure when you make a change from one problem to another. This is an example of an adaptive method. Actual adaptive methods are much more complicated than this.

9. Use the Runge-Kutta method to find good graphs on some interval $(0, r)$ of the solutions to the following initial value problems. Determine whether the solution exists for all t or if it only exists locally. You might want to adapt the step size as discussed above in order to account for large values of the slope.

(a) $y' = \sin(t)y^2 + y^3 - y^5$, $y(0) = 1$

(b) $y' = \sin(y) + \cos(y)$, $y(0) = 1$

(c) $y' = \tan\left(1 + y^2\right)$, $y(0) = 0$

(d) $y' = 1 + y^2$, $y(0) = 0$

(e) $y' = 1 + y^4$, $y(0) = 0$

(f) $y' = 1 + y^4 - \frac{1}{10}y^5$, $y(0) = 0$. Actually, this one should have a solution on $[0, T]$ for any $T > 0$. Explain why. Then use Runge-Kutta for step size .001 and a large number of iterations to see a graph.

10. The Runge-Kutta method pertains to an interval of the form $[0, r]$. Show there is an **easy way** to obtain the solution on $[-r, 0]$. Thus there is no lack of generality in specializing to $[0, r]$. Similarly, show that if you are interested in an interval $[a, b]$ this can always be considered by changing to the interval $[0, b - a]$. **Hint:** Consider changing the independent variable.

11. Suppose you have an equation $\mathbf{x}' = \mathbf{f}(t, \mathbf{x})$, $\mathbf{x}(0) = \mathbf{x}_0$ and for some constant $\lambda > 0$,

$$(\lambda \mathbf{x} - \mathbf{f}(t, \mathbf{x})) \cdot \mathbf{x} \geq 0$$

and also $(t, \mathbf{x}) \to \mathbf{f}(t, \mathbf{x})$ has all partial derivatives and they are continuous. Show that the initial value problem has a solution on $[0, T]$ where T is completely arbitrary.

12. A multi-step method uses not just the value of y_n, t_n to compute y_{n+1} but other earlier values as well. The idea is to approximate the derivative with a polynomial and then integrate the polynomial. You know

$$(t_{k-3}, y_{k-3}), (t_{k-2}, y_{k-2}), (t_{k-1}, y_{k-1}), (t_k, y_k).$$

Then the derivatives y' would be respectively

$$f(t_{k-3}, y_{k-3}), f(t_{k-2}, y_{k-2}), f(t_{k-1}, y_{k-1}), f(t_k, y_k). \qquad (*)$$

If a polynomial $p(t)$ is obtained which has these values at the points $0, h, 2h, 3h$, then to get y_{n+1} one could use an integral, $y_{n+1} = y_n + \int_{3h}^{4h} p(t)\, dt$. It would be a degree three polynomial, $a + bt + ct^2 + dt^3$.

$$\int_{3h}^{4h} \left(a + bt + ct^2 + dt^3\right) dt = \frac{175}{4} dh^4 + \frac{37}{3} ch^3 + \frac{7}{2} bh^2 + ah \qquad (**)$$

Thus it only remains to find a, b, c, d such that the values in $*$ are obtained at the points $0, h, 2h, 3h$. Show that the correct values are

$$a = f_{k-3}, b = \frac{1}{6h}\left(2f_k - 9f_{k-1} + 18f_{k-2} - 11f_{k-3}\right),$$

$$c = -\frac{1}{2h^2}\left(f_k - 4f_{k-1} + 5f_{k-2} - 2f_{k-3}\right),$$

$$d = \frac{1}{6h^3}\left(f_k - 3f_{k-1} + 3f_{k-2} - f_{k-3}\right)$$

$f_{k-3} \equiv f(t_{k-3}, y_{k-3})$ and so forth. Inserting these into $**$, show that you get

$$\frac{h}{24}\left(55f_k - 59f_{k-1} + 37f_{k-2} - 9f_{n-3}\right)$$

Thus

$$y_{n+1} = y_n + \frac{h}{24}\left(55f_k - 59f_{k-1} + 37f_{k-2} - 9f_{n-3}\right)$$

Use the Runge-Kutta method for $k = 1, 2, 3$ and then continue with this method for $k = 4, \cdots, n$. Write the directions in MATLAB and test on some examples. This multistep method is called the Adams-Bashforth method. Try this on $y' = y^2, y(0) = 1$. For this equation, you know the solution so compare with the result given by the Adams-Bashforth method and Runge-Kutta method.

13. Give an estimate of the maximum interval on which a solution exists for the following equations. Use either analytical methods or graphing the numerical solution. You might want to adapt the step size to help determine this. Thus if the solution blows up, you might expect to see the method be unable to take you beyond some value which will give an estimate of the length of the interval of existence.

(a) $y' = (1 + t) y^2, y(0) = 1$ $y(0) = 1$

(b) $y' = (1 + t)(1 + y^2), y(0) = 1$ (d) $y' = \frac{1 + ty^4}{1 + y^2}, y(0) = 1$

(c) $y' = 1 + t(y + y^2 + y^4),$ (e) $y' = \tan(y), y(0) = 1.$

14. Recall Gronwall's inequality. Suppose $u(t) \geq 0$ and for all $t \in [0, T]$,

$$u(t) \leq u_0 + \int_0^t ku(s) \, ds. \tag{11.17}$$

where k is some constant. Then $u(t) \leq u_0 e^{kt}$. Suppose now that k is replaced with $k(s) \geq 0$ in (11.17), $s \to k(s)$ continuous. Suppose $K'(t) = k(t), K(0) = 0$. Show that

$$u(t) \leq u_0 e^{K(t)}$$

You can assume what you like about $k(s)$. Let it be continuous for example to make things go easily and also assume $u(\cdot)$ is continuous.

15. To see why the Runge-Kutta method might be a good idea, show that if $p(x) = a + bx + cx^2 + dx^3$, then

$$\int_x^{x+h} p(t) \, dt = \frac{h}{6}\left(p(x) + 4p\left(x + \frac{h}{2}\right) + p(x + h)\right)$$

Thus the right side integrates any cubic polynomial exactly. Use this along with Taylor's theorem to verify that for y' a smooth function,

$$y(x + h) - y(x) = \frac{h}{6}\left(y'(x) + 4y'\left(x + \frac{h}{2}\right) + y'(x + h)\right) + O(h^5)$$

This is used to obtain the error estimate for the Runge-Kutta method. The details are excruciating.

16. Excruciating details summarized: Show

$$y(t) + y'\left(t + \left(\frac{h}{2}\right)\right)\left(\frac{h}{2}\right) - y\left(t + \left(\frac{h}{2}\right)\right)$$

$$= \frac{1}{2}y''(t)\left(\frac{h}{2}\right)^2 + \frac{1}{3}y'''(t)\left(\frac{h}{2}\right)^3 + O\left(\left(\frac{h}{2}\right)^4\right)$$

$$y(t) + hy'\left(t + \frac{h}{2}\right) = y(t + h) + y'''(t)\left(-\frac{1}{24}h^3\right) + O(h^4)$$

Let $J \equiv f_y\left(t + \frac{h}{2}, y\left(t + \frac{h}{2}\right)\right)$. Using the definition of the k_i and the above show that

$$k_1 = y'(t), k_2 = y'\left(t + \frac{h}{2}\right) + J\left(-\left(y''(t)\frac{h^2}{8} + y'''(t)\frac{h^3}{48}\right)\right) + O(h^4),$$

$$k_3 = y'\left(t + \frac{h}{2}\right) + J\left(\begin{array}{c}\frac{1}{2}y''(t)\left(\frac{h}{2}\right)^2 + \frac{1}{3}y'''(t)\left(\frac{h}{2}\right)^3 \\ +J\left(-\left(y''(t)\frac{h^3}{16}\right)\right)\end{array}\right) + O(h^4)$$

Also show that $f_y(t + h, y(t + h)) = J + O(h)$. Then you verify

$$k_4 = y'(t + h) + J\left(y'''(t)\left(-\frac{1}{24}h^3\right) + Jy''(t)\left(\frac{h}{2}\right)^3\right) + O(h^4)$$

It follows then that things cancel out when you form $k_1 + 2k_2 + 2k_3 + k_4$ to yield

$$y'(t) + 4y'\left(t + \frac{h}{2}\right) + y'(t + h)$$

Then the final result follows from the formula of the preceding problem.

17. In this problem is another way to show the Picard iterates converge. Let $M_{\mathbf{f}} \equiv \max\{|\mathbf{f}(s, \mathbf{x}_0)| : s \in [a, b]\}$ and letting $\mathbf{x}_k(t)$ be the kth Picard iterate, show that for any $t \in [a, b]$,

$$|\mathbf{x}_1(t) - \mathbf{x}_0| \leq \left|\int_c^t |\mathbf{f}(s, \mathbf{x}_0)|\, ds\right| \leq M_{\mathbf{f}}|t - c|. \tag{11.18}$$

Now using this estimate and the Lipschitz condition for $\mathbf{f}$, explain why

$$|\mathbf{x}_2(t) - \mathbf{x}_1(t)| \leq \left|\int_c^t |\mathbf{f}(s, \mathbf{x}_1(s)) - \mathbf{f}(s, \mathbf{x}_0)|\, ds\right|$$

$$\leq \left|\int_c^t K|\mathbf{x}_1(s) - \mathbf{x}_0|\, ds\right| \leq KM_{\mathbf{f}}\left|\int_c^t |s - c|\, ds\right| \leq KM_{\mathbf{f}}\frac{|t - c|^2}{2}. \tag{11.19}$$

Continuing in this way verify that for $k \geq 2$,

$$|\mathbf{x}_k(t) - \mathbf{x}_{k-1}(t)| \leq M_{\mathbf{f}}K^{k-1}\frac{|t - c|^k}{k!}. \tag{11.20}$$

Next verify that for each $t \in [a, b]$, $\{\mathbf{x}_k(t)\}_{k=1}^{\infty}$ is a Cauchy sequence. Letting $\mathbf{x}(t) \equiv \lim_{k \to \infty} \mathbf{x}_k(t)$, show that

$$|\mathbf{x}(t) - \mathbf{x}_l(t)| \leq M_{\mathbf{f}}\sum_{r=l}^{\infty}\frac{(K(b - a))^r}{(r + 1)!} < \varepsilon$$

whenever l is large enough. Thus $\|\mathbf{x} - \mathbf{x}_l\| \to 0$. Explain why $\mathbf{x}$ is continuous. Now explain why $\lim_{l \to \infty} \int_c^t \mathbf{f}(s, \mathbf{x}_l(s))\, ds = \int_c^t \mathbf{f}(s, \mathbf{x}(s))\, ds$. Thus the Picard iterates converge and the function to which they converge is the solution to the initial value problem. Show uniqueness from Gronwall's inequality.

Chapter 12

Equilibrium Points and Limit Cycles

12.1 Stability with Graphing

Existence of solutions to the initial value problem

$$\begin{pmatrix} x \\ y \end{pmatrix}' = \begin{pmatrix} f_1(x,y) \\ f_2(x,y) \end{pmatrix}, \ \begin{pmatrix} x(0) \\ y(0) \end{pmatrix} = \begin{pmatrix} x_0 \\ y_0 \end{pmatrix}$$

were discussed earlier. However, there is another aspect to this which is very significant. If you have an initial condition $\mathbf{x}_0$ near an equilibrium point, a point $\mathbf{x}$ where $\mathbf{f}(\mathbf{x}) = \mathbf{0}$, what is the behavior of the solutions starting at $\mathbf{x}_0$. Sometimes one can get a good idea by having a computer graph the vector field along with some solution curves and then using Observation 11.7.3 to gain understanding about what happens as $t \to \infty$.

Example 12.1.1 *Here is a differential equation*

$$\begin{pmatrix} x \\ y \end{pmatrix}' = \begin{pmatrix} x - xy \\ xy - y \end{pmatrix}$$

The equilibrium points are $(0,0), (1,1)$. Are they stable or not?

To get an idea, here is a graph done by a computer of the direction vectors,

$$\begin{pmatrix} x - xy \\ xy - y \end{pmatrix}$$

The idea is that the computer graphs vectors corresponding to various choices of (x, y). For example, at $(1, 2)$, the computer would graph the vector $(-1, 0)$. At $(2, 3)$ it would graph $(-4, 4)$. In order to convey the information of which are the largest vectors, it scales everything appropriately, according to the length of the longest vector, and then places direction vectors having the correct direction and magnitude at many points on the plane. Those arrows give directions of motion and also indicate magnitude by how long they are. The syntax is on the right.

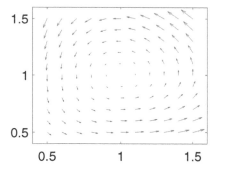

```
>>[a,b]=meshgrid(.5:.1:1.5,.5:.1:1.5);
u=a-a.*b;
v=a.*b-b;
figure
quiver(a,b,u,v)
```

Note that in MATLAB, you should type a.*b and not a*b. In this case, it will not work with a*b because a,b are vectors. Also, be sure to place quiver(a,b,u,v) on a new line. Remember that to do this, you type shift enter.

The problem with this graph is that the vectors become very small and it is not clear where they are pointing. One can modify the vector field so that the resulting vectors still point in the same direction but they don't shrink away. You can also graph numerical solutions to the differential equation for various initial conditions on the same axes as follows.

```
>>[a,b]=meshgrid(.2:.2:2.5,.2:.2:2.5);
u=(a-a.*b); v=(a.*b-b);
r=(u.*u+v.*v).^(1/2);
figure
quiver(a,b,u./r,v./r,'autoscalefactor',.5)
f=@(t,x)[x(1)-x(1)*x(2);x(1)*x(2)-x(2)];
hold on
for z=.5:.5:2
[ts,ys]=ode45(f,[0:.03:10],[.5;z]);
plot(ys(:,1),ys(:,2),'linewidth',2)
end
```

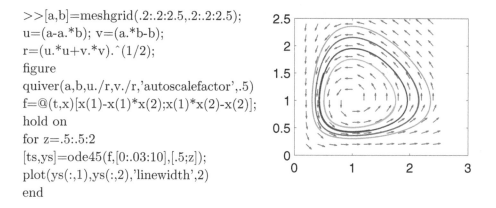

Note in the ode45 line "[0:.03:10]". This says that all the step sizes will be no more than .03 in length and insures that the graph will be smooth. Otherwise, you don't get such a nice result.

Example 12.1.2 *Here is a differential equation*

$$\begin{pmatrix} x \\ y \end{pmatrix}' = \begin{pmatrix} y + xy + y^3 \\ xy^2 - 3y - 2x \end{pmatrix}$$

Then $(0,0)$ is an equilibrium point. How do the solutions behave for initial conditions near this equilibrium point?

To get an idea of what is happening, here is a graph done by a computer of the direction vectors indicated by the right side of the differential equation along with solutions.

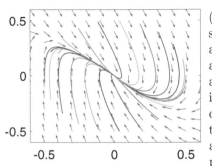

You see that in this case, it appears that $(0,0)$ is stable in the sense that solutions which start near $(0,0)$ end up moving towards $(0,0)$ and it appears that in fact, they even have $(0,0)$ as a limit as t increases. Note how the picture appears to contradict Observation 11.7.3. This is because the computer can't draw the plane curves coming from the initial value problem thin enough to harmonize with this observation and if it could, you could not see them. The point $(0,0)$ is said to be stable because initial conditions near it lead to solutions which converge to this point in the limit. When the phase-portrait looks like this, the point is called a node, a stable node. Here is another example.

Example 12.1.3 *Here is a differential equation*

$$\left(\begin{array}{c} x \\ y \end{array} \right)' = \left(\begin{array}{c} -1y + .01x^2 \\ -1x + .01y^2 \end{array} \right)$$

Then $(0,0)$ is an equilibrium point. How do the solutions behave for initial conditions near this equilibrium point?

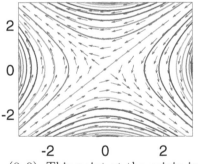

Note how if the initial data is not in just the right location, the solution will move away from $(0,0)$.

The line $y = x$ is close to something called the stable manifold for $(0,0)$ because initial conditions on this set of points will be drawn toward $(0,0)$. However, we do not refer to $(0,0)$ as being a stable equilibrium point because for points not on this stable manifold but near $(0,0)$, the solution curve does not stay near $(0,0)$. This point at the origin is called a saddle point.

Here is another example.

Example 12.1.4 *Here is a differential equation*

$$\left(\begin{array}{c} x \\ y \end{array} \right)' = \left(\begin{array}{c} y + x^2 y \\ -(y + 2x) + x^3 \end{array} \right)$$

Then $(0,0)$ is an equilibrium point. How do the solutions behave for initial conditions near this equilibrium point?

In this case, the point $(0,0)$ is said to be asymptotically stable. If you start near it, the solutions converge to it. In addition, because it looks the way it does, it is called a focus. Note that if you don't start near it, the solutions will not be drawn to it. There is quite a lot which goes into the study of geometric properties of solutions of ordinary differential equations. However, in this book, we will mainly be interested in whether an equilibrium point is stable or unstable and also will consider the case of periodic orbits which are limit sets. The picture follows with its MATLAB syntax.

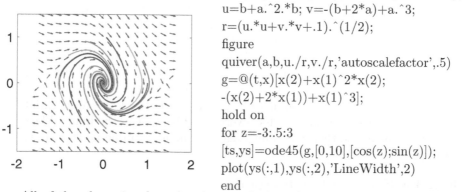

```
>>[a,b]=meshgrid(-1.5:.2:1.5,-1.5:.2:1.5);
u=b+a.^2.*b; v=-(b+2*a)+a.^3;
r=(u.*u+v.*v+.1).^(1/2);
figure
quiver(a,b,u./r,v./r,'autoscalefactor',.5)
g=@(t,x)[x(2)+x(1)^2*x(2);
-(x(2)+2*x(1))+x(1)^3];
hold on
for z=-3:.5:3
[ts,ys]=ode45(g,[0,10],[cos(z);sin(z)]);
plot(ys(:,1),ys(:,2),'LineWidth',2)
end
```

    All of the above involves drawing conclusions from pictures drawn by a computer. All of these considerations are also in two dimensions. Are there analytic methods and criteria which will allow one to conclude various properties of the solution curves which don't depend on pictures? The next few sections will consider this question. To do this, a very interesting approach due to Putzer will be presented and used for linear systems. After this, things will be extended to almost linear systems. In all of this, we will only consider autonomous equations.

12.2   Minimal Polynomial

The Cayley-Hamilton theorem says that any $n \times n$ matrix A has the property that $q(A) = 0$ where $q(x)$ is the characteristic polynomial of the matrix. It is proved in Theorem D.6.4 in the appendix.

Definition 12.2.1 *Let A be an $n \times n$ matrix. The characteristic polynomial is defined as*

$$q_A(t) \equiv \det(tI - A)$$

and the solutions to $q_A(t) = 0$ are called eigenvalues. For A a matrix and $p(t) = t^n + a_{n-1}t^{n-1} + \cdots + a_1 t + a_0$, denote by $p(A)$ the matrix defined by

$$p(A) \equiv A^n + a_{n-1}A^{n-1} + \cdots + a_1 A + a_0 I.$$

The explanation for the last term is that A^0 is interpreted as I, the identity matrix.

    Then the Cayley-Hamilton theorem says that $q_A(A) = 0$ with the convention just discussed that $A^0 \equiv I$. It was shown by Putzer [18] in 1966 that this theorem can be used to produce the fundamental matrix. This is a very significant thing because the other ways of doing it tend to involve a much harder theorem called the Jordan canonical form. This Putzer method is discussed first and then it is used to consider questions of stability. Note that if $\{\lambda_1, \cdots, \lambda_n\}$ are the eigenvalues listed according to algebraic multiplicity, the characteristic polynomial is of the form

$$q_A(t) = \prod_{i=1}^{n} (t - \lambda_i)$$

Thus

$$\prod_{i=1}^{n} (A - \lambda_i I) = q_A(A) = 0$$

The minimal polynomial $p(t)$ is the monic polynomial of the form $t^m + b_{m-1} t^{m-1} + \cdots + b_1 t + b_0$ such that $p(A) = 0$ and m is as small as possible. Thus $m \leq n$.

Proposition 12.2.2 *The minimal polynomial divides the characteristic polynomial.*

Proof: By the Euclidean algorithm,

$$q(t) = l(t) p(t) + r(t)$$

where $r(t)$ has degree smaller than that of $p(t)$ or else is 0. However, $q(A) = p(A) = 0$ and so $r(A) = 0$ which cannot happen unless $r(t) = 0$ since otherwise, $p(t)$ would not be the minimal polynomial. ∎

It follows that

$$p(t) = \prod_{i=1}^{m} (t - \lambda_i), \ m \leq n, \ \text{repeats in } \lambda_i \text{ are possible} \tag{*}$$

Proposition 12.2.3 *If the minimal polynomial of A is given in $*$, then the eigenvalues of A are $\lambda_1, \cdots, \lambda_m$.*

Proof: We need to show they are each eigenvalues, and then we need to show that there are no other eigenvalues. Note that A commutes with A. Thus we can take the factors in $\prod_{i=1}^{m} (A - \lambda_i I)$ in any order desired.

First, there is a vector $\mathbf{x}$ such that $\prod_{i \neq j} (A - \lambda_i I) \mathbf{x} \neq \mathbf{0}$. Otherwise $p(t)$ was not minimal. Then

$$(A - \lambda_j I) \left(\prod_{i \neq j} (A - \lambda_i I) \mathbf{x} \right) = \prod_{i=1}^{m} (A - \lambda_i I) \mathbf{x} = \mathbf{0}.$$

Thus each λ_j is an eigenvalue. Are there any others? Suppose μ is an eigenvalue. Then $(A - \mu I) \mathbf{x} = \mathbf{0}$ for some $\mathbf{x} \neq \mathbf{0}$.

$$\mathbf{0} = \prod_{i=1}^{m} (A - \lambda_i I) \mathbf{x} = \prod_{i=1}^{m} (A - \mu I + (\mu - \lambda_i) I) \mathbf{x} = \prod_{i=1}^{m} (\mu - \lambda_i) \mathbf{x}$$

which is impossible unless μ is one of the λ_i. ∎

12.3   Putzer's Method and Linear Systems

Here is a discussion of linear systems of the form

$$\mathbf{x}' = A\mathbf{x} + \mathbf{f}(t)$$

where A is an $n \times n$ matrix and $\mathbf{f}$ is a vector valued function having all entries continuous. Of course it is a very special case of the general considerations above but this section will give a more algebraic description of the fundamental matrix. This also leads to important theorems about stability of equilibrium points without having to include quite so much difficult linear algebra. Recall Lemma 9.1.2 which is the product rule for multiplication of matrices,

$$(M(t) N(t))' = M'(t) N(t) + M(t) N'(t)$$

Here these are matrices which have entries which are differentiable functions.

Using Gronwall's inequality, here is a theorem on uniqueness of solutions to the initial value problem,

$$\mathbf{x}' = A\mathbf{x} + \mathbf{f}(t), \ \mathbf{x}(a) = \mathbf{x}_a, \tag{12.1}$$

in which A is an $n \times n$ matrix and $\mathbf{f}$ is a continuous function having values in $\mathbb{C}^n$. We state it here for convenience but it is a special case of Theorem 11.4.1

Theorem 12.3.1 *Suppose $\mathbf{x}$ and $\mathbf{y}$ satisfy (12.1). Then $\mathbf{x}(t) = \mathbf{y}(t)$ for all t.*

Recall also the definition of the fundamental matrix.

Definition 12.3.2 *Let A be an $n \times n$ matrix. We say $\Phi(t)$ is the fundamental matrix for A if*

$$\Phi'(t) = A\Phi(t), \ \Phi(0) = I, \tag{12.2}$$

Recall that $\Phi(t)^{-1}$ exists for all $t \in \mathbb{R}$.

This section is devoted to an algebraic method for finding this fundamental matrix.

The next theorem gives a formula for the fundamental matrix (12.2). It is known as Putzer's method [1],[18]. It is a remarkably clever approach to finding the fundamental matrix and computes the fundamental matrix in terms of the eigenvalues of the matrix A.

Theorem 12.3.3 *Let A be an $n \times n$ matrix whose eigenvalues are*

$$\{\lambda_1, \cdots, \lambda_m\}, m \leq n,$$

listed according to multiplicity as roots of the minimal polynomial. Let the minimal polynomial be

$$p(t) \equiv \prod_{i=1}^{m} (t - \lambda_i), \ m \leq n$$

in which there may be repeated λ_i. Define

$$P_k(A) \equiv \prod_{i=1}^{k}(A - \lambda_i I), \ P_0(A) \equiv I,$$

and let the scalar valued functions, $r_k(t)$ be defined as, $r_0(t) \equiv 0$ and for $k \geq 1$, r_k is obtained as the solution to the following initial value problem

$$\begin{pmatrix} r_1'(t) \\ r_2'(t) \\ \vdots \\ r_m'(t) \end{pmatrix} = \begin{pmatrix} \lambda_1 r_1(t) \\ \lambda_2 r_2(t) + r_1(t) \\ \vdots \\ \lambda_m r_m(t) + r_{m-1}(t) \end{pmatrix}, \ \begin{pmatrix} r_1(0) \\ r_2(0) \\ \vdots \\ r_m(0) \end{pmatrix} = \begin{pmatrix} 1 \\ 0 \\ \vdots \\ 0 \end{pmatrix}$$

Note the system amounts to a list of single first order linear differential equations. Now define

$$\Phi(t) \equiv \sum_{k=0}^{m-1} r_{k+1}(t) P_k(A). \qquad (*)$$

Then

$$\Phi'(t) = A\Phi(t), \ \Phi(0) = I. \qquad (12.3)$$

Furthermore, if $\Phi(t)$ is a solution to (12.3) for all t, then it follows $\Phi(t)^{-1}$ exists for all t and $\Phi(t)$ is the unique fundamental matrix for A.

Proof: The first part of this follows from a computation. First note that since $p(t)$ is the minimal polynomial, $P_m(A) = 0$. Now for the computation: Let $\Phi(t)$ satisfy $*$.

$$\Phi'(t) = \sum_{k=0}^{m-1} r_{k+1}'(t) P_k(A) = \sum_{k=0}^{m-1} (\lambda_{k+1} r_{k+1}(t) + r_k(t)) P_k(A) =$$

$$\sum_{k=0}^{m-1} \lambda_{k+1} r_{k+1}(t) P_k(A) + \sum_{k=1}^{m} r_k(t) P_k(A) - \overbrace{r_m(t) P_m(A)}^{=0}$$

$$= \sum_{k=0}^{m-1} \lambda_{k+1} r_{k+1}(t) P_k(A) + \sum_{k=1}^{m} r_k(t)(A - \lambda_k I) P_{k-1}(A)$$

$$= \sum_{k=0}^{m-1} \lambda_{k+1} r_{k+1}(t) P_k(A) + \sum_{k=0}^{m-1} r_{k+1}(t)(A - \lambda_{k+1} I) P_k(A)$$

$$= \sum_{k=0}^{m-1} \lambda_{k+1} r_{k+1}(t) P_k(A) + A \sum_{k=0}^{m-1} r_{k+1}(t) P_k(A) - \sum_{k=0}^{m-1} \lambda_{k+1} r_{k+1} P_k(A) = A\Phi(t)$$

This shows $\Phi'(t) = A\Phi(t)$. That $\Phi(0) = I$ follows from

$$\Phi(0) = \sum_{k=0}^{m-1} r_{k+1}(0) P_k(A) = r_1(0) P_0(A) = I.$$

Then it follows from Proposition 10.2.9 and Lemma 10.2.6 that the inverse exists and that this is the unique fundamental matrix. ■

Theorem 12.3.4 *Let $\mathbf{f}$ be continuous on $[0, T]$ and let A be an $n \times n$ matrix and $\mathbf{x}_0$ a vector in $\mathbb{C}^n$. Then there exists a unique solution $\mathbf{x}$ to*

$$\mathbf{x}'(t) = A\mathbf{x}(t) + \mathbf{f}(t), \ \mathbf{x}(0) = \mathbf{x}_0$$

given by the variation of constants formula,

$$\mathbf{x}(t) = \Phi(t)\mathbf{x}_0 + \Phi(t) \int_0^t \Phi(s)^{-1} \mathbf{f}(s)\, ds \tag{12.4}$$

for $\Phi(t)$ the fundamental matrix for A. Also, $\Phi(t)^{-1} = \Phi(-t)$ and $\Phi(t+s) = \Phi(t)\Phi(s)$ for all t, s and the above variation of constants formula can also be written as

$$\begin{aligned}
\mathbf{x}(t) &= \Phi(t)\mathbf{x}_0 + \int_0^t \Phi(t-s)\mathbf{f}(s)\, ds \tag{12.5} \\
&= \Phi(t)\mathbf{x}_0 + \int_0^t \Phi(s)\mathbf{f}(t-s)\, ds \tag{12.6}
\end{aligned}$$

As shown earlier, Theorem 12.3.4 is general enough to include all constant coefficient linear differential equations of any order.

What if you could only find the eigenvalues approximately? Then Putzer's method applied to the approximate eigenvalues would end up giving a fairly good solution because there are no eigenvectors mentioned. You are not left stranded if you can't find the eigenvalues exactly, as you are if you use the methods usually taught.

12.4   Stability and Eigenvalues

To begin with, notice that for A an $m \times n$ matrix, $\mathbf{x} \to A\mathbf{x}$ is continuous. To see this, observe that, from the Cauchy-Schwarz inequality, (1.1) on Page 13 $|A\mathbf{x} - A\mathbf{y}|^2 =$

$$\sum_{i=1}^m \left| \sum_{j=1}^n A_{ij}(x_j - y_j) \right|^2 \leq \sum_{i=1}^m \left(\left(\sum_{j=1}^n |A_{ij}|^2 \right)^{1/2} \left(\sum_{j=1}^n |x_j - y_j|^2 \right)^{1/2} \right)^2$$

$$= \left(\sum_{i=1}^m \sum_{j=1}^n |A_{ij}|^2 \right) |\mathbf{x} - \mathbf{y}|^2$$

Conclusions may be drawn about stability of an equilibrium point for $\mathbf{x}' = A\mathbf{x} + \mathbf{g}(\mathbf{x})$ from knowledge of the eigenvalues of A and this idea will be developed here. First of all, here is a fundamental estimate for the entries of a fundamental matrix.

Lemma 12.4.1 *Let the functions, r_k be given in the statement of Theorem 12.3.3 and suppose that A is an $n \times n$ matrix whose eigenvalues are*

$$\{\lambda_1, \cdots, \lambda_m\}, m \leq n,$$

listed according to multiplicity as roots of the minimal polynomial. Suppose that these eigenvalues are ordered such that

$$\operatorname{Re}(\lambda_1) \leq \operatorname{Re}(\lambda_2) \leq \cdots \leq \operatorname{Re}(\lambda_m)$$

Also, if $0 > -\delta > \operatorname{Re}(\lambda_m)$ is given, there exists a constant, C such that for each $k = 0, 1, \cdots, m$,

$$|r_k(t)| \leq Ce^{-\delta t} \tag{12.7}$$

For the ordering of the eigenvalues as above, one can give a description of $r_k(t)$ in the Putzer description of the fundamental matrix. $r_k(t) =$

$$e^{\lambda_k t}\left(p_k(t) + \sum_{i=1}^{m_k} e^{\alpha_i t} q_i(t)\right), \quad \deg(p_k(t)) \leq k - 1, \operatorname{Re}\alpha_i < 0 \tag{12.8}$$

and $\deg(p_k(t)) \leq \deg(p_{k+1}(t))$. Strict inequality holds if and only if $\lambda_{k+1} = \lambda_k$. Also, $p_1(t) = 1$.

Proof: $r_1(t) = e^{\lambda_1 t}$. Next

$$r_2'(t) = \lambda_2 r_2(t) + e^{\lambda_1 t}, \quad r_2(0) = 0$$

thus

$$r_2(t) = \frac{1}{\lambda_1 - \lambda_2}e^{\lambda_1 t} + \frac{1}{\lambda_2 - \lambda_1}e^{\lambda_2 t}$$

or in case $\lambda_1 = \lambda_2$,

$$r_2(t) = te^{\lambda_2 t}$$

Thus (12.8) has just been shown in case of $k = 1, 2$. Suppose true for k. By induction and the variation of constants formula,

$$r_{k+1}(t) = e^{\lambda_{k+1} t}\int_0^t e^{-\lambda_{k+1} s}r_k(s)\,ds$$

$$= e^{\lambda_{k+1} t}\int_0^t e^{-\lambda_{k+1} s}e^{\lambda_k s}\left(p_k(s) + \sum_{i=1}^{m_k} e^{\alpha_i s}q_i(s)\right)ds, \quad \operatorname{Re}\alpha_i < 0$$

where p_k has degree at most $k - 1$. In case, $\lambda_{k+1} = \lambda_k$, you get

$$r_{k+1}(t) = e^{\lambda_{k+1} t}\left(p_{k+1}(t) + \sum_{i=1}^{\hat{m}_k} e^{\alpha_i t}\hat{q}_i(t)\right), \quad \operatorname{Re}\alpha_i < 0$$

where the degree of p_{k+1} is one more than the degree of $p_k(t)$. If $\lambda_{k+1} \neq \lambda_k$, then integrating by parts multiple times, you obtain the same formula but this time the degree of $p_{k+1}(t)$ equals the degree of $p_k(t)$ and the α_i might change.

As to the estimate for the case where $\operatorname{Re}\lambda_i < -\delta$ for all i,

$$|r_k(t)| = \left|\left(p_k(t) + \sum_{i=1}^{m_k} e^{\alpha_i t}q_i(t)\right)e^{\lambda_k t}\right| < C_k e^{-\delta t}$$

for some C_k since $\left|p_k\left(t\right) + \sum_{i=1}^{m_k} e^{\alpha_i t} q_i\left(t\right)\right| e^{\mathrm{Re}\left(\lambda_k\right)t} e^{\delta t}$ must be bounded because $\mathrm{Re}\left(\lambda_k\right) < -\delta$ and polynomial growth is slower than exponential growth. Now let C be the maximum of the C_k. ∎

With the lemma, the following sloppy estimate is available for a fundamental matrix. Recall that

$$\Phi\left(t\right) \equiv \sum_{k=0}^{n-1} r_{k+1}\left(t\right) P_k\left(A\right)$$

Theorem 12.4.2 *Let A be an $n \times n$ matrix and let $\Phi\left(t\right)$ be the fundamental matrix for A. That is,*

$$\Phi'\left(t\right) = A\Phi\left(t\right), \ \Phi\left(0\right) = I.$$

Suppose also the eigenvalues of A are $\{\lambda_1, \cdots, \lambda_m\}, m \leq n$ where these eigenvalues are ordered such that

$$\mathrm{Re}\left(\lambda_1\right) \leq \mathrm{Re}\left(\lambda_2\right) \leq \cdots \leq \mathrm{Re}\left(\lambda_m\right) < 0.$$

Then if $0 > -\delta > \mathrm{Re}\left(\lambda_n\right),$ is given, there exists a constant, K related to the earlier constant C such that for all $t > 0$

$$\left|\Phi\left(t\right)\mathbf{x}\right| \leq K e^{-\delta t} \left|\mathbf{x}\right|. \tag{12.9}$$

Proof: From the above formula,

$$
\begin{aligned}
\left|\Phi\left(t\right)\mathbf{x}\right| &= \left|\sum_{k=0}^{m} r_{k+1}\left(t\right) P_k\left(A\right)\mathbf{x}\right| \leq \sum_{k=0}^{m} \left|r_{k+1}\left(t\right)\right| \left|P_k\left(A\right)\mathbf{x}\right| \\
&\leq \sum_{k=0}^{m} C e^{-\delta t} \left\|P_k\left(A\right)\right\| \left|\mathbf{x}\right| \leq C e^{-\delta t} \max_k \left\{\left\|P_k\left(A\right)\right\|\right\} n \left|\mathbf{x}\right| \equiv K e^{-\delta t} \left|\mathbf{x}\right|
\end{aligned}
$$

Here $\left\|P_k\left(A\right)\right\|,$ the operator norm of $P_k\left(A\right)$ is defined as

$$\sup_{\left|\mathbf{x}\right|=1} \left\{\left|P_k\left(A\right)\mathbf{x}\right|\right\} = \max_{\left|\mathbf{x}\right|=1} \left\{\left|P_k\left(A\right)\mathbf{x}\right|\right\}$$

which is finite by Theorem E.2.21 because the set $\{\mathbf{x} : \left|\mathbf{x}\right| = 1\}$ is a closed and bounded set. ∎

For a little more on the operator norm of a matrix, see Problem 31 on Page 329.

Definition 12.4.3 *Let $\mathbf{f} : U \to \mathbb{R}^n$ where U is an open subset of $\mathbb{R}^n$ such that $\mathbf{a} \in U$ and $\mathbf{f}\left(\mathbf{a}\right) = \mathbf{0}$. A point $\mathbf{a}$ where $\mathbf{f}\left(\mathbf{a}\right) = \mathbf{0}$ is called an equilibrium point. Then $\mathbf{a}$ is asymptotically stable if for any $\varepsilon > 0$ there exists $r > 0$ such that whenever $\left|\mathbf{x}_0 - \mathbf{a}\right| < r$ and $\mathbf{x}\left(t\right)$ the solution to the initial value problem,*

$$\mathbf{x}' = \mathbf{f}\left(\mathbf{x}\right), \ \mathbf{x}\left(0\right) = \mathbf{x}_0,$$

it follows

$$\lim_{t \to \infty} \mathbf{x}\left(t\right) = \mathbf{a}, \ \left|\mathbf{x}\left(t\right) - \mathbf{a}\right| < \varepsilon$$

A differential equation of the form $\mathbf{x}' = \mathbf{f}\left(\mathbf{x}\right)$ is called autonomous as opposed to a non-autonomous equation of the form $\mathbf{x}' = \mathbf{f}\left(t, \mathbf{x}\right).$ The equilibrium point $\mathbf{a}$ is *stable if for every $\varepsilon > 0$ there exists $\delta > 0$ such that if $|\mathbf{x}_0 - \mathbf{a}| < \delta$, then if $\mathbf{x}$ is the solution of*

$$\mathbf{x}' = \mathbf{f}(\mathbf{x}), \ \mathbf{x}(0) = \mathbf{x}_0, \tag{12.10}$$

then $|\mathbf{x}(t) - \mathbf{a}| < \varepsilon$ for all $t > 0$. Sometimes we say somewhat more informally, that $\mathbf{0}$ is stable if whenever $|\mathbf{x}_0 - \mathbf{a}|$ starts off sufficiently small, less than δ, $|\mathbf{x}(t) - \mathbf{a}|$ remains less than δ.

Obviously asymptotic stability implies stability.

An ordinary differential equation is called almost linear if it is of the form

$$\mathbf{x}' = A\mathbf{x} + \mathbf{g}(\mathbf{x})$$

where A is an $n \times n$ matrix and

$$\lim_{\mathbf{x} \to 0} \frac{\mathbf{g}(\mathbf{x})}{|\mathbf{x}|} = \mathbf{0}.$$

We assume $\mathbf{g}$ has continuous partial derivatives. The above is often stated by saying that $\mathbf{g}$ is $\mathbf{o}(\mathbf{x})$.

Now the stability of an equilibrium point of an autonomous system,

$$\mathbf{x}' = \mathbf{f}(\mathbf{x})$$

can always be reduced to the consideration of the stability of $\mathbf{0}$ for an almost linear system. Here is why. If you are considering the equilibrium point $\mathbf{a}$ for $\mathbf{x}' = \mathbf{f}(\mathbf{x})$, you could define a new variable $\mathbf{y}$ by

$$\mathbf{a} + \mathbf{y} = \mathbf{x}.$$

Then asymptotic stability would involve $|\mathbf{y}(t)| < \varepsilon$ and $\lim_{t \to \infty} \mathbf{y}(t) = \mathbf{0}$ while stability would only require $|\mathbf{y}(t)| < \varepsilon$. Then since $\mathbf{a}$ is an equilibrium point, $\mathbf{y}$ solves the following initial value problem.

$$\mathbf{y}' = \mathbf{f}(\mathbf{a} + \mathbf{y}) - \mathbf{f}(\mathbf{a}), \ \mathbf{y}(0) = \mathbf{y}_0,$$

where $\mathbf{y}_0 = \mathbf{x}_0 - \mathbf{a}$.

Let $A = D\mathbf{f}(\mathbf{a})$. Then from the definition of the derivative of a function,

$$\mathbf{y}' = A\mathbf{y} + \mathbf{g}(\mathbf{y}), \ \mathbf{y}(0) = \mathbf{y}_0 \tag{12.11}$$

where

$$\lim_{\mathbf{y} \to 0} \frac{\mathbf{g}(\mathbf{y})}{|\mathbf{y}|} = \mathbf{0}, \ A = D\mathbf{f}(\mathbf{a})$$

Thus there is never any loss of generality in considering only the equilibrium point $\mathbf{0}$ for an almost linear system.* Therefore, from now on I will only consider the case of almost linear systems and the equilibrium point $\mathbf{0}$.

---

*This is no longer true when you study partial differential equations as ordinary differential equations in infinite dimensional spaces.

If you don't know what the derivative is for a function of many variables, just say that you are considering an almost linear system of equations. All this does is point out that this is always the case assuming $\mathbf{f}$ is smooth.

Theorem 12.4.4 *Consider the almost linear system of equations,*

$$\mathbf{y}' = A\mathbf{y} + \mathbf{g}(\mathbf{y}) \tag{12.12}$$

where

$$\lim_{\mathbf{x} \to 0} \frac{\mathbf{g}(\mathbf{x})}{|\mathbf{x}|} = 0$$

and $\mathbf{g}$ is a C^1 function. Suppose that for all λ an eigenvalue of A, $\operatorname{Re}\lambda < 0$. Then $\mathbf{0}$ is asymptotically stable. In fact, there is $\delta, \varepsilon > 0$ and a positive constant λ such that for all $\mathbf{y}_0 \in B(\mathbf{0},\delta)$, $|\mathbf{y}(t,\mathbf{y}_0)| \le \varepsilon e^{-\lambda t}$.

Proof: By Theorem 12.4.2 there exist constants $\delta > 0$ and K such that for $\Phi(t)$ the fundamental matrix for A,

$$|\Phi(t)\mathbf{x}| \le Ke^{-\delta t}|\mathbf{x}|.$$

Let $\varepsilon > 0$ be given and let r be small enough that $Kr < \varepsilon$ and for

$$|\mathbf{x}| < (K+1)r, |\mathbf{g}(\mathbf{x})| < \eta|\mathbf{x}|$$

where η is so small that $K\eta < \delta$, and let $|\mathbf{y}_0| < r$. Then by the variation of constants formula, the solution to (12.12), at least for small t satisfies

$$\mathbf{y}(t) = \Phi(t)\mathbf{y}_0 + \int_0^t \Phi(t-s)\mathbf{g}(\mathbf{y}(s))\,ds.$$

The following estimate holds.

$$
\begin{aligned}
|\mathbf{y}(t)| &\le Ke^{-\delta t}|\mathbf{y}_0| + \int_0^t Ke^{-\delta(t-s)}\eta|\mathbf{y}(s)|\,ds \\
&< Ke^{-\delta t}r + \int_0^t Ke^{-\delta(t-s)}\eta|\mathbf{y}(s)|\,ds.
\end{aligned}
$$

Therefore,

$$e^{\delta t}|\mathbf{y}(t)| < Kr + \int_0^t K\eta e^{\delta s}|\mathbf{y}(s)|\,ds.$$

By Gronwall's inequality,

$$e^{\delta t}|\mathbf{y}(t)| < Kre^{K\eta t}$$

and so

$$|\mathbf{y}(t)| < Kre^{(K\eta-\delta)t} < \varepsilon e^{(K\eta-\delta)t}$$

Therefore, $|\mathbf{y}(t)| < Kr < \varepsilon$ for all t and so from Corollary 11.5.4, the solution to (12.12) exists for all $t \ge 0$, and since $K\eta - \delta < 0$, $\lim_{t\to\infty}|\mathbf{y}(t)| = 0$. The computation shows that for all $|\mathbf{y}_0| < r$, $|\mathbf{y}(t,\mathbf{y}_0)| \le \varepsilon e^{(K\eta-\delta)t}$. Let $-\lambda = K\eta - \delta$. ∎

Corollary 12.4.5 *Consider the almost linear system of equations,*

$$\mathbf{x}' = A\mathbf{x} + \mathbf{g}(\mathbf{x}) \tag{12.13}$$

where

$$\lim_{\mathbf{x} \to 0} \frac{\mathbf{g}(\mathbf{x})}{|\mathbf{x}|} = 0$$

and $\mathbf{g}$ *is a* C^1 *function. Suppose that for all* λ *an eigenvalue of* A, $\mathrm{Re}\,\lambda > 0$. *Then* $\mathbf{0}$ *unstable in the sense that there is* $\delta > 0$ *such that if* $\mathbf{x}_0 \in B(\mathbf{x}_0, \delta) \setminus \{\mathbf{0}\}$, *then*

$$\lim_{t \to -\infty} \mathbf{x}(t, \mathbf{x}_0) = \mathbf{0}.$$

In fact, there exists $\delta > 0$ *such that if* $|\mathbf{x}_0| < \delta$ *but* $\mathbf{x}_0 \neq \mathbf{0}$, *then* $|\mathbf{x}(t, \mathbf{x}_0)| \geq \delta$ *for all* t *large enough.*

Proof: This follows from Theorem 12.4.4 above. Let $s = -t, \mathbf{y}(s) = \mathbf{x}(t)$. Then the equation in terms of s and $\mathbf{y}$ is an almost linear system of the form

$$\mathbf{y}' = (-A)\mathbf{y} + \hat{\mathbf{g}}(\mathbf{y})$$

All eigenvalues of $-A$ have negative real parts and so for $\mathbf{x}_0$ close to $\mathbf{0}$, it follows that

$$\lim_{t \to -\infty} \mathbf{x}(t) = \lim_{s \to \infty} \mathbf{y}(s) = \mathbf{0}$$

Choose $\delta > 0$ small enough that if $|\mathbf{x}_0| < \delta$, then

$$|\mathbf{x}(-t, \mathbf{x}_0)| \leq \varepsilon e^{-\lambda t}, \lambda > 0$$

Then if $\lim_{n \to \infty} t_n = \infty$, and if $\mathbf{x}(t_n, \mathbf{x}_0) \in B(\mathbf{x}_0, \delta)$, then

$$|\mathbf{x}_0| = |\mathbf{x}(-t_n, \mathbf{x}(t_n, \mathbf{x}_0))| \leq \varepsilon e^{-\lambda t_n}$$

which implies that $\mathbf{x}_0 = \mathbf{0}$. Thus, for all t large enough, $|\mathbf{x}(t, \mathbf{x}_0)| \geq \delta$. ∎

It can be proved that if the matrix, A has eigenvalues such that the real parts are either positive or negative and it also has some whose real parts are positive, then $\mathbf{0}$ is not stable for the almost linear system. In fact there exists a set containing $\mathbf{0}$ such that if the initial condition is not on that set, then the solution to the differential equation fails to stay in some open ball containing $\mathbf{0}$ but if the initial condition is on this set, then the solution does converge to $\mathbf{0}$. However, this requires more work to establish and is shown later. Here is what happens when there is an eigenvalue which has positive real part.

Theorem 12.4.6 *Consider the almost linear system*

$$\mathbf{x}' = A\mathbf{x} + \mathbf{g}(\mathbf{x})$$

where

$$\lim_{\mathbf{x} \to \mathbf{0}} \frac{\mathbf{g}(\mathbf{x})}{|\mathbf{x}|} = \mathbf{0}$$

and $\mathbf{g}$ *is* C^1. *Then if* A *has an eigenvalue* λ *with positive real part, it follows that* $\mathbf{0}$ *is not stable.*

The above theorem is proved in Section 12.8. See also Problem 32 on Page 329. The above theorems are summarized in the following table for stability of $\mathbf{0}$ in an almost linear system $\mathbf{x}' = A\mathbf{x} + \mathbf{g}(\mathbf{x})$. Here $\sigma(A)$ denotes the eigenvalues of A.

$\mathrm{Re}(\lambda) < 0$ for all $\lambda \in \sigma(A)$	asymptotically stable
$\mathrm{Re}(\lambda) > 0$ for some $\lambda \in \sigma(A)$	unstable

Of course more can be said about the geometry of the solutions near the equilibrium point which can often be obtained by using a numerical method to graph these solutions, as discussed earlier.

Here are some examples.

Example 12.4.7 *Here is a differential equation*

$$\begin{pmatrix} x \\ y \end{pmatrix}' = \begin{pmatrix} x^2 - 1 \\ x^2 + 2y^2 - 9 \end{pmatrix}$$

Determine the equilibrium points of this system and whether they are stable.

First, what are the equilibrium points? Some simple manipulations will show that these are $(1, 2), (-1, 2), (1, -2), (-1, -2)$. Now we need to determine whether these are stable. As mentioned above, one should write in terms of new variables so that the equilibrium point corresponds to $(0, 0)$ for the new variables. Begin with $(1, 2)$. Let

$$x = 1 + u, \ y = 2 + v$$

Now substitute this into the equation.

$$\begin{pmatrix} u \\ v \end{pmatrix}' = \begin{pmatrix} (1 + u)^2 - 1 \\ (1 + u)^2 + 2(2 + v)^2 - 9 \end{pmatrix} = \begin{pmatrix} u^2 + 2u \\ u^2 + 2u + 2v^2 + 8v \end{pmatrix}$$

$$= \begin{pmatrix} 2 & 0 \\ 2 & 8 \end{pmatrix} \begin{pmatrix} u \\ v \end{pmatrix} + \begin{pmatrix} u^2 \\ u^2 + 2v^2 \end{pmatrix}$$

The last term is C^1 and is $o((u, v))$. The eigenvalues are positive for the matrix in the above and so the origin $(0, 0)$ is unstable for this system. It follows that $(1, 2)$ is unstable for the original system. Now consider the point $(-1, -2)$. Similar to the above, let $x = -1 + u, y = -2 + v$ and write the system in terms of the new variables.

$$\begin{pmatrix} u \\ v \end{pmatrix}' = \begin{pmatrix} (-1 + u)^2 - 1 \\ (-1 + u)^2 + 2(-2 + v)^2 - 9 \end{pmatrix} = \begin{pmatrix} u^2 - 2u \\ u^2 - 2u + 2v^2 - 8v \end{pmatrix}$$

$$= \begin{pmatrix} -2 & 0 \\ -2 & -8 \end{pmatrix} \begin{pmatrix} u \\ v \end{pmatrix} + \begin{pmatrix} u^2 \\ u^2 + 2v^2 \end{pmatrix}$$

This time, the eigenvalues are all negative and so the point $(-1, -2)$ is stable. Next consider $(1, -2)$. Again, let $x = 1 + u, y = -2 + v$. The new system is

$$\begin{pmatrix} u \\ v \end{pmatrix}' = \begin{pmatrix} (1 + u)^2 - 1 \\ (1 + u)^2 + 2(-2 + v)^2 - 9 \end{pmatrix} = \begin{pmatrix} u^2 + 2u \\ u^2 + 2u + 2v^2 - 8v \end{pmatrix}$$

$$= \begin{pmatrix} 2 & 0 \\ 2 & -8 \end{pmatrix} \begin{pmatrix} u \\ v \end{pmatrix} + \begin{pmatrix} u^2 \\ u^2 + 2v^2 \end{pmatrix}$$

This time, the eigenvalues have one positive and the other negative. Hence, $(1, -2)$ is not stable because there is a positive eigenvalue.

Example 12.4.8 *A damped vibrating pendulum can be described by the nonlinear system*

$$\begin{pmatrix} x' \\ y' \end{pmatrix} = \begin{pmatrix} y \\ -\frac{g}{l}\sin x - \frac{k}{l}y \end{pmatrix}$$

Then $(0,0)$ *is an equilibrium point. Is it stable?*

You can replace $\sin x$ with its power series as follows.

$$\begin{pmatrix} x' \\ y' \end{pmatrix} = \begin{pmatrix} y \\ -\frac{g}{l}\left(x - \frac{x^3}{3!} + o\left(x^4\right)\right) - \frac{k}{l}y \end{pmatrix}$$

$$= \begin{pmatrix} 0 & 1 \\ -\frac{g}{l} & -\frac{k}{l} \end{pmatrix}\begin{pmatrix} x \\ y \end{pmatrix} + \overbrace{\begin{pmatrix} 0 \\ \frac{g}{l}\frac{x^3}{3!} + o\left(x^4\right) \end{pmatrix}}^{\mathbf{g}(x,y)}$$

The eigenvalues of the matrix are both negative so $(0,0)$ is stable, in fact asymptotically stable. To see they are both negative without doing any work, recall that the trace is their sum and the determinant is their product. Now here the determinant is positive so the two eigenvalues have the same sign. The trace is negative so they are both negative.

Another thing to note is the use of MATLAB in finding eigenvalues. There are ways to do this numerically and MATLAB uses these. Here is an example.

>> A=[1 1 2;2 1 -1;1 2 3];
eig(A)

then press enter and it gives numerical eigenvalues.

$$4.0000 + 0.0000i, 0.5000 + 0.8660i, 0.5000 - 0.8660i$$

This is enough to determine stability. If A were the matrix in one of these almost linear systems, the origin would not be stable.

You can also do this in Scientific Notebook. Enter the matrix by choosing an appropriate template from the tool bar. Then click on the matrix and then compute, matrices, and eigenvalues. The following is what resulted. In this case, you can find them exactly in terms of square roots. I told it to find them numerically by writing 2.0 instead of 2.

$$\begin{pmatrix} 1 & 1 & 2.0 \\ 2 & 1 & -1 \\ 1 & 2 & 3 \end{pmatrix}, \text{ eigenvalues: } 4.0, 0.5 + 0.866\,03i, 0.5 - 0.866\,03i$$

This simple criterion about sign of real parts of eigenvalues can be used in many cases to determine whether the equilibrium points are stable. However, it does not always work. Here is an important example. This is sometimes called the predator prey equations because it is used to model the numbers of predators versus the number of prey. Maybe you can figure out which of x, y is the predator and which is the prey. If not, it was discussed earlier.

Example 12.4.9 *The Lotka-Volterra equations are*

$$\begin{pmatrix} x \\ y \end{pmatrix}' = \begin{pmatrix} \alpha x - \beta xy \\ \delta xy - \gamma y \end{pmatrix}$$

where $\alpha, \delta, \gamma, \beta$ are positive constants. Determine the equilibrium points and whether they are stable.

The equilibrium points are $(0,0), \left(\frac{\gamma}{\delta}, \frac{\alpha}{\beta}\right)$. We need to consider their stability. As to the first,

$$\begin{pmatrix} x \\ y \end{pmatrix}' = \begin{pmatrix} \alpha & 0 \\ 0 & \gamma \end{pmatrix} \begin{pmatrix} x \\ y \end{pmatrix} + \begin{pmatrix} -\beta xy \\ \delta xy \end{pmatrix}$$

the last term being $o\left((x,y)\right)$ and C^1. The eigenvalues are positive so $(0,0)$ is not stable. Consider the other point. Let $x = \frac{\gamma}{\delta} + u$, $y = \frac{\alpha}{\beta} + v$ so that $(0,0)$ for (u,v) corresponds to $\left(\frac{\gamma}{\delta}, \frac{\alpha}{\beta}\right)$ for (x,y). Then in terms of these new variables, the equations are

$$\begin{pmatrix} u \\ v \end{pmatrix}' = \begin{pmatrix} \alpha \left(\frac{\gamma}{\delta} + u\right) - \beta \left(\frac{\gamma}{\delta} + u\right) \left(\frac{\alpha}{\beta} + v\right) \\ \delta \left(\frac{\gamma}{\delta} + u\right) \left(\frac{\alpha}{\beta} + v\right) - \gamma \left(\frac{\alpha}{\beta} + v\right) \end{pmatrix} = \begin{pmatrix} -uv\beta - v\beta\frac{\gamma}{\delta} \\ uv\delta + u\frac{\alpha}{\beta}\delta \end{pmatrix}$$

$$= \begin{pmatrix} 0 & -\frac{\gamma\beta}{\delta} \\ \frac{\alpha\delta}{\beta} & 0 \end{pmatrix} \begin{pmatrix} u \\ v \end{pmatrix} + \begin{pmatrix} -\beta uv \\ \delta uv \end{pmatrix}$$

the last term being C^1 and $o\left((u,v)\right)$. This time the eigenvalues are pure imaginary so the only real parts are 0. Thus the above criterion does not work to determine stability of this second equilibrium point.

These equations are of the form

$$x' = xy\left(\frac{\alpha}{y} - \beta\right), \ y' = xy\left(\delta - \frac{\gamma}{x}\right)$$

More generally, following [19], consider

$$x' = g(y) R(x,y), \ y' = f(x) R(x,y) \tag{12.14}$$

Then along any solution, you have

$$F(x) - G(y) = C, \ \ F'(x) = f(x), G'(y) = g(y)$$

To see this, differentiate both sides with respect to t, Then you get

$$f(x) x' - g(y) y' = f(x) g(y) R(x,y) - g(y) f(x) R(x,y) = 0$$

and so the function $F(x) - G(y)$ is constant along solution curves of the system $x' = g(y) R(x,y), \ y' = f(x) R(x,y)$. Now for the Lotka-Volterra system,

$$x' = xy\left(\frac{\alpha}{y} - \beta\right), \ y' = xy\left(\delta - \frac{\gamma}{x}\right)$$

Therefore,

$$\delta x - \gamma \ln|x| - (\alpha \ln|y| - \beta y) = C$$

along a solution curve. Then the level curves correspond to solution curves of the equations. Let's just pick some values for $\delta, \gamma, \alpha, \beta$. Let them all be equal to 2 for simplicity.

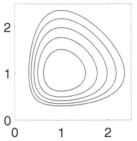

Then this is equivalent to $2(x+y) - \ln(x^2y^2) = C$ for various values of C. The equilibrium point $(1,1)$ corresponds to $C = 4$ so we graph various level curves of the form $2(x+y) - \ln(x^2y^2) = 4+r$ for small r.

This appears to show that in fact, the equilibrium point is stable because if you begin near it, the solution stays near it. Note however, that it is not asymptotically stable. The solution just goes around the equilibrium point without wandering too far. This is called a periodic orbit. Such a stable point is called a center. Much more can be said about this kind of behavior.

Example 12.4.10 *The case of the nonlinear pendulum can be considered as the following first order system.*

$$x' = y$$
$$y' = -\sin x$$

Then $(0,0)$ is clearly an equilibrium point. Is it stable?

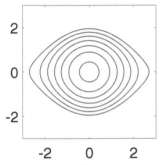

Multiply the top equation by $\sin x$ and bottom by y. Then add. This yields $-\cos x + \frac{y^2}{2} = C$ which defines some level curves which enclose $(0,0)$ as shown in the picture. Thus $(0,0)$ is also a center and is stable in the same way as the Lotka-Volterra system.

It is nice when you can resolve stability issues by simply looking at eigenvalues of a matrix, but often you can't do this. This is what is illustrated by these two important examples. Something else is needed and this is the topic of the next section.

12.5 Lyapunov Functions

One interesting technique for determining stability is the method of Lyapunov[†] functions. These are defined next. For simplicity, we will let all the functions be defined on all of $\mathbb{R}^n$ to avoid fussing with details. However, if it is clear that the solutions of the equation will have values in some open subset of $\mathbb{R}^n$, then one only needs to have the Lyapunov function defined on this open subset.

Definition 12.5.1 *A function W which is C^1 is a Lyapunov function for the differential equation*

$$\mathbf{x}' = \mathbf{f}(\mathbf{x})$$

if it has the following properties:

[†]Aleksandr Mikhailovich Lyapunov (1857-1918), worked on stability theory in differential equations as well as mathematical physics and probability theory. He is most famous for the Lyapunov functions of this section, but he also gave an improved proof of the central limit theorem in probability. Lyapunov's brother Sergey was a fairly important composer who wrote some wonderful music for piano.

1. $\nabla W(\mathbf{x}) \cdot \mathbf{f}(\mathbf{x}) \leq 0$

2. $W(\mathbf{x}) \geq 0$ *and* $W(\mathbf{0}) = 0$

3. If B is a set and $\sup_{\mathbf{x} \in B} W(\mathbf{x}) < \infty$, *then* $\sup_{\mathbf{x} \in B} |\mathbf{x}| < \infty$.

The reason this is interesting is that if you look at a solution to the differential equation $\mathbf{x}' = \mathbf{f}(\mathbf{x})$ and consider the time rate of change of $W(\mathbf{x})$, the chain rule gives

$$\frac{d}{dt} W(\mathbf{x}) = \nabla W(\mathbf{x}) \cdot \mathbf{x}' = \nabla W(\mathbf{x}(t)) \cdot \mathbf{f}(\mathbf{x}(t)) \leq 0$$

so that $t \to W(\mathbf{x}(t))$ is decreasing along every solution to the equation and we can conclude this without first knowing the solution. This is really nice even from the point of view of existence of a global solution for $t \geq 0$ because if $\mathbf{f}$ is C^1, then 3. above implies that $\mathbf{x}(t)$ remains in a bounded set and so Lemma 11.5.1 implies that one can consider $\mathbf{f}$ to satisfy a Lipschitz condition and so there exists a global solution on all of $[0, \infty)$ to the differential equation. This is the following Proposition.

Proposition 12.5.2 *Let $\mathbf{f}$ be a C^1 function defined on $\mathbb{R}^n$ and suppose there is a Lyapunov function for the differential equation*

$$\mathbf{x}' = \mathbf{f}(\mathbf{x})$$

Then for any $\mathbf{x}_0 \in \mathbb{R}^n$, there exists a unique solution to the initial value problem

$$\mathbf{x}' = \mathbf{f}(\mathbf{x}), \quad \mathbf{x}(0) = \mathbf{x}_0 \tag{12.15}$$

which is valid on $[0, \infty)$.

As earlier, $\mathbf{x}(t, \mathbf{x}_0)$ denotes the solution to the above initial value problem (12.15).

Next is the definition of a limit set.

Definition 12.5.3 *Let $\mathbf{x}_0 \in \mathbb{R}^n$ be given and define the set Λ^+ to be those $\mathbf{z} \in \mathbb{R}^n$ such that there is a sequence $\{t_n\}$ such that $t_n \to \infty$ and $\mathbf{z} = \lim_{n \to \infty} \mathbf{x}(t_n, \mathbf{x}_0)$.*

For example, a stable equilibrium point could be a limit set, but there are other examples of limit sets. Sometimes these are periodic orbits. These are discussed later. The following is due to LaSalle and gives an important property of Lyapunov functions pertaining to the limit set.

Proposition 12.5.4 *Let the initial value problem (12.15) have a Lyapunov function W. Let the limit set Λ^+ correspond to some initial condition $\mathbf{x}_0$. Then if $\mathbf{z} \in \Lambda^+$ it follows that $\mathbf{x}(t, \mathbf{z}) \in \Lambda^+$ for all $t \geq 0$. That is, Λ^+ is positively invariant. Also*

$$\Lambda^+ \subseteq \{\mathbf{z} : \nabla W(\mathbf{z}) \cdot \mathbf{f}(\mathbf{z}) = 0\} = \left\{\mathbf{z} : \frac{d}{dt} W(\mathbf{x}(t, \mathbf{z})) |_{t=0} = 0\right\}.$$

In addition, the limit set Λ^+ exists.

Proof: Let $\mathbf{x}(t_n, \mathbf{x}_0) \to \mathbf{z}$. Then by continuity on initial data, Theorem 11.6.2 and also Proposition 11.7.2,

$$\mathbf{x}(t, \mathbf{z}) = \lim_{n \to \infty} \mathbf{x}(t, \mathbf{x}(t_n, \mathbf{x}_0)) = \lim_{n \to \infty} \mathbf{x}(t + t_n, \mathbf{x}_0)$$

and letting $s_n \equiv t + t_n$, it follows that $\mathbf{x}(t, \mathbf{z}) \in \Lambda^+$.

Now consider the second claim about the limit set. Let $\mathbf{z} \in \Lambda^+$. By Property 1. of Lyapunov functions and Property 2 which says that $W(\mathbf{x}) \geq 0$, it follows that

$$\lim_{t \to \infty} W(\mathbf{x}(t, \mathbf{x}_0)) = \beta$$

Indeed, you have a function of t which is decreasing and bounded below so of course the limit exists. Therefore, $W(\mathbf{z}) = \beta$ for all $z \in \Lambda^+$. From the first part and the chain rule,

$$0 = \lim_{t \downarrow 0} \frac{\beta - \beta}{t} = \lim_{t \downarrow 0} \frac{W(\mathbf{x}(t, \mathbf{z})) - W(\mathbf{z})}{t} = \nabla W(\mathbf{z}) \cdot \mathbf{x}'(t, \mathbf{z}) = \nabla W(\mathbf{z}) \cdot \mathbf{f}(\mathbf{z})$$

Note that Property 3. implies the global existence of the solution and that it stays in a bounded set. This also implies the existence of a limit set thanks to the Heine Borel theorem, Theorem E.2.10 which says that a bounded set has a convergent subsequence. Starting with an initial condition $\mathbf{x}_0, W(\mathbf{x}(t, \mathbf{x}_0))$ is bounded since it is no larger than $W(\mathbf{x}_0)$. Therefore, $\{\mathbf{x}(t, \mathbf{x}_0)\}_{t \geq 0}$ is also bounded. Thus by the Heine Borel theorem, there is a sequence $\{t_n\}$ such that $t_n \to \infty$ and $\mathbf{x}(t_n, \mathbf{x}_0)$ is a convergent sequence. ∎

This is convenient because it gives a formula for the limit set or at least something which contains the limit set. It also says that the limit set is invariant. This can sometimes be used to rule out the excess points as shown in the following examples.

Example 12.5.5 *Consider the following system of nonlinear equations*

$$x' = 2y^3 - x$$
$$y' = -(x + y)$$

Find a Lyapunov function and limit set.

You need to have $(W_x, W_y) \cdot (2y^3 - x, -(x + y))$. After experimenting somewhat you find that $W(x, y) = x^2 + y^4$ works. It equals

$$(2x, 4y^3) \cdot (2y^3 - x, -(x + y))$$
$$= 4xy^3 - 2x^2 + (-4xy^3) + (-4y^4) = -(2x^2 + 4y^4)$$

Then the limit set Λ^+ equals $\{(x, y) : -(2x^2 + 4y^4) = 0\}$. Thus there is only one point in Λ^+ and it is $(0, 0)$. This implies that

$$\lim_{t \to \infty} (x(t, x_0), y(t, y_0)) = (0, 0)$$

for any choice of initial data (x_0, y_0). If not, there would exist a subsequence $\{t_n\}$ such that $(x(t_n, x_0), y(t_n, y_0))$ is always further than $\varepsilon > 0$ for some $\varepsilon > 0$. However, it is a bounded subsequence and so there is a further subsequence, still denoted as t_n such that $(x(t_n, x_0), y(t_n, y_0)) \to (z_1, z_2)$ where $|(z_1, z_2) - (0, 0)| \geq \varepsilon$ but this is impossible because (z_1, z_2) by definition is in the limit set.

The next example is a little harder because you can't immediately identify the limit set. However, you can eliminate extraneous points by using the fact that the limit set is positively invariant.

Example 12.5.6 *Consider the following system of nonlinear equations*

$$x' = -y - x^3$$
$$y' = x^5$$

Find a Lyapunov function and limit set.

First note that the equation is of the form

$$\left(\begin{array}{c} x \\ y \end{array} \right)' = \left(\begin{array}{cc} 0 & -1 \\ 0 & 0 \end{array} \right) \left(\begin{array}{c} x \\ y \end{array} \right) + \left(\begin{array}{c} -x^3 \\ x^5 \end{array} \right)$$

The eigenvalues are both 0 for the matrix so the method of looking at the signs of the matrix in the linearized system won't yield anything. Therefore, we try to find a Lyapunov function. You need to have $(W_x, W_y) \cdot (-y - x^3, x^5) \leq 0$. You try a few things to get something to work out. This is rather ad hoc but that is what you have to do with these. Try

$$\frac{1}{3} x^6 + y^2 = W(x, y)$$

Then this seems to work because the above dot product becomes

$$(2x^5, 2y) \cdot (-y - x^3, x^5) = -2yx^5 - 2x^8 + 2yx^5 = -2x^8 \leq 0$$

Thus this satisfies the three criteria for a Lyapunov function. Hence, from the above theorem, the limit set is contained in

$$\{(x, y) : (2x^5, 2y) \cdot (-y - x^3, x^5) = 0\}$$

From the above computation, this shows that the limit set is contained in

$$\{(x, y) : -2x^8 = 0\}$$

so we have to have $x = 0$. The limit set for a given initial condition must be contained in the y axis. Recall also from Proposition 12.5.4 that it must be invariant. Now consider the differential equation with initial condition equal to $(0, y_0)$ where $y_0 \neq 0$. Then the first equation shows that x does not remain zero. Hence the limit set can't contain any points $(0, y)$ for $y \neq 0$. It follows that there is only one point in the limit set and it is $(0, 0)$.

Now start with an arbitrary initial condition (x_0, y_0) and let $(x(t), y(t))$ be the resulting solution. Then $W(x(t), y(t))$ is bounded. Therefore, $(x(t), y(t))$ is bounded. It follows from the Heine Borel theorem, Theorem E.2.10 that there is a sequence $t_n \to \infty$ such that $(x(t_n), y(t_n)) \to (0, 0)$, the only possible point in the limit set. Therefore, in fact, $\lim_{t\to\infty}(x(t), y(t)) = (0,0)$ because this shows that if you have any sequence $\{t_n\}$ converging to ∞,

$$(x(t_n), y(t_n)) \to (0,0)$$

because if not, an application of the Heine Borel theorem would yield a subsequence which would converge to something other than $(0,0)$ which is impossible since this is the only possible point in a limit set. Thus $(0,0)$ is asymptotically stable in the sense that $(x(t, x_0), y(t, y_0)) \to (0,0)$ regardless of the choice of (x_0, y_0).

Example 12.5.7 *Consider the equation*

$$\left(\begin{array}{c} x \\ y \end{array}\right)' = \left(\begin{array}{c} -x^3 + 2y^3 \\ -2xy^2 \end{array}\right)$$

Find limit set and Lyapunov function.

We need to have

$$(W_x, W_y) \cdot \left(-x^3 + 2y^3, -2xy^2\right) \leq 0$$

Let's look for $W = ax^2 + bxy + cy^2$ and see if the a, b, c can be chosen so that everything works.

$$(2ax + by, bx + 2cy) \cdot \left(-x^3 + 2y^3, -2xy^2\right)$$

$$= 2by^4 - 2ax^4 + 4axy^3 - bx^3y - 4cxy^3 - 2bx^2y^2$$

Let's let $a = c$. Then this reduces to

$$2by^4 - 2ax^4 - bx^3y - 2bx^2y^2$$

clearly we can get what is wanted by letting $b = 0$. Thus it appears that $W(x,y) = x^2 + y^2$ will work fine. Then

$$(W_x, W_y) \cdot \left(-x^3 + 2y^3, -2xy^2\right) = (2x, 2y) \cdot \left(-x^3 + 2y^3, -2xy^2\right)$$
$$= -2x^4 + 4xy^3 - 4xy^3 = -2x^4$$

Thus the limit set is $x = 0$ as in the previous example. Thus the limit set is in the y axis. However, if you start with an initial condition $(0, y_0)$ where $y_0 \neq 0$, the first equation will cause $x(t)$ to leave the y axis contrary to this set being invariant, and so the limit set is actually only $(0,0)$. Now the same reasoning as in the above example shows that $(0,0)$ is asymptotically stable in the sense that $(x(t, x_0), y(t, y_0)) \to (0,0)$.

12.6 Periodic Orbits, Poincare Bendixon Theorem

The fundamental result in this subject, at least in the plane, is the Poincare Bendixon theorem.[‡] A proof of this theorem is in Appendix H. A periodic orbit is a set of points

$$\{\mathbf{x}(t, \mathbf{x}_0), t \geq 0\}$$

[‡]Ivar Otto Bendixson (1861-1935) was a Swedish mathematician. He is most famous for the Poincare Bendixon theorem presented here. He also did work in topology.

such that for some $T > 0$, $\mathbf{x}(t + T, \mathbf{x}_0) = \mathbf{x}(t, \mathbf{x}_0)$ for all $t \geq 0$. The number T is called a period. Thus the point $\mathbf{x}(t, \mathbf{x}_0)$ goes around and around always returning to the point from where it started.

Theorem 12.6.1 *Let D be the closure of a bounded region of the plane such that $\mathbf{f}$ is a C^1 function which has no zeros in D, and suppose that $\mathbf{x}(t, \mathbf{x}_0)$ stays in D for all $t \geq 0$ if $\mathbf{x}_0 \in D$, where this is the solution to*

$$\mathbf{x}' = \mathbf{f}(\mathbf{x}), \ \mathbf{x}(0) = \mathbf{x}_0$$

Then letting $\Lambda_+ = \cup_{t \geq 0}\mathbf{x}(t, \mathbf{x}_0)$, it follows that Λ_+ is either a periodic orbit or $t \to \mathbf{x}(t, \mathbf{x}_0)$ spirals in toward a periodic orbit.

It is a plausible result. Say you have that every initial condition which starts off in a bounded closed set stays in that set and there are no equilibrium points. Thus $t \to \mathbf{x}(t, \mathbf{x}_0)$ just keeps moving. Then from this theorem, there must be a periodic orbit somewhere such that either this function traces out a periodic orbit or it gets close to one. For example, consider the system

$$\left(\begin{array}{c} x \\ y \end{array} \right)' = \left(\begin{array}{c} x + y - x\left(x^2 + 2y^2\right) \\ -x + y - y\left(2x^2 + y^2\right) \end{array} \right)$$

From looking at the eigenvalues of the matrix in the almost linear system, you will see that they are both positive. Hence every solution near $(0, 0)$ but not equal to $(0, 0)$ must fail to remain near $(0, 0)$. Also, you can see from the graph of the direction field that the conditions of the above theorem are satisfied. Therefore, there should exist a periodic orbit from the above theorem. The following shows a graph of this direction field along with graphs of solutions to the equations.

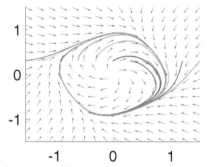

As explained in the appendix, the following very interesting theorem can be obtained from the above.

Theorem 12.6.2 *If you have a periodic orbit of a solution to an autonomous two dimensional differential equation, $\mathbf{x}' = \mathbf{f}(\mathbf{x})$, then it must go around some equilibrium point. If there is only one equilibrium point inside the periodic orbit, then it cannot be a saddle point.*

12.7   Van der Pol Equation

An important example of an equation which has periodic solutions even though it is very nonlinear is the Van der Pol equation. In this case, the existence of a region which absorbs the graphs of solutions starting in it is not as clear.

Example 12.7.1 *Consider the following equation, $k > 0$*

$$x'' + k\left(x^2 - 1\right)x' + x = 0$$

Show that it has a non constant periodic solution.

This explanation follows [19]. To consider this question, write as a first order system.

$$x' = y$$
$$y' = -\left(k\left(x^2 - 1\right)y + x\right)$$

Now the argument will be based on the following picture in which we will adjust the values of h and r to get a region which has the needed properties to use the Poincare Bendixon theorem.

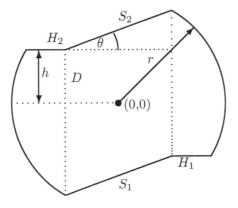

In the picture, r will always be large, certainly larger than 1. Also the two vertical dotted lines will be symmetric as shown and the distance between them is at least 2. Thus all points on the circular part of the boundary will have $|x| > 1$.

Multiply the top by x and the bottom by y. Then add.

$$\frac{d}{dt}\left(\frac{x^2}{2} + \frac{y^2}{2}\right) = xx' + yy' = -k\left(x^2 - 1\right)y^2 < 0 \text{ if } |x| > 1 \qquad (12.16)$$

Thus the distance to $(0,0)$ gets smaller out on the circular parts of the closed region D. That is, on this part of the boundary of D, $(x(t), y(t))$ is moving closer to $(0,0)$ as t increases.

Now consider the horizontal segments H_1 and H_2. Consider the top one $y = h$. From the differential equations, you get the following for such points.

$$y' = -\left(k\left(x^2 - 1\right)h - |x|\right)$$

Is this negative? You would need $|x| - kh\left(x^2 - 1\right) < 0$ which requires that

$$\frac{|x|}{x^2 - 1} < kh$$

Require that the two vertical dotted lines in the picture be at least 4 apart. Say the left vertical line is $x = -a, a \geq 2$. In fact, to be definite, let them be exactly 4 apart. Thus the left line is $x = -2$ and the right is $x = 2$. Then the function of x on the left in the above is increasing for $x < -2$ and so the above condition will be satisfied if it is required that $x = 2$ and

$$\frac{2}{4-1} = \frac{2}{3} < kh$$

Hence choose r large enough and let $h > \frac{2}{3k}$. This will force $y' < 0$ on the top horizontal segment. Similarly, for h this large, $y' > 0$ on the bottom horizontal segment.

Making r still larger, this causes an increase in θ and makes the slope m of the segment S_2 as large as desired. Now this line has equation $y = m(x + 2) + h$ and so a downward pointing normal vector to this line segment is $(m, -1)$. Then if the angle between this vector and the direction vector (x', y') is acute, this will ensure that (x', y') will point into D. Using the equations, the dot product between these two is

$$x'm - y' = x'm + \left(k\left(x^2 - 1\right)y + x\right) = ym + \left(k\left(x^2 - 1\right)y + x\right)$$

$$= \left(m(x+2) + h\right)m + \left(k\left(x^2 - 1\right)\left(m(x+2) + h\right) + x\right)$$

$$= \left(m(x+2) + h\right)m + \left(k\left(x^2 - 1\right)\left(m(x+2) + h\right) + x\right)$$

There are two parts of this line, the part where $-2 \leq x \leq -\frac{3}{2}$ and the part where $-\frac{3}{2} \leq x \leq 2$. On the second part, the term $m^2(x+2) \geq m^2\frac{1}{2}$ and so for large enough m, obtained by simply enlarging r sufficiently, the m^2 term dominates and the expression is positive. On the first part of this line where $-2 \leq x \leq -\frac{3}{2}$, the expression is at least as large as

$$hm + \left(k\left(\frac{9}{4} - 1\right)h - 2\right) = hm + \left(\frac{5}{4}kh - 2\right)$$

Thus, making r still larger if needed, m gets increasingly large and this is also positive. The situation is similar on S_1. Thus, choosing r large enough and h large enough, the direction of motion of any solution curve on the boundary of D is toward the inside of D.

There is exactly one point in D where $\mathbf{f}(\mathbf{x}) = \mathbf{0}$ and it is the origin $(0, 0)$. As shown above,

$$\frac{d}{dt}\left(\frac{x^2}{2} + \frac{y^2}{2}\right) = xx' + yy' = -k\left(x^2 - 1\right)y^2$$

Suppose $x^2 + y^2 = 1/4$. Then $|x| \leq 1/2$ and

$$\frac{d}{dt}\left(\frac{x^2}{2} + \frac{y^2}{2}\right) = k\left(1 - x^2\right)y^2 \geq k\left(1 - \frac{1}{4}\right)y^2 > 0$$

unless $y = 0$. But if $y = 0$, the equation $y' = -\left(k\left(x^2 - 1\right)y + x\right)$ requires that $y' = \pm\frac{1}{2}$ and $x' = 0$. Thus, even at this point, motion is away from the inside of the disk $x^2 + y^2 \leq 1$. It follows that we can modify D by leaving out the disk centered at $(0, 0)$ which has radius $1/2$ and the resulting region in the plane is perfect for application of the Poincare Bendixon theorem. Thus there exists a periodic orbit in this modified D such that the solution for any initial condition (x_0, y_0) starting in B is either a periodic orbit or is spirals toward one. Now here is the vector field for this system in which $k = 1$.

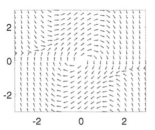

Note that also, the only restriction on r was that it be sufficiently large. Neither is it necessary to take out the disk of radius $1/2$. One could have removed any smaller disk centered at $(0,0)$ with the same result. Thus every initial condition $(x_0, y_0) \neq (0,0)$ has this property that it is either a periodic orbit or spirals toward one. In fact there is only one and this is illustrated by the following phase portrait in which $k = 1$. Again, this is only an illustration. The solution curves cannot merge, contrary to what is indicated in the picture.

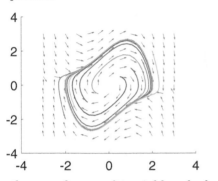

An open region is an open set which is connected. Such an open region D in the plane is called simply connected if the inside[§] of every simple closed curve contained in D is also contained in D. Then there is a very interesting theorem which gives a condition which excludes the possibility of a periodic orbit, a non constant solution such that for some positive $T, \mathbf{x}(t+T) = \mathbf{x}(t)$ for all $t \geq 0$. It is called the Bendixon Dulac theorem. This theorem requires the use of Green's theorem from multivariable calculus. For proofs of this theorem, see [10] or [16] or for a very general version, [2].

Theorem 12.7.2 *Consider the system of equations*

$$\begin{pmatrix} x \\ y \end{pmatrix}' = \begin{pmatrix} f(x,y) \\ g(x,y) \end{pmatrix}$$

where f and g are C^1 functions on a simply connected open region D. Suppose there exists a C^1 function $R : D \to \mathbb{R}$ such that $(fR)_x + (gR)_y$ is either always positive on D or negative on D. Then the above system of equations has no periodic solution other than an equilibrium point.

Proof: Suppose there is such a periodic solution. Let T be the smallest number such that for all $t \geq 0, \mathbf{x}(t) = \mathbf{x}(t+T)$. Let C be the oriented curve consisting of the values of $\mathbf{x}(t)$ for $t \in [0,T]$. Thus C is a simple closed curve. In fact, it is a C^1 curve. This is sufficient to apply Green's theorem. Then letting U be the inside of this curve,

$$0 \neq \int_U (fR)_x + (gR)_y \, dA =$$

$$\int_U \left((fR)_x - (-gR)_y\right) dA = \pm \int_C -gR dx + fR dy$$

[§]The Jordan curve theorem states that every simple closed curve is the boundary of exactly two regions, an inside which is bounded and an outside which is unbounded. It was first proved successfully by Veblen around 1911 although it was first formulated by Jordan several decades earlier. A proof may be found in Appendix G.

$$\equiv \pm \int_0^T \left(-g\left(x\left(t\right),y\left(t\right)\right)R\left(x\left(t\right),y\left(t\right)\right)\right)x'\left(t\right)$$
$$+f\left(x\left(t\right),y\left(t\right)\right)R\left(x\left(t\right),y\left(t\right)\right)y'\left(t\right)dt$$
$$= \pm \int_0^T R\left(x\left(t\right),y\left(t\right)\right)\left(-g\left(x\left(t\right),y\left(t\right)\right)\right)x'\left(t\right)+f\left(x\left(t\right),y\left(t\right)\right)y'\left(t\right)dt$$
$$= \pm \int_0^T R\left(x\left(t\right),y\left(t\right)\right)\left(-y'\left(t\right)x'\left(t\right)+x'\left(t\right)y'\left(t\right)\right)dt = 0$$

This is a contradiction. ∎

Example 12.7.3 *Consider the following system of equations defined on $\mathbb{R}^2$.*

$$\left(\begin{array}{c} x \\ y \end{array}\right)' = \left(\begin{array}{c} 2xy^3 - y \\ 3x^2y^2 - x \end{array}\right)$$

Could this have a periodic orbit which is contained in the set $\{(x,y) : y > 0\}$?

The answer is no. Let $R = 1$ and consider $f_x + g_y$ which is $y\left(2y^2 + 6x^2\right) > 0$ on this set. Hence by the above theorem, there is no periodic orbit in the upper half plane. Similarly, there is no periodic orbit in the lower half plane $y < 0$.

Example 12.7.4 *Consider the following system of equations on $\mathbb{R}^2$.*

$$\left(\begin{array}{c} x \\ y \end{array}\right)' = \left(\begin{array}{c} ax + f\left(y\right) \\ by + g\left(x\right) \end{array}\right)$$

Could this have a periodic orbit if $a + b \neq 0$?

The answer is no. Again, just let $R = 1$. You have $\left(ax + f\left(y\right)\right)_x + \left(by + g\left(x\right)\right)_y = a + b \neq 0$. The only possible periodic orbits are equilibrium points.

12.8   Stable Manifold

This section will be somewhat more technical and demanding than what comes earlier. You will need to understand the linear algebra in the appendix on block diagonal matrices in order to complete this material. You need to be able to reduce a matrix to a block diagonal matrix with a similarity transformation where the blocks correspond to particular eigenvalues of the matrix. Thanks to Putzer's method presented earlier, it will be somewhat simpler than the usual approach which is typically based on the Jordan canonical form but it still draws on a lot of linear algebra. You will also need the abstract material on fixed points in a complete normed linear space in Section 11.3. It is also convenient to introduce some new ideas in discussing this question of stability in the case of eigenvalues with positive real parts.

In this section, $\mathbf{x}\left(t, \mathbf{x}_0\right)$ will denote the solution to the initial value problem

$$\mathbf{x}' = A\mathbf{x} + \mathbf{g}\left(\mathbf{x}\right), \ \mathbf{x}\left(0\right) = \mathbf{x}_0$$

and we are concerned with $t \geq 0$.

What of the case where you have some eigenvalues with negative real parts and some with positive real parts? In this case, there is a special set of possible initial data such that for the initial condition in this special set, the solution to the initial value problem remains bounded and for initial data not in this set, the solution will not be bounded.

To begin with, assume A takes a special form. One can easily reduce to this case using standard linear algebra presented in the appendix in the section on block diagonal matrices.

$$A = \begin{pmatrix} A_- & 0 \\ 0 & A_+ \end{pmatrix} \tag{12.17}$$

where A_- and A_+ are square matrices of size $k \times k$ and $(n-k) \times (n-k)$ respectively. Also assume A_- has eigenvalues whose real parts are all less than $-\alpha$ while A_+ has eigenvalues whose real parts are all larger than α. That is, there are no eigenvalues having zero real part. Assume also that each of A_- and A_+ is upper triangular.

Also, I will use the following convention. For $\mathbf{v} \in \mathbb{C}^n$,

$$\mathbf{v} = \begin{pmatrix} \mathbf{v}_- \\ \mathbf{v}_+ \end{pmatrix}$$

where $\mathbf{v}_-$ consists of the first k entries of $\mathbf{v}$.

Recall Lemma 12.4.1. It was

Lemma 12.8.1 *Let the functions, r_k be given in the statement of Theorem 12.3.3 and suppose that A is an $n \times n$ matrix whose eigenvalues are $\{\lambda_1, \cdots, \lambda_n\}$. Suppose that these eigenvalues are ordered such that*

$$\mathrm{Re}\,(\lambda_1) \leq \mathrm{Re}\,(\lambda_2) \leq \cdots \leq \mathrm{Re}\,(\lambda_n)$$

Then for all $1 \leq k \leq n$,

$$|r_k(t)| \leq \frac{1}{(k-1)!} t^{k-1} e^{\mathrm{Re}(\lambda_k)t}$$

Also, if $0 > -\delta > \mathrm{Re}\,(\lambda_n)$ is given, there exists a constant, C such that for each $k = 0, 1, \cdots, n$,

$$|r_k(t)| \leq C e^{-\delta t} \tag{12.18}$$

Also, regardless of the ordering of the $\mathrm{Re}\,\lambda_i$, one can give a description of $r_k(t)$.

$$r_k(t) = p_k(t) e^{\lambda_k t}, \ \deg(p_k(t)) \leq k-1 \tag{12.19}$$

and $\deg(p_k(t)) \leq \deg(p_{k+1}(t))$. Strict inequality holds if and only if $\lambda_{k+1} = \lambda_k$. Also, $p_1(t) = 1$.

Also recall how this was used to obtain the fundamental matrix.

$$\Phi(t) \equiv \sum_{k=0}^{n-1} r_{k+1}(t) P_k(A)$$

where $P_k(A)$ was a polynomial in A.

Lemma 12.8.2 *Let A be of the form given in (12.17) as explained above and let $\Phi_+ (t)$ and $\Phi_- (t)$ be the fundamental matrices corresponding to A_+ and A_- respectively. Then there exist positive constants, α and γ such that*

$$|\Phi_+ (t) \mathbf{y}| \leq C e^{\alpha t} \text{ for all } t < 0 \qquad (12.20)$$

$$|\Phi_- (t) \mathbf{y}| \leq C e^{-(\alpha + \gamma)t} \text{ for all } t > 0. \qquad (12.21)$$

Also for any nonzero $\mathbf{x} \in \mathbb{C}^{n-k}$,

$$|\Phi_+ (t) \mathbf{x}| \text{ is unbounded.} \qquad (12.22)$$

Proof: The first two claims are easy consequences of (12.19). It suffices to pick α and γ such that $-(\alpha + \gamma)$ is larger than the real parts of all eigenvalues of A_- and α is smaller than all eigenvalues of A_+. Then note that exponential growth always exceeds polynomial growth. It remains to verify (12.22). From the Putzer formula for $\Phi_+ (t)$,

$$\Phi_+ (t) \mathbf{x} = \sum_{k=0}^{n-1} r_{k+1} (t) P_k (A) \mathbf{x}$$

where $P_0 (A) \equiv I$. Now each r_k is a polynomial (possibly a constant) times an exponential $e^{\alpha t}$ for $\alpha > 0$ and from Lemma 12.4.1 $\lambda_k = \lambda_{k+1}$ if and only if the degrees of the polynomials change. Thus no cancellation is possible and you have the sum of terms of the form $e^{\alpha t} p (t) \mathbf{x}$ where $\alpha > 0$ where $p (t) \neq 0$, the first being $e^{\lambda_1 t} \mathbf{x}$. Thus (12.22) must hold. ∎

It is clear that the fundamental matrix for the original system is of the form

$$\begin{pmatrix} \Phi_- (t) & 0 \\ 0 & \Phi_+ (t) \end{pmatrix}$$

Lemma 12.8.3 *Consider the initial value problem for the almost linear system*

$$\mathbf{x}' (t) = A\mathbf{x} (t) + \mathbf{g} (\mathbf{x} (t)), \ \mathbf{x} (0) = \mathbf{x}_0,$$

and $\mathbf{g}$ satisfies: For all $\eta > 0$, there exists δ such that if $|\mathbf{x}|, |\mathbf{y}| < \delta$, then

$$|\mathbf{g} (\mathbf{x}) - \mathbf{g} (\mathbf{y})| \leq \eta |\mathbf{x} - \mathbf{y}|, \qquad (12.23)$$

and[¶] A is of the special form

$$A = \begin{pmatrix} A_- & 0 \\ 0 & A_+ \end{pmatrix}$$

in which A_- is a $k \times k$ matrix which has eigenvalues for which the real parts are all negative and A_+ is a $(n - k) \times (n - k)$ matrix for which the real parts of all the

---

[¶]In fact this condition on $\mathbf{g}(\mathbf{x})$ can be proved by simply assuming that $\mathbf{g}$ is C^1 but I am just stating here what will be needed. By assumption, it is easy to prove that $\frac{\partial \mathbf{g}}{\partial x_i} (0) = \mathbf{0}$ which implies by continuity that these partial derivatives are small for $\mathbf{x}$ near $\mathbf{0}$. Then one uses Lemma 11.5.1 to get the desired condition.

eigenvalues are positive. Then $\mathbf{0}$ *is not stable. More precisely, there exists a set of points* $(\mathbf{a}_-, \psi(\mathbf{a}_-))$ *for* $\mathbf{a}_-$ *small such that for* $\mathbf{x}_0$ *on this set,*

$$\lim_{t \to \infty} \mathbf{x}(t, \mathbf{x}_0) = \mathbf{0}$$

and for small $\mathbf{x}_0$ *not on this set, there exists a* $\delta > 0$ *such that* $|\mathbf{x}(t, \mathbf{x}_0)|$ *cannot remain less than* δ *for all positive* t. *This set of points is called the stable manifold.*

Proof: First, suppose $\mathbf{x}(s, \mathbf{a})$ is a solution to the initial value problem with $\mathbf{x}(0, \mathbf{a}) = \mathbf{a}$ such that $|\mathbf{x}(t, \mathbf{a})| < \delta$ for all $t \geq 0$. Then

$$|\mathbf{g}_+(\mathbf{x}(s, \mathbf{a}))| = |\mathbf{g}_+(\mathbf{x}(s, \mathbf{a})) - \mathbf{g}_+(\mathbf{0})| \leq \eta |\mathbf{x}(s, \mathbf{a})|$$

$$\left| - \int_t^\infty \Phi_+(t - s) \mathbf{g}_+(\mathbf{x}(s, \mathbf{a})) \, ds \right| \leq \int_t^\infty |\Phi_+(t - s) \mathbf{g}_+(\mathbf{x}(s, \mathbf{a}))| \, ds$$

$$\leq \int_t^\infty C e^{\alpha(t - s)} \eta \delta \, ds = \eta \delta \int_0^\infty C e^{-\alpha u} \, du = C \frac{\eta \delta}{\alpha} \qquad (**)$$

Always assume that η is small enough that $\eta C / \alpha < 1/12$, and $|\mathbf{a}| < \delta/2$. Also, if $|\mathbf{x}(t, \mathbf{a})| < \delta$ for all t, then

$$\left| \int_0^t \Phi_-(t - s) \mathbf{g}_-(\mathbf{x}(s, \mathbf{a})) \, ds \right| \leq \int_0^t C e^{-(\alpha + \gamma)(t - s)} \eta \delta \, ds \leq C \frac{1}{\alpha + \gamma} \eta \delta$$

so this term is also less than $\delta/12$.

Consider the initial value problem for the almost linear equation,

$$\mathbf{x}' = A\mathbf{x} + \mathbf{g}(\mathbf{x}), \ \mathbf{x}(0) \equiv \mathbf{a} \equiv \begin{pmatrix} \mathbf{a}_- \\ \mathbf{a}_+ \end{pmatrix}.$$

Then $\mathbf{x}(t, \mathbf{a})$ is a bounded solution, $|\mathbf{x}(t, \mathbf{a})|$ bounded by δ, if and only if

$$\mathbf{x}(t, \mathbf{a}) = \begin{pmatrix} \Phi_-(t) & 0 \\ 0 & \Phi_+(t) \end{pmatrix} \begin{pmatrix} \mathbf{a}_- \\ \mathbf{a}_+ \end{pmatrix}$$

$$+ \int_0^t \begin{pmatrix} \Phi_-(t - s) & 0 \\ 0 & \Phi_+(t - s) \end{pmatrix} \mathbf{g}(\mathbf{x}(s, \mathbf{a})) \, ds \qquad (12.24)$$

and $|\mathbf{x}(t, \mathbf{a})| \leq \delta$ for all $t \geq 0$.

Writing (12.24) differently yields

$$\mathbf{x}(t, \mathbf{a}) = \begin{pmatrix} \Phi_-(t) & 0 \\ 0 & \Phi_+(t) \end{pmatrix} \begin{pmatrix} \mathbf{a}_- \\ \mathbf{a}_+ \end{pmatrix} + \begin{pmatrix} \int_0^t \Phi_-(t - s) \mathbf{g}_-(\mathbf{x}(s, \mathbf{a})) \, ds \\ 0 \end{pmatrix}$$

$$+ \begin{pmatrix} 0 \\ \int_0^t \Phi_+(t - s) \mathbf{g}_+(\mathbf{x}(s, \mathbf{a})) \, ds \end{pmatrix}$$

$$= \begin{pmatrix} \Phi_-(t) & 0 \\ 0 & \Phi_+(t) \end{pmatrix} \begin{pmatrix} \mathbf{a}_- \\ \mathbf{a}_+ \end{pmatrix} + \begin{pmatrix} \int_0^t \Phi_-(t - s) \mathbf{g}_-(\mathbf{x}(s, \mathbf{a})) \, ds \\ 0 \end{pmatrix}$$

$$+ \begin{pmatrix} 0 \\ \int_0^\infty \Phi_+ (t - s) \, \mathbf{g} (\mathbf{x} (s, \mathbf{a})) \, ds - \int_t^\infty \Phi_+ (t - s) \, \mathbf{g} (\mathbf{x} (s, \mathbf{a})) \, ds \end{pmatrix}.$$

These improper integrals converge thanks to the assumption that $\mathbf{x} (t)$ is bounded and the estimates (12.20) and (12.21). Continuing the rewriting,

$$\begin{pmatrix} \mathbf{x}_- (t) \\ \mathbf{x}_+ (t) \end{pmatrix} = \begin{pmatrix} \left(\Phi_- (t) \, \mathbf{a}_- + \int_0^t \Phi_- (t - s) \, \mathbf{g}_- (\mathbf{x} (s, \mathbf{a})) \, ds \right) \\ \Phi_+ (t) \left(\mathbf{a}_+ + \int_0^\infty \Phi_+ (-s) \, \mathbf{g}_+ (\mathbf{x} (s, \mathbf{a})) \, ds \right) \end{pmatrix}$$

$$+ \begin{pmatrix} 0 \\ - \int_t^\infty \Phi_+ (t - s) \, \mathbf{g}_+ (\mathbf{x} (s, \mathbf{a})) \, ds \end{pmatrix}.$$

It follows from Lemma 12.8.2 and $**$ that if $\mathbf{x} (t, \mathbf{a})$ is a solution to the initial value problem which is bounded by δ if and only if $\mathbf{a}_+ + \int_0^\infty \Phi_+ (-s) \, \mathbf{g}_+ (\mathbf{x} (s, \mathbf{a})) \, ds = \mathbf{0}$. Otherwise you would not have $\mathbf{x}_+ (t)$ bounded. Indeed, by $**$ the last term would be bounded and the lemma would imply that $\Phi_+ (t) \left(\mathbf{a}_+ + \int_0^\infty \Phi_+ (-s) \, \mathbf{g}_+ (\mathbf{x} (s, \mathbf{a})) \, ds \right)$ is unbounded. In particular, $|\mathbf{x} (s, \mathbf{a})|$ could not remain less than δ.

It follows that $\mathbf{x} (t, \mathbf{a})$ is a solution to the initial value problem which is bounded by δ if and only if

$$\mathbf{x} (t, \mathbf{a}) = \Phi (t) \begin{pmatrix} \mathbf{a}_- \\ \mathbf{0} \end{pmatrix} + \begin{pmatrix} \int_0^t \Phi_- (t - s) \, \mathbf{g} (\mathbf{x} (s, \mathbf{a})) \, ds \\ - \int_t^\infty \Phi_+ (t - s) \, \mathbf{g} (\mathbf{x} (s, \mathbf{a})) \, ds \end{pmatrix} \qquad (12.25)$$

So what if you have the above condition holding with

$$\mathbf{a}_+ + \int_0^\infty \Phi_+ (-s) \, \mathbf{g}_+ (\mathbf{x} (s, \mathbf{a})) \, ds = \mathbf{0}?$$

Then you can go backwards in the above argument and obtain $\mathbf{x} (t, \mathbf{a})$ is a solution to the initial value problem. In addition, the above estimates $\eta C / \alpha < 1/12$, $|\mathbf{a}| < \delta / 2$ imply that $|\mathbf{x} (t, \mathbf{a})|$ is bounded by δ.

For $\mathbf{x} \in E_\gamma$, used in Lemma 11.3.8, and $|\mathbf{a}_-| < \frac{\delta}{2C}$,

$$F\mathbf{x} (t) \equiv \left(\Phi_- (t) \, \mathbf{a}_- + \int_0^t \Phi_- (t - s) \, \mathbf{g}_- (\mathbf{x} (s)) \, ds - \int_t^\infty \Phi_+ (t - s) \, \mathbf{g}_+ (\mathbf{x} (s)) \, ds \right).$$

I need to find a fixed point of F. Letting $\|\mathbf{x}\|_\gamma \leq \delta$, it follows that $\sup_{t \geq 0} |e^{\gamma t} \mathbf{x} (t)| < \delta$ From estimates of Lemma 12.8.2,

$$e^{\gamma t} |F\mathbf{x} (t)| \leq e^{\gamma t} |\Phi_- (t) \, \mathbf{a}_-| + e^{\gamma t} \int_0^t C e^{-(\alpha + \gamma)(t - s)} \eta \, |\mathbf{x} (s)| \, ds$$

$$+ e^{\gamma t} \int_t^\infty C e^{\alpha (t - s)} \eta \, |\mathbf{x} (s)| \, ds$$

$$\leq e^{\gamma t} C \frac{\delta}{2C} e^{-(\alpha + \gamma) t} + e^{\gamma t} \|\mathbf{x}\|_\gamma C \eta \int_0^t e^{-(\alpha + \gamma)(t - s)} e^{-\gamma s} ds$$

$$+ e^{\gamma t} C \eta \int_t^\infty e^{\alpha (t - s)} e^{-\gamma s} ds \, \|\mathbf{x}\|_\gamma$$

$$< \frac{\delta}{2} + \delta C \eta \int_0^t e^{-\alpha (t - s)} ds + C \eta \delta \int_t^\infty e^{(\alpha + \gamma)(t - s)} ds$$

$$< \frac{\delta}{2} + \delta C \eta \frac{1}{\alpha} + \frac{\delta C \eta}{\alpha + \gamma} \leq \delta \left(\frac{1}{2} + \frac{C \eta}{\alpha} \right) < \delta \left(\frac{2}{3} \right)$$

Thus F maps every $\mathbf{x} \in E_\gamma$ having $||\mathbf{x}||_\gamma \leq \delta$ to $F\mathbf{x}$ where $||F\mathbf{x}||_\gamma \leq \frac{2\delta}{3}$.

Now let $\mathbf{x}, \mathbf{y} \in E_\gamma$ where $||\mathbf{x}||_\gamma, ||\mathbf{y}||_\gamma < \delta$. Then

$$
\begin{aligned}
e^{\gamma t} |F\mathbf{x}(t) - F\mathbf{y}(t)| \quad &\leq \quad e^{\gamma t} \int_0^t |\Phi_-(t-s)| \, \eta e^{-\gamma s} e^{\gamma s} |\mathbf{x}(s) - \mathbf{y}(s)| \, ds \\
&\quad + e^{\gamma t} \int_t^\infty |\Phi_+(t-s)| \, e^{-\gamma s} e^{\gamma s} \eta |\mathbf{x}(s) - \mathbf{y}(s)| \, ds
\end{aligned}
$$

$$
\leq C\eta \, ||\mathbf{x} - \mathbf{y}||_\gamma \left(\int_0^t e^{-\alpha(t-s)} ds + \int_t^\infty e^{(\alpha+\gamma)(t-s)} ds \right)
$$

$$
\leq C\eta \left(\frac{1}{\alpha} + \frac{1}{\alpha+\gamma} \right) ||\mathbf{x} - \mathbf{y}||_\gamma < \frac{2C\eta}{\alpha} ||\mathbf{x} - \mathbf{y}||_\gamma < \frac{1}{3} ||\mathbf{x} - \mathbf{y}||_\gamma \, .
$$

It follows from Theorem 11.3.6, for each $\mathbf{a}_-$ such that $|\mathbf{a}_-| < \frac{\delta}{2C}$, there exists a unique solution to (12.25) in E_γ, the unique fixed point of F.

Thus, in particular, this fixed point $\mathbf{x}(\cdot)$ is bounded by δ. It follows from the argument given above that for $\mathbf{a}$ the initial condition of this fixed point of F,

$$
\mathbf{a}_+ = - \int_0^\infty \Phi_+(-s) \, \mathbf{g}_+(\mathbf{x}(s, \mathbf{a})) \, ds
$$

Define $\psi(\mathbf{a}_-) \equiv \mathbf{a}_+$, the initial condition of the fixed point of F which is determined by $\mathbf{a}_-$. Then for an initial condition $\mathbf{a} \equiv (\mathbf{a}_-, \psi(\mathbf{a}_-))$ and $|\mathbf{a}_-|$ sufficiently small, the solution $\mathbf{x}(t, \mathbf{a})$ having this initial condition satisfies

$$
|\mathbf{x}(t, \mathbf{a})| \leq \delta e^{-\gamma t}.
$$

However, as shown above, small initial data which are not of this form have the property that $|\mathbf{x}(t, \mathbf{a})|$ cannot remain less than δ. ∎

The following theorem is the main result. It involves a use of linear algebra and the above lemma.

Recall that if S is an $n \times n$ matrix and $||S|| \equiv \max\{|S_{ij}|, i, j\}$, then

$$
\begin{aligned}
|S\mathbf{x}| \quad &\equiv \quad \left(\sum_{i=1}^n \left| \sum_{i=1}^n S_{ij} x_j \right|^2 \right)^{1/2} \leq \left(\sum_{i=1}^n \left| \sum_{i=1}^n ||S|| \, |\mathbf{x}| \right|^2 \right)^{1/2} \\
&= \quad \left(\sum_{i=1}^n (n \, ||S|| \, |\mathbf{x}|)^2 \right)^{1/2} \leq n \, ||S|| \, |\mathbf{x}|
\end{aligned}
$$

Theorem 12.8.4 *Consider the initial value problem for the almost linear system*

$$
\mathbf{x}' = A\mathbf{x} + \mathbf{g}(\mathbf{x}), \quad \mathbf{x}(0) = \mathbf{x}_0
$$

in which $\mathbf{g}$ is C^1 and $\mathbf{o}(\mathbf{x})$ or more generally, for every $\eta > 0$ there exists $\delta > 0$ such that if $|\mathbf{x}|, |\mathbf{y}| < \delta$, then

$$
|\mathbf{g}(\mathbf{x}) - \mathbf{g}(\mathbf{y})| < \eta |\mathbf{x} - \mathbf{y}|,
$$

and where there are $k < n$ eigenvalues of A which have negative real parts and $n - k$ eigenvalues of A which have positive real parts. Then $\mathbf{0}$ is not stable. More precisely, there exists a set of points $(\mathbf{a}, \boldsymbol{\psi}(\mathbf{a}))$ for $\mathbf{a}$ small and in a k dimensional subspace such that for $\mathbf{x}_0$ on this set,

$$\lim_{t \to \infty} \mathbf{x}(t, \mathbf{x}_0) = \mathbf{0}$$

and for $\mathbf{x}_0$ not on this set, there exists a $\delta > 0$ such that $|\mathbf{x}(t, \mathbf{x}_0)|$ cannot remain less than δ for all positive t.

Proof: This involves nothing more than a reduction to the situation of Lemma 12.8.3. From Theorem D.10.2 on Page 469 A is similar to a matrix of the form described in Lemma 12.8.3. Thus $A = S^{-1} \begin{pmatrix} A_- & 0 \\ 0 & A_+ \end{pmatrix} S$. Letting $\mathbf{y} = S\mathbf{x}$, it follows

$$\mathbf{y}' = \begin{pmatrix} A_- & 0 \\ 0 & A_+ \end{pmatrix} \mathbf{y} + \mathbf{g}\left(S^{-1}\mathbf{y}\right)$$

Now $|\mathbf{x}| = |S^{-1}S\mathbf{x}| \le n \|S^{-1}\| \, |\mathbf{y}|$ and $|\mathbf{y}| = |SS^{-1}\mathbf{y}| \le n \|S\| \, |\mathbf{x}|$. Therefore,

$$\frac{1}{n\|S\|} |\mathbf{y}| \le |\mathbf{x}| \le n \|S^{-1}\| \, |\mathbf{y}|.$$

It follows all conclusions of Lemma 12.8.3 are valid for this theorem. In particular $\mathbf{x}(t) \to \mathbf{0}$ if and only if $\mathbf{y}(t) \to \mathbf{0}$ and the two have the same stability properties. ∎

The points $S^{-1}(\mathbf{a}, \boldsymbol{\psi}(\mathbf{a}))$ for small $\mathbf{a} \in \mathbb{C}^k$ is called the stable manifold. Actually, it is even more interesting to consider something called the center unstable manifold because it turns out that all solutions are in a sense attracted to it. We will not get into this in this book. An equilibrium point for an almost linear system in which there are no eigenvalues with zero real parts is called hyperbolic.

As to stability, what if you have eigenvalues which equal 0 along with eigenvalues which have positive real parts and eigenvalues which have negative real part? Will the $\mathbf{0}$ equilibrium point still be unstable? The answer is yes. This is an easy corollary to the above theorem. First note that all of the above works just as well if $\mathbf{g}(\mathbf{x}(t))$ is replaced with $e^{-\varepsilon t}\mathbf{g}(e^{\varepsilon t}\mathbf{x}(t))$ and you are considering a solution involving this instead of $\mathbf{g}(\mathbf{x})$.

Corollary 12.8.5 *Consider the initial value problem for the almost linear system*

$$\mathbf{x}' = A\mathbf{x} + \mathbf{g}(\mathbf{x}), \quad \mathbf{x}(0) = \mathbf{x}_0 \tag{12.26}$$

in which $\mathbf{g}$ is C^1 and $o(\mathbf{x})$ or more generally, for every $\eta > 0$ there exists $\delta > 0$ such that if $|\mathbf{x}|, |\mathbf{y}| < \delta$, then

$$|\mathbf{g}(\mathbf{x}) - \mathbf{g}(\mathbf{y})| < \eta |\mathbf{x} - \mathbf{y}|,$$

and where A has some eigenvalue which has positive real part. Then $\mathbf{0}$ is not stable.

Proof: There are $k < n$ eigenvalues of A which have negative real parts and $n-k$ eigenvalues of A which have nonnegative real parts. Consider $\mathbf{y}(t) \equiv (e^{-\varepsilon t}\mathbf{x}(t))$ where ε is very small and $\mathbf{x}$ is the solution to (12.26). Then

$$\mathbf{y}'(t) = -\varepsilon e^{-\varepsilon t}\mathbf{x}(t) + e^{-\varepsilon t}\mathbf{x}'(t)$$

$$= -\varepsilon e^{-\varepsilon t} \mathbf{x}(t) + e^{-\varepsilon t} (A\mathbf{x} + \mathbf{g}(\mathbf{x}))$$

Writing in terms of $\mathbf{y}$,

$$\mathbf{y}'(t) = (A - \varepsilon I)\mathbf{y}(t) + e^{-\varepsilon t}\mathbf{g}(e^{\varepsilon t}\mathbf{y}(t))$$

Then for small positive ε, the eigenvalues of $A - \varepsilon I$ have real parts which are either positive or negative and at least one which is positive. By Theorem 12.8.4, $\mathbf{0}$ is not stable for $\mathbf{y}$. Thus for $\delta > 0$, if $|\mathbf{x}_0| < \delta$, $|\mathbf{y}(t, \mathbf{x}_0)| = |e^{-\varepsilon t}\mathbf{x}(t, \mathbf{x}_0)|$ does not stay less than δ. Therefore, $|\mathbf{x}(t, \mathbf{x}_0)|$ also does not stay less than δ because it is larger than $|\mathbf{y}(t, \mathbf{x}_0)|$. $\blacksquare$

This corollary completes the proof of Theorem 12.4.6.

12.9 Exercises

1. Recall the Lotka-Volterra equations in which the parameters are taken as 1

$$\begin{pmatrix} x \\ y \end{pmatrix}' = \begin{pmatrix} x - xy \\ xy - y \end{pmatrix}$$

and how the stable point is at $(1, 1)$ which is a center. Change it slightly. Instead, consider the system which is slightly perturbed.

$$\begin{pmatrix} x \\ y \end{pmatrix}' = \begin{pmatrix} x - xy + .1 \\ xy - y \end{pmatrix}$$

Use MATLAB to graph a solution to the above system. Show that solutions spiral in towards the equilibrium point. **Hint:** Use something like the following.

g=@(t,y) [y(1)-y(1).*y(2)+.1;y(1).*y(2)-y(2)];
[a,b]=meshgrid(0:.2:3,0:.2:2.4);
u=(a-a.*b+.1); v=(a.*b-b); r=(u.^2+v.^2).^(1/2)
hold on
quiver(a,b,u./r ,v./r,'autoscalefactor',.5)
[t,y]=ode45(g,[0:.03:60],[2;2]);
plot(y(:,1),y(:,2),'linewidth',1.5,'color','red')

Notice also the way the vectors are normalized. Since u,v are vectors, you have to do .*,.^and so forth to ensure that operations are done on the entries of the vector which is what you desire to happen. This is one of the peculiarities of MATLAB that it allows you to do this. Also note that in the ode part, the step size is .03 to be sure that the curves look nice and smooth. Be sure you place a semicolon after the ode line so that you don't get a monstrous table of values.

2. In the above problem, what is the equilibrium point of the system in which the .1 was added in? For this equilibrium point (a, b), let $x = a + u, y = b + v$ and write the system in terms of u, v to get a system which has $(0, 0)$ as its

equilibrium point. Now show this is an almost linear system $\mathbf{x}' = A\mathbf{x} + \mathbf{g}(\mathbf{x})$ in which the eigenvalues of the matrix A are complex and have negative real parts. What does this say about the stability of the equilibrium point?

3. A more general model of predator prey interactions is the following system of equations in which all constants are positive but ε, δ are very small relative to the other constants.

$$x' = ax - bxy - \varepsilon x^2$$
$$y' = -cy + dxy - \delta y^2$$

Find an equilibrium point other than $(0, 0)$. Determine its stability by consideration of eigenvalues. In doing so, recall that the determinant is the product of the eigenvalues and the trace is the sum of the eigenvalues. You don't really care about finding the eigenvalues exactly, just whether their real parts are negative. Graph solutions to the system when $a = b = c = d = 1$ and $\varepsilon = \delta = .1$. Use a modification of the steps given in the above problem. Note that you couldn't see whether the equilibrium point was stable from looking at the direction field of vectors although the graph of the solutions to the equations do show that this is the case.

4. Now graph the vector field associated with the Lotka-Volterra equations. From this vector field, explain why the solutions which start off in $[0, 4] \times [0, 4]$ appear to stay in this bounded region. What does the Poincare Bendixon theorem allow you to conclude based on this?

```
[a,b]=meshgrid(0:.2:4,0:.2:4);
u=(a-a.*b); v=(a.*b-b);
figure
quiver(a,b,u./(u.^2+v.^2+.1).^(1/2),v./(u.^2+v.^2+.1).^(1/2))
```

5. Pictures are not entirely satisfactory. The Lotka-Volterra equations are

$$x' = ax - bxy$$
$$y' = cxy - dy$$

where a, b, c, d are positive constants. This problem is a review of what is in the text. Do the following:

$$x' = (xy)\left(\frac{a}{y} - b\right)$$
$$y' = (xy)\left(c - \frac{d}{x}\right)$$

Now consider antiderivatives of $\left(\frac{a}{y} - b\right), \left(c - \frac{d}{x}\right)$

$$a \ln y - by, \quad cx - d \ln x$$

Show that, from the equations,

$$[(a \ln y - by) - (cx - d \ln x)]' = 0$$

Therefore,

$$\ln\left(y^a x^d\right) - (cx + by) = \text{constant}$$

For $x, y > 0$ this function on the left is differentiable. Show that it has the form of a mountain which has its summit in the set where $x, y > 0$. Explain why the level curves of this function of two variables give the solution curves of the Lotka-Volterra system and show that the maximum of this mountain occurs at the non- zero equilibrium point of the above equations. Obtain a graph of this function of two variables using MATLAB. You type something like

```
[x,y]=meshgrid(.1:.5:30,.1:.5:30);
z=log(y.^2.*x.^3)-(.5.*x+.6.*y); surf(x,y,z)
```

6. In the Lotka-Volterra equations above, which gives the number of predators, x or y? Explain why these are reasonable equations for modeling the interaction between predators and prey.

7. Consider the following system of equations which models two competing species.

$$\begin{pmatrix} x \\ y \end{pmatrix}' = \begin{pmatrix} x - \frac{1}{3}xy - \frac{1}{3}x^2 \\ y - xy - y^2 \end{pmatrix} = \begin{pmatrix} x\frac{(3-(x+y))}{3} \\ y\left(1 - (x+y)\right) \end{pmatrix} \qquad (*)$$

Show by graphing solutions of these equations that y must become extinct and that x eventually becomes 3 in the limit.

```
[a,b]=meshgrid(0:.2:5,0:.2:5);u=a-((1/3).*a.*b+(1/3).*a.^2);
v=b-(a.*b+b.^2);r=(u.^2+v.^2+.1).^(1/2);
figure
quiver(a,b,u./r,v./r)
f=@(t,x)[x(1)-((1/3)*x(1)*x(2)+(1/3)*x(1)^2);x(2)-(x(2)*x(1)+x(2)^2)];
hold on
for z=0:.3:5   [t,y]=ode45(f,[0:.04:30],[z;3]);
plot(y(:,1),y(:,2),'linewidth',1.6,'color','red')
end
```

Thus, if the maximum possible population for x is 3 and the maximum possible for y is 1, then eventually, x will approach its maximum and y will die out. Why is 3 the maximum possible for x and why is 1 the maximum possible for y? How is this system of equations similar to the logistic equation in one dimension? Explain why it really does model competing species.

8. In the above problem, divide the first quadrant into regions according to whether x', y' are positive. Thus you will consider the regions in the first quadrant determined by $y + x = 3, x + y = 1$. Sketch a graph of these regions and and label them according to whether x', y' are positive. Explain why the above result happens. More generally, you could have the following system.

$$\begin{pmatrix} x \\ y \end{pmatrix}' = \begin{pmatrix} \alpha x\frac{(K-(x+y))}{K} \\ \beta y\frac{(L-(x+y))}{L} \end{pmatrix}$$

where $K > L, \alpha, \beta > 0$. Explain why eventually y will become extinct. Show that any solution to the equations which begins in the region between $x + y = K$ and $x + y = L$ will stay in this region.

9. As pointed out, the nonlinear pendulum

$$x' = y$$
$$y' = -\sin x$$

has a center at the origin. Modify it slightly by introducing a friction term.

$$x' = y$$
$$y' = -\alpha y - \sin x$$

where α is a small positive number. Show that the origin is asymptotically stable for this system. Give graphs for various values of α. To do this, try something like the following.

hold on
for a=0:.2:2
ga=@(t,y)[y(2);-(a*y(2)+sin(y(1)))];
[t,y]=ode45(ga,[0:.05:30],[-.2;-.2]);
plot(y(:,1),y(:,2))
end

Explain what the resulting graph shows. Note that it is graphing solutions to different differential equations and so there is nothing which says that the curves cannot cross each other.

10. If you want something even prettier, try this

hold on
for a=0:.2:2
for b=0:.3:1
ga=@(t,y)[y(2);-(a*y(1)+sin(y(2)))];
[t,y]=ode45(ga,[0:.05:30],[-b;-b]);
plot(y(:,1),y(:,2))
end
end

what is the meaning of what you are getting in the resulting graph? Note that the case where $a = 0$ is included.

11. The pictures are nice, but can you prove that the origin is asymptotically stable for that frictionally damped pendulum?

12. Show that $(0, 0)$ is a center for the system

$$\begin{pmatrix} x \\ y \end{pmatrix}' = \begin{pmatrix} -4y^3 \\ 2x \end{pmatrix}$$

Graph several solutions to this. **Hint:** Try something like this in MATLAB:

```
hold on
for a=.5:.2:2
f=@(t,x)[-4*x(2)^3;2*x(1)];
[ts,xs]=ode45(f,[0,10],[a;a]);
plot(xs(:,1),xs(:,2))
end
```

Find a Lyapunov function for this equation.

13. Now consider the system

$$\begin{pmatrix} x \\ y \end{pmatrix}' = \begin{pmatrix} -4y^3 - x \\ 2x \end{pmatrix}$$

Is $(0,0)$ stable? Consider graphing solutions

```
hold on
for a=-2:.2:2
f=@(t,x)[-4*x(2)^3-x(1);2*x(1)];
[ts,ys]=ode45(f,[0,10],[a;a]);
plot(ys(:,1),ys(:,2))
end
```

Now recall that the trajectories of autonomous systems cannot cross. Also look for a Lyapunov function. Consider the limit set. Why can't you determine stability by looking at eigenvalues of the linearization of the system?

14. Show the following system has a center at $(0,0)$.

$$\begin{pmatrix} x \\ y \end{pmatrix}' = \begin{pmatrix} -(y + x^2 y) \\ x + xy^2 \end{pmatrix}$$

Find a Lyapunov function. You could also use MATLAB to graph the trajectories. for small initial conditions.

15. Show that $(0,0)$ is stable for the system

$$\begin{pmatrix} x \\ y \end{pmatrix}' = \begin{pmatrix} -(y + x^2 y + x) \\ x + xy^2 \end{pmatrix}$$

Why can't you determine stability by looking at eigenvalues? **Hint:** You might find a Lyapunov function. Maybe try $x^2 + y^2 + x^2 y^2$ and consider a limit set. You might also graph the trajectories and draw conclusions that way using the fact that they don't cross.

16. Recall the Van der Pol equation. As a first order system, it has the form

$$\begin{pmatrix} x \\ y \end{pmatrix}' = \begin{pmatrix} y \\ -(x^2 - 1)^2 y + x \end{pmatrix}$$

where the dependent variable of interest is x. Then use the following to obtain a graph of a trajectory which shows fairly convincingly that there must be a limit cycle. Recall that the graphs of solutions cannot cross.

[a,b]=meshgrid(-3:.2:3,-3:.2:3); p=b; q=-((a.^2-1).*b+a);
r=(p.^2+q.^2).^(1/2); u=p./r; v=q./r; hold on
quiver(a,b,u,v,'autoscalefactor',.5)
f=@(t,x)[x(2);-((x(1)^2-1)*x(2)+x(1))];[t,y]=ode45(f,[0,10],[3;3]);
plot(y(:,1),y(:,2),'linewidth',1.5)

17. Consider the system of equations

$$\left(\begin{array}{c} x \\ y \end{array}\right)' = \left(\begin{array}{c} x^3 - \sin y + x \\ y + \ln\left(1 + x^2\right) - \sin\left(\cos\left(x^2\right)\right) \end{array}\right)$$

Show that it has no periodic solution.

18. Consider the system of equations

$$\left(\begin{array}{c} x \\ y \end{array}\right)' = \left(\begin{array}{c} ax^3 + f(y) \\ by^5 + g(x) \end{array}\right), ab > 0$$

Show that if it has a periodic solution, then the graph of this periodic solution must contain $(0,0)$ on the bounded region determined by this graph. That is, the graph must go around the origin.

19. Determine stability of the following systems at the given equilibrium points. When you have an equilibrium point $\mathbf{a}$, you should replace $\mathbf{x}$ with $\mathbf{a} + \mathbf{u}$ and then look at the resulting system in terms of $\mathbf{u}$ which has an equilibrium point at $\mathbf{0}$. Then use the above theory about eigenvalues to try and determine stability.

(a)
$$x' = yx^2 + 2y$$
$$y' = y^3 + 3y - x$$
$$(0,0)$$

(b)
$$x' = yx^2 - 2yx + 3y$$
$$y' = y^2 + 3y - x + 1$$
$$(1,0)$$

(c)
$$x' = x^2 + 5x - 6y + 6$$
$$y' = 4x - 4y + xy^2 - 2xy + 4$$
$$(0,1), (1,2)$$

(d)
$$x' = x^2 - x - 2y$$
$$y' = x - 4y + xy$$
$$(2,1), (0,0), (3,3)$$

(e)
$$x' = y^2 - 4y + x$$
$$y' = 2x - 5y + xy$$
$$(-5,-1), (0,0), (3,3)$$

(f)
$$x' = y^2 + 2y - 5x$$
$$y' = xy - 2y - x$$
$$(0,0), (3,3), \left(\frac{8}{5}, -4\right)$$

(g)
$$x' = 2y - 2z + xy$$
$$y' = y^2 - 5y - 2x + 4z$$
$$z' = z - 2y - x + xz$$
$$\left(1,1,\frac{3}{2}\right), (-2,1,0), (0,0,0)$$

(h)
$$x' = x^2 + 2y - 2z$$
$$y' = 16z - 13y - 6x + xy$$
$$z' = 13z - 10y - 5x + xz$$
$$(0,0,0), (2,-4,-2), \left(1,-\frac{1}{2},0\right)$$

20. In Theorem 12.7.2, use Corollary E.2.32 of the appendix to show that it suffices to say that $(Rf)_x + (Rg)_y \neq 0$.

21. Consider
$$\begin{pmatrix} x \\ y \end{pmatrix}' = \begin{pmatrix} \mu x + y - x\left(x^2 + 2y^2\right) \\ -x + \mu y - y\left(2x^2 + y^2\right) \end{pmatrix}$$
It has an equilibrium point at $(0,0)$. Show that this equilibrium point is stable if $\mu < 0$ and unstable if $\mu > 0$. Argue from the right side that there is a large ball centered at $(0,0)$ such that every solution starting at a point in this ball remains in the ball. Show that the only equilibrium point is $(0,0)$ regardless of μ. Explain why for $\mu > 0$ there is a limit cycle which attracts solutions. Thus $(0,0)$ looses its stability as μ increases through 0 and this stability becomes attached to an appropriate limit cycle. Graph direction fields for $\mu = -.2, -.1, 0, .1, .2$. This phenomenon is called Hopf bifurcation.[‖]

22. Suppose $A(\mu)$ is a 2×2 matrix and consider the almost linear system
$$\begin{pmatrix} x \\ y \end{pmatrix}' = A(\mu) \begin{pmatrix} x \\ y \end{pmatrix} + \begin{pmatrix} f(x,y) \\ g(x,y) \end{pmatrix}$$
where $A(0)$ has eigenvalues with 0 real part while the eigenvalues of $A(\mu)$ are distinct and have negative real parts for $\mu < 0$ and positive real parts for $\mu > 0$. Suppose also that there is only one equilibrium point which is at $(0,0)$ and that there exists a circle of radius R called C which is centered at $(0,0)$ such that if (x,y) is on C, then
$$\left(A(\mu) \begin{pmatrix} x \\ y \end{pmatrix} + \begin{pmatrix} f(x,y) \\ g(x,y) \end{pmatrix} \right) \cdot \begin{pmatrix} x \\ y \end{pmatrix} < 0$$
Show that $(0,0)$ is stable for $\mu < 0$ and unstable for $\mu > 0$ and for $\mu > 0$, then for each $(x_0, y_0) \in B\left((0,0), R\right)$ but not equal to $(0,0)$,
$$t \to (x(t, x_0, y_0), y(t, x_0, y_0))$$
must either be a periodic orbit or it must spiral towards a limit cycle.

23. This and some of the other problems are in [5]. Rayleigh's equation is
$$y'' - \mu\left(1 - \frac{1}{3}(y')^2\right)y' + y = 0, \quad \mu > 0$$

   (a) Write as a first order system and use MATLAB to graph a vector field for $\mu = 1$.

   (b) By consideration of eigenvalues, determine stability for the origin in this first order system independent of μ.

   (c) Determine whether there exists a limit cycle which attracts solutions beginning in some region.

[‖] Eberhard Frederich Ferdinand Hopf, (1902-1983) is probably best known for the Hopf bifurcation because of his work on it in the early 1940s, but he also made major contributions to elliptic partial differential equations, in particular the Hopf maximum principal.

(d) If there is a limit cycle, graph y for initial conditions close to a point on the limit cycle and estimate the period of a resulting periodic solution.

24. Certain chemical reactions lead to first order systems of equations. One such is of the form
$$\begin{pmatrix} x \\ y \end{pmatrix}' = \begin{pmatrix} 1 - (b+1)x + x^2y/4 \\ bx - x^2y/4 \end{pmatrix}$$

(a) Find the equilibrium point.

(b) Determine stability of this equilibrium point.

(c) Graph a solution which converges to a limit cycle for $b = 2$. **Hint:** Consider the equation in terms of u, v which are displacements away from the equilibrium point. In other words, modify things to that the equilibrium point is at the origin.

25. The Fitzhugh-Nagumo equations model neural impulses. These equations are, following [5]
$$x' = 3\left(x + y - \frac{1}{3}x^3 - k\right), \quad y' = -\frac{1}{3}(x + .8y - .7)$$

Here k is a parameter which is the external stimulus.

(a) Show there is only one equilibrium point.

(b) Let (a, b) be the equilibrium point for k. Then let $x = a + u, y = b + v$. Find the equation in terms of u, v. Next determine stability according to the values of a.

(c) What values of k correspond to these values of a?

26. Graph the direction field for the equations of the above problem for $k = 1$. Explain why there must exist a limit cycle. Now graph solutions to the equations to see the limit cycle. You can use the following syntax.

[a,b]=meshgrid(-5:.4:5,-5:.4:5);u=3*(a+b-(a.^3)/3-1);

v=-3^(-1)*(a+.8*b+(-.7));r=(u.^2+v.^2).^(1/2);

figure quiver(a,b,u./r,v./r,'autoscalefactor',.5)

f=@(t,x)[3*(x(1)+x(2)-((1/3)*x(1)^3+1)); (-1/3)*(x(1)+.8*x(2)-.7)];

hold on for z=-3:.3:3

[t,x]=ode45(f,[0,10],[z;z]); plot(x(:,1),x(:,2)) end

You put in the new lines where needed. Now do the same thing for $k = 0$ which is outside of the region where the limit cycle and instability of the equilibrium point occurs.

27. The Lorenz equations are
$$\begin{pmatrix} x \\ y \\ z \end{pmatrix}' = \begin{pmatrix} 10(y - x) \\ rx - (y + xz) \\ -\frac{8}{3}z + xy \end{pmatrix}$$

actually, there are more parameters and this problem picked values for two of them to correspond to reasonable values for the earth's atmosphere. Determine stability of $(0, 0, 0)$ for various values of r.

28. For $r = 30$ solve the Lorenz equations numerically and graph $(x(t), y(t)), t \in [0, 20]$ and the initial condition of the form $[3, 3, 3]$. Then graph

$$(t, x(t)), (t, y(t)), (t, z(t)).$$

Recall this is a value of r for which the origin is not stable. Also graph $(x(t), y(t))$ for many initial conditions. Pick initial conditions of the form $[z; z; z]$ where z takes many values between -5 and 5.

29. Here are some equations. Explain why there is no periodic solution.

$$\begin{pmatrix} x \\ y \end{pmatrix}' = \begin{pmatrix} -3 & -4 \\ 2 & 3 \end{pmatrix} \begin{pmatrix} x \\ y \end{pmatrix} + \begin{pmatrix} -(3x + 4y)^3 \\ (2x + 3y)^3 \end{pmatrix}$$

30. Recall the Lotka-Volterra equations in which the parameters are equal to 1.

$$\begin{pmatrix} x \\ y \end{pmatrix}' = \begin{pmatrix} x - xy \\ xy - y \end{pmatrix}$$

There were periodic solutions around the equilibrium point $(1, 1)$ which were in the first quadrant. Show that there are no periodic solutions which are contained in the second, third or fourth quadrants. Also there are no small periodic solutions around $(0, 0)$.

31. As pointed out in the chapter, for A an $m \times n$ matrix,

$$\|A\| \equiv \sup_{|\mathbf{x}| = 1} |A\mathbf{x}| = \max_{|\mathbf{x}| = 1} |A\mathbf{x}| < \infty$$

Explain why for any $\mathbf{x} \in \mathbb{C}^n, |A\mathbf{x}| \leq \|A\| \, |\mathbf{x}|$. Also show that $\|AB\| \leq \|A\| \, \|B\|$.

32. Suppose

$$\mathbf{x}' = A\mathbf{x} + \mathbf{f}(\mathbf{x})$$

is an almost linear system and $\operatorname{Re} \lambda_m > 0, \operatorname{Re} \lambda_1 \leq \operatorname{Re} \lambda_2 \leq \cdots \leq \operatorname{Re} \lambda_m$, $m \leq n$ (There could be repeated eigenvalues). Show that in this case, $\mathbf{0}$ is unstable directly by using the variation of constants formula for the solution.

   (a) If $\mathbf{x}_0$ is given, explain why the solution is

$$\mathbf{x}(t) = \sum_{k=0}^{m-1} r_{k+1}(t) P_k(A) \mathbf{x}_0 + \sum_{k=0}^{m-1} \int_0^t r_{k+1}(t - s) P_k(A) \mathbf{f}(\mathbf{x}(s)) \, ds \tag{*}$$

   where $r_k(t) = p_k(t) e^{\lambda_k t}$.

   (b) Explain why $\mathbf{x}_0$ can be chosen such that $P_{m-1}(A) \mathbf{x}_0 \neq \mathbf{0}$.

(c) If $|\mathbf{x}(s)| \leq \varepsilon$ for all $s > 0$, where ε is sufficiently small in comparison to $P_k(A)\mathbf{x}_0$, argue that the highest order terms from the integral are dominated by the highest order terms of $r_m(t)P_{m-1}(A)\mathbf{x}_0$. Thus it can't happen that $|\mathbf{x}(s)| < \varepsilon$ for all s. Hence $\mathbf{0}$ cannot be stable.

   i. Let $\delta < \frac{\operatorname{Re}\lambda_m}{5}$ and let ε be small enough that if $|\mathbf{x}| < \varepsilon$, then $|\mathbf{f}(\mathbf{x})| \leq \delta|\mathbf{x}|$.

   ii. Use the part a. and reduce attention to those terms which have exponential $e^{\lambda_m t}$ times t^p where p is the largest of all exponents in any of the polynomials occurring in the definition of $r_k(t)$. Recall these are of the form $p_k(t)e^{\lambda_k t}$ and the degrees of the p_k are increasing in k.

   iii. Let $C = \sup\{|P_{m-1}(A)\mathbf{x}| : |\mathbf{x}| = 1\}$ and choose $\mathbf{x}_0, |\mathbf{x}_0| \in \left(\frac{\varepsilon}{2}, \varepsilon\right)$ with
$$|P_{m-1}(A)(\mathbf{x}_0/|\mathbf{x}_0|)| > C/2$$
so
$$|P_{m-1}(A)(\mathbf{x}_0)| > C|\mathbf{x}_0|/2 > \frac{C\varepsilon}{4}$$

   iv. Argue that $\int_0^t (t-s)^p e^{\lambda_m(t-s)}ds$ is a polynomial of degree p times $e^{\lambda_m t}$ and the coefficient of the t^p term is $(1/\lambda_m)$.

   v. Now obtain a contradiction from an assumption that $|\mathbf{x}(s)| < \varepsilon$ for all $s > 0$ by comparing the highest order terms.

33. The Putzer method, as presented here, came up with the fundamental matrix using the minimal polynomial. That is, $P_k(A) = \prod_{i=1}^k (A - \lambda_i I)$ where the minimal polynomial was $\prod_{i=1}^m (t - \lambda_i)$, $m \leq n$. There could be repeats in the λ_i. The characteristic polynomial was of the form $\prod_{i=1}^n (t - \lambda_i)$ where the eigenvalues are listed according to multiplicity as roots of the characteristic polynomial. Show that the same scheme will yield the fundamental matrix using the characteristic polynomial. It will be $\sum_{k=0}^{n-1} r_{k+1}(t) P_k(A)$ where $P_k(A)$ is defined as above except that individual eigenvalues may now be listed more times. Thus there are more terms in the sum.

Part V

Partial Differential Equations

Chapter 13

Boundary Value Problems, Fourier Series

13.1 Boundary Value Problems

The initial value problem can always be formulated as

$$\mathbf{y}' = A\mathbf{y} + \mathbf{f}, \ \mathbf{y}(a) = \mathbf{y}_0.$$

These are very nice problems because they always have a unique solution. A boundary value problem is different. They don't always have solutions.

Definition 13.1.1 *A two-point boundary value problem is to find a solution* y *to a differential equation*

$$y'' + p(x)y' + q(x)y = g(x), \ x \in [a,b]$$

which also satisfies boundary conditions which are given at the two end points.

Examples of boundary values would be to give the value of y at the end points or the value of y' at the end points or some combination of y and y' at the end points.

Example 13.1.2 *Find the solutions to the equation* $y'' + y = \sin x$, *and boundary conditions*

$$y(0) = 0, y(\pi) = 0$$

The general solution to the differential equation is easily seen to be $A\cos(x) + B\sin(x) - \frac{1}{2}x\cos(x)$. You have find A, B such that the boundary conditions are satisfied. Substituting $t = 0$ yields $A = 0$. Then you also need $B\sin(\pi) - \frac{1}{2}\pi\cos(\pi) = 0$ which is impossible. Therefore, there is no solution to this boundary value problem.

This is a very significant issue because there are numerical methods for solving boundary value problems. These methods will give you an answer even if there isn't one, but if it isn't there, you won't find it. Of course, just because you can't find it does not necessarily mean it isn't there. This is the interesting thing about math. In the absence of good existence and uniqueness theorems, you sometimes don't know what you are getting.

Example 13.1.3 *Find the solutions to the equation $y'' + y = \sin x$, and boundary conditions $y(0) = 0$, $y\left(\frac{\pi}{2}\right) = 0$.*

It is the same equation, but the end points are different. As in the above example, if it has a solution, then it is of the form $B \sin x - \frac{1}{2}x \cos(x)$ Now let $x = \pi/2$ and you find $B - \frac{1}{4}\pi 0 = 0$. Thus a solution to this boundary value problem is $y = -\frac{1}{2}x \cos(x)$

In this example, there was exactly one solution. Next consider

Example 13.1.4 *Find the solutions to the equation $y'' + y = \sin x$, and boundary conditions*

$$y(0) = 0, \ y'\left(\frac{\pi}{2}\right) = \frac{\pi}{4}$$

The general solution to the differential equation is easily seen to be $A \cos(x) + B \sin(x) - \frac{1}{2}x \cos(x)$. You have find A, B such that the boundary conditions are satisfied. Substituting $t = 0$ yields $A = 0$. Thus if there is a solution it is of the form $y = B \sin(x) - \frac{1}{2}x \cos(x)$. Then $y'(x) = B \cos x - \frac{1}{2}\cos x + \frac{1}{2}x \sin x$. Then you also need $\frac{\pi}{4} = y'\left(\frac{\pi}{2}\right) = B \cos\left(\frac{\pi}{2}\right) + 0 + \frac{\pi}{4}$ which happens for any value of B. Therefore, for any $B, y = B \sin(x) - \frac{1}{2}x \cos(x)$ is a solution to this two point boundary value problem.

This is an example of a boundary value problem which has infinitely many solutions. Notice how all three examples involved the same differential equation, just different boundary conditions.

It turns out that for two point boundary value problems it is always this way. Either there are no solutions, exactly one or there are infinitely many. This may look familiar. Recall that it was this way for systems of linear equations. There are profound reasons why this similarity takes place but they are not for a book like this.

13.2 Eigenvalue Problems

I suppose these are best discussed through the example which will be featured most prominently.

Example 13.2.1 *Find the values of λ such that there exist* $\boxed{\textbf{\textit{nonzero}}}$ *solutions to the boundary value problem*

$$\begin{aligned} y'' + \lambda y &= 0 \\ y(0) &= y(L) = 0 \end{aligned}$$

Along with any pair of boundary conditions which satisfy the conditions

$$y(0)\, y'(0) = 0, \ y(L)\, y'(L) = 0$$

Multiply by y and integrate from 0 to L.

$$\int_0^L y'' y\, dx + \lambda \int_0^L y^2 dx = 0 \tag{13.1}$$

Integrate by parts.

$$y'y|_0^L - \int_0^L (y')^2\, dx + \lambda \int_0^L y^2 dx = 0$$

Consider now the boundary term. It equals 0 by assumption. Therefore,

$$- \int_0^L (y')^2\, dx + \lambda \int_0^L y^2 dx = 0$$

If $\lambda < 0$, this equation could not be true and have $y \neq 0$ because it would imply $\int_0^L y^2 dx = 0$ so $y = 0$. Therefore, for any such example, $\lambda \geq 0$.

Case 1: Now consider some cases each of which have the property that yy' equals 0 at the end points of the interval $[0, L]$. First suppose $y = 0$ at the ends of the interval. To save notation, write $\lambda = \mu^2$. Then you want

$$\begin{aligned} y'' + \mu^2 y &= 0 \\ y(0) &= y(L) = 0 \end{aligned}$$

The solution to the differential equation is

$$C_1 \sin \mu x + C_2 \cos \mu x$$

Insert the boundary conditions. This yields

$$C_2 = 0, \ \ C_1 \sin(\mu L) = 0$$

Therefore, for some nonnegative integer n, you must have $\mu L = n\pi$. You can't have $n = 0$ since then $y = 0$ and this is not allowed. Therefore, n is a positive integer and the eigenvalues are

$$\lambda = \frac{n^2 \pi^2}{L^2}, \ n = 1, 2, \cdots$$

The corresponding eigenfunctions are

$$\sin\left(\frac{n\pi}{L} x\right), \ n = 1, 2, \cdots$$

Case 2: Next consider the case where $y' = 0$ at the ends. Thus you want nonzero y and λ such that

$$\begin{aligned} y'' + \mu^2 y &= 0 \\ y'(0) &= y'(L) = 0 \end{aligned}$$

The solution to the differential equation is

$$y = C_1 \sin \mu x + C_2 \cos \mu x$$

Then

$$y' = C_1 \mu \cos \mu x - C_2 \mu \sin \mu x$$

Insert the boundary conditions. At 0 this requires that

$$C_1 \mu = 0$$

At the right end point this requires

$$C_2 \mu \sin(\mu L) = 0$$

One case is for $\mu = 0$. This would result in an eigenfunction

$$y = 1$$

which is a nonzero function. Of course any nonzero multiple of this is also an eigenfunction. If μ is not zero, then you need

$$\mu L = n\pi, \ n = 1, 2, \cdots$$

so

$$\lambda = \frac{n^2 \pi^2}{L^2}, \ n = 0, 1, 2, \cdots$$

The eigenfunctions in this case are

$$1, \ \cos\left(\frac{n\pi}{L} x\right), \ n = 1, 2, \cdots$$

Case 3: Next consider the case where $y(0) = 0$ and $y'(L) = 0$. Thus you want nonzero y and λ such that

$$\begin{aligned} y'' + \mu^2 y &= 0 \\ y(0) &= y'(L) = 0 \end{aligned}$$

In this case, you would have

$$y = C_1 \sin \mu x + C_2 \cos \mu x$$

and on inserting the left boundary condition, this requires that $C_2 = 0$. Now consider the right boundary condition. You can't have $\mu = 0$ in this case, because if you did, you would have $y = 0$ which is not allowed. Hence you have

$$y'(L) = C_1 \mu \cos(\mu L) = 0$$

since $\mu \neq 0$, you must have

$$\mu L = (2n - 1)\pi \text{ for } n = 1, 2, \cdots$$

Therefore, in this case the eigenvalues are

$$\lambda = \frac{(2n - 1)^2 \pi^2}{L^2}, \ n = 1, 2, \cdots$$

and the eigenfunctions are

$$\sin\left(\frac{(2n - 1)\pi}{L} x\right), \ n = 1, 2, \cdots$$

13.3 Fourier Series

A Fourier series is a series which is intended to somehow approximate a given periodic function by an infinite sum of the form

$$a_0 + \sum_{k=1}^{\infty} a_k \cos\left(\frac{k\pi}{L}x\right) + \sum_{k=1}^{\infty} b_k \sin\left(\frac{k\pi}{L}x\right)$$

First of all, what is a periodic function?

Definition 13.3.1 *A function $f : \mathbb{R} \to \mathbb{R}$ is called **periodic** of period T if for all $x \in \mathbb{R}$,*

$$f(x+T) = f(x).$$

An example of a periodic function having period $2L$ is $x \to \sin\left(\frac{k\pi}{L}x\right)$ and $x \to \cos\left(\frac{k\pi}{L}x\right)$. If you want to approximate a function with these periodic functions, then it is necessary that it be periodic of period $2L$. Otherwise it would not be reasonable to expect to be able to approximate the function in any useful way with these periodic functions.

Before doing anything else, here are some important trig. identities.

$$\sin a \cos b = \frac{1}{2}\left(\sin(a+b) + \sin(a-b)\right) \tag{13.2}$$

$$\cos a \cos b = \frac{1}{2}\left(\cos(a-b) + \cos(a+b)\right) \tag{13.3}$$

$$\sin a \sin b = \frac{1}{2}\left(\cos(a-b) - \cos(a+b)\right) \tag{13.4}$$

These follow right away from the standard trig. identities for the sum of two angles. Here is a lemma which gives an orthogonality condition.

Lemma 13.3.2 *The following formulas hold. For m,n positive integers,*

$$\int_{-L}^{L} \frac{1}{\sqrt{L}}\sin\left(\frac{m\pi}{L}x\right)\frac{1}{\sqrt{L}}\sin\left(\frac{n\pi}{L}x\right)dx = \begin{cases} 0 & \text{if } m \neq n \\ 1 & \text{if } m = n \end{cases}$$

$$\int_{-L}^{L} \frac{1}{\sqrt{L}}\cos\left(\frac{m\pi}{L}x\right)\frac{1}{\sqrt{L}}\cos\left(\frac{n\pi}{L}x\right)dx = \begin{cases} 0 & \text{if } m \neq n \\ 1 & \text{if } m = n \end{cases}$$

$$\int_{-L}^{L} \sin\left(\frac{m\pi}{L}x\right)\cos\left(\frac{n\pi}{L}x\right)dx = 0$$

Proof: Consider the first of these formulas. From one of the above trig. identities,

$$\int_{-L}^{L} \sin\left(\frac{m\pi}{L}x\right)\sin\left(\frac{n\pi}{L}x\right)dx =$$

$$\frac{1}{2}\int_{-L}^{L} \cos\left(\left(\frac{m\pi}{L} - \frac{n\pi}{L}\right)x\right) - \cos\left(\left(\frac{m\pi}{L} + \frac{n\pi}{L}\right)x\right)dx$$

If $m \neq n$, this clearly integrates to 0. If $m = n$, you have

$$\frac{1}{2} \int_{-L}^{L} \left(1 - \cos\left(\frac{2n}{L}x\right)\right) dx = L$$

Thus

$$\int_{-L}^{L} \frac{1}{\sqrt{L}} \sin\left(\frac{n\pi}{L}x\right) \frac{1}{\sqrt{L}} \sin\left(\frac{n\pi}{L}x\right) dx = 1$$

The second formula works out the same way. Consider the third.

$$\int_{-L}^{L} \sin\left(\frac{m\pi}{L}x\right) \cos\left(\frac{n\pi}{L}x\right) dx =$$

$$\frac{1}{2} \int_{-L}^{L} \left(\sin\left(\left(\frac{m\pi}{L} + \frac{n\pi}{L}\right)x\right) + \sin\left(\left(\frac{m\pi}{L} - \frac{n\pi}{L}\right)x\right)\right) dx$$

It is easy to see that this integral is always 0 regardless the choice of m, n. ∎

Now suppose you succeed in approximating f with a Fourier series in some meaningful way.

$$f(x) \approx a_0 \frac{1}{\sqrt{2L}} + \sum_{k=1}^{\infty} a_k \frac{1}{\sqrt{L}} \cos\left(\frac{k\pi}{L}x\right) + \sum_{k=1}^{\infty} b_k \frac{1}{\sqrt{L}} \sin\left(\frac{k\pi}{L}x\right) \qquad (13.5)$$

What should be the formula for a_k and b_k? Multiply both sides by $\frac{1}{\sqrt{L}} \sin\left(\frac{m\pi}{L}x\right)$ and then integrate the resulting infinite sum by saying the integral of the sum is the sum of the integrals. Since the sum involves a limit, this is nothing but a formal and highly speculative piece of pseudo mathematical nonsense but we will not let a little thing like that get in the way. Thus

$$\int_{-L}^{L} f(x) \frac{1}{\sqrt{L}} \sin\left(\frac{m\pi}{L}x\right) dx = a_0 \int_{-L}^{L} \frac{1}{\sqrt{2L}} \sin\left(\frac{m\pi}{L}x\right) dx +$$

$$\sum_{k=1}^{\infty} a_k \int_{-L}^{L} \frac{1}{\sqrt{L}} \cos\left(\frac{k\pi}{L}x\right) \frac{1}{\sqrt{L}} \sin\left(\frac{m\pi}{L}x\right) dx$$

$$+ \sum_{k=1}^{\infty} b_k \int_{-L}^{L} \frac{1}{\sqrt{L}} \sin\left(\frac{k\pi}{L}x\right) \frac{1}{\sqrt{L}} \sin\left(\frac{m\pi}{L}x\right) dx$$

All these integrals equal 0 but one and that is the one involving the sine and $k = m$. This is by the above lemma. Therefore,

$$\int_{-L}^{L} f(x) \frac{1}{\sqrt{L}} \sin\left(\frac{m\pi}{L}x\right) dx = b_m \frac{1}{L} \int_{-L}^{L} \sin^2\left(\frac{m\pi}{L}x\right) dx = b_m$$

It seems likely therefore, that b_m should be defined as

$$b_m = \int_{-L}^{L} f(x) \frac{1}{\sqrt{L}} \sin\left(\frac{m\pi}{L}x\right) dx \qquad (13.6)$$

Next do the same thing after multiplying by $\frac{1}{\sqrt{L}}\cos\left(\frac{m\pi}{L}x\right)$. Another use of the same lemma implies that the appropriate choice for a_m is

$$a_m = \int_{-L}^{L} f(x)\frac{1}{\sqrt{L}}\cos\left(\frac{m\pi}{L}x\right)dx \tag{13.7}$$

Finally integrate both sides of (13.5). This yields

$$\int_{-L}^{L} f(x)\frac{1}{\sqrt{2L}}dx = a_0 \tag{13.8}$$

and so the appropriate description of a_0 is given above. Thus the Fourier series is of the form

$$\int_{-L}^{L} f(y)\frac{1}{\sqrt{2L}}dy\frac{1}{\sqrt{2L}} + \sum_{m=1}^{\infty}\left(\int_{-L}^{L} f(y)\frac{1}{\sqrt{L}}\cos\left(\frac{m\pi}{L}y\right)dy\right)\frac{1}{\sqrt{L}}\cos\left(\frac{m\pi}{L}x\right)$$

$$+ \sum_{m=1}^{\infty}\left(\int_{-L}^{L} f(y)\frac{1}{\sqrt{L}}\sin\left(\frac{m\pi}{L}y\right)dy\right)\frac{1}{\sqrt{L}}\sin\left(\frac{m\pi}{L}x\right)$$

Combining the $\sqrt{L}$ terms, this yields

$$= \frac{1}{2L}\int_{-L}^{L} f(y)\,dy + \sum_{m=1}^{\infty}\left(\frac{1}{L}\int_{-L}^{L} f(y)\cos\left(\frac{m\pi}{L}y\right)dy\right)\cos\left(\frac{m\pi}{L}x\right)$$

$$+ \sum_{m=1}^{\infty}\left(\frac{1}{L}\int_{-L}^{L} f(y)\sin\left(\frac{m\pi}{L}y\right)dy\right)\sin\left(\frac{m\pi}{L}x\right)$$

This is so far completely speculative, but this was they often did things back in the time when Fourier came up with the idea back in the early 1800s. Here is the definition of the Fourier series in which we combine the various constant terms to make it easier to remember.

Definition 13.3.3 *Let f be a function defined on $\mathbb{R}$ which is $2L$ periodic and Riemann integrable on every closed interval of length $2L$. Then the Fourier series is defined as*

$$a_0 + \sum_{k=1}^{\infty} a_k\cos\left(\frac{k\pi}{L}x\right) + \sum_{k=1}^{\infty} b_k\sin\left(\frac{k\pi}{L}x\right)$$

where a_0, a_m, b_m are given as

$$a_0 = \frac{1}{2L}\int_{-L}^{L} f(y)\,dy,\ a_m = \frac{1}{L}\int_{-L}^{L} f(y)\cos\left(\frac{m\pi}{L}y\right)dy$$

$$b_m = \frac{1}{L}\int_{-L}^{L} f(y)\sin\left(\frac{m\pi}{L}y\right)dy$$

We will refer to a_0, a_n, b_n as Fourier coefficients.

13.4   Mean Square Approximation

When you have two functions defined on an interval $[a, b]$, how do you measure the distance between them? It turns out there are infinitely many ways to do this. One way is to say the distance between f and g, denoted as $\|f - g\|$ is defined as

$$\|f - g\| = \sup \{|f(x) - g(x)|, t \in [a, b]\}$$

To say that two functions are close in this sense is to say that for each x you have $f(x)$ close to $g(x)$. The two functions are said to be uniformly close if they are close in this norm. This norm is also called the uniform norm.

   This is a good way to define distance between functions, but it turns out that a more useful way in many situations is the following. You define

$$\|f - g\| \equiv \left(\int_a^b |f(x) - g(x)|^2 \, dx \right)^{1/2}$$

Then $\|f - g\|$ is called the mean square norm with the above definition. You should verify that if two functions are close in the uniform norm, then they must be close in the mean square norm, but not the other way around. Often the mean square norm is denoted as $|f - g|$. So why is this a norm and what is meant by a norm? First here is a simple lemma.

Lemma 13.4.1 *Suppose f, g are Riemann integrable functions. Define*

$$(f, g) \equiv \int_a^b f(x) g(x) \, dx.$$

Then the following are satisfied.

$$(f, g) = (g, f)$$

$$(f, f) \geq 0$$

For a, b real numbers,

$$\begin{aligned} (af + bg, h) &= a(f, h) + b(g, h) \\ (f, ag + bh) &= a(f, g) + b(f, h) \end{aligned}$$

The following inequality called the Cauchy-Schwarz inequality holds.

$$|(f, g)| \leq |f| \, |g| \equiv (f, f)^{1/2} (g, g)^{1/2}$$

where $|f|$ denotes the mean square distance defined above.

   Proof: All of the above are completely obvious except for the last one. As to that one, note that from the first obvious properties, for $t \in \mathbb{R}$

$$0 \leq (tf + g, tf + g) = t^2 (f, f) + 2t (f, g) + (g, g)$$

If $(f, f) = 0$ there is nothing to prove because you must have $(f, g) = 0$ since otherwise, the above inequality would be violated for suitable choice of t. It follows that the above is a quadratic polynomial whose graph opens up and which has at most one real zero. Hence by the quadratic formula,

$$4 (f, g)^2 - 4 (f, f) (g, g) \le 0$$

which reduces to the Cauchy-Schwarz inequality. ∎

Now the mean square norm amounts to nothing more than $|f| = (f, f)^{1/2}$.

Proposition 13.4.2 *The mean square norm* $\|f\| = |f| = (f, f)^{1/2}$ *satisfies the following axioms.*

1. $\|f\| \ge 0$

2. *If a is a number,* $\|af\| = |a| \, \|f\|$

3. $\|f + g\| \le \|f\| + \|g\|$

Proof: The only one which is not completely obvious is the last. Then by the definition of the norm and the properties of $(\cdot, \cdot)$,

$$\|f + g\|^2 \equiv (f + g, f + g) = \|f\|^2 + \|g\|^2 + 2 (f, g)$$

$$
\begin{aligned}
&\le \ \|f\|^2 + \|g\|^2 + 2 \, |(f, g)| \\
&\le \ \|f\|^2 + \|g\|^2 + 2 \, \|f\| \, \|g\| \\
&= \ (\|g\| + \|f\|)^2
\end{aligned}
$$

Now taking the square root of both sides yields the desired inequality. ∎

The reason this is important is that if you have f close to g and h close to g, then you have f close to h. Indeed,

$$\|f - h\| \le \|f - g\| + \|g - h\|$$

and if both of the terms on the right are small, then the term on the left is also.

There are $2n + 1$ functions, $\frac{1}{\sqrt{2L}}, \frac{1}{\sqrt{L}} \cos\left(\frac{k\pi}{L} x\right), \frac{1}{\sqrt{L}} \sin\left(\frac{j\pi}{L} x\right)$ for $k, j \in 1, 2, \cdots, n$. Denote these functions as $\{\phi_k\}_{k=1}^{2n+1}$ to save on notation. It was shown above that $(\phi_k, \phi_j) = \delta_{jk}$ which is 1 if $k = j$ and 0 if $k \ne j$. Then for f a Riemann integrable function on $[-L, L]$, our problem is to choose α_k to minimize

$$\left| f - \sum_{k=1}^{2n+1} \alpha_k \phi_k \right|^2 = (A + B, A + B)$$

where $A = f - \sum_{k=1}^{2n+1} (f, \phi_k) \phi_k, B = \sum_{k=1}^{2n+1} ((f, \phi_k) - \alpha_k) \phi_k$. Thus the above is

$$|A|^2 + 2 (A, B) + |B|^2 \qquad (*)$$

Consider the middle term.

$$
\begin{aligned}
\left(f - \sum_{k=1}^{2n+1} (f, \phi_k) \phi_k, \phi_j \right) &= (f, \phi_j) - \sum_k (f, \phi_k) (\phi_k, \phi_j) \\
&= (f, \phi_j) - (f, \phi_j) = 0
\end{aligned}
$$

and so the middle term of $*$ equals 0 because

$$\left(f - \sum_{k=1}^{2n+1} (f, \phi_k)\, \phi_k, \sum_{k=1}^{2n+1} a_k \phi_k \right) = 0$$

for any choice of a_k which includes the case of (A, B). Thus $*$ implies

$$\left| f - \sum_{k=1}^{2n+1} \alpha_k \phi_k \right|^2 = \left| f - \sum_{k=1}^{2n+1} (f, \phi_k)\, \phi_k \right|^2 + \left| \sum_{k=1}^{2n+1} ((f, \phi_k) - \alpha_k)\, \phi_k \right|^2$$

From the definition of the norm, the second term is

$$\sum_{j,k} ((f, \phi_k) - \alpha_k)\, ((f, \phi_j) - \alpha_j)\, (\phi_k, \phi_j) = \sum_k ((f, \phi_k) - \alpha_k)^2$$

Hence

$$\left| f - \sum_{k=1}^{2n+1} \alpha_k \phi_k \right|^2 = \left| f - \sum_{k=1}^{2n+1} (f, \phi_k)\, \phi_k \right|^2 + \sum_{k=1}^{2n+1} ((f, \phi_k) - \alpha_k)^2 \qquad (**)$$

which shows that the left side is minimized exactly when $\alpha_k = (f, \phi_k)$. It is clear then that corresponding to $\frac{1}{\sqrt{L}} \cos\left(\frac{k\pi x}{L}\right)$, you would have

$$\alpha_k = \int_{-L}^{L} f(x)\, \frac{1}{\sqrt{L}} \cos\left(\frac{k\pi x}{L}\right) dx$$

and so, the term in the Fourier series which corresponds to this would be

$$\left(\int_{-L}^{L} f(y)\, \frac{1}{\sqrt{L}} \cos\left(\frac{k\pi y}{L}\right) dy \right) \frac{1}{\sqrt{L}} \cos\left(\frac{k\pi x}{L}\right)$$

$$= \left(\frac{1}{L} \int_{-L}^{L} f(y) \cos\left(\frac{k\pi y}{L}\right) dy \right) \cos\left(\frac{k\pi x}{L}\right)$$

which is exactly what was determined earlier.

In $**$, let each $\alpha_k = 0$. Then this equation implies

$$|f|^2 \geq \sum_{k=1}^{2n+1} (f, \phi_k)^2$$

which is Bessel's inequality. In particular, in the case of most interest here, this inequality is

$$|f|^2 \geq \frac{1}{2L} \left(\int_{-L}^{L} f(x)\, dx \right)^2 + \frac{1}{L} \sum_{k=1}^{n} \left(\int_{-L}^{L} f(x) \sin\left(\frac{k\pi x}{L}\right) dx \right)^2$$

$$+ \frac{1}{L} \sum_{k=1}^{n} \left(\int_{-L}^{L} f(x) \cos\left(\frac{k\pi x}{L}\right) dx \right)^2 \qquad (**)$$

It follows that the sequence of partial sums in the sum on the right in ** converges and so

$$\lim_{k\to\infty} \int_{-L}^{L} f(x) \cos\left(\frac{k\pi}{L}x\right) dx = 0$$

$$\lim_{k\to\infty} \int_{-L}^{L} f(x) \sin\left(\frac{k\pi}{L}x\right) dx = 0 \tag{13.9}$$

The two limits in (13.9) are special cases of the Riemann-Lebesgue lemma. These are the considerations which make it possible to consider the pointwise convergence properties of Fourier series. In particular, the following lemma is used.

Lemma 13.4.3 *Suppose f is a Riemann integrable function defined on $[-L, L]$. Then*

$$\lim_{k\to\infty} \int_{-L}^{L} f(x) \sin\left(\left(k+\frac{1}{2}\right)\frac{\pi}{L}x\right) dx = 0$$

Proof: It equals

$$\lim_{k\to\infty}\left[\int_{-L}^{L} f(x)\cos\left(\frac{\pi x}{2L}\right)\sin\left(\frac{k\pi}{L}x\right) dx + \int_{-L}^{L} f(x)\sin\left(\frac{\pi x}{2L}\right)\cos\left(\frac{k\pi}{L}x\right) dx\right]$$

and each of these converge to 0 thanks to (13.9). ∎

Fourier series are really all about mean square convergence. If you are interested in pointwise approximation with trig. polynomials, there are better ways to do it than with Fourier series. However, the pointwise convergence is also very interesting and this is discussed in the next section.

13.5 Pointwise Convergence of Fourier Series

For each x the Fourier series yields an infinite series. One wonders whether it converges to $f(x)$. It is completely obvious that this is not necessarily the case. This is because the Fourier series is completely unchanged if f is changed at any finite set of points. Therefore, to obtain any sort of meaningful convergence, one must assume something about the function. The following is an elementary theorem which is a special case of a more substantial real analysis result.

Definition 13.5.1 *The one sided limits are*

$$f(x+) \equiv \lim_{h\to 0+} f(x+h), \ \ f(x-) \equiv \lim_{h\to 0+} f(x-h)$$

.

Theorem 13.5.2 *Suppose f is a periodic function of period $2L$ such that f has only finitely many jump discontinuities on the interval $[-L, L)$. Suppose there exists a constant K such that for all x,*

$$|f(x+) - f(x+y)| < Ky$$

for all sufficiently small positive y. Also

$$|f(x-) - f(x-y)| < Ky$$

for all y sufficiently small. Then

$$\frac{f(x+) + f(x-)}{2} = a_0 + \sum_{k=1}^{\infty} a_k \cos\left(\frac{k\pi}{L}x\right) + \sum_{k=1}^{\infty} b_k \sin\left(\frac{k\pi}{L}x\right)$$

In words, this says that the Fourier series converges to the midpoint of the jump. A picture which represents a part of the graph of f is as follows.

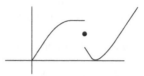

You note that the dot is at the midpoint of the jump. The condition in the theorem is there to rule out excessive steepness of the graph of the function. In fact, one can do a lot better than what it says in this theorem. You should see [2] for two more general treatments of this theorem.

One way to satisfy the condition on not having excessive steepness is to have f be piecewise continuous such that if a, b are successive discontinuities, then redefining f on $[a, b]$ to equal $f(a+)$ at the left and $f(b-)$ at the right, the new function has a continuous derivative on $[a, b]$.

Theorem 13.5.2 is the convergence theorem. I am going to give a discussion of this convergence theorem. If you are not interested in understanding why it works, ignore the proof. It is included in case someone would be interested.

13.5.1   Explanation of Pointwise Convergence Theorem

Proof of the convergence theorem: The convergence of sums has to do with the limit of the sequence of partial sums. Let

$$S_n f(x) = a_0 + \sum_{k=1}^{n} a_k \cos\left(\frac{k\pi}{L}x\right) + \sum_{k=1}^{n} b_k \sin\left(\frac{k\pi}{L}x\right)$$

From the definition, this equals

$$\frac{1}{L}\int_{-L}^{L} \frac{f(y)}{2}\,dy + \sum_{k=1}^{n} \frac{1}{L}\int_{-L}^{L} f(y)\cos\left(\frac{k\pi}{L}y\right)dy \cos\left(\frac{k\pi}{L}x\right)$$

$$+ \sum_{k=1}^{n} \frac{1}{L}\int_{-L}^{L} f(y)\sin\left(\frac{k\pi}{L}y\right)dy \sin\left(\frac{k\pi}{L}x\right)$$

This simplifies to

$$\frac{1}{L}\int_{-L}^{L} \frac{f(y)}{2}\,dy+$$

$$\sum_{k=1}^{n} \frac{1}{L}\int_{-L}^{L} f(y)\cos\left(\frac{k\pi}{L}y\right)\cos\left(\frac{k\pi}{L}x\right) + f(y)\sin\left(\frac{k\pi}{L}y\right)\sin\left(\frac{k\pi}{L}x\right)dy$$

which equals

$$\frac{1}{L}\int_{-L}^{L}\frac{f\left(y\right)}{2}dy+\sum_{k=1}^{n}\frac{1}{L}\int_{-L}^{L}f\left(y\right)\cos\left(\frac{k\pi}{L}\left(x-y\right)\right)dy$$

Simplifying this a little more yields

$$\int_{-L}^{L}\frac{1}{L}\left(\frac{1}{2}+\sum_{k=1}^{n}\cos\left(\frac{k\pi}{L}\left(x-y\right)\right)\right)f\left(y\right)dy$$

$$\equiv\int_{-L}^{L}D_{n}\left(x-y\right)f\left(y\right)dy$$

Here $D_{n}\left(t\right)$ is called the Dirichlet kernel. In order to consider the convergence of the partial sums, it is necessary to study the properties of the Dirichlet kernel.

Lemma 13.5.3 *The Dirichlet kernel is periodic of period $2L$.*

$$\int_{-L}^{L}D_{n}\left(t\right)dt=2\int_{0}^{L}D_{n}\left(t\right)dt=1.$$

There is also a formula for this kernel,

$$D_{n}\left(t\right)=\frac{\sin\left(\left(n+\frac{1}{2}\right)\frac{\pi}{L}t\right)}{2L\sin\left(\frac{\pi}{2L}t\right)}$$

Proof: As indicated above,

$$D_{n}\left(t\right)=\frac{1}{L}\left(\frac{1}{2}+\sum_{k=1}^{n}\cos\left(\frac{k\pi}{L}t\right)\right)$$

and so, it is obvious that

$$\int_{-L}^{L}D_{n}\left(t\right)dt=1. \tag{13.10}$$

From the above formula, it follows that $D_{n}\left(t\right)=D_{n}\left(-t\right)$ and $D_{n}\left(x+2L\right)=D_{n}\left(x\right)$. Since $D_{n}\left(t\right)=D_{n}\left(-t\right),$

$$\int_{-L}^{L}D_{n}\left(t\right)dt=2\int_{0}^{L}D_{n}\left(t\right)dt$$

It remains to find a formula. Use (13.2)

$$\sin\left(\frac{\pi}{2L}t\right)D_{n}\left(t\right)=\frac{1}{L}\left(\frac{1}{2}\sin\left(\frac{\pi}{2L}t\right)+\sum_{k=1}^{n}\sin\left(\frac{\pi}{2L}t\right)\cos\left(\frac{k\pi}{L}t\right)\right)$$

$$=\frac{1}{L}\left(\frac{1}{2}\sin\left(\frac{\pi}{2L}t\right)+\frac{1}{2}\sum_{k=1}^{n}\sin\left(\left(\frac{k\pi}{L}+\frac{\pi}{2L}\right)t\right)-\sin\left(\left(\frac{k\pi}{L}-\frac{\pi}{2L}\right)t\right)\right)$$

$$= \frac{1}{2L} \left[\sin\left(\frac{\pi}{2L}t\right) + \sum_{k=1}^{n} \sin\left(\left(k+\frac{1}{2}\right)\frac{\pi}{L}t\right) - \sum_{k=1}^{n} \sin\left(\left(k-\frac{1}{2}\right)\frac{\pi}{L}t\right) \right]$$

$$= \frac{1}{2L} \left[\sin\left(\frac{\pi}{2L}t\right) + \sum_{k=1}^{n} \sin\left(\left(k+\frac{1}{2}\right)\frac{\pi}{L}t\right) - \sum_{k=0}^{n-1} \sin\left(\left(k+\frac{1}{2}\right)\frac{\pi}{L}t\right) \right]$$

$$= \frac{1}{2L} \sin\left(\left(n+\frac{1}{2}\right)\frac{\pi}{L}t\right)$$

Thus the desired formula is

$$D_n(t) = \frac{\sin\left(\left(n+\frac{1}{2}\right)\frac{\pi}{L}t\right)}{2L \sin\left(\frac{\pi}{2L}t\right)} \quad \blacksquare$$

Here is a graph of the first seven of these Dirichlet kernels, $n \geq 1$ for $L = \pi$.

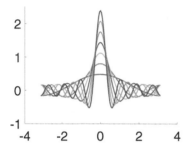

Next, it follows from the above that

$$S_n f(x) = \int_{-L}^{L} D_n(x-y) f(y)\, dy.$$

Change the variables. Let $u = x - y$. Then this reduces to

$$\int_{-L+x}^{L+x} D_n(u) f(x-u)\, du$$

Since D_n and f are both periodic of period $2L$, this equals

$$\int_{-L}^{L} D_n(y) f(x-y)\, dy$$

Therefore, since $\int_{-L}^{L} D_n(y)\, dy = 1$,

$$\left| \frac{f(x+) + f(x-)}{2} - S_n f(x) \right| = \left| \frac{f(x+) + f(x-)}{2} - \int_{-L}^{L} D_n(y) f(x-y)\, dy \right|$$

$$= \left| \int_{-L}^{L} \left(\frac{f(x+) + f(x-)}{2} - f(x-y) \right) D_n(y)\, dy \right|$$

$$= \left| \int_{0}^{L} (f(x+) + f(x-)) D_n(y)\, dy - \int_{0}^{L} (f(x-y) + f(x+y)) D_n(y)\, dy \right|$$

$$\leq \left| \int_{0}^{L} \frac{f(x+) - f(x+y)}{2L \sin\left(\frac{\pi}{2L}y\right)} \sin\left(\left(n+\frac{1}{2}\right)\frac{\pi}{L}y\right) dy \right| +$$

$$\left| \int_{0}^{L} \frac{f(x-) - f(x-y)}{2L \sin\left(\frac{\pi}{2L}y\right)} \sin\left(\left(n+\frac{1}{2}\right)\frac{\pi}{L}y\right) dy \right|$$

Both of these converge to 0 thanks to Lemma 13.4.3. To use this lemma, it is only necessary to verify that the functions

$$y \to \frac{f(x-) - f(x-y)}{2L \sin\left(\frac{\pi}{2L}y\right)}, \quad y \to \frac{f(x+) - f(x+y)}{2L \sin\left(\frac{\pi}{2L}y\right)}$$

are each Riemann integrable on $[-L, L]$.

I will show this now. Each is continuous except for finitely many points of discontinuity. The only remaining issue is whether the functions are bounded as $y \to 0$. However, there exists a constant K such that

$$\left| \frac{f(x+) - f(x+y)}{2L \sin\left(\frac{\pi}{2L}y\right)} \right| \leq \frac{K|y|}{\left| 2L \sin\left(\frac{\pi}{2L}y\right) \right|}$$

and this expression converges to K/π, so the function is Riemann integrable. The other function is similar. ∎

Example 13.5.4 *Let* $f(x) = |x|$ *for* $x \in [-1, 1)$ *and let* f *be periodic of period 2. Find the Fourier series of* f.

Here you need $L = 1$. Then

$$a_0 = \frac{1}{2} \int_{-1}^{1} |x|\, dx = \frac{1}{2}$$

$$a_k = \int_{-1}^{1} |x| \cos(k\pi x)\, dx = \frac{2}{\pi^2 k^2} \left((-1)^k - 1 \right)$$

Note that $a_k = 0$ if k is even and it equals $-4/\left(\pi^2 k^2\right)$ when k is odd.

Since the function is even, the $b_k = 0$. Therefore, the Fourier series equals

$$\frac{1}{2} - \sum_{k=1}^{\infty} \frac{4}{\pi^2 (2k-1)^2} \cos(2k-1)\pi x$$

Now here is the graph of the function between -1 and 1 along with the sum up to 2 in the Fourier series. You will notice that after only three terms the Fourier series appears to be very close to the function on the interval $[-1, 1]$. This also shows how the Fourier series approximates the periodic extension of this function off this interval.

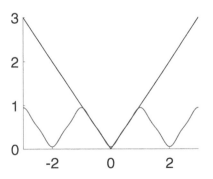

Notice that if you take $x = 0$ in the above, the theorem on pointwise convergence of Fourier series implies that the sum converges to the value of the function which is 0. Therefore,

$$\frac{1}{2} = \sum_{k=1}^{\infty} \frac{4}{\pi^2 (2k-1)^2}$$

It follows that

$$\frac{\pi^2}{8} = \sum_{k=1}^{\infty} \frac{1}{(2k-1)^2}.$$

This is a remarkable assertion.

Now here is another example for which the Fourier series will have to struggle harder to approximate the function.

Example 13.5.5 *Let $f(x) = 1$ on $(0, 2]$ and $f(x) = -1$ on $(-2, 0]$ and $f(x + 4) = f(x)$.*

First note that $L = 2$. In this case, the function is odd and so all the $a_k = 0$.

$$b_k = \frac{1}{2} \int_{-2}^{2} f(x) \sin \left(\frac{k\pi x}{2} \right) dx = \int_{0}^{2} \sin \left(\frac{k\pi x}{2} \right) dx$$

Then $b_k = \frac{2}{\pi k} \left(1 - (-1)^k \right)$. Thus for k even, this is 0. For k odd, this is $\frac{4}{\pi k}$. It follows the Fourier series is

$$\sum_{k=1}^{\infty} \frac{4}{\pi (2k - 1)} \sin \left(\frac{(2k - 1) \pi x}{2} \right)$$

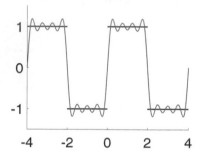

In the picture, is a graph of the addition of the first four terms of the Fourier series along with part of the function. Notice the way the Fourier series is struggling to do the impossible, approximate uniformly a discontinuous function with one which is very smooth. That little blip near the jump in the function will never go away by taking more terms in the sum.

Note that if you take $x = 1$ the series must converge to 1. Therefore,

$$1 = \sum_{k=1}^{\infty} \frac{4}{\pi (2k - 1)} (-1)^{k-1}$$

It follows that

$$\frac{\pi}{4} = \sum_{k=1}^{\infty} \frac{(-1)^{k-1}}{2k - 1}$$

This is another remarkable assertion.

13.5.2   Mean Square Convergence

It is the case that if f is Riemann integrable and $2L$ periodic, then the Fourier series converges to the function f in the mean square sense. That is

$$\lim_{n \to \infty} \int_{-L}^{L} |f(x) - S_n f(x)|^2 \, dx = 0$$

I will show this now, leaving out a few details which will be reasonable to believe. Suppose that f is continuous and periodic with period $2L$. The Cesaro means of f are defined as follows.

$$\sigma_n f(x) \equiv \frac{1}{n+1} \sum_{k=0}^{n} S_k f(x), \quad S_0 f(x) = a_0 \equiv \frac{1}{2L} \int_{-L}^{L} f(x) \, dx$$

Thus, from what was shown above,

$$\sigma_n f(x) = \frac{1}{n+1} \sum_{k=0}^{n} \int_{-L}^{L} D_k(x-y) f(y) \, dy$$

$$= \int_{-L}^{L} \left(\frac{1}{n+1} \sum_{k=0}^{n} D_k(x-y) \right) f(y) \, dy$$

Then the Fejer kernel is

$$F_n(t) = \frac{1}{n+1} \sum_{k=0}^{n} D_k(t) \qquad (*)$$

We compute this now. Recall that

$$D_n(t) = \frac{\sin\left(\left(n+\frac{1}{2}\right)\frac{\pi}{L}t\right)}{2L \sin\left(\frac{\pi}{2L}t\right)}$$

Thus

$$\sin^2\left(\frac{\pi}{2L}t\right) F_n(t) = \frac{1}{2L}\frac{1}{n+1} \sum_{k=0}^{n} \sin\left(\frac{\pi}{2L}t\right) \sin\left(\left(k+\frac{1}{2}\right)\frac{\pi}{L}t\right)$$

$$= \frac{1}{2L(n+1)}\frac{1}{2}\sum_{k=0}^{n}\left[\cos\left(\left(k+\frac{1}{2}\right)\frac{\pi}{L}t - \left(\frac{\pi}{2L}t\right)\right) - \cos\left(\frac{\pi}{2L}t + \left(k+\frac{1}{2}\right)\frac{\pi}{L}t\right)\right]$$

$$= \frac{1}{2L}\frac{1}{n+1}\frac{1}{2}\sum_{k=0}^{n}\left[\cos\left(\frac{\pi}{L}kt\right) - \cos\left(\frac{\pi}{L}t(k+1)\right)\right]$$

$$= \frac{1}{4L(n+1)}\left(1 - \cos\left(\frac{\pi}{L}t(n+1)\right)\right)$$

Thus

$$F_n(t) = \frac{1}{4L(n+1)}\frac{\left(1 - \cos\left(\frac{\pi}{L}t(n+1)\right)\right)}{\sin^2\left(\frac{\pi}{2L}t\right)} \qquad (**)$$

Here are graphs of $F_n(t)$ for $n = 1, 2, \cdots, 7$ for $L = \pi$. Notice how they are nonnegative and are large on a small interval containing 0. As you increase n, the bump in the middle gets taller.

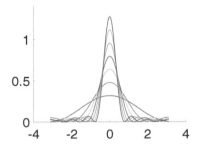

There are certain properties which are obvious. First of all, $\int_{-L}^{L} F_n(t) \, dt = 1$. This follows from $*$ and (13.10) which pertained to the Dirichlet kernel. Another which is obvious from the above formula $**$ is that $F_n(t) \geq 0$. Finally, for any small $\delta > 0$, if $|t| > \delta$ then

$$F_n(t) \leq \frac{1}{4L(n+1)}\frac{\left(1 - \cos\left(\frac{\pi}{L}t(n+1)\right)\right)}{\sin^2\left(\frac{\pi}{2L}\delta\right)} \qquad (***)$$

It follows from periodicity that for $M \geq \max_x |f(x)|$

$$|f(x) - \sigma_n f(x)| = \left| \int_{-L}^{L} (f(x) - f(t)) F_n(x-t)\, dt \right|$$

$$= \left| \int_{-L}^{L} (f(x) - f(x-u)) F_n(u)\, du \right|$$

$$\leq \int_{L \geq |u| \geq \delta} |f(x) - f(x-u)| F_n(u)\, du + 2M \int_{|u| < \delta} |f(x) - f(x-u)| F_n(u)\, du$$

$$\leq 2M \frac{1}{4L(n+1)} \frac{2}{\sin^2\left(\frac{\pi}{2L}\delta\right)} + 2M \int_{|u| < \delta} |f(x) - f(x-u)| F_n(u)\, du$$

Now if $\varepsilon > 0$, there is $\delta > 0$ such that if $|u| < \delta$, then for all x, $|f(x) - f(x-u)| < \varepsilon/2$. Thus for such a choice of δ and $***$,

$$|f(x) - \sigma_n f(x)| \leq 2M \frac{1}{4L(n+1)} \frac{2}{\sin^2\left(\frac{\pi}{2L}\delta\right)} + \frac{\varepsilon}{2}$$

and so, if n is large enough, you get

$$|f(x) - \sigma_n f(x)| < \varepsilon,$$

this for any x. Thus the convergence of $\sigma_n f(x)$ to $f(x)$ is uniform. It follows that

$$\lim_{n \to \infty} \int_{-L}^{L} |f(x) - \sigma_n f(x)|^2\, dx = 0$$

This shows the following interesting result.

Proposition 13.5.6 *If f is continuous and $2L$ periodic, then the Cesaro means converge uniformly to f and also they converge to f in the mean square sense.*

From this, it is not hard to establish that the Cesaro means converge in mean square to any $2L$ periodic function f which is Riemann integrable on intervals of length $2L$. To do this, you argue that, given a Riemann integrable function which is $2L$ periodic, there exists a continuous function which is close to it in the mean square norm. Then apply the above proposition to this continuous function and get a Cesaro mean close to it in mean square which is close to the original function in mean square sense.

The Cesaro means are trig polynomials of the form

$$a_0 + \sum_{k=1}^{n} a_k \cos\left(\frac{k\pi x}{L}\right) + b_k \sin\left(\frac{k\pi x}{L}\right).$$

One of these can be made as close as desired to f in the mean square sense. Hence the corresponding Fourier series is even closer, by the above section on mean square approximation. Thus, for every $\varepsilon > 0$ there exists N such that if $n > N$, then

$$\int_{-L}^{L} |S_n f(x) - f(x)|^2\, dx < \varepsilon$$

which says the Fourier series converge in the mean square sense to f.

Note that the above proposition also shows an improved result about pointwise convergence. The function f did not need to have any control on its derivative and yet the Cesaro means converged uniformly to the function. If the function were piecewise continuous, the Cesaro means would converge to the mid point of the jump with no condition on the derivatives from left or right. This is easy to show but is as far as this will be taken here. If you want uniform approximation using trigonometric series, you should not be using the Fourier series. You should use the Cesaro means.

13.6 Integrating and Differentiating Fourier Series

Suppose that f is $2L$ periodic and piecewise continuous. This is defined next.

Definition 13.6.1 *Let f be a bounded function defined on $[a, b]$. It is called piecewise continuous if there is a partition of $[a, b]$, $\{x_0, \cdots, x_n\}$ and for each k, a continuous function g_k such that $f(x) = g_k(x)$ for all $x \in (x_{k-1}, x_k)$.*

It turns out that you can integrate a Fourier series term by term. This is generally true but I will show it here for piecewise continuous $2L$ periodic functions. Let f be such a function equal to a continuous function on $[x_i, x_{i+1}]$ for $i \leq n$. Then consider

$$G(x) \equiv \int_{-L}^{x} (f(t) - a_0)\, dt$$

where a_0 is the Fourier coefficient

$$a_0 = \frac{1}{2L} \int_{-L}^{L} f(t)\, dt$$

Thus $G(-L) = G(L) = 0$ and if we continue using G to denote the $2L$ periodic extension, it follows from Theorem 13.5.2 that the Fourier series of G

$$A_0 + \sum_{n=1}^{\infty} A_n \cos\left(\frac{n\pi x}{L}\right) + \sum_{n=1}^{\infty} B_n \sin\left(\frac{n\pi x}{L}\right)$$

converges to G at every point. This is because $|f(t) - a_0|$ is bounded by some M due to the assumption that it is piecewise continuous and the observation that

$$|G(x) - G(\hat{x})| \leq \left| \int_{\hat{x}}^{x} |f(t) - a_0| \right| \leq M |x - \hat{x}|$$

Then plugging in π to the Fourier series for G we get

$$0 = A_0 + \sum_{n=1}^{\infty} A_n (-1)^n, \quad A_0 = -\sum_{n=1}^{\infty} A_n (-1)^n \tag{*}$$

Next consider $A_n, n > 0$.

$$LA_n = \int_{-L}^{L} \int_{-L}^{x} (f(t) - a_0)\, dt \cos\left(\frac{n\pi x}{L}\right) dx$$

$$= \sum_{k=0}^{n-1} \int_{x_k}^{x_{k+1}} \left(\int_{-L}^{x_k} (f(t) - a_0)\, dt + \int_{x_k}^{x} (f(t) - a_0)\, dt \right) \cos\left(\frac{n\pi x}{L}\right) dx$$

$$= \sum_{k=0}^{n-1} \int_{x_k}^{x_{k+1}} \int_{-L}^{x_k} (f(t) - a_0)\, dt \cos\left(\frac{n\pi x}{L}\right) dx$$

$$+ \sum_{k=0}^{n-1} \left(\begin{array}{c} \frac{L}{n\pi} \sin\left(\frac{n\pi x}{L}\right) \int_{x_k}^{x} (f(t) - a_0)\, dt \Big|_{x_k}^{x_{k+1}} \\ - \int_{x_k}^{x_{k+1}} \frac{L}{n\pi} \sin\left(\frac{n\pi x}{L}\right) (f(x) - a_0)\, dx \end{array} \right)$$

$$= \sum_{k=0}^{n-1} \int_{x_k}^{x_{k+1}} G(x_k) \cos\left(\frac{n\pi x}{L}\right) dx + \sum_{k=0}^{n-1} \frac{L}{n\pi} \sin\left(\frac{n\pi x_{k+1}}{L}\right) (G(x_{k+1}) - G(x_k))$$

$$- \sum_{k=0}^{n-1} \int_{x_k}^{x_{k+1}} \frac{L}{n\pi} \sin\left(\frac{n\pi x}{L}\right) (f(x) - a_0)\, dx$$

Now do an integration on the first sum. This yields

$$\frac{L}{n\pi} \sum_{k=0}^{n-1} G(x_k) \left(\sin\left(\frac{n\pi x_{k+1}}{L}\right) - \sin\left(\frac{n\pi x_k}{L}\right) \right)$$

$$+ \frac{L}{n\pi} \sum_{k=0}^{n-1} \sin\left(\frac{n\pi x_{k+1}}{L}\right) (G(x_{k+1}) - G(x_k))$$

$$- \sum_{k=0}^{n-1} \int_{x_k}^{x_{k+1}} \frac{L}{n\pi} \sin\left(\frac{n\pi x}{L}\right) (f(x) - a_0)\, dx$$

The sums simplify and the result one obtains is

$$\frac{L}{n\pi} \sum_{k=0}^{n-1} G(x_{k+1}) \sin\left(\frac{n\pi x_{k+1}}{L}\right) - G(x_k) \sin\left(\frac{n\pi x_k}{L}\right)$$

$$- \sum_{k=0}^{n-1} \int_{x_k}^{x_{k+1}} \frac{L}{n\pi} \sin\left(\frac{n\pi x}{L}\right) (f(x) - a_0)\, dx$$

The series telescopes and the result is 0 because $G(L) = G(-L) = 0$. Thus the result of it all is

$$LA_n = -\sum_{k=0}^{n-1} \int_{x_k}^{x_{k+1}} \frac{L}{n\pi} \sin\left(\frac{n\pi x}{L}\right) (f(x) - a_0)\, dx$$

$$= \int_{-L}^{L} -\frac{L}{n\pi} \sin\left(\frac{n\pi x}{L}\right) (f(x) - a_0)\, dx$$

Thus

$$A_n = -\frac{L}{n\pi} \frac{1}{L} \int_{-L}^{L} \sin\left(\frac{n\pi x}{L}\right) (f(x) - a_0)\, dx = -\frac{L}{n\pi} b_n$$

Similar computations will show that for $n > 0$,

$$B_n = \frac{L}{n\pi} \frac{1}{L} \int_{-L}^{L} \cos\left(\frac{n\pi x}{L}\right) (f(x) - a_0)\, dx = \frac{L}{n\pi} a_n$$

where a_n, b_n are, respectively, the cosine and sine Fourier coefficients of f. Thus we have from $*$,

$$G(x) = \int_{-L}^{x} (f(t) - a_0)\, dt = -\sum_{n=1}^{\infty} \left(-\frac{\overbrace{\frac{L}{n\pi}}^{A_n}}{} b_n \right) (-1)^n +$$

$$\sum_{n=1}^{\infty} -\frac{L}{n\pi} b_n \cos \left(\frac{n\pi x}{L} \right) + \sum_{n=1}^{\infty} \frac{L}{n\pi} a_n \sin \left(\frac{n\pi x}{L} \right)$$

Hence

$$\int_{-L}^{x} (f(t) - a_0)\, dt = \sum_{n=1}^{\infty} \frac{L b_n}{n\pi} \left(\cos \left(\frac{-n\pi L}{L} \right) - \cos \left(\frac{n\pi x}{L} \right) \right)$$

$$+ \sum_{n=1}^{\infty} \frac{L}{n\pi} a_n \sin \left(\frac{n\pi x}{L} \right)$$

Thus

$$\int_{-L}^{x} f(t)\, dt = \int_{-L}^{x} a_0\, dt + \sum_{n=1}^{\infty} b_n \int_{-L}^{x} \sin \left(\frac{n\pi t}{L} \right) dt$$

$$+ \sum_{n=1}^{\infty} a_n \int_{-L}^{x} \cos \left(\frac{n\pi t}{L} \right) dt$$

This proves the following theorem.

Theorem 13.6.2 *Let f be piecewise continuous and $2L$ periodic. Then for every $x \in [-L, L]$,*

$$\int_{-L}^{x} f(t)\, dt = \int_{-L}^{x} a_0\, dt + \sum_{n=1}^{\infty} b_n \int_{-L}^{x} \sin \left(\frac{n\pi t}{L} \right) dt$$

$$+ \sum_{n=1}^{\infty} a_n \int_{-L}^{x} \cos \left(\frac{n\pi t}{L} \right) dt$$

where a_0, a_k, b_k are the Fourier coefficients for f.

Note that there is nothing which says that the Fourier series of f converges to f! This is a wonderful result.

You can't expect to be able to differentiate Fourier series. See the exercises. However, there is something which can be said. Suppose for $x \in [-L, L)$

$$f(x) = f(-L) + \int_{-L}^{x} f'(t)\, dt$$

and that f' is piecewise continuous and $2L$ periodic. Let f denote the $2L$ periodic extension of the above f. Then let the formal Fourier series for f' be

$$a_0 + \sum_{n=1}^{\infty} a_n \cos \left(\frac{n\pi x}{L} \right) + \sum_{n=1}^{\infty} b_n \sin \left(\frac{n\pi x}{L} \right)$$

Then by Theorem 13.6.2,

$$\int_{-L}^{x} f'(t)\,dt = \int_{-L}^{x} a_0\,dt + \sum_{n=1}^{\infty} a_n \frac{L}{n\pi} \sin\left(\frac{n\pi x}{L}\right)$$

$$+ \sum_{n=1}^{\infty} b_n \frac{L}{n\pi}\left((-1)^n - \cos\left(\frac{n\pi x}{L}\right)\right)$$

Then $a_0 \equiv \frac{1}{2L}\int_{-L}^{L} f'(t)\,dt = \frac{1}{2L}(f(L) - f(-L)) = 0$.

$$a_n \equiv \frac{1}{L}\int_{-L}^{L} f'(t)\cos\left(\frac{n\pi t}{L}\right) dt = \frac{1}{L} f(t)\cos\left(\frac{n\pi t}{L}\right)\Big|_{-L}^{L}$$

$$+ \frac{1}{L}\frac{n\pi}{L}\int_{-L}^{L} f(t)\sin\left(\frac{n\pi t}{L}\right) dt$$

$$= \frac{1}{L}\frac{n\pi}{L}\int_{-L}^{L} f(t)\sin\left(\frac{n\pi t}{L}\right) dt = B_n \frac{n\pi}{L}$$

where B_n is the Fourier coefficient for $f(t)$. Similarly,

$$b_n = \frac{1}{L}\int_{-L}^{L} f'(t)\sin\left(\frac{n\pi t}{L}\right) dt = -\frac{1}{L}\frac{n\pi}{L}\int_{-L}^{L} f(t)\cos\left(\frac{n\pi t}{L}\right) dt = -\frac{n\pi}{L}A_n$$

where A_n is the nth cosine Fourier coefficient for f. Thus

$$\int_{-L}^{x} f'(t)\,dt = \sum_{n=1}^{\infty} B_n \frac{n\pi}{L}\frac{L}{n\pi}\sin\left(\frac{n\pi x}{L}\right)$$

$$+ \sum_{n=1}^{\infty}\left(-\frac{n\pi}{L}A_n\right)\frac{L}{n\pi}\left((-1)^n - \cos\left(\frac{n\pi x}{L}\right)\right)$$

$$f(x) - f(-L) = \sum_{n=1}^{\infty} B_n \sin\left(\frac{n\pi x}{L}\right) + \sum_{n=1}^{\infty} A_n \cos\left(\frac{n\pi x}{L}\right) - \sum_{n=1}^{\infty} A_n (-1)^n$$

$$f(x) = \sum_{n=1}^{\infty} B_n \sin\left(\frac{n\pi x}{L}\right) + \sum_{n=1}^{\infty} A_n \cos\left(\frac{n\pi x}{L}\right) + \left(f(-L) - \sum_{n=1}^{\infty} A_n (-1)^n\right)$$

Thus that constant on the end is A_0. It follows that

$$f(x) = A_0 + \sum_{n=1}^{\infty} B_n \sin\left(\frac{n\pi x}{L}\right) + \sum_{n=1}^{\infty} A_n \cos\left(\frac{n\pi x}{L}\right)$$

and $-\frac{n\pi}{L}A_n = b_n$, $B_n\frac{n\pi}{L} = a_n$ and so

$$f'(x) = \sum_{n=1}^{\infty} a_n \cos\left(\frac{n\pi x}{L}\right) + \sum_{n=1}^{\infty} b_n \sin\left(\frac{n\pi x}{L}\right)$$

$$= \sum_{n=1}^{\infty} B_n \frac{n\pi}{L}\cos\left(\frac{n\pi x}{L}\right) + \sum_{n=1}^{\infty} A_n \left(-\frac{n\pi}{L}\right)\sin\left(\frac{n\pi x}{L}\right)$$

$$= \sum_{n=1}^{\infty} B_n \frac{d}{dx}\sin\left(\frac{n\pi x}{L}\right) + \sum_{n=1}^{\infty} A_n \frac{d}{dx}\cos\left(\frac{n\pi x}{L}\right)$$

This proves the following.

Theorem 13.6.3 *Let f denote the $2L$ periodic extension of the function f given on $[-L, L)$ by*

$$f(x) = f(-L) + \int_{-L}^{x} f'(t)\, dt$$

and suppose f' is $2L$ periodic and piecewise continuous. Then for each $x \in [-L, L]$,

$$f(x) = A_0 + \sum_{n=1}^{\infty} B_n \sin\left(\frac{n\pi x}{L}\right) + \sum_{n=1}^{\infty} A_n \cos\left(\frac{n\pi x}{L}\right)$$

where the A_k, B_k are the Fourier coefficients of f and the Fourier series for f' is

$$\sum_{n=1}^{\infty} B_n \frac{d}{dx} \sin\left(\frac{n\pi x}{L}\right) + \sum_{n=1}^{\infty} A_n \frac{d}{dx} \cos\left(\frac{n\pi x}{L}\right)$$

13.7   Odd and Even Extensions

Often, as in the above examples and in the applications which follow, the function you are finding the Fourier series for is either even or odd. One way this often occurs is when the function of interest is defined on an interval $[0, L]$ and it is only its values on this interval which are of interest. Then you could consider either the even or the odd extension of this function to $[-L, L]$ and then extend it to be a $2L$ periodic function. For example, consider the following pictures.

The first of these is an even extension to $[-L, L]$ and the second is an odd extension to $[-L, L]$. In the first case where there is an even extension, the Fourier coefficients are $b_k = 0$

$$a_0 = \frac{1}{2L} \int_{-L}^{L} f(x)\, dx = \frac{1}{L} \int_{0}^{L} f(x)\, dx$$

$$a_k = \frac{1}{L} \int_{-L}^{L} f(x) \cos\left(\frac{k\pi x}{L}\right) dx = \frac{2}{L} \int_{0}^{L} f(x) \cos\left(\frac{k\pi x}{L}\right) dx$$

In the second case where you are dealing with the odd extension, each $a_k = 0$ and

$$b_k = \frac{1}{L} \int_{-L}^{L} f(x) \sin\left(\frac{k\pi x}{L}\right) dx = \frac{2}{L} \int_{0}^{L} f(x) \sin\left(\frac{k\pi x}{L}\right) dx$$

Example 13.7.1 *Let $f(x) = x$ on $[0, 1]$. Find the Fourier series of its even extension.*

Its even extension is nothing more than the function of Example 13.5.4. This is

$$\frac{1}{2} - \sum_{k=1}^{\infty} \frac{4}{\pi^2 (2k-1)^2} \cos (2k-1) \pi x$$

Example 13.7.2 *Let $f(x) = x$ on $[0,1]$. Find the Fourier series of its odd extension which is periodic of period 2.*

This would be the function $f(x) = x$ on $(-1, 1)$ extended to be periodic of period 2. Thus $L = 1$. Since it is an odd function, all the $a_k = 0$ and from the above,

$$b_k = 2 \int_0^1 x \sin (k\pi x)\, dx = \frac{2}{\pi k} (-1)^{k+1}$$

Then the Fourier series is

$$\sum_{k=1}^{\infty} \frac{2}{\pi k} (-1)^{k+1} \sin (k\pi x)$$

The graph of the sum of the first five terms is given.

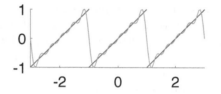

Note the difficulty in handling the jump with the little bump right before the discontinuity. This illustrates that if you are only interested in the function on $[0, 1]$, it would be better to use the even extension than the odd extension. However, in the applications, you don't get to choose.

Also, note that, unlike power series, Fourier series are attempting to approximate a function on a whole interval, not just near a single point. This is much more interesting.

There is a general sort of problem called a Sturm-Liouville problem discussed in Problem 13. It turns out that there are general theorems about convergence of expansions in terms of eigenfunctions to such problems [22]. However, you can often see that convergence in the mean square sense will hold from observing that the Fourier series for the eigenfunctions will converge because it is the restriction of the Fourier series of an even or odd extension as discussed in this section. For many other considerations on Sturm-Liouville problems, the old book by Ince [15] is very good. These problems have been intensively studied since around 1830.

13.8   Exercises

1. Let $f(x)$ be the even extension of $\sin x$. Find the Fourier series and at $x = \pi/2$ write a series which says that the Fourier series converges to the function at this point. Note that here $L = \pi$ and so $n\pi/L = n$.

2. Let $f(x)$ be the odd 2π periodic extension of $y = x^2$. Find what the sum converges to at $x = \pi/2$. Again $L = \pi$ and so $n\pi/L = \pi$.

3. Let $f(x)$ be the even 2π periodic extension of $y = x^2$. Find the Fourier coefficients and obtain an interesting series by letting $x = \pi$.

4. In Example 13.5.5 the Fourier series was found for the function f which is 1 on $[0, 2]$ and -1 on $(-2, 0)$.

$$\sum_{k=1}^{\infty} \frac{4}{\pi (2k-1)} \sin \left(\frac{(2k-1)\pi x}{2} \right)$$

This function has a jump so it is not differentiable at 0,2,4, etc. However, it is differentiable at most points, other than a few jumps. Furthermore, the Fourier series converges to the function at these points. Can you differentiate the Fourier series term by term and get something which converges to the derivative of the function? What does this show about interchange of limits?

5. In one of the problems above, you found that the Fourier series for the 2π periodic extension of $y = x^2$ is $\frac{\pi^2}{3} + \sum_{k=1}^{\infty} 4 \frac{(-1)^k}{k^2} \cos(kx)$. The derivative of this function, $y = 2x$ is sure piecewise continuous. Find the Fourier series expansion for $y = 2x$ without any effort.

6. Find a Fourier series which converges to the 2π periodic extension of

$$\int_{-\pi}^{x} \left(t^2 - \frac{\pi^2}{3} \right) dt = \frac{1}{3}x^3 - \frac{1}{3}\pi^2 x.$$

7. Suppose f is periodic with period $2L$. Does it follow that f' is also periodic of period $2L$? Explain.

8. Here are some boundary value problems. Find nonzero solutions if there are any or determine that there are none.

(a) $y'' + \frac{1}{4}\pi^2 y = 0$,
    $y(0) = 0, y(2) = 0$

(b) $y'' + \left(\frac{7\pi}{5}\right)^2 y = 0$,
    $y(0) = 0, y\left(\frac{5}{2}\right) = 0$

(c) $y'' + \left(\frac{5\pi}{3}\right)^2 y = 0$,
    $y(0) = 0, y\left(\frac{3}{2}\right) = 0$

(d) $y'' + \left(\frac{2\pi}{7}\right)^2 y = 0$,
    $y(0) = 0, y\left(\frac{7}{2}\right) = 0$

(e) $y'' + \left(\frac{1}{2}\pi\right)^2 y = 0$,
    $y(0) = 0, y(1) = 0$

(f) $y'' + \pi^2 y = 0$,
    $y(0) = 0, y(1) = 0$

(g) $y'' + \left(\frac{4\pi}{2}\right)^2 y = 0$,
    $y(0) = 0, y\left(\frac{2}{2}\right) = 0$

(h) $y'' + \frac{9}{25}\pi^2 y = 0$,
    $y(0) = 0, y\left(\frac{5}{2}\right) = 0$

9. Here are some boundary value problems. Find nonzero solutions if there are any or determine that there are none.

(a) $y'' + \left(\frac{11\pi}{2}\right)^2 y = 0,$
    $y(0) = 0, y'\left(\frac{2}{2}\right) = 0$

(b) $y'' + \left(\frac{4\pi}{5}\right)^2 y = 0,$
    $y(0) = 0, y'\left(\frac{5}{2}\right) = 0$

(c) $y'' + \left(\frac{2\pi}{7}\right)^2 y = 0,$
    $y(0) = 0, y'\left(\frac{7}{2}\right) = 0$

(d) $y'' + \left(\frac{7\pi}{3}\right)^2 y = 0,$

$y(0) = 0, y'\left(\frac{3}{2}\right) = 0$

(e) $y'' + \left(\frac{\pi}{13}\right)^2 y = 0,$
    $y(0) = 0, y'\left(\frac{13}{2}\right) = 0$

(f) $y'' + \left(\frac{2\pi}{5}\right)^2 y = 0,$
    $y(0) = 0, y'\left(\frac{5}{2}\right) = 0$

(g) $y'' + \left(\frac{5\pi}{11}\right)^2 y - 0,$
    $y(0) = 0, y'\left(\frac{11}{2}\right) = 0$

10. In boundary value problems like the above, why is it that there is either no nonzero solution or infinitely many?

11. In the study of buckling beams, you have an equation

$$y^{(4)}(x) + \lambda y''(x) = 0, x \in [0, L]$$

along with boundary conditions like

$$y(0) = y'(0) = 0, \text{ clamped at left end}$$
$$y(L) = 0 = y''(L), \text{ hinged at right end}$$

where λ increases with the axial force and depends on geometrical and physical properties of the beam. The idea is to find values of λ for which there is a nonzero solution to the differential equation and the boundary conditions. Assume all boundary conditions considered have $y(0) = y(L) = 0$ and at each end, either y' or y'' is equal to 0. Thus one considers beams for which each end is either clamped or hinged. Show that if λ is such that there exists a nonzero solution, then $\lambda > 0$. **Hint:** You show this by multiplying the equation by y and integrating by parts.

12. Letting $\lambda = \delta^2$, in the above problem, show that there exist infinitely many values for δ and corresponding nonzero solutions to the boundary value problem for the following situation.

    (a) $y(0) = y'(0) = 0, y(L) = y'(L) = 0$
    (b) $y(0) = y''(0) = 0, y(L) = y'(L) = 0$
    (c) $y(0) = y''(0) = 0, y(L) = y''(L) = 0$

13. A **Sturm-Liouville problem** involves the differential equation for an unknown function of x which is denoted here by y,

$$(p(x) y'(x))' + (\lambda q(x) + r(x)) y = 0, \ x \in [a, b]$$

and it is assumed that $p(t), q(t) \geq 0$ and are nonzero except for finitely many points in $[a, b]$ for any t along with boundary conditions,

$$C_1 y(a) + C_2 y'(a) = 0$$
$$C_3 y(b) + C_4 y'(b) = 0$$

where
$$C_1^2 + C_2^2 > 0, \text{ and } C_3^2 + C_4^2 > 0.$$

There is an immense theory connected to these important problems. The constant λ is called an eigenvalue. Show that if y is a solution to the above problem corresponding to $\lambda = \lambda_1$ and if z is a solution corresponding to $\lambda = \lambda_2 \neq \lambda_1$, then

$$\int_a^b q(x) y(x) z(x) \, dx = 0. \tag{13.11}$$

Hint: Do something like this:

$$(p(x) y')' z + (\lambda_1 q(x) + r(x)) yz = 0,$$

$$(p(x) z')' y + (\lambda_2 q(x) + r(x)) zy = 0.$$

Now subtract and either use integration by parts or show

$$(p(x) y')' z - (p(x) z')' y = ((p(x) y') z - (p(x) z') y)'$$

and then integrate. Use the boundary conditions to show that $y'(a) z(a) - z'(a) y(a) = 0$ and $y'(b) z(b) - z'(b) y(b) = 0$. The formula (13.11) is called an orthogonality relation and it makes possible an expansion in terms of certain functions called eigenfunctions.

14. Here is a really nice result. Suppose you have y, z are both solutions of the differential equation

$$(p(x) y'(x))' + q(x) y(x) = 0$$

Show that $p(x) W(y, z)(x) = C$ a constant. Here $W(y, z)$ is the Wronskian.

15. In the above problem, change the variables as follows. Let $z(x) = p(x) \frac{y'(x)}{y(x)}$ and determine the equation which results for z. This kind of equation is called a Riccati equation. In particular, show that

$$z' + \frac{1}{p(x)} z^2 + q(x) = 0$$

This kind of equation is like a Bernouli equation with exponent 2, but with another function added in. For more on this, see [17].

16. Suppose in the equation of Problem 14 you have two solutions u, v whose Wronskian is nonzero so they are independent solutions. Suppose that a, b are consecutive zeros of u and that $p(x) > 0$ on $[a, b]$. Show that v has exactly one zero in (a, b). This is called the Sturm separation theorem. **Hint:** Use the result of the above mentioned problem and argue that $v(a) \neq 0$ and that you can assume that $v(a) > 0$ and that u is positive on the open interval (a, b).

17. Letting $[a, b] = [-\pi, \pi]$, consider an example of a Sturm-Liouville problem which is of the form

$$y'' + \lambda y = 0, \; y(-\pi) = 0, \; y(\pi) = 0.$$

Show that if $\lambda = n^2$ and $y_n(x) = \sin(nx)$ for n a positive integer, then y_n is a solution to this regular Sturm-Liouville problem. In this case, $q(x) = 1$ and so from Problem 13, it must be the case that

$$\int_{-\pi}^{\pi} \sin(nx) \sin(mx)\, dx = 0$$

if $n \neq m$. Show directly using integration by parts that the above equation is true.

18. Sometimes one encounters an eigenvalue problem of the form

$$x^2 y'' + x y' + (\lambda x^2 - n^2) y = 0, \quad C_1 y(L) + C_2 y'(L) = 0$$

not both C_i equal zero and y bounded near 0. Discover an orthogonality relation between this solution and one for which λ is changed to μ. **Hint:** You might divide by x.

19. Let $x \to J_n(x)$ be a solution to the Bessel equation

$$x^2 y'' + x y' + (x^2 - n^2) y = 0$$

and suppose α is a positive number. Let $z(x) \equiv J_n(\alpha x)$. Find a differential equation satisfied by z. You should show that it satisfies $x^2 z'' + xz + (\alpha^2 x^2 - n^2) z = 0$.

20. Let α, β be two zeros of the Bessel function $J_n(x)$. It was shown in Proposition 8.10.1 on Page 211 that there are infinitely many of these zeros. Now consider the two functions $x \to J_n\left(\frac{\alpha}{L}x\right), x \to J_n\left(\frac{\beta}{L}x\right)$. Show that

$$\int_0^L J_n\left(\frac{\alpha}{L}x\right) J_n\left(\frac{\beta}{L}x\right) x\, dx = 0$$

21. Consider

$$x^2 y'' + x y' + (\delta^2 x^2 - n^2) y = 0, y(L) = 0, y \text{ bounded near } 0$$

Show that there are only certain values of δ which work and they are of the form $\delta^2 = (\alpha/L)^2$ where α is some zero of a solution to Bessel's equation.

22. Show that the only eigenvalues λ for

$$x^2 y'' + x y' + (\lambda x^2 - n^2) y = 0, \quad y(L) = 0$$

are positive.

23. Recall that for n an integer, the general solution to Bessel's equation is $C_1 J_n(x) + C_2 Y_n(x)$ where Y_n is unbounded at 0. Using the above problem, characterize all eigenvalues λ of the eigenvalue problem

$$x^2 y'' + x y' + (\lambda x^2 - n^2) y = 0, y(L) = 0, y \text{ bounded near } 0.$$

and describe all solutions to this boundary value problem in terms of Bessel functions. **Hint:** Rule out Y_n to begin with. Then consider $z\left(\sqrt{\lambda}x\right) = y(x)$ for y a solution to the above Sturm-Liouville equation.

24. A Sturm-Liouville eigenvalue problem involves the equation

$$(p(x) y'(x))' + (\lambda q(x) + r(x)) y = 0, x \in (a, b)$$

The Liouville transformation is

$$z = (p(x) q(x))^{1/4} y, \; t = \int_c^x \left(\frac{q(s)}{p(s)} \right)^{1/2} ds, c \in (a, b)$$

Then determine the equation solved by z. **Hint:** This is a little involved. First verify that the left side reduces to

$$\frac{d}{dt} \left(p \frac{d}{dx} \left((pq)^{-1/4} \right) z + (pq)^{1/4} \frac{dz}{dt} \right) \sqrt{\frac{q}{p}} + (\lambda q + r) (pq)^{-1/4} z = 0$$

Next verify that the $z'(t)$ terms all cancel. That way, in the above, you can neglect these terms in using the product rule. This leads to

$$\left(\frac{\frac{d}{dx} \left(-\frac{1}{4} p^{-1/4} q^{-5/4} \frac{d}{dx} (pq) \right)}{(p^{-1/4} q^{3/4})} + \frac{r (pq)^{-1/4}}{(p^{-1/4} q^{3/4})} \right) z + z'' + \lambda z = 0$$

Now argue that the equation is of the form

$$z'' + (\lambda + m(t)) z = 0$$

where $m(t)$ is a function which depends on p, q.

25. Consider the eigenvalue problem for Bessel's equation,

$$x^2 y'' + x y' + (\lambda x^2 - n^2) y = 0, \quad y(L) = 0$$

Show it can be written in self adjoint form as

$$(xy')' + \left(\lambda x - \frac{n^2}{x} \right) y = 0$$

Thus in this case, $q(x) = x$ and $r(x) = -n^2/x$. What is the form of the equation if Liouville's transformation is applied to this Bessel eigenvalue problem? **Hint:** Just use the specific description of what was obtained above and that $r(x) = -n^2/x, p(x) = q(x) = x$, and so $t = x$. You should get something like

$$z'' + \lambda z + \left(\frac{1 - 4n^2}{4x^2} \right) z = 0$$

26. In the above problem, let $\lambda = 1$ and let $n = 1/2$ and use to find the general solution to the Bessel equation in which $\nu = 1/2$. Show, using the above, that this general solution is of the form

$$C_1 x^{-1/2} \cos x + C_2 x^{-1/2} \sin x.$$

27. Show that the polynomial $q(x)$ of degree n which minimizes

$$\int_{-1}^{1} |f(x) - p(x)|^2 \, dx$$

out of all polynomials p of degree n is the nth partial sum of the Fourier series taken with respect to the Legendre polynomials $q(x) = S_n f(x)$, where $S_n f(x) \equiv \sum_{k=0}^{n} c_k p_k(x), c_k = \int_{-1}^{1} q_k(x) f(x) \, dx$.

28. Recall the normalized Legendre polynomials

$$q_n(x) = \frac{\sqrt{2n+1}}{\sqrt{2}} p_n(x)$$

which have the property that

$$\int_{-1}^{1} q_j(x) q_k(x) \, dx = \delta_{jk} = \begin{cases} 1 \text{ if } j = k \\ 0 \text{ if } j \neq k \end{cases}$$

If f is a Riemann integrable function, show that

$$\lim_{n \to \infty} \int_{-1}^{1} f(x) q_n(x) \, dx = 0$$

29. Show that if f is any continous function on $[-1, 1]$, then the Fourier series in terms of Legendre polynomials converges to f in the mean square sense. This means that for $S_n f(x) \equiv \sum_{k=0}^{n} c_k p_k(x), c_k = \int_{-1}^{1} q_k(x) f(x) \, dx$, it follows that

$$\lim_{n \to \infty} \int_{-1}^{1} |f(x) - S_n f(x)|^2 \, dx = 0$$

Hint: You should use Problem 24 on Page 222.

30. In Section 8.4 it was shown that there are no continuous, nonzero solutions to Legendre's equation

$$\left((1 - x^2) y' \right)' + \lambda y = 0$$

defined on $[-1, 1]$ unless $\lambda = n(n+1)$ for n an integer. Use the above problem to give a shorter proof.

31. One of the applications of Fourier series is to obtain solutions to linear differential equations which have a periodic right side. This is done by expanding the right side which is a forcing function in a Fourier series, solving the simple equation which corresponds to each term and then adding these solutions to obtain what is hoped to be a representation of the solution. Find a particular solution for each of the following. Let

$$y'' + 3y = f(t),$$

where $f(t)$ is the step function which is periodic of period 2 and equals -1 on $[-1, 0)$ and 1 on $(0, 1]$. Here are the steps. First find a Fourier series for f. Say $\sum_{n=1}^{\infty} b_n \sin(n\pi x)$. Then let y_n be the solution to

$$y_n'' + 3y_n = \sin(n\pi t)$$

and then hopefully, on neglecting mathematical issues, the solution to the original problem is

$$y(t) = \sum_{n=1}^{\infty} b_n y_n(t)$$

32. Explain why the above procedure should give a particular solution if mathematical issues related to interchange of limit operations are ignored.

33. This problem is tedious but maybe it is better to do it all at once than to repeat seemingly endless virtually identical problems. In this problem, a is positive and b is a nonzero real number while n is a nonnegative integer. Find the real and imaginary parts of a solution y to

$$y'' + 2ay' + by = \exp\left(i\frac{n\pi t}{L}\right)$$

using the method of undetermined coefficients. Show that the real part is

$$\frac{2\pi L^3 an \sin\frac{\pi}{L}nt - \pi^2 L^2 n^2 \cos\frac{\pi}{L}nt}{L^4 b^2 + 4\pi^2 L^2 a^2 n^2 - 2\pi^2 L^2 b n^2 + \pi^4 n^4}$$

and the imaginary part of the solution is

$$\frac{\left(L^4 b - \pi^2 L^2 n^2\right)\sin\frac{\pi}{L}nt - 2\pi L^3 an \cos\frac{\pi}{L}nt}{L^4 b^2 + 4\pi^2 L^2 a^2 n^2 - 2\pi^2 L^2 b n^2 + \pi^4 n^4}$$

Explain why the real part is a particular solution to

$$y'' + 2ay' + by = \cos\left(\frac{n\pi t}{L}\right)$$

and the imaginary part is a particular solution to

$$y'' + 2ay' + by = \sin\left(\frac{n\pi t}{L}\right)$$

In case n is 0, a solution is $1/b$.

34. Using the above problem, describe the solution after a long time to the equation

$$y'' + 2y' + 2y = f(t)$$

where $f(t)$ is a periodic function which has the following Fourier series. Note that the transient terms will disappear due to the fact that $a = 1$ is positive. Note that with the above problem, you could do many other examples in which a and b are not given as here.

(a) $\sum_{n=1}^{\infty} \frac{1}{n^2} \cos\left(\frac{n\pi t}{3}\right) + \sum_{n=1}^{\infty} \frac{1}{1+n^2} \sin\left(\frac{n\pi t}{3}\right) + 3$

(b) $\sum_{n=1}^{\infty} e^{-n} \cos\left(\frac{n\pi t}{2}\right) + \sum_{n=1}^{\infty} \frac{1}{n^4} \sin\left(\frac{n\pi t}{2}\right) + 1$

(c) $\sum_{n=1}^{\infty} \frac{1}{n^3} \cos\left(\frac{n\pi t}{4}\right) + \sum_{n=1}^{\infty} \frac{1}{n^3+1} \sin\left(\frac{n\pi t}{4}\right) - 2$

35. Suppose you have an undamped equation

$$y'' + 4y = f(t)$$

where f is periodic. Suppose in the Fourier expansion of $f(t)$ there is a nonzero term which is of the form $b \sin(2t)$. Say it describes the transverse vibrations of a bridge in the center. What will likely happen to this bridge?

36. Consider the functions $y_n(x) = \sin(n\pi x)$ on the interval $[0, 2]$. Show that these functions satisfy $\int_0^2 y_n(x) y_m(x) \, dx$ is 1 if $n = m$ and zero if $n \neq m$. Now consider using them to expand the function $f(x) = x$ in a Fourier series. Thus you would have

$$\sum_{n=1}^{\infty} b_n \sin(n\pi x)$$

where

$$b_n = \int_0^2 x \sin(n\pi x) \, dx$$

Graph the sum of the first seven terms in this Fourier series expansion along with the function it is supposedly approximating. What does this tell you about being able to approximate with orthogonal functions? Now do the same problem with the orthonormal functions $\sin\left(n\frac{\pi}{2}x\right)$.

37. Recall that a sequence of functions defined on $[a, b]$ $\{f_n\}$ converges to f in the mean square sense if

$$\lim_{n\to\infty} \int_a^b |f_n(x) - f(x)|^2 \, dx = 0$$

consider the function $f_n(x)$ for $x \in [0, 1]$ defined as follows. $f_n(x) = \sqrt{n}$ on $(0, 1/n)$ and $f_n(x) = 0$ for x not on this interval. Show that $\lim_{n\to\infty} f_n(x) = 0$ for each x but f_n fails to converge to 0 in the mean square sense. Now let $f_n(x) = 1$ for $x \in \{1, 1/2, 1/2^2, \cdots, 1/2^n\}$ but it equals zero at all other points. Show that f_n converges to 0 in the mean square sense but not at every point.

38. Using Example 13.5.5 and the convergence theorem for Fourier series, explain why

$$1 = \sum_{k=1}^{\infty} \frac{4}{\pi(2k-1)} \sin\left(\frac{(2k-1)\pi\alpha}{2}\right) \quad \text{for all } \alpha \in (0, 2).$$

Chapter 14

Some Partial Differential Equations

14.1 Laplacian in Orthogonal Curvilinear Coordinates

The Laplacian in rectangular coordinates is

$$\Delta u = u_{xx} + u_{yy} + u_{zz}$$

It is the divergence of the gradient.

$$\nabla \cdot (\nabla u)$$

However, in other coordinate systems, it is not obvious how to express the Laplacian of a function.

I am going to give a geometrical approach to finding the Laplacian in various coordinate systems. For a more complete treatment of this important problem, see good physics books or the appendix of [16]. This geometrical approach is very handy however and works well for the coordinate systems of most interest which are orthogonal curvilinear coordinates. See [10], [4].

Recall spherical coordinates from calculus

$$x = \rho \sin \phi \cos \theta$$
$$y = \rho \sin \phi \sin \theta$$
$$z = \rho \cos \phi$$

Here ϕ is the angle measured from positive z axis, and θ is the same as the angle in polar coordinates. Denote this function as follows.

$$\mathbf{r}(\rho, \phi, \theta) = (\rho \sin \phi \cos \theta)\mathbf{i} + (\rho \sin \phi \sin \theta)\mathbf{j} + (\rho \cos \phi)\mathbf{k}$$

You can also check that

$$\mathbf{r}_\rho \times \mathbf{r}_\phi \cdot \mathbf{r}_\theta = \rho^2 \sin(\phi) \geq 0$$

Recall that $\rho^2 \sin(\theta)\, d\rho d\phi d\theta$ is the increment of volume in spherical coordinates.

This is an orthogonal system because

$$\mathbf{r}_\rho \cdot \mathbf{r}_\theta = \mathbf{r}_\rho \cdot \mathbf{r}_\phi = \mathbf{r}_\theta \cdot \mathbf{r}_\phi = 0$$

You should check that this is in fact the case. More generally, we have for $\mathbf{x} = x\mathbf{i} + y\mathbf{j} + z\mathbf{k}$ and coordinates (u_1, u_2, u_3),

$$\mathbf{x} = \mathbf{r}(u_1, u_2, u_3) = \mathbf{r}(\mathbf{u})$$

is an orthogonal system of coordinates if

$$\mathbf{r}_{u_1} \cdot \mathbf{r}_{u_2} = \mathbf{r}_{u_1} \cdot \mathbf{r}_{u_3} = \mathbf{r}_{u_2} \cdot \mathbf{r}_{u_3} = 0$$

We will assume that one can invert $\mathbf{r}(\mathbf{u})$ to solve $\mathbf{u} = \mathbf{r}^{-1}(\mathbf{x})$ so we assume that $\mathbf{r}$ is one to one and it and its inverse have continuous partial derivatives. The inverse or implicit function theorem allows one to justify this. We will also assume that the axes in the u_i system are oriented such that $\mathbf{r}_{u_1} \times \mathbf{r}_{u_2} \cdot \mathbf{r}_{u_3} > 0$. From calculus, the increment of volume in this system is $dV = (\mathbf{r}_{u_1} \times \mathbf{r}_{u_2} \cdot \mathbf{r}_{u_3})\, du_1 du_2 du_3$.

First of all, note that, from the chain rule,

$$\frac{\partial u_i}{\partial x_1}\frac{\partial x_1}{\partial u_j} + \frac{\partial u_i}{\partial x_2}\frac{\partial x_2}{\partial u_j} + \frac{\partial u_i}{\partial x_3}\frac{\partial x_3}{\partial u_j} = \frac{\partial u_i}{\partial u_j} = \left\{ \begin{array}{l} 1 \text{ if } i = j \\ 0 \text{ if } i \neq j \end{array} \right. \equiv \delta_{ij}$$

In short,

$$\nabla u_i \cdot \mathbf{r}_{u_j} = \delta_{ij}$$

The vector $\mathbf{r}_{u_1}$ is in the direction of motion of the curve $u_1 \to \mathbf{r}(u_1, u_2, u_3)$, similar for the other u_i. By the orthogonality condition, ∇u_i must be in the direction of $\mathbf{r}_{u_i}$. It is parallel to it and their dot product is positive from the above, and so they are indeed in the same direction. Now define a scaling factor

$$h_i \equiv |\mathbf{r}_{u_j}|$$

Thus by the orthogonality condition,

$$|\nabla u_i|\,|\mathbf{r}_{u_j}| = |\nabla u_i|\, h_i = 1 \tag{14.1}$$

so $|\nabla u_i| = 1/h_i$.

Also, the volume element is

$$dV = (\mathbf{r}_{u_1} \times \mathbf{r}_{u_2} \cdot \mathbf{r}_{u_3})\, du_1 du_2 du_3 = h_1 h_2 h_3 du_1 du_2 du_3 \tag{14.2}$$

because of orthogonality of the vectors $\mathbf{r}_{u_i}$. Define the orthonormal set of vectors

$$\mathbf{e}_i \equiv \mathbf{r}_{u_i}/h_i, i = 1, 2, 3$$

Thus ∇u_i points in the direction of $\mathbf{e}_i$ and has magnitude $1/h_i$ so it follows that

$$\nabla u_i = \frac{1}{h_i}\frac{\mathbf{r}_{u_i}}{h_i} = \frac{1}{h_i}\mathbf{e}_i \tag{*}$$

Now let f be some C^1 function defined on an open set U in three dimensional space. This function is defined at a geometric point of space which can be identified using either the coordinates u_i or the rectangular coordinates x, y, z. Thus we write $f(u_1, u_2, u_3) = f(x, y, z)$. Then the gradient of f at some point is equal to the following thanks to the chain rule and $*$.

$$\nabla f = \sum_i \frac{\partial f}{\partial u_i} \frac{\partial u_i}{\partial x} \mathbf{i} + \sum_i \frac{\partial f}{\partial u_i} \frac{\partial u_i}{\partial y} \mathbf{j} + \sum_i \frac{\partial f}{\partial u_i} \frac{\partial u_i}{\partial z} \mathbf{k} = \sum_i \frac{\partial f}{\partial u_i} \nabla u_i = \sum_i \frac{1}{h_i} \frac{\partial f}{\partial u_i} \mathbf{e}_i$$

Example 14.1.1 *What is the gradient in spherical coordinates?*

We need the vectors $\mathbf{e}_i$ and the scaling factors which are just $|\mathbf{r}_r|, |\mathbf{r}_\phi|, |\mathbf{r}_\theta|$ respectively. Here are the $\mathbf{e}_i$.

$$\mathbf{e}_1 = \begin{pmatrix} \sin\phi\cos\theta \\ \sin\phi\sin\theta \\ \cos\phi \end{pmatrix}, \mathbf{e}_2 = \frac{1}{\rho}\begin{pmatrix} \rho\cos\phi\cos\theta \\ \rho\cos\phi\sin\theta \\ -\rho\sin\phi \end{pmatrix},$$

$$\mathbf{e}_3 = \frac{1}{\rho\sin\phi}\begin{pmatrix} -\rho\sin\phi\sin\theta \\ \rho\sin\phi\cos\theta \\ 0 \end{pmatrix}$$

The scaling factors being $1, \rho, \rho\sin\phi$ respectively corresponding to ρ, ϕ, θ. Then the gradient in spherical coordinates is

$$\frac{\partial f}{\partial \rho}\mathbf{e}_1 + \frac{1}{\rho}\frac{\partial f}{\partial \phi}\mathbf{e}_2 + \frac{1}{\rho\sin\phi}\frac{\partial f}{\partial \theta}\mathbf{e}_3$$

Next consider the divergence. Let $\mathbf{F}$ be a vector field. Thus it can be written as

$$\mathbf{F} = F_1\mathbf{e}_1 + F_2\mathbf{e}_2 + F_3\mathbf{e}_3.$$

If $\mathbf{F} = \nabla f$ then from what was shown above, it is of the form

$$\sum_i \frac{1}{h_i}\frac{\partial f}{\partial u_i}\mathbf{e}_i = \frac{1}{h_1}\frac{\partial f}{\partial u_1}\mathbf{e}_1 + \frac{1}{h_2}\frac{\partial f}{\partial u_2}\mathbf{e}_2 + \frac{1}{h_3}\frac{\partial f}{\partial u_3}\mathbf{e}_3$$

In any case $\mathbf{F} \cdot \mathbf{e}_i = F_i$.

From the divergence theorem, it is approximately the flux out of a very small box illustrated below. This box is not drawn correctly. It is actually much smaller, but needs to be blown up to illustrate. Also the sides of this box are in reality curved, but the idea is to let it shrink.

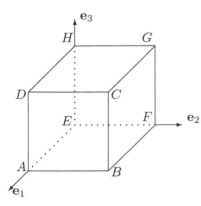

This is a geometrical argument on the level of what is usually done in calculus to obtain the volume increment in another coordinate system, not a rigorous one.

The sides of the box are, respectively, of length $\Delta u_1 h_1, \Delta u_2 h_2$, and $\Delta u_3 h_3$. This is because $\mathbf{e}_i = \mathbf{r}_{u_i}/h_i$ so if you have a change in u_1 called Δu_1, this would correspond approximately to

$$\mathbf{r}(u_1 + \Delta u_1, u_2, u_3) - \mathbf{r}(u_1, u_2, u_3)$$
$$\approx \mathbf{r}_{u_1}(u_1, u_2, u_3)\Delta u_1 = h_1\Delta u_1\mathbf{e}_1$$

Similar considerations apply to the other sides.

Consider the flux out of the two sides $ABCD$ and $EFGH$. $\mathbf{e}_1$ is perpendicular to the side $ABCD$ and pointing away from the box. $-\mathbf{e}_1$ is pointing away from the box on side $EFGH$ and perpendicular to this face. Hence the flux out of these two sides is of the form

$$F_1 h_2 h_3 \Delta u_2 \Delta u_3 - F_1 h_2 h_3 \Delta u_2 \Delta u_3 = (F_1 h_2 h_3 - F_1 h_2 h_3) \Delta u_2 \Delta u_3$$

$$\approx \frac{\partial}{\partial u_1} (F_1 h_2 h_3) \Delta u_1 \Delta u_2 \Delta u_3$$

when Δu_i gets increasingly small. Similar considerations apply to the other faces. Thus you get the flux is approximately

$$\frac{\partial}{\partial u_1} (F_1 h_2 h_3) \Delta u_1 \Delta u_2 \Delta u_3 + \frac{\partial}{\partial u_2} (F_2 h_1 h_3) \Delta u_1 \Delta u_2 \Delta u_3 + \frac{\partial}{\partial u_3} (F_3 h_1 h_2) \Delta u_1 \Delta u_2 \Delta u_3$$

with the approximation getting increasingly good as $\Delta u_i \to 0$. This flux is equal to the integral of the divergence of $\mathbf{F}$ over the interior of the box. This is by the divergence theorem. Therefore, if you divide by the volume of the box, which is approximately $h_1 h_2 h_3 \Delta u_1 \Delta u_2 \Delta u_3$, you get

$$\frac{1}{h_1 h_2 h_3} \left(\frac{\partial}{\partial u_1} (F_1 h_2 h_3) + \frac{\partial}{\partial u_2} (F_2 h_1 h_3) + \frac{\partial}{\partial u_3} (F_3 h_1 h_2) \right)$$

where the approximation gets better if the size of the box shrinks. It follows that $\nabla \cdot \mathbf{F} =$

$$\frac{1}{h_1 h_2 h_3} \left(\frac{\partial}{\partial u_1} (F_1 h_2 h_3) + \frac{\partial}{\partial u_2} (F_2 h_1 h_3) + \frac{\partial}{\partial u_3} (F_3 h_1 h_2) \right)$$

This is the formula for the divergence. As the case of most interest, we have in mind the divergence of a gradient. As shown above,

$$\nabla f = \frac{1}{h_1} \frac{\partial f}{\partial u_1} \mathbf{e}_1 + \frac{1}{h_2} \frac{\partial f}{\partial u_2} \mathbf{e}_2 + \frac{1}{h_3} \frac{\partial f}{\partial u_3} \mathbf{e}_3$$

Therefore, $\nabla \cdot \nabla f =$

$$\frac{1}{h_1 h_2 h_3} \left(\frac{\partial}{\partial u_1} \left(\frac{1}{h_1} \frac{\partial f}{\partial u_1} h_2 h_3 \right) + \frac{\partial}{\partial u_2} \left(\frac{1}{h_2} \frac{\partial f}{\partial u_2} h_1 h_3 \right) + \frac{\partial}{\partial u_3} \left(\frac{1}{h_3} \frac{\partial f}{\partial u_3} h_1 h_2 \right) \right)$$

Example 14.1.2 *Laplacian in spherical coordinates.*

Here the new coordinates are $(u_1, u_2, u_3) = (\rho, \phi, \theta)$ and as shown above,

$$h_1 = 1, h_2 = \rho, h_3 = \rho \sin \phi.$$

Therefore, the Laplacian is $\Delta f = \nabla \cdot \nabla f =$

$$\frac{1}{\rho^2 \sin \phi} \left(\frac{\partial}{\partial \rho} \left(\rho^2 \sin \phi \frac{\partial f}{\partial \rho} \right) + \frac{\partial}{\partial \phi} \left(\frac{\rho \sin \phi}{\rho} \frac{\partial f}{\partial \phi} \right) + \frac{\partial}{\partial \theta} \left(\frac{\rho}{\rho \sin \phi} \frac{\partial f}{\partial \theta} \right) \right)$$

$$= \frac{1}{\rho^2} \frac{\partial}{\partial \rho} \left(\rho^2 \frac{\partial f}{\partial \rho} \right) + \frac{1}{\rho^2 \sin \phi} \frac{\partial}{\partial \phi} \left(\sin (\phi) \frac{\partial f}{\partial \phi} \right) + \frac{1}{\rho^2 \sin^2 \phi} \frac{\partial^2 f}{\partial \theta^2}$$

Example 14.1.3 *Laplacian in cylindrical coordinates.*

$$x = r \cos \theta$$
$$y = r \sin \theta$$
$$z = z$$

The $\mathbf{e}_i$ are

$$\begin{pmatrix} \cos \theta \\ \sin \theta \\ 0 \end{pmatrix}, \begin{pmatrix} -r \sin \theta \\ r \cos \theta \\ 0 \end{pmatrix}, \begin{pmatrix} 0 \\ 0 \\ 1 \end{pmatrix}$$

Here we are listing them as ρ, θ, z. You can see that the orientation is consistent with the above. Then the magnification factors are $1, r, 1$. Thus the Laplacian in cylindrical coordinates is $\Delta f =$

$$\frac{1}{r} \left(\frac{\partial}{\partial r} \left(r \frac{\partial f}{\partial r} \right) + \frac{\partial}{\partial \theta} \left(\frac{1}{r} \frac{\partial f}{\partial \theta} \right) + \frac{\partial}{\partial z} \left(r \frac{\partial f}{\partial z} \right) \right) = \frac{1}{r} \frac{\partial}{\partial r} \left(r \frac{\partial f}{\partial r} \right) + \frac{1}{r^2} \frac{\partial^2 f}{\partial \theta^2} + \frac{\partial^2 f}{\partial z^2}.$$

As to polar coordinates, there is no real difference. In two dimensions, the formula would be similar,

$$\frac{1}{h_1 h_2} \left(\frac{\partial}{\partial u_1} \left(\frac{1}{h_1} \frac{\partial f}{\partial u_1} h_2 \right) + \frac{\partial}{\partial u_2} \left(\frac{1}{h_2} \frac{\partial f}{\partial u_2} h_1 \right) \right)$$

also, the magnification factors are $1, r$ for r, θ in this order. Therefore, $\Delta f =$

$$\frac{1}{r} \left(\frac{\partial}{\partial r} \left(r \frac{\partial f}{\partial r} \right) + \frac{\partial}{\partial \theta} \left(\frac{1}{r} \frac{\partial f}{\partial \theta} \right) \right) = \frac{1}{r} \frac{\partial}{\partial r} \left(r \frac{\partial f}{\partial r} \right) + \frac{1}{r^2} \frac{\partial^2 f}{\partial \theta^2}$$

14.2   Heat and Wave Equations

14.2.1   Heat Equation

Fourier's law of heat conduction is that the heat flux $\mathbf{J}$ is proportional to the temperature gradient ∇u where here u is the temperature. Specifically it says that

$$\mathbf{J} = -k \nabla u$$

So what is the "heat flux"? Hopefully, you saw flux integrals in calculus but here is a short review. If you have a surface S and a field of unit normals on S denoted as $\mathbf{n}$, then the rate at which the heat crosses S in the direction of $\mathbf{n}$ is

$$\int_S \mathbf{J} \cdot \mathbf{n} dS$$

where this is an integral over the surface. Now consider a ball B with boundary S in a heat conducting material. Then the heat in B is given by

$$\int_B \rho c u dV$$

where ρ is the density and c the specific heat. Then if no heat is being produced by some chemical reaction for example, it follows that the time rate of change of the total heat in B is equal to the rate at which heat flows into B. Thus

$$\frac{d}{dt}\left(\int_B \rho cu dV\right) = -\int_S \mathbf{J} \cdot \mathbf{n} dS$$

where $\mathbf{n}$ is the outer normal from B. This is why there is a minus sign on the right. You want the rate at which heat enters B. Then from the divergence theorem,

$$\frac{d}{dt}\left(\int_B \rho cu dV\right) = -\int_B \nabla \cdot \mathbf{J} dV$$

The integral is a sort of a sum, here over the spacial variables and so it makes sense to formally take the time derivative into the integral* and write, using the Fourier law of heat conduction

$$\int_B \frac{\partial (\rho cu)}{\partial t} dV = \int_B \nabla \cdot (k\nabla u)\, dV$$

This must hold for any ball B and so the only way this could take place is to have

$$\frac{\partial (\rho cu)}{\partial t} = \nabla \cdot (k\nabla u)$$

We now let k, c, ρ all be constants and obtain

$$\frac{\partial u}{\partial t} = \frac{k}{\rho c}\Delta u$$

Of course these things are typically not constant, especially k but if we don't assume this, we can't solve the equation.

In one dimension, this reduces to

$$u_t = \alpha^2 u_{xx}$$

and this is the equation in what follows. There are other issues besides the equation to consider.

You have a rod of length L. The heat equation for the temperature u in the rod is

$$u_t = \alpha^2 u_{xx}$$

In addition to this, there are boundary conditions given on u at the ends of the rod. For example, you could have

$$u(0,t) = u(L,t) = 0$$

and there is also an initial temperature given

$$u(x,0) = f(x)$$

---

*This is horrible mathematics because it exchanges two limit operations. However, when modeling, one doesn't worry about rigorous math.

Then the idea is to find the unknown function $u(t, x)$. Here t is time and x is the coordinate of a point on the rod. The constant α^2 varies from material to material. It is different for iron than for aluminum for example. Here you have $x \in [0, L]$ and $t > 0$.

This is a rectangular shape and so it is reasonable to look for a nonzero solution to the above partial differential equation and boundary condition in the form

$$u(x, t) = a(t)b(x)$$

Then

$$a'(t)b(x) = \alpha^2 a(t)b''(x)$$

One can separate the variables as follows.

$$\frac{a'(t)}{\alpha^2 a(t)} = \frac{b''(x)}{b(x)} \tag{14.3}$$

Both sides must equal to some constant c since otherwise they could not be equal. One way to see this is to differentiate both sides with respect to t. Then

$$\left(\frac{a'(t)}{\alpha^2 a(t)}\right)' = 0 \text{ and so } \frac{a'(t)}{\alpha^2 a(t)} = c,$$

a constant. Consider the side involving x.

$$b''(x) - cb(x) = 0, \ b(0) = b(L) = 0$$

Of course you can't have $b(x) = 0$ since if it were 0, you would have $u(x, t) = 0$. Therefore, from Example 13.2.1, $-c = \frac{n^2 \pi^2}{L^2}$ where n is a positive integer and

$$b(x) = \sin\left(\frac{n\pi x}{L}\right)$$

Of course there is such a function for each n a positive integer. Having picked such a positive integer, (14.3) now forces $a(t)$ to satisfy the equation

$$a'(t) + \frac{n^2 \pi^2 \alpha^2}{L^2}a(t) = 0$$

Therefore,

$$a(t) = a_n e^{-\frac{n^2 \pi^2 \alpha^2}{L^2}t}$$

It follows that for each n, there exists a solution to the partial differential equation along with the boundary conditions which is of the form

$$u_n(x, t) = a_n e^{-\frac{n^2 \pi^2 \alpha^2}{L^2}t} \sin\left(\frac{n\pi x}{L}\right)$$

Now if you have solutions to the differential equation along with the boundary condition and you add them together, you have another solution to these things. Therefore, it is not unreasonable to hope that this would also be true for an infinite sum of such solutions. Therefore, we look for a solution to the partial differential equation which is of the form

$$u\left(x,t\right) = \sum_{n=1}^{\infty} a_n e^{-\frac{n^2\pi^2\alpha^2}{L^2}t} \sin\left(\frac{n\pi x}{L}\right)$$

At least formally, such a thing would solve everything but the initial condition. Now you choose a_n in such a way that when $t = 0$,

$$f\left(x\right) = \sum_{n=1}^{\infty} a_n \sin\left(\frac{n\pi x}{L}\right)$$

for $x \in [0, L]$. This is now a Fourier series problem.

Another point of view is to look for eigenfunctions. b such that

$$b''\left(x\right) + \lambda b\left(x\right) = 0, \ b\left(0\right) = b\left(L\right) = 0$$

This is because if you had such an eigenfunction, you could replace the $b''\left(x\right)$ with $-\lambda b\left(x\right)$. From Example 13.2.1 on Page 334, $\lambda = \frac{n^2\pi^2}{L^2}$ where n is a positive integer and

$$b\left(x\right) = \sin\left(\frac{n\pi x}{L}\right)$$

Denote by b_n this eigenfunction. Then look for a solution to the whole problem which is in the form

$$u\left(x,t\right) = \sum_{k=1}^{\infty} a_k\left(t\right) b_k\left(x\right)$$

Then proceeding formally,

$$\sum_{k=1}^{\infty} a_k'\left(t\right) b_k\left(x\right) = \alpha^2 \sum_{k=1}^{\infty} a_k\left(t\right) b_k''\left(x\right) = \sum_{k=1}^{\infty} a_k\left(t\right)\left(-\frac{k^2\pi^2}{L^2}\alpha^2\right) b_k\left(x\right)$$

It follows that you should have

$$a_k'\left(t\right) + \frac{k^2\pi^2\alpha^2}{L^2}a_k\left(t\right) = 0$$

so this results in

$$u\left(x,t\right) = \sum_{k=1}^{\infty} a_k e^{-\frac{k^2\pi^2\alpha^2}{L^2}t} \sin\left(\frac{k\pi x}{L}\right),$$

the same as before. Then you just try and find the a_k to satisfy the initial condition.

Here is a summary of the method. This method is general and will work for all the examples discussed here.

Procedure 14.2.1 *To find the solution to an equation*

$$u_t = \alpha^2 u_{xx}, \ \textit{zero boundary conditions, Initial condition}$$

you do the following.

1. *First find eigenfunctions, nonzero solutions to*

$$y'' + \lambda^2 y = 0, \;\; boundary \; conditions$$

There will typically be infinitely many of these $\{y_n(x)\}_{n=1}^{\infty}$ corresponding to eigenvalues λ_n where $\lim_{n \to \infty} \lambda_n = \infty$.

2. *Your solution will then be of the form*

$$u(x,t) = \sum_{n=1}^{\infty} b_n(t) y_n(x)$$

3. *Choose $b_n(t)$ to satisfy the equation $b'_n(t) = -\lambda_n^2 b_n(t)$ in order that the terms of the sum satisfy the partial differential equation. Thus*

$$b_n(t) = b_n \exp\left(-t\lambda_n^2\right)$$

Then the solution to the problem is

$$u(x,t) = \sum_{n=1}^{\infty} b_n \exp\left(-t\lambda_n^2\right) y_n(x)$$

where b_n is chosen such that $\sum_{n=1}^{\infty} b_n y_n(x)$ is the Fourier series expansion for the initial condition.

Example 14.2.2 *Find the solution to the initial boundary value problem*

$$
\begin{aligned}
u_t &= .1u_{xx}, \;\; u(0,t) = u(2,t) = 0 \\
u(x,0) &= 1 - (1-x)^2
\end{aligned}
$$

where

$$f(x) = \begin{cases} x & if \; x \in [0,1] \\ 1-x & if \; x \in [1,2] \end{cases}$$

From the above discussion,

$$u(x,t) = \sum_{k=1}^{\infty} a_k e^{-\frac{.1k^2\pi^2}{2^2}t} \sin\left(\frac{k\pi x}{2}\right)$$

the eigenfunctions being $\sin\left(\frac{k\pi x}{2}\right)$, and to satisfy the initial condition, you need

$$a_k = \frac{2}{2} \int_0^2 \left(1 - (1-x)^2\right) \sin\left(\frac{k\pi x}{2}\right) dx$$

After some tedious computations, this yields

$$a_k = \frac{16}{\pi^3 k^3}\left(1 - (-1)^k\right)$$

Thus when k is even, this is 0 and when k is odd, it equals $\frac{32}{\pi^3 k^3}$. Thus

$$u(x,t) = \sum_{k=1}^{\infty} \frac{32}{\pi^3 (2k-1)^3} e^{-\frac{(2k-1)^2\pi^2}{4}(.1)t} \sin\left(\frac{(2k-1)\pi x}{2}\right)$$

The next example has to do with the same equation but with one end insulated and the other held at a temperature of 0. The physical modeling of this equation shows that to consider an insulated boundary, say at L, you let $u_x(L, t) = 0$.

Example 14.2.3 *Solve the problem*

$$u_t = .1u_{xx}, \ u(0, t) = u_x(2, t) = 0$$
$$u(x, 0) = 1 - (1 - x)^2$$

To do this, first look for eigenfunctions. Find solutions to

$$y'' + \lambda y = 0, \ y(0) = 0, y'(2) = 0$$

Then the eigenfunctions are in Example 13.2.1. They are

$$\sin\left(\frac{(2n - 1)\pi x}{4}\right), \ n = 1, 2, \cdots$$

It follows that the solution desired is of the form

$$\sum_{n=1}^{\infty} b_n(t) \sin\left(\frac{(2n - 1)\pi x}{4}\right)$$

and one needs

$$b_n'(t) = -\frac{1}{10}\left(\frac{(2n - 1)\pi}{4}\right)^2 b_n(t)$$

so

$$b_n(t) = b_n \exp\left(-\frac{1}{10}\left(\frac{(2n - 1)\pi}{4}\right)^2 t\right)$$

Then the Fourier series expansion of the solution is

$$\sum_{n=1}^{\infty} b_n \exp\left(-\frac{1}{10}\left(\frac{(2n - 1)\pi}{4}\right)^2 t\right) \sin\left(\frac{(2n - 1)\pi x}{4}\right)$$

where b_n is an appropriate Fourier coefficient chosen to satisfy the initial condition. Thus

$$b_n = \frac{2}{2}\int_0^2 \sin\left(\frac{(2n - 1)\pi x}{4}\right)\left(1 - (1 - x)^2\right) dx$$

$$= \frac{32}{\pi^3(2n - 1)^3}\left(2(-1)^n \pi n - (-1)^n \pi + 4\right)$$

Then the solution to this problem is

$$\sum_{n=1}^{\infty}\left(\frac{32}{\pi^3(2n - 1)^3}\left(2(-1)^n \pi n - (-1)^n \pi + 4\right)\right) e^{-\frac{1}{10}\left(\frac{(2n-1)\pi}{4}\right)^2 t} \sin\left(\frac{(2n - 1)\pi x}{4}\right)$$

A graph of this function of two variables in which the sum is taken up to 8 for $(t, x) \in [0, 12] \times [0, 2]$ is:

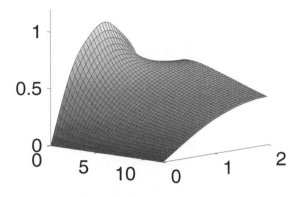

14.2.2   The Wave Equation

The next example is of a different sort of equation, the wave equation. This equation is of the form

$$u_{tt} = c^2 u_{xx}$$

It models the transverse displacements of a vibrating string. Here is a picture to discuss why this is an appropriate equation. It is important to note that it is a string, not a beam. This means that it cannot support itself in the sense that there is no internal stiffness. It is also very important to note that the transverse displacements are assumed to be very small. Thus the picture drawn below is blown up in the vertical direction.

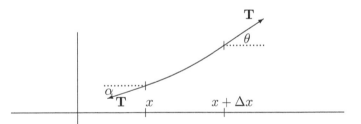

Let ρ be the length density of this string which is assumed constant. This means that the mass of the segment of string shown is just ρ (length of the segment of string). Since the transverse displacements are very small, this is essentially $\rho \Delta x$. The force acting on the segment of string shown is $T \sin \theta - T \sin \alpha$, where T is the magnitude of the vector $\mathbf{T}$. We assume also that the magnitude of the tension in the string is also a constant due to the assumption that the displacements are small. Let $u(t, x)$ denote the vertical displacement from horizontal. For Δx small enough, the acceleration $u_{tt}(t, x)$ should be essentially constant on the interval $[x, x + \Delta x]$. Then by Newton's second law,

$$\rho \Delta x u_{tt}(t, x) = T(\sin \theta - \sin \alpha)$$

Since the displacement is very small, we can assume that there is really no difference in replacing $\sin \theta$, $\sin \alpha$ with $\tan \theta$, $\tan \alpha$ respectively. But $\tan \theta$ is just the slope of the tangent line at $(t, x + \Delta x)$. Thus

$$\rho \Delta x u_{tt} (t, x) = T (u_x (t, x + \Delta x) - u_x (t, x))$$

Divide by Δx and let $\Delta x \to 0$ to obtain

$$\rho u_{tt} = T u_{xx}, \quad u_{tt} = \frac{T}{\rho} u_{xx}.$$

This is the wave equation for a vibrating string.

Since it is second order in t you need two initial conditions, one on the velocity and the other on the displacement in order to get a unique solution. However, other than this, the procedure is essentially the same.

Example 14.2.4 *Find the solution to the initial boundary value problem*

$$\begin{aligned}
u_{tt} &= \alpha^2 u_{xx}, \quad u(0, t) = u(2, t) = 0, \\
u(x, 0) &= 1 - (1 - x)^2 \\
u_t(x, 0) &= 0
\end{aligned}$$

The eigenfunctions are solutions to

$$y''(x) + \lambda y(x) = 0, \; y(0) = 0 = y(2)$$

This is discussed in Example 13.2.1. The eigenfunctions are

$$\sin \left(\frac{n \pi x}{2} \right)$$

and the eigenvalues are $\lambda = \frac{n^2 \pi^2}{4}$.

Then you look for a solution to the equation with boundary conditions of the form

$$a(t) \sin \left(\frac{n \pi x}{2} \right)$$

Thus you need

$$a''(t) \sin \left(\frac{n \pi x}{2} \right) = -\alpha^2 \frac{n^2 \pi^2}{4} a(t) \sin \left(\frac{n \pi x}{2} \right)$$

Hence

$$a'' + \alpha^2 \frac{n^2 \pi^2}{4} a = 0$$

and so, since you know the general solution to this equation, it is

$$a(t) = a_n \cos \left(\alpha \frac{n \pi}{2} t \right) + b_n \sin \left(\alpha \frac{n \pi}{2} t \right)$$

It follows that the solution to the full problem will be of the form

$$u(x, t) = \sum_{n=1}^{\infty} \left(a_n \cos \left(\alpha \frac{n \pi}{2} t \right) + b_n \sin \left(\alpha \frac{n \pi}{2} t \right) \right) \sin \left(\frac{n \pi x}{2} \right)$$

Now you need to find a_n and b_n to get the initial conditions. Letting $t = 0$, you need to have

$$1 - (1 - x)^2 = \sum_{n=1}^{\infty} a_n \sin\left(\frac{n\pi x}{2}\right)$$

and this is something which was done earlier. You need

$$a_n = \frac{16}{\pi^3 n^3}\left(1 - (-1)^n\right)$$

Next, what about b_n? Differentiate both sides. Thus

$$u_t(x, t) = \sum_{n=1}^{\infty}\left(a_n\left(-\alpha\frac{n\pi}{2}\right)\sin\left(\alpha\frac{n\pi}{2}t\right) + b_n\left(\alpha\frac{n\pi}{2}\right)\cos\left(\alpha\frac{n\pi}{2}t\right)\right)\sin\left(\frac{n\pi x}{2}\right)$$

Of course this operation is complete garbage because it involves the interchange of limit operations without any justification. However, we do it anyway. In fact it is all right. You can do the formal manipulations and then you can rigorously verify that what you end up with really is a solution to the problem in some sense. Now plug in $t = 0$. Then you need

$$0 = \sum_{n=0}^{\infty} b_n\left(\alpha\frac{n\pi}{2}\right)\sin\left(\frac{m\pi x}{2}\right)$$

Clearly you should take $b_n = 0$. Therefore, the desired solution is

$$u(x, t) = \sum_{n=1}^{\infty}\left(\frac{32}{\pi^3(2n-1)^3}\cos\left(\alpha\frac{(2n-1)\pi}{2}t\right)\right)\sin\left(\frac{n\pi x}{2}\right)$$

Let's let $\alpha^2 = .09$. Then the specific solution is

$$u(x, t) = \sum_{n=1}^{\infty}\left(\frac{32}{\pi^3(2n-1)^3}\cos\left(.3\frac{(2n-1)\pi}{2}t\right)\right)\sin\left(\frac{n\pi x}{2}\right)$$

Note that from calculus, the series makes perfect sense because in fact, it converges absolutely.

Example 14.2.5 *Solve the initial boundary value problem*

$$\begin{aligned}
u_{tt} &= \alpha^2 u_{xx}, \quad u(0, t) = u(4, t) = 0,\\
u(x, 0) &= f(x)\\
u_t(x, 0) &= 0
\end{aligned}$$

where

$$f(x) = \begin{cases} 1 - (x - 2)^2 & \text{on } [1, 3]\\ 0 & \text{on the rest of } [0, 4]\end{cases}$$

By similar reasoning to the above example,

$$u\left(x,t\right) = \sum_{n=1}^{\infty}\left(a_n\cos\left(\alpha\frac{n\pi}{4}t\right) + b_n\sin\left(\alpha\frac{n\pi}{4}t\right)\right)\sin\left(\frac{n\pi x}{4}\right)$$

Then, as above, $b_n = 0$ and a_n must be chosen such that

$$f\left(x\right) = \sum_{n=1}^{\infty}a_n\sin\left(\frac{n\pi x}{4}\right)$$

Thus

$$a_n = \frac{2}{4}\int_1^3\left(1 - (x-2)^2\right)\sin\left(\frac{n\pi x}{4}\right)dx$$

Then after doing the hard work, you end up with

$$a_n = -16\frac{4\cos\frac{3}{4}n\pi + n\pi\sin\frac{3}{4}n\pi - 4\cos\frac{1}{4}n\pi + n\pi\sin\frac{1}{4}n\pi}{n^3\pi^3}$$

Then the solution is

$$u\left(x,t\right) = \sum_{n=1}^{\infty}\left(-16\frac{4\cos\frac{3}{4}n\pi + n\pi\sin\frac{3}{4}n\pi - 4\cos\frac{1}{4}n\pi + n\pi\sin\frac{1}{4}n\pi}{n^3\pi^3}\right)\cdot$$
$$\cos\left(\alpha\frac{n\pi}{4}t\right)\sin\left(\frac{n\pi x}{4}\right)$$

Let $\alpha = .5$ to give a specific example. Here is a graph of the function of two variables in which the sum is taken up to $n = 6$. The t axis goes from 0 to 10 and if you fix t and imagine a cross section, it will be $x \to u(x,t)$.

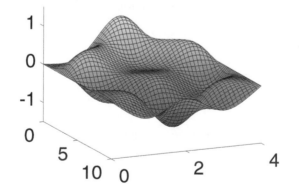

14.3   Nonhomogeneous Problems

For the sake of completeness, here is a brief discussion of what can be done if you have a nonhomogeneous equation of the form $u_t = au_{xx} + f$ along with an initial condition

$$u\left(x,0\right) = g\left(x\right)$$

and boundary conditions. As before, there are eigenfunctions y_n satisfying the boundary conditions and

$$y_n'' = -\lambda_n^2 y_n, \quad \lim_{n \to \infty} \lambda_n = \infty$$

such that also

$$\int_0^L y_n(x) \, y_m(x) \, dx = \delta_{nm} = \begin{cases} 1 \text{ if } n = m \\ 0 \text{ if } n \neq m \end{cases}$$

and it is assumed that one can obtain a valid Fourier series expansion in terms of these eigenfunctions of all the functions of interest. Note how, for the sake of simplicity, it is assumed that

$$\int_0^L y_n^2(x) \, dx = 1$$

You multiply by an appropriate constant to make it this way. Thus, if the eigenfunctions are multiples of $\sin\left(\frac{n\pi}{L}x\right)$, you choose the multiple to satisfy the above equation. Let

$$f(x, t) = \sum_{n=0}^{\infty} f_n(t) \, y_n(x)$$

Thus it is desired to have

$$\sum_{n=0}^{\infty} b_n'(t) \, y_n(x) = -a \sum_{n=0}^{\infty} \lambda_n b_n(t) \, y_n(x) + \sum_{n=0}^{\infty} f_n(t) \, y_n(x)$$

and this is achieved if

$$b_n'(t) = -a\lambda_n b_n(t) + f_n(t)$$

which is a familiar equation, the solution being

$$b_n(t) = e^{-a\lambda_n t} b_n(0) + \int_0^t e^{-a\lambda_n(t-s)} f_n(s) \, ds$$

Then the solution is

$$u(x, t) = \sum_{n=0}^{\infty} \left(e^{-a\lambda_n t} b_n(0) + \int_0^t e^{-a\lambda_n(t-s)} f_n(s) \, ds \right) y_n(x)$$

where $b_n(0)$ needs to be chosen to satisfy the initial condition. Thus it is required that

$$b_n(0) = \int_0^L g(u) \, y_n(u) \, du$$

In what was done earlier, y_n was typically something like $(2/L)^{1/2} \sin\left(\frac{n\pi x}{L}\right)$. Then the solution is

$$u(x, t) = \sum_{n=0}^{\infty} \left(e^{-a\lambda_n t} \left(\int_0^L g(u) \, y_n(u) \, du \right) + \int_0^t e^{-a\lambda_n(t-s)} f_n(s) \, ds \right) y_n(x)$$

Example 14.3.1 *Find the solution to*

$$u_t(x,t) = u_{xx}(x,t) + f(x,t)$$
$$u(0,t) = 0 = u(2,t)$$
$$u(x,0) = x$$

where $f(x,t) = xt$.

First find the eigenfunctions and eigenvalues for the equation

$$y'' + \lambda y = 0, y(0) = 0 = y(2)$$

You must have λ strictly positive. The eigenvalues are $\lambda = \left(\frac{n\pi}{2}\right)^2$, and the eigenfunctions are $\sin\left(\frac{n\pi x}{2}\right)$. Also, there is a Fourier series expansion for $f(x,t)$ as follows.

$$f(x,t) = \sum_{n=1}^{\infty} f_n(t) \sin\left(\frac{n\pi x}{2}\right)$$

where

$$f_n(t) = \frac{2}{2} \int_0^2 f(x,t) \sin\left(\frac{n\pi x}{2}\right) dx$$

Thus

$$f_n(t) = \int_0^2 (xt) \sin\left(\frac{n\pi x}{2}\right) dx = \frac{1}{\pi^2 n^2}\left(4\pi nt(-1)^{n+1}\right)$$

Now the solution is

$$u(x,t) = \sum_{n=0}^{\infty} \left(\begin{array}{c} e^{-\left(\frac{n\pi}{2}\right)^2 t}\left(\int_0^L u \sin\left(\frac{n\pi u}{2}\right) du\right) + \\ \int_0^t e^{-\left(\frac{n\pi}{2}\right)^2(t-s)}\left(\frac{1}{\pi^2 n^2}\left(4\pi ns(-1)^{n+1}\right)\right) ds \end{array} \right) \sin\left(\frac{n\pi x}{2}\right)$$

Once you know how to solve this kind of problem, it becomes routine, if long, to find solutions to problems like this.

Example 14.3.2 *Find the solution to the initial-boundary value problem*

$$u_t(x,t) = u_{xx}(x,t) + f(x,t)$$
$$u(0,t) = 0, u(L,t) = g(t)$$
$$u(x,0) = h(x)$$

In this case, you massage the problem to get one which is like one you do know how to do which involves zero boundary conditions. Let

$$w(x,t) = u(x,t) - \frac{x}{L}g(t)$$

then

$$w_t = u_t - \frac{x}{L}g'(t) = u_{xx} + f - \frac{x}{L}g'(t) = w_{xx} + f(x,t) - \frac{x}{L}g'(t)$$
$$w(0,t) = u(0,t) = 0, \quad w(L,t) = u(L,t) - g(t) = 0$$
$$w(x,0) = u(x,0) - \frac{x}{L}g(0) = h(x) - \frac{x}{L}g(0)$$

and now you solve for w using the above procedure. There are seemingly endless variations of this but all amount to the following.

Procedure 14.3.3 *To solve*

$$u_t = Au + f$$

nonzero boundary conditions

initial condition $u(x,0) = l(x)$

You let $w = u - k(x,t)$ *where* k *is a known function chosen such that the boundary conditions on* w *involve* w *or its partial* x *derivatives set equal to 0. Then adjust to consider the equation solved for* w *which is of the form*

$$w_t = Aw + \hat{f}$$

zero boundary conditions

modified initial condition $w(x,0) = \hat{l}(x)$

This is then of the right form which can be solved. Obtain eigenfunctions $\{y_n\}$

$$Ay_n = -\lambda_n^2 y_n$$

Find the eigenfunction expansion for f *in terms of these.*

$$\sum_{n=0}^{\infty} f_n(t) y_n(x)$$

Then you need

$$w(x,t) = \sum_{n=0}^{\infty} b_n(t) y_n(x)$$

where

$$b_n'(t) = -\lambda_n^2 b_n(t) + f_n(t)$$

and $b_n(0)$ *is an appropriate Fourier coefficient chosen to satisfy the initial condition. Find* w *and then* $u(x,t) = w(x,t) + k(x,t)$.

In case the problem is second order in time, you do something similar except that the differential equation for b_n will now be second order in time and you will need to adjust both $b_n(0)$ and $b_n'(0)$ to achieve appropriate initial conditions.

Example 14.3.4 *Solve the following*

$$u_{tt} = u_{xx}, \quad u(x,0) = 0, u_t(x,0) = 0$$

$$u(0,t) = 0, \quad u(L,t) = \sin t$$

Initially the string is at rest and then something starts moving the right side up and down. What happens?

Following the procedure, let

$$w(x,t) = u(x,t) - \frac{x}{L}\sin(t)$$

this works because w has zero boundary conditions. Then

$$w_{tt} = u_{tt} + \frac{x}{L}\sin t = u_{xx} + \frac{x}{L}\sin t = w_{xx} + \frac{x}{L}\sin t$$

$$w(0,t) = w(L,t) = 0$$

$$w(x,0) = 0, \quad w_t(x,0) = -\frac{x}{L}\cos(t)$$

The eigenfunctions are $\sin\left(\frac{n\pi}{L}x\right)$. Then the expansion for $(x/L)\sin(t)$ is

$$\sum_{n=1}^{\infty} \left(\frac{2}{L} \int_0^L \left(\frac{x}{L}\sin t \right) \sin\left(\frac{n\pi}{L}x\right) dx \right) \sin\left(\frac{n\pi}{L}x\right)$$

$$= \sum_{n=1}^{\infty} \left(2\frac{(-1)^{n+1}}{\pi n}\sin(t) \right) \sin\left(\frac{n\pi}{L}x\right)$$

then the solution is

$$w(x,t) = \sum_{n=1}^{\infty} b_n(t)\sin\left(\frac{n\pi}{L}x\right)$$

where

$$b_n''(t) = -\frac{n^2\pi^2}{L^2}b_n(t) + 2\frac{(-1)^{n+1}}{\pi n}\sin(t) \qquad (*)$$

The Fourier series expansion for $w_t(x,t) = -\frac{x}{L}\cos(t)$ in terms of these eigenfunctions is

$$\sum_{n=1}^{\infty} \left(\frac{2}{L} \int_0^L \left(-\frac{x}{L}\cos(t) \right) \sin\left(\frac{n\pi x}{L}\right) dx \right) \sin\left(\frac{n\pi x}{L}\right)$$

$$= \sum_{n=1}^{\infty} \left(2\frac{(-1)^n}{\pi n}\cos t \right) \sin\left(\frac{n\pi x}{L}\right)$$

Now the solution to $*$ is $b_n(t) =$

$$\left(\cos\left(\frac{\pi}{L}nt\right) \right) b_n(0) + \frac{1}{\pi}\frac{L}{n} \left(\sin\left(\frac{\pi}{L}nt\right) \right) b_n'(0) + 2(-1)^{n+1}L^2\frac{\sin t}{\pi^3 n^3 - \pi L^2 n}$$

Clearly the initial condition for w gives $b_n(0) = 0$. It remains to find $b_n'(0)$. The solution is

$$w(x,t) = \sum_{n=1}^{\infty} \left(\frac{1}{\pi}\frac{L}{n} \left(\sin\left(\frac{\pi}{L}nt\right) \right) b_n'(0) + 2(-1)^{n+1}L^2\frac{\sin t}{\pi^3 n^3 - \pi L^2 n} \right) \sin\left(\frac{n\pi}{L}x\right)$$

Now

$$w_t(x,t) = \sum_{n=1}^{\infty} \left(\left(\cos\frac{\pi}{L}nt \right) b_n'(0) + 2(-1)^{n+1}L^2\left(\frac{\cos t}{\pi^3 n^3 - \pi L^2 n} \right) \right) \sin\left(\frac{n\pi}{L}x\right)$$

and so the initial condition for w_t requires

$$\sum_{n=1}^{\infty} \left(b_n'(0) + \frac{2(-1)^{n+1}L^2}{\pi^3 n^3 - \pi L^2 n} \right) \sin\left(\frac{n\pi}{L}x\right) = \sum_{n=1}^{\infty} \left(2\frac{(-1)^n}{\pi n} \right) \sin\left(\frac{n\pi x}{L}\right)$$

and so

$$b_n'(0) = 2\frac{(-1)^n}{\pi n} - \frac{2(-1)^{n+1} L^2}{\pi^3 n^3 - \pi L^2 n} = 2(-1)^n \pi \frac{n}{\pi^2 n^2 - L^2}$$

Thus

$$w(x,t) = \sum_{n=1}^{\infty} \left(\begin{array}{c} 2(-1)^n L \frac{\sin \frac{\pi}{L} nt}{\pi^2 n^2 - L^2} \\ +2(-1)^{n+1} L^2 \frac{\sin t}{\pi^3 n^3 - \pi L^2 n} \end{array} \right) \sin \left(\frac{n\pi}{L} x \right)$$

Therefore, $u(x,t) = w(x,t) + \frac{x}{L} \sin(t)$.

14.4 Laplace Equation

The Laplace equation is $\Delta u = 0$. In two dimensions and in rectangular coordinates,

$$\Delta u = u_{xx} + u_{yy} = 0$$

Here u is a function of the two variables x, y. Note first that Δ is a linear operator. That is, for a, b scalars and u, v functions,

$$\Delta(au + bv) = a\Delta u + b\Delta v$$

Because of this, if you have several solutions to the Laplace equation $u_1, \cdots, u_m$, and if you have scalars c_i, then

$$\Delta \left(\sum_{i=1}^{n} c_i u_i \right) = \sum_{i=1}^{n} c_i \Delta u_i = \sum_{i=1}^{n} c_i 0 = 0.$$

It is understood that the point (x, y) is contained in some region in the plane. One looks for a solution to the equation which also satisfies boundary conditions on the boundary of the region. When these conditions involve given values for the function u it is called the Dirichlet problem. When it involves giving values for the normal derivative of u defined by $\nabla u \cdot \mathbf{n}$ for $\mathbf{n}$ the unit outer normal, it is called a Neuman problem. In this short introduction this region will be either a circular disk or a rectangle. These are called boundary value problems.

14.4.1 Rectangles

First consider the rectangle. Here is a typical problem. The boundary conditions are as shown in the picture, zero on the top bottom and left side and $f(y)$ on the right.

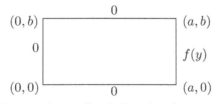

You can solve this the usual way. Look for eigenfunctions. These need to correspond to the two opposite sides where the boundary condition is 0. Thus the eigenfunctions are the nonzero solutions to

$$f''(y) + \lambda f(y) = 0, \ f(0) = 0 = f(b)$$

It follows the eigenfunctions are

$$\sin\left(\frac{n\pi}{b}y\right), \ n = 1, 2, \cdots$$

and the eigenvalues are $\frac{\pi^2}{b^2}n^2, n = 1, 2, \cdots$. Next you need to find some $g(x)$ such that $g(x)\sin\left(\frac{n\pi}{b}y\right)$ solves the boundary conditions and the equation. The boundary conditions are automatic. Now consider the equation. You need

$$g''(x)\sin\left(\frac{n\pi}{b}y\right) + g(x)\left(-\frac{\pi^2}{b^2}n^2\sin\left(\frac{\pi}{b}ny\right)\right) = 0$$

Thus, you need

$$g''(x) - \frac{\pi^2}{b^2}n^2 g(x) = 0$$

You know the solution is

$$C_1 e^{\frac{n\pi}{b}x} + C_2 e^{-\frac{n\pi}{b}x}$$

Now it turns out that in this application, it is much more convenient to write the general solution as

$$a_n \cosh\left(\frac{n\pi}{b}x\right) + b_n \sinh\left(\frac{n\pi}{b}x\right)$$

This gives the same general solution. The above functions are linear combinations of the known solutions and so things in the above form are solutions. Furthermore, the ratio of the two solutions is not constant so their Wronskian does not vanish. Hence it is the general solution. Now you try and get the solution to the boundary value problem in the form

$$u(x, y) = \sum_{n=1}^{\infty}\left(a_n \cosh\left(\frac{n\pi}{b}x\right) + b_n \sinh\left(\frac{n\pi}{b}x\right)\right)\sin\left(\frac{n\pi}{b}y\right)$$

when $x = 0$, you get $\sum_{n=1}^{\infty} a_n \sin\left(\frac{n\pi}{b}y\right) = 0$ and so each $a_n = 0$. When $x = a$, you need

$$f(y) = \sum_{n=1}^{\infty} b_n \sinh\left(\frac{n\pi}{b}a\right)\sin\left(\frac{n\pi}{b}y\right)$$

Hence you need

$$b_n \sinh\left(\frac{n\pi}{b}a\right) = \frac{2}{b}\int_0^b f(t)\sin\left(\frac{n\pi}{b}t\right)dt b_n = \frac{2}{b\sinh\left(\frac{n\pi}{b}a\right)}\int_0^b f(t)\sin\left(\frac{n\pi}{b}t\right)dt$$

Therefore, with this formula for b_n

$$u(x, y) = \sum_{n=1}^{\infty} b_n \sinh\left(\frac{n\pi}{b}x\right)\sin\left(\frac{n\pi}{b}y\right)$$

This shows how to solve a more general problem in which you have functions given on the edges. You solve the problem for the situation in which there is something nonzero on exactly one edge with 0 on the others and then you add these solutions together.

Example 14.4.1 *Find the solution to the boundary value problem*

$$u_{xx} + u_{yy} = 0$$

where the boundary conditions and rectangle are as expressed in the following picture.

First find the solution which has $\sin y$ on the right and zero on the other edges. This was done in the above. It is

$$u_1(x, y) = \sum_{n=1}^{\infty} b_n \sinh(n\pi x) \sin(n\pi y)$$

where b_n is as given above. Thus

$$b_n = \frac{2}{\sinh(n\pi a)} \int_0^1 \sin(t) \sin(n\pi t)\, dt = \frac{2}{\sinh n\pi a} \frac{n\pi \sin 1 (-1)^n}{n^2 \pi^2 - 1}$$

Hence this partial solution is

$$u_1(x, y) = \sum_{n=1}^{\infty} \frac{2}{\sinh n\pi a} \frac{n\pi \sin 1 (-1)^n}{n^2 \pi^2 - 1} \sinh(n\pi x) \sin(n\pi y)$$

Next find the solution to the equation which has $1 - (x - 1)^2$ on the top and zero on the other sides. This is just like what was done earlier except that you would switch a and b. You find the eigenfunctions for the two opposite zero boundary conditions. These are

$$\sin\left(\frac{n\pi}{2}x\right), \quad n = 1, 2, \cdots$$

with eigenvalues $\frac{n^2\pi^2}{4}$. Next you look for solutions to the equation which involve

$$a(y) \sin\left(\frac{n\pi}{2}x\right)$$

Thus

$$a''(y) \sin\left(\frac{n\pi}{2}x\right) + a(y)\left(-\frac{\pi^2}{4}n^2 \sin\frac{\pi}{2}nx\right) = 0$$

Hence

$$a''(y) - \frac{\pi^2}{4}n^2 a(y) = 0$$

and so

$$a(y) = a_n \cosh\left(\frac{n\pi}{2}y\right) + b_n \sinh\left(\frac{n\pi}{2}y\right)$$

Then the general solution is

$$u_2(x, y) = \sum_{n=1}^{\infty} \left(a_n \cosh\left(\frac{n\pi}{2}y\right) + b_n \sinh\left(\frac{n\pi}{2}y\right) \right) \sin\left(\frac{n\pi}{2}x\right)$$

When $y = 0$, you are supposed to get 0 for the boundary condition. Hence $a_n = 0$. When $y = b$ you need

$$1 - (1 - x)^2 = \sum_{n=1}^{\infty} \left(b_n \sinh\left(\frac{n\pi}{2}\right) \right) \sin\left(\frac{n\pi}{2}x\right)$$

Therefore, you need

$$b_n \sinh\left(\frac{n\pi}{2}\right) = \int_0^2 \left(1 - (1 - s)^2\right) \sin\left(\frac{n\pi}{2}s\right) ds = 8\frac{2 - 2(-1)^n}{n^3\pi^3}$$

Then this solution is of the form

$$u_2(x, y) = \sum_{n=1}^{\infty} \left(8\frac{2 - 2(-1)^n}{n^3\pi^3} \frac{1}{\sinh\left(\frac{n\pi}{2}\right)} \right) \sinh\left(\frac{n\pi}{2}y\right) \sin\left(\frac{n\pi}{2}x\right)$$

$$= \sum_{n=1}^{\infty} \frac{32}{(2n-1)^3\pi^3} \frac{1}{\sinh\left(\frac{(2n-1)\pi}{2}\right)} \sinh\left(\frac{(2n-1)\pi y}{2}\right) \sin\left(\frac{(2n-1)\pi}{2}x\right)$$

Therefore, the solution to the boundary value problem is the sum of these two solutions.

$$u(x, y) = \sum_{n=1}^{\infty} \frac{2}{\sinh n\pi a} \frac{n\pi \sin 1 \,(-1)^n}{n^2\pi^2 - 1} \sinh(n\pi x) \sin(n\pi y) +$$

$$\sum_{n=1}^{\infty} \frac{32}{(2n-1)^3\pi^3} \frac{1}{\sinh\left(\frac{(2n-1)\pi}{2}\right)} \sinh\left(\frac{(2n-1)\pi y}{2}\right) \sin\left(\frac{(2n-1)\pi}{2}x\right)$$

You can probably see how to consider given functions in place of 0 on the remaining two sides.

14.4.2　Circular Disks

This is more interesting than the above because it is more often the case that you encounter it in real situations. Most pipes are circular for example. The Laplacian in rectangular coordinates is

$$\Delta = u_{xx} + u_{yy}$$

However, rectangular coordinates are not natural for considering circles. For example, the boundaries of a rectangle are obtained by letting one of the variables be constant. If you want something like this to happen for a circular shape, you should consider polar coordinates. For example, the boundary of a circular disk is obtained by letting $r = c$ a constant. Recall the relation between polar and rectangular coordinates. $\theta \in [0, 2\pi), r > 0$,

$$
\begin{aligned}
x &= r\cos\theta \\
y &= r\sin\theta
\end{aligned}
\tag{14.4}
$$

You have a scalar field u and it is a function of a point in two dimensional space. This point can be described in terms of either polar coordinates or rectangular coordinates. Thus

$$u(x, y) = u(r, \theta)$$

there (x, y) and (r, θ) pertain to the same point in two dimensions. As discussed above in Section 14.1, the Laplacian in polar coordinates is

$$u_{rr} + \frac{1}{r}u_r + \frac{1}{r^2}u_{\theta\theta} = \frac{1}{r}\frac{\partial}{\partial r}\left(r\frac{\partial u}{\partial r}\right) + \frac{1}{r^2}\frac{\partial^2 u}{\partial \theta^2}$$

Example 14.4.2 *Find the solution to*

$$\Delta u = u_{rr} + \frac{1}{r}u_r + \frac{1}{r^2}u_{\theta\theta} = 0$$

on the disc of radius R if on the boundary of this disk,

$$u(R, \theta) = f(\theta)$$

where $f(0) = f(2\pi)$. This last condition is necessary because $\theta = 0$ and $\theta = 2\pi$ correspond to the same point on the boundary of this disk. Note how everything is in terms of the variables r, θ and that in terms of these variables, the circular disk is actually a rectangle.

Use the method of separation of variables. Look for a solution to the equation which is of the form $R(r)\Theta(\theta)$.

$$r^2 R''(r)\Theta(\theta) + rR'(r)\Theta(\theta) + R(r)\Theta''(\theta) = 0$$

So divide by $R\Theta$. This leads to

$$r^2\frac{R''}{R} + r\frac{R'}{R} + \frac{\Theta''}{\Theta} = 0$$

Hence

$$\frac{\Theta''}{\Theta} = -\lambda = r^2\frac{R''}{R} + r\frac{R'}{R}$$

for some constant λ. First consider Θ. You must have $\Theta(0) = \Theta(2\pi)$. Also

$$\Theta'' + \lambda\Theta = 0$$

Multiply both sides by Θ and integrate. This leads to

$$\Theta'(\theta)\Theta(\theta)\,|_0^{2\pi} - \int_0^{2\pi}(\Theta')^2\,d\theta + \lambda\int_0^{2\pi}\Theta^2\,d\theta = 0$$

The boundary terms disappear because you must also have $\Theta'(2\pi) = \Theta'(0)$. Therefore, to have a solution, it is necessary that $\lambda \geq 0$. If $\lambda = 0$, you need to have $\Theta' = 0$ and so $\Theta(\theta) = C$ a constant. Otherwise, you need $\lambda = \mu^2, \mu > 0$. Then the solution to the equation is

$$C_1 \cos \mu\theta + C_2 \sin \mu\theta$$

and you need to have $\Theta(0) = \Theta(2\pi)$. Therefore, it is required that $\mu 2\pi$ is an integer multiple of 2π so $\mu = n$ for n an integer. Thus the eigenvalues are the nonnegative integers and you get

$$\Theta_n(\theta) = (a_n \cos(n\theta) + b_n \sin(n\theta)), \quad n = 0, 1, 2, \cdots$$

It follows that for each of these n,

$$r^2 R_n'' + r R_n' - n^2 R_n = 0$$

This is an Euler equation and you look for solutions in the form $R(r) = r^\alpha$. Then to find α, you insert this into the equation.

$$r^2 \alpha(\alpha - 1) r^{\alpha-2} + r\alpha r^{\alpha-1} - n^2 r^\alpha = 0$$

and so you get the indicial equation

$$\alpha(\alpha - 1) + \alpha - n^2 = (\alpha - n)(\alpha + n) = 0$$

Therefore, the solutions are of the form

$$c_n r^n + d_n r^{-n}$$

We can immediately conclude that $d_n = 0$ because it makes no sense to have the solution to the differential equation be unbounded as $r \to 0$. Recall the theorem from calculus that on a closed and bounded set, a continuous function achieves its maximum and minimum. If u is going to be continuous, which we certainly expect it to be, then this cannot be harmonized with $d_n \neq 0$. Thus this has found many solutions to the partial differential equation which are of the form

$$r^n (a_n \cos(n\theta) + b_n \sin(n\theta))$$

The solution to the equation will then be an infinite sum of the functions of the above form. Thus combining the c_n with a_n and b_n,

$$u(r, \theta) = \sum_{n=0}^\infty r^n (a_n \cos(n\theta) + b_n \sin(n\theta))$$

If you want to achieve the boundary condition, then you need to have $R^n a_n =$

$$\frac{1}{\pi} \int_0^{2\pi} \cos(n\theta) f(\theta) \, d\theta, \quad a_0 = \frac{1}{2\pi} \int_0^{2\pi} f(\theta) \, d\theta, \quad R^n b_n = \frac{1}{\pi} \int_0^{2\pi} \sin(n\theta) f(\theta) \, d\theta$$

If you like, you can simplify this and write an interesting formula for the solution to this problem.

$$u(r, \theta) = \frac{1}{2\pi} \int_0^{2\pi} f(\theta) \, d\theta +$$

$$\frac{1}{\pi} \sum_{n=1}^{\infty} \frac{r^n}{R^n} \left(\left(\int_0^{2\pi} \cos(n\alpha) f(\alpha) \, d\alpha \right) \cos(n\theta) + \left(\int_0^{2\pi} \sin(n\alpha) f(\alpha) \, d\alpha \right) \sin(n\theta) \right)$$

$$= \frac{1}{2\pi} \int_0^{2\pi} f(\alpha) \, d\alpha + \frac{1}{\pi} \sum_{n=1}^{\infty} \frac{r^n}{R^n} \int_0^{2\pi} f(\alpha) \cos(n(\theta - \alpha)) \, d\alpha$$

In fact, it can be proved that the infinite sum and the integral can be interchanged. This is thanks to the term $(r/R)^n$ which yields absolute convergence. There is no problem if it were a finite sum and thanks to this term, the tail of the series is negligible. Thus one can reduce to the finite sum case and make the interchange. Thus the above implies

$$u(r, \theta) = \int_0^{2\pi} \frac{1}{\pi} \left(\frac{1}{2} + \sum_{n=1}^{\infty} \frac{r^n}{R^n} \cos(n(\theta - \alpha)) \right) f(\alpha) \, d\alpha$$

You can find a formula for this.

$$\sum_{n=1}^{\infty} \frac{r^n}{R^n} \cos(nt) = \text{Re} \sum_{n=1}^{\infty} \left(\frac{r}{R} e^{it} \right)^n = \text{Re} \left(\frac{\frac{r}{R} e^{it}}{1 - \frac{r}{R} e^{it}} \right) = \text{Re} \left(\frac{r e^{it}}{R - r e^{it}} \right)$$

$$= \frac{R r \cos(t) - r^2}{(R - r \cos t)^2 + r^2 \sin^2(t)} = \frac{R r \cos(t) - r^2}{R^2 - 2 (\cos t) R r + r^2}$$

Then

$$\frac{1}{2} + \sum_{n=1}^{\infty} \frac{r^n}{R^n} \cos(nt) = \frac{1}{2} + \frac{R r \cos(t) - r^2}{R^2 - 2 (\cos t) R r + r^2} = \frac{1}{2} \frac{R^2 - r^2}{R^2 - 2 (\cos t) R r + r^2}$$

Thus

$$u(r, \theta) = \frac{1}{2\pi} \int_0^{2\pi} \left(\frac{R^2 - r^2}{R^2 - 2 (\cos(\theta - \alpha)) R r + r^2} \right) f(\alpha) \, d\alpha$$

Note that this shows that if $r = 0$ so you are at the center, then

$$u(r, \theta) = \frac{1}{2\pi} \int_0^{2\pi} f(\alpha) \, d\alpha$$

so the value at the center is the average of the boundary values. This proves the following fundamental result.

Theorem 14.4.3 *The solution to the problem*

$$\Delta u = u_{rr} + \frac{1}{r} u_r + \frac{1}{r^2} u_{\theta\theta} = 0$$

on the disc of radius R where on the boundary of this disk,

$$u(R, \theta) = f(\theta), \ f(0) = f(2\pi)$$

is given by the formula

$$u(r, \theta) = \frac{1}{2\pi} \int_0^{2\pi} \left(\frac{R^2 - r^2}{R^2 - 2 (\cos(\theta - \alpha)) R r + r^2} \right) f(\alpha) \, d\alpha$$

14.5 Exercises

1. Solve the following initial boundary value problems.

   (a) $u_t = u_{xx}, u(x,0) = 1, u(0,t) = 0, u(4,t) = 0$

   (b) $u_t = 2u_{xx}, u(x,0) = 1, u_x(0,t) = 0, u(3,t) = 0$

   (c) $u_t = 3u_{xx}, u(x,0) = 1 - x, u(0,t) = 0, u(2,t) = 0$

   (d) $u_t = 4u_{xx}, u(x,0) = 1 - x, u(0,t) = 0, u(2,t) = 0$

   (e) $u_t = 5u_{xx}, u(x,0) = 1 - x, u(0,t) = 0, u_x(1,t) = 0$

   (f) $u_t = u_{xx}, u(x,0) = x + 1, u_x(0,t) = 0, u(2,t) = 0$

   (g) $u_t = 3u_{xx}, u(x,0) = x, u_x(0,t) = 0, u(1,t) = 0$

   (h) $u_t = 3u_{xx}, u(x,0) = x^2, u(0,t) = 0, u(5,t) = 0$

   (i) $u_t = 4u_{xx}, u(x,0) = 1, u(0,t) = 0, u_x(1,t) = 0$

   (j) $u_t = u_{xx}, u(x,0) = x, u(0,t) = 0, u_x(4,t) = 0$

   (k) $u_t = 2u_{xx}, u(x,0) = 1, u_x(0,t) = 0, u_x(5,t) = 0$

   (l) $u_t = 2u_{xx}, u(x,0) = x, u_x(0,t) = 0, u_x(4,t) = 0$

   (m) $u_t = 2u_{xx}, u(x,0) = 1 - x, u_x(0,t) = 0, u_x(3,t) = 0$

2. Find the solution to the initial boundary value problem

$$
\begin{aligned}
u_{tt} &= 3u_{xx}, u(0,t) = 0, u(5,t) = 0, \\
u(x,0) &= 3x(x-5), u_t(x,0) = x^2
\end{aligned}
$$

3. Find the solution to the initial boundary value problem

$$
\begin{aligned}
u_{tt} &= 4u_{xx}, u(0,t) = 0, u(5,t) = 0, \\
u(x,0) &= 3x(x-5), u_t(x,0) = x + 1
\end{aligned}
$$

4. Find the solution to the initial boundary value problem

$$
\begin{aligned}
u_{tt} &= 4u_{xx}, u(0,t) = 0, u(2,t) = 0, \\
u(x,0) &= -x(x-2), u_t(x,0) = x^2
\end{aligned}
$$

5. Describe how to solve the initial boundary value problem

$$
\begin{aligned}
u_{tt} + 2u_t &= 2u_{xx}, u(0,t) = 0, u(5,t) = 0, \\
u(x,0) &= -x(x-5), u_t(x,0) = x^2
\end{aligned}
$$

Hint: You might consider defining $w = e^{2t}u$ and see what equation is solved by w.

6. Find the solution to the initial boundary value problem

$$
\begin{aligned}
u_t - 2u &= u_{xx}, u(0,t) = 0, u(5,t) = 0, \\
u(x,0) &= x
\end{aligned}
$$

Hint: It is like before. You get eigenfunctions and match coefficients.

7. Find the solution to the initial boundary value problem

$$
\begin{aligned}
u_t &= u_{xx} + (\cos x), \, u\,(0,t) = 0, \, u\,(2,t) = 0, \\
u\,(x,0) &= 1 - x
\end{aligned}
$$

8. Find the solution to the initial boundary value problem

$$
\begin{aligned}
u_t &= 2u_{xx} + (x-1), \, u\,(0,t) = 0, \, u\,(4,t) = 0, \\
u\,(x,0) &= 1
\end{aligned}
$$

9. Find the solution to the initial boundary value problem

$$
\begin{aligned}
u_t &= 5u_{xx} + (x-1), \, u\,(0,t) = 0, \, u\,(1,t) = 0, \\
u\,(x,0) &= x + 1
\end{aligned}
$$

10. Find the solution to the initial boundary value problem

$$
\begin{aligned}
u_t &= 2u_{xx}, \, u\,(0,t) = 0, \, u\,(5,t) = 0, \\
u\,(x,0) &= \begin{array}{l} x \text{ for } x \in \left[0, \frac{5}{4}\right] \\ 0 \text{ if } x \in \left(\frac{5}{4}, 5\right] \end{array}
\end{aligned}
$$

11. Find the solution to the initial boundary value problem

$$
\begin{aligned}
u_t &= 3u_{xx}, \, u\,(0,t) = 0, \, u\,(3,t) = 0, \\
u\,(x,0) &= \begin{array}{l} 1 \text{ for } x \in [0,1] \\ 0 \text{ if } x \in (1,3] \end{array}
\end{aligned}
$$

12. Find the solution to the initial boundary value problem

$$
\begin{aligned}
u_t &= 4u_{xx}, \, u\,(0,t) = 0, \, u\,(2,t) = 0, \\
u\,(x,0) &= \begin{array}{l} x \text{ for } x \in \left[0, \frac{2}{3}\right] \\ 1 - \frac{1}{2}x \text{ if } x \in \left(\frac{2}{3}, 2\right] \end{array}
\end{aligned}
$$

13. Find the solution to the initial boundary value problem

$$
\begin{aligned}
u_t &= 5u_{xx}, \, u\,(0,t) = 0, \, u\,(1,t) = 0, \\
u\,(x,0) &= \begin{array}{l} x \text{ for } x \in \left[0, \frac{1}{2}\right] \\ 1 - x \text{ if } x \in \left(\frac{1}{2}, 1\right] \end{array}
\end{aligned}
$$

14. Find the solution to the initial boundary value problem

$$
\begin{aligned}
u_t &= 3u_{xx}, \, u\,(0,t) = 0, \, u\,(5,t) = 0, \\
u\,(x,0) &= \begin{array}{l} x \text{ for } x \in \left[0, \frac{5}{2}\right] \\ 5 - x \text{ if } x \in \left(\frac{5}{2}, 5\right] \end{array}
\end{aligned}
$$

15. Find the solution to the initial boundary value problem

$$u_t = 2u_{xx}, u_x(0,t) = 0, u(2,t) = 0,$$

$$u(x,0) = \begin{array}{l} x \text{ for } x \in [0,\frac{1}{2}] \\ \frac{2}{3} - \frac{1}{3}x \text{ if } x \in (\frac{1}{2},2] \end{array}$$

16. Find the solution to the initial boundary value problem

$$u_t = 5u_{xx}, u_x(0,t) = 0, u(1,t) = 0,$$

$$u(x,0) = \begin{array}{l} x \text{ for } x \in [0,\frac{1}{2}] \\ 1 - x \text{ if } x \in (\frac{1}{2},1] \end{array}$$

17. Find the solution to the initial boundary value problem

$$u_t = 5u_{xx}, u_x(0,t) = 0, u(2,t) = 0,$$

$$u(x,0) = \begin{array}{l} x \text{ for } x \in [0,\frac{2}{3}] \\ 1 - \frac{1}{2}x \text{ if } x \in (\frac{2}{3},2] \end{array}$$

18. Consider the following initial boundary value problem,

$$u_t = u_{xx}, \ u(0,t) = 0, u(2,t) + u_x(2,t) = 0,$$
$$u(x,0) = f(x)$$

Determine the appropriate equation for the eigenfunctions and show that there exists a sequence of strictly positive eigenvalues converging to ∞. Also explain why the solution u if it exists, must have a limit $\lim_{t\to\infty} u(x,t) = w(x)$ and that this limit satisfies $w(x) = 0$.

19. Consider the following initial boundary value problem,

$$u_t = u_{xx}, \ u_x(0,t) = 0, u(2,t) + u_x(2,t) = 0,$$
$$u(x,0) = f(x)$$

Determine the appropriate equation for the eigenfunctions and show that there exists a sequence of strictly positive eigenvalues converging to ∞. Also explain why the solution u if it exists, must have a limit $\lim_{t\to\infty} u(x,t) = w(x)$ and that this limit satisfies $w''(x) = w(x) = 0$.

20. Consider the following initial boundary value problem,

$$u_t = u_{xx}, \ u_x(0,t) = 0, u_x(2,t) = 0,$$
$$u(x,0) = f(x)$$

Determine the appropriate equation for the eigenfunctions and show that there exists a sequence of strictly positive eigenvalues converging to ∞. Also explain why the solution u if it exists, must have a limit $\lim_{t\to\infty} u(x,t) = \frac{1}{2}\int_0^2 f(x)\,dx$.

21. Recall that on the circular disk of radius R centered at the origin, denoted here as D_R

$$u(r, \theta) = \int_0^{2\pi} \frac{1}{\pi} \left(\frac{1}{2} + \sum_{n=1}^{\infty} \frac{r^n}{R^n} \cos(n(\theta - \alpha)) \right) f(\alpha) \, d\alpha$$

gave the solution to $\Delta u = 0$ and $f(\alpha)$ a given function on the boundary where $f(0) = f(2\pi)$. Show, using the divergence theorem from calculus that there is at most one smooth solution to this problem. Then explain why

$$\int_0^{2\pi} \frac{1}{\pi} \left(\frac{1}{2} + \sum_{n=1}^{\infty} \frac{r^n}{R^n} \cos(n(\theta - \alpha)) \right) d\alpha = 1$$

22. Recall that on a simple computation was done which showed that

$$\frac{1}{\pi} \left(\frac{1}{2} + \sum_{n=1}^{\infty} \frac{r^n}{R^n} \cos(n(\theta - \alpha)) \right) = \frac{1}{2\pi} \frac{R^2 - r^2}{R^2 - 2(\cos(\theta - \alpha))Rr + r^2}$$

Therefore,

$$\int_0^{2\pi} \frac{1}{2\pi} \frac{R^2 - r^2}{R^2 - 2(\cos(\theta - \alpha))Rr + r^2} d\alpha = 1$$

Explain why it is also the case that

$$\frac{1}{2\pi} \frac{R^2 - r^2}{R^2 - 2(\cos(\theta - \alpha))Rr + r^2} \geq 0$$

and if $|\theta - \alpha| \geq \delta > 0$, then

$$\lim_{r \to R-} \frac{1}{2\pi} \frac{R^2 - r^2}{R^2 - 2(\cos(\theta - \alpha))Rr + r^2} = 0$$

uniformly for such α.

23. The solution to Laplace's equation on the disk D_R which has boundary values $f(\alpha)$ was derived and it is

$$u(r, \theta) = \int_0^{2\pi} \frac{1}{2\pi} \frac{R^2 - r^2}{R^2 - 2(\cos(\theta - \alpha))Rr + r^2} f(\alpha) \, d\alpha$$

Show that

$$\lim_{r \to R-} u(r, \theta) = f(\theta)$$

This shows how the boundary values are obtained.

24. Recall that $u(r, \theta) =$

$$\frac{1}{2\pi} \int_0^{2\pi} f(\theta) \, d\theta + \frac{1}{\pi} \sum_{n=1}^{\infty} \frac{r^n}{R^n} \left(\left(\int_0^{2\pi} \cos(n\alpha) f(\alpha) \, d\alpha \right) \cos(n\theta) \right)$$

$$+ \left(\int_0^{2\pi} \sin{(n\alpha)} f(\alpha) \, d\alpha \right) \sin{(n\theta)} \right)$$

Explain why if f is a 2π periodic continuous function, it follows that there is a trigonometric polynomial which is uniformly close to $f(\theta)$ for $\theta \in [0, 2\pi]$. **Hint:** From the above problem, convergence to $f(\theta)$ as $r \to R-$ takes place. Note that from the argument, this actually happens uniformly thanks to the uniform continuity of f. Now argue that the tail $\sum_{n=N}^{\infty}$ of the above series is uniformly small if N is large.

25. Let

$$f(x) = \begin{cases} x & \text{if } x \in [0, 1] \\ 2 - x & \text{if } x \in [1, 2] \end{cases}$$

Solve the following initial boundary value problems

(a) $u_t = a^2 u_{xx}, u(x, 0) = f(x), u(0, t) = 0 = u(2, t)$

(b) $u_t = a^2 u_{xx}, u(x, 0) = f(x), u_x(0, t) = 0 = u(2, t)$

(c) $u_t = a^2 u_{xx}, u(x, 0) = f(x), u_x(0, t) = 0 = u_x(2, t)$

Appendix A

MATLAB Syntax Summarized

In this short chapter is a review and summary of some MATLAB syntax. I am including this because I think that the most difficult part of the use of computer algebra systems involves getting the syntax right and hope that this short chapter will be useful. A few generalities are that ";" placed at the end of a statement causes MATLAB to know about the statement but to defer any action on it.

A.1 Matrices in MATLAB

To enter the matrix

$$\begin{pmatrix} 1 & 2 \\ 3 & 4 \end{pmatrix}$$

in MATLAB and then find its inverse, you enter the following. The semicolon says to start a new row.

$$>>a=[1\ 2;3\ 4];\ \text{inv(a)}$$

Then press "enter". It will present you with the inverse. If you want its eigenvalues, you do the following.

$$>>a=[1\ 2;3\ 4];\ \text{eig(a)}$$

If you want its eigenvalues and eigenvectors, do this:

$$>>a=[1\ 2;3\ 4];\ [V,D]=\text{eig(a)}$$

It will give the eigenvectors as columns in the matrix V. The eigenvalues will be the diagonal entries in D. It will find independent eigenvectors if it can to place as columns in V. However, it might be that there is not a basis of eigenvectors. Then the columns of V can be repeated. You adjust the above for other sized matrices. To enter a 3×1 column vector, you do [1;2;3].

When you have x, y two $1 \times n$ matrices. You write $x. * y$ to obtain the $1 \times n$ matrix which has in the jth position $x_j y_j$. Other uses of this are similar.

A.2   Summation in MATLAB

Suppose you want to sum $\sum_{k=1}^{5} (k+1)$. You would do the following:

> syms k;
> symsum(k+1,k,0,10)

then press enter. You need the symbolic math toolbox to do this.

If you were summing terms in a Fourier series to get a function, here is what will work. Say you want to consider this function

$$\sum_{k=1}^{7} \frac{4}{(2k-1)\pi} \sin\left((2k-1)\frac{\pi x}{2}\right)$$

on $[-4, 4]$ and graph it. Then you could do it like this.

```
x=-4:.05:4;
y=(4/pi)*sin(pi*x/2);
hold on
for k=2:7
y=y+(4/((2*k-1)*pi))*sin((2*k-1)*pi*x/2);
end
plot(x,y)
```

I am not saying this is the only way to do it, but it seems to work well and does not need the symbolic math toolbox.

Suppose you want to graph $\sum_{k=1}^{7} \frac{1}{k} \sin(kx)\cos(ky)$. Then you could do it like this.

```
>>[x,y]=meshgrid(0:.1:3*pi,0:.1:4*pi);
z=sin(x).*cos(y);
hold on
for k=2:7
z=z+(1/k)*sin(k*x).*cos(k*y);
end
surf(x,y,z)
```

Incidentally, in Scientific Notebook, you would just type in math mode:

$$\sum_{k=1}^{7} \frac{1}{k} \sin(kx)\cos(ky)$$

and then click on compute, Plot3D and then rectangular and then modify in the dialog box to give whatever interval is desired and how many points are used. You could do this, or even more simply, click on the little icon in the tool bar which says to graph in three dimensions. It has a picture of three axes. However, the graphs in MATLAB look better and they can be exported as eps files and there is more flexibility in specifying the size of the fonts and so forth, especially in two dimensions.

A.3   Graphing Slope Fields and Vector Fields

To graph a slope field, do the following:

```
>>[a,b]=meshgrid(-1:.2:3,-1:.2:3);
c=(b-1).*b.*(2-b); d=(c.*c+1).^(1/2);
u=1./d; v=c./d;
quiver(a,b,u,v,'ShowArrowHead','off')
```

This will graph a slope field for the differential equation $y' = (y - 1) y (2 - y)$.

To graph a vector field in which it is designed to expose the direction of the vectors, do the following:

```
>>[a,b]=meshgrid(-1.5:.2:1.5,-1.5:.2:1.5);
u=b+a.^2.*b; v=-(b+2*a)+a.^3; r=(u.*u+v.*v+.1).^(1/2);
figure
quiver(a,b,u./r,v./r,'autoscalefactor',.5)
```

This will give a direction field for the vector field $\left(y + x^2y, - (y + 2x) + x^3\right)$.

A.4   Using MATLAB to Find a Numerical Solution of ODE

If you want MATLAB to find a solution to an initial value problem, do this:

```
>> f=@(t,x) [x-x^3];
>> hold on
for z=-2:.5:2
[t,x]=ode45(f,[0,4],z);
plot(t,x)
end
```

This will actually do several of them, one for each initial condition z, the differential equation being $y' = y - y^3$. If you just want one, you would do

```
>> f=@(t,x) [x-x^3];
[t,x]=ode45(f,[0,4],3);
plot(t,x)
```

If you want values, at various points, you would do

```
>>  f=@(t,x) [x-x^3];
S=ode45(f,[0,4],3);
deval(S,[1,2,3,4])
```

This would give the values of the solution S at the points 1,2,3,4. If you want the solution to more decimal places or less, you would type

```
vpa(deval(S,[1,2,3,4]),8)
```

You adjust the 8 to be the number of digits you want to appear.

If your ordinary differential equation is a system, you do something similar.

> >>f=@(t,y)[t^2*y(1)+sin(t)*y(2);(t+1)*y(1)+2*y(2)];
> [s,x]=ode45(f,[0,2],[0;1]);
> S=ode45(f,[0,2],[0;1]);
> plot(s,x)
> vpa(deval(S,[0,.2,.4,.6,.8,1,1.2,1.4,1.6,1.8,2]),3)

The vpa command along with the " ,3" says to display 3 digits. This will solve the system

$$\begin{pmatrix} x \\ y \end{pmatrix}' = \begin{pmatrix} t^2x + \sin(t)\,y \\ (1+t)\,x + 2y \end{pmatrix}, \quad \begin{pmatrix} x(0) \\ y(0) \end{pmatrix} = \begin{pmatrix} 0 \\ 1 \end{pmatrix}$$

and give a graph of $t \to x(t)$ and $t \to y(t)$ on $[0, 2]$. If you want to graph the space curve $t \to \begin{pmatrix} x(t) \\ y(t) \end{pmatrix}$, you would type

$$\text{plot}(x(:,1),x(:,2))$$

instead of plot(s,x). In this example, it won't be very interesting. You can also force ode45 to consider smaller step sizes as follows. You replace [0,2] with [0:.03:2]. The .03 makes sure that the step size is no more than .03.

A.5   Laplace Transforms and MATLAB

Here is the syntax for solving an initial value problem using the Laplace transform and MATLAB. Say you want to solve

$$\begin{pmatrix} x \\ y \\ z \end{pmatrix}' = \begin{pmatrix} 3 & 3 & 2 \\ -1 & 0 & -1 \\ 0 & -1 & 1 \end{pmatrix} \begin{pmatrix} x \\ y \\ z \end{pmatrix} + \begin{pmatrix} 1 \\ 1 \\ \sin t \end{pmatrix}, \quad \begin{pmatrix} x(0) \\ y(0) \\ z(0) \end{pmatrix} = \begin{pmatrix} 0 \\ 1 \\ 0 \end{pmatrix}$$

> >>syms s t; a=[0;1;0]; b=s*[1 0 0;0 1 0;0 0 1]-[3 3 2;-1 0 -1;0 -1 1];
> c=[1;1;sin(t)]; simplify(ilaplace(inv(b)*(a+laplace(c))))

If you wanted a numerical solution, which might be all you could get if the characteristic polynomial does not factor exactly, you would replace with

> >> syms s t; a=[0;1;0]; b=s*[1 0 0;0 1 0;0 0 1]-[3 3 2;-1 0 -1;0 -1 1];
> c=[1;1;sin(t)]; vpa(ilaplace(inv(b)*(a+laplace(c))),4)

This would display four digits in the numbers which occur in an approximate solution.

A.6   Plotting in MATLAB

To plot do the following:

$$>> \text{x=-3:.01:3;}$$
$$\text{y=sin(10.*x)+x.\^2;}$$
$$\text{plot(x,y)}$$
$$\text{ylim([-8,8]); xlim([-3,3])}$$

This will plot the graph of $y = \sin(10x) + x^2$ on $[-3, 3]$ and chop off anything for which $|y| > 8$. If you want to plot surfaces, you do something else which is similar. Here is a plot of a parametric surface.

To plot a parametric surface, do this:

$$>>\text{[u,v]=meshgrid(1:.2:4,0:.2:24);}$$
$$\text{x=3.*u.*cos(v); y=3.*u.*sin(v); z=4.*v;}$$
$$\text{surf(x,y,z)}$$

This will be a graph of $(3u \cos v, 3u \sin v, 4v)$ for (u, v) in $[1, 4] \times [0, 24]$. Note the use of .* rather than * because u,v are vectors and you want to do the operation on the components of these vectors. Here is syntax for graphing a cylinder intersecting a sphere.

```
[s,t]=meshgrid(0:.02*pi:2*pi,0:.02*pi:pi);
[u,v]=meshgrid(0:.02*pi:2*pi,-1.4:.2:1.4);
hold on
surf(sin(t).*cos(s),sin(t).*sin(s),cos(t),'edgecolor','none')
alpha .7
surf(.5*cos(u),.5*sin(u),v,'edgecolor','none')
axis equal
```

The line alpha .7 makes the sphere partially transparent. You can adust this number to make it more or less transparent. It needs to be between 0 and 1.

Suppose you want to graph three surfaces. You do something like this.

$$>>\text{[x,y]=meshgrid(-3:.1:3,-3:.1:3);}$$
$$\text{z=sin(.2.*x.*y); w=cos(.2.*x.*y); u=.3.*x+.5.*y;}$$
$$\text{hold on}$$
$$\text{mesh(x,y,z)}$$
$$\text{mesh(x,y,w)}$$
$$\text{mesh(x,y,u)}$$

To plot a curve in three dimensions, do this:

$$>> \text{t=0:.2:10;}$$
$$\text{x=cos(t); y=sin(t); z=t;}$$
$$\text{plot3(x,y,z)}$$

This will graph the space curve $(\cos t, \sin t, t)$ for $t \in [0, 10]$.

You can also graph contour lines in MATLAB. To do this, you do the following:

```
>> [x,y]=meshgrid(-4:.1:4,-4:.1:4);
z=y-(y.^3/3+x.^3/3);
contour(x,y,z,[-.5,-1,-.3],'color', 'red','LineWidth',3);
axis equal
```

This will graph the level curves corresponding to $z = y - \left(y^3/3 + x^3/3\right)$ for $z = -.5, -1, -.3$ and they will be red and thick. If you just pick one number, then you are getting an implicit plot.

To do a graph of a surface with contour lines but no mesh lines other than the contours, do this.

```
[x,y]=meshgrid(-4:.05:4,-4:.05:4); z=x.^2-y.^2;
hold on
contour3(x,y,z,[-20,-16,-12,-8,-4,0,4,8,12,16,20],'linewidth',.4,'color','black')
surf(x,y,z,'edgecolor','none')
hold off (If you want no axes, type: axis off )
To get transparency, you type alpha(x) where x is a number in [0, 1].
```

Those numbers in contour3 give the values of z corresponding to a contour. If you replace contour3 with contour, then you will get these contour lines in the plane.

Appendix B

Calculus Review

B.1 The Limit of A Sequence

The definition of the limit of a sequence was defined by Bolzano.[*]

Definition B.1.1 *A sequence $\{a_n\}_{n=1}^{\infty}$ converges to a,*

$$\lim_{n \to \infty} a_n = a \ or \ a_n \to a$$

if and only if for every $\varepsilon > 0$ there exists n_ε such that whenever $n \geq n_\varepsilon$,

$$|a_n - a| < \varepsilon.$$

In words the definition says that given any measure of closeness ε, the terms of the sequence are eventually all this close to a. Note the similarity with the concept of limit. Here, the word "eventually" refers to n being sufficiently large. The above definition is always the definition of what is meant by the limit of a sequence. If the a_n are complex numbers or vectors the definition remains the same. If $a_n = x_n + iy_n$ and $a = x + iy$, $|a_n - a| = \sqrt{(x_n - x)^2 + (y_n - y)^2}$. Recall the way you measure distance between two complex numbers.

Theorem B.1.2 *If $\lim_{n \to \infty} a_n = a$ and $\lim_{n \to \infty} a_n = a_1$ then $a_1 = a$.*

Proof: Suppose $a_1 \neq a$. Then let $0 < \varepsilon < |a_1 - a|/2$ in the definition of the limit. It follows there exists n_ε such that if $n \geq n_\varepsilon$, then $|a_n - a| < \varepsilon$ and $|a_n - a_1| < \varepsilon$. Therefore, for such n,

$$
\begin{aligned}
|a_1 - a| \ &\leq \ |a_1 - a_n| + |a_n - a| \\
&< \ \varepsilon + \varepsilon < |a_1 - a|/2 + |a_1 - a|/2 = |a_1 - a|,
\end{aligned}
$$

a contradiction. ∎

---

[*]Bernhard Bolzano lived from 1781 to 1848. He was a Catholic priest and held a position in philosophy at the University of Prague. He had strong views about the absurdity and waste of war, educational reform, and the need for individual conscience. These views offended many important people and led to the loss of his university position in 1819. He understood the need for absolute rigor in mathematics and sought to replace vague notions of infinitesimals with careful definitions. He also gave a proof of the fundamental theorem of algebra and the intermediate value theorem of calculus. His ideas are the ones which have survived and are the basis for math analysis.

Example B.1.3 *Let $a_n = \frac{1}{n^2+1}$.*

Then it seems clear that

$$\lim_{n\to\infty} \frac{1}{n^2+1} = 0.$$

In fact, this is true from the definition. Let $\varepsilon > 0$ be given. Let $n_\varepsilon \geq \sqrt{\varepsilon^{-1}}$. Then if

$$n > n_\varepsilon \geq \sqrt{\varepsilon^{-1}},$$

it follows that $n^2 + 1 > \varepsilon^{-1}$ and so

$$0 < \frac{1}{n^2+1} = a_n < \varepsilon..$$

Thus $|a_n - 0| < \varepsilon$ whenever n is this large.

Note the definition was of no use in finding a candidate for the limit. This had to be produced based on other considerations. The definition is for verifying beyond any doubt that something is the limit. It is also what must be referred to in establishing theorems which are good for finding limits.

Example B.1.4 *Let $a_n = n^2$*

Then in this case $\lim_{n\to\infty} a_n$ does not exist. Sometimes this situation is also referred to by saying $\lim_{n\to\infty} a_n = \infty$.

Example B.1.5 *Let $a_n = (-1)^n$.*

In this case, $\lim_{n\to\infty} (-1)^n$ does not exist. This follows from the definition. Let $\varepsilon = 1/2$. If there exists a limit, l, then eventually, for all n large enough, $|a_n - l| < 1/2$. However, $|a_n - a_{n+1}| = 2$ and so,

$$2 = |a_n - a_{n+1}| \leq |a_n - l| + |l - a_{n+1}| < 1/2 + 1/2 = 1$$

which cannot hold. Therefore, there can be no limit for this sequence.

Theorem B.1.6 *Suppose $\{a_n\}$ and $\{b_n\}$ are sequences and that*

$$\lim_{n\to\infty} a_n = a \text{ and } \lim_{n\to\infty} b_n = b.$$

Also suppose x and y are real numbers. Then

$$\lim_{n\to\infty} xa_n + yb_n = xa + yb \tag{2.1}$$

$$\lim_{n\to\infty} a_n b_n = ab \tag{2.2}$$

If $b \neq 0$,

$$\lim_{n\to\infty} \frac{a_n}{b_n} = \frac{a}{b}. \tag{2.3}$$

Proof: The first of these claims is left for you to do. To do the second, let $\varepsilon > 0$ be given and choose n_1 such that if $n \geq n_1$ then

$$|a_n - a| < 1.$$

Then for such n, the triangle inequality implies

$$
\begin{aligned}
|a_n b_n - ab| &\leq |a_n b_n - a_n b| + |a_n b - ab| \\
&\leq |a_n| |b_n - b| + |b| |a_n - a| \\
&\leq (|a| + 1) |b_n - b| + |b| |a_n - a|.
\end{aligned}
$$

Now let n_2 be large enough that for $n \geq n_2$,

$$|b_n - b| < \frac{\varepsilon}{2(|a| + 1)}, \text{ and } |a_n - a| < \frac{\varepsilon}{2(|b| + 1)}.$$

Such a number exists because of the definition of limit. Therefore, let

$$n_\varepsilon > \max(n_1, n_2).$$

For $n \geq n_\varepsilon$,

$$
\begin{aligned}
|a_n b_n - ab| &\leq (|a| + 1) |b_n - b| + |b| |a_n - a| \\
&< (|a| + 1) \frac{\varepsilon}{2(|a| + 1)} + |b| \frac{\varepsilon}{2(|b| + 1)} \leq \varepsilon.
\end{aligned}
$$

This proves (2.2). Next consider (2.3).

Let $\varepsilon > 0$ be given and let n_1 be so large that whenever $n \geq n_1$,

$$|b_n - b| < \frac{|b|}{2}.$$

Thus for such n,

$$
\left| \frac{a_n}{b_n} - \frac{a}{b} \right| = \left| \frac{a_n b - a b_n}{b b_n} \right| \leq \frac{2}{|b|^2} [|a_n b - ab| + |ab - ab_n|]
$$

$$
\leq \frac{2}{|b|} |a_n - a| + \frac{2|a|}{|b|^2} |b_n - b|.
$$

Now choose n_2 so large that if $n \geq n_2$, then

$$|a_n - a| < \frac{\varepsilon |b|}{4}, \text{ and } |b_n - b| < \frac{\varepsilon |b|^2}{4(|a| + 1)}.$$

Letting $n_\varepsilon > \max(n_1, n_2)$, it follows that for $n \geq n_\varepsilon$,

$$
\begin{aligned}
\left| \frac{a_n}{b_n} - \frac{a}{b} \right| &\leq \frac{2}{|b|} |a_n - a| + \frac{2|a|}{|b|^2} |b_n - b| \\
&< \frac{2}{|b|} \frac{\varepsilon |b|}{4} + \frac{2|a|}{|b|^2} \frac{\varepsilon |b|^2}{4(|a| + 1)} < \varepsilon. \blacksquare
\end{aligned}
$$

Another very useful theorem for finding limits is the squeezing theorem.

Theorem B.1.7 *Suppose* $\lim_{n\to\infty} a_n = a = \lim_{n\to\infty} b_n$ *and* $a_n \leq c_n \leq b_n$ *for all n large enough. Then* $\lim_{n\to\infty} c_n = a$.

Proof: Let $\varepsilon > 0$ be given and let n_1 be large enough that if $n \geq n_1$,

$$|a_n - a| < \varepsilon/2 \text{ and } |b_n - a| < \varepsilon/2.$$

Then for such n,

$$|c_n - a| \leq |a_n - a| + |b_n - a| < \varepsilon.$$

The reason for this is that if $c_n \geq a$, then

$$|c_n - a| = c_n - a \leq b_n - a \leq |a_n - a| + |b_n - a|$$

because $b_n \geq c_n$. On the other hand, if $c_n \leq a$, then

$$|c_n - a| = a - c_n \leq a - a_n \leq |a - a_n| + |b - b_n|. \ \blacksquare$$

As an example, consider the following.

Example B.1.8 *Let*

$$c_n \equiv (-1)^n \frac{1}{n}$$

and let $b_n = \frac{1}{n}$, *and* $a_n = -\frac{1}{n}$. *Then you may easily show that*

$$\lim_{n\to\infty} a_n = \lim_{n\to\infty} b_n = 0.$$

Since $a_n \leq c_n \leq b_n$, *it follows* $\lim_{n\to\infty} c_n = 0$ *also.*

Theorem B.1.9 $\lim_{n\to\infty} r^n = 0$. *Whenever* $|r| < 1$.

Proof: If $0 < r < 1$ if follows $r^{-1} > 1$. Why? Letting $\alpha = \frac{1}{r} - 1$, it follows

$$r = \frac{1}{1 + \alpha}.$$

Therefore, by the binomial theorem,

$$0 < r^n = \frac{1}{(1 + \alpha)^n} \leq \frac{1}{1 + \alpha n}.$$

Therefore, $\lim_{n\to\infty} r^n = 0$ if $0 < r < 1$. Now in general, if $|r| < 1$, $|r^n| = |r|^n \to 0$ by the first part. $\ \blacksquare$

Definition B.1.10 *Let* $\{a_n\}$ *be a sequence and let* $n_1 < n_2 < n_3, \cdots$ *be any strictly increasing list of integers such that* n_1 *is at least as large as the first number in the domain of the function. Then if* $b_k \equiv a_{n_k}$, $\{b_k\}$ *is called a subsequence of* $\{a_n\}$.

An important theorem is the one which states that if a sequence converges, so does every subsequence.

Theorem B.1.11 *Let* $\{x_n\}$ *be a sequence with* $\lim_{n\to\infty} x_n = x$ *and let* $\{x_{n_k}\}$ *be a subsequence. Then* $\lim_{k\to\infty} x_{n_k} = x$.

Proof: Let $\varepsilon > 0$ be given. Then there exists n_ε such that if $n > n_\varepsilon$, then $|x_n - x| < \varepsilon$. Suppose $k > n_\varepsilon$. Then $n_k \geq k > n_\varepsilon$ and so

$$|x_{n_k} - x| < \varepsilon$$

showing $\lim_{k\to\infty} x_{n_k} = x$ as claimed. $\ \blacksquare$

B.2   Cauchy Sequences

Definition B.2.1 $\{a_n\}$ *is a **Cauchy sequence** if for all $\varepsilon > 0$, there exists n_ε such that whenever $n, m \geq n_\varepsilon$,*

$$|a_n - a_m| < \varepsilon.$$

A sequence is Cauchy means the terms are "bunching up to each other" as m, n get large.

Theorem B.2.2 *The set of terms in a Cauchy sequence in $\mathbb{R}$ is bounded above and below.*

Proof: Let $\varepsilon = 1$ in the definition of a Cauchy sequence and let $n > n_1$. Then from the definition, $|a_n - a_{n_1}| < 1$. It follows that for all $n > n_1, |a_n| < 1 + |a_{n_1}|$. Therefore, for all n, $|a_n| \leq 1 + |a_{n_1}| + \sum_{k=1}^{n_1} |a_k|$. ∎

Theorem B.2.3 *If a sequence $\{a_n\}$ in $\mathbb{R}$ converges, then the sequence is a Cauchy sequence. Hence every subsequence converges. Also, in contrast to an arbitrary sequence, if any subsequence of a Cauchy sequence converges, then the Cauchy sequence converges.*

Proof: Let $\varepsilon > 0$ be given and suppose $a_n \to a$. Then from the definition of convergence, there exists n_ε such that if $n > n_\varepsilon$, it follows that $|a_n - a| < \frac{\varepsilon}{2}$. Therefore, if $m, n \geq n_\varepsilon + 1$, it follows that

$$|a_n - a_m| \leq |a_n - a| + |a - a_m| < \frac{\varepsilon}{2} + \frac{\varepsilon}{2} = \varepsilon$$

showing that, since $\varepsilon > 0$ is arbitrary, $\{a_n\}$ is a Cauchy sequence.

Now suppose $\lim_{k \to \infty} a_{n_k} = a$ and $\{a_n\}$ is a Cauchy sequence. Let $\varepsilon > 0$ be given. There exists N_1 such that if $m, n \geq N_1$ then $|a_n - a_m| < \varepsilon/2$. There also exists N_2 such that if $k \geq N_2$, then $|a - a_{n_k}| < \varepsilon/2$. Pick

$$k > \max(N_1, N_2) \equiv N$$

then if $n \geq N$

$$|a - a_n| \leq |a - a_{n_k}| + |a_{n_k} - a_n|$$

Since the list of n_k is strictly increasing, it follows that $n_k > N$ and so the above is smaller than $\varepsilon/2 + \varepsilon/2$. ∎

Example B.2.4 *Consider the sequence $(-1)^n = a_n$. This is not a Cauchy sequence so it does not converge. However, it does have a convergent subsequence. Just consider a_{2n}. These are always equal to 1 so this subsequence does converge.*

B.3   Continuity and the Limit of a Sequence

There is a very useful way of thinking of continuity in terms of limits of sequences found in the following theorem. In words, it says a function is continuous if it takes convergent sequences to convergent sequences whenever possible.

Theorem B.3.1 *A function $f : D(f) \to \mathbb{R}$ is continuous at $x \in D(f)$ if and only if, whenever $x_n \to x$ with $x_n \in D(f)$, it follows $f(x_n) \to f(x)$.*

Proof: Suppose first that f is continuous at x and let $x_n \to x$. Let $\varepsilon > 0$ be given. By continuity, there exists $\delta > 0$ such that if $|y - x| < \delta$, then $|f(x) - f(y)| < \varepsilon$. However, there exists n_δ such that if $n \geq n_\delta$, then $|x_n - x| < \delta$ and so for all n this large,

$$|f(x) - f(x_n)| < \varepsilon$$

which shows $f(x_n) \to f(x)$.

Now suppose the condition about taking convergent sequences to convergent sequences holds at x. Suppose f fails to be continuous at x. Then there exists $\varepsilon > 0$ and $x_n \in D(f)$ such that $|x - x_n| < \frac{1}{n}$, yet

$$|f(x) - f(x_n)| \geq \varepsilon.$$

But this is clearly a contradiction because, although $x_n \to x$, $f(x_n)$ fails to converge to $f(x)$. It follows f must be continuous after all. ∎

The above theorem has an easy corollary. Nothing changes if $D(f) \subseteq \mathbb{R}^n$ meaning that f is a function of many variables. You simply write $\mathbf{x}$ rather than x and measure the distance as

$$|\mathbf{x} - \mathbf{y}| \equiv \left(\sum_{i=1}^n (x_i - y_i)^2 \right)^{1/2} = ((\mathbf{x} - \mathbf{y}) \cdot (\mathbf{x} - \mathbf{y}))^{1/2}$$

Recall from multivariable calculus that the triangle inequality holds for this measure of distance. That is,

$$|\mathbf{x} + \mathbf{y}| \leq |\mathbf{x}| + |\mathbf{y}|$$

If you don't remember this fact from calculus or if it was not mentioned, simply define your distance as follows:

$$\|\mathbf{x}\| \equiv \max \{|\mathbf{x}_i| : i = 1, 2, \cdots, n\}$$

The triangle inequality is obvious for this. Indeed,

$$\begin{aligned} \|\mathbf{x} + \mathbf{y}\| &= \max_i \{|x_i + y_i|\} \leq \max_i \{|x_i| + |y_i|\} \\ &\leq \max_i |x_i| + \max_i |y_i| = \|\mathbf{x}\| + \|\mathbf{y}\| \end{aligned}$$

Both measures of distance mean that $\mathbf{x}_k \to \mathbf{x}$ if and only if the corresponding components converge in $\mathbb{R}$ and this is all that matters.

Then the definition of continuity is the same as for a function of one variable except that you use bold face to indicate that the point is in $\mathbb{R}^n$ rather than $\mathbb{R}$.

Then

Corollary B.3.2 *Nothing changes in the above theorem if $D(f) \subseteq \mathbb{R}^n$ rather than $\mathbb{R}$.*

B.4 Integrals

The integral is needed in order to consider the basic questions of differential equations. Consider the following initial value problem, differential equation and initial condition.

$$A'(x) = e^{x^2}, \ A(0) = 0.$$

So what is the solution to this initial value problem and does it even have a solution? More generally, for which functions, f does there exist a solution to the initial value problem, $y'(x) = f(x), y(0) = y_0$? The solution to these sorts of questions depend on the integral. Since this is usually not done well in beginning calculus courses, I will give a presentation of the theory of the integral. I assume the reader is familiar with the usual techniques for finding antiderivatives and integrals such as partial fractions, integration by parts and integration by substitution. These topics are usually done very well in beginning calculus courses.

The integral depends on completeness of $\mathbb{R}$.

Definition B.4.1 *One of the equivalent definitions of completeness of $\mathbb{R}$ is that if S is any nonempty subset of $\mathbb{R}$ which is bounded above, then there exists a least upper bound for S and if S is bounded below, then there exists a greatest lower bound for S. The least upper bound of S is denoted as $\sup(S)$ or sometimes as $l.u.b.(S)$ while the greatest lower bound of S is denoted as $\inf(S)$ sometimes as $g.l.b.(S)$. If there is no upper bound for S we say $\sup(S) = \infty$. If there is no lower bound, we say $\inf(S) = -\infty$.*

The words mean exactly what they say. $\sup(S)$ is a number with the property that $s \leq \sup(S)$ for all $s \in S$ and out of all such "upper bounds" it is the smallest. $\inf(S)$ has the property that $\inf(S) \leq s$ for all $s \in S$ and if $l \leq s$ for all $s \in S$, then $l \leq \inf(S)$. In words, it is the largest lower bound and $\sup(S)$ is the smallest upper bound. Here the meaning of small and large are as follows. To say that x is smaller than y means that $x \leq y$ which also says that y is larger than x.

A consequence of this axiom is the nested interval lemma, Lemma B.4.2.

Lemma B.4.2 *Let $I_k = [a^k, b^k]$ and suppose that for all $k = 1, 2, \cdots$,*

$$I_k \supseteq I_{k+1}.$$

Then there exists a point, $c \in \mathbb{R}$ which is an element of every I_k. If

$$\lim_{k \to \infty} b^k - a^k = 0$$

then there is exactly one point in all of these intervals.

Proof: Since $I_k \supseteq I_{k+1}$, this implies

$$a^k \leq a^{k+1}, \ b^k \geq b^{k+1}. \tag{2.4}$$

Consequently, letting $k \leq l$,

$$a^l \leq a^l \leq b^l \leq b^k. \tag{2.5}$$

Thus

$$c \equiv \sup\left\{a^l : l = 1, 2, \cdots\right\} = \sup\left\{a^l : l = k, k+1, \cdots\right\} \leq b^k$$

because b^k is an upper bound for all the a^l. Then $c \geq a^l$ for all l. In other words $x \geq a^k$ for all k. Also $c \leq b^k$ for all k. Therefore, $c \in [a^k, b^k]$ for all k.

If the length of these intervals converges to 0, then there can be at most one point in their intersection since otherwise, you would have two different points c, d and the length of the kth interval would then be at least as large as $|d - c|$ but both of these points would need to be in intervals having smaller length than this which can't happen. ∎

Corollary B.4.3 *Suppose* $\{x_n\}$ *is a sequence contained in* $[a, b]$. *Then there exists* $x \in [a, b]$ *and a subsequence* $\{x_{n_k}\}$ *such that* $\lim_{k \to \infty} x_{n_k} = x$.

Proof: Consider a sequence of closed intervals contained in $[a, b], I_1, I_2, \cdots$ where I_{k+1} is one half of I_k and each I_k contains x_n for infinitely many values of n. Thus $I_1 = [a, b]$, I_2 is either $\left[a, \frac{a+b}{2}\right]$ or $\left[\frac{a+b}{2}, b\right]$, depending on which one contains x_n for infinitely many values of n. If both intervals have this property, just pick one. Let $x_{n_k} \in I_k$ and let $x_{n_{k+1}} \in I_{k+1}$ with $n_{k+1} > n_k$. This is possible to do because each I_k contains x_n for infinitely many values of n. Then by the nested interval lemma, there exists a unique point x contained in all of these intervals and $|x - x_{n_k}| < (b - a)/2^k$. ∎

This has an obvious corollary in the case that $[a, b]$ is replaced with $Q = \prod_{i=1}^n [a_i, b_i]$ the Cartesian product of closed intervals.

Corollary B.4.4 *Suppose* $Q = \prod_{i=1}^n [a_i, b_i]$ *and let* $\{\mathbf{x}_k\}$ *be a sequence which is contained in* Q. *Then there exists* $\mathbf{x} \in Q$ *and a subsequence* $\{\mathbf{x}_{k_j}\}$ *such that* $\lim_{j \to \infty} \mathbf{x}_{k_j} = \mathbf{x}$.

Proof: Let $\mathbf{x}_k = \left(x_k^1, x_k^2, \cdots, x_k^n\right)$. By Corollary B.4.3 there exists a subsequence still denoted by $\{\mathbf{x}_k\}$ such that $\lim_{k \to \infty} x_k^1 = x_1 \in [a_1, b_1]$. Then by the same corollary, there is a further subsequence, still denoted as $\{\mathbf{x}_k\}$ such that in addition to this, $x_k^2 \to x_2 \in [a_2, b_2]$. Continuing this way, taking subsequences, we eventually obtain a subsequence, still denoted as $\{\mathbf{x}_k\}$ such that

$$\lim_{k \to \infty} x_k^i = x_i \in [a_i, b_i]$$

for each $i \leq n$. However, from the way we measure distance in $\mathbb{R}^n$ as

$$|\mathbf{x} - \mathbf{y}| = \left(\sum_i (x_i - y_i)^2\right)^{1/2}$$

The above limits say exactly that $\lim_{k \to \infty} |\mathbf{x}_k - \mathbf{x}| = 0$ and that $\mathbf{x} \in Q$. ∎

The next corollary is the extreme value theorem from calculus.

Corollary B.4.5 *If* $f : [a, b] \to \mathbb{R}$ *is continuous, then there exists* $x_M \in [a, b]$ *such that*

$$f(x_M) = \sup \{f(x) : x \in [a, b]\}$$

and there exists $x_m \in [a, b]$ *such that*

$$f(x_m) = \inf \{f(x) : x \in [a, b]\}$$

Proof: From the definition of $\inf\{f(x) : x \in [a, b]\}$, there exists $x_n \in [a, b]$ such that

$$f(x_n) \leq \inf\{f(x) : x \in [a, b]\} + 1/n$$

That is, $\lim_{n \to \infty} f(x_n) = \inf\{f(x) : x \in [a, b]\}$. This is called a minimizing sequence. Therefore, there is a subsequence $\{x_{n_k}\}$ which converges to $x \in [a, b]$. By continuity of f it follows that

$$\inf\{f(x) : x \in [a, b]\} = \lim_{k \to \infty} f(x_{n_k}) = f(x)$$

The case where f achieves its maximum is similar. You just use a maximizing sequence. ∎

You might write down a corresponding generalization in case

$$f : Q = \prod_{i=1}^{n} [a_i, b_i] \to \mathbb{R}$$

which is based on Corollary B.4.4.

Corollary B.4.6 *If $\{x_n\}$ is a Cauchy sequence, then it converges.*

Proof: The Cauchy sequence is contained in some closed interval $[a, b]$ thanks to Theorem B.2.2. By Corollary B.4.3, there is a subsequence of the Cauchy sequence which converges to some $x \in [a, b]$. Then by Theorem B.2.3, the original Cauchy sequence converges to x. ∎

Actually, the convergence of every Cauchy sequence is equivalent to completeness and so it gives another way of defining completeness in contexts where no order is available.

The Riemann integral pertains to bounded functions which are defined on a bounded interval. Let $[a, b]$ be a closed interval. A set of points in $[a, b]$, $\{x_0, \cdots, x_n\}$ is a partition if

$$a = x_0 < x_1 < \cdots < x_n = b.$$

Such partitions are denoted by P or Q.

Definition B.4.7 *A function $f : [a, b] \to \mathbb{R}$ is bounded if the set of values of f is contained in some interval. Thus*

$$\sup\{f(x) : x \in [a, b]\} < \infty, \quad \inf\{f(x) : x \in [a, b]\} > -\infty$$

. Letting P denote a partition,

$$\|P\| \equiv \max\{|x_{i+1} - x_i| : i = 0, \cdots, n - 1\}.$$

A Riemann sum for a bounded f corresponding to a partition $P = \{x_0, \cdots, x_n\}$ is a sum of the form

$$\sum_P f \equiv \sum_{i=1}^{n} f(y_i)(x_i - x_{i-1})$$

where $y_i \in [x_{i-1}, x_i]$. Then there are really many different Riemann sums corresponding to a given partition, depending on which y_i is chosen.

For example, suppose f is a function with positive values. The above Riemann sum involves adding areas of rectangles. Here is a picture:

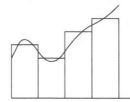

The area under the curve is close to the sum of the areas of these rectangles and one would imagine that this would become an increasingly good approximation if you included more and narrower rectangles.

Definition B.4.8 *A bounded function defined on an interval $[a, b]$ is Riemann integrable means that there exists a number I such that for every $\varepsilon > 0$, there exists a $\delta > 0$ such that whenever $\|P\| < \delta$, and $\sum_P f$ is some Riemann sum corresponding to this partition, it follows that*

$$\left| \sum_P f - I \right| < \varepsilon$$

This is written as

$$\lim_{\|P\| \to 0} \sum_P f = I$$

and when this number exists, it is denoted by

$$I = \int_a^b f(x) \, dx$$

One of the big theorems is on the existence of the integral whenever f is a continuous function. This requires a technical lemma which follows.

Lemma B.4.9 *Let $f : [a, b] \to \mathbb{R}$ be continuous. Then for every $\varepsilon > 0$ there exists a $\delta > 0$ such that if $|x - y| < \delta, x, y \in [a, b]$, it follows that $|f(x) - f(y)| < \varepsilon$.*

Proof: If not, then there exists $\varepsilon > 0$ and $x_n, y_n, |x_n - y_n| < 1/n$ but

$$|f(x_n) - f(y_n)| \geq \varepsilon.$$

By Corollary B.4.3, there exists a subsequence $\{x_{n_k}\}$ and a point $x \in [a, b]$ such that $\lim_{k \to \infty} x_{n_k} = x$. Then it follows that also $\lim_{k \to \infty} y_{n_k} = x$ also because

$$|y_{n_k} - x| \leq |y_{n_k} - x_{n_k}| + |x_{n_k} - x|$$

and both of the terms on the right converge to 0. But then by continuity of f,

$$0 = f(x) - f(x) = \lim_{k \to \infty} (f(x_{n_k}) - f(y_{n_k}))$$

which is impossible because $|f(x_{n_k}) - f(y_{n_k})| \geq \varepsilon$ for all k. ∎

The lemma has an obvious generalization to the case where $f : Q = \prod_{i=1}^n [a_i, b_i] \to \mathbb{R}$.

Corollary B.4.10 *Let $f : Q = \prod_{i=1}^{n} [a_i, b_i] \to \mathbb{R}$ be continuous. Then for every $\varepsilon > 0$ there exists a $\delta > 0$ such that whenever $|\mathbf{x} - \mathbf{y}| < \delta, \mathbf{x}, \mathbf{y} \in Q$, it follows that $|f(\mathbf{x}) - f(\mathbf{y})| < \varepsilon$.*

Proof: The proof is identical except, instead of Corollary B.4.3, you use Corollary B.4.4 and you use bold face or something else to indicate that the points are in $\mathbb{R}^n$. ∎

With this preparation, here is the major result on the existence of the integral of a continuous function.

Theorem B.4.11 *Let $f : [a, b] \to \mathbb{R}$ be continuous. Then $\int_a^b f(x)\, dx$ exists. In fact, there exists a sequence δ_m converging to 0 such that if $\|P\| < \delta_m$, and if $\sum_P f$ is a Riemann sum, then*

$$\left| \sum_P f - \int_a^b f\, dx \right| < \frac{2}{m} (b - a)$$

δ_m is defined to be such that if $|x - y| < \delta_m$, then $|f(x) - f(y)| < \frac{1}{m}$ and the sequence is decreasing.

Proof: Consider a partition P given by $a = x_0 < x_1 < \cdots < x_n = b$. Then you could add in another point as follows:

$$a = x_0 < x_1 < \cdots < x_{i-1} < x^* < x_i < \cdots < x_n = b$$

Denote this one by Q. Then if you have a Riemann sum,

$$\sum_P f = \sum_{j=1}^{n} f(y_j)(x_j - x_{j-1})$$

You could write this sum in the following form.

$$\sum_{j=1}^{i-1} f(y_j)(x_j - x_{j-1}) + f(y_i)(x^* - x_{i-1}) + f(y_i)(x_i - x^*) + \sum_{j=i+1}^{n} f(y_j)(x_j - x_{j-1})$$

In fact, you could continue adding in points and doing the same trick and thereby write the original sum in terms of any partition containing P. If R is a partition containing P and if δ_m corresponds to $\varepsilon = 1/m$ in the above Lemma with $\cdots > \delta_m > \delta_{m+1} \cdots$ B.4.9, then one can conclude that if $\|P\| < \delta_m$, then

$$\left| \sum_P f - \sum_R f \right| \leq \frac{1}{m}(b - a)$$

Now if $\|P\|, \|Q\| < \delta_m$, let $R = P \cup Q$. Then

$$\left| \sum_P f - \sum_Q f \right| \leq \left| \sum_P f - \sum_R f \right| + \left| \sum_R f - \sum_Q f \right|$$

$$\leq \frac{1}{m}(b - a) + \frac{1}{m}(b - a) = \frac{2}{m}(b - a)$$

Let $M \geq \max \{|f(x)| : x \in [a, b]\}$. Then all Riemann sums are in the interval

$$[-M(b-a), M(b-a)]$$

Now let

$$S_n \equiv \left\{ \sum_P f : \|P\| < \delta_n \right\}$$

Then $S_n \supseteq S_{n+1}$ for all n thanks to the fact that the δ_n are decreasing. Let

$$I_n = [\inf(S_n), \sup(S_n)]$$

These are nested intervals contained in $[-M(b-a), M(b-a)]$ and so there exists I contained in them all. However, from the above computation,

$$\sup(S_n) - \inf(S_n) \leq \frac{2}{n}(b-a)$$

and so there is only one such I. Hence for any $\varepsilon > 0$ given, there exists $\delta > 0$ such that if $\|P\| < \delta$, then

$$\left| \sum_S f - I \right| < \varepsilon \quad \blacksquare$$

We say that a bounded function f defined on an interval $[a, b]$ is Riemann integrable if the above integral exists. This is written as $f \in R([a, b])$. The above just showed that every continuous function is Riemann integrable.

Not all bounded functions are Riemann integrable. For example, let $x \in [0, 1]$ and

$$f(x) \equiv \begin{cases} 1 \text{ if } x \in \mathbb{Q} \\ 0 \text{ if } x \in \mathbb{R} \setminus \mathbb{Q} \end{cases} \tag{2.6}$$

This has no Riemann integral because you can pick a sequence of partitions P_n, such that $\|P_n\| < 1/n$ and each partition point is rational. Then for your Riemann sums, take the value of the function at the left end point. The resulting Riemann sum will always equal 1. But you could just as easily pick your point y_i in the Riemann sum to equal an irrational number and these Riemann sums will all equal 0. Therefore, the condition for integrability is violated for $\varepsilon = 1/4$.

B.5   Gamma Function

This belongs to a larger set of ideas concerning improper integrals. I will just give enough of an introduction to this to present the very important gamma function. The Riemann integral only is defined for bounded functions which are defined on a bounded interval. If this is not the case, then the integral has not been defined. Of course, just because the function is bounded does not mean the integral exists as mentioned above, but if it is not bounded, then there is no hope for it at all. However, one can consider limits of Riemann integrals. The following definition is sufficient to deal with the gamma function in the generality needed in this book.

Definition B.5.1 *We say that f defined on $[0, \infty)$ is improper Riemann integrable if it is Riemann integrable on $[\delta, R]$ for each $R > 1 > \delta > 0$ and the following limits exist.*

$$\int_0^\infty f(t)\, dt \equiv \lim_{\delta \to 0+} \int_\delta^1 f(t)\, dt + \lim_{R \to \infty} \int_1^R f(t)\, dt$$

The gamma function is defined by

$$\Gamma(\alpha) \equiv \int_0^\infty e^{-t} t^{\alpha - 1}\, dt$$

whenever $\alpha > 0$.

Lemma B.5.2 *The limits in the above definition exists for each $\alpha > 0$.*

Proof: Note first that as $\delta \to 0+$, the Riemann integrals

$$\int_\delta^1 e^{-t} t^{\alpha - 1} dt$$

increase. Thus $\lim_{\delta \to 0+} \int_\delta^1 e^{-t} t^{\alpha - 1} dt$ either is $+\infty$ or it will converge to the least upper bound thanks to completeness of $\mathbb{R}$. However,

$$\int_\delta^1 t^{\alpha - 1} dt \leq \frac{1}{\alpha}$$

so the limit of these integrals exists. Also $e^{-t} t^{\alpha - 1} \leq C e^{-(t/2)}$ for suitable C if $t > 1$. This is obvious if $\alpha - 1 < 0$ and in the other case it is also clear because exponential growth exceeds polynomial growth. Thus

$$\int_1^R e^{-t} t^{\alpha - 1} dt \leq \int_1^R C e^{-(t/2)} dt \leq 2C e^{(-1/2)} - 2C e^{(-R/2)} \leq 2C e^{(-1/2)}$$

Thus these integrals also converge as $R \to \infty$. It follows that $\Gamma(\alpha)$ makes sense. ∎

This gamma function has some fundamental properties described in the following proposition. In case the improper integral exists, we can obviously compute it in the form

$$\lim_{\delta \to 0+} \int_\delta^{1/\delta} f(t)\, dt$$

which is used in what follows. Thus also the usual algebraic properties of the Riemann integral are inherited by the improper integral.

Proposition B.5.3 *For n a positive integer, $n! = \Gamma(n+1)$. In general, $\Gamma(1) = 1, \Gamma(\alpha + 1) = \alpha \Gamma(\alpha)$*

Proof: First of all, $\Gamma(1) = \lim_{\delta \to 0} \int_\delta^{\delta^{-1}} e^{-t} dt = \lim_{\delta \to 0} \left(e^{-\delta} - e^{-(\delta^{-1})} \right) = 1$. Next, for $\alpha > 0$,

$$\Gamma(\alpha + 1) = \lim_{\delta \to 0} \int_\delta^{\delta^{-1}} e^{-t} t^\alpha dt = \lim_{\delta \to 0} \left[-e^{-t} t^\alpha |_\delta^{\delta^{-1}} + \alpha \int_\delta^{\delta^{-1}} e^{-t} t^{\alpha - 1} dt \right]$$

$$= \lim_{\delta \to 0} \left(e^{-\delta} \delta^{\alpha} - e^{-(\delta^{-1})} \delta^{-\alpha} + \alpha \int_{\delta}^{\delta^{-1}} e^{-t} t^{\alpha-1} dt \right) = \alpha \Gamma(\alpha)$$

Now it is defined that $0! = 1$ and so $\Gamma(1) = 0!$. Suppose that $\Gamma(n+1) = n!$, what of $\Gamma(n+2)$? Is it $(n+1)!$? if so, then by induction, the proposition is established. From what was just shown,

$$\Gamma(n+2) = \Gamma(n+1)(n+1) = n!(n+1) = (n+1)!$$

and so this proves the proposition. ∎

The properties of the gamma function also allow for a fairly easy proof about differentiating under the integral in a Laplace transform. First is a definition.

Definition B.5.4 *A function ϕ has exponential growth on $[0, \infty)$ if there are positive constants λ, C such that $|\phi(t)| \leq Ce^{\lambda t}$ for all t.*

Theorem B.5.5 *Let $f(s) = \int_0^{\infty} e^{-st} \phi(t) \, dt$ where $t \to \phi(t) e^{-st}$ is improper Riemann integrable for all s large enough and ϕ has exponential growth. Then for s large enough, $f^{(k)}(s)$ exists and equals $\int_0^{\infty} (-t)^k e^{-st} \phi(t) \, dt$.*

Proof: Suppose true for some $k \geq 0$. By definition it is so for $k = 0$. Then always assuming $s > \lambda, |h| < s - \lambda$, where $|\phi(t)| \leq Ce^{\lambda t}, \lambda \geq 0$,

$$\frac{f^{(k)}(s+h) - f^{(k)}(s)}{h} = \int_0^{\infty} (-t)^k \frac{e^{-(s+h)t} - e^{-st}}{h} \phi(t) \, dt$$

$$= \int_0^{\infty} (-t)^k e^{-st} \left(\frac{e^{-ht} - 1}{h} \right) \phi(t) \, dt = \int_0^{\infty} (-t)^k e^{-st} \left((-t) e^{\theta(h,t)} \right) \phi(t) \, dt$$

where $\theta(h, t)$ is between $-ht$ and 0, this by the mean value theorem. Thus by mean value theorem again,

$$\left| \frac{f^{(k)}(s+h) - f^{(k)}(s)}{h} - \int_0^{\infty} (-t)^{k+1} e^{-st} \phi(t) \, dt \right|$$

$$\leq \int_0^{\infty} |t|^{k+1} Ce^{\lambda t} e^{-st} \left| e^{\theta(h,t)} - 1 \right| dt \leq \int_0^{\infty} t^{k+1} Ce^{\lambda t} e^{-st} e^{\alpha(h,t)} |ht| \, dt$$

$$\leq \int_0^{\infty} t^{k+2} Ce^{\lambda t} e^{-st} |h| e^{t|h|} dt = C|h| \int_0^{\infty} t^{k+2} e^{-(s-(\lambda+|h|))t} dt$$

Let $u = (s - (\lambda + |h|)) t, du = (s - (\lambda + |h|)) \, dt$. Then the above equals

$$C|h| \int_0^{\infty} \left(\frac{u}{s - (\lambda + |h|)} \right)^{k+2} e^{-u} \frac{1}{(s - (\lambda + |h|))} du$$

$$= \frac{C|h|}{(s - (\lambda + |h|))^{k+3}} \int_0^{\infty} e^{-u} u^{k+2} du = \frac{C|h|}{(s - (\lambda + |h|))^{k+3}} \Gamma(k+3)$$

Thus, as $h \to 0$, this converges to 0 and so this proves the theorem. ∎

Appendix C

Series

C.1 Infinite Series of Numbers

This chapter gives a review of series. This is material encountered in calculus. It is included here in case there is a need for review.

C.1.1 Basic Considerations

In Definition B.1.1 on Page 401, the notion of limit of a sequence is discussed. There is a very closely related concept called an infinite series which is dealt with in this section.

Definition C.1.1 *Define*

$$\sum_{k=m}^{\infty} a_k \equiv \lim_{n \to \infty} \sum_{k=m}^{n} a_k$$

whenever the limit exists and is finite. In this case the series is said to converge. If it does not converge, it is said to diverge. The sequence $\{\sum_{k=m}^{n} a_k\}_{n=m}^{\infty}$ in the above is called the sequence of partial sums.

From this definition, it should be clear that infinite sums do not always make sense. Sometimes they do and sometimes they do not, depending on the behavior of the partial sums. As an example, consider $\sum_{k=1}^{\infty} (-1)^k$. The partial sums corresponding to this symbol alternate between -1 and 0. Therefore, there is no limit for the sequence of partial sums. It follows the symbol just written is meaningless and the infinite sum diverges.

Example C.1.2 *Find the infinite sum $\sum_{n=1}^{\infty} \frac{1}{n(n+1)}$.*

Note $\frac{1}{n(n+1)} = \frac{1}{n} - \frac{1}{n+1}$ and so $\sum_{n=1}^{N} \frac{1}{n(n+1)} = \sum_{n=1}^{N} \left(\frac{1}{n} - \frac{1}{n+1} \right) = -\frac{1}{N+1} + 1$. Therefore,

$$\lim_{N \to \infty} \sum_{n=1}^{N} \frac{1}{n(n+1)} = \lim_{N \to \infty} \left(-\frac{1}{N+1} + 1 \right) = 1.$$

Proposition C.1.3 *Let* $a_k \geq 0$. *Then* $\{\sum_{k=m}^{n} a_k\}_{n=m}^{\infty}$ *is an increasing sequence. If this sequence is bounded above, then* $\sum_{k=m}^{\infty} a_k$ *converges and its value equals*

$$\sup\left\{\sum_{k=m}^{n} a_k : n = m, m+1, \cdots\right\}.$$

When the sequence is not bounded above, $\sum_{k=m}^{\infty} a_k$ *diverges.*

Proof: It follows $\{\sum_{k=m}^{n} a_k\}_{n=m}^{\infty}$ is an increasing sequence because

$$\sum_{k=m}^{n+1} a_k - \sum_{k=m}^{n} a_k = a_{n+1} \geq 0.$$

If it is bounded above, let

$$l \equiv \sup\left\{\sum_{k=m}^{n} a_k : n = m, m+1, \cdots\right\}$$

be the least upper bound. Then for every $\varepsilon > 0$ there exists p such that $l - \varepsilon < \sum_{k=m}^{p} a_k \leq l$ since otherwise l would not be the least upper bound of the partial sums. Then since each $a_k \geq 0$, it follows for all $n \geq p$, $l - \varepsilon < \sum_{k=m}^{p} a_k \leq \sum_{k=m}^{n} a_k \leq l$, so by definition, the partial sums converge to l. It follows the sequence of partial sums converges to $\sup\{\sum_{k=m}^{n} a_k : n = m, m+1, \cdots\}$. If the sequence of partial sums is not bounded, then it is not a Cauchy sequence and so it does not converge. See Theorem B.2.3 on Page 405. ∎

In the case where $a_k \geq 0$, the above proposition shows there are only two alternatives available. Either the sequence of partial sums is bounded above or it is not bounded above. In the first case convergence occurs and in the second case, the infinite series diverges. For this reason, people will sometimes write $\sum_{k=m}^{\infty} a_k < \infty$ to denote the case where convergence occurs and $\sum_{k=m}^{\infty} a_k = \infty$ for the case where divergence occurs. Be very careful you never think this way in the case where it is not true that all $a_k \geq 0$. For example, the partial sums of $\sum_{k=1}^{\infty} (-1)^k$ are bounded because they are all either -1 or 0 but the series does not converge.

One of the most important examples of a convergent series is the geometric series. This series is $\sum_{n=0}^{\infty} r^n$. The study of this series depends on simple high school algebra and Theorem B.1.9 on Page 404. Let $S_n \equiv \sum_{k=0}^{n} r^k$. Then

$$S_n = \sum_{k=0}^{n} r^k, \ rS_n = \sum_{k=0}^{n} r^{k+1} = \sum_{k=1}^{n+1} r^k.$$

Therefore, subtracting the second equation from the first yields $(1-r) S_n = 1 - r^{n+1}$, and so a formula for S_n is available. In fact, if $r \neq 1$,

$$S_n = \frac{1 - r^{n+1}}{1 - r}.$$

By Theorem B.1.9, $\lim_{n \to \infty} S_n = \frac{1}{1-r}$ in the case when $|r| < 1$. Now if $|r| \geq 1$, the limit clearly does not exist because S_n fails to be a Cauchy sequence (Why?). This shows the following.

Theorem C.1.4 *The geometric series $\sum_{n=0}^{\infty} r^n$ converges and equals $\frac{1}{1-r}$ if $|r| < 1$ and diverges if $|r| \geq 1$.*

If the series do converge, the following holds about combinations of infinite series.

Theorem C.1.5 *If $\sum_{k=m}^{\infty} a_k$ and $\sum_{k=m}^{\infty} b_k$ both converge and x, y are numbers, then*

$$\sum_{k=m}^{\infty} a_k = \sum_{k=m+j}^{\infty} a_{k-j} \tag{3.1}$$

$$\sum_{k=m}^{\infty} x a_k + y b_k = x \sum_{k=m}^{\infty} a_k + y \sum_{k=m}^{\infty} b_k \tag{3.2}$$

$$\left| \sum_{k=m}^{\infty} a_k \right| \leq \sum_{k=m}^{\infty} |a_k| \tag{3.3}$$

where in the last inequality, the last sum equals $+\infty$ if the partial sums are not bounded above.

Proof: The above theorem is really only a restatement of Theorem B.1.6 about limits of sequences on Page 402 and the above definitions of infinite series. Formula (3.3) follows from the observation that, from the triangle inequality,

$$\left| \sum_{k=m}^{n} a_k \right| \leq \sum_{k=m}^{\infty} |a_k|,$$

and so

$$\left| \sum_{k=m}^{\infty} a_k \right| = \lim_{n \to \infty} \left| \sum_{k=m}^{n} a_k \right| \leq \sum_{k=m}^{\infty} |a_k|. \blacksquare$$

Example C.1.6 *Find $\sum_{n=0}^{\infty} \left(\frac{5}{2^n} + \frac{6}{3^n} \right)$.*

From the above theorem and Theorem C.1.4,

$$\sum_{n=0}^{\infty} \left(\frac{5}{2^n} + \frac{6}{3^n} \right) = 5 \sum_{n=0}^{\infty} \frac{1}{2^n} + 6 \sum_{n=0}^{\infty} \frac{1}{3^n} = 5 \frac{1}{1 - (1/2)} + 6 \frac{1}{1 - (1/3)} = 19.$$

The following criterion is useful in checking convergence.

Theorem C.1.7 *The sum $\sum_{k=m}^{\infty} a_k$ converges if and only if for all $\varepsilon > 0$, there exists n_ε such that if $q \geq p \geq n_\varepsilon$, then*

$$\left| \sum_{k=p}^{q} a_k \right| < \varepsilon. \tag{3.4}$$

Proof: Suppose first that the series converges. Then $\{\sum_{k=m}^{n} a_k\}_{n=m}^{\infty}$ is a Cauchy sequence by Theorem B.2.3 on Page 405. Therefore, there exists $n_\varepsilon > m$ such that if $q \geq p - 1 \geq n_\varepsilon > m$,

$$\left| \sum_{k=m}^{q} a_k - \sum_{k=m}^{p-1} a_k \right| = \left| \sum_{k=p}^{q} a_k \right| < \varepsilon. \tag{3.5}$$

Next suppose (3.4) holds. Then from (3.5), it follows upon letting p be replaced with $p + 1$ that $\{\sum_{k=m}^{n} a_k\}_{n=m}^{\infty}$ is a Cauchy sequence and so, by the completeness axiom, it converges. By the definition of infinite series, this shows the infinite sum converges as claimed. ∎

Definition C.1.8 *A series $\sum_{k=m}^{\infty} a_k$ converges absolutely if $\sum_{k=m}^{\infty} |a_k|$ converges. If the series does converge but does not converge absolutely, then it is said to converge conditionally.*

Theorem C.1.9 *If $\sum_{k=m}^{\infty} a_k$ converges absolutely, then it converges. Also if each $a_k \geq 0$ and the partial sums are bounded above, then $\sum_{k=m}^{\infty} a_k$ converges.*

Proof: Let $\varepsilon > 0$ be given. Then by assumption and Theorem C.1.7, there exists n_ε such that whenever $q \geq p \geq n_\varepsilon$, $\sum_{k=p}^{q} |a_k| < \varepsilon$. Therefore, from the triangle inequality, $\varepsilon > \sum_{k=p}^{q} |a_k| \geq \left| \sum_{k=p}^{q} a_k \right|$. By Theorem C.1.7, $\sum_{k=m}^{\infty} a_k$ converges and this proves the first part. The second claim follows from Proposition C.1.3. ∎

In fact, the above theorem is really another version of the completeness axiom. Thus its validity implies completeness. You might try to show this.

Theorem C.1.10 *(comparison test) Suppose $\{a_n\}$ and $\{b_n\}$ are sequences of non-negative real numbers and suppose for all n sufficiently large, $a_n \leq b_n$. Then*

1. *If $\sum_{n=k}^{\infty} b_n$ converges, then $\sum_{n=m}^{\infty} a_n$ converges.*

2. *If $\sum_{n=k}^{\infty} a_n$ diverges, then $\sum_{n=m}^{\infty} b_n$ diverges.*

Proof: Consider the first claim. From the assumption there exists n^* such that $n^* > \max(k, m)$ and for all $n \geq n^*$ $b_n \geq a_n$. Then if $p \geq n^*$,

$$\sum_{n=m}^{p} a_n \leq \sum_{n=m}^{n^*} a_n + \sum_{n=n^*+1}^{k} b_n \leq \sum_{n=m}^{n^*} a_n + \sum_{n=k}^{\infty} b_n.$$

Thus the sequence $\{\sum_{n=m}^{p} a_n\}_{p=m}^{\infty}$ is bounded above and increasing. Therefore, it converges by Theorem C.1.9. The second claim is left as an exercise. ∎

Example C.1.11 *Determine the convergence of $\sum_{n=1}^{\infty} \frac{1}{n^2}$.*

For $n > 1$,

$$\frac{1}{n^2} \leq \frac{1}{n(n-1)}.$$

Now

$$\sum_{n=2}^{p} \frac{1}{n(n-1)} = \sum_{n=2}^{p} \left[\frac{1}{n-1} - \frac{1}{n} \right] = 1 - \frac{1}{p}$$

which converges to 1 as $p \to \infty$. Therefore, letting $a_n = \frac{1}{n^2}$ and $b_n = \frac{1}{n(n-1)}$, it follows by the comparison test, the given series converges.

A convenient way to implement the comparison test is to use the limit comparison test. This is considered next.

Theorem C.1.12 *Let $a_n, b_n > 0$ and suppose for all n large enough,*

$$0 < a < \frac{a_n}{b_n} < b < \infty.$$

Then $\sum a_n$ and $\sum b_n$ converge or diverge together.

Proof: Let n^* be such that $n \geq n^*$, then

$$\frac{a_n}{b_n} > a \text{ and } \frac{a_n}{b_n} < b$$

and so for all such n,

$$ab_n < a_n < bb_n$$

and so the conclusion follows from the comparison test. ∎

Corollary C.1.13 *If $a_n, b_n > 0$ and $\lim_{n \to \infty} \frac{a_n}{b_n} = r \in (0, \infty)$, then $\sum a_n, \sum b_n$ converge or diverge together.*

Proof: The assumption implies that for all n large enough,

$$\frac{r}{2} < \frac{a_n}{b_n} < 2r$$

The conclusion now follows from Theorem C.1.12. ∎

Example C.1.14 *Determine the convergence of $\sum_{k=1}^{\infty} \frac{1}{\sqrt{n^4+2n+7}}$.*

This series converges by the limit comparison test. Compare with the series of Example C.1.11.

$$\lim_{n \to \infty} \frac{\left(\frac{1}{n^2} \right)}{\left(\frac{1}{\sqrt{n^4+2n+7}} \right)} = \lim_{n \to \infty} \frac{\sqrt{n^4 + 2n + 7}}{n^2} = \lim_{n \to \infty} \sqrt{1 + \frac{2}{n^3} + \frac{7}{n^4}} = 1.$$

Therefore, the series converges with the series of Example C.1.11. How did I know what to compare with? I noticed that $\sqrt{n^4 + 2n + 7}$ is essentially like $\sqrt{n^4} = n^2$ for large enough n. You see, the higher order term n^4 dominates the other terms in $n^4 + 2n + 7$. Therefore, reasoning that $1/\sqrt{n^4 + 2n + 7}$ is a lot like $1/n^2$ for large n, it was easy to see what to compare with. Of course this is not always easy and there is room for acquiring skill through practice.

To really exploit this limit comparison test, it is desirable to get lots of examples of series, some which converge and some which do not. The tool for obtaining these examples here will be the following wonderful theorem known as the Cauchy condensation test. It is an extremely simple idea. You group the terms in blocks and use the assumption that the terms of the series are decreasing.

Theorem C.1.15 *Let $a_n \geq 0$ and suppose the terms of the sequence $\{a_n\}$ are decreasing. Thus $a_n \geq a_{n+1}$ for all n. Then the two series*

$$\sum_{n=1}^{\infty} a_n, \quad \sum_{n=0}^{\infty} 2^n a_{2^n}$$

converge or diverge together.

Proof: This follows from the inequality of the following claim.
Claim:

$$\sum_{k=0}^{n} 2^k a_{2^k} \geq \sum_{k=1}^{2^n} a_k \geq \frac{1}{2} \sum_{k=0}^{n} 2^k a_{2^k}.$$

Proof of the Claim: The proof is by induction.
It is clear that the above inequality is true if $n = 1$ because it says

$$a_1 + 2a_2 \geq a_1 + a_2 \geq \frac{1}{2}a_1 + a_2,$$

which is obvious. Suppose the inequality is true for n. Then by induction,

$$\sum_{k=0}^{n+1} 2^k a_{2^k} \; = \; 2^{n+1} a_{2^{n+1}} + \sum_{k=0}^{n} 2^k a_{2^k} \geq 2^{n+1} a_{2^{n+1}} + \sum_{k=1}^{2^n} a_k$$

$$\geq \; 2^{n+1} a_{2^{n+1}} + \frac{1}{2}\sum_{k=0}^{n} 2^k a_{2^k} \geq \frac{1}{2}\sum_{k=0}^{n+1} 2^k a_{2^k},$$

the last inequality because $2^{n+1} a_{2^{n+1}} \geq \frac{1}{2} 2^{n+1} a_{2^{n+1}}$. ∎

Example C.1.16 *Determine the convergence of $\sum_{k=1}^{\infty} \frac{1}{k^p}$ where p is a positive number. These are called the p series.*

Let $a_n = \frac{1}{n^p}$. Then $a_{2^n} = \left(\frac{1}{2^p}\right)^n$. From the Cauchy condensation test the two series

$$\sum_{n=1}^{\infty} \frac{1}{n^p} \text{ and } \sum_{n=0}^{\infty} 2^n \left(\frac{1}{2^p}\right)^n = \sum_{n=0}^{\infty} \left(2^{(1-p)}\right)^n$$

converge or diverge together. If $p > 1$, the last series above is a geometric series having common ratio less than 1, and so it converges. If $p \leq 1$, it is still a geometric series but in this case the common ratio is either 1 or greater than 1, so the series diverges. It follows that the p series converges if $p > 1$ and diverges if $p \leq 1$. In particular, $\sum_{n=1}^{\infty} n^{-1}$ diverges while $\sum_{n=1}^{\infty} n^{-2}$ converges. The following table summarizes the above results.

The p series table

	$p \leq 1$	$p > 1$
$\sum \frac{1}{n^p}$	diverges	converges

Example C.1.17 *Determine the convergence of $\sum_{k=1}^{\infty} \frac{1}{\sqrt{n^2+100n}}$.*

Use the limit comparison test.

$$\lim_{n\to\infty} \frac{\left(\frac{1}{n}\right)}{\left(\frac{1}{\sqrt{n^2+100n}}\right)} = 1,$$

and so this series diverges with $\sum_{k=1}^{\infty} \frac{1}{k}$.

Example C.1.18 *Determine the convergence of $\sum_{k=2}^{\infty} \frac{1}{k \ln k}$.*

Use the Cauchy condensation test. The above series does the same thing in terms of convergence as the series

$$\sum_{n=1}^{\infty} 2^n \frac{1}{2^n \ln(2^n)} = \sum_{n=1}^{\infty} \frac{1}{n \ln 2}$$

and this series diverges by limit comparison with the series $\sum \frac{1}{n}$.

Sometimes it is good to be able to say a series does not converge. The nth term test gives such a condition which is sufficient for this. Sometimes this is called the divergence test. It is really a corollary of Theorem C.1.7.

Theorem C.1.19 *If $\sum_{n=m}^{\infty} a_n$ converges, then $\lim_{n\to\infty} a_n = 0$.*

Proof: By Theorem C.1.7, $\lim_{n\to\infty} a_n = \lim_{n\to\infty} \sum_{k=n}^{n} a_k = 0.\blacksquare$

It is very important to observe that this theorem goes only in one direction. That is, you **cannot conclude** the series converges if $\lim_{n\to\infty} a_n = 0$. If this happens, you don't know anything from this information. Recall $\lim_{n\to\infty} n^{-1} = 0$ but $\sum_{n=1}^{\infty} n^{-1}$ diverges. The following picture is descriptive of the situation.

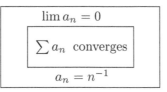

C.2 More Tests for Convergence

C.2.1 Convergence Because of Cancellation

So far, the tests for convergence have been applied to nonnegative terms only. Sometimes, a series converges, not because the terms of the series get small fast enough, but because of cancellation taking place between positive and negative terms. A discussion of this involves some simple algebra.

Let $\{a_n\}$ and $\{b_n\}$ be sequences and let

$$A_n \equiv \sum_{k=1}^{n} a_k, \quad A_{-1} \equiv A_0 \equiv 0.$$

Then if $p < q$

$$\sum_{n=p}^{q} a_n b_n = \sum_{n=p}^{q} b_n \left(A_n - A_{n-1} \right) = \sum_{n=p}^{q} b_n A_n - \sum_{n=p}^{q} b_n A_{n-1}$$

$$= \sum_{n=p}^{q} b_n A_n - \sum_{n=p-1}^{q-1} b_{n+1} A_n = b_q A_q - b_p A_{p-1} + \sum_{n=p}^{q-1} A_n \left(b_n - b_{n+1} \right)$$

This formula is called the partial summation formula. It is just like integration by parts.

Theorem C.2.1 *(Dirichlet's test) Suppose A_n is bounded and $\lim_{n \to \infty} b_n = 0$, with $b_n \geq b_{n+1}$. Then*

$$\sum_{n=1}^{\infty} a_n b_n$$

converges.

Proof: This follows quickly from Theorem C.1.7. Indeed, letting $|A_n| \leq C$, and using the partial summation formula above along with the assumption that the b_n are decreasing,

$$\left| \sum_{n=p}^{q} a_n b_n \right| = \left| b_q A_q - b_p A_{p-1} + \sum_{n=p}^{q-1} A_n \left(b_n - b_{n+1} \right) \right|$$

$$\leq C \left(|b_q| + |b_p| \right) + C \sum_{n=p}^{q-1} \left(b_n - b_{n+1} \right) = C \left(|b_q| + |b_p| \right) + C \left(b_p - b_q \right)$$

and by assumption, this last expression is small whenever p and q are sufficiently large. ■

Definition C.2.2 *If $b_n > 0$ for all n, a series of the form*

$$\sum_{k} \left(-1 \right)^k b_k \ \ or \ \ \sum_{k} \left(-1 \right)^{k-1} b_k$$

*is known as an **alternating series**.*

The following corollary is known as the alternating series test.

Corollary C.2.3 *(alternating series test) If $\lim_{n \to \infty} b_n = 0$, with $b_n \geq b_{n+1}$, then $\sum_{n=1}^{\infty} \left(-1 \right)^n b_n$ converges.*

Proof: Let $a_n = \left(-1 \right)^n$. Then the partial sums of $\sum_n a_n$ are bounded, and so Theorem C.2.1 applies. ■

In the situation of Corollary C.2.3 there is a convenient error estimate available.

Theorem C.2.4 *Let $b_n > 0$ for all n such that $b_n \geq b_{n+1}$ for all n and $\lim_{n \to \infty} b_n = 0$ and consider either $\sum_{n=1}^{\infty} \left(-1 \right)^n b_n$ or $\sum_{n=1}^{\infty} \left(-1 \right)^{n-1} b_n$. Then*

$$\left| \sum_{n=1}^{\infty} \left(-1 \right)^n b_n - \sum_{n=1}^{N} \left(-1 \right)^n b_n \right| \leq |b_{N+1}|, \left| \sum_{n=1}^{\infty} \left(-1 \right)^{n-1} b_n - \sum_{n=1}^{N} \left(-1 \right)^{n-1} b_n \right| \leq |b_{N+1}|$$

Example C.2.5 *How many terms must I take in the sum $\sum_{n=1}^{\infty}(-1)^n \frac{1}{n^2+1}$ to be closer than $\frac{1}{10}$ to $\sum_{n=1}^{\infty}(-1)^n \frac{1}{n^2+1}$?*

From Theorem C.2.4, I need to find n such that $\frac{1}{n^2+1} \leq \frac{1}{10}$ and then $n-1$ is the desired value. Thus $n = 3$, and so

$$\left| \sum_{n=1}^{\infty}(-1)^n \frac{1}{n^2+1} - \sum_{n=1}^{2}(-1)^n \frac{1}{n^2+1} \right| \leq \frac{1}{10}$$

C.2.2 Ratio and Root Tests

A favorite test for convergence is the ratio test. This is discussed next. It is the at the other extreme. This test is completely oblivious to any sort of cancelation. It only gives absolute convergence or spectacular divergence.

Theorem C.2.6 *Suppose $|a_n| > 0$ for all n and suppose*

$$\lim_{n\to\infty} \frac{|a_{n+1}|}{|a_n|} = r.$$

Then

$$\sum_{n=1}^{\infty} a_n \begin{cases} \text{diverges if } r > 1 \\ \text{converges absolutely if } r < 1 \\ \text{test fails if } r = 1 \end{cases}.$$

Proof: Suppose $r < 1$. Then there exists n_1 such that if $n \geq n_1$, then

$$0 < \left| \frac{a_{n+1}}{a_n} \right| < R$$

where $r < R < 1$. Then

$$|a_{n+1}| < R |a_n|$$

for all such n. Therefore,

$$|a_{n_1+p}| < R |a_{n_1+p-1}| < R^2 |a_{n_1+p-2}| < \cdots < R^p |a_{n_1}| \tag{3.6}$$

and so if $m > n$, then $|a_m| < R^{m-n_1} |a_{n_1}|$. By the comparison test and the theorem on geometric series, $\sum |a_n|$ converges. This proves the convergence part of the theorem.

To verify the divergence part, note that if $r > 1$, then (3.6) can be turned around for some $R > 1$. Showing $\lim_{n\to\infty} |a_n| = \infty$. Since the nth term fails to converge to 0, it follows the series diverges.

To see the test fails if $r = 1$, consider $\sum n^{-1}$ and $\sum n^{-2}$. The first series diverges while the second one converges but in both cases, $r = 1$. (Be sure to check this last claim.) ∎

The ratio test is very useful for many different examples but it is somewhat unsatisfactory mathematically. One reason for this is the assumption that $a_n \neq 0$, necessitated by the need to divide by a_n, and the other reason is the possibility that the limit might not exist. The next test, called the root test removes both of these objections.

Theorem C.2.7 *Suppose $|a_n|^{1/n} < R < 1$ for all n sufficiently large. Then*

$$\sum_{n=1}^{\infty} a_n \text{ converges absolutely.}$$

If there are infinitely many values of n such that $|a_n|^{1/n} \geq 1$, then

$$\sum_{n=1}^{\infty} a_n \text{ diverges.}$$

Proof: Suppose first that $|a_n|^{1/n} < R < 1$ for all n sufficiently large. Say this holds for all $n \geq n_R$. Then for such n,

$$\sqrt[n]{|a_n|} < R.$$

Therefore, for such n,

$$|a_n| \leq R^n,$$

and so the comparison test with a geometric series applies and gives absolute convergence as claimed.

Next suppose $|a_n|^{1/n} \geq 1$ for infinitely many values of n. Then for those values of n, $|a_n| \geq 1$, and so the series fails to converge because of Theorem C.1.19. ■

Corollary C.2.8 *Suppose $\lim_{n\to\infty} |a_n|^{1/n}$ exists and equals r. Then*

$$\sum_{k=m}^{\infty} a_k \begin{cases} \text{converges absolutely if } r < 1 \\ \text{test fails if } r = 1 \\ \text{diverges if } r > 1 \end{cases}$$

Proof: The first and last alternatives follow from Theorem C.2.7. To see the test fails if $r = 1$, consider the two series $\sum_{n=1}^{\infty} \frac{1}{n}$ and $\sum_{n=1}^{\infty} \frac{1}{n^2}$ both of which have $r = 1$ but having different convergence properties. ■

C.3   Double Series

Sometimes it is required to consider double series which are of the form

$$\sum_{k=m}^{\infty} \sum_{j=m}^{\infty} a_{jk} \equiv \sum_{k=m}^{\infty} \left(\sum_{j=m}^{\infty} a_{jk} \right).$$

In other words, first sum on j yielding something which depends on k and then sum these. The major consideration for these double series is the question of when

$$\sum_{k=m}^{\infty} \sum_{j=m}^{\infty} a_{jk} = \sum_{j=m}^{\infty} \sum_{k=m}^{\infty} a_{jk}.$$

In other words, when does it make no difference which subscript is summed over first? In the case of finite sums there is no issue here. You can always interchange the order because addition is commutative. However, there are limits involved with infinite sums and the interchange in order of summation involves taking limits in a different order. Therefore, it is not always true that it is permissible to interchange the two sums. A general rule of thumb is this: If something involves changing the order in which two limits are taken, you may not do it without agonizing over the question. In general, limits foul up algebra and also introduce things which are counter intuitive. Here is an example. This example is a little technical. It is placed here just to prove conclusively there is a question which needs to be considered.

Example C.3.1 *Consider the following picture which depicts some of the ordered pairs (m, n) where m, n are positive integers.*

$$0. \quad 0. \quad 0. \quad 0. \quad 0. \quad c. \quad 0. \quad -c.$$
$$0. \quad 0. \quad 0. \quad 0. \quad c. \quad 0. \quad -c. \quad 0.$$
$$0. \quad 0. \quad 0. \quad c. \quad 0. \quad -c. \quad 0. \quad 0.$$
$$0. \quad 0. \quad c. \quad 0. \quad -c. \quad 0. \quad 0. \quad 0.$$
$$0. \quad c. \quad 0. \quad -c. \quad 0. \quad 0. \quad 0. \quad 0.$$
$$b. \quad 0. \quad -c. \quad 0. \quad 0. \quad 0. \quad 0. \quad 0.$$
$$0. \quad a. \quad 0. \quad 0. \quad 0. \quad 0. \quad 0. \quad 0.$$

The numbers next to the point are the values of a_{mn}. You see $a_{nn} = 0$ for all n, $a_{21} = a, a_{12} = b, a_{mn} = c$ for (m, n) on the line $y = 1 + x$ whenever $m > 1$, and $a_{mn} = -c$ for all (m, n) on the line $y = x - 1$ whenever $m > 2$.

Then $\sum_{m=1}^{\infty} a_{mn} = a$ if $n = 1$, $\sum_{m=1}^{\infty} a_{mn} = b - c$ if $n = 2$ and if $n > 2, \sum_{m=1}^{\infty} a_{mn} = 0$. Therefore, $\sum_{n=1}^{\infty} \sum_{m=1}^{\infty} a_{mn} = a + b - c$. Next observe that $\sum_{n=1}^{\infty} a_{mn} = b$ if $m = 1$, $\sum_{n=1}^{\infty} a_{mn} = a + c$ if $m = 2$, and $\sum_{n=1}^{\infty} a_{mn} = 0$ if $m > 2$. Therefore, $\sum_{m=1}^{\infty} \sum_{n=1}^{\infty} a_{mn} = b + a + c$, and so the two sums are different. Moreover, you can see that by assigning different values of a, b, and c, you can get an example for any two different numbers desired.

This happens because, as indicated above, limits are taken in two different orders. An infinite sum always involves a limit and this illustrates why you must always remember this. This example in no way violates the commutative law of addition which has nothing to do with limits. However, it turns out that if $a_{ij} \geq 0$ for all i, j, then you can always interchange the order of summation. This is shown next and is based on the following lemma. First, some notation should be discussed.

Definition C.3.2 *Let $f(a, b) \in [-\infty, \infty]$ for $a \in A$ and $b \in B$ where A, B are sets which means that $f(a, b)$ is either a number, ∞, or $-\infty$. The symbol $+\infty$ is interpreted as a point out at the end of the number line which is larger than every real number and read as plus infinity. Of course there is no such number. That is why it is called ∞. The symbol $-\infty$ is interpreted similarly. Then $\sup_{a \in A} f(a, b)$ means $\sup(S_b)$ where $S_b \equiv \{f(a, b) : a \in A\}$.*

Unlike limits, you can take the sup in different orders.

Lemma C.3.3 *Let $f(a,b) \in [-\infty, \infty]$ for $a \in A$ and $b \in B$ where A, B are sets. Then*

$$\sup_{a \in A} \sup_{b \in B} f(a,b) = \sup_{b \in B} \sup_{a \in A} f(a,b).$$

Proof: Note that for all a, b, $f(a,b) \leq \sup_{b \in B} \sup_{a \in A} f(a,b)$ and therefore, for all a, $\sup_{b \in B} f(a,b) \leq \sup_{b \in B} \sup_{a \in A} f(a,b)$. Therefore,

$$\sup_{a \in A} \sup_{b \in B} f(a,b) \leq \sup_{b \in B} \sup_{a \in A} f(a,b).$$

Repeat the same argument interchanging a and b, to get the conclusion of the lemma. ∎

Lemma C.3.4 *If $\{A_n\}$ is an increasing sequence in $[-\infty, \infty]$, then*

$$\sup\{A_n\} = \lim_{n \to \infty} A_n.$$

Proof: Let $\sup(\{A_n : n \in \mathbb{N}\}) = r$. In the first case, suppose $r < \infty$. Then letting $\varepsilon > 0$ be given, there exists n such that $A_n \in (r - \varepsilon, r]$. Since $\{A_n\}$ is increasing, it follows if $m > n$, then $r - \varepsilon < A_n \leq A_m \leq r$ and so $\lim_{n \to \infty} A_n = r$ as claimed. In the case where $r = \infty$, then if a is a real number, there exists n such that $A_n > a$. Since $\{A_k\}$ is increasing, it follows that if $m > n$, $A_m > a$. But this is what is meant by $\lim_{n \to \infty} A_n = \infty$. The other case is that $r = -\infty$. But in this case, $A_n = -\infty$ for all n and so $\lim_{n \to \infty} A_n = -\infty$. ∎

Theorem C.3.5 *Let $a_{ij} \geq 0$. Then $\sum_{i=1}^{\infty} \sum_{j=1}^{\infty} a_{ij} = \sum_{j=1}^{\infty} \sum_{i=1}^{\infty} a_{ij}$.*

Proof: First note there is no trouble in defining these sums because the a_{ij} are all nonnegative. If a sum diverges, it only diverges to ∞ and so ∞ is the value of the sum. Next note that

$$\sum_{j=r}^{\infty} \sum_{i=r}^{\infty} a_{ij} \geq \sup_{n} \sum_{j=r}^{\infty} \sum_{i=r}^{n} a_{ij}$$

because for all j, $\sum_{i=r}^{\infty} a_{ij} \geq \sum_{i=r}^{n} a_{ij}$. Therefore,

$$\sum_{j=r}^{\infty} \sum_{i=r}^{\infty} a_{ij} \geq \sup_{n} \sum_{j=r}^{\infty} \sum_{i=r}^{n} a_{ij} = \sup_{n} \lim_{m \to \infty} \sum_{j=r}^{m} \sum_{i=r}^{n} a_{ij}$$

$$= \sup_{n} \lim_{m \to \infty} \sum_{i=r}^{n} \sum_{j=r}^{m} a_{ij} = \sup_{n} \sum_{i=r}^{n} \lim_{m \to \infty} \sum_{j=r}^{m} a_{ij}$$

$$= \sup_{n} \sum_{i=r}^{n} \sum_{j=r}^{\infty} a_{ij} = \lim_{n \to \infty} \sum_{i=r}^{n} \sum_{j=r}^{\infty} a_{ij} = \sum_{i=r}^{\infty} \sum_{j=r}^{\infty} a_{ij}$$

Interchanging the i and j in the above argument proves the theorem. ∎

The following is the fundamental result on double sums.

Theorem C.3.6 *Let a_{ij} be a number and suppose*

$$\sum_{i=r}^{\infty}\sum_{j=r}^{\infty}|a_{ij}| < \infty .$$

Then

$$\sum_{i=r}^{\infty}\sum_{j=r}^{\infty}a_{ij} = \sum_{j=r}^{\infty}\sum_{i=r}^{\infty}a_{ij}$$

and every infinite sum encountered in the above equation converges.

Proof: By Theorem C.3.5

$$\sum_{j=r}^{\infty}\sum_{i=r}^{\infty}|a_{ij}| = \sum_{i=r}^{\infty}\sum_{j=r}^{\infty}|a_{ij}| < \infty$$

Therefore, for each j, $\sum_{i=r}^{\infty}|a_{ij}| < \infty$ and for each i, $\sum_{j=r}^{\infty}|a_{ij}| < \infty$. By Theorem C.1.9 on Page 418, $\sum_{i=r}^{\infty}a_{ij}$, $\sum_{j=r}^{\infty}a_{ij}$ both converge, the first one for every j and the second for every i. Also,

$$\sum_{j=r}^{\infty}\left|\sum_{i=r}^{\infty}a_{ij}\right| \le \sum_{j=r}^{\infty}\sum_{i=r}^{\infty}|a_{ij}| < \infty$$

and

$$\sum_{i=r}^{\infty}\left|\sum_{j=r}^{\infty}a_{ij}\right| \le \sum_{i=r}^{\infty}\sum_{j=r}^{\infty}|a_{ij}| < \infty$$

so by Theorem C.1.9 again, $\sum_{j=r}^{\infty}\sum_{i=r}^{\infty}a_{ij}$, $\sum_{i=r}^{\infty}\sum_{j=r}^{\infty}a_{ij}$ both exist. It only remains to verify they are equal. Note $0 \le (|a_{ij}| + a_{ij}) \le |a_{ij}|$. Therefore, by Theorem C.3.5 and Theorem C.1.5 on Page 417

$$\sum_{j=r}^{\infty}\sum_{i=r}^{\infty}|a_{ij}| + \sum_{j=r}^{\infty}\sum_{i=r}^{\infty}a_{ij} = \sum_{j=r}^{\infty}\sum_{i=r}^{\infty}(|a_{ij}| + a_{ij}) = \sum_{i=r}^{\infty}\sum_{j=r}^{\infty}(|a_{ij}| + a_{ij})$$

$$= \sum_{i=r}^{\infty}\sum_{j=r}^{\infty}|a_{ij}| + \sum_{i=r}^{\infty}\sum_{j=r}^{\infty}a_{ij} = \sum_{j=r}^{\infty}\sum_{i=r}^{\infty}|a_{ij}| + \sum_{i=r}^{\infty}\sum_{j=r}^{\infty}a_{ij}$$

and so $\sum_{j=r}^{\infty}\sum_{i=r}^{\infty}a_{ij} = \sum_{i=r}^{\infty}\sum_{j=r}^{\infty}a_{ij}$ as claimed. ∎

One of the most important applications of this theorem is to the problem of multiplication of series.

Definition C.3.7 *Let $\sum_{i=r}^{\infty}a_i$ and $\sum_{i=r}^{\infty}b_i$ be two series. For $n \ge r$, define*

$$c_n \equiv \sum_{k=r}^{n}a_k b_{n-k+r}.$$

The series $\sum_{n=r}^{\infty}c_n$ is called the Cauchy product of the two series.

It is not hard to see where this comes from. Formally write the following in the case $r = 0$:

$$(a_0 + a_1 + a_2 + a_3 \cdots)(b_0 + b_1 + b_2 + b_3 \cdots)$$

and start multiplying in the usual way. This yields

$$a_0 b_0 + (a_0 b_1 + b_0 a_1) + (a_0 b_2 + a_1 b_1 + a_2 b_0) + \cdots$$

and you see the expressions in parentheses above are just the c_n for $n = 0, 1, 2, \cdots$. Therefore, it is reasonable to conjecture that

$$\sum_{i=r}^{\infty} a_i \sum_{j=r}^{\infty} b_j = \sum_{n=r}^{\infty} c_n$$

and of course there would be no problem with this in the case of finite sums but in the case of infinite sums, it is necessary to prove a theorem. The following is a special case of Merten's theorem.

Theorem C.3.8 *Suppose $\sum_{i=r}^{\infty} a_i$ and $\sum_{j=r}^{\infty} b_j$ both converge absolutely.*[*] *Then*

$$\left(\sum_{i=r}^{\infty} a_i \right) \left(\sum_{j=r}^{\infty} b_j \right) = \sum_{n=r}^{\infty} c_n$$

where

$$c_n = \sum_{k=r}^{n} a_k b_{n-k+r}.$$

Proof: Let $p_{nk} = 1$ if $r \leq k \leq n$ and $p_{nk} = 0$ if $k > n$. Then

$$c_n = \sum_{k=r}^{\infty} p_{nk} a_k b_{n-k+r}.$$

Also,

$$\sum_{k=r}^{\infty} \sum_{n=r}^{\infty} p_{nk} |a_k| |b_{n-k+r}| = \sum_{k=r}^{\infty} |a_k| \sum_{n=r}^{\infty} p_{nk} |b_{n-k+r}| = \sum_{k=r}^{\infty} |a_k| \sum_{n=k}^{\infty} |b_{n-k+r}|$$

$$= \sum_{k=r}^{\infty} |a_k| \sum_{n=k}^{\infty} \left| b_{n-(k-r)} \right| = \sum_{k=r}^{\infty} |a_k| \sum_{m=r}^{\infty} |b_m| < \infty.$$

Therefore, by Theorem C.3.6

$$\sum_{n=r}^{\infty} c_n = \sum_{n=r}^{\infty} \sum_{k=r}^{n} a_k b_{n-k+r} = \sum_{n=r}^{\infty} \sum_{k=r}^{\infty} p_{nk} a_k b_{n-k+r}$$

$$= \sum_{k=r}^{\infty} a_k \sum_{n=r}^{\infty} p_{nk} b_{n-k+r} = \sum_{k=r}^{\infty} a_k \sum_{n=k}^{\infty} b_{n-k+r} = \sum_{k=r}^{\infty} a_k \sum_{m=r}^{\infty} b_m \ \blacksquare$$

_____

[*]Actually, it is only necessary to assume one of the series converges and the other converges absolutely. This is known as Merten's theorem and may be read in the 1974 book by Apostol listed in the bibliography.

C.4 Taylor's Formula

There is a fundamental result for approximating functions. It is nearly always left out in calculus courses, many of the books not even bothering to give a proof. Therefore, here is a good proof. This is from Ellis and Gulek, a good calculus book which did prove the theorems.

Theorem C.4.1 *Suppose f has $n+1$ derivatives on an interval (a,b) and let $c \in (a,b)$. Then if $x \in (a,b)$, there exists ξ between c and x such that*

$$f(x) = f(c) + \sum_{k=1}^{n} \frac{f^{(k)}(c)}{k!}(x-c)^k + \frac{f^{(n+1)}(\xi)}{(n+1)!}(x-c)^{n+1}.$$

(In this formula, the symbol $\sum_{k=1}^{0} a_k$ will denote the number 0.)

Proof: If $n = 0$ then the theorem is true because it is just the mean value theorem. Suppose the theorem is true for $n-1, n \geq 1$. It can be assumed $x \neq c$ because if $x = c$ there is nothing to show. Then there exists K such that

$$f(x) - \left(f(c) + \sum_{k=1}^{n} \frac{f^{(k)}(c)}{k!}(x-c)^k + K(x-c)^{n+1} \right) = 0 \qquad (3.7)$$

In fact,

$$K = \frac{-f(x) + \left(f(c) + \sum_{k=1}^{n} \frac{f^{(k)}(c)}{k!}(x-c)^k \right)}{(x-c)^{n+1}}.$$

It remains to find K. Define $F(t)$ for t in the closed interval determined by x and c by

$$F(t) \equiv f(x) - \left(f(t) + \sum_{k=1}^{n} \frac{f^{(k)}(c)}{k!}(x-t)^k + K(x-t)^{n+1} \right).$$

The c in (3.7) got replaced by t.

Therefore, $F(c) = 0$ by the way K was chosen and also $F(x) = 0$. By the mean value theorem or Rolle's theorem, there exists t_1 between x and c such that $F'(t_1) = 0$. Therefore,

$$\begin{aligned}
0 &= f'(t_1) - \sum_{k=1}^{n} \frac{f^{(k)}(c)}{k!} k(x-t_1)^{k-1} - K(n+1)(x-t_1)^n \\
&= f'(t_1) - \left(f'(c) + \sum_{k=1}^{n-1} \frac{f^{(k+1)}(c)}{k!}(x-t_1)^k \right) - K(n+1)(x-t_1)^n \\
&= f'(t_1) - \left(f'(c) + \sum_{k=1}^{n-1} \frac{f'^{(k)}(c)}{k!}(x-t_1)^k \right) - K(n+1)(x-t_1)^n
\end{aligned}$$

By induction applied to f', there exists ξ between x and t_1 such that the above simplifies to $0 =$

$$\frac{f'^{(n)}(\xi)(x-t_1)^n}{n!} - K(n+1)(x-t_1)^n = \frac{f^{(n+1)}(\xi)(x-t_1)^n}{n!} - K(n+1)(x-t_1)^n$$

therefore,

$$K = \frac{f^{(n+1)}(\xi)}{(n+1)\,n!} = \frac{f^{(n+1)}(\xi)}{(n+1)!}$$

and the formula is true for n. ∎

The term $\frac{f^{(n+1)}(\xi)}{(n+1)!}(x-c)^{n+1}$, is called the remainder, and this particular form of the remainder is called the Lagrange form of the remainder.

Appendix D

Review of Linear Algebra

D.1   Systems of Equations

A system of linear equations is of the form

$$\sum_j a_{ij}x_j = b_i, \ i = 1, \cdots, m \tag{4.1}$$

Your want to find the solutions which are values for the x_j which make the equations true. Note that the set of solutions is unchanged if the order of any two of these equations is reversed. Also the set of solutions is unchanged if any equation is multiplied by a nonzero number. The most interesting observation is that the solution set is unchanged when one equation is replaced by a multiple of another equation added to it.

Why exactly does the replacement of one equation with a multiple of another added to it not change the solution set? Suppose you have two of these equations,

$$E_1 = f_1, E_2 = f_2 \tag{4.2}$$

where E_1 and E_2 are expressions involving the variables. The claim is that if a is a number, then (4.2) has the same solution set as

$$E_1 = f_1, \ E_2 + aE_1 = f_2 + af_1. \tag{4.3}$$

Why is this?

If (x, y) solves (4.2) then it solves the first equation in (4.3). Also, it satisfies $aE_1 = af_1$ and so, since it also solves $E_2 = f_2$ it must solve the second equation in (4.3). If (x, y) solves (4.3) then it solves the first equation of (4.2). Also $aE_1 = af_1$ and it is given that the second equation of (4.3) is verified. Therefore, $E_2 = f_2$ and it follows (x, y) is a solution of the second equation in (4.2). This shows the solutions to (4.2) and (4.3) are exactly the same which means they have the same solution set.

It is foolish to write the variables every time you do these operations. It is easier to write the system (4.1) as the following augmented matrix.

$$\begin{pmatrix} a_{11} & \cdots & a_{1n} & b_1 \\ \vdots & & \vdots & \vdots \\ a_{m1} & \cdots & a_{mn} & b_m \end{pmatrix}$$

Now the operations described above where you switch the order of the equations, multiply an equation by a nonzero number, or replace an equation by a multiple of another equation added to it are equivalent to the following row operations which are done on this augmented matrix.

Definition D.1.1 *The row operations consist of the following*

1. *Switch two rows.*

2. *Multiply a row by a nonzero number.*

3. *Replace a row by a multiple of another row added to it.*

It is important to observe that any row operation can be "undone" by another inverse row operation. For example, if $\mathbf{r}_1, \mathbf{r}_2$ are two rows, and $\mathbf{r}_2$ is replaced with $\mathbf{r}_2' = \alpha \mathbf{r}_1 + \mathbf{r}_2$ using row operation 3, then you could get back to where you started by replacing the row $\mathbf{r}_2'$ with $-\alpha$ times $\mathbf{r}_1$ and adding to $\mathbf{r}_2'$. In the case of operation 2, you would simply multiply the row that was changed by the inverse of the scalar which multiplied it in the first place, and in the case of row operation 1, you would just make the same switch again and you would be back to where you started. In each case, the row operation which undoes what was done is called the **inverse row operation.**

Also, you have a solution $\mathbf{x} = (x_1, \cdots, x_n)$ to system of equations given above if and only if

$$x_1 \begin{pmatrix} a_{11} \\ \vdots \\ a_{m1} \end{pmatrix} + \cdots + x_n \begin{pmatrix} a_{1n} \\ \vdots \\ a_{mn} \end{pmatrix} = \begin{pmatrix} b_1 \\ \vdots \\ b_m \end{pmatrix}$$

Recall from linear algebra and the way we multiply matrices, this is the same as saying that $\mathbf{x}$ satisfies

$$\begin{pmatrix} a_{11} & \cdots & a_{1n} \\ \vdots & & \vdots \\ a_{m1} & \cdots & a_{mn} \end{pmatrix} \begin{pmatrix} x_1 \\ \vdots \\ x_n \end{pmatrix} = \begin{pmatrix} b_1 \\ \vdots \\ b_m \end{pmatrix}, \quad A\mathbf{x} = \mathbf{b}$$

The important thing to observe is that when a row operation is done to the augmented matrix $(A|\mathbf{b})$ you get a new augmented matrix which corresponds to a system of equations which has the same solution set as the original system.

D.2 Matrices

Recall the following definition of matrix multiplication and addition.

Definition D.2.1 *Letting M_{ij} denote the entry of M in the ith row and j^{th} column, let A be an $m \times n$ matrix and let B be an $n \times p$ matrix. Then AB is an $m \times p$ matrix and*

$$(AB)_{ij} = \sum_k A_{ik} B_{kj}$$

If A, B are both $m \times n$, then $(A + B)$ is also $m \times n$ and $(A + B)_{ij} = A_{ij} + B_{ij}$. If α is a scalar, then $(\alpha A)_{ij} = \alpha A_{ij}$.

Recall the following properties of matrix arithmetic which follow right away from the above.

$$A + B = B + A, \tag{4.4}$$

the commutative law of addition,

$$(A + B) + C = A + (B + C), \tag{4.5}$$

the associative law for addition,

$$A + 0 = A, \tag{4.6}$$

the existence of an additive identity,

$$A + (-A) = 0, \tag{4.7}$$

the existence of an additive inverse. Also, for α, β scalars, the following also hold.

$$\alpha (A + B) = \alpha A + \alpha B, \tag{4.8}$$

$$(\alpha + \beta) A = \alpha A + \beta A, \tag{4.9}$$

$$\alpha (\beta A) = \alpha \beta (A), \tag{4.10}$$

$$1A = A. \tag{4.11}$$

The above properties, (4.4)-(4.11) are known as the vector space axioms and the fact that the $m \times n$ matrices satisfy these axioms is what is meant by saying this set of matrices with addition and scalar multiplication as defined above forms a vector space.

There are also properties which are related to matrix multiplication.

Proposition D.2.2 *If all multiplications and additions make sense, the following hold for matrices, A, B, C and a, b scalars.*

$$A (aB + bC) = a (AB) + b (AC) \tag{4.12}$$

$$(B + C) A = BA + CA \tag{4.13}$$

$$A (BC) = (AB) C \tag{4.14}$$

Proof: Using the above definition of matrix multiplication,

$$(A(aB + bC))_{ij} = \sum_k A_{ik}(aB + bC)_{kj} = \sum_k A_{ik}(aB_{kj} + bC_{kj})$$

$$= a\sum_k A_{ik}B_{kj} + b\sum_k A_{ik}C_{kj} = a(AB)_{ij} + b(AC)_{ij} = (a(AB) + b(AC))_{ij}$$

showing that $A(B + C) = AB + AC$ as claimed. Formula (4.13) is entirely similar.

Consider (4.14), the associative law of multiplication. Before reading this, review the definition of matrix multiplication in terms of entries of the matrices.

$$\begin{aligned}(A(BC))_{ij} &= \sum_k A_{ik}(BC)_{kj} = \sum_k A_{ik}\sum_l B_{kl}C_{lj} \\ &= \sum_l (AB)_{il}C_{lj} = ((AB)C)_{ij}.\blacksquare\end{aligned}$$

Recall also that AB is sometimes not equal to BA. For example,

$$\begin{pmatrix} 1 & 2 \\ 3 & 4 \end{pmatrix}\begin{pmatrix} 0 & 1 \\ 1 & 0 \end{pmatrix} \neq \begin{pmatrix} 0 & 1 \\ 1 & 0 \end{pmatrix}\begin{pmatrix} 1 & 2 \\ 3 & 4 \end{pmatrix}$$

Definition D.2.3 *Let A be an $m \times n$ matrix. Then A^T denotes the $n \times m$ matrix which is defined as follows.*

$$\left(A^T\right)_{ij} = A_{ji}$$

The transpose of a matrix has the following important property.

Lemma D.2.4 *Let A be an $m \times n$ matrix and let B be a $n \times p$ matrix. Then*

$$(AB)^T = B^T A^T \tag{4.15}$$

and if α and β are scalars,

$$(\alpha A + \beta B)^T = \alpha A^T + \beta B^T \tag{4.16}$$

Proof: From the definition,

$$\left((AB)^T\right)_{ij} = (AB)_{ji} = \sum_k A_{jk}B_{ki} = \sum_k \left(B^T\right)_{ik}\left(A^T\right)_{kj} = \left(B^T A^T\right)_{ij}$$

(4.16) is left as an exercise. $\blacksquare$

Definition D.2.5 *A real $n \times n$ matrix A is said to be symmetric if $A = A^T$. It is said to be skew symmetric if $A^T = -A$. There is a special matrix called I and defined by*

$$I_{ij} = \delta_{ij}$$

where δ_{ij} is the Kronecker symbol defined by

$$\delta_{ij} = \begin{cases} 1 & \text{if } i = j \\ 0 & \text{if } i \neq j \end{cases}$$

It is called the identity matrix because it is a multiplicative identity in the following sense.

The following lemma follows from the above definition.

Lemma D.2.6 *Suppose A is an $m \times n$ matrix and I_n is the $n \times n$ identity matrix. Then $AI_n = A$. If I_m is the $m \times m$ identity matrix, it also follows that $I_m A = A$.*

Definition D.2.7 *An $n \times n$ matrix A has an inverse A^{-1} if and only if there exists a matrix, denoted as A^{-1} such that $AA^{-1} = A^{-1}A = I$ where $I = (\delta_{ij})$ for*

$$\delta_{ij} \equiv \begin{cases} 1 & \text{if } i = j \\ 0 & \text{if } i \neq j \end{cases}$$

Such a matrix is called invertible.

If it acts like an inverse, then it is the inverse. This is the message of the following proposition.

Proposition D.2.8 *Suppose $AB = BA = I$. Then $B = A^{-1}$.*

Proof: From the definition B is an inverse for A. Could there be another one B'?

$$B' = B'I = B'(AB) = (B'A)B = IB = B.$$

Thus, the inverse, if it exists, is unique. ∎

Recall the definition of the special vectors $\mathbf{e}_k$.

$$\mathbf{e}_k = \begin{pmatrix} 0 & \cdots & 0 & 1 & 0 & \cdots & 0 \end{pmatrix}^T$$

where the 1 is in the kth position from the left.

Definition D.2.9 *Let A be a matrix. $N(A) = \{\mathbf{x} : A\mathbf{x} = \mathbf{0}\}$. This may also be referred to as $\ker(A)$.*

There is a fundamental result in the case where $m < n$. In this case, the matrix A looks like the following.

Theorem D.2.10 *Let A be an $m \times n$ matrix where $m < n$. Then $N(A)$ contains nonzero vectors.*

Proof: It is clear that the theorem is true if A is $1 \times n$ with $n > 1$. Suppose it is true whenever A is $m-1 \times k$ for $m-1 < k$. Say A is $m \times n$ with $m < n, m > 1$,

$$A = \begin{pmatrix} \mathbf{a}_1 & \cdots & \mathbf{a}_n \end{pmatrix}, \ \mathbf{a}_i \text{ the } i^{th} \text{ column}$$

If $\mathbf{a}_1 = \mathbf{0}$, consider the vector $\mathbf{x} = \mathbf{e}_1$. $A\mathbf{e}_1 = \mathbf{0}$. If $\mathbf{a}_1 \neq \mathbf{0}$, do row operations to obtain a matrix $\hat{A}$ such that the solutions of $A\mathbf{y} = \mathbf{0}$ are the same as the solutions of $\hat{A}\mathbf{y} = \mathbf{0}$, and $\hat{A} =$

$$\begin{pmatrix} 1 & \mathbf{a}^T \\ \mathbf{0} & B \end{pmatrix}$$

Now B is $m - 1 \times n - 1$ where $m - 1 < n - 1$ and so by induction, there is a nonzero vector $\mathbf{x}$ such that $B\mathbf{x} = \mathbf{0}$. Consider the nonzero vector $\mathbf{z} = \begin{pmatrix} 0 \\ \mathbf{x} \end{pmatrix}$. Then $\hat{A}\mathbf{z} = \mathbf{0}$ and so $A\mathbf{z} = \mathbf{0}$. ■

D.3   Subspaces and Spans

To save on notation, $\mathbb{F}$ will denote a field. In this course, it will be either $\mathbb{R}$ or $\mathbb{C}$, the real or the complex numbers.

Definition D.3.1 *Let $\{\mathbf{x}_1, \cdots, \mathbf{x}_p\}$ be vectors in $\mathbb{F}^n$. A linear combination is any expression of the form*

$$\sum_{i=1}^{p} c_i \mathbf{x}_i$$

where the c_i are scalars. The set of all linear combinations of these vectors is called span $(\mathbf{x}_1, \cdots, \mathbf{x}_n)$. *If $V \subseteq \mathbb{F}^n$, then V is called a subspace if whenever α, β are scalars and $\mathbf{u}$ and $\mathbf{v}$ are vectors of V, it follows $\alpha \mathbf{u} + \beta \mathbf{v} \in V$. That is, it is "closed under the algebraic operations of vector addition and scalar multiplication". A linear combination of vectors is said to be trivial if all the scalars in the linear combination equal zero. A set of vectors is said to be linearly independent if the only linear combination of these vectors which equals the zero vector is the trivial linear combination. Thus $\{\mathbf{x}_1, \cdots, \mathbf{x}_n\}$ is called linearly independent if whenever*

$$\sum_{k=1}^{p} c_k \mathbf{x}_k = \mathbf{0}$$

it follows that all the scalars c_k equal zero. A set of vectors, $\{\mathbf{x}_1, \cdots, \mathbf{x}_p\}$, is called linearly dependent if it is not linearly independent. Thus the set of vectors is linearly dependent if there exist scalars $c_i, i = 1, \cdots, n$, not all zero such that $\sum_{k=1}^{p} c_k \mathbf{x}_k = \mathbf{0}$.

Proposition D.3.2 *Let $V \subseteq \mathbb{F}^n$. Then V is a subspace if and only if it is a vector space itself with respect to the same operations of scalar multiplication and vector addition.*

Proof: Suppose first that V is a subspace. All algebraic properties involving scalar multiplication and vector addition hold for V because these things hold for $\mathbb{F}^n$. Is $\mathbf{0} \in V$? Yes it is. This is because $0\mathbf{v} \in V$ and $0\mathbf{v} = \mathbf{0}$. By assumption, for α a scalar and $\mathbf{v} \in V, \alpha \mathbf{v} \in V$. Therefore, $-\mathbf{v} = (-1)\mathbf{v} \in V$. Thus V has the additive identity and additive inverse. By assumption, V is closed with respect to the two operations. Thus V is a vector space. If $V \subseteq \mathbb{F}^n$ is a vector space, then by definition, if α, β are scalars and $\mathbf{u}, \mathbf{v}$ vectors in V, it follows that $\alpha \mathbf{v} + \beta \mathbf{u} \in V$. ∎

Thus, from the above, subspaces of $\mathbb{F}^n$ are just subsets of $\mathbb{F}^n$ which are themselves vector spaces.

Lemma D.3.3 *A set of vectors $\{\mathbf{x}_1, \cdots, \mathbf{x}_p\}$ is linearly independent if and only if none of the vectors can be obtained as a linear combination of the others.*

Proof: Suppose first that $\{\mathbf{x}_1, \cdots, \mathbf{x}_p\}$ is linearly independent. If $\mathbf{x}_k = \sum_{j \neq k} c_j \mathbf{x}_j$, then

$$\mathbf{0} = 1\mathbf{x}_k + \sum_{j \neq k} (-c_j)\, \mathbf{x}_j,$$

a nontrivial linear combination, contrary to assumption. This shows that if the set is linearly independent, then none of the vectors is a linear combination of the others.

Now suppose no vector is a linear combination of the others. Is $\{\mathbf{x}_1, \cdots, \mathbf{x}_p\}$ linearly independent? If it is not, there exist scalars c_i, not all zero such that

$$\sum_{i=1}^{p} c_i \mathbf{x}_i = \mathbf{0}.$$

Say $c_k \neq 0$. Then you can solve for $\mathbf{x}_k$ as

$$\mathbf{x}_k = \sum_{j \neq k} (-c_j)\, /c_k \mathbf{x}_j$$

contrary to assumption. ∎

The following is called the exchange theorem.

Theorem D.3.4 *(Exchange Theorem) Let $\{\mathbf{x}_1, \cdots, \mathbf{x}_r\}$ be a linearly independent set of vectors such that each $\mathbf{x}_i$ is in $span(\mathbf{y}_1, \cdots, \mathbf{y}_s)$. Then $r \leq s$.*

Proof : Suppose not. Then $r > s$. By assumption, there exist scalars a_{ji} such that

$$\mathbf{x}_i = \sum_{j=1}^{s} a_{ji} \mathbf{y}_j$$

The matrix whose ji^{th} entry is a_{ji} has more columns than rows. Therefore, by Theorem D.2.10 there exists a **nonzero** vector $\mathbf{b} \in \mathbb{F}^r$ such that $A\mathbf{b} = \mathbf{0}$. Thus

$$0 = \sum_{i=1}^{r} a_{ji} b_i, \text{ each } j.$$

Then

$$\sum_{i=1}^{r} b_i \mathbf{x}_i = \sum_{i=1}^{r} b_i \sum_{j=1}^{s} a_{ji} \mathbf{y}_j = \sum_{j=1}^{s} \left(\sum_{i=1}^{r} a_{ji} b_i \right) \mathbf{y}_j = \mathbf{0}$$

contradicting the assumption that $\{\mathbf{x}_1, \cdots, \mathbf{x}_r\}$ is linearly independent. ∎

Definition D.3.5 *A finite set of vectors, $\{\mathbf{x}_1, \cdots, \mathbf{x}_r\}$ is a basis for $\mathbb{F}^n$ if*

$$\text{span}(\mathbf{x}_1, \cdots, \mathbf{x}_r) = \mathbb{F}^n$$

and $\{\mathbf{x}_1, \cdots, \mathbf{x}_r\}$ is linearly independent.

Corollary D.3.6 *Let $\{\mathbf{x}_1, \cdots, \mathbf{x}_r\}$ and $\{\mathbf{y}_1, \cdots, \mathbf{y}_s\}$ be two bases* of $\mathbb{F}^n$. Then $r = s = n$.*

Proof: From the exchange theorem, $r \leq s$ and $s \leq r$. Now note the vectors,

$$\mathbf{e}_i = \overbrace{(0, \cdots, 0, 1, 0 \cdots, 0)}^{1 \text{ is in the } i^{th} \text{ slot}}$$

for $i = 1, 2, \cdots, n$ are a basis for $\mathbb{F}^n$. ∎

Lemma D.3.7 *Let $\{\mathbf{v}_1, \cdots, \mathbf{v}_r\}$ be a set of vectors. Then $V \equiv \text{span}(\mathbf{v}_1, \cdots, \mathbf{v}_r)$ is a subspace.*

Proof: Suppose α, β are two scalars and let $\sum_{k=1}^{r} c_k \mathbf{v}_k$ and $\sum_{k=1}^{r} d_k \mathbf{v}_k$ are two elements of V. What about

$$\alpha \sum_{k=1}^{r} c_k \mathbf{v}_k + \beta \sum_{k=1}^{r} d_k \mathbf{v}_k?$$

Is it also in V?

$$\alpha \sum_{k=1}^{r} c_k \mathbf{v}_k + \beta \sum_{k=1}^{r} d_k \mathbf{v}_k = \sum_{k=1}^{r} (\alpha c_k + \beta d_k) \mathbf{v}_k \in V$$

so the answer is yes. ∎

Definition D.3.8 *A finite set of vectors, $\{\mathbf{x}_1, \cdots, \mathbf{x}_r\}$ is a basis for a subspace V of $\mathbb{F}^n$ if $\text{span}(\mathbf{x}_1, \cdots, \mathbf{x}_r) = V$ and $\{\mathbf{x}_1, \cdots, \mathbf{x}_r\}$ is linearly independent.*

Corollary D.3.9 *Let $\{\mathbf{x}_1, \cdots, \mathbf{x}_r\}$ and $\{\mathbf{y}_1, \cdots, \mathbf{y}_s\}$ be two bases for V. Then $r = s$.*

Proof: From the exchange theorem, $r \leq s$ and $s \leq r$. ∎

Definition D.3.10 *Let V be a subspace of $\mathbb{F}^n$. Then $\dim(V)$ read as the dimension of V is the number of vectors in a basis.*

Of course you should wonder right now whether an arbitrary subspace even has a basis. In fact it does and this is in the next theorem. First, here is an interesting lemma.

Lemma D.3.11 *Suppose $\mathbf{v} \notin \text{span}(\mathbf{u}_1, \cdots, \mathbf{u}_k)$ and $\{\mathbf{u}_1, \cdots, \mathbf{u}_k\}$ is linearly independent. Then $\{\mathbf{u}_1, \cdots, \mathbf{u}_k, \mathbf{v}\}$ is also linearly independent.*

---

*This is the plural form of basis. We could say basiss but it would involve an inordinate amount of hissing as in "The sixth shiek's sixth sheep is sick." This is the reason that bases is used instead of basiss.

Proof: Suppose $\sum_{i=1}^{k} c_i \mathbf{u}_i + d\mathbf{v} = \mathbf{0}$. It is required to verify that each $c_i = 0$ and that $d = 0$. But if $d \neq 0$, then you can solve for $\mathbf{v}$ as a linear combination of the vectors, $\{\mathbf{u}_1, \cdots, \mathbf{u}_k\}$,

$$\mathbf{v} = -\sum_{i=1}^{k} \left(\frac{c_i}{d}\right) \mathbf{u}_i$$

contrary to assumption. Therefore, $d = 0$. But then $\sum_{i=1}^{k} c_i \mathbf{u}_i = 0$ and the linear independence of $\{\mathbf{u}_1, \cdots, \mathbf{u}_k\}$ implies each $c_i = 0$ also. ∎

Theorem D.3.12 *Let V be a nonzero subspace of $\mathbb{F}^n$. Then V has a basis.*

Proof: Let $\mathbf{v}_1 \in V$ where $\mathbf{v}_1 \neq 0$. If span $\{\mathbf{v}_1\} = V$, stop. $\{\mathbf{v}_1\}$ is a basis for V. Otherwise, there exists $\mathbf{v}_2 \in V$ which is not in span $\{\mathbf{v}_1\}$. By Lemma D.3.11 $\{\mathbf{v}_1, \mathbf{v}_2\}$ is a linearly independent set of vectors. If span $\{\mathbf{v}_1, \mathbf{v}_2\} = V$ stop, $\{\mathbf{v}_1, \mathbf{v}_2\}$ is a basis for V. If span $\{\mathbf{v}_1, \mathbf{v}_2\} \neq V$, then there exists $\mathbf{v}_3 \notin$ span $\{\mathbf{v}_1, \mathbf{v}_2\}$ and $\{\mathbf{v}_1, \mathbf{v}_2, \mathbf{v}_3\}$ is a larger linearly independent set of vectors. Continuing this way, the process must stop before $n + 1$ steps because if not, it would be possible to obtain $n + 1$ linearly independent vectors, contrary to the exchange theorem. ∎

In words the following corollary states that any linearly independent set of vectors can be enlarged to form a basis.

Corollary D.3.13 *Let V be a subspace of $\mathbb{F}^n$ and let $\{\mathbf{v}_1, \cdots, \mathbf{v}_r\}$ be a linearly independent set of vectors in V. Then either it is a basis for V or there exist vectors, $\mathbf{v}_{r+1}, \cdots, \mathbf{v}_s$ such that $\{\mathbf{v}_1, \cdots, \mathbf{v}_r, \mathbf{v}_{r+1}, \cdots, \mathbf{v}_s\}$ is a basis for V.*

Proof: This follows immediately from the proof of Theorem D.3.12. You do exactly the same argument except you start with $\{\mathbf{v}_1, \cdots, \mathbf{v}_r\}$ rather than $\{\mathbf{v}_1\}$. ∎

It is also true that any spanning set of vectors can be restricted to obtain a basis.

Theorem D.3.14 *Let V be a subspace of $\mathbb{F}^n$ and suppose span $(\mathbf{u}_1 \cdots, \mathbf{u}_p) = V$ where the $\mathbf{u}_i$ are nonzero vectors. Then there exist vectors $\{\mathbf{v}_1 \cdots, \mathbf{v}_r\}$ such that $\{\mathbf{v}_1 \cdots, \mathbf{v}_r\} \subseteq \{\mathbf{u}_1 \cdots, \mathbf{u}_p\}$ and $\{\mathbf{v}_1 \cdots, \mathbf{v}_r\}$ is a basis for V.*

Proof: Let r be the smallest positive integer with the property that for some set $\{\mathbf{v}_1 \cdots, \mathbf{v}_r\} \subseteq \{\mathbf{u}_1 \cdots, \mathbf{u}_p\}$,

$$\text{span}(\mathbf{v}_1 \cdots, \mathbf{v}_r) = V.$$

Then $r \leq p$ and it must be the case that $\{\mathbf{v}_1 \cdots, \mathbf{v}_r\}$ is linearly independent because if it were not so, one of the vectors, say $\mathbf{v}_k$ would be a linear combination of the others. But then you could delete this vector from $\{\mathbf{v}_1 \cdots, \mathbf{v}_r\}$ and the resulting list of $r - 1$ vectors would still span V contrary to the definition of r. ∎

D.4   Application to Matrices

The following is a theorem of major significance.

Theorem D.4.1 *Suppose A is an $n \times n$ matrix. Then A is one to one (injective) if and only if A is onto (surjective). Also, if B is an $n \times n$ matrix and $AB = I$, then it follows $BA = I$.*

Proof: First suppose A is one to one. Consider the vectors, $\{Ae_1, \cdots, Ae_n\}$ where e_k is the column vector which is all zeros except for a 1 in the kth position. This set of vectors is linearly independent because, since multiplication by A is linear, if

$$\sum_{k=1}^{n} c_k Ae_k = \mathbf{0}, \text{ then } A\left(\sum_{k=1}^{n} c_k e_k\right) = \mathbf{0}$$

and so, since A is one to one,

$$\sum_{k=1}^{n} c_k e_k = \mathbf{0}$$

which implies each $c_k = 0$ because the e_k are clearly linearly independent.

Therefore, $\{Ae_1, \cdots, Ae_n\}$ must be a basis for $\mathbb{F}^n$ because if not there would exist a vector, $\mathbf{y} \notin \text{span}(Ae_1, \cdots, Ae_n)$ and then by Lemma D.3.11, $\{Ae_1, \cdots, Ae_n, \mathbf{y}\}$ would be an independent set of vectors having $n + 1$ vectors in it, contrary to the exchange theorem. It follows that for $\mathbf{y} \in \mathbb{F}^n$ there exist constants, c_i such that

$$\mathbf{y} = \sum_{k=1}^{n} c_k Ae_k = A\left(\sum_{k=1}^{n} c_k e_k\right)$$

showing that, since $\mathbf{y}$ was arbitrary, A is onto.

Next suppose A is onto. Say $A\mathbf{x} = \mathbf{0}$ where $\mathbf{x} \neq \mathbf{0}$. But this would say that the columns of A are linearly dependent. If A were onto, then the columns of A would be a spanning set of $\mathbb{F}^n$. However, one column is a linear combination of the others and so there would exist a spanning set of fewer than n vectors contradicting the above exchange theorem.

Now suppose $AB = I$. Why is $BA = I$? Since $AB = I$ it follows B is one to one since otherwise, there would exist, $\mathbf{x} \neq \mathbf{0}$ such that $B\mathbf{x} = \mathbf{0}$ and then $AB\mathbf{x} = A\mathbf{0} = \mathbf{0} \neq I\mathbf{x}$. Therefore, from what was just shown, B is also onto. In addition to this, A must be one to one because if $A\mathbf{y} = \mathbf{0}$, then $\mathbf{y} = B\mathbf{x}$ for some $\mathbf{x}$ and then $\mathbf{x} = AB\mathbf{x} = A\mathbf{y} = \mathbf{0}$ showing $\mathbf{y} = \mathbf{0}$. Now from what is given to be so, it follows $(AB)A = A$ and so using the associative law for matrix multiplication,

$$A(BA) - A = A(BA - I) = 0.$$

But this means $(BA - I)\mathbf{x} = \mathbf{0}$ for all $\mathbf{x}$ since otherwise, A would not be one to one. Hence $BA = I$ as claimed. ∎

This theorem shows that if an $n \times n$ matrix B acts like an inverse when multiplied on one side of A, it follows that $B = A^{-1}$ and it will act like an inverse on both sides of A.

D.5   Mathematical Theory of Determinants

D.5.1   The Function sgn

The following lemma will be essential in the definition of the determinant.

Lemma D.5.1 *There exists a function,* sgn_n *which maps each ordered list of numbers from* $\{1, \cdots, n\}$ *to one of the three numbers,* $0, 1,$ *or* -1 *which also has the following properties.*

$$\text{sgn}_n(1, \cdots, n) = 1 \tag{4.17}$$

$$\text{sgn}_n(i_1, \cdots, p, \cdots, q, \cdots, i_n) = -\text{sgn}_n(i_1, \cdots, q, \cdots, p, \cdots, i_n) \tag{4.18}$$

In words, the second property states that if two of the numbers are switched, the value of the function is multiplied by -1. *Also, in the case where* $n > 1$ *and* $\{i_1, \cdots, i_n\} = \{1, \cdots, n\}$ *so that every number from* $\{1, \cdots, n\}$ *appears in the ordered list,* $(i_1, \cdots, i_n),$

$$\text{sgn}_n(i_1, \cdots, i_{\theta-1}, n, i_{\theta+1}, \cdots, i_n) \equiv$$

$$(-1)^{n-\theta} \text{sgn}_{n-1}(i_1, \cdots, i_{\theta-1}, i_{\theta+1}, \cdots, i_n) \tag{4.19}$$

where $n = i_\theta$ *in the ordered list,* $(i_1, \cdots, i_n).$

Proof: Define $\text{sign}(x) = 1$ if $x > 0, -1$ if $x < 0$ and 0 if $x = 0$. If $n = 1$, there is only one list and it is just the number 1. Thus one can define $\text{sgn}_1(1) \equiv 1$. For the general case where $n > 1$, simply define

$$\text{sgn}_n(i_1, \cdots, i_n) \equiv \text{sign}\left(\prod_{r<s}(i_s - i_r)\right)$$

This delivers either $-1, 1,$ or 0 by definition. What about the other claims? Suppose you switch i_p with i_q where $p < q$ so two numbers in the ordered list $(i_1, \cdots, i_n)$ are switched. Denote the new ordered list of numbers as $(j_1, \cdots, j_n)$. Thus $j_p = i_q$ and $j_q = i_p$ and if $r \notin \{p, q\}$, $j_r = i_r$. See the following illustration

$$\frac{i_1}{1} \quad \frac{i_2}{2} \quad \cdots \quad \frac{i_p}{p} \quad \cdots \quad \frac{i_q}{q} \quad \cdots \quad \frac{i_n}{n}$$

$$\frac{i_1}{1} \quad \frac{i_2}{2} \quad \cdots \quad \frac{i_q}{p} \quad \cdots \quad \frac{i_p}{q} \quad \cdots \quad \frac{i_n}{n}$$

$$\frac{j_1}{1} \quad \frac{j_2}{2} \quad \cdots \quad \frac{j_p}{p} \quad \cdots \quad \frac{j_q}{q} \quad \cdots \quad \frac{j_n}{n}$$

Then

$$\text{sgn}_n(j_1, \cdots, j_n) \equiv \text{sign}\left(\prod_{r<s}(j_s - j_r)\right)$$

$$= \text{sign}\left((i_p - i_q) \overbrace{\prod_{p<j<q}(i_j - i_q) \prod_{p<j<q}(i_p - i_j)}^{\text{one of } p,q} \overbrace{\prod_{r<s, r,s \notin \{p,q\}}(i_s - i_r)}^{\text{neither } p \text{ nor } q}\right)$$

with "both p,q" labeling $(i_p - i_q)$.

The last product consists of the product of terms which were in the unswitched product $\prod_{r<s}(i_s - i_r)$ so produces no change in sign, while the two products in

the middle both introduce $q - p - 1$ minus signs. Thus their product produces no change in sign. The first factor is of opposite sign to the $i_q - i_p$ which occurred in $\text{sgn}_n (i_1, \cdots, i_n)$. Therefore, this switch introduced a minus sign and

$$\text{sgn}_n (j_1, \cdots, j_n) = - \text{sgn}_n (i_1, \cdots, i_n)$$

Now consider the last claim. In computing $\text{sgn}_n (i_1, \cdots, i_{\theta-1}, n, i_{\theta+1}, \cdots, i_n)$ there will be the product of $n - \theta$ negative terms

$$(i_{\theta+1} - n) \cdots (i_n - n)$$

and the other terms in the product for computing $\text{sgn}_n (i_1, \cdots, i_{\theta-1}, n, i_{\theta+1}, \cdots, i_n)$ are those which are required to compute $\text{sgn}_{n-1} (i_1, \cdots, i_{\theta-1}, i_{\theta+1}, \cdots, i_n)$ multiplied by terms of the form $(n - i_j)$ which are nonnegative. It follows that

$$\text{sgn}_n (i_1, \cdots, i_{\theta-1}, n, i_{\theta+1}, \cdots, i_n) = (-1)^{n-\theta} \text{sgn}_{n-1} (i_1, \cdots, i_{\theta-1}, i_{\theta+1}, \cdots, i_n)$$

It is obvious that if there are repeats in the list, the function gives 0. ∎

Lemma D.5.2 *Every ordered list of distinct numbers from $\{1, 2, \cdots, n\}$ can be obtained from every other such ordered list by a finite number of switches. Also, sgn_n is unique.*

Proof: This is obvious if $n = 1$ or 2. Suppose then that it is true for sets of $n - 1$ elements. Take two ordered lists of numbers, P_1, P_2. Make one switch in both to place n at the end. Call the result P_1^n and P_2^n. Then using induction, there are finitely many switches in P_1^n so that it will coincide with P_2^n. Now switch the n in what results to where it was in P_2.

To see sgn_n is unique, if there exist two functions, f and g both satisfying (4.17) and (4.18), you could start with $f(1, \cdots, n) = g(1, \cdots, n) = 1$ and applying the same sequence of switches, eventually arrive at $f(i_1, \cdots, i_n) = g(i_1, \cdots, i_n)$. If any numbers are repeated, then (4.18) gives both functions are equal to zero for that ordered list. ∎

Definition D.5.3 *When you have an ordered list of distinct numbers from*

$$\{1, 2, \cdots, n\},$$

say

$$(i_1, \cdots, i_n),$$

this ordered list is called a permutation. The symbol for all such permutations is S_n. The number $\text{sgn}_n (i_1, \cdots, i_n)$ is called the sign of the permutation.

A permutation can also be considered as a function from the set

$$\{1, 2, \cdots, n\} \text{ to } \{1, 2, \cdots, n\}$$

as follows. Let $f(k) = i_k$. Permutations are of fundamental importance in certain areas of math. For example, it was by considering permutations that Galois was able to give a criterion for solution of polynomial equations by radicals, but this is a different direction than what is being attempted here.

In what follows sgn will often be used rather than sgn_n because the context supplies the appropriate n.

D.5.2   Determinants

Definition D.5.4 *Let f be a function which has the set of ordered lists of numbers from $\{1, \cdots, n\}$ as its domain. Define*

$$\sum_{(k_1, \cdots, k_n)} f(k_1 \cdots k_n)$$

to be the sum of all the $f(k_1 \cdots k_n)$ for all possible choices of ordered lists $(k_1, \cdots, k_n)$ of numbers of $\{1, \cdots, n\}$. For example,

$$\sum_{(k_1, k_2)} f(k_1, k_2) = f(1,2) + f(2,1) + f(1,1) + f(2,2).$$

D.5.3   Definition of Determinants

Definition D.5.5 *Let $(a_{ij}) = A$ denote an $n \times n$ matrix. The determinant of A, denoted by $\det(A)$ is defined by*

$$\det(A) \equiv \sum_{(k_1, \cdots, k_n)} \text{sgn}(k_1, \cdots, k_n) a_{1k_1} \cdots a_{nk_n}$$

where the sum is taken over all ordered lists of numbers from $\{1, \cdots, n\}$. Note it suffices to take the sum over only those ordered lists in which there are no repeats because if there are, $\text{sgn}(k_1, \cdots, k_n) = 0$ and so that term contributes 0 to the sum.

D.5.4   Permuting Rows or Columns

Let A be an $n \times n$ matrix, $A = (a_{ij})$ and let $(r_1, \cdots, r_n)$ denote an ordered list of n numbers from $\{1, \cdots, n\}$. Let $A(r_1, \cdots, r_n)$ denote the matrix whose k^{th} row is the r_k row of the matrix A. Thus

$$\det(A(r_1, \cdots, r_n)) = \sum_{(k_1, \cdots, k_n)} \text{sgn}(k_1, \cdots, k_n) a_{r_1 k_1} \cdots a_{r_n k_n} \tag{4.20}$$

and

$$A(1, \cdots, n) = A.$$

Proposition D.5.6 *Let*

$$(r_1, \cdots, r_n)$$

be an ordered list of numbers from $\{1, \cdots, n\}$. Then

$$\text{sgn}(r_1, \cdots, r_n) \det(A)$$

$$= \sum_{(k_1, \cdots, k_n)} \text{sgn}(k_1, \cdots, k_n) a_{r_1 k_1} \cdots a_{r_n k_n} \tag{4.21}$$

$$= \det(A(r_1, \cdots, r_n)). \tag{4.22}$$

Proof: Let $(1, \cdots, n) = (1, \cdots, r, \cdots s, \cdots, n)$ so $r < s$.

$$\det\left(A\left(1, \cdots, r, \cdots, s, \cdots, n\right)\right) = \tag{4.23}$$

$$\sum_{(k_1, \cdots, k_n)} \operatorname{sgn}\left(k_1, \cdots, k_r, \cdots, k_s, \cdots, k_n\right) a_{1k_1} \cdots a_{rk_r} \cdots a_{sk_s} \cdots a_{nk_n},$$

and renaming the variables, calling k_s, k_r and k_r, k_s, this equals

$$= \sum_{(k_1, \cdots, k_n)} \operatorname{sgn}\left(k_1, \cdots, k_s, \cdots, k_r, \cdots, k_n\right) a_{1k_1} \cdots a_{rk_s} \cdots a_{sk_r} \cdots a_{nk_n}$$

$$= \sum_{(k_1, \cdots, k_n)} -\operatorname{sgn}\left(k_1, \cdots, \overbrace{k_r, \cdots, k_s}^{\text{These got switched}}, \cdots, k_n\right) a_{1k_1} \cdots a_{sk_r} \cdots a_{rk_s} \cdots a_{nk_n}$$

$$= -\det\left(A\left(1, \cdots, s, \cdots, r, \cdots, n\right)\right). \tag{4.24}$$

Consequently,

$$\det\left(A\left(1, \cdots, s, \cdots, r, \cdots, n\right)\right) =$$

$$-\det\left(A\left(1, \cdots, r, \cdots, s, \cdots, n\right)\right) = -\det\left(A\right)$$

Now letting $A\left(1, \cdots, s, \cdots, r, \cdots, n\right)$ play the role of A, and continuing in this way, switching pairs of numbers,

$$\det\left(A\left(r_1, \cdots, r_n\right)\right) = (-1)^p \det\left(A\right)$$

where it took p switches to obtain $(r_1, \cdots, r_n)$ from $(1, \cdots, n)$. By Lemma D.5.1, this implies

$$\det\left(A\left(r_1, \cdots, r_n\right)\right) = (-1)^p \det\left(A\right) = \operatorname{sgn}\left(r_1, \cdots, r_n\right) \det\left(A\right)$$

and proves the proposition in the case when there are no repeated numbers in the ordered list, $(r_1, \cdots, r_n)$. However, if there is a repeat, say the rth row equals the sth row, then the reasoning of (4.23) and (4.24) shows that $\det A\left(r_1, \cdots, r_n\right) = 0$ and also $\operatorname{sgn}\left(r_1, \cdots, r_n\right) = 0$ so the formula holds in this case also. ∎

Observation D.5.7 *There are $n!$ ordered lists of distinct numbers from the integers $\{1, \cdots, n\}$.*

To see this, consider n slots placed in order. There are n choices for the first slot. For each of these choices, there are $n - 1$ choices for the second. Thus there are $n(n - 1)$ ways to fill the first two slots. Then for each of these ways there are $n - 2$ choices left for the third slot. Continuing this way, there are $n!$ ordered lists of distinct numbers from $\{1, \cdots, n\}$ as stated in the observation.

D.5.5 A Symmetric Definition

With the above, it is possible to give a more symmetric description of the determinant from which it will follow that $\det(A) = \det(A^T)$.

Corollary D.5.8 *The following formula for* $\det(A)$ *is valid.*

$$\det(A) = \frac{1}{n!} \cdot$$

$$\sum_{(r_1,\cdots,r_n)} \sum_{(k_1,\cdots,k_n)} \text{sgn}(r_1,\cdots,r_n)\,\text{sgn}(k_1,\cdots,k_n)\,a_{r_1 k_1}\cdots a_{r_n k_n}. \tag{4.25}$$

And also $\det(A^T) = \det(A)$ *where* A^T *is the transpose of A. (Recall that for* $A^T = (a_{ij}^T)$, $a_{ij}^T = a_{ji}$.)

Proof: From Proposition D.5.6, if the r_i are distinct,

$$\det(A) = \sum_{(k_1,\cdots,k_n)} \text{sgn}(r_1,\cdots,r_n)\,\text{sgn}(k_1,\cdots,k_n)\,a_{r_1 k_1}\cdots a_{r_n k_n}.$$

Summing over all ordered lists, $(r_1,\cdots,r_n)$ where the r_i are distinct, (If the r_i are not distinct, $\text{sgn}(r_1,\cdots,r_n) = 0$ and so there is no contribution to the sum.)

$$n!\det(A) =$$

$$\sum_{(r_1,\cdots,r_n)} \sum_{(k_1,\cdots,k_n)} \text{sgn}(r_1,\cdots,r_n)\,\text{sgn}(k_1,\cdots,k_n)\,a_{r_1 k_1}\cdots a_{r_n k_n}.$$

This proves the corollary since the formula gives the same number for A as it does for A^T. ∎

D.5.6 Alternating Property of the Determinant

Corollary D.5.9 *If two rows or two columns in an* $n \times n$ *matrix A, are switched, the determinant of the resulting matrix equals* (-1) *times the determinant of the original matrix. If A is an* $n \times n$ *matrix in which two rows are equal or two columns are equal then* $\det(A) = 0$. *Suppose the* i^{th} *row of A equals*

$$(xa_1 + yb_1, \cdots, xa_n + yb_n).$$

Then

$$\det(A) = x\det(A_1) + y\det(A_2)$$

where the ith *row of* A_1 *is* $(a_1,\cdots,a_n)$ *and the* i^{th} *row of* A_2 *is* $(b_1,\cdots,b_n)$, *all other rows of* A_1 *and* A_2 *coinciding with those of A. In other words,* det *is a linear function of each row A. The same is true with the word "row" replaced with the word "column".*

Proof: By Proposition D.5.6 when two rows are switched, the determinant of the resulting matrix is (-1) times the determinant of the original matrix. By Corollary D.5.8 the same holds for columns because the columns of the matrix equal the rows of the transposed matrix. Thus if A_1 is the matrix obtained from A by switching two columns,

$$\det (A) = \det \left(A^T \right) = - \det \left(A_1^T \right) = - \det (A_1).$$

If A has two equal columns or two equal rows, then switching them results in the same matrix. Therefore, $\det (A) = - \det (A)$ and so $\det (A) = 0$.

It remains to verify the last assertion.

$$\det (A) \equiv \sum_{(k_1,\cdots,k_n)} \operatorname{sgn} (k_1, \cdots, k_n) a_{1k_1} \cdots (x a_{k_i} + y b_{k_i}) \cdots a_{nk_n}$$

$$= x \sum_{(k_1,\cdots,k_n)} \operatorname{sgn} (k_1, \cdots, k_n) a_{1k_1} \cdots a_{k_i} \cdots a_{nk_n}$$

$$+ y \sum_{(k_1,\cdots,k_n)} \operatorname{sgn} (k_1, \cdots, k_n) a_{1k_1} \cdots b_{k_i} \cdots a_{nk_n}$$

$$\equiv x \det (A_1) + y \det (A_2).$$

The same is true of columns because $\det \left(A^T \right) = \det (A)$ and the rows of A^T are the columns of A. ∎

D.5.7　Linear Combinations and Determinants

Linear combinations have been discussed already. However, here is a review and some new terminology.

Definition D.5.10 *A vector* $\mathbf{w}$, *is a linear combination of the vectors* $\{\mathbf{v}_1, \cdots, \mathbf{v}_r\}$ *if there exists scalars,* $c_1, \cdots c_r$ *such that* $\mathbf{w} = \sum_{k=1}^r c_k \mathbf{v}_k$. *This is the same as saying*

$$\mathbf{w} \in \operatorname{span} (\mathbf{v}_1, \cdots, \mathbf{v}_r).$$

The following corollary is also of great use.

Corollary D.5.11 *Suppose A is an $n \times n$ matrix and some column (row) is a linear combination of r other columns (rows). Then $\det (A) = 0$.*

Proof: Let $A = \left(\begin{array}{ccc} \mathbf{a}_1 & \cdots & \mathbf{a}_n \end{array} \right)$ be the columns of A and suppose the condition that one column is a linear combination of r of the others is satisfied. Then by using Corollary D.5.9 the determinant of A is zero if and only if the determinant of the matrix B, which has this special column placed in the last position, equals zero. Thus $\mathbf{a}_n = \sum_{k=1}^r c_k \mathbf{a}_k$ and so

$$\det (B) = \det \left(\begin{array}{ccccc} \mathbf{a}_1 & \cdots & \mathbf{a}_r & \cdots & \mathbf{a}_{n-1} \quad \sum_{k=1}^r c_k \mathbf{a}_k \end{array} \right).$$

By Corollary D.5.9

$$\det (B) = \sum_{k=1}^r c_k \det \left(\begin{array}{ccccc} \mathbf{a}_1 & \cdots & \mathbf{a}_r & \cdots & \mathbf{a}_{n-1} \quad \mathbf{a}_k \end{array} \right) = 0.$$

because there are two equal columns. The case for rows follows from the fact that $\det (A) = \det \left(A^T \right)$. ∎

D.5.8 Determinant of a Product

Recall the following definition of matrix multiplication.

Definition D.5.12 *If A and B are $n \times n$ matrices, $A = (a_{ij})$ and $B = (b_{ij})$, $AB = (c_{ij})$ where*

$$c_{ij} \equiv \sum_{k=1}^{n} a_{ik} b_{kj}.$$

One of the most important rules about determinants is that the determinant of a product equals the product of the determinants.

Theorem D.5.13 *Let A and B be $n \times n$ matrices. Then*

$$\det(AB) = \det(A)\det(B).$$

Proof: Let c_{ij} be the ij^{th} entry of AB. Then by Proposition D.5.6,

$$\det(AB) =$$

$$\sum_{(k_1,\cdots,k_n)} \operatorname{sgn}(k_1,\cdots,k_n) c_{1k_1} \cdots c_{nk_n}$$

$$= \sum_{(k_1,\cdots,k_n)} \operatorname{sgn}(k_1,\cdots,k_n) \left(\sum_{r_1} a_{1r_1} b_{r_1 k_1} \right) \cdots \left(\sum_{r_n} a_{nr_n} b_{r_n k_n} \right)$$

$$= \sum_{(r_1\cdots,r_n)} \sum_{(k_1,\cdots,k_n)} \operatorname{sgn}(k_1,\cdots,k_n) b_{r_1 k_1} \cdots b_{r_n k_n} (a_{1r_1} \cdots a_{nr_n})$$

$$= \sum_{(r_1\cdots,r_n)} \operatorname{sgn}(r_1\cdots r_n) a_{1r_1} \cdots a_{nr_n} \det(B) = \det(A)\det(B). \blacksquare$$

D.5.9 Cofactor Expansions

Lemma D.5.14 *Suppose a matrix is of the form*

$$M = \begin{pmatrix} A & * \\ \mathbf{0} & a \end{pmatrix} \tag{4.26}$$

or

$$M = \begin{pmatrix} A & \mathbf{0} \\ * & a \end{pmatrix} \tag{4.27}$$

where a is a number and A is an $(n-1) \times (n-1)$ matrix and $*$ denotes either a column or a row having length $n-1$ and the $\mathbf{0}$ denotes either a column or a row of length $n-1$ consisting entirely of zeros. Then $\det(M) = a\det(A)$.

Proof: Denote M by (m_{ij}). Thus in the first case, $m_{nn} = a$ and $m_{ni} = 0$ if $i \neq n$ while in the second case, $m_{nn} = a$ and $m_{in} = 0$ if $i \neq n$. From the definition of the determinant,

$$\det(M) \equiv \sum_{(k_1,\cdots,k_n)} \text{sgn}_n(k_1,\cdots,k_n) m_{1k_1} \cdots m_{nk_n}$$

Letting θ denote the position of n in the ordered list, $(k_1,\cdots,k_n)$ then using Lemma D.5.1, $\det(M)$ equals

$$\sum_{(k_1,\cdots,k_n)} (-1)^{n-\theta} \text{sgn}_{n-1} \left(k_1,\cdots,k_{\theta-1}, \overset{\theta}{k_{\theta+1}},\cdots, \overset{n-1}{k_n} \right) m_{1k_1} \cdots m_{nk_n}$$

Now suppose (4.27). Then if $k_n \neq n$, the term involving m_{nk_n} in the above expression equals zero. Therefore, the only terms which survive are those for which $\theta = n$ or in other words, those for which $k_n = n$. Therefore, the above expression reduces to

$$a \sum_{(k_1,\cdots,k_{n-1})} \text{sgn}_{n-1}(k_1,\cdots k_{n-1}) m_{1k_1} \cdots m_{(n-1)k_{n-1}} = a \det(A).$$

To get the assertion in the situation of (4.26) use Corollary D.5.8 and (4.27) to write

$$\det(M) = \det(M^T) = \det \left(\begin{pmatrix} A^T & \mathbf{0} \\ * & a \end{pmatrix} \right) = a \det(A^T) = a \det(A). \blacksquare$$

In terms of the theory of determinants, arguably the most important idea is that of Laplace expansion along a row or a column. This will follow from the above definition of a determinant.

Definition D.5.15 *Let $A = (a_{ij})$ be an $n \times n$ matrix. Then a new matrix called the cofactor matrix, $\text{cof}(A)$ is defined by $\text{cof}(A) = (c_{ij})$ where to obtain c_{ij} delete the ith row and the jth column of A, take the determinant of the $(n-1) \times (n-1)$ matrix which results, (This is called the ij^{th} minor of A.) and then multiply this number by $(-1)^{i+j}$. To make the formulas easier to remember, $\text{cof}(A)_{ij}$ will denote the ij^{th} entry of the cofactor matrix.*

The following is the main result. Earlier this was given as a definition and the outrageous totally unjustified assertion was made that the same number would be obtained by expanding the determinant along any row or column. The following theorem proves this assertion.

Theorem D.5.16 *Let A be an $n \times n$ matrix where $n \geq 2$. Then*

$$\det(A) = \sum_{j=1}^{n} a_{ij} \text{cof}(A)_{ij} = \sum_{i=1}^{n} a_{ij} \text{cof}(A)_{ij}. \tag{4.28}$$

The first formula consists of expanding the determinant along the ith row and the second expands the determinant along the jth column.

Proof: Let $(a_{i1}, \cdots, a_{in})$ be the ith row of A. Let B_j be the matrix obtained from A by leaving every row the same except the ith row which in B_j equals

$$(0, \cdots, 0, a_{ij}, 0, \cdots, 0).$$

Then by Corollary D.5.9,

$$\det(A) = \sum_{j=1}^{n} \det(B_j)$$

Denote by A^{ij} the $(n-1) \times (n-1)$ matrix obtained by deleting the ith row and the jth column of A. Thus $\operatorname{cof}(A)_{ij} \equiv (-1)^{i+j} \det(A^{ij})$. At this point, recall that from Proposition D.5.6, when two rows or two columns in a matrix M, are switched, this results in multiplying the determinant of the old matrix by -1 to get the determinant of the new matrix. Therefore, by Lemma D.5.14,

$$\det(B_j) = (-1)^{n-j} (-1)^{n-i} \det \left(\begin{pmatrix} A^{ij} & * \\ \mathbf{0} & a_{ij} \end{pmatrix} \right)$$

$$= (-1)^{i+j} \det \left(\begin{pmatrix} A^{ij} & * \\ \mathbf{0} & a_{ij} \end{pmatrix} \right) = a_{ij} \operatorname{cof}(A)_{ij}.$$

Therefore,

$$\det(A) = \sum_{j=1}^{n} a_{ij} \operatorname{cof}(A)_{ij}$$

which is the formula for expanding $\det(A)$ along the ith row. Also,

$$\det(A) = \det(A^T) = \sum_{j=1}^{n} a_{ij}^T \operatorname{cof}(A^T)_{ij}$$

$$= \sum_{j=1}^{n} a_{ji} \operatorname{cof}(A)_{ji}$$

which is the formula for expanding $\det(A)$ along the ith column. ∎

D.5.10 Formula for the Inverse

Note that this gives an easy way to write a formula for the inverse of an $n \times n$ matrix.

Theorem D.5.17 *A^{-1} exists if and only if $\det(A) \neq 0$. If $\det(A) \neq 0$, then $A^{-1} = \left(a_{ij}^{-1} \right)$ where*

$$a_{ij}^{-1} = \det(A)^{-1} \operatorname{cof}(A)_{ji}$$

for $\operatorname{cof}(A)_{ij}$ the ij^{th} cofactor of A.

Proof: By Theorem D.5.16 and letting $(a_{ir}) = A$, if $\det(A) \neq 0$,

$$\sum_{i=1}^{n} a_{ir} \operatorname{cof}(A)_{ir} \det(A)^{-1} = \det(A) \det(A)^{-1} = 1.$$

Now consider

$$\sum_{i=1}^{n} a_{ir} \operatorname{cof}(A)_{ik} \det(A)^{-1}$$

when $k \neq r$. Replace the kth column with the rth column to obtain a matrix B_k whose determinant equals zero by Corollary D.5.9. However, expanding this matrix along the kth column yields

$$0 = \det(B_k) \det(A)^{-1} = \sum_{i=1}^{n} a_{ir} \operatorname{cof}(A)_{ik} \det(A)^{-1}$$

Summarizing,

$$\sum_{i=1}^{n} a_{ir} \operatorname{cof}(A)_{ik} \det(A)^{-1} = \delta_{rk}.$$

Using the other formula in Theorem D.5.16, and similar reasoning,

$$\sum_{j=1}^{n} a_{rj} \operatorname{cof}(A)_{kj} \det(A)^{-1} = \delta_{rk}$$

This proves that if $\det(A) \neq 0$, then A^{-1} exists with $A^{-1} = \left(a_{ij}^{-1}\right)$, where

$$a_{ij}^{-1} = \operatorname{cof}(A)_{ji} \det(A)^{-1}.$$

Now suppose A^{-1} exists. Then by Theorem D.5.13,

$$1 = \det(I) = \det\left(AA^{-1}\right) = \det(A) \det\left(A^{-1}\right)$$

so $\det(A) \neq 0$. ∎

The next corollary points out that if an $n \times n$ matrix A has a right or a left inverse, then it has an inverse.

Corollary D.5.18 *Let A be an $n \times n$ matrix and suppose there exists an $n \times n$ matrix B such that $BA = I$. Then A^{-1} exists and $A^{-1} = B$. Also, if there exists C an $n \times n$ matrix such that $AC = I$, then A^{-1} exists and $A^{-1} = C$.*

Proof: Since $BA = I$, Theorem D.5.13 implies

$$\det B \det A = 1$$

and so $\det A \neq 0$. Therefore from Theorem D.5.17, A^{-1} exists. Therefore,

$$A^{-1} = (BA) A^{-1} = B \left(AA^{-1}\right) = BI = B.$$

The case where $CA = I$ is handled similarly. ∎

The conclusion of this corollary is that left inverses, right inverses, and inverses are all the same in the context of $n \times n$ matrices.

Theorem D.5.17 says that to find the inverse, take the transpose of the cofactor matrix and divide by the determinant. The transpose of the cofactor matrix is called the adjugate or sometimes the classical adjoint of the matrix A. It is an abomination to call it the adjoint although you do sometimes see it referred to in this way. In words, A^{-1} is equal to one over the determinant of A times the adjugate matrix of A.

D.5.11 Cramer's Rule

In case you are solving a system of equations, $A\mathbf{x} = \mathbf{y}$ for $\mathbf{x}$, it follows that if A^{-1} exists,

$$\mathbf{x} = \left(A^{-1}A\right)\mathbf{x} = A^{-1}\left(A\mathbf{x}\right) = A^{-1}\mathbf{y}$$

thus solving the system. Now in the case that A^{-1} exists, there is a formula for A^{-1} given above. Using this formula,

$$x_i = \sum_{j=1}^{n} a_{ij}^{-1} y_j = \sum_{j=1}^{n} \frac{1}{\det(A)} \operatorname{cof}(A)_{ji} \, y_j.$$

By the formula for the expansion of a determinant along a column,

$$x_i = \frac{1}{\det(A)} \det \begin{pmatrix} * & \cdots & y_1 & \cdots & * \\ \vdots & & \vdots & & \vdots \\ * & \cdots & y_n & \cdots & * \end{pmatrix},$$

where here the ith column of A is replaced with the column vector $(y_1 \cdots, y_n)^T$, and the determinant of this modified matrix is taken and divided by $\det(A)$. This formula is known as Cramer's rule.

D.5.12 Upper Triangular Matrices

Definition D.5.19 *A matrix M, is upper triangular if $M_{ij} = 0$ whenever $i > j$. Thus such a matrix equals zero below the main diagonal, the entries of the form M_{ii} as shown.*

$$\begin{pmatrix} * & * & \cdots & * \\ 0 & * & \ddots & \vdots \\ \vdots & \ddots & \ddots & * \\ 0 & \cdots & 0 & * \end{pmatrix}$$

A lower triangular matrix is defined similarly as a matrix for which all entries above the main diagonal are equal to zero.

With this definition, here is a simple corollary of Theorem D.5.16.

Corollary D.5.20 *Let M be an upper (lower) triangular matrix. Then $\det(M)$ is obtained by taking the product of the entries on the main diagonal.*

D.6   Cayley-Hamilton Theorem

Definition D.6.1 *Let A be an $n \times n$ matrix. The characteristic polynomial is defined as*

$$q_A(t) \equiv \det(tI - A)$$

and the solutions to $q_A(t) = 0$ are called eigenvalues. For A a matrix and $p(t) = t^n + a_{n-1}t^{n-1} + \cdots + a_1 t + a_0$, denote by $p(A)$ the matrix defined by

$$p(A) \equiv A^n + a_{n-1}A^{n-1} + \cdots + a_1 A + a_0 I.$$

The explanation for the last term is that A^0 is interpreted as I, the identity matrix.

The Cayley-Hamilton theorem states that every matrix satisfies its characteristic equation, that equation defined by $q_A(t) = 0$. It is one of the most important theorems in linear algebra.[†] The proof in this section is not the most general proof, but works well when the field of scalars is $\mathbb{R}$ or $\mathbb{C}$. The following lemma will help with its proof.

Lemma D.6.2 *Suppose for all $|\lambda|$ large enough,*

$$A_0 + A_1 \lambda + \cdots + A_m \lambda^m = 0,$$

where the A_i are $n \times n$ matrices. Then each $A_i = 0$.

Proof: Multiply by λ^{-m} to obtain

$$A_0 \lambda^{-m} + A_1 \lambda^{-m+1} + \cdots + A_{m-1}\lambda^{-1} + A_m = 0.$$

Now let $|\lambda| \to \infty$ to obtain $A_m = 0$. With this, multiply by λ to obtain

$$A_0 \lambda^{-m+1} + A_1 \lambda^{-m+2} + \cdots + A_{m-1} = 0.$$

Now let $|\lambda| \to \infty$ to obtain $A_{m-1} = 0$. Continue multiplying by λ and letting $\lambda \to \infty$ to obtain that all the $A_i = 0$. ∎
    With the lemma, here is a simple corollary.

Corollary D.6.3 *Let A_i and B_i be $n \times n$ matrices and suppose*

$$A_0 + A_1 \lambda + \cdots + A_m \lambda^m = B_0 + B_1 \lambda + \cdots + B_m \lambda^m$$

for all $|\lambda|$ large enough. Then $A_i = B_i$ for all i. If $A_i = B_i$ for each A_i, B_i then one can substitute an $n \times n$ matrix M for λ and the identity will continue to hold.

Proof: Subtract and use the result of the lemma. The last claim is obvious by matching terms. ∎
    With this preparation, here is a relatively easy proof of the Cayley-Hamilton theorem.

[†]A special case was first proved by Hamilton in 1853. The general case was announced by Cayley some time later and a proof was given by Frobenius in 1878.

Theorem D.6.4 *Let A be an $n \times n$ matrix and let $q(\lambda) \equiv \det(\lambda I - A)$ be the characteristic polynomial. Then $q(A) = 0$.*

Proof: Let $C(\lambda)$ equal the transpose of the cofactor matrix of $(\lambda I - A)$ for $|\lambda|$ large. (If $|\lambda|$ is large enough, then λ cannot be in the finite list of eigenvalues of A and so for such λ, $(\lambda I - A)^{-1}$ exists.) Therefore, by Theorem D.5.17

$$C(\lambda) = q(\lambda)(\lambda I - A)^{-1}.$$

Say

$$q(\lambda) = a_0 + a_1\lambda + \cdots + \lambda^n$$

Note that each entry in $C(\lambda)$ is a polynomial in λ having degree no more than $n - 1$. For example, you might have something like

$$C(\lambda) = \begin{pmatrix} \lambda^2 - 6\lambda + 9 & 3 - \lambda & 0 \\ 2\lambda - 6 & \lambda^2 - 3\lambda & 0 \\ \lambda - 1 & \lambda - 1 & \lambda^2 - 3\lambda + 2 \end{pmatrix}$$

$$= \begin{pmatrix} 9 & 3 & 0 \\ -6 & 0 & 0 \\ -1 & -1 & 2 \end{pmatrix} + \lambda \begin{pmatrix} -6 & -1 & 0 \\ 2 & -3 & 0 \\ 1 & 1 & -3 \end{pmatrix} + \lambda^2 \begin{pmatrix} 1 & 0 & 0 \\ 0 & 1 & 0 \\ 0 & 0 & 1 \end{pmatrix}$$

Therefore, collecting the terms in the general case,

$$C(\lambda) = C_0 + C_1\lambda + \cdots + C_{n-1}\lambda^{n-1}$$

for C_j some $n \times n$ matrix. Then

$$C(\lambda)(\lambda I - A) = \left(C_0 + C_1\lambda + \cdots + C_{n-1}\lambda^{n-1}\right)(\lambda I - A) = q(\lambda)I$$

Then multiplying out the middle term, it follows that for all $|\lambda|$ sufficiently large,

$$a_0 I + a_1 I\lambda + \cdots + I\lambda^n = C_0\lambda + C_1\lambda^2 + \cdots + C_{n-1}\lambda^n$$

$$- \left[C_0 A + C_1 A\lambda + \cdots + C_{n-1}A\lambda^{n-1}\right]$$

$$= -C_0 A + (C_0 - C_1 A)\lambda + (C_1 - C_2 A)\lambda^2 + \cdots + (C_{n-2} - C_{n-1}A)\lambda^{n-1} + C_{n-1}\lambda^n$$

Then, using Corollary D.6.3, one can replace λ on both sides with A. Then the right side is seen to equal 0. Hence the left side, $q(A)I$ is also equal to 0. ∎

The following theorem is of fundamental importance and ties together many of the ideas presented above.

Theorem D.6.5 *Let A be an $n \times n$ matrix. Then the following are equivalent.*

1. $\det(A) = 0$.

2. A, A^T are not one to one.

3. A is not onto.

Proof: Suppose $\det(A) = 0$. Then A cannot be one to one because if it were, then it would be onto as well thanks to Theorem D.4.1. Hence you would have the existence of A^{-1} because, there would exist $\mathbf{b}_i$ such that $A\mathbf{b}_i = \mathbf{e}_i$ and so $B \equiv \begin{pmatrix} \mathbf{b}_1 & \cdots & \mathbf{b}_n \end{pmatrix}$ satisfies $AB = I$ and so $B = A^{-1}$. But then

$$1 = \det\left(AA^{-1}\right) = \det(A)\det\left(A^{-1}\right) = 0.$$

Now $\det(A) = \det\left(A^T\right)$ and so the same reasoning implies A^T is not one to one. This verifies that 1. implies 2..

Now suppose 2. Then since A^T is not one to one, it follows there exists $\mathbf{x} \neq \mathbf{0}$ such that

$$A^T\mathbf{x} = \mathbf{0}.$$

Taking the transpose of both sides yields

$$\mathbf{x}^T A = \mathbf{0}^T$$

where the $\mathbf{0}^T$ is a $1 \times n$ matrix or row vector. Now if $A\mathbf{y} = \mathbf{x}$, then

$$|\mathbf{x}|^2 = \mathbf{x}^T(A\mathbf{y}) = \left(\mathbf{x}^T A\right)\mathbf{y} = \mathbf{0}\mathbf{y} = 0$$

contrary to $\mathbf{x} \neq \mathbf{0}$. Consequently there can be no $\mathbf{y}$ such that $A\mathbf{y} = \mathbf{x}$ and so A is not onto. This shows that 2. implies 3..

Finally, suppose 3. If 1. does not hold, then $\det(A) \neq 0$ but then from Theorem D.5.17 A^{-1} exists and so for every $\mathbf{y} \in \mathbb{F}^n$ there exists a unique $\mathbf{x} \in \mathbb{F}^n$ such that $A\mathbf{x} = \mathbf{y}$. In fact $\mathbf{x} = A^{-1}\mathbf{y}$. Thus A would be onto contrary to 3.. This shows 3. implies 1.. ■

Corollary D.6.6 *Let A be an $n \times n$ matrix. Then the following are equivalent.*

1. $\det(A) \neq 0$.

2. A and A^T are one to one.

3. A is onto.

Proof: This follows immediately from the above theorem.

D.7   Eigenvalues and Eigenvectors of a Matrix

D.7.1   Definition of Eigenvectors and Eigenvalues

In this section, the field of scalars $\mathbb{F}$ will be $\mathbb{C}$. The following is the definition of eigenvalues and eigenvectors.

Definition D.7.1 *Let M be an $n \times n$ matrix and let $\mathbf{x} \in \mathbb{C}^n$ be a **nonzero vector** for which*

$$M\mathbf{x} = \lambda\mathbf{x} \tag{4.29}$$

for some scalar λ. Then **x** *is called an* **eigenvector** *and λ is called an* **eigenvalue** *(characteristic value) of the matrix M.*

$$\boxed{\text{\textit{Note: } \textbf{\textit{Eigenvectors \underline{are} \underline{never} \underline{equal} \underline{to} \underline{zero}!}}}}$$

The set of all eigenvalues of an $n \times n$ matrix M, is denoted by $\sigma(M)$ and is referred to as the **spectrum** *of M.*

The following corollary of Theorem D.6.5 tells how to obtain eigenvalues and eigenvectors.

Corollary D.7.2 *Let M be an $n \times n$ matrix, the entries in a field of scalars $\mathbb{F}$, for us $\mathbb{C}$. Then λ is an eigenvalue of M if and only if $\det(M - \lambda I) = 0$. Thus $\det(M - \lambda I) = 0$ if and only if there exists $\mathbf{x} \neq \mathbf{0}$ such that $(M - \lambda I)\mathbf{x} = \mathbf{0}$.*

Note that the above corollary is valid if $M - \lambda I$ is replaced with $\lambda I - M$.

If you have an eigenvalue $\lambda \in \mathbb{F}$, for us $\mathbb{F} = \mathbb{C}$, then to get the eigenvectors for this λ, you find nonzero solutions $\mathbf{x}$ to $(M - \lambda I)\mathbf{x} = \mathbf{0}$. These will exist and you can find them using row operations as is usually the case. If λ is complex, the computations must of course involve complex arithmetic. Other than this, there is no change. For a general field, you simply do the computations in that field.

D.7.2 Triangular Matrices

Although it is usually hard to solve the eigenvalue problem, there is a kind of matrix for which this is not the case. These are the upper or lower triangular matrices. I will illustrate by examples.

Example D.7.3 *Let $A = \begin{pmatrix} 1 & 2 & 4 \\ 0 & 4 & 7 \\ 0 & 0 & 6 \end{pmatrix}$. Find its eigenvalues.*

You need to solve

$$
\begin{aligned}
0 &= \det\left(\begin{pmatrix} 1 & 2 & 4 \\ 0 & 4 & 7 \\ 0 & 0 & 6 \end{pmatrix} - \lambda \begin{pmatrix} 1 & 0 & 0 \\ 0 & 1 & 0 \\ 0 & 0 & 1 \end{pmatrix}\right) \\
&= \det\begin{pmatrix} 1-\lambda & 2 & 4 \\ 0 & 4-\lambda & 7 \\ 0 & 0 & 6-\lambda \end{pmatrix} = (1-\lambda)(4-\lambda)(6-\lambda).
\end{aligned}
$$

Thus the eigenvalues are just the diagonal entries of the original matrix. You can see it would work this way with any such matrix. These matrices are called **upper triangular.** Stated precisely, a matrix A is upper triangular if $A_{ij} = 0$ for all $i > j$. Similarly, it is easy to find the eigenvalues for a lower triangular matrix, one which has all zeros above the main diagonal.

D.7.3   Defective and Nondefective Matrices

In this book, we are tacitly assuming that the field of scalars is $\mathbb{C}$ so that the characteristic polynomial can always be factored.

Definition D.7.4 *By the fundamental theorem of algebra, there exists a factorization of the characteristic equation in the form*

$$(\lambda - \lambda_1)^{r_1} (\lambda - \lambda_2)^{r_2} \cdots (\lambda - \lambda_m)^{r_m} = 0$$

*where r_i is some integer no smaller than 1. Thus the eigenvalues are $\lambda_1, \lambda_2, \cdots, \lambda_m$. The **algebraic multiplicity** of λ_j is defined to be r_j.*

Example D.7.5 *Consider the matrix*

$$A = \begin{pmatrix} 1 & 1 & 0 \\ 0 & 1 & 1 \\ 0 & 0 & 1 \end{pmatrix} \tag{4.30}$$

What is the algebraic multiplicity of the eigenvalue $\lambda = 1$?

In this case the characteristic equation is

$$\det (A - \lambda I) = (1 - \lambda)^3 = 0$$

or equivalently,

$$\det (\lambda I - A) = (\lambda - 1)^3 = 0.$$

Therefore, λ is of algebraic multiplicity 3.

Definition D.7.6 *The **geometric multiplicity** of an eigenvalue is the dimension of the eigenspace,*

$$N (A - \lambda I).$$

Example D.7.7 *Find the geometric multiplicity of $\lambda = 1$ for the matrix in (4.30).*

We need to solve

$$\begin{pmatrix} 0 & 1 & 0 \\ 0 & 0 & 1 \\ 0 & 0 & 0 \end{pmatrix} \begin{pmatrix} x \\ y \\ z \end{pmatrix} = \begin{pmatrix} 0 \\ 0 \\ 0 \end{pmatrix}.$$

The solutions are of the form

$$t \begin{pmatrix} 1 & 0 & 0 \end{pmatrix}^T, \, t \in \mathbb{F}.$$

It follows the geometric multiplicity of $\lambda = 1$ is 1.

Definition D.7.8 *An $n \times n$ matrix is called **defective** if the geometric multiplicity is not equal to the algebraic multiplicity for some eigenvalue. Sometimes such an eigenvalue for which the geometric multiplicity is not equal to the algebraic multiplicity is called a defective eigenvalue. If the geometric multiplicity for an eigenvalue equals the algebraic multiplicity, the eigenvalue is sometimes referred to as nondefective.*

Here is another more interesting example of a defective matrix.

Example D.7.9 *Let*

$$A = \begin{pmatrix} 2 & -2 & -1 \\ -2 & -1 & -2 \\ 14 & 25 & 14 \end{pmatrix}.$$

Find the eigenvectors and eigenvalues.

In this case the eigenvalues are $3, 6, 6$ where we have listed 6 twice because it is a zero of algebraic multiplicity two, the characteristic equation being

$$(\lambda - 3)(\lambda - 6)^2 = 0.$$

Then a computation shows that the geometric multiplicity of 6 is 1.

Consider the eigenvectors for $\lambda = 6$. This requires you to solve

$$\left(\begin{pmatrix} 2 & -2 & -1 \\ -2 & -1 & -2 \\ 14 & 25 & 14 \end{pmatrix} - 6 \begin{pmatrix} 1 & 0 & 0 \\ 0 & 1 & 0 \\ 0 & 0 & 1 \end{pmatrix} \right) \begin{pmatrix} x \\ y \\ z \end{pmatrix} = \begin{pmatrix} 0 \\ 0 \\ 0 \end{pmatrix}$$

and the augmented matrix for this system of equations is

$$\begin{pmatrix} -4 & -2 & -1 & 0 \\ -2 & -7 & -2 & 0 \\ 14 & 25 & 8 & 0 \end{pmatrix}$$

Then from row operations,

$$\begin{pmatrix} 1 & 0 & \frac{1}{8} & 0 \\ 0 & 1 & \frac{1}{4} & 0 \\ 0 & 0 & 0 & 0 \end{pmatrix}$$

and so the eigenvectors for $\lambda = 6$ are of the form

$$t \begin{pmatrix} -\frac{1}{8} & -\frac{1}{4} & 1 \end{pmatrix}^T \text{ or more simply } t \begin{pmatrix} -1 & -2 & 8 \end{pmatrix}^T$$

where $t \in \mathbb{F}$.

Note that in this example the eigenspace for the eigenvalue $\lambda = 6$ is of dimension 1 because there is only one parameter. However, this eigenvalue is of multiplicity two as a root to the characteristic equation. Thus this eigenvalue is a defective eigenvalue. However, the eigenvalue 3 is nondefective. The matrix is defective because it has a defective eigenvalue.

The word, defective, seems to suggest there is something wrong with the matrix. This is in fact the case. Defective matrices are a lot of trouble in applications and we wish they never occurred. However, they do occur as the above example shows. The reason these matrices are so horrible to work with is that it is impossible to obtain a basis of eigenvectors.

In terms of algebra, this lack of a basis of eigenvectors says that it is impossible to obtain a diagonal matrix which is similar to the given matrix. See below for this definition.

Although there may be repeated roots to the characteristic equation, it is not known whether the matrix is defective. However, there is an important theorem which holds when considering eigenvectors which correspond to distinct eigenvalues.

Theorem D.7.10 *Suppose* $M\mathbf{v}_i = \lambda_i\mathbf{v}_i, i = 1, \cdots, r$, $\mathbf{v}_i \neq 0$, *and that if* $i \neq j$, *then* $\lambda_i \neq \lambda_j$. *Then the set of eigenvectors,* $\{\mathbf{v}_1, \cdots, \mathbf{v}_r\}$ *is linearly independent.*

Proof. Suppose the claim of the lemma is not true. Then there exists a subset of this set of vectors

$$\{\mathbf{w}_1, \cdots, \mathbf{w}_r\} \subseteq \{\mathbf{v}_1, \cdots, \mathbf{v}_k\}$$

such that

$$\sum_{j=1}^{r} c_j\mathbf{w}_j = \mathbf{0} \tag{4.31}$$

where each $c_j \neq 0$. Say $M\mathbf{w}_j = \mu_j\mathbf{w}_j$ where

$$\{\mu_1, \cdots, \mu_r\} \subseteq \{\lambda_1, \cdots, \lambda_k\},$$

the μ_j being distinct eigenvalues of M. Out of all such subsets, let this one be such that r is as small as possible. Then necessarily, $r > 1$ because otherwise, $c_1\mathbf{w}_1 = \mathbf{0}$ which would imply $\mathbf{w}_1 = \mathbf{0}$, which is not allowed for eigenvectors.

Now apply M to both sides of (4.31).

$$\sum_{j=1}^{r} c_j\mu_j\mathbf{w}_j = \mathbf{0}. \tag{4.32}$$

Next pick $\mu_k \neq 0$ and multiply both sides of (4.31) by μ_k. Such a μ_k exists because $r > 1$. Thus

$$\sum_{j=1}^{r} c_j\mu_k\mathbf{w}_j = \mathbf{0} \tag{4.33}$$

Subtract the sum in (4.33) from the sum in (4.32) to obtain

$$\sum_{j=1}^{r} c_j \left(\mu_k - \mu_j\right) \mathbf{w}_j = \mathbf{0}$$

Now one of the constants $c_j \left(\mu_k - \mu_j\right)$ equals 0, when $j = k$. Therefore, r was not as small as possible after all. ∎

D.7.4 Diagonalization

First of all, here is what it means for two matrices to be similar.

Definition D.7.11 *Let A, B be two $n \times n$ matrices. Then they are **similar** if and only if there exists an invertible matrix S such that*

$$A = S^{-1}BS$$

Proposition D.7.12 *Define for $n \times n$ matrices $A \sim B$ if A is similar to B. Then*

$$A \sim A,$$
$$\text{If } A \sim B \text{ then } B \sim A$$
$$\text{If } A \sim B \text{ and } B \sim C \text{ then } A \sim C$$

Proof: It is clear that $A \sim A$ because you could just take $S = I$. If $A \sim B$, then for some S invertible,

$$A = S^{-1}BS$$

and so

$$SAS^{-1} = B$$

But then

$$\left(S^{-1}\right)^{-1} AS^{-1} = B$$

which shows that $B \sim A$.

Now suppose $A \sim B$ and $B \sim C$. Then there exist invertible matrices S, T such that

$$A = S^{-1}BS, \ B = T^{-1}CT.$$

Therefore,

$$A = S^{-1}T^{-1}CTS = (TS)^{-1} C (TS)$$

showing that A is similar to C. ∎

For your information, when $\sim$ satisfies the above conditions, it is called a similarity relation. Similarity relations are very significant in mathematics.

When a matrix is similar to a diagonal matrix, the matrix is said to be diagonalizable. I think this is one of the worst monstrosities for a word that I have ever seen. Nevertheless, it is commonly used in linear algebra. It turns out to be the same as nondefective. The following is the precise definition.

Definition D.7.13 *Let A be an $n \times n$ matrix. Then A is **diagonalizable** if there exists an invertible matrix S such that*

$$S^{-1}AS = D$$

where D is a diagonal matrix. This means D has a zero as every entry except for the main diagonal. More precisely, $D_{ij} = 0$ unless $i = j$. Such matrices look like the following.

$$\begin{pmatrix} * & & 0 \\ & \ddots & \\ 0 & & * \end{pmatrix}$$

where $*$ might not be zero.

The most important theorem about diagonalizability[‡] is the following major result.

Theorem D.7.14 *An $n \times n$ matrix is diagonalizable if and only if $\mathbb{F}^n$ has a basis of eigenvectors of A. Furthermore, you can take the matrix S described above, to be given as*

$$S = \left(\begin{array}{cccc} \mathbf{v}_1 & \mathbf{v}_2 & \cdots & \mathbf{v}_n \end{array} \right)$$

where here the v_k are the eigenvectors in the basis for $\mathbb{F}^n$. If A is diagonalizable, the eigenvalues of A are the diagonal entries of the diagonal matrix.

Proof: Suppose there exists a basis of eigenvectors $\{\mathbf{v}_k\}$ where $A\mathbf{v}_k = \lambda_k \mathbf{v}_k$. Then let S be given as above. It follows S^{-1} exists because these vectors are linearly independent so $N(S) = \{0\}$ which implies S is one to one which implies $\det(S) \neq 0$ which implies S^{-1} exists. Let S^{-1} be of the form

$$S^{-1} = \left(\begin{array}{c} \mathbf{w}_1^T \\ \mathbf{w}_2^T \\ \vdots \\ \mathbf{w}_n^T \end{array} \right)$$

where $\mathbf{w}_k^T \mathbf{v}_j = \delta_{kj}$. Then

$$\left(\begin{array}{ccc} \lambda_1 & & 0 \\ & \ddots & \\ 0 & & \lambda_n \end{array} \right) = \left(\begin{array}{c} \mathbf{w}_1^T \\ \mathbf{w}_2^T \\ \vdots \\ \mathbf{w}_n^T \end{array} \right) \left(\begin{array}{cccc} \lambda_1 \mathbf{v}_1 & \lambda_2 \mathbf{v}_2 & \cdots & \lambda_n \mathbf{v}_n \end{array} \right)$$

$$= \left(\begin{array}{c} \mathbf{w}_1^T \\ \mathbf{w}_2^T \\ \vdots \\ \mathbf{w}_n^T \end{array} \right) \left(\begin{array}{cccc} A\mathbf{v}_1 & A\mathbf{v}_2 & \cdots & A\mathbf{v}_n \end{array} \right)$$

$$= S^{-1} A S$$

Next suppose A is diagonalizable so that $S^{-1} A S = D$. Let

$$S = \left(\begin{array}{cccc} \mathbf{v}_1 & \mathbf{v}_2 & \cdots & \mathbf{v}_n \end{array} \right)$$

where the columns are the $\mathbf{v}_k$ and

$$D = \left(\begin{array}{ccc} \lambda_1 & & 0 \\ & \ddots & \\ 0 & & \lambda_n \end{array} \right)$$

[‡]This word has 9 syllables.

Then

$$AS = SD = \begin{pmatrix} \mathbf{v}_1 & \mathbf{v}_2 & \cdots & \mathbf{v}_n \end{pmatrix} \begin{pmatrix} \lambda_1 & & 0 \\ & \ddots & \\ 0 & & \lambda_n \end{pmatrix}$$

and so

$$\begin{pmatrix} A\mathbf{v}_1 & A\mathbf{v}_2 & \cdots & A\mathbf{v}_n \end{pmatrix} = \begin{pmatrix} \lambda_1\mathbf{v}_1 & \lambda_2\mathbf{v}_2 & \cdots & \lambda_n\mathbf{v}_n \end{pmatrix}$$

showing the $\mathbf{v}_i$ are eigenvectors of A and the λ_k are eigenvectors. Now the $\mathbf{v}_k$ form a basis for $\mathbb{F}^n$ because the matrix S having these vectors as columns is given to be invertible. ∎

In other words, to diagonalize A you get a basis of eigenvectors $\{\mathbf{v}_1, \cdots, \mathbf{v}_n\}$ and let $S = \begin{pmatrix} \mathbf{v}_1 & \cdots & \mathbf{v}_n \end{pmatrix}$. Then $S^{-1}AS = D$ a diagonal matrix which has the eigenvalues down the main diagonal listed according to multiplicity. Note also that for n a positive integer,

$$A^n = \overbrace{SDS^{-1}SDS^{-1}SDS^{-1}\cdots SDS^{-1}}^{50 \text{ times}}$$

The interior $S^{-1}S$ cancel and so this reduces to

$$A^n = SD^nS^{-1}$$

and it is easy to compute D^m. More generally, you can define functions of the matrix using power series in this way.

D.8   Schur's Theorem

Consider the following system of equations for $x_1, x_2, \cdots, x_n$

$$\sum_{j=1}^{n} a_{ij}x_j = 0, \ i = 1, 2, \cdots, m \tag{4.34}$$

where $m < n$. Then the following theorem is a fundamental observation.

Theorem D.8.1 *Let the system of equations be as just described in (4.34) where $m < n$. Then letting*

$$\mathbf{x}^T \equiv (x_1, x_2, \cdots, x_n) \in \mathbb{F}^n,$$

there exists $\mathbf{x} \neq \mathbf{0}$ such that the components satisfy each of the equations of (4.34). Here $\mathbb{F}$ is a field of scalars. Think $\mathbb{R}$ or $\mathbb{C}$ for example.

Proof: The above system is of the form

$$A\mathbf{x} = \mathbf{0}$$

where A is an $m \times n$ matrix with $m < n$. Therefore, from Theorem D.2.10, there exists $\mathbf{x} \neq \mathbf{0}$ such that $A\mathbf{x} = \mathbf{0}$. ∎

Definition D.8.2 *A set of vectors in* $\mathbb{F}^n, \mathbb{F} = \mathbb{R}$ *or* $\mathbb{C}$, $\{\mathbf{x}_1, \cdots, \mathbf{x}_k\}$ *is called an* ***orthonormal*** *set of vectors if*

$$\overline{\mathbf{x}_i^T}\mathbf{x}_j = \delta_{ij} \equiv \begin{cases} 1 \;\; if \; i = j \\ 0 \;\; if \; i \neq j \end{cases}$$

Theorem D.8.3 *Let* $\mathbf{v}_1$ *be a unit vector* ($|\mathbf{v}_1| = 1$) *in* $\mathbb{F}^n$, $n > 1$. *Then there exist vectors* $\{\mathbf{v}_2, \cdots, \mathbf{v}_n\}$ *such that*

$$\{\mathbf{v}_1, \cdots, \mathbf{v}_n\}$$

is an orthonormal set of vectors.

Proof: The equation for $\mathbf{x}$, $\overline{\mathbf{v}_1}^T\mathbf{x} = 0$ has a nonzero solution $\mathbf{x}$ by Theorem D.8.1. Pick such a solution and divide by its magnitude to get $\mathbf{v}_2$ a unit vector such that $\overline{\mathbf{v}_1}^T \cdot \mathbf{v}_2 = 0$. Now suppose $\mathbf{v}_1, \cdots, \mathbf{v}_k$ have been chosen such that $\{\mathbf{v}_1, \cdots, \mathbf{v}_k\}$ is an orthonormal set of vectors. Then consider the equations

$$\overline{\mathbf{v}_j}^T\mathbf{x} = 0 \;\; j = 1, 2, \cdots, k$$

This amounts to the situation of Theorem D.8.1 in which there are more variables than equations. Therefore, by this theorem, there exists a nonzero $\mathbf{x}$ solving all these equations. Divide by its magnitude and this gives $\mathbf{v}_{k+1}$. ∎

Definition D.8.4 *If* U *is an* $n \times n$ *matrix whose columns form an orthonormal set of vectors, then* U *is called an* ***orthogonal matrix*** *if it is real and a* ***unitary matrix*** *if it is complex. Note that from the way we multiply matrices,*

$$U^T U = U U^T = I$$

in case U *is orthogonal. Thus* $U^{-1} = U^T$. *If* U *is only unitary, then from the dot product in* $\mathbb{C}^n$, *we replace the above with*

$$U^* U = U U^* = I.$$

Where the $*$ *indicates to take the conjugate of the transpose.*

Note the product of orthogonal or unitary matrices is orthogonal or unitary because

$$\begin{aligned} (U_1 U_2)^T (U_1 U_2) &= U_2^T U_1^T U_1 U_2 = I \\ (U_1 U_2)^* (U_1 U_2) &= U_2^* U_1^* U_1 U_2 = I. \end{aligned}$$

Two matrices A and B are similar if there is some invertible matrix S such that $A = S^{-1}BS$. Note that similar matrices have the same characteristic equation because by Theorem D.5.13 which says the determinant of a product is the product of the determinants,

$$\det(\lambda I - A) = \det(\lambda I - S^{-1}BS) = \det\left(S^{-1}(\lambda I - B)S\right)$$

$$= \det\left(S^{-1}\right)\det(\lambda I - B)\det(S) = \det\left(S^{-1}S\right)\det(\lambda I - B) = \det(\lambda I - B)$$

With this preparation, here is Schur's theorem.

Theorem D.8.5 *Let A be a real or complex $n \times n$ matrix. Then there exists a unitary matrix U such that*

$$U^* A U = T, \tag{4.35}$$

where T is an upper triangular matrix having the eigenvalues of A on the main diagonal, listed according to multiplicity as zeros of the characteristic equation. If A has all real entries and eigenvalues, then U can be chosen to be orthogonal.

Proof: The theorem is clearly true if A is a 1×1 matrix. Just let $U = 1$ the 1×1 matrix which has 1 down the main diagonal and zeros elsewhere. Suppose it is true for $(n-1) \times (n-1)$ matrices and let A be an $n \times n$ matrix. Then let $\mathbf{v}_1$ be a unit eigenvector for A. There exists λ_1 such that

$$A\mathbf{v}_1 = \lambda_1 \mathbf{v}_1, \ |\mathbf{v}_1| = 1.$$

By Theorem D.8.3 there exists $\{\mathbf{v}_1, \cdots, \mathbf{v}_n\}$, an orthonormal set in $\mathbb{F}^n$. Let U_0 be a matrix whose i^{th} column is $\mathbf{v}_i$. Then from the above, it follows U_0 is unitary. Then from the way you multiply matrices $U_0^* A U_0$ is of the form

$$\begin{pmatrix} \lambda_1 & * & \cdots & * \\ 0 & & & \\ \vdots & & A_1 & \\ 0 & & & \end{pmatrix}$$

where A_1 is an $n - 1 \times n - 1$ matrix. The above matrix is similar to A so it has the same eigenvalues and indeed the same characteristic equation. Now by induction there exists an $(n-1) \times (n-1)$ unitary matrix $\widetilde{U}_1$ such that

$$\widetilde{U}_1^* A_1 \widetilde{U}_1 = T_{n-1},$$

an upper triangular matrix. Consider

$$U_1 \equiv \begin{pmatrix} 1 & \mathbf{0} \\ \mathbf{0} & \widetilde{U}_1 \end{pmatrix}$$

From the way we multiply matrices, this is a unitary matrix and

$$U_1^* U_0^* A U_0 U_1 = \begin{pmatrix} 1 & \mathbf{0} \\ \mathbf{0} & \widetilde{U}_1^* \end{pmatrix} \begin{pmatrix} \lambda_1 & * \\ \mathbf{0} & A_1 \end{pmatrix} \begin{pmatrix} 1 & \mathbf{0} \\ \mathbf{0} & \widetilde{U}_1 \end{pmatrix} = \begin{pmatrix} \lambda_1 & * \\ \mathbf{0} & T_{n-1} \end{pmatrix} \equiv T$$

where T is upper triangular. Then let $U = U_0 U_1$. Then $U^* A U = T$. If A is real having real eigenvalues, all of the above can be accomplished using the real dot product and using real eigenvectors. Thus U can be orthogonal. ∎

D.9   Direct Sums

Here is a very useful theorem due to Sylvester. The field of scalars will be $\mathbb{F}$ and the entries of A, B come from $\mathbb{F}$. We will have in mind $\mathbb{F} = \mathbb{C}$ but the results hold more generally. You really only need to be able to completely factor something called the minimal polynomial which will be described below.

Theorem D.9.1 *Let A, B be matrices such that BA makes sense. Then*

$$\dim\left(\ker\left(BA\right)\right) \leq \dim\left(\ker\left(B\right)\right) + \dim\left(\ker\left(A\right)\right).$$

Proof: If $\mathbf{x} \in \ker\left(BA\right)$, then $A\mathbf{x} \in \ker\left(B\right)$ and so $A\left(\ker\left(BA\right)\right) \subseteq \ker\left(B\right)$. The following picture may help.

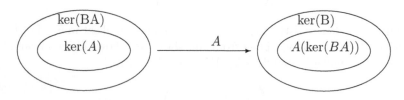

Now let $\{\mathbf{x}_1, \cdots, \mathbf{x}_n\}$ be a basis of $\ker\left(A\right)$ and let $\{A\mathbf{y}_1, \cdots, A\mathbf{y}_m\}$ be a basis for $A\left(\ker\left(BA\right)\right)$. Take any $\mathbf{z} \in \ker\left(BA\right)$. Then $A\mathbf{z} = \sum_{i=1}^{m} a_i A\mathbf{y}_i$ and so

$$A\left(\mathbf{z} - \sum_{i=1}^{m} a_i\mathbf{y}_i\right) = \mathbf{0}$$

which means $\mathbf{z} - \sum_{i=1}^{m} a_i\mathbf{y}_i \in \ker\left(A\right)$ and so there are scalars b_i such that

$$\mathbf{z} - \sum_{i=1}^{m} a_i\mathbf{y}_i = \sum_{j=1}^{n} b_i\mathbf{x}_i.$$

It follows $\operatorname{span}\left(\mathbf{x}_1, \cdots, \mathbf{x}_n, \mathbf{y}_1, \cdots, \mathbf{y}_m\right) \supseteq \ker\left(BA\right)$ and so by the first part (see the picture)

$$\dim\left(\ker\left(BA\right)\right) \leq n + m \leq \dim\left(\ker\left(A\right)\right) + \dim\left(\ker\left(B\right)\right) \quad \blacksquare$$

Of course this result holds for any finite product of square matrices by induction. One way this is quite useful is in the case where you have a finite product $\prod_{i=1}^{l} L_i$ all of which are square matrices of the same size. Then

$$\dim\left(\ker\prod_{i=1}^{l} L_i\right) \leq \sum_{i=1}^{l} \dim\left(\ker L_i\right)$$

and so if you can find a linearly independent set of vectors in $\ker\left(\prod_{i=1}^{l} L_i\right)$ of size

$$\sum_{i=1}^{l} \dim\left(\ker L_i\right),$$

then it must be a basis for $\ker\left(\prod_{i=1}^{l} L_i\right)$.

Definition D.9.2 *Let $\{V_i\}_{i=1}^r$ be subspaces of some vector space V. We have in mind $V = \mathbb{F}^p$ but this is not necessary. Then*

$$\sum_{i=1}^r V_i$$

denotes all sums of the form $\sum_{i=1}^r \mathbf{v}_i$ where $\mathbf{v}_i \in V_i$. If whenever

$$\sum_{i=1}^r \mathbf{v}_i = \mathbf{0}, \ \mathbf{v}_i \in V_i, \tag{4.36}$$

it follows that $\mathbf{v}_i = \mathbf{0}$ for each i, then a special notation is used to denote $\sum_{i=1}^r V_i$. This notation is

$$V_1 \oplus \cdots \oplus V_r$$

and it is called a direct sum of subspaces.

Lemma D.9.3 *If $V = V_1 \oplus \cdots \oplus V_r$ and if $\beta_i = \{\mathbf{v}_1^i, \cdots, \mathbf{v}_{m_i}^i\}$ is a basis for V_i, then a basis for V is $\{\beta_1, \cdots, \beta_r\}$.*

Proof: Suppose $\sum_{i=1}^r \sum_{j=1}^{m_i} c_{ij} \mathbf{v}_j^i = 0$. then since it is a direct sum, it follows for each i,

$$\sum_{j=1}^{m_i} c_{ij} \mathbf{v}_j^i = 0$$

and now since $\{\mathbf{v}_1^i, \cdots, \mathbf{v}_{m_i}^i\}$ is a basis, each $c_{ij} = 0$. ∎

Here is a useful lemma.

Lemma D.9.4 *Let L_i a square matrix for $i = 1, \cdots, p$ each of the same size, and suppose for $i \neq j, L_i L_j = L_j L_i$ and also L_i is one to one on $\ker(L_j)$ whenever $i \neq j$. Then*

$$\ker\left(\prod_{i=1}^p L_i\right) = \ker(L_1) \oplus + \cdots + \oplus \ker(L_p)$$

Here $\prod_{i=1}^p L_i$ is the product of all the matrices. A symbol like $\prod_{j \neq i} L_j$ is the product of all of them except for L_i.

Proof: Note that since the matrices commute, $L_j : \ker(L_i) \mapsto \ker(L_i)$. Here is why. If $L_i \mathbf{y} = \mathbf{0}$ so that $\mathbf{y} \in \ker(L_i)$, then

$$L_i L_j \mathbf{y} = L_j L_i \mathbf{y} = L_j \mathbf{0} = \mathbf{0}$$

and so $L_j : \ker(L_i) \mapsto \ker(L_i)$. Next observe that it is obvious that, since the operators commute,

$$\sum_{i=1}^p \ker(L_p) \subseteq \ker\left(\prod_{i=1}^p L_i\right)$$

Suppose

$$\sum_{i=1}^p \mathbf{v}_i = \mathbf{0}, \ \mathbf{v}_i \in \ker(L_i),$$

but some $\mathbf{v}_i \neq \mathbf{0}$. Then do $\prod_{j \neq i} L_j$ to both sides. Since the matrices commute, this results in

$$\prod_{j \neq i} L_j \mathbf{v}_i = \mathbf{0}$$

which contradicts the assumption that these L_j are one to one and the above observation that they map $\ker(L_i)$ to $\ker(L_i)$. Thus if

$$\sum_i \mathbf{v}_i = \mathbf{0}, \; \mathbf{v}_i \in \ker(L_i)$$

then each $\mathbf{v}_i = \mathbf{0}$. It follows that

$$\ker(L_1) \oplus + \cdots + \oplus \ker(L_p) \subseteq \ker\left(\prod_{i=1}^{p} L_i\right) \tag{*}$$

From Sylvester's theorem and the observation about direct sums in Lemma D.9.3,

$$\sum_{i=1}^{p} \dim(\ker(L_i)) = \dim(\ker(L_1) \oplus + \cdots + \oplus \ker(L_p))$$

$$\leq \dim\left(\ker\left(\prod_{i=1}^{p} L_i\right)\right) \leq \sum_{i=1}^{p} \dim(\ker(L_i))$$

which implies all these are equal. Now in general, if W is a subspace of V, a finite dimensional vector space and the two have the same dimension, then $W = V$. This is because W has a basis and if $\mathbf{v}$ is not in the span of this basis, then $\mathbf{v}$ adjoined to the basis of W would be a linearly independent set so the dimension of V would then be strictly larger than the dimension of W. It follows from * that

$$\ker(L_1) \oplus + \cdots + \oplus \ker(L_p) = \ker\left(\prod_{i=1}^{p} L_i\right) \quad \blacksquare$$

Here is a situation in which the above holds. $\ker(A - \lambda_i I)^r$ is sometimes called a generalized eigenspace in case λ_i is an eigenvalue.

Theorem D.9.5 *Let A be an $n \times n$ matrix and suppose $\{\lambda_1, \cdots, \lambda_k\}$ are distinct scalars. Define for r_i a positive integer,*

$$V_i = \ker(A - \lambda_i I)^{r_i} \tag{4.37}$$

Then

$$\ker\left(\prod_{i=1}^{p}(A - \lambda_i I)^{r_i}\right) = V_i \oplus \cdots \oplus V_p. \tag{4.38}$$

Proof: It is obvious the linear transformations $(A - \lambda_i I)^{r_i}$ commute. Now here is a claim.

Claim : Let $\mu \neq \lambda_i$. Then $(A - \mu I)^m : V_i \mapsto V_i$ and is one to one and onto for every m a positive integer.

Proof: It is clear $(A - \mu I)^m$ maps V_i to V_i because $\mathbf{v} \in V_i$ is equivalent to $(A - \lambda_i I)^{r_i} \mathbf{v} = \mathbf{0}$. Consequently,

$$(A - \lambda_i I)^{r_i} (A - \mu I)^m \mathbf{v} = (A - \mu I)^m (A - \lambda_i I)^{r_i} \mathbf{v} = (A - \mu I)^m \mathbf{0} = \mathbf{0}$$

which shows that $(A - \mu I)^m \mathbf{v} \in V_i$.

It remains to verify $(A - \mu I)^m$ is one to one. This will be done by showing that $(A - \mu I)$ is one to one. Let $\mathbf{w} \in V_i$ and suppose $(A - \mu I) \mathbf{w} = 0$ so that $A\mathbf{w} = \mu \mathbf{w}$. Then for $m \equiv r_i$, $(A - \lambda_i I)^m \mathbf{w} = \mathbf{0}$ and so by the binomial theorem,

$$(\mu - \lambda_i)^m \mathbf{w} = \sum_{l=0}^{m} \binom{m}{l} (-\lambda_i)^{m-l} \mu^l \mathbf{w} = \sum_{l=0}^{m} \binom{m}{l} (-\lambda_i)^{m-l} A^l \mathbf{w}$$

$$= (A - \lambda_i I)^m \mathbf{w} = (A - \lambda_i I)^{r_i} \mathbf{w} = \mathbf{0}.$$

Therefore, since $\mu \neq \lambda_i$, it follows $\mathbf{w} = 0$ and this verifies $(A - \mu I)$ is one to one. Thus $(A - \mu I)^m$ is also one to one on V_i. Letting $\{\mathbf{u}_1^i, \cdots, \mathbf{u}_{r_k}^i\}$ be a basis for V_i, it follows

$$\{(A - \mu I)^m \mathbf{u}_1^i, \cdots, (A - \mu I)^m \mathbf{u}_{r_k}^i\}$$

is also a basis and so $(A - \mu I)^m$ is also onto. The desired result now follows from Lemma D.9.4. ∎

By the Cayley-Hamilton theorem for A a $n \times n$ complex matrix, it satisfies its characteristic polynomial which was a polynomial of degree n, denoted as $q(\lambda)$. Let the minimal polynomial $p(\lambda)$ be the monic polynomial (leading coefficient is 1) of smallest degree such that $p(A) = 0$. In all of this, A^0 is defined as I. Thus the minimal polynomial is of the form

$$p(\lambda) = \lambda^m + a_{m-1}\lambda^{m-1} + \cdots + a_1\lambda + a_0, m \leq n$$

and

$$p(A) \equiv A^m + a_{m-1}A^{m-1} + \cdots + a_1 A + a_0 I = 0$$

where the 0 is the $n \times n$ 0 matrix.

Lemma D.9.6 *The minimal polynomial divides the characteristic polynomial* $q(\lambda) = \det(\lambda I - A)$ *and is unique.*

Proof: By division of polynomials, Lemma 4.1.2,

$$q(\lambda) = p(\lambda) k(\lambda) + r(\lambda)$$

where either $r(\lambda) = 0$ or else it has smaller degree than $p(\lambda)$. However, in the second case, you would have $0 = \overset{=0}{p(A)}k(A) + r(A) = r(A)$ which is impossible because $p(\lambda)$ has the smallest degree with the property that $p(A) = 0$. As to this polynomial being unique, suppose $\hat{p}(\lambda)$ is another one. Then the same argument just used shows that $p(\lambda)$ divides $\hat{p}(\lambda)$ and $\hat{p}(\lambda)$ divides $p(\lambda)$. Since both are monic polynomials of the same degree, they must be the same polynomial. ∎

From now on, assume the field of scalars is $\mathbb{C}$ so that all polynomials have all roots.

Corollary D.9.7 *Let the minimal polynomial be $\prod_{i=1}^{m} (\lambda - \lambda_i)^{r_i}$. Then λ_i has the property that there exists a nonzero $\mathbf{v}$ such that $(A - \lambda_i I) \mathbf{v} = 0$. Furthermore the λ_i are the only scalars with this property. Thus these are **the eigenvalues of** A.*

Proof: For the first claim, just factor out $(A - \lambda_i I)$

$$(A - \lambda_i I) \prod_{j \neq i}^{m} (A - \lambda_j I)^{r_j} (A - \lambda_i I)^{r_i - 1} \equiv (A - \lambda_i I) B$$

If $B\mathbf{v} = \mathbf{0}$ for all $\mathbf{v}$, then this would contradict that $p(\lambda)$ is the minimal polynomial. Hence there exists $\mathbf{v}$ such that $B\mathbf{v} \neq \mathbf{0}$ and then $(A - \lambda_i I)(B\mathbf{v}) = \mathbf{0}$ showing that $B\mathbf{v}$ is an eigenvector. Thus each λ_i is an eigenvalue for A. It remains to show that these are the **only** eigenvalues for A.

Next suppose

$$(A - \mu I) \mathbf{v} = \mathbf{0}$$

for some μ and $\mathbf{v} \neq \mathbf{0}$. Then

$$
\begin{aligned}
\mathbf{0} &= \prod_{k=1}^{m} (A - \lambda_k I) \mathbf{v} = \prod_{k=1}^{m-1} (A - \lambda_k I) \left(\overbrace{A\mathbf{v}}^{=\mu\mathbf{v}} - \lambda_m \mathbf{v} \right) \\
&= (\mu - \lambda_m) \left(\prod_{k=1}^{m-1} (A - \lambda_k I) \right) \mathbf{v} \\
&= (\mu - \lambda_m) \left(\prod_{k=1}^{m-2} (A - \lambda_k I) \right) (A\mathbf{v} - \lambda_{m-1}\mathbf{v}) \\
&= (\mu - \lambda_m)(\mu - \lambda_{m-1}) \left(\prod_{k=1}^{m-2} (A - \lambda_k I) \right) \mathbf{v}
\end{aligned}
$$

continuing this way yields $= \prod_{k=1}^{m} (\mu - \lambda_k) \mathbf{v}$, a contradiction unless $\mu = \lambda_k$ for some k. ∎

Theorem D.9.8 *Letting the minimal polynomial of A a complex $n \times n$ matrix be*

$$p(\lambda) = \prod_{i=1}^{m} (\lambda - \lambda_i)^{r_i}$$

and

$$V_i \equiv \ker (A - \lambda_i I)^{r_i}$$

it follows that

$$\mathbb{C}^n = V_1 \oplus \cdots \oplus V_p$$

That is, the vector space equals the direct sum of its generalized eigenspaces.

Proof: Since $\mathbb{C}^n = \ker \left(\prod_{k=1}^{p} (A - \lambda_k I)^{r_k} \right)$, the conclusion follows from Theorem D.9.5. ∎

D.10   Block Diagonal Matrices

Now with this, it is time to consider the notion of block diagonal matrices.

Definition D.10.1 *Let A and B be two $n \times n$ matrices. Then A is similar to B, written as $A \sim B$ when there exists an invertible matrix S such that $A = S^{-1}BS$.*

Theorem D.10.2 *Let A be an $n \times n$ matrix. Letting $\lambda_1, \lambda_2, \cdots, \lambda_r$ be the distinct eigenvalues of A, arranged in some order, there exist square matrices $P_1, \cdots, P_r$ such that A is similar to the block diagonal matrix*

$$
P = \begin{pmatrix} P_1 & \cdots & 0 \\ \vdots & \ddots & \vdots \\ 0 & \cdots & P_r \end{pmatrix}
$$

in which P_k has the single eigenvalue λ_k. Denoting by r_k the size of P_k it follows that r_k equals the dimension of the generalized eigenspace for λ_k which is $V_{\lambda_i} =$

$$
\ker (A - \lambda_i I)^{r_i} , \text{ for minimal polynomial } p(\lambda) = \prod_{i=1}^{m} (\lambda - \lambda_i)^{r_i}
$$

Furthermore, if S is the matrix satisfying

$$
S^{-1}AS = P,
$$

then S is of the form

$$
\begin{pmatrix} B_1 & \cdots & B_r \end{pmatrix}
$$

where $B_k = \begin{pmatrix} \mathbf{u}_1^k & \cdots & \mathbf{u}_{r_k}^k \end{pmatrix}$ in which the columns, $\{\mathbf{u}_1^k, \cdots, \mathbf{u}_{r_k}^k\} = D_k$ constitute a basis for V_{λ_k}.

Proof: By Lemma D.9.3,

$$
\mathbb{C}^n = V_{\lambda_1} \oplus \cdots \oplus V_{\lambda_k}
$$

and a basis for $\mathbb{C}^n$ is $\{D_1, \cdots, D_r\}$ where D_k is a basis for V_{λ_k}, $\ker (A - \lambda_k I)^{r_k}$.
Let

$$
S = \begin{pmatrix} B_1 & \cdots & B_r \end{pmatrix}
$$

where the B_i are the matrices described in the statement of the theorem. Then S^{-1} must be of the form

$$
S^{-1} = \begin{pmatrix} C_1 & \cdots & C_r \end{pmatrix}^T
$$

where $C_i B_i = I_{r_i \times r_i}$. Also, if $i \neq j$, then $C_i A B_j = 0$ the last claim holding because $A : V_{\lambda_j} \mapsto V_{\lambda_j}$ so the columns of $A B_j$ are linear combinations of the columns of B_j

and each of these columns is orthogonal to the rows of C_i since $C_i B_j = 0$ if $i \neq j$. Therefore,

$$S^{-1}AS = \begin{pmatrix} C_1 \\ \vdots \\ C_r \end{pmatrix} A \begin{pmatrix} B_1 & \cdots & B_r \end{pmatrix} = \begin{pmatrix} C_1 \\ \vdots \\ C_r \end{pmatrix} \begin{pmatrix} AB_1 & \cdots & AB_r \end{pmatrix}$$

$$= \begin{pmatrix} C_1 AB_1 & & 0 \\ & \ddots & \\ 0 & & C_r AB_r \end{pmatrix} \equiv \begin{pmatrix} P_1 & & 0 \\ & \ddots & \\ 0 & & P_r \end{pmatrix}$$

and $C_{r_k} AB_{r_k} \equiv P_r$ is an $r_k \times r_k$ matrix.

What about the eigenvalues of P_k? The only eigenvalue of A restricted to V_{λ_k} is λ_k because if $A\mathbf{x} = \mu\mathbf{x}$ for some $\mathbf{x} \in V_{\lambda_k}$ and $\mu \neq \lambda_k$, then

$$(A - \lambda_k I)^{r_k} \mathbf{x} = (A - \mu I + (\mu - \lambda_k) I)^{r_k} \mathbf{x}$$

$$= \sum_{j=0}^{r_k} \binom{r_k}{j} (\mu - \lambda_k)^{r_k - j} (A - \mu I)^j \mathbf{x} = (\mu - \lambda_k)^{r_k} \mathbf{x} \neq \mathbf{0}$$

contrary to the assumption that $\mathbf{x} \in V_{\lambda_k}$. Now if μ is an eigenvalue for P_k, $P_k \mathbf{z} = \mu\mathbf{z}, \mathbf{z} \neq \mathbf{0}$, then

$$AS \begin{pmatrix} \mathbf{0} \\ \mathbf{z} \\ \mathbf{0} \end{pmatrix} = S \begin{pmatrix} \mathbf{0} \\ P_k \mathbf{z} \\ \mathbf{0} \end{pmatrix} = \mu S \begin{pmatrix} \mathbf{0} \\ \mathbf{z} \\ \mathbf{0} \end{pmatrix}$$

From the definition of S, the vector on the right in the above equation is nonzero, in V_{λ_k} and so from what was just observed, $\mu = \lambda_k$. ∎

Appendix E

Theory of Functions of Many Variables

This section contains the proofs of theorems which are needed to understand the more theoretical sections of this book. When one considers questions about existence and uniqueness and geometric theory such as stability of equilibrium points, you end up having to consider some hard analysis, not the sort of thing encountered in a watered down calculus course. The symbol $|\mathbf{x}|$ will refer to the norm of $\mathbf{x} \in \mathbb{R}^p$, defined as

$$\sqrt{\sum_{j=1}^{p} |x_j|^2} = |\mathbf{x}|$$

but we could also use the norm $\|\cdot\| = \|\cdot\|_{\infty}$ defined as

$$\|\mathbf{x}\|_{\infty} = \max\left\{|x_i|, i = 1, \cdots, p\right\}$$

From linear algebra or calculus, for $\|\cdot\|$ a norm,

$$\|\mathbf{x}\| \geq 0 \text{ and } \|\mathbf{x}\| = 0 \text{ if and only if } \mathbf{x} = \mathbf{0}$$

$$\text{For } \alpha \text{ a scalar, } \|\alpha\mathbf{x}\| = |\alpha| \|\mathbf{x}\|$$

$$\|\mathbf{x} + \mathbf{y}\| \leq \|\mathbf{x}\| + \|\mathbf{y}\|$$

You should verify that the above axioms for a norm hold for each of the two norms given above. Note that the above two norms are equivalent in the sense that

$$\|\mathbf{x}\| \leq |\mathbf{x}| \leq \sqrt{p}\|\mathbf{x}\| \qquad (*)$$

Thus in what follows, it will not matter which norm is being used.

E.1 Closed and Open Sets

The definition of open and closed sets is next.

Definition E.1.1 *Let U be a set of points in a $\mathbb{R}^n$. A point, $\mathbf{p} \in U$ is said to be an interior point of U if whenever $||\mathbf{x} - \mathbf{p}||$ is sufficiently small, it follows $\mathbf{x} \in U$ also. The set of points $\mathbf{x}$ which are closer to $\mathbf{p}$ than δ is denoted by*

$$B(\mathbf{p}, \delta) \equiv \{\mathbf{x} \in V : ||\mathbf{x} - \mathbf{p}|| < \delta\}.$$

This symbol, $B(\mathbf{p}, \delta)$ is called an open ball of radius δ. Thus a point $\mathbf{p}$ is an interior point of U if there exists $\delta > 0$ such that $\mathbf{p} \in B(\mathbf{p}, \delta) \subseteq U$. An open set is one for which every point of the set is an interior point. Closed sets are those which are complements of open sets. Thus H is closed means H^C is open.

Note that the definition of open and closed does not depend on the choice of the norm because of $*$ which implies that $B(\mathbf{x}, r) \subseteq B_\infty(\mathbf{x}, r) \subseteq B(\mathbf{x}, \sqrt{p}r)$. It turns out that all norms are equivalent but we will not need this in this book. Thus when we mention open set or closed set, it makes no difference whether this is taken with respect to $|\cdot|$ or $||\cdot||$.

Theorem E.1.2 *The intersection of any finite collection of open sets is open. The union of any collection of open sets is open. The intersection of any collection of closed sets is closed and the union of any finite collection of closed sets is closed.*

Proof: To see that any union of open sets is open, note that every point $\mathbf{p}$ of the union is in at least one of the open sets U. Therefore, it is an interior point of U and hence an interior point of the entire union.

Now let $\{U_1, \cdots, U_m\}$ be some open sets and suppose $\mathbf{p} \in \cap_{k=1}^m U_k$. Then there exists $r_k > 0$ such that $B(\mathbf{p}, r_k) \subseteq U_k$. Let $0 < r \leq \min(r_1, r_2, \cdots, r_m)$. Then $B(\mathbf{p}, r) \subseteq \cap_{k=1}^m U_k$ and so the finite intersection is open. Note that if the finite intersection is empty, there is nothing to prove because it is certainly true in this case that every point in the intersection is an interior point because there aren't any such points.

Suppose $\{H_1, \cdots, H_m\}$ is a finite set of closed sets. Then $\cup_{k=1}^m H_k$ is closed if its complement is open. However, from DeMorgan's laws, Problem 1 on Page 14,

$$\left(\cup_{k=1}^m H_k\right)^C = \cap_{k=1}^m H_k^C,$$

a finite intersection of open sets which is open by what was just shown.

Next let $\mathcal{C}$ be some set consisting of closed sets. Then

$$(\cap \mathcal{C})^C = \cup \{H^C : H \in \mathcal{C}\},$$

a union of open sets which is therefore open by the first part of the proof. Thus $\cap \mathcal{C}$ is closed. This proves the theorem. ∎

Next there is the concept of a limit point which gives another way of characterizing closed sets.

Definition E.1.3 *Let A be any nonempty set and let $\mathbf{x}$ be a point. Then $\mathbf{x}$ is said to be a limit point of A if for every $r > 0, B(\mathbf{x}, r)$ contains a point of A which is not equal to $\mathbf{x}$.*

Example E.1.4 *Consider $A = B(\mathbf{x}, \delta)$, an open ball in a $\mathbb{R}^n$. Then every point of $B(\mathbf{x}, \delta)$ is a limit point. (There are more general situations than $\mathbb{R}^n$ in which this assertion is false but these are of no concern in this book.)*

If $\mathbf{z} \in B(\mathbf{x}, \delta)$, consider $\mathbf{z} + \frac{1}{k}(\mathbf{x} - \mathbf{z}) \equiv \mathbf{w}_k$ for $k \in \mathbb{N}$. Then

$$\|\mathbf{w}_k - \mathbf{x}\| = \left\|\mathbf{z} + \frac{1}{k}(\mathbf{x} - \mathbf{z}) - \mathbf{x}\right\|$$

$$= \left\|\left(1 - \frac{1}{k}\right)\mathbf{z} - \left(1 - \frac{1}{k}\right)\mathbf{x}\right\| = \frac{k-1}{k}\|\mathbf{z} - \mathbf{x}\| < \delta$$

and also

$$\|\mathbf{w}_k - \mathbf{z}\| \leq \frac{1}{k}\|\mathbf{x} - \mathbf{z}\| < \delta/k$$

so $\mathbf{w}_k \to \mathbf{z}$. Furthermore, the $\mathbf{w}_k$ are distinct. Thus $\mathbf{z}$ is a limit point of A as claimed. This is because every ball containing $\mathbf{z}$ contains infinitely many of the $\mathbf{w}_k$ and since they are all distinct, they can't all be equal to $\mathbf{z}$.

A mapping $\mathbf{f} : \{k, k+1, k+2, \cdots\} \to \mathbb{R}^p$ is called a sequence. We usually write it in the form $\{\mathbf{a}_j\}$ where it is understood that $\mathbf{a}_j \equiv \mathbf{f}(j)$. In the same way as for sequences of real numbers, one can define what it means for convergence to take place.

Definition E.1.5 *A sequence, $\{\mathbf{a}_k\}$ is said to **converge** to $\mathbf{a}$ if for every $\varepsilon > 0$ there exists n_ε such that if $n > n_\varepsilon$, then $|\mathbf{a} - \mathbf{a}_n| < \varepsilon$. The usual notation for this is $\lim_{n \to \infty} \mathbf{a}_n = \mathbf{a}$ although it is often written as $\mathbf{a}_n \to \mathbf{a}$.*

One can also define a subsequence in the same way as in the case of real valued sequences.

Definition E.1.6 *$\{\mathbf{a}_{n_k}\}$ is a **subsequence** of $\{\mathbf{a}_n\}$ if $n_1 < n_2 < \cdots$.*

Nothing changes if you use $\|\cdot\|$ instead of $|\cdot|$ thanks to the equivalence of these norms. The following theorem says the limit, if it exists, is unique. Thus the limit is well defined.

Theorem E.1.7 *If a sequence, $\{\mathbf{a}_n\}$ converges to $\mathbf{a}$ and to $\mathbf{b}$ then $\mathbf{a} = \mathbf{b}$.*

Proof: There exists n_ε such that if $n > n_\varepsilon$ then $|\mathbf{a}_n - \mathbf{a}| < \frac{\varepsilon}{2}$ and if $n > n_\varepsilon$, then $|\mathbf{a}_n - \mathbf{b}| < \frac{\varepsilon}{2}$. Then pick such an n.

$$|\mathbf{a} - \mathbf{b}| < |\mathbf{a} - \mathbf{a}_n| + |\mathbf{a}_n - \mathbf{b}| < \frac{\varepsilon}{2} + \frac{\varepsilon}{2} = \varepsilon.$$

Since ε is arbitrary, this proves the theorem. ∎

Then the following is about limit points.

Theorem E.1.8 *Let A be a nonempty set in $\mathbb{R}^p$. A point $\mathbf{a}$ is a limit point of A if and only if there exists a sequence of distinct points of A, $\{\mathbf{a}_k\}$ which converges to $\mathbf{a}$. Also a nonempty set A is closed if and only if it contains all its limit points.*

Proof: Suppose first $\mathbf{a}$ is a limit point of A. There exists $\mathbf{a}_1 \in B(\mathbf{a}, 1) \cap A$ such that $\mathbf{a}_1 \neq \mathbf{a}$. Now supposing distinct points, $\mathbf{a}_1, \cdots, \mathbf{a}_n$ have been chosen such that none are equal to $\mathbf{a}$ and for each $k \leq p$, $\mathbf{a}_k \in B(\mathbf{a}, 1/k)$, let

$$0 < r_{n+1} < \min\left\{\frac{1}{n+1}, \|\mathbf{a} - \mathbf{a}_1\|, \cdots, \|\mathbf{a} - \mathbf{a}_n\|\right\}.$$

Then there exists $\mathbf{a}_{n+1} \in B(\mathbf{a}, r_{n+1}) \cap A$ with $\mathbf{a}_{n+1} \neq \mathbf{a}$. Because of the definition of r_{n+1}, $\mathbf{a}_{n+1}$ is not equal to any of the other $\mathbf{a}_k$ for $k < n+1$. Also since $\|\mathbf{a} - \mathbf{a}_m\| < 1/m$, it follows $\lim_{m \to \infty} \mathbf{a}_m = \mathbf{a}$. Conversely, if there exists a sequence of distinct points of A converging to $\mathbf{a}$, then $B(\mathbf{a}, r)$ contains all $\mathbf{a}_n$ for n large enough. Thus $B(\mathbf{a}, r)$ contains infinitely many points of A since all are distinct. Thus at least one of them is not equal to $\mathbf{a}$. This establishes the first part of the theorem.

Now consider the second claim. If A is closed then it is the complement of an open set. Since A^C is open, it follows that if $\mathbf{a} \in A^C$, then there exists $\delta > 0$ such that $B(\mathbf{a}, \delta) \subseteq A^C$ and so no point of A^C can be a limit point of A. In other words, every limit point of A must be in A. Conversely, suppose A contains all its limit points. Then A^C does not contain any limit points of A. It also contains no points of A. Therefore, if $\mathbf{a} \in A^C$, since it is not a limit point of A, there exists $\delta > 0$ such that $B(\mathbf{a}, \delta)$ contains no points of A different than $\mathbf{a}$. However, $\mathbf{a}$ itself is not in A because $\mathbf{a} \in A^C$. Therefore, $B(\mathbf{a}, \delta)$ is entirely contained in A^C. Since $\mathbf{a} \in A^C$ was arbitrary, this shows every point of A^C is an interior point and so A^C is open. ∎

E.2   Compactness

The following is the definition of compactness. It is a very useful notion which can be used to prove existence theorems.

Definition E.2.1 *A set, $K \subseteq \mathbb{R}^p$, is said to be sequentially compact if whenever $\{\mathbf{a}_n\} \subseteq K$ is a sequence, there exists a subsequence, $\{\mathbf{a}_{n_k}\}$ such that this subsequence converges to a point of K. Let $\mathcal{C}$ be a set of open sets in $\mathbb{R}^p$. It is called an open cover for K if $\cup\mathcal{C} \supseteq K$. Then K is called compact if every such open cover has the property that there are finitely many sets of $\mathcal{C}$, $U_1, \cdots, U_m$ such that $K \subseteq \cup_{i=1}^m U_i$.*

Lemma E.2.2 *If K is a nonempty compact set, then it must be closed.*

Proof: If it is not closed, then it has a limit point $\mathbf{k}$ not in the set. Thus, K is not a finite set. For each $\mathbf{x} \in K$, there is $B(\mathbf{x}, r_{\mathbf{x}})$ such that $B(\mathbf{x}, r_{\mathbf{x}}) \cap B(\mathbf{k}, r_{\mathbf{x}}) = \emptyset$. Since K is compact, finitely many of these balls $B(\mathbf{x}_i, r_{\mathbf{x}_i})$ $i = 1, \cdots, m$ must cover K. But then one could consider $r = \min\{r_{\mathbf{x}_i}, i = 1 \cdots m\} > 0$ and $B(\mathbf{k}, r)$ would contain points of K which are not covered by $\{B(\mathbf{x}_i, r_{\mathbf{x}_i})\}$ because it is a limit point. This is a contradiction. ∎

Recall the nested interval lemma, B.4.2.

Lemma E.2.3 *Let $I_k = [a^k, b^k]$ and suppose that for all $k = 1, 2, \cdots$,*

$$I_k \supseteq I_{k+1}.$$

Then there exists a point, $c \in \mathbb{R}$ which is an element of every I_k. If

$$\lim_{k \to \infty} b^k - a^k = 0$$

then there is exactly one point in all of these intervals.

This generalizes right away to the following version in $\mathbb{R}^p$.

Definition E.2.4 *The **diameter** of a set S, is defined as*

$$\text{diam}(S) \equiv \sup\{|\mathbf{x} - \mathbf{y}| : \mathbf{x}, \mathbf{y} \in S\}.$$

Thus $\text{diam}(S)$ is just a careful description of what you would think of as the diameter. It measures how stretched out the set is.

Here is a multidimensional version of the nested interval lemma.

Lemma E.2.5 *Let $I_k = \prod_{i=1}^{p} [a_i^k, b_i^k] \equiv \{\mathbf{x} \in \mathbb{R}^p : x_i \in [a_i^k, b_i^k]\}$ and suppose that for all $k = 1, 2, \cdots,$*

$$I_k \supseteq I_{k+1}.$$

Then there exists a point $\mathbf{c} \in \mathbb{R}^p$ which is an element of every I_k. If

$$\lim_{k \to \infty} \text{diam}(I_k) = 0,$$

then the point $\mathbf{c}$ is unique.

Proof: For each $i = 1, \cdots, p$, $[a_i^k, b_i^k] \supseteq [a_i^{k+1}, b_i^{k+1}]$ and so, by the nested interval lemma, there exists a point $c_i \in [a_i^k, b_i^k]$ for all k. Then letting $\mathbf{c} \equiv (c_1, \cdots, c_p)$ it follows $\mathbf{c} \in I_k$ for all k. If the condition on the diameters holds, then the lengths of the intervals $\lim_{k \to \infty} [a_i^k, b_i^k] = 0$ and so by the same lemma, each c_i is unique. Hence $\mathbf{c}$ is unique. ∎

Theorem E.2.6 *Let*

$$I = \prod_{k=1}^{p} [a_k, b_k] \subseteq \mathbb{R}^p$$

Then I is sequentially compact and compact.

Proof: First consider compactness. Let I_0 be the given set and suppose it is not compact. Then there is a set of open sets $\mathcal{C}$ which admits no finite sub-cover of I_0. Now we describe a nested sequence of such products recursively as follows. I_n cannot be covered with finitely many sets of $\mathcal{C}$ (like I_0) and if $I_n = \prod_{k=1}^{p} [a_k^n, b_k^n]$, let c_k^n be $\frac{a_k^n + b_k^n}{2}$ and consider all products which are of the form $\prod_{k=1}^{p} [\alpha_k, \beta_k]$ where $\alpha_k \leq \beta_k$ and one of α_k, β_k is c_k^n. Thus there are 2^p of these products $\{J_k\}_{k=1}^{2^n}$ and $\text{diam}(J_k) = 2^{-1} \text{diam}(I_n)$. If each can be covered by finitely many sets of $\mathcal{C}$ then so can I_n contrary to assumption. Hence one of these cannot be covered with finitely many sets of $\mathcal{C}$. Call it I_{n+1}. Then iterating the relation on the diameter, it follows that $\text{diam}(I_{n+1}) \leq 2^{-n} \text{diam}(I_0)$ and so the diameters of these nested products converges to 0. By the above nested interval lemma, Lemma E.2.5, it follows that there exists a unique $\mathbf{c}$ in $\cap_n I_n$. Then $\mathbf{c}$ is contained in some $U \in \mathcal{C}$ and so, since the diameters of the I_n converge to 0, for large enough n, it follows that $I_n \subseteq U$ also, contrary to assumption. Thus I is compact.

Next let $\{\mathbf{x}_n\}$ be a sequence in I. Using the same nested sequence bisection method described above, we can choose a nested sequence $\{I_k\}$ such that nested $\lim_{k \to \infty} \text{diam}(I_k) = 0$ and I_k contains $\mathbf{x}_k$ for infinitely many indices k. Letting $\mathbf{c} = \cap_{k=1}^\infty I_k$ one can choose an increasing sequence $\{n_k\}$ such that $\mathbf{x}_{n_k} \in I_k$. Then it follows that $\lim_{k \to \infty} \mathbf{x}_{n_k} = \mathbf{c}$. ∎

A useful corollary of this theorem is the following.

Corollary E.2.7 *Let $\{\mathbf{x}_k\}_{k=1}^\infty$ be a bounded sequence. Then it has a convergent subsequence.*

Proof: The given sequence is contained in some I as in Theorem E.2.6 which was shown to be a sequentially compact set. Hence the given sequence has a convergent subsequence. ∎

In $\mathbb{R}^p$, the two versions of compactness are equivalent. In fact, this holds more generally for metric spaces.

Lemma E.2.8 *Let $K \neq \emptyset$ be sequentially compact in $\mathbb{R}^p$ and let $\mathcal{O}$ be an open cover. Then K is compact.*

Proof: First, I claim there is a number $\delta > 0$, called a Lebesgue number such that $B(\mathbf{k}, \delta)$ is contained in some set of $\mathcal{O}$ for any $\mathbf{k} \in K$. If not, there would be a sequence $\{\mathbf{k}_n\}$ of points of K such that $B(\mathbf{k}_n, 1/n)$ is not contained in any single set of of $\mathcal{O}$. By sequential compactness, there is a subsequence, still denoted as $\{\mathbf{k}_n\}$ which converges to $\mathbf{k} \in K$. Now $B(\mathbf{k}, \delta) \subseteq O \in \mathcal{O}$ for some open set O. Consider n large enough that $1/n < \delta/5$ and also $\mathbf{k}_n \in B(\mathbf{k}, \delta/5)$. Then

$$B(\mathbf{k}_n, 1/n) \subseteq B(\mathbf{k}_n, \delta/5) \subseteq B(\mathbf{k}, 2\delta/5) \subseteq B(\mathbf{k}, \delta) \subseteq O$$

contrary to the construction of the $\mathbf{k}_n$. This shows the claim.

Now pick $\mathbf{k}_1 \in K$. If $B(\mathbf{k}_1, \delta)$ covers K, stop. Otherwise pick $\mathbf{k}_2 \in K \setminus B(\mathbf{k}_1, \delta)$. If $B(\mathbf{k}_1, \delta), B(\mathbf{k}_2, \delta)$ covers K, stop. Otherwise pick $\mathbf{k}_3$ not covered. Continue this way obtaining a sequence of points any pair further apart than δ. Therefore, this process must stop since otherwise, there would be no subsequence which could converge to a point of K and K would fail to be sequentially compact. Therefore, there is some n such that $\{B(\mathbf{k}_i, \delta)\}_{i=1}^n$ is an open cover. However, from the choice of $\delta, B(\mathbf{k}_i, \delta) \subseteq O_i \in \mathcal{O}$ and so $\{O_1, \cdots, O_n\}$ is an open cover. ∎

Conversely, we have the following.

Lemma E.2.9 *If $K \neq \emptyset$ is compact set in $\mathbb{R}^p$, then it is sequentially compact.*

Proof: Let $\{\mathbf{k}_n\}$ be a sequence in K and it is desired to show it has a convergent subsequence. Suppose it does not. For $\mathbf{k} \in K$ there must be a ball $B(\mathbf{k}, \delta_k)$ which contains $\mathbf{k}_k$ for only finitely many k since otherwise, one could extract a convergent subsequence by Corollary E.2.7 which would be in K since K is closed (Lemma E.2.2). Since K is compact, there are finitely many of these balls which cover K. But now, there are only finitely many of the indices accounted for, a contradiction. Hence K is sequentially compact. ∎

This proves most of the following theorem. First, a set is bounded if it is contained in $I = \prod_{k=1}^p [a_k, b_k]$ for some choice of intervals $[a_k, b_k]$.

Theorem E.2.10 *A set $K \subseteq \mathbb{R}^p$ is compact if and only if it is sequentially compact if and only if it is closed and bounded.*

Proof: The first equivalence was established in the lemmas. Suppose then that K is compact. Why is it bounded? If not, then there exist $\mathbf{x}_n \in K$ such that $\|\mathbf{x}_n\| \geq n$. This sequence can have no subsequence which converges. Therefore, K cannot be sequentially compact. Hence it is not compact either. Thus K is bounded. It was shown earlier that K is closed, Lemma E.2.2. Thus if K is compact, then it is closed and bounded.

Now suppose K is closed and bounded. Let $I \supseteq K$ where $I = \prod_{k=1}^{p} [a_k, b_k]$. Now let $\{\mathbf{k}_n\} \subseteq K$. By Theorem E.2.6 it has a subsequence converging to $\mathbf{k} \in I$. But K is closed and so $\mathbf{k} \in K$. Thus K is sequentially compact and hence is compact by the first part of this theorem. ■

Since $\mathbb{C}^p$ is just $\mathbb{R}^{2p}$, closed and bounded sets are compact in $\mathbb{C}^p$ also as a special case of the above.

E.2.1 Continuous Functions

You saw continuous functions in beginning calculus. It is no different for vector valued functions.

Definition E.2.11 *A function $\mathbf{f} : D(\mathbf{f}) \subseteq \mathbb{R}^p \to \mathbb{R}^q$ is continuous at $\mathbf{x} \in D(\mathbf{f})$ if for each $\varepsilon > 0$ there exists $\delta > 0$ such that whenever $\mathbf{y} \in D(\mathbf{f})$ and*

$$|\mathbf{y} - \mathbf{x}| < \delta$$

it follows that

$$|\mathbf{f}(\mathbf{x}) - \mathbf{f}(\mathbf{y})| < \varepsilon.$$

$\mathbf{f}$ *is continuous if it is continuous at every point of $D(\mathbf{f})$.*

By equivalence of norms, this is equivalent to the same statement with $\|\cdot\|$ in place of $|\cdot|$.

Proposition E.2.12 *If $\mathbf{f} : D(\mathbf{f}) \to \mathbb{R}^q$ as above, then if it is continuous on $D(\mathbf{f})$, it follows that for any open set V in $\mathbb{R}^q, \mathbf{f}^{-1}(V)$ is open in $D(\mathbf{f})$ which means that if $\mathbf{x} \in \mathbf{f}^{-1}(V)$, then there exists $\delta > 0$ such that $\mathbf{f}(B(\mathbf{x}, \delta) \cap D(\mathbf{f})) \subseteq V$.*

Proof: It comes from the definition. $\mathbf{f}(\mathbf{x}) \in V$ and V is open. Hence, there is $\varepsilon > 0$ such that $B(f(\mathbf{x}), \varepsilon) \subseteq V$. Then the existence of the δ in the above claim is nothing more than the definition of continuity. ■

Note that the converse is also true. It is nothing more than specializing to let V be $B(\mathbf{f}(\mathbf{x}), \varepsilon)$.

Now here are some basic properties of continuous functions.

Theorem E.2.13 *The following assertions are valid.*

1. *The function $a\mathbf{f} + b\mathbf{g}$ is continuous at $\mathbf{x}$ when $\mathbf{f}, \mathbf{g}$ are continuous at $\mathbf{x} \in D(\mathbf{f}) \cap D(\mathbf{g})$ and $a, b \in \mathbb{R}$.*

2. *If and f and g are each real valued functions continuous at $\mathbf{x}$, then fg is continuous at $\mathbf{x}$. If, in addition to this, $g(\mathbf{x}) \neq 0$, then f/g is continuous at $\mathbf{x}$.*

3. *If $\mathbf{f}$ is continuous at $\mathbf{x}$, $\mathbf{f}(\mathbf{x}) \in D(\mathbf{g}) \subseteq \mathbb{R}^p$, and $\mathbf{g}$ is continuous at $\mathbf{f}(\mathbf{x})$, then $\mathbf{g} \circ \mathbf{f}$ is continuous at $\mathbf{x}$.*

4. *If $\mathbf{f} = (f_1, \cdots, f_q) : D(\mathbf{f}) \to \mathbb{R}^q$, then $\mathbf{f}$ is continuous if and only if each f_k is a continuous real valued function.*

5. *The function $f : \mathbb{R}^p \to \mathbb{R}$, given by $f(\mathbf{x}) = |\mathbf{x}|$ is continuous.*

Proof: Begin with (1). Let $\varepsilon > 0$ be given. By assumption, there exist $\delta_1 > 0$ such that whenever $|\mathbf{x} - \mathbf{y}| < \delta_1$, it follows $|\mathbf{f}(\mathbf{x}) - \mathbf{f}(\mathbf{y})| < \frac{\varepsilon}{2(|a|+|b|+1)}$ and there exists $\delta_2 > 0$ such that whenever $|\mathbf{x} - \mathbf{y}| < \delta_2$, it follows that $|\mathbf{g}(\mathbf{x}) - \mathbf{g}(\mathbf{y})| < \frac{\varepsilon}{2(|a|+|b|+1)}$. Then let $0 < \delta \leq \min(\delta_1, \delta_2)$. If $|\mathbf{x} - \mathbf{y}| < \delta$, then everything happens at once. Therefore, using the triangle inequality

$$|a\mathbf{f}(\mathbf{x}) + b\mathbf{f}(\mathbf{x}) - (a\mathbf{g}(\mathbf{y}) + b\mathbf{g}(\mathbf{y}))|$$

$$\leq |a||\mathbf{f}(\mathbf{x}) - \mathbf{f}(\mathbf{y})| + |b||\mathbf{g}(\mathbf{x}) - \mathbf{g}(\mathbf{y})|$$

$$< |a|\left(\frac{\varepsilon}{2(|a|+|b|+1)}\right) + |b|\left(\frac{\varepsilon}{2(|a|+|b|+1)}\right) < \varepsilon.$$

Now begin on (2). There exists $\delta_1 > 0$ such that if $|\mathbf{y} - \mathbf{x}| < \delta_1$, then $|f(\mathbf{x}) - f(\mathbf{y})| < 1$. Therefore, for such $\mathbf{y}$,

$$|f(\mathbf{y})| < 1 + |f(\mathbf{x})|.$$

It follows that for such $\mathbf{y}$,

$$|fg(\mathbf{x}) - fg(\mathbf{y})| \leq |f(\mathbf{x})g(\mathbf{x}) - g(\mathbf{x})f(\mathbf{y})| + |g(\mathbf{x})f(\mathbf{y}) - f(\mathbf{y})g(\mathbf{y})|$$

$$\leq |g(\mathbf{x})||f(\mathbf{x}) - f(\mathbf{y})| + |f(\mathbf{y})||g(\mathbf{x}) - g(\mathbf{y})|$$
$$\leq (1 + |g(\mathbf{x})| + |f(\mathbf{y})|)[|g(\mathbf{x}) - g(\mathbf{y})| + |f(\mathbf{x}) - f(\mathbf{y})|].$$

Now let $\varepsilon > 0$ be given. There exists δ_2 such that if $|\mathbf{x} - \mathbf{y}| < \delta_2$, then

$$|g(\mathbf{x}) - g(\mathbf{y})| < \frac{\varepsilon}{2(1 + |g(\mathbf{x})| + |f(\mathbf{y})|)},$$

and there exists δ_3 such that if $|\mathbf{x} - \mathbf{y}| < \delta_3$, then

$$|f(\mathbf{x}) - f(\mathbf{y})| < \frac{\varepsilon}{2(1 + |g(\mathbf{x})| + |f(\mathbf{y})|)}$$

Now let $0 < \delta \leq \min(\delta_1, \delta_2, \delta_3)$. Then if $|\mathbf{x} - \mathbf{y}| < \delta$, all the above hold at once and

$$|fg(\mathbf{x}) - fg(\mathbf{y})| \leq$$

$$(1 + |g(\mathbf{x})| + |f(\mathbf{y})|) [|g(\mathbf{x}) - g(\mathbf{y})| + |f(\mathbf{x}) - f(\mathbf{y})|]$$

$$< (1 + |g(\mathbf{x})| + |f(\mathbf{y})|) \left(\frac{\varepsilon}{2(1 + |g(\mathbf{x})| + |f(\mathbf{y})|)} + \frac{\varepsilon}{2(1 + |g(\mathbf{x})| + |f(\mathbf{y})|)} \right) = \varepsilon.$$

This proves the first part of (2). To obtain the second part, let δ_1 be as described above and let $\delta_0 > 0$ be such that for $|\mathbf{x} - \mathbf{y}| < \delta_0$,

$$|g(\mathbf{x}) - g(\mathbf{y})| < |g(\mathbf{x})| / 2$$

and so by the triangle inequality,

$$- |g(\mathbf{x})| / 2 \leq |g(\mathbf{y})| - |g(\mathbf{x})| \leq |g(\mathbf{x})| / 2$$

which implies $|g(\mathbf{y})| \geq |g(\mathbf{x})| / 2$, and $|g(\mathbf{y})| < 3 |g(\mathbf{x})| / 2$.

Then if $|\mathbf{x} - \mathbf{y}| < \min(\delta_0, \delta_1)$,

$$\left| \frac{f(\mathbf{x})}{g(\mathbf{x})} - \frac{f(\mathbf{y})}{g(\mathbf{y})} \right| = \left| \frac{f(\mathbf{x}) g(\mathbf{y}) - f(\mathbf{y}) g(\mathbf{x})}{g(\mathbf{x}) g(\mathbf{y})} \right|$$

$$\leq \frac{|f(\mathbf{x}) g(\mathbf{y}) - f(\mathbf{y}) g(\mathbf{x})|}{\left(\frac{|g(\mathbf{x})|^2}{2} \right)}$$

$$= \frac{2 |f(\mathbf{x}) g(\mathbf{y}) - f(\mathbf{y}) g(\mathbf{x})|}{|g(\mathbf{x})|^2}$$

$$\leq \frac{2}{|g(\mathbf{x})|^2} [|f(\mathbf{x}) g(\mathbf{y}) - f(\mathbf{y}) g(\mathbf{y}) + f(\mathbf{y}) g(\mathbf{y}) - f(\mathbf{y}) g(\mathbf{x})|]$$

$$\leq \frac{2}{|g(\mathbf{x})|^2} [|g(\mathbf{y})| |f(\mathbf{x}) - f(\mathbf{y})| + |f(\mathbf{y})| |g(\mathbf{y}) - g(\mathbf{x})|]$$

$$\leq \frac{2}{|g(\mathbf{x})|^2} \left[\frac{3}{2} |\mathbf{g}(\mathbf{x})| |f(\mathbf{x}) - f(\mathbf{y})| + (1 + |f(\mathbf{x})|) |g(\mathbf{y}) - g(\mathbf{x})| \right]$$

$$\leq \frac{2}{|g(\mathbf{x})|^2} (1 + 2 |f(\mathbf{x})| + 2 |g(\mathbf{x})|) [|f(\mathbf{x}) - f(\mathbf{y})| + |g(\mathbf{y}) - g(\mathbf{x})|]$$

$$\equiv M [|f(\mathbf{x}) - f(\mathbf{y})| + |g(\mathbf{y}) - g(\mathbf{x})|]$$

where

$$M \equiv \frac{2}{|g(\mathbf{x})|^2} (1 + 2 |f(\mathbf{x})| + 2 |g(\mathbf{x})|)$$

Now let δ_2 be such that if $|\mathbf{x} - \mathbf{y}| < \delta_2$, then

$$|f(\mathbf{x}) - f(\mathbf{y})| < \frac{\varepsilon}{2} M^{-1}$$

and let δ_3 be such that if $|\mathbf{x} - \mathbf{y}| < \delta_3$, then

$$|g(\mathbf{y}) - g(\mathbf{x})| < \frac{\varepsilon}{2} M^{-1}.$$

Then if $0 < \delta \leq \min(\delta_0, \delta_1, \delta_2, \delta_3)$, and $|\mathbf{x} - \mathbf{y}| < \delta$, everything holds and

$$\left| \frac{f(\mathbf{x})}{g(\mathbf{x})} - \frac{f(\mathbf{y})}{g(\mathbf{y})} \right| \leq M [|f(\mathbf{x}) - f(\mathbf{y})| + |g(\mathbf{y}) - g(\mathbf{x})|]$$

$$< M \left[\frac{\varepsilon}{2} M^{-1} + \frac{\varepsilon}{2} M^{-1} \right] = \varepsilon.$$

This completes the proof of the second part of (2). Note that in these proofs no effort is made to find some sort of "best" δ. The problem is one which has a yes or a no answer. Either it is or it is not continuous.

Now begin on (3). If $\mathbf{f}$ is continuous at $\mathbf{x}$, $\mathbf{f}(\mathbf{x}) \in D(\mathbf{g}) \subseteq \mathbb{R}^p$, and $\mathbf{g}$ is continuous at $\mathbf{f}(\mathbf{x})$, then $\mathbf{g} \circ \mathbf{f}$ is continuous at $\mathbf{x}$. Let $\varepsilon > 0$ be given. Then there exists $\eta > 0$ such that if $|\mathbf{y} - \mathbf{f}(\mathbf{x})| < \eta$ and $\mathbf{y} \in D(\mathbf{g})$, it follows that $|\mathbf{g}(\mathbf{y}) - \mathbf{g}(\mathbf{f}(\mathbf{x}))| < \varepsilon$. It follows from continuity of $\mathbf{f}$ at $\mathbf{x}$ that there exists $\delta > 0$ such that if $|\mathbf{x} - \mathbf{z}| < \delta$ and $\mathbf{z} \in D(\mathbf{f})$, then $|\mathbf{f}(\mathbf{z}) - \mathbf{f}(\mathbf{x})| < \eta$. Then if $|\mathbf{x} - \mathbf{z}| < \delta$ and $\mathbf{z} \in D(\mathbf{g} \circ \mathbf{f}) \subseteq D(\mathbf{f})$, all the above hold and so

$$|\mathbf{g}(\mathbf{f}(\mathbf{z})) - \mathbf{g}(\mathbf{f}(\mathbf{x}))| < \varepsilon.$$

This proves part (3).

Part (4) says: If $\mathbf{f} = (f_1, \cdots, f_q) : D(\mathbf{f}) \to \mathbb{R}^q$, then $\mathbf{f}$ is continuous if and only if each f_k is a continuous real valued function. Then

$$|f_k(\mathbf{x}) - f_k(\mathbf{y})| \leq |\mathbf{f}(\mathbf{x}) - \mathbf{f}(\mathbf{y})| \equiv \left(\sum_{i=1}^{q} |f_i(\mathbf{x}) - f_i(\mathbf{y})|^2 \right)^{1/2}$$
$$\leq \sum_{i=1}^{q} |f_i(\mathbf{x}) - f_i(\mathbf{y})|. \tag{5.1}$$

Suppose first that $\mathbf{f}$ is continuous at $\mathbf{x}$. Then there exists $\delta > 0$ such that if $|\mathbf{x} - \mathbf{y}| < \delta$, then $|\mathbf{f}(\mathbf{x}) - \mathbf{f}(\mathbf{y})| < \varepsilon$. The first part of the above inequality then shows that for each $k = 1, \cdots, q$, $|f_k(\mathbf{x}) - f_k(\mathbf{y})| < \varepsilon$. This shows the only if part. Now suppose each function f_k is continuous. Then if $\varepsilon > 0$ is given, there exists $\delta_k > 0$ such that whenever $|\mathbf{x} - \mathbf{y}| < \delta_k$

$$|f_k(\mathbf{x}) - f_k(\mathbf{y})| < \varepsilon/q.$$

Now let $0 < \delta \leq \min(\delta_1, \cdots, \delta_q)$. For $|\mathbf{x} - \mathbf{y}| < \delta$, the above inequality holds for all k and so the last part of (5.1) implies

$$|\mathbf{f}(\mathbf{x}) - \mathbf{f}(\mathbf{y})| \leq \sum_{i=1}^{q} |f_i(\mathbf{x}) - f_i(\mathbf{y})| < \sum_{i=1}^{q} \frac{\varepsilon}{q} = \varepsilon.$$

This proves part (4).

To verify part (5), let $\varepsilon > 0$ be given and let $\delta = \varepsilon$. Then if $|\mathbf{x} - \mathbf{y}| < \delta$, the triangle inequality implies

$$|f(\mathbf{x}) - f(\mathbf{y})| = ||\mathbf{x}| - |\mathbf{y}|| \leq |\mathbf{x} - \mathbf{y}| < \delta = \varepsilon.$$

This proves part (5) and completes the proof of the theorem. ∎

E.2.2   Convergent Sequences

The following says that subsequences converge to the same thing that a convergent sequence converges to. We leave the proof to you.

Theorem E.2.14 *Let $\{\mathbf{x}_n\}$ be a vector valued sequence with $\lim_{n\to\infty} \mathbf{x}_n = \mathbf{x}$ and let $\{\mathbf{x}_{n_k}\}$ be a subsequence. Then $\lim_{k\to\infty} \mathbf{x}_{n_k} = \mathbf{x}$.*

The following is the definition of a Cauchy sequence in $\mathbb{R}^p$.

Definition E.2.15 $\{\mathbf{a}_n\}$ *is a Cauchy sequence if for all $\varepsilon > 0$, there exists n_ε such that whenever $n, m \geq n_\varepsilon$,*

$$|\mathbf{a}_n - \mathbf{a}_m| < \varepsilon.$$

One has the same set of Cauchy sequences if $|\cdot|$ is replaced with $\|\cdot\|$ thanks to equivalence of norms.

A sequence is Cauchy, means the terms are "bunching up to each other" as m, n get large.

Theorem E.2.16 *The set of terms in a Cauchy sequence in $\mathbb{R}^p$ is bounded in the sense that for all n, $|\mathbf{a}_n| < M$ for some $M < \infty$.*

Proof: Let $\varepsilon = 1$ in the definition of a Cauchy sequence and let $n > n_1$. Then from the definition,

$$|\mathbf{a}_n - \mathbf{a}_{n_1}| < 1.$$

It follows that for all $n > n_1$,

$$|\mathbf{a}_n| < 1 + |\mathbf{a}_{n_1}|.$$

Therefore, for all n,

$$|\mathbf{a}_n| \leq 1 + |\mathbf{a}_{n_1}| + \sum_{k=1}^{n_1} |\mathbf{a}_k|. \blacksquare$$

Theorem E.2.17 *If a sequence $\{\mathbf{a}_n\}$ in $\mathbb{R}^p$ converges, then the sequence is a Cauchy sequence. Also, if some subsequence of a Cauchy sequence converges, then the original sequence converges.*

Proof: Let $\varepsilon > 0$ be given and suppose $\mathbf{a}_n \to \mathbf{a}$. Then from the definition of convergence, there exists n_ε such that if $n > n_\varepsilon$, it follows that

$$|\mathbf{a}_n - \mathbf{a}| < \frac{\varepsilon}{2}$$

Therefore, if $m, n \geq n_\varepsilon + 1$, it follows that

$$|\mathbf{a}_n - \mathbf{a}_m| \leq |\mathbf{a}_n - \mathbf{a}| + |\mathbf{a} - \mathbf{a}_m| < \frac{\varepsilon}{2} + \frac{\varepsilon}{2} = \varepsilon$$

showing that, since $\varepsilon > 0$ is arbitrary, $\{\mathbf{a}_n\}$ is a Cauchy sequence. It remains to that the last claim.

Suppose then that $\{\mathbf{a}_n\}$ is a Cauchy sequence and $\mathbf{a} = \lim_{k\to\infty} \mathbf{a}_{n_k}$ where $\{\mathbf{a}_{n_k}\}_{k=1}^\infty$ is a subsequence. Let $\varepsilon > 0$ be given. Then there exists K such that if $k, l \geq K$, then $|\mathbf{a}_k - \mathbf{a}_l| < \frac{\varepsilon}{2}$. Then if $k > K$, it follows $n_k > K$ because $n_1, n_2, n_3, \cdots$ is strictly increasing as the subscript increases. Also, there exists

K_1 such that if $k > K_1, |\mathbf{a}_{n_k} - \mathbf{a}| < \frac{\varepsilon}{2}$. Then letting $n > \max(K, K_1)$, pick $k > \max(K, K_1)$. Then

$$|\mathbf{a} - \mathbf{a}_n| \leq |\mathbf{a} - \mathbf{a}_{n_k}| + |\mathbf{a}_{n_k} - \mathbf{a}_n| < \frac{\varepsilon}{2} + \frac{\varepsilon}{2} = \varepsilon.$$

Therefore, the sequence converges. ∎

There is no change replacing $|\cdot|$ with $\|\cdot\|$ thanks to equivalence of the norms.

Theorem E.2.18 *Every Cauchy sequence in $\mathbb{R}^p$ converges.*

Proof: Let $\{\mathbf{a}_k\}$ be a Cauchy sequence. Since it is Cauchy, it must be bounded (Why?). Thus there is some box $\prod_{i=1}^p [a_i, b_i]$ containing all the terms of $\{\mathbf{a}_k\}$. Therefore, by Theorem E.2.6, a subsequence converges to a point of $\prod_{i=1}^p [a_i, b_i]$. By Theorem E.2.17, the original sequence converges. ∎

As mentioned above, sequential compactness and compactness are equivalent in $\mathbb{R}^p$. The following is a very important property pertaining to compact sets. It is a surprising result.

Proposition E.2.19 *Suppose $\mathcal{F}$ is a nonempty collection of nonempty compact sets with the finite intersection property. This means that the intersection of any finite subset of $\mathcal{F}$ is nonempty. Then $\cap \mathcal{F} \neq \emptyset$.*

Proof: First I show each compact set is closed. Let K be a nonempty compact set and suppose $p \notin K$. Then for each $x \in K$, there are open sets U_x, V_x such that $x \in V_x$ and $p \in U_x$ and $U_x \cap V_x = \emptyset$. Just let these be balls centered at p, x respectively having radius equal to one third the distance from x to p. Then since V is compact, there are finitely many V_x which cover K say $V_{x_1}, \cdots, V_{x_n}$. Then let $U = \cap_{i=1}^n U_{x_i}$. It follows $p \in U$ and U has empty intersection with K. In fact U has empty intersection with $\cup_{i=1}^n V_{x_i}$. Since U is an open set and $p \in K^C$ is arbitrary, it follows K^C is an open set.

Consider now the claim about the intersection. If this were not so,

$$\cup \left\{ F^C : F \in \mathcal{F} \right\} = X$$

and so, in particular, picking some $F_0 \in \mathcal{F}$,

$$\left\{ F^C : F \in \mathcal{F} \right\}$$

would be an open cover of F_0. A point in F_0 is not in F_0^C so it must be in one of the above sets. Since F_0 is compact, some finite subcover, $F_1^C, \cdots, F_m^C$ exists. But then

$$F_0 \subseteq \cup_{k=1}^m F_k^C$$

which means $\cap_{k=0}^m F_k = \emptyset$, contrary to the finite intersection property. ∎

E.2.3   Continuity and the Limit of a Sequence

Just as in the case of a function of one variable, there is a very useful way of thinking of continuity in terms of limits of sequences found in the following theorem. In words, it says a function is continuous if it takes convergent sequences to convergent sequences whenever possible.

Theorem E.2.20 *A function* $\mathbf{f} : D(\mathbf{f}) \to \mathbb{R}^q$ *is continuous at* $\mathbf{x} \in D(\mathbf{f})$ *if and only if, whenever* $\mathbf{x}_n \to \mathbf{x}$ *with* $\mathbf{x}_n \in D(\mathbf{f})$, *it follows* $\mathbf{f}(\mathbf{x}_n) \to \mathbf{f}(\mathbf{x})$.

Proof: Suppose first that $\mathbf{f}$ is continuous at $\mathbf{x}$ and let $\mathbf{x}_n \to \mathbf{x}$. Let $\varepsilon > 0$ be given. By continuity, there exists $\delta > 0$ such that if $|\mathbf{y} - \mathbf{x}| < \delta$, then $|\mathbf{f}(\mathbf{x}) - \mathbf{f}(\mathbf{y})| < \varepsilon$. However, there exists n_δ such that if $n \geq n_\delta$, then $|\mathbf{x}_n - \mathbf{x}| < \delta$, and so for all n this large,

$$|\mathbf{f}(\mathbf{x}) - \mathbf{f}(\mathbf{x}_n)| < \varepsilon$$

which shows $\mathbf{f}(\mathbf{x}_n) \to \mathbf{f}(\mathbf{x})$.

Now suppose the condition about taking convergent sequences to convergent sequences holds at $\mathbf{x}$. Suppose $\mathbf{f}$ fails to be continuous at $\mathbf{x}$. Then there exists $\varepsilon > 0$ and $\mathbf{x}_n \in D(f)$ such that $|\mathbf{x} - \mathbf{x}_n| < \frac{1}{n}$, yet

$$|\mathbf{f}(\mathbf{x}) - \mathbf{f}(\mathbf{x}_n)| \geq \varepsilon.$$

But this is clearly a contradiction because, although $\mathbf{x}_n \to \mathbf{x}$, $\mathbf{f}(\mathbf{x}_n)$ fails to converge to $\mathbf{f}(\mathbf{x})$. It follows $\mathbf{f}$ must be continuous after all. ∎

E.2.4   The Extreme Value Theorem and Uniform Continuity

Here is a proof of the extreme value theorem.

Theorem E.2.21 *Let* $C \subseteq \mathbb{R}^p$ *be closed and bounded and let* $f : C \to \mathbb{R}$ *be continuous. Then* f *achieves its maximum and its minimum on* C. *This means there exist* $\mathbf{x}_1, \mathbf{x}_2 \in C$ *such that for all* $\mathbf{x} \in C$,

$$f(\mathbf{x}_1) \leq f(\mathbf{x}) \leq f(\mathbf{x}_2).$$

Proof: Let $M = \sup\{f(\mathbf{x}) : \mathbf{x} \in C\}$. Let $\mathbf{x}_n \in C$ such that $M = \lim_{n \to \infty} f(\mathbf{x}_n)$. Thus for any subsequence, $M = \lim_{k \to \infty} f(\mathbf{x}_{n_k})$. By Theorem E.2.10, $\{\mathbf{x}_n\}$ has a convergent subsequence $\mathbf{x}_{n_k}$ such that $\lim_{k \to \infty} \mathbf{x}_{n_k} = \mathbf{x} \in C$. Then by continuity and Theorem E.2.20,

$$M = \lim_{k \to \infty} f(\mathbf{x}_{n_k}) = f(\mathbf{x}).$$

The case of the minimum value is similar. Just replace sup with inf. ∎

As in the case of a function of one variable, there is a concept of uniform continuity.

Definition E.2.22 *A function* $\mathbf{f} : D(\mathbf{f}) \to \mathbb{R}^q$ *is uniformly continuous if for every* $\varepsilon > 0$ *there exists* $\delta > 0$ *such that whenever* $\mathbf{x}, \mathbf{y}$ *are points of* $D(\mathbf{f})$ *such that* $|\mathbf{x} - \mathbf{y}| < \delta$, *it follows* $|\mathbf{f}(\mathbf{x}) - \mathbf{f}(\mathbf{y})| < \varepsilon$.

Theorem E.2.23 *Let* $\mathbf{f} : K \to \mathbb{R}^q$ *be continuous at every point of* K *where* K *is a closed and bounded set in* $\mathbb{R}^p$. *Then* $\mathbf{f}$ *is uniformly continuous.*

Proof: Suppose not. Then there exists $\varepsilon > 0$ and sequences $\{\mathbf{x}_j\}$ and $\{\mathbf{y}_j\}$ of points in K such that

$$|\mathbf{x}_j - \mathbf{y}_j| < \frac{1}{j}$$

but $|\mathbf{f}(\mathbf{x}_j) - \mathbf{f}(\mathbf{y}_j)| \geq \varepsilon$. Then Theorem E.2.10 which says K is sequentially compact, there is a subsequence $\{\mathbf{x}_{n_k}\}$ of $\{\mathbf{x}_j\}$ which converges to a point $\mathbf{x} \in K$. Then since $|\mathbf{x}_{n_k} - \mathbf{y}_{n_k}| < \frac{1}{k}$, it follows that $\{\mathbf{y}_{n_k}\}$ also converges to $\mathbf{x}$. Therefore,

$$\varepsilon \leq \lim_{k \to \infty} |\mathbf{f}(\mathbf{x}_{n_k}) - \mathbf{f}(\mathbf{y}_{n_k})| = |\mathbf{f}(\mathbf{x}) - \mathbf{f}(\mathbf{x})| = 0,$$

a contradiction. Therefore, $\mathbf{f}$ is uniformly continuous as claimed. ∎

E.2.5   Connected Sets

Stated informally, connected sets are those which are in one piece. In order to define what is meant by this, I will first consider what it means for a set to **not** be in one piece. This is called **separated**. Connected sets are defined in terms of **not** being separated. This is why theorems about connected sets sometimes seem a little tricky.

Definition E.2.24 *Let A be a nonempty subset $\mathbb{R}^n$. Then $\overline{A}$ is defined to be the intersection of all closed sets which contain A. This is called the closure of A. Note the whole space, $\mathbb{R}^n$ is one such closed set which contains A.*

Lemma E.2.25 *Let A be a nonempty set in $\mathbb{R}^n$. Then $\overline{A}$ is a closed set and*

$$\overline{A} = A \cup A'$$

where A' denotes the set of limit points of A.

Proof: First of all, denote by $\mathcal{C}$ the set of closed sets which contain A. Then

$$\overline{A} = \cap \mathcal{C}$$

and this will be closed if its complement is open. However,

$$\overline{A}^C = \cup \left\{ H^C : H \in \mathcal{C} \right\}.$$

Each H^C is open and so the union of all these open sets must also be open. This is because if $\mathbf{x}$ is in this union, then it is in at least one of them. Hence it is an interior point of that one. But this implies it is an interior point of the union of them all which is an even larger set. Thus $\overline{A}$ is closed.

The interesting part is the next claim. First note that from the definition, $A \subseteq \overline{A}$ so if $\mathbf{x} \in A$, then $\mathbf{x} \in \overline{A}$. Now consider $\mathbf{y} \in A'$ but $\mathbf{y} \notin A$. If $\mathbf{y} \notin \overline{A}$, a closed set, then there exists $B(\mathbf{y}, r) \subseteq \overline{A}^C$. Thus $\mathbf{y}$ cannot be a limit point of A, a contradiction. Therefore,

$$A \cup A' \subseteq \overline{A}$$

Next suppose $\mathbf{x} \in \overline{A}$ and suppose $\mathbf{x} \notin A$. Then if $B(\mathbf{x}, r)$ contains no points of A different than $\mathbf{x}$, since $\mathbf{x}$ itself is not in A, it would follow that $B(\mathbf{x}, r) \cap A = \emptyset$ and so recalling that open balls are open, $B(\mathbf{x}, r)^C$ is a closed set containing A so from the definition, it also contains $\overline{A}$ which is contrary to the assertion that $\mathbf{x} \in \overline{A}$. Hence if $\mathbf{x} \notin A$, then $\mathbf{x} \in A'$ and so

$$A \cup A' \supseteq \overline{A} \quad ∎$$

Now is a definition about what it means to not be connected. This is called separated.

Definition E.2.26 *A set, S in $\mathbb{R}^n$, is separated if there exist sets A, B such that*

$$S = A \cup B, \ A, B \neq \emptyset, \ and \ \overline{A} \cap B = \overline{B} \cap A = \emptyset.$$

In this case, the sets A and B are said to separate S. A set is connected if it is not separated. Remember $\overline{A}$ denotes the closure of the set A.

Note that the concept of connected sets is defined in terms of what it is not. This makes it somewhat difficult to understand. One of the most important theorems about connected sets is the following.

Theorem E.2.27 *Suppose $\mathcal{U}$ is a set of connected sets and that there exists a point p which is in all of these connected sets. Then $K \equiv \cup \mathcal{U}$ is connected.*

Proof: Suppose

$$K = A \cup B$$

where $\bar{A} \cap B = \bar{B} \cap A = \emptyset, A \neq \emptyset, B \neq \emptyset$. Let $U \in \mathcal{U}$. Then

$$U = (U \cap A) \cup (U \cap B)$$

and this would separate U if both sets in the union are nonempty since the limit points of $U \cap B$ are contained in the limit points of B. It follows that every set of $\mathcal{U}$ is contained in one of A or B. Suppose then that some $U \subseteq A$. Then all $U \in \mathcal{U}$ must be contained in A because if one is contained in B, this would violate the assumption that they all have a point p in common. Thus K is connected after all because this requires $B = \emptyset$. Alternatively, p is in one of these sets. Say $p \in A$. Then by the above argument every U must be in A because if not, the above would be a separation of U. Thus $B = \emptyset$. ∎

The intersection of connected sets is not necessarily connected as is shown by the following picture.

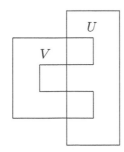

Theorem E.2.28 *Let $\mathbf{f} : X \to \mathbb{R}^m$ be continuous where X is connected. Then $\mathbf{f}(X)$ is also connected.*

Proof: To do this you show $\mathbf{f}(X)$ is not separated. Suppose to the contrary that $\mathbf{f}(X) = A \cup B$ where A and B separate $\mathbf{f}(X)$. Then consider the sets $\mathbf{f}^{-1}(A)$ and $\mathbf{f}^{-1}(B)$. If $\mathbf{z} \in \mathbf{f}^{-1}(B)$, then $\mathbf{f}(\mathbf{z}) \in B$ and so $\mathbf{f}(\mathbf{z})$ is not a limit point of A. Therefore, there exists an open set, U containing $\mathbf{f}(\mathbf{z})$ such that $U \cap A = \emptyset$. But then, the continuity of $\mathbf{f}$ implies that $\mathbf{f}^{-1}(U)$ is an open set containing $\mathbf{z}$ such that $\mathbf{f}^{-1}(U) \cap \mathbf{f}^{-1}(A) = \emptyset$. Therefore, $\mathbf{f}^{-1}(B)$ contains no limit points of $\mathbf{f}^{-1}(A)$.

Similar reasoning implies $\mathbf{f}^{-1}(A)$ contains no limit points of $\mathbf{f}^{-1}(B)$. It follows that X is separated by $\mathbf{f}^{-1}(A)$ and $\mathbf{f}^{-1}(B)$, contradicting the assumption that X was connected. ∎

An arbitrary set can be written as a union of maximal connected sets called connected components. This is the concept of the next definition.

Definition E.2.29 *Let S be a set and let $\mathbf{p} \in S$. Denote by $C_{\mathbf{p}}$ the union of all connected subsets of S which contain $\mathbf{p}$. This is called the connected component determined by $\mathbf{p}$.*

Theorem E.2.30 *Let $C_{\mathbf{p}}$ be a connected component of a set S. Then $C_{\mathbf{p}}$ is a connected set and if $C_{\mathbf{p}} \cap C_{\mathbf{q}} \neq \emptyset$, then $C_{\mathbf{p}} = C_{\mathbf{q}}$.*

Proof: Let $\mathcal{C}$ denote the connected subsets of S which contain $\mathbf{p}$. By Theorem E.2.27, $\cup \mathcal{C} = C_{\mathbf{p}}$ is connected. If $\mathbf{x} \in C_{\mathbf{p}} \cap C_{\mathbf{q}}$, then from Theorem E.2.27, $C_{\mathbf{p}} \supseteq C_{\mathbf{p}} \cup C_{\mathbf{q}}$ and so $C_{\mathbf{p}} \supseteq C_{\mathbf{q}}$. The inclusion goes the other way by the same reason. ∎

This shows the connected components of a set are equivalence classes and partition the set.

A set, I is an interval in $\mathbb{R}$ if and only if whenever $x, y \in I$ then $(x, y) \subseteq I$. The following theorem is about the connected sets in $\mathbb{R}$.

Theorem E.2.31 *A set C in $\mathbb{R}$ is connected if and only if C is an interval.*

Proof: Let C be connected. If C consists of a single point, p, there is nothing to prove. The interval is just $[p, p]$. Suppose $p < q$ and $p, q \in C$. You need to show $(p, q) \subseteq C$. If

$$x \in (p, q) \setminus C$$

let $C \cap (-\infty, x) \equiv A$, and $C \cap (x, \infty) \equiv B$. Then $C = A \cup B$ and the sets A and B separate C contrary to the assumption that C is connected.

Conversely, let I be an interval. Suppose I is separated by A and B. Pick $x \in A$ and $y \in B$. Suppose without loss of generality that $x < y$. Now define the set,

$$S \equiv \{t \in [x, y] : [x, t] \subseteq A\}$$

and let l be the least upper bound of S. Then $l \in \overline{A}$ so $l \notin B$ which implies $l \in A$. But if $l \notin \overline{B}$, then for some $\delta > 0$,

$$(l, l + \delta) \cap B = \emptyset$$

contradicting the definition of l as an upper bound for S. Therefore, $l \in \overline{B}$ which implies $l \notin A$ after all, a contradiction. It follows I must be connected. ∎

This yields a generalization of the intermediate value theorem from one variable calculus.

Corollary E.2.32 *Let E be a connected set in $\mathbb{R}^n$ and suppose $f : E \to \mathbb{R}$ and that $y \in (f(e_1), f(e_2))$ where $e_i \in E$. Then there exists $e \in E$ such that $f(e) = y$.*

Proof: From Theorem E.2.28, $f(E)$ is a connected subset of $\mathbb{R}$. By Theorem E.2.31 $f(E)$ must be an interval. In particular, it must contain y. This proves the corollary. ∎

The following theorem is a very useful description of the open sets in $\mathbb{R}$.

Theorem E.2.33 *Let U be an open set in $\mathbb{R}$. Then there exist countably many disjoint open sets $\{(a_i, b_i)\}_{i=1}^{\infty}$ such that $U = \cup_{i=1}^{\infty} (a_i, b_i)$.*

Proof: Let $p \in U$ and let $z \in C_p$, the connected component determined by p. Since U is open, there exists, $\delta > 0$ such that $(z - \delta, z + \delta) \subseteq U$. It follows from Theorem E.2.27 that

$$(z - \delta, z + \delta) \subseteq C_p.$$

This shows C_p is open. By Theorem E.2.31, this shows C_p is an open interval, (a, b) where $a, b \in [-\infty, \infty]$. There are therefore at most countably many of these connected components because each must contain a rational number and the rational numbers are countable. Denote by $\{(a_i, b_i)\}_{i=1}^{\infty}$ the set of these connected components. ∎

Definition E.2.34 *A set E in $\mathbb{R}^n$ is arcwise connected if for any two points, $\mathbf{p}, \mathbf{q} \in E$, there exists a closed interval, $[a, b]$ and a continuous function, $\gamma : [a, b] \to E$ such that $\gamma(a) = p$ and $\gamma(b) = q$.*

An example of an arcwise connected metric space would be any subset of $\mathbb{R}^n$ which is the continuous image of an interval. Arcwise connected is not the same as connected. A well known example is the following.

$$\left\{ \left(x, \sin \frac{1}{x} \right) : x \in (0, 1] \right\} \cup \{(0, y) : y \in [-1, 1]\} \tag{5.2}$$

You can verify that this set of points in $\mathbb{R}^2$ is not arcwise connected but is connected.

Lemma E.2.35 *In $\mathbb{R}^n$, $B(\mathbf{z}, r)$ is arcwise connected.*

Proof: This is easy from the convexity of the set. If $\mathbf{x}, \mathbf{y} \in B(\mathbf{z}, r)$, then let $\gamma(t) = \mathbf{x} + t(\mathbf{y} - \mathbf{x})$ for $t \in [0, 1]$.

$$\begin{aligned} \|\mathbf{x} + t(\mathbf{y} - \mathbf{x}) - \mathbf{z}\| &= \|(1 - t)(\mathbf{x} - \mathbf{z}) + t(\mathbf{y} - \mathbf{z})\| \\ &\leq (1 - t)\|\mathbf{x} - \mathbf{z}\| + t\|\mathbf{y} - \mathbf{z}\| \\ &< (1 - t)r + tr = r \end{aligned}$$

showing $\gamma(t)$ stays in $B(\mathbf{z}, r)$. ∎

Proposition E.2.36 *If $X \neq \emptyset$ is arcwise connected, then it is connected.*

Proof: Let $p \in X$. Then by assumption, for any $x \in X$, there is an arc joining p and x. This arc is connected because it is the continuous image of an interval which is connected. Since x is arbitrary, every x is in a connected subset of X which contains p. Hence $C_p = X$ and so X is connected. ∎

Theorem E.2.37 *Let U be an open subset of a $\mathbb{R}^n$. Then U is arcwise connected if and only if U is connected. Also the connected components of an open set are open sets.*

Proof: By Proposition E.2.36 it is only necessary to verify that if U is connected and open in the context of this theorem, then U is arcwise connected. Pick $\mathbf{p} \in U$. Say $\mathbf{x} \in U$ satisfies $\mathcal{P}$ if there exists a continuous function, $\boldsymbol{\gamma} : [a, b] \to U$ such that $\gamma(a) = \mathbf{p}$ and $\boldsymbol{\gamma}(\mathbf{b}) = \mathbf{x}$.

$$A \equiv \{\mathbf{x} \in U \text{ such that } \mathbf{x} \text{ satisfies } \mathcal{P}.\}$$

If $\mathbf{x} \in A$, then Lemma E.2.35 implies $B(\mathbf{x}, r) \subseteq U$ is arcwise connected for small enough r. Thus letting $\mathbf{y} \in B(\mathbf{x}, r)$, there exist intervals, $[a, b]$ and $[c, d]$ and continuous functions having values in U, $\boldsymbol{\gamma}, \boldsymbol{\eta}$ such that $\boldsymbol{\gamma}(a) = \mathbf{p}, \boldsymbol{\gamma}(b) = \mathbf{x}, \boldsymbol{\eta}(c) = \mathbf{x}$, and $\eta(d) = \mathbf{y}$. Then let $\boldsymbol{\gamma}_1 : [a, b + d - c] \to U$ be defined as

$$\gamma_1(t) \equiv \begin{cases} \boldsymbol{\gamma}(t) & \text{if } t \in [a, b] \\ \boldsymbol{\eta}(t + c - b) & \text{if } t \in [b, b + d - c] \end{cases}$$

Then it is clear that $\boldsymbol{\gamma}_1$ is a continuous function mapping $\mathbf{p}$ to $\mathbf{y}$ and showing that $B(\mathbf{x}, r) \subseteq A$. Therefore, A is open. $A \neq \emptyset$ because since U is open there is an open set, $B(\mathbf{p}, \delta)$ containing $\mathbf{p}$ which is contained in U and is arcwise connected.

Now consider $B \equiv U \setminus A$. I claim this is also open. If B is not open, there exists a point $\mathbf{z} \in B$ such that every open set containing $\mathbf{z}$ is not contained in B. Therefore, letting $B(\mathbf{z}, \delta)$ be such that $\mathbf{z} \in B(\mathbf{z}, \delta) \subseteq U$, there exist points of A contained in $B(\mathbf{z}, \delta)$. But then, a repeat of the above argument shows $z \in A$ also. Hence B is open and so if $B \neq \emptyset$, then $U = B \cup A$ and so U is separated by the two sets B and A contradicting the assumption that U is connected.

It remains to verify the connected components are open. Let $\mathbf{z} \in C_{\mathbf{p}}$ where $C_{\mathbf{p}}$ is the connected component determined by $\mathbf{p}$. Then picking $B(\mathbf{z}, \delta) \subseteq U, C_{\mathbf{p}} \cup B(\mathbf{z}, \delta)$ is connected and contained in U and so it must also be contained in $C_{\mathbf{p}}$. Thus $\mathbf{z}$ is an interior point of $C_{\mathbf{p}}$. ∎

As an application, consider the following corollary.

Corollary E.2.38 *Let $f : \Omega \to \mathbb{Z}$ be continuous where Ω is a connected open set in $\mathbb{R}^n$. Then f must be a constant.*

Proof: Suppose not. Then it achieves two different values, k and $l \neq k$. Then $\Omega = f^{-1}(l) \cup f^{-1}(\{m \in \mathbb{Z} : m \neq l\})$ and these are disjoint nonempty open sets which separate Ω. To see they are open, note

$$f^{-1}(\{m \in \mathbb{Z} : m \neq l\}) = f^{-1}\left(\cup_{m \neq l}\left(m - \frac{1}{6}, m + \frac{1}{6}\right)\right)$$

which is the inverse image of an open set while $f^{-1}(l) = f^{-1}\left(\left(l - \frac{1}{6}, l + \frac{1}{6}\right)\right)$ also an open set. ∎

Appendix F

Implicit Function Theorem

The implicit function theorem is one of the greatest theorems in mathematics. There are many versions of this theorem which are of far greater generality than the one given here. The proof given here is like one found in one of Caratheodory's books on the calculus of variations. It is not as elegant as some of the others which are based on a contraction mapping principle but it may be more accessible. However, it is an advanced topic. Don't waste your time with it unless you have first read and understood the material on rank and determinants found in the chapter on the mathematical theory of determinants. You will also need to use the extreme value theorem for a function of n variables and the chain rule of multivariable calculus as well as everything about matrix multiplication.

Definition F.0.39 *Suppose U is an open set in $\mathbb{R}^n \times \mathbb{R}^m$ and $(\mathbf{x}, \mathbf{y})$ will denote a typical point of $\mathbb{R}^n \times \mathbb{R}^m$ with $\mathbf{x} \in \mathbb{R}^n$ and $\mathbf{y} \in \mathbb{R}^m$. Let $\mathbf{f} : U \to \mathbb{R}^p$ be in $C^1(U)$. Then define*

$$
D_1\mathbf{f}(\mathbf{x}, \mathbf{y}) \equiv \begin{pmatrix} f_{1,x_1}(\mathbf{x}, \mathbf{y}) & \cdots & f_{1,x_n}(\mathbf{x}, \mathbf{y}) \\ \vdots & & \vdots \\ f_{p,x_1}(\mathbf{x}, \mathbf{y}) & \cdots & f_{p,x_n}(\mathbf{x}, \mathbf{y}) \end{pmatrix},
$$

$$
D_2\mathbf{f}(\mathbf{x}, \mathbf{y}) \equiv \begin{pmatrix} f_{1,y_1}(\mathbf{x}, \mathbf{y}) & \cdots & f_{1,y_m}(\mathbf{x}, \mathbf{y}) \\ \vdots & & \vdots \\ f_{p,y_1}(\mathbf{x}, \mathbf{y}) & \cdots & f_{p,y_m}(\mathbf{x}, \mathbf{y}) \end{pmatrix}.
$$

Theorem F.0.40 *(implicit function theorem) Suppose U is an open set in $\mathbb{R}^n \times \mathbb{R}^m$. Let $\mathbf{f} : U \to \mathbb{R}^n$ be in $C^1(U)$ and suppose*

$$\mathbf{f}(\mathbf{x}_0, \mathbf{y}_0) = \mathbf{0}, \ D_1\mathbf{f}(\mathbf{x}_0, \mathbf{y}_0)^{-1} \ exists. \tag{6.1}$$

Then there exist positive constants, δ, η, such that for every $\mathbf{y} \in B(\mathbf{y}_0, \eta)$ there exists a unique $\mathbf{x}(\mathbf{y}) \in B(\mathbf{x}_0, \delta)$ such that

$$\mathbf{f}(\mathbf{x}(\mathbf{y}), \mathbf{y}) = \mathbf{0}. \tag{6.2}$$

Furthermore, the mapping, $\mathbf{y} \to \mathbf{x}(\mathbf{y})$ is in $C^1(B(\mathbf{y}_0, \eta))$.

Proof: Let

$$\mathbf{f}(\mathbf{x}, \mathbf{y}) = \begin{pmatrix} f_1(\mathbf{x}, \mathbf{y}) \\ f_2(\mathbf{x}, \mathbf{y}) \\ \vdots \\ f_n(\mathbf{x}, \mathbf{y}) \end{pmatrix}.$$

Define for $(\mathbf{x}^1, \cdots, \mathbf{x}^n) \in \overline{B(\mathbf{x}_0, \delta)}^n$ and $\mathbf{y} \in B(\mathbf{y}_0, \eta)$ the following matrix.

$$J(\mathbf{x}^1, \cdots, \mathbf{x}^n, \mathbf{y}) \equiv \begin{pmatrix} f_{1,x_1}(\mathbf{x}^1, \mathbf{y}) & \cdots & f_{1,x_n}(\mathbf{x}^1, \mathbf{y}) \\ \vdots & & \vdots \\ f_{n,x_1}(\mathbf{x}^n, \mathbf{y}) & \cdots & f_{n,x_n}(\mathbf{x}^n, \mathbf{y}) \end{pmatrix}. \tag{*}$$

Then by the assumption of continuity of all the partial derivatives and the extreme value theorem, there exists $r > 0$ and $\delta_0, \eta_0 > 0$ such that if $\delta \leq \delta_0$ and $\eta \leq \eta_0$, it follows that for all $(\mathbf{x}^1, \cdots, \mathbf{x}^n) \in \overline{B(\mathbf{x}_0, \delta)}^n$ and $\mathbf{y} \in \overline{B(\mathbf{y}_0, \eta)}$,

$$\left| \det\left(J(\mathbf{x}^1, \cdots, \mathbf{x}^n, \mathbf{y}) \right) \right| > r > 0. \tag{6.3}$$

and $\overline{B(\mathbf{x}_0, \delta_0)} \times \overline{B(\mathbf{y}_0, \eta_0)} \subseteq U$. By continuity of all the partial derivatives and the extreme value theorem, it can also be assumed there exists a constant, K such that for all $(\mathbf{x}, \mathbf{y}) \in \overline{B(\mathbf{x}_0, \delta_0)} \times \overline{B(\mathbf{y}_0, \eta_0)}$ and $i = 1, 2, \cdots, n$, the ith row of $D_2\mathbf{f}(\mathbf{x}, \mathbf{y})$, given by $D_2 f_i(\mathbf{x}, \mathbf{y})$ satisfies

$$|D_2 f_i(\mathbf{x}, \mathbf{y})| < K, \tag{6.4}$$

and for all $(\mathbf{x}^1, \cdots, \mathbf{x}^n) \in \overline{B(\mathbf{x}_0, \delta_0)}^n$ and $\mathbf{y} \in \overline{B(\mathbf{y}_0, \eta_0)}$ the ith row of the matrix,

$$J(\mathbf{x}^1, \cdots, \mathbf{x}^n, \mathbf{y})^{-1}$$

which equals $\mathbf{e}_i^T \left(J(\mathbf{x}^1, \cdots, \mathbf{x}^n, \mathbf{y})^{-1} \right)$ satisfies

$$\left| \mathbf{e}_i^T \left(J(\mathbf{x}^1, \cdots, \mathbf{x}^n, \mathbf{y})^{-1} \right) \right| < K. \tag{6.5}$$

(Recall that $\mathbf{e}_i$ is the column vector consisting of all zeros except for a 1 in the ith position.)

To begin with it is shown that for a given $\mathbf{y} \in B(\mathbf{y}_0, \eta)$ there is at most one $\mathbf{x} \in B(\mathbf{x}_0, \delta)$ such that $\mathbf{f}(\mathbf{x}, \mathbf{y}) = \mathbf{0}$.

Pick $\mathbf{y} \in B(\mathbf{y}_0, \eta)$ and suppose there exist $\mathbf{x}, \mathbf{z} \in \overline{B(\mathbf{x}_0, \delta)}$ such that $\mathbf{f}(\mathbf{x}, \mathbf{y}) = \mathbf{f}(\mathbf{z}, \mathbf{y}) = \mathbf{0}$. Consider f_i and let

$$h(t) \equiv f_i(\mathbf{x} + t(\mathbf{z} - \mathbf{x}), \mathbf{y}).$$

Then $h(1) = h(0)$ and so by the mean value theorem, $h'(t_i) = 0$ for some $t_i \in (0, 1)$. Therefore, from the chain rule and for this value of t_i,

$$h'(t_i) = \sum_{j=1}^{n} \frac{\partial}{\partial x_j} f_i(\mathbf{x} + t_i(\mathbf{z} - \mathbf{x}), \mathbf{y})(z_j - x_j) = 0. \tag{6.6}$$

Then denote by $\mathbf{x}^i$ the vector, $\mathbf{x} + t_i (\mathbf{z} - \mathbf{x})$. It follows from (6.6) that

$$J \left(\mathbf{x}^1, \cdots, \mathbf{x}^n, \mathbf{y}\right) (\mathbf{z} - \mathbf{x}) = \mathbf{0}$$

and so from (6.3) $\mathbf{z} - \mathbf{x} = \mathbf{0}$. (The matrix, in the above is invertible since its determinant is nonzero.) Now it will be shown that if η is chosen sufficiently small, then for all $\mathbf{y} \in B (\mathbf{y}_0, \eta)$, there exists a unique $\mathbf{x} (\mathbf{y}) \in B (\mathbf{x}_0, \delta)$ such that $\mathbf{f} (\mathbf{x} (\mathbf{y}), \mathbf{y}) = \mathbf{0}$.

Claim: If η is small enough, then the function, $\mathbf{x} \to h_{\mathbf{y}} (\mathbf{x}) \equiv |\mathbf{f} (\mathbf{x}, \mathbf{y})|^2$ achieves its minimum value on $\overline{B (\mathbf{x}_0, \delta)}$ at a point of $B (\mathbf{x}_0, \delta)$. (The existence of a point in $\overline{B (\mathbf{x}_0, \delta)}$ at which $h_{\mathbf{y}}$ achieves its minimum follows from the extreme value theorem.)

Proof of claim: Suppose this is not the case. Then there exists a sequence $\eta_k \to 0$ and for some $\mathbf{y}_k$ having $|\mathbf{y}_k - \mathbf{y}_0| < \eta_k$, the minimum of $h_{\mathbf{y}_k}$ on $\overline{B (\mathbf{x}_0, \delta)}$ occurs on a point $\mathbf{x}_k$ such that $|\mathbf{x}_0 - \mathbf{x}_k| = \delta$. Now taking a subsequence, still denoted by k, it can be assumed that $\mathbf{x}_k \to \mathbf{x}$ with $|\mathbf{x} - \mathbf{x}_0| = \delta$ and $\mathbf{y}_k \to \mathbf{y}_0$. This follows from the fact that $\left\{\mathbf{x} \in \overline{B (\mathbf{x}_0, \delta)} : |\mathbf{x} - \mathbf{x}_0| = \delta\right\}$ is a closed and bounded set and is therefore sequentially compact. Let $\varepsilon > 0$. Then for k large enough, the continuity of $\mathbf{y} \to h_{\mathbf{y}} (\mathbf{x}_0)$ implies $h_{\mathbf{y}_k} (\mathbf{x}_0) < \varepsilon$ because $h_{\mathbf{y}_0} (\mathbf{x}_0) = 0$ since $\mathbf{f} (\mathbf{x}_0, \mathbf{y}_0) = \mathbf{0}$. Therefore, from the definition of $\mathbf{x}_k$, it is also the case that $h_{\mathbf{y}_k} (\mathbf{x}_k) < \varepsilon$. Passing to the limit yields $h_{\mathbf{y}_0} (\mathbf{x}) \leq \varepsilon$. Since $\varepsilon > 0$ is arbitrary, it follows that $h_{\mathbf{y}_0} (\mathbf{x}) = 0$ which contradicts the first part of the argument in which it was shown that for $\mathbf{y} \in B (\mathbf{y}_0, \eta)$ there is at most one point, $\mathbf{x}$ of $\overline{B (\mathbf{x}_0, \delta)}$ where $\mathbf{f} (\mathbf{x}, \mathbf{y}) = \mathbf{0}$. Here two have been obtained, $\mathbf{x}_0$ and $\mathbf{x}$. This proves the claim.

Choose $\eta < \eta_0$ and also small enough that the above claim holds and let $\mathbf{x} (\mathbf{y})$ denote a point of $B (\mathbf{x}_0, \delta)$ at which the minimum of $h_{\mathbf{y}}$ on $\overline{B (\mathbf{x}_0, \delta)}$ is achieved. Since $\mathbf{x} (\mathbf{y})$ is an interior point, you can consider $h_{\mathbf{y}} (\mathbf{x} (\mathbf{y}) + t\mathbf{v})$ for $|t|$ small and conclude this function of t has a zero derivative at $t = 0$. Now

$$h_{\mathbf{y}} (\mathbf{x} (\mathbf{y}) + t\mathbf{v}) = \sum_{i=1}^n f_i^2 (\mathbf{x} (\mathbf{y}) + t\mathbf{v}, \mathbf{y})$$

and so from the chain rule,

$$\frac{d}{dt} h_{\mathbf{y}} (\mathbf{x} (\mathbf{y}) + t\mathbf{v}) = \sum_{i=1}^n \sum_{j=1}^n 2 f_i (\mathbf{x} (\mathbf{y}) + t\mathbf{v}, \mathbf{y}) \frac{\partial f_i (\mathbf{x} (\mathbf{y}) + t\mathbf{v}, \mathbf{y})}{\partial x_j} v_j.$$

Therefore, letting $t = 0$, it is required that for every $\mathbf{v}$,

$$\sum_{i=1}^n \sum_{j=1}^n 2 f_i (\mathbf{x} (\mathbf{y}), \mathbf{y}) \frac{\partial f_i (\mathbf{x} (\mathbf{y}), \mathbf{y})}{\partial x_j} v_j = 0.$$

In terms of matrices this reduces to

$$0 = 2\mathbf{f} (\mathbf{x} (\mathbf{y}), \mathbf{y})^T D_1 \mathbf{f} (\mathbf{x} (\mathbf{y}), \mathbf{y}) \mathbf{v}$$

for every vector $\mathbf{v}$. Therefore,

$$\mathbf{0} = \mathbf{f} (\mathbf{x} (\mathbf{y}), \mathbf{y})^T D_1 \mathbf{f} (\mathbf{x} (\mathbf{y}), \mathbf{y})$$

From (6.3), it follows $\mathbf{f} (\mathbf{x} (\mathbf{y}), \mathbf{y}) = \mathbf{0}$. This proves the existence of the function $\mathbf{y} \to \mathbf{x} (\mathbf{y})$ such that $\mathbf{f} (\mathbf{x} (\mathbf{y}), \mathbf{y}) = \mathbf{0}$ for all $\mathbf{y} \in B (\mathbf{y}_0, \eta)$.

It remains to verify this function is a C^1 function. To do this, let $\mathbf{y}_1$ and $\mathbf{y}_2$ be points of $B(\mathbf{y}_0, \eta)$. Then as before, consider the ith component of $\mathbf{f}$ and consider the same argument using the mean value theorem to write

$$0 = f_i(\mathbf{x}(\mathbf{y}_1), \mathbf{y}_1) - f_i(\mathbf{x}(\mathbf{y}_2), \mathbf{y}_2)$$

$$= f_i(\mathbf{x}(\mathbf{y}_1), \mathbf{y}_1) - f_i(\mathbf{x}(\mathbf{y}_2), \mathbf{y}_1) + f_i(\mathbf{x}(\mathbf{y}_2), \mathbf{y}_1) - f_i(\mathbf{x}(\mathbf{y}_2), \mathbf{y}_2) \qquad (6.7)$$

$$= D_1 f_i(\mathbf{x}^i, \mathbf{y}_1)(\mathbf{x}(\mathbf{y}_1) - \mathbf{x}(\mathbf{y}_2)) + D_2 f_i(\mathbf{x}(\mathbf{y}_2), \mathbf{y}^i)(\mathbf{y}_1 - \mathbf{y}_2).$$

where $\mathbf{y}^i$ is a point on the line segment joining $\mathbf{y}_1$ and $\mathbf{y}_2$. Thus from (6.4) and the Cauchy-Schwarz inequality,

$$\left| D_2 f_i(\mathbf{x}(\mathbf{y}_2), \mathbf{y}^i)(\mathbf{y}_1 - \mathbf{y}_2) \right| \le K |\mathbf{y}_1 - \mathbf{y}_2|.$$

Therefore, letting $M(\mathbf{y}^1, \cdots, \mathbf{y}^n) \equiv M$ denote the matrix having the ith row equal to $D_2 f_i(\mathbf{x}(\mathbf{y}_2), \mathbf{y}^i)$, it follows

$$|M(\mathbf{y}_1 - \mathbf{y}_2)| \le \left(\sum_i K^2 |\mathbf{y}_1 - \mathbf{y}_2|^2 \right)^{1/2} = \sqrt{m} K |\mathbf{y}_1 - \mathbf{y}_2|. \qquad (6.8)$$

Also, from (6.7),

$$J(\mathbf{x}^1, \cdots, \mathbf{x}^n, \mathbf{y}_1)(\mathbf{x}(\mathbf{y}_1) - \mathbf{x}(\mathbf{y}_2)) = -M(\mathbf{y}_1 - \mathbf{y}_2) \qquad (6.9)$$

and so from (6.8), (6.5), $|\mathbf{x}(\mathbf{y}_1) - \mathbf{x}(\mathbf{y}_2)| =$

$$= \left| J(\mathbf{x}^1, \cdots, \mathbf{x}^n, \mathbf{y}_1)^{-1} M(\mathbf{y}_1 - \mathbf{y}_2) \right|$$

$$= \left(\sum_{i=1}^n \left| \mathbf{e}_i^T J(\mathbf{x}^1, \cdots, \mathbf{x}^n, \mathbf{y}_1)^{-1} M(\mathbf{y}_1 - \mathbf{y}_2) \right|^2 \right)^{1/2}$$

$$\le \left(\sum_{i=1}^n K^2 |M(\mathbf{y}_1 - \mathbf{y}_2)|^2 \right)^{1/2} \le \left(\sum_{i=1}^n K^2 \left(\sqrt{m} K |\mathbf{y}_1 - \mathbf{y}_2| \right)^2 \right)^{1/2}$$

$$= K^2 \sqrt{mn} |\mathbf{y}_1 - \mathbf{y}_2|$$

Now let $\mathbf{y}_2 = \mathbf{y}, \mathbf{y}_1 = \mathbf{y} + h\mathbf{e}_k$ for small h. Then M depends on h and

$$\lim_{h \to 0} M(h) = D_2 \mathbf{f}(\mathbf{x}(\mathbf{y}), \mathbf{y})$$

thanks to the continuity of $\mathbf{y} \to \mathbf{x}(\mathbf{y})$ just shown. Also,

$$\frac{\mathbf{x}(\mathbf{y} + h\mathbf{e}_k) - \mathbf{x}(\mathbf{y})}{h} = -J(\mathbf{x}^1(h), \cdots, \mathbf{x}^n(h), \mathbf{y} + h\mathbf{e}_k)^{-1} M(h) \mathbf{e}_k$$

Passing to a limit and using the formula for the inverse of a matrix in terms of the cofactor matrix, and the continuity of $\mathbf{y} \to \mathbf{x}(\mathbf{y})$ shown above, this yields

$$\frac{\partial \mathbf{x}}{\partial y_k} = -D_1 \mathbf{f}(\mathbf{x}(\mathbf{y}), \mathbf{y})^{-1} D_2 f_i(\mathbf{x}(\mathbf{y}), \mathbf{y}) \mathbf{e}_k$$

Then continuity of $\mathbf{y} \to \mathbf{x}(\mathbf{y})$ and the assumed continuity of the partial derivatives of $\mathbf{f}$ shows that each partial derivative of $\mathbf{y} \to \mathbf{x}(\mathbf{y})$ exists and is continuous. ∎

Appendix G

The Jordan Curve Theorem

This short chapter is devoted to giving an elementary proof of the Jordan curve theorem. This is normally done in courses on algebraic topology or it can be done using degree theory. I am following lecture notes from a topology course given by Fernley at BYU in the 1970s. The ideas used in this presentation are elementary and also lead to more general notions in algebraic topology. In addition to this, these techniques are very useful in complex analysis. This is a very strange thing to put in an ordinary differential equations book. It is here because I think that it is at least as interesting as the Poincare Bendixon theorem which is where it is used and I am so very tired of people referring to theorems with little or no explanation as though mathematics is some sort of religion based on faith in what authority figures have determined.

Definition G.0.41 *A grating G is a **finite** set of horizontal and vertical lines, each of which separate the plane. The grating divides the plane into two dimensional domains the closures of which are called 2 cells of G. The 1 cells of G are the edges of the 2 cells and the 0 cells of G are the end points of the 1 cells.*

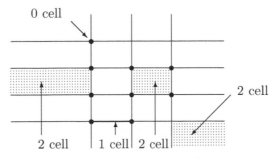

For $k = 0, 1, 2$, one speaks of k chains. For $\{a_j\}_{j=1}^n$ a set of k cells, the k chain is denoted as a formal sum

$$C = a_1 + a_2 + \cdots + a_n$$

where the sum is taken modulo 2. The sums are just formal expressions like the above. Thus for a a k cell, $a + a = 0, 0 + a = a$, the summation sign is commutative. In other words, if a k cell is repeated an even number of times in the formal sum, it disappears resulting in 0 defined by $0 + a = a + 0 = a$. For a a k cell, $|a|$ denotes the points of the plane which are contained in a. For a k chain, C as above,

$$|C| \equiv \{x : x \in |a_j| \text{ for some } a_j\}$$

so $|C|$ is the union of the k cells in the sum remembering that when a k cell occurs twice, it is gone and does not contribute to $|C|$.

The following picture illustrates the above definition. The following is a picture of the 2 cells in a 2 chain. The dotted lines indicate the lines in the grating.

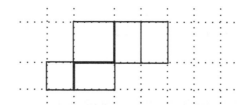

Now the following is a picture of the 1 chain consisting of the sum of the 1 cells which are the edges of the above 2 cells. Remember when a 1 cell is added to itself, it disappears from the chain. Thus if you add up the 1 cells which are the edges of the above 2 cells, lots of them cancel off. In fact all the edges which are shared between two 2 cells disappear. The following is what results.

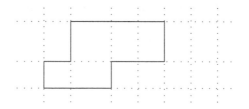

Definition G.0.42 *Next the boundary operator is defined. This is denoted by ∂. ∂ takes k cells to $k-1$ chains. If a is a 2 cell, then ∂a consists of the edges of a. If a is a 1 cell, then ∂a consists of the ends of the 1 cell. If a is a 0 cell, then $\partial a \equiv 0$. This extends in a natural way to k chains. For*

$$C = a_1 + a_2 + \cdots + a_n,$$

$$\partial C \equiv \partial a_1 + \partial a_2 + \cdots + \partial a_n$$

A k chain C is called a cycle if $\partial C = 0$.

In the second of the above pictures, you have a 1 cycle. Here is a picture of another one in which the boundary of another 2 cell has been included over on the right.

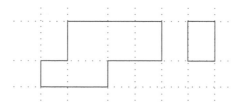

This 1 cycle shown above is the boundary of exactly two 2 chains. What are they? C_1 consists of the 2 cells in the first picture above along with the 2 cell whose boundary is the 1 cycle over on the right. C_2 is all the other 2 cells of the grating. You see this clearly works. Could you make that 2 cell on the right be in C_2? No, you couldn't do it. This is because the 1 cells which are shown would disappear, being listed twice.

This illustrates the fundamental lemma of the plane which comes next.

Lemma G.0.43 *If C is a bounded 1 cycle ($\partial C = 0$), then there are exactly two 2 chains D_1, D_2 such that*
$$C = \partial D_1 = \partial D_2.$$

Proof: The lemma is vacuously true unless there are at least two vertical lines and at least two horizontal lines in the grating G. It is also obviously true if there are exactly two vertical lines and two horizontal lines in G. Suppose the theorem is true for n lines in G. Then as just mentioned, there is nothing to prove unless there are either 2 or more vertical lines and two or more horizontal lines. Suppose without loss of generality there are at least as many vertical lines are there are horizontal lines and that this number is at least 3. If it is only two, there is nothing left to show. Let l be the second vertical line from the left. Let $\{e_1, \cdots, e_m\}$ be the 1 cells of C with the property that $|e_j| \subseteq l$. Note that e_j occurs only once in C since if it occurred twice, it would disappear because of the rule for addition. Pick one of the 2 cells adjacent to e_j, b_j and add in ∂b_j which is a 1 cycle. Thus

$$C + \sum_j \partial b_j$$

is a bounded 1 cycle and it has the property that it has no 1 cells contained in l. Thus you could eliminate l from the grating G and all the 1 cells of the above 1 chain are edges of the grating $G \setminus \{l\}$. By induction, there are exactly two 2 chains D_1, D_2 composed of 2 cells of $G \setminus \{l\}$ such that for $i = 1, 2$,

$$\partial D_i = C + \sum_j \partial b_j \tag{7.1}$$

Since none of the 2 cells of D_i have any edges on l, one can add l back in and regard D_1 and D_2 as 2 chains in G. Therefore, adding $\sum_j \partial b_j$ to both sides of the above yields

$$C = \partial D_i + \sum_j \partial b_j = \partial \left(D_i + \sum_j b_j \right), i = 1, 2.$$

and this shows there exist two 2 chains which have C as the boundary. If $\partial D_i' = C$, then

$$\partial D_i' + \sum_j \partial b_j = \partial \left(D_i' + \sum_j b_j \right) = C + \sum_j \partial b_j$$

and by induction, there are exactly two 2 chains which $D_i' + \sum_j b_j$ can equal. Thus adding $\sum_j b_j$ there are exactly two 2 chains which D_i' can equal.

Here is another proof which is not by induction. This proof also gives an algorithm for identifying the two 2 chains. The 1 cycle is bounded and so every 1 cell in it is part of the boundary of a 2 cell which is bounded. For the unbounded 2 cells on the left, label them all as A. Now starting from the left and moving toward the right, toggle between A and B every time you hit a vertical 1 cell of C. This will label every 2 cell with either A or B. Next, starting at the top, label all the unbounded 2 cells as A and move down and toggle between A and B every time you encounter a horizontal 1 cell of C. This also labels every 2 cell as either A or B. Suppose there is a contradiction in the labeling. Pick the first column in which a contradiction occurs and then pick the top contradictory 2 cell in this column. There are various cases which can occur, each leading to the existence of a vertex of C which is contained in an odd number of 1 cells of C, thus contradicting the conclusion that C is a 1 cycle. In the following picture, AB will mean the labeling from the left to right gives A and the labeling from top to bottom yields B with similar modification for AA and BB.

$$
\begin{array}{c|c} AA & BB \\ \hline BB & AB \end{array}
\qquad
\begin{array}{c:c} AA & AA \\ \hline AA & AB \end{array}
$$

A solid line indicates the corresponding 1 cell is in C. It is there because a change took place either from top to bottom or from left to right. Note that in both of those situations the vertex right in the middle of the crossed lines will occur in ∂C and so C is not a 1 cycle. There are 8 similar pictures you can draw and in each case this happens. The vertex in the center gets added in an odd number of times. You can also notice that if you start with the contradictory 2 cell and move counter clockwise, crossing 1 cells as you go and starting with B, you must end up at A as a result of crossing 1 cells of C and this requires crossing either one or three of these 1 cells of C.

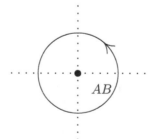

Thus that center vertex is a boundary point of C and so C is not a 1 cycle after all. Similar considerations would hold if the contradictory 2 cell were labeled BA. Thus there can be no contradiction in the two labeling schemes. They label the 2 cells in G either A or B in an unambiguous manner.

The labeling algorithm encounters every 1 cell of C (in fact of G) and gives a label to every 2 cell of G. Define the two 2 chains as A and B where A consists of those labeled as A and B those labeled as B. The 1 cells which cause a change to take place in the labeling are exactly those in C and each is contained in one 2 cell from A and one 2 cell from B. Therefore, each of these 1 cells of C appears in ∂A and ∂B which shows $C \subseteq \partial A$ and $C \subseteq \partial B$. On the other hand, if l is a 1 cell in ∂A, then it can only occur in a single 2 cell of A and so the 2 cell adjacent to that one along l must be in B and so l is one of the 1 cells of C by definition. As to uniqueness, in moving from left to right, you must assign adjacent 2 cells joined at a 1 cell of C to different 2 chains or else the 1 cell would not appear when you take the boundary of either A or B since it would be added in twice. Thus there are exactly two 2 chains with the desired property. ■

The next lemma is interesting because it gives the existence of a continuous curve joining two points.

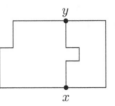

Lemma G.0.44 *Let C be a bounded 1 chain such that $\partial C = x + y$. Then both x, y are contained in a continuous curve which is a subset of $|C|$.*

Proof: There are an odd number of 1 cells of C which have x at one end. Otherwise $\partial C \neq x + y$. Begin at x and move along an edge leading away from x. Continue till there is no new edge to travel along. You must be at y since otherwise, you would have found another boundary point of C. This point would be in either one or three one cells of C. It can't be x because x is contained in either one or three one cells of C. Thus, there is always a way to leave x if the process returns to it. IT follows that there is a continuous curve in $|C|$ joining x to y. ■

The next lemma gives conditions under which you can go around a couple of closed sets. It is called Alexander's lemma. The following picture is a rough illustration of the situation. Roughly, it says that if you can miss F_1 and you can miss F_2 in going from x to y, then you can miss both F_1 and F_2 by climbing around F_1.

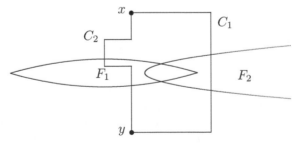

Lemma G.0.45 *Let F_1 be compact and F_2 closed. Suppose C_1, C_2 are two bounded 1 chains in a grating which has no unbounded two cells having nonempty intersection with F_1. Suppose $\partial C_i = x + y$ where $x, y \notin F_1 \cup F_2$. Suppose C_2 does not intersect F_2 and C_1 does not intersect F_1. Also suppose the 1 cycle $C_1 + C_2$ bounds a 2 chain D for which $|D| \cap F_1 \cap F_2 = \emptyset$. Then there exists a 1 chain C such*

that $\partial C = x + y$ and $|\partial C| \cap (F_1 \cup F_2) = \emptyset$. In particular x, y cannot be in different components of the complement of $F_1 \cup F_2$.

Proof: Let $a_1, a_2, \cdots, a_m$ be the 2 cells of D which intersect the compact set F_1. Consider

$$C \equiv C_2 + \sum_k \partial a_k.$$

This is a 1 chain and $\partial C = x + y$ because $\partial \partial a_k = 0$. Then $|a_k| \cap F_2 = \emptyset$. This is because $|a_k| \cap F_1 \neq \emptyset$ and none of the 2 cells of D intersect both F_1 and F_2 by assumption. Therefore, C is a bounded 1 chain which avoids intersecting F_2.

Does it also avoid F_1? Suppose to the contrary that l is a one cell of C which does intersect F_1. If $|l| \subseteq |C_1 + C_2|$, then it would be an edge of some 2 cell of D and would have to be a 1 cell of C_2 since it intersects F_1 so it would have been added twice, once from C_2 and once from

$$\sum_k \partial a_k$$

and therefore could not be a summand in C. Therefore, $|l|$ is not in $|C_1 + C_2|$. It follows l must be an edge of some $a_k \in D$ and it is not a 1 cell of $C_1 + C_2$. Therefore, if b is the 2 cell adjacent to a_k, it must follow $b \in D$ since otherwise l would be a 1 cell of $C_1 + C_2$ the boundary of D and this was just ruled out. But now it would follow that l would occur twice in the above sum so l cannot be a summand of C. Therefore C misses F_1 also.

Here is another argument. Suppose $|l| \cap F_1 \neq \emptyset$. $l \in C = C_2 + \sum_k \partial a_k$. First note that $l \notin C_1$ since $|C_1| \cap F_1 = \emptyset$.

Case 1: $l \in C_2$.

In this case it is in $C_1 + C_2$ because, as just noted it is not in C_1. Therefore, there exists $a \in D$ such that l is an edge of a and is not in the two cell adjacent to a. But this would require l to disappear since it would occur in both C_2 and $\sum_k \partial a_k$. Hence $l \notin C_2$.

Case 2: In this case $l \notin C_2$. Then l is the edge of some $a \in D$ which intersects F_1. Letting b be the two cell adjacent to a sharing l, then b cannot be in D since otherwise l would occur twice in the above sum and would then disappear. Hence $b \notin D$ and $l \in \partial D = C_1 + C_2$ but this cannot happen because $l \notin C_1$ and in this case $l \notin C_2$ either. ∎

Lemma G.0.46 *Let C be a bounded 1 cycle such that $|C| \cap H = \emptyset$ where H is a connected set. Also let D, E be the 2 chains with $\partial D = C = \partial E$. Then either $|H| \subseteq |E|$ or $|H| \subseteq |D|$.*

Proof: If p is a limit point of $|E|$ and $p \in |D|$, then p must be contained in an edge of some 2 cell of D since otherwise it could not be a limit point, being contained in an open set whose intersection with $|E|$ is empty. If p is a point of an even number of edges of 2 cells of D, then it is likewise an interior point of $|D|$ which cannot be a limit point of $|E|$. Therefore, if p is a limit point of $|E|$, it must be the case that $p \in |C|$. A similar observation holds for the case where $p \in |E|$ and is a limit point of $|D|$. Thus if $H \cap |D|$ and $H \cap |E|$ are both nonempty, then they separate the connected set H and so H must be a subset of one of $|D|$ or $|E|$. ∎

Definition G.0.47 *A Jordan arc is a set of points of the form $\Gamma \equiv \mathbf{r}\left([a,b]\right)$ where $\mathbf{r}$ is a one to one map from $[a,b]$ to the plane. For $p,q \in \Gamma$, say $p < q$ if $p = \mathbf{r}\left(t_1\right), q = \mathbf{r}\left(t_2\right)$ for $t_1 < t_2$. Also let pq denote the arc $\mathbf{r}\left([t_1,t_2]\right)$.*

Theorem G.0.48 *Let Γ be a Jordan arc. Then its complement is connected.*

Proof: Suppose this is not so. Then there exists x,y points in Γ^C which are in different components of Γ^C. Let G be a grating having x,y as points of intersection of a horizontal line and a vertical line of G and let p,q be the points at the ends of the Jordan arc. Also let G be such that no unbounded two cell has nonempty intersection with Γ. Let $p = \mathbf{r}\left(a\right)$ and $q = \mathbf{r}\left(b\right)$. Now let $z = \mathbf{r}\left(\frac{a+b}{2}\right)$ and consider the two arcs pz and zq. If $\partial C = x + y$ then it is required $|C| \cap \Gamma \neq \emptyset$ since otherwise these two points would not be in different components. Suppose there exists $C_1, \partial C_1 = x + y$ and $|C_1| \cap zq = \emptyset$ and $C_2, \partial C_2 = x + y$ but $|C_2| \cap pz = \emptyset$. Then $C_1 + C_2$ is a 1 cycle and so by Lemma G.0.43 there are exactly two 2 chains whose boundaries are $C_1 + C_2$. Since $z \notin |C_i|$, it follows $z = pz \cap zq$ can only be in one of these 2 chains because it is a single point. Then by Lemma G.0.45, Alexander's lemma, there exists C a 1 chain with $\partial C = x + y$ and $|C| \cap (pz \cup zq) = \emptyset$ so by Lemma G.0.44 x, y are not in different components of Γ^C contrary to the assumption they are in different components. Hence one of pz, zq has the property that every 1 chain, $\partial C = x + y$ goes through it. Say every such 1 chain goes through zq. Then let zq play the role of pq and conclude every 1 chain C such that $\partial C = x + y$ goes through either zw or wq there

$$w = \mathbf{r}\left(\left(\frac{a+b}{2} + b\right)\frac{1}{2}\right)$$

Thus, continuing this way, there is a sequence of Jordan arcs $p_k q_k$ where $\mathbf{r}\left(t_k\right) = q_k$ and $\mathbf{r}\left(s_k\right) = p_k$ with $|t_k - s_k| < \frac{b-a}{2^k}$, $[s_k, t_k] \subseteq [a,b]$ such that every C with $\partial C = x + y$ has nonempty intersection with $p_k q_k$. The intersection of these arcs is $\mathbf{r}\left(s\right)$ where $s = \cap_{k=1}^{\infty}[s_k, t_k]$. Then all such C must go through $\mathbf{r}\left(s\right)$ because such C with $\partial C = x + y$ must intersect $p_k q_k$ for each k and their intersection is $\mathbf{r}\left(s\right)$. But now there is an obvious contradiction to having every 1 chain whose boundary is $x + y$ intersecting $\mathbf{r}\left(s\right)$.

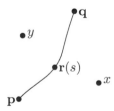

Pick a 1 chain whose boundary is $x + y$. Let D be the two chain of at most four 2 cells consisting of those two cells which have $\mathbf{r}\left(s\right)$ on some edge. Then $\partial\left(C + \partial D\right) = \partial C = x + y$ but $\mathbf{r}\left(s\right) \notin |C + \partial D|$. Therefore, this contradiction shows Γ^C must be connected after all. ∎

The other important observation about a Jordan arc is that it has no interior points. This will follow later from a harder result but it is also easy to prove.

Lemma G.0.49 *Let* $\Gamma = \mathbf{r}\left([a,b]\right)$ *be a Jordan arc where* $\mathbf{r}$ *is as above, one to one, onto and continuous. Then* Γ *has no interior points.*

Proof: Suppose to the contrary that Γ has an interior point p. Then for some $r > 0$,

$$B\left(p,r\right) \subseteq \Gamma.$$

Consider the circles of radius $\delta < r$ centered at p. Denoting as C_δ one of these, it follows the C_δ are disjoint. Therefore, since $\mathbf{r}$ is one to one, the sets $\mathbf{r}^{-1}\left(C_\delta\right)$ are also disjoint. Now $\mathbf{r}$ is continuous and one to one mapping to a compact set. Therefore, $\mathbf{r}^{-1}$ is also continuous. It follows $\mathbf{r}^{-1}\left(C_\delta\right)$ is connected and compact. Thus by Theorem E.2.31 each of these sets is a closed interval of positive length since $\mathbf{r}$ is one to one. It follows there exist disjoint open nonempty intervals consisting of the interiors of $\mathbf{r}^{-1}\left(C_\delta\right), \{I_\delta\}_{\delta < r}$. This is a contradiction to the density of $\mathbb{Q}$ and the fact that $\mathbb{Q}$ is at most countable. ■

Definition G.0.50 *Let* $\mathbf{r}$ *map* $[a,b]$ *to the plane such that* $\mathbf{r}$ *is one to one on* $[a,b)$ *and* $(a,b]$ *but* $\mathbf{r}\left(a\right) = \mathbf{r}\left(b\right)$. *Then* $J = \mathbf{r}\left([a,b]\right)$ *is called a simple closed curve. It is also called a Jordan curve. Also since the term "boundary" has been given a specialized meaning relative to chains of various sizes, we say* x *is in the frontier of* S *if every open ball containing* x *contains points of* S *as well as points of* S^C.

Note that if J is a Jordan curve, then it is the union of two Jordan arcs whose intersection is two distinct points of J. You could pick $z \in (a,b)$ and consider $\mathbf{r}\left([a,z]\right)$ and $\mathbf{r}\left([z,b]\right)$ as the two Jordan arcs.

The next lemma gives a probably more convenient way of thinking about a Jordan curve. It says essentially that a Jordan curve is a wriggly circle. First consider the following simple lemma.

Lemma G.0.51 *Let* K *be a compact set in* $\mathbb{R}^n$ *and let* $\mathbf{f} : K \to \mathbb{R}^m$ *be continuous and one to one. Then* $\mathbf{f}^{-1} : \mathbf{f}\left(K\right) \to K$ *is also continuous.*

Proof: Suppose $\{\mathbf{f}\left(k_n\right)\}$ is a convergent sequence in $\mathbf{f}\left(K\right)$ converging to $\mathbf{f}\left(k\right)$. Does it follow that $k_n \to k$? If not, there exists a subsequence $\{k_{n_k}\}$ which converges as $k \to \infty$ to $l \neq k$. Then by continuity of $\mathbf{f}$ it follows $\mathbf{f}\left(k_{n_k}\right) \to \mathbf{f}\left(l\right)$. Hence $\mathbf{f}\left(l\right) = \mathbf{f}\left(k\right)$ which violates the condition that $\mathbf{f}$ is one to one.

Lemma G.0.52 *J is a simple closed curve if and only if there exists a mapping* $\boldsymbol{\theta} : S^1 \to J$ *where* S^1 *is the unit circle*

$$\left\{(x,y) : x^2 + y^2 = 1\right\},$$

such that $\boldsymbol{\theta}$ *is one to one and continuous.*

Proof: Suppose that J is a simple closed curve so there is a parametrization $\mathbf{r}$ and an interval $[a,b]$ such that $\mathbf{r}$ is continuous and one to one on $[a,b)$ and $(a,b]$ with $\mathbf{r}\left(a\right) = \mathbf{r}\left(b\right)$. Let $C_0 = \mathbf{r}\left((a,b)\right), C_\delta = \mathbf{r}\left([a+\delta, b-\delta]\right)$, and let S^1 denote the unit circle. Let l be a linear one to one map from $[a, b]$ onto $[0, 2\pi]$. Consider the following diagram.

$$[a, b] \overset{l}{\to} [0, 2\pi]$$
$$\downarrow \mathbf{r} \qquad \downarrow \mathbf{R}$$
$$C \qquad S^1$$

where $\mathbf{R}(\theta) \equiv (\cos\theta, \sin\theta)$. Then clearly $\mathbf{R}$ is continuous. It is also the case that, from the above lemma, $\mathbf{r}^{-1}$ is continuous on C_δ. Therefore, since $\delta > 0$ is arbitrary, $\boldsymbol{\theta} \equiv \mathbf{R} \circ l \circ \mathbf{r}^{-1}$ is a one to one and onto mapping from C_0 to $S^1 \setminus (1, 0)$. Also, letting $\mathbf{p} = \mathbf{r}(a) = \mathbf{r}(b)$, it follows that $\boldsymbol{\theta}(\mathbf{p}) = (1, 0)$. It remains to verify that $\boldsymbol{\theta}$ is continuous at $\mathbf{p}$. Suppose then that $\mathbf{r}(x_n) \to \mathbf{p} = \mathbf{r}(a) = \mathbf{r}(b)$. If $\boldsymbol{\theta}(\mathbf{r}(x_n))$ fails to converge to $(1, 0) = \boldsymbol{\theta}(\mathbf{p})$, then there is a subsequence, still denoted as x_n and $\varepsilon > 0$ such that $|\boldsymbol{\theta}(\mathbf{r}(x_n)) - \boldsymbol{\theta}(\mathbf{p})| \geq \varepsilon$. In particular $x_n \notin \{a, b\}$. By the above lemma, $\mathbf{r}^{-1}$ is continuous on $\mathbf{r}([a, b))$ since this is true for $\mathbf{r}([a, b - \eta])$ for each $\eta > 0$. Since $\mathbf{p} = \mathbf{r}(a)$, it follows that

$$|\boldsymbol{\theta}(\mathbf{r}(x_n)) - \boldsymbol{\theta}(\mathbf{r}(a))| = |\mathbf{R} \circ l(x_n) - \mathbf{R} \circ l(a)| \geq \varepsilon$$

Hence there is some $\delta > 0$ such that $|x_n - a| \geq \delta_1$. Similarly, $|x_n - b| \geq \delta_2 > 0$. Letting $\delta = \min(\delta_1, \delta_2)$, it follows that $x_n \in [a + \delta, b - \delta]$. Taking a convergent subsequence, still denoted as $\{x_n\}$, there exists $x \in [a + \delta, b - \delta]$ such that $x_n \to x$. However, this implies that $\mathbf{r}(x_n) \to \mathbf{r}(x)$ and so $\mathbf{r}(x) = \mathbf{r}(a) = \mathbf{p}$, a contradiction to the fact that $\mathbf{r}$ is one to one on $[a, b)$.

Next suppose J is the image of the unit circle as just explained. Then let $\mathbf{R}: [0, 2\pi] \to S^1$ be defined as $\mathbf{R}(t) \equiv (\cos(t), \sin(t))$. Then consider $\mathbf{r}(t) \equiv \boldsymbol{\theta}(\mathbf{R}(t))$. $\mathbf{r}$ is one to one on $[0, 2\pi)$ and $(0, 2\pi]$ with $\mathbf{r}(0) = \mathbf{r}(2\pi)$ and is continuous, being the composition of continuous functions. ∎

Before the proof of the Jordan curve theorem, recall Theorem E.2.37 which says that the connected components of an open sets are open and that an open connected set is arcwise connected. If J is a Jordan curve then it is the continuous image of the compact set S^1 and so J is also compact. Therefore, its complement is open and the connected components of J^C are connected. The following lemma is a fairly obvious conclusion of this. A square curve is a continuous curve which consists entirely of line segments which are either horizontal or vertical.

Lemma G.0.53 *Let U be a connected open set and let x, y be points of U. Then there is a square curve which joins x and y.*

Proof: Let V denote those points of U which can be joined to x by a square curve. Then if $z \in V$, there exists $B(z, r) \subseteq U$. It is clear that every point of $B(z, r)$ can be joined to z by a square curve. Also V^C must be open since if $z \in V^C, B(z, r) \subseteq U$ for some r. Then if any $w \in B(z, r)$ is in V, one could join w to z by a square curve and conclude that $z \in V$ after all. The fact that both V, V^C are both open would result in a contradiction unless both $x, y \in V$ since otherwise, U is separated by V, V^C. ∎

Theorem G.0.54 *Let J be a Jordan curve in the plane. Then J^C consists of exactly two components, a bounded component, and an unbounded component, and J is the frontier of both of these components. Furthermore, J has empty interior.*

Proof: To begin with consider the claim there are no more than two components. Suppose this is not so. Then there exist x, y, z each of which is in a different component of J^C. Let $J = H \cup K$ where H and K are two Jordan arcs joined at the points a and b. If the Jordan curve is $\mathbf{r}([c,d])$ where $\mathbf{r}(c) = \mathbf{r}(d)$ as described above, you could take $H = \mathbf{r}\left(\left[c, \frac{c+d}{2}\right]\right)$ and $K = \mathbf{r}\left(\left[\frac{c+d}{2}, d\right]\right)$. Thus the points on the Jordan curve illustrated in the following picture could be

$$a = \mathbf{r}(c), b = \mathbf{r}\left(\frac{c+d}{2}\right)$$

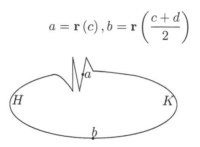

First we show that there is at most two components in J^C. Suppose to the contrary that there exists x, y, z, each in a different component. By the Jordan arc theorem above, and the above lemma about square curves, there exists a square curve C_{xyH} such that $\partial C_{xyH} = x + y$ and $|C_{x,yH}| \cap H = \emptyset$. Using the same notation in relation to the other points, there exist square curves in the following list.

$$\begin{aligned}
C_{xyH}, \ \partial C_{xyH} &= x + y, \ C_{yzH}, \ \partial C_{yzH} = y + z \\
C_{xyK}, \ \partial C_{xyH} &= x + y, \ C_{yzK}, \ \partial C_{yzK} = y + z
\end{aligned}$$

Let these square curves be part of a grating which includes all vertices of all these square curves and contains the compact set J in the bounded two cells. First note that $C_{xyH} + C_{xyK}$ is a one cycle and that

$$|C_{xyH} + C_{xyK}| \cap (H \cap K) = \emptyset$$

Also note that $H \cap K = \{a, b\}$ since $\mathbf{r}$ is one to one on $[c, d)$ and $(c, d]$. Therefore, there exist unique two chains D, E such that $\partial D = \partial E = C_{xyH} + C_{xyK}$. Now if one of these two chains contains both a, b then then the other two chain does not contain either a nor b. Then by Alexander's lemma, Lemma G.0.45, there would exist a square curve C such that $|C| \cap (H \cup K) = |C| \cap J = \emptyset$ and $\partial C = x + y$ which is assumed not to happen. Therefore, one of the two chains contains a and the other contains b. Say $a \in |D|$ and $b \in |E|$. Similarly there exist unique two chains P, Q such that

$$\partial P = \partial Q = C_{yzH} + C_{yzK}$$

where $a \in |P|$ and $b \in |Q|$. Now consider

$$\partial(D + Q) = C_{xyH} + C_{xyK} + C_{yzH} + C_{yzK}$$

$$= (C_{xyH} + C_{yzH}) + (C_{xyK} + C_{yzK})$$

This is a one cycle because its boundary is $x + y + y + z + x + y + y + z = 0$. By Lemma G.0.43, the fundamental lemma of the plane, there are exactly two two chains whose boundaries equal this one cycle. Therefore, $D + Q$ must be one of them. Also $b \in |Q|$

and is not in $|D|$. Hence $b \in |D + Q|$. Similarly $a \in |D + Q|$. It follows that the other two chain whose boundary equals the above one cycle contains neither a nor b. In addition to this, $C_{xyH} + C_{yzH}$ misses H and $C_{xyK} + C_{yzK}$ misses K. Both of these one chains have boundary equal to $x + z$. By Alexander's lemma, there exists a one chain C which misses both H and K (all of J) such that $\partial C = x + z$ which contradicts the assertion that x, z are in different components. This proves the assertion that there are only two components to J^C.

Next, why are there at least two components in J^C? Suppose there is only one and let a, b be the points of J described above and H, K also as above. Let Q be a small square 1 cycle which encloses a on its inside such that b is not inside Q. Thus a is on the inside of Q and b is on the outside of Q as shown in the picture.

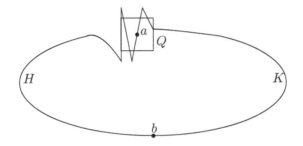

Now let G be a grating which has the corners of Q as points of intersection of horizontal and vertical lines and also has all the 2 cells so small that none of them can intersect both of the disjoint compact sets $H \cap |Q|$ and $|Q| \cap K$. Let P be the 1 cells contained in Q which have nonempty intersection with H. Some of them must have nonempty intersection with H because if not, then H would fail to be connected, having points inside Q, a, and points outside Q, b, but no points on Q. Similarly some of these one cells in Q have nonempty intersection with K. Let $\partial P = x_1 + \cdots + x_m$. Then it follows each $x_k \notin H$. Could $\partial P = 0$? Suppose $\partial P = 0$. If l is a one cell of P, then since its ends are not in ∂P, the two adjacent one cells to l which are in Q must also intersect H. Moving counterclockwise around Q, it would follow that all the one cells contained in Q would intersect H. However, at least one must intersect K because if not, a is a point of K inside the square Q while b is a point of K outside Q thus separating K which is a connected set. However, this contradicts the choice of the grating. Therefore, $\partial P \neq 0$. Now this violates the assumption that no 2 cell of G can intersect both of those disjoint compact sets $H \cap |Q|$ and $|Q| \cap K$. Starting with a one cell of Q which does not intersect H, move counter clockwise till you obtain the first one which intersects H. This will produce a point of ∂P. Then the next point of ∂P will occur when the first one cell of P which does not intersect H is encountered. Thus a pair of points in ∂P are obtained. Now you are in the same position as before, continue moving counter clockwise and obtaining pairs of points of ∂P till there are no more one cells of Q which intersect H. You must have encountered an even number of points for ∂P.

Since it is assumed there is only one component of J^C, it follows upon refining G if necessary, there exist 1 chains B_k contained in J^C such that $\partial B_k = x_1 + x_k$ **and**

it is the existence of these B_k which will give the desired contradiction.
Let $B = B_2 + \cdots + B_m$. Then $P + B$ is a 1 cycle which misses K. It is a one cycle
because m is even.

$$\partial(P + B) = \sum_{k=2}^{m=2l} x_1 + x_k + \sum_{k=1}^{2l} x_k = \sum_{k=1}^{2l} x_k + \sum_{k=2}^{2l} x_k + x_k = 0$$

It misses K because B misses J and all the 1 cells of P in the original grating G
intersect $H \cap |Q|$ so they cannot intersect K. Also $P + Q + B$ is a 1 cycle which
misses H. This is because B misses J and every 1 cell of P which intersects H
disappears because $P + Q$ causes them to be added twice. Since H and K are
connected, it follows from Lemma G.0.46, G.0.43 that $P + B$ bounds a 2 chain D
which is contained entirely in K^C(the one which does not contain K). Similarly
$P + Q + B$ bounds a 2 chain E which is contained in H^C(the 2 chain which does
not contain H). Thus $D + E$ is a 2 chain which does not contain either a or b. (D
misses K and E misses H and $\{a, b\} = H \cap K$) However,

$$\partial(D + E) = P + B + P + Q + B = Q$$

and so $D + E$ is one of the 2 chains of Lemma G.0.43 which have Q as boundary.
However, Q bounds the 2 chain of 2 cells which are inside Q which contains a and
the 2 chain of 2 cells which are outside Q which contains b. This is a contradiction
because neither of these 2 chains miss both a and b and this shows there are two
components of J^C.

In the above argument, if each pair $\{x_1, x_i\}$ can be joined by a square curve B_i
which lies in J^C, then the contradiction was obtained. Therefore, there must exist
a pair $\{x_1, x_i\}$ which can't be joined by any square curve in J^C and this requires
these points to be in different components by Lemma G.0.53 above. Since they are
both on Q and Q could be as small as desired, this shows a is in the frontier of both
components of J^C. Furthermore, a was arbitrary so every point of J is a frontier
point of both the components of J^C. These are the only frontier points because the
components of J^C are open.

By Lemma G.0.52, J is the continuous image of the compact set S^1 so it follows
J is bounded. The unbounded component of J^C is the one which contains the
connected set $B(0, R)^C$ where $J \subseteq B(0, R)$. Thus there are two components for
J^C, the unbounded one which contains $B(0, R)^C$ and the bounded one which must
be contained in $B(0, R)$. This proves the theorem. ∎

Appendix H

Poincare Bendixon Theorem

This section contains an explanation of the Poincare Bendixon theorem which is used to determine the existence of periodic solutions to a nonlinear system of differential equations in the plane. This is a very remarkable result, but the proof is a little technical and will depend on the implicit function theorem as well as the Jordan curve theorem.

We have a planar system of equations

$$\mathbf{x}' = \mathbf{f}(\mathbf{x}), \ \mathbf{x} = (x, y), \ \mathbf{x}_0 = (x_0, y_0) = \mathbf{x}(0)$$

the solution of this being $\mathbf{x}(t, \mathbf{x}_0)$. Then we make the following definition.

Definition H.0.55 *For a given* $\mathbf{x}_0, \Lambda^+$ *will denote*

$$\left\{ \mathbf{z} : \mathbf{z} = \lim_{n \to \infty} \mathbf{x}(t_n, \mathbf{x}_0) \text{ for some } t_n \to \infty \right\}$$

Also let Λ_+ *be given by*

$$\cup_{t \geq 0} \mathbf{x}(t, \mathbf{x}_0)$$

In all these considerations, we assume that $\mathbf{f}$ is C^1 at least and that there is a bounded open set D_0 such that for D its closure and if $\mathbf{x}_0 \in D$, then $\mathbf{x}(t, \mathbf{x}_0)$ remains in D for all $t \geq 0$ and that therefore, it is reasonable to assume that $\mathbf{f}$ is a Lipschitz continuous function so that there are no issues about whether the solution to the initial value problem exists or is unique.

First here is a very interesting lemma.

Lemma H.0.56 *Suppose* $\mathbf{x}(t + T, \mathbf{x}_1) = \mathbf{x}(t, \mathbf{x}_1)$ *for* $T, t \geq 0$ *if and only if* $\mathbf{x}(T, \mathbf{x}_1) = \mathbf{x}_1$.

Proof: $\implies$ Suppose that for all $t \geq 0$, $\mathbf{x}(t + T, \mathbf{x}_1) = \mathbf{x}(t, \mathbf{x}(T, \mathbf{x}_1))$ and also this equals $\mathbf{x}(t, \mathbf{x}_1)$ which is the first hypothesis. By Proposition 11.7.2, $\mathbf{x}_1 = \mathbf{x}(T, \mathbf{x}_1)$.

$\impliedby$ If $\mathbf{x}(T, \mathbf{x}_1) = \mathbf{x}_1$, then by uniqueness, $\mathbf{x}(t + T, \mathbf{x}_1) = \mathbf{x}(t, \mathbf{x}(T, \mathbf{x}_1)) = \mathbf{x}(t, \mathbf{x}_1)$. ∎

Lemma H.0.57 *Assume* $\mathbf{f}(\mathbf{x}) \neq \mathbf{0}$ *for every* $\mathbf{x} \in \Lambda^+$ *for* Λ^+ *the limit set coming from the initial condition* $\mathbf{x}_0$. *Then if there exists* $\mathbf{x}_1 \in \Lambda^+ \cap \Lambda_+$, *it follows that* $\Lambda_+ = \Lambda^+$ *and is a periodic orbit meaning that* $\mathbf{x}(t+T, \mathbf{x}_0) = \mathbf{x}(t, \mathbf{x}_0)$ *for all* $t \geq 0$.

Proof: Let $\mathbf{x}_1 = \mathbf{x}(t_1, \mathbf{x}_0)$. Then, since $\mathbf{x}_1 \in \Lambda^+$, for every $\varepsilon > 0$, there exists $\mathbf{x}(\hat{t}_2, \mathbf{x}_0) \in B(\mathbf{x}_1, \varepsilon)$ for $\hat{t}_2$ as large as desired. In particular this is true for some $\hat{t}_2$ larger than t_1. Assume that $t \to \mathbf{x}(t, \mathbf{x}_1)$ is one to one on every closed interval $[0, \hat{t}]$. This will lead to a contradiction by showing that in fact, there exists T such that $\mathbf{x}(T, \mathbf{x}_1) = \mathbf{x}_1$ and this is the way the argument will develop.

Now we consider the size of ε just mentioned. $\mathbf{f}(\mathbf{x}_1) \neq \mathbf{0}$ and so there exist constants a, b, c such that $\mathbf{x}_1$ is on the line

$$\{\mathbf{x} : (a, b) \cdot \mathbf{x} = c\},$$

where

$$\mathbf{f}(\mathbf{x}_1) \cdot (a, b) > 0$$

the last inequality being equivalent to saying that the direction vector $\mathbf{x}'(0, \mathbf{x}_1)$ makes an angle of less than $90°$ with the vector (a, b). In geometric terms, it crosses the line. By continuity of $\mathbf{f}$, let ε be small enough that for all $\mathbf{z} \in B(\mathbf{x}_1, \varepsilon)$,

$$\mathbf{f}(\mathbf{z}) \cdot (a, b) > 0$$

Now consider the function

$$g(t, \mathbf{z}) \equiv (a, b) \cdot \mathbf{x}(t, \mathbf{z}) - c,$$

Since $\mathbf{f}$ is C^1, it follows from the discussion after Theorem 11.6.2 that $\mathbf{x}$ in the above equation is C^1. Also

$$\begin{aligned} g(0, \mathbf{x}_1) &\equiv ax(0, \mathbf{x}_1) + by(0, \mathbf{x}_1) - c \\ &= (a, b) \cdot \mathbf{x}_1 - c = 0 \end{aligned}$$

and

$$\begin{aligned} g_t(0, \mathbf{x}_1) &= ax_t(0, \mathbf{x}_1) + by_t(0, \mathbf{x}_1) = (a, b) \cdot \mathbf{x}'(0, \mathbf{x}_1) \\ &= (a, b) \cdot \mathbf{f}(\mathbf{x}_1) > 0 \end{aligned}$$

By the implicit function theorem, there exists $\delta > 0$ very small, $\delta < 1$ and ε at least as small as what was just chosen such that for all $\mathbf{z} \in B(\mathbf{x}_1, \varepsilon)$, there exists a unique $\hat{t} \in (-\delta, \delta)$ such that

$$g(\hat{t}, \mathbf{x}(\hat{t}, \mathbf{z})) = 0.$$

Now apply this for $\mathbf{z} = \mathbf{x}(\hat{t}_2, \mathbf{x}_0) \in B(\mathbf{x}_1, \varepsilon)$ where $\hat{t}_2 > t_1 + 1$. Then

$$ax(\hat{t}, \mathbf{x}(\hat{t}_2, \mathbf{x}_0)) + by(\hat{t}, \mathbf{x}(\hat{t}_2, \mathbf{x}_0)) = c$$

Thus by Proposition 11.7.2,

$$ax(\hat{t} + \hat{t}_2, \mathbf{x}_0) + by(\hat{t} + \hat{t}_2, \mathbf{x}_0) = (a, b) \cdot \mathbf{x}(\hat{t} + \hat{t}_2, \mathbf{x}_0) = c$$

Thus $(x(t_1, \mathbf{x}_0), y(t_1, \mathbf{x}_0)) = \mathbf{x}_1$ is on the line l which has equation $ax + by = c$ and there is $t_2 \equiv \hat{t}_2 + \hat{t} > t_1$ such that $(x(t_2, \mathbf{x}_0), y(t_2, \mathbf{x}_0))$ is on the line l. It is also being assumed that

$$(x(t_1, \mathbf{x}_0), y(t_1, \mathbf{x}_0)) \neq (x(t_2, \mathbf{x}_0), y(t_2, \mathbf{x}_0))$$

and that in fact, $t \to \mathbf{x}(t, \mathbf{x}_1)$ is one to one on every closed interval.

How many intersections can there be between $\mathbf{x}(t, \mathbf{x}_0)$ and l inside $B(\mathbf{x}_1, \varepsilon)$? For $t \in [t_1, \hat{t}]$ there can be only finitely many. If there are infinitely many, say $\{t_n\}$, then there is a subsequence, still denoted as t_n such that $t_n \to \tilde{t} \in [t_1, \hat{t}]$ and also $\mathbf{x}(t_n, \mathbf{x}_0) \to \mathbf{z} \in l$. Since both $\{t_n\}$ and $\{\mathbf{x}(t_n, \mathbf{x}_0)\}$ would be in a compact set. Then $\mathbf{z} = \mathbf{x}(\tilde{t}, \mathbf{x}_0)$

$$\mathbf{x}'(\tilde{t}, \mathbf{x}_0) = \lim_{n \to \infty} \frac{\mathbf{x}(t_n, \mathbf{x}_0) - \mathbf{x}(\tilde{t}, \mathbf{x}_0)}{t_n - \tilde{t}} = \mathbf{f}(\mathbf{x}(\tilde{t}, \mathbf{x}_0))$$

But also, these points are on l and so

$$a(x(t_n, \mathbf{x}_0) - x(\tilde{t}, \mathbf{x}_0)) + b(y(t_n, \mathbf{x}_0) - y(\tilde{t}, \mathbf{x}_0)) = 0$$
$$a\frac{x(t_n, \mathbf{x}_0) - x(\tilde{t}, \mathbf{x}_0)}{t_n - \tilde{t}} + b\frac{y(t_n, \mathbf{x}_0) - y(\tilde{t}, \mathbf{x}_0)}{t_n - \tilde{t}} = 0$$

so, passing to a limit, you get

$$(a, b) \cdot \mathbf{x}'(\tilde{t}, \mathbf{x}_0) = (a, b) \cdot \mathbf{f}(\mathbf{x}(\tilde{t}, \mathbf{x}_0)) = 0$$

which contradicts the definition of ε above. Thus there are finitely many points of intersection. Let t_2 be the smallest which is larger than t_1 such that $\mathbf{x}(t_2, \mathbf{x}_0) \in l$ and $\mathbf{x}(t_2, \mathbf{x}_0) \in B(\mathbf{x}_1, \varepsilon)$.

Then one of the following pictures must hold.

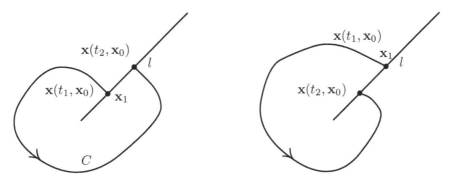

Let's assume the first of the two holds. The reasoning is the same if the second one holds. The thing to observe is that if $\hat{t} > t_1$ and if $\mathbf{x}(\hat{t}, \mathbf{x}_0)$ is on the line l inside $B(\mathbf{x}_1, \varepsilon)$, then $\mathbf{x}(\hat{t}, \mathbf{x}_0)$ is further from $\mathbf{x}_1$ than $\mathbf{x}(t_2, \mathbf{x}_0)$. This is because the line segment from $\mathbf{x}_1$ to $\mathbf{x}(t_2, \mathbf{x}_0)$ along with the curve C is a simple closed curve, that is, the one to one continuous image of a circle. This due to the assumption that $t \to \mathbf{x}(t, \mathbf{x}_1)$ is one to one on every closed interval. Therefore, by the Jordan curve theorem, this segment along with the curve C divides the plane into two connected open sets of which it is the complete boundary. If $\mathbf{x}\left(\hat{t}, \mathbf{x}_0\right)$ is closer to $\mathbf{x}_1$ than $\mathbf{x}\left(t_2, \mathbf{x}_0\right), t \to \mathbf{x}\left(t, \mathbf{x}_0\right)$ would have to cross either C or the segment between $\mathbf{x}_1$ and $\mathbf{x}\left(t_2, \mathbf{x}_0\right)$. The first is impossible thanks to the assumption that $t \to \mathbf{x}\left(t, \mathbf{x}_1\right)$ is one to one. The second is also impossible.

Indeed, that illustrated in the picture in which $\mathbf{x}\left(\hat{t}, \mathbf{x}_0\right) = \hat{\mathbf{x}}$ cannot happen because it would require the vector valued function $t \to \mathbf{x}\left(t, \mathbf{x}_0\right)$ to cross the segment in the opposite direction and this is manifested by

$$\mathbf{x}'\left(\hat{t}, \mathbf{x}_0\right) \cdot (a, b) = \mathbf{f}\left(\mathbf{x}\left(\hat{t}, \mathbf{x}_0\right)\right) \cdot (a, b) < 0$$

and inside $B\left(\mathbf{x}_1, \varepsilon\right)$ the last dot product does not change sign and is always positive.

This yields a contradiction to the assumption that

$$\left(x\left(t_1, \mathbf{x}_0\right), y\left(t_1, \mathbf{x}_0\right)\right) \neq \left(x\left(t_2, \mathbf{x}_0\right), y\left(t_2, \mathbf{x}_0\right)\right)$$

because all of the above will apply to ε smaller than the ε just used and less than the length of the segment between $\mathbf{x}_1$ and $\mathbf{x}\left(t_2, \mathbf{x}_0\right)$. This would yield a $\hat{t}_2 > t_1$ such that $\mathbf{x}\left(\hat{t}_2, \mathbf{x}_0\right)$ intersects the segment in a point closer to $\mathbf{x}_1$ than $\mathbf{x}\left(t_2, \mathbf{x}_0\right)$. Since $\mathbf{x}\left(\hat{t}_2, \mathbf{x}_0\right)$ intersects l, $\hat{t}_2$ must be larger than t_2 but it was just argued that for such $\hat{t}_2$, $\mathbf{x}\left(\hat{t}_2, \mathbf{x}_0\right)$ must be further away from $\mathbf{x}_1$ than $\mathbf{x}\left(t_2, \mathbf{x}_0\right)$. Hence

$$\mathbf{x}\left(t_1, \mathbf{x}_0\right) = \mathbf{x}\left(t_2, \mathbf{x}_0\right)$$

for $t_2 > t_1$ and so for $\hat{T} = t_2 - t_1$, $\mathbf{x}\left(\hat{T}, \mathbf{x}_1\right) = \mathbf{x}_1$. If $t \geq 0$,

$$\mathbf{x}\left(t + \hat{T}, \mathbf{x}_1\right) = \mathbf{x}\left(t, \mathbf{x}\left(\hat{T}, \mathbf{x}_1\right)\right) = \mathbf{x}\left(t, \mathbf{x}_1\right)$$

so $\cup_{t \geq 0}\mathbf{x}\left(t, \mathbf{x}_1\right)$ is a periodic orbit. This contradicts the assumption that $\mathbf{x}\left(\cdot, \mathbf{x}_1\right)$ is one to one on every closed interval which was the premise for drawing this conclusion. Hence there exist $T > 0, \hat{t} \geq 0$ such that $\mathbf{x}\left(\hat{t} + T, \mathbf{x}_1\right) = \mathbf{x}\left(\hat{t}, \mathbf{x}_1\right)$. By Lemma H.0.56, $\mathbf{x}\left(T, \mathbf{x}_1\right) = \mathbf{x}_1$ and for all $t \geq 0, \mathbf{x}\left(t + T, \mathbf{x}_1\right) = \mathbf{x}\left(t, \mathbf{x}_1\right)$.

Next observe that if $\mathbf{x}\left(t, \mathbf{x}_0\right) \in \Lambda^+$, then for all $s > t, \mathbf{x}\left(s, \mathbf{x}_0\right) \in \Lambda^+$. This is because there exists $t_n \to \infty$ such that $\mathbf{x}\left(t, \mathbf{x}_0\right) = \lim_{n \to \infty} \mathbf{x}\left(t_n, \mathbf{x}_0\right)$. Then by Proposition 11.7.2 and the continuous dependence on initial data,

$$\begin{aligned} \mathbf{x}\left(s, \mathbf{x}_0\right) &= \mathbf{x}\left(s - t + t, \mathbf{x}_0\right) = \mathbf{x}\left(s - t, \mathbf{x}\left(t, \mathbf{x}_0\right)\right) \\ &= \lim_{n \to \infty} \mathbf{x}\left(s - t, \mathbf{x}\left(t_n, \mathbf{x}_0\right)\right) = \lim \mathbf{x}\left(s - t + t_n, \mathbf{x}_0\right) \end{aligned}$$

showing that $\mathbf{x}\left(s, \mathbf{x}_0\right) \in \Lambda^+$. Thus $\Lambda^+ \subseteq \cup_{t > 0}\left\{\mathbf{x}\left(t, \mathbf{x}_1\right)\right\} = \mathbf{x}\left([0, T], \mathbf{x}_1\right)$, a closed and bounded set. Also, for $t \in [0, T], \mathbf{x}\left(t, \mathbf{x}_1\right) = \mathbf{x}\left(t, \mathbf{x}\left(t_1, \mathbf{x}_0\right)\right) = \mathbf{x}\left(t + t_1, \mathbf{x}_0\right) \in \Lambda^+$ due to the above which has just shown that Λ^+ is invariant. Thus

$$\Lambda^+ = \cup_{t \geq 0}\left\{\mathbf{x}\left(t, \mathbf{x}_1\right)\right\}. \tag{8.1}$$

Is $\mathbf{x}_0 \in \Lambda^+$? More generally, is $\mathbf{x}\left(t, \mathbf{x}_0\right) \in \Lambda^+$ for all $t \geq 0$? Let m be a positive integer and let $mT > t + t_1$. Then

$$\begin{aligned} \mathbf{x}\left(t_1, \mathbf{x}\left(mT - (t_1), \mathbf{x}_1\right)\right) &= \mathbf{x}\left(mT, \mathbf{x}_1\right) = \mathbf{x}_1 \\ \mathbf{x}\left(t_1, \mathbf{x}_0\right) &= \mathbf{x}_1 \end{aligned}$$

However, if $\mathbf{x}_0 \notin \Lambda^+$, then $\mathbf{x}(mT - t_1, \mathbf{x}_1) \neq \mathbf{x}_0$ and so this contradicts Proposition 11.7.2. Therefore, $\mathbf{x}_0 \in \Lambda^+$ and

$$\mathbf{x}_0 = \mathbf{x}(mT - t_1, \mathbf{x}_1) = \mathbf{x}(mT - t_1, \mathbf{x}(t_1, \mathbf{x}_0)) = \mathbf{x}(mT, \mathbf{x}_0) \in \Lambda^+$$

Therefore, from what was shown above in (8.1), $\mathbf{x}_0 = \mathbf{x}(t_0, \mathbf{x}_1)$ for some $t_0 \in [0, T]$. Hence for $t \geq 0$,

$$\begin{aligned} \mathbf{x}(t, \mathbf{x}_0) &= \mathbf{x}(t, \mathbf{x}(t_0, \mathbf{x}_1)) = \mathbf{x}(t, \mathbf{x}(t_0 + T, \mathbf{x}_1)) \\ &= \mathbf{x}(t + T, \mathbf{x}(t_0, \mathbf{x}_1)) = \mathbf{x}(t + T, \mathbf{x}_0) \end{aligned}$$

Therefore, $\mathbf{x}_0 \in \Lambda^+$ and $\Lambda^+ = \Lambda_+$ and is a periodic orbit. $\blacksquare$

What about the case where there is no point of D in both Λ^+ and Λ_+? That is, $\mathbf{x}(t, \mathbf{x}_0) \notin \Lambda^+$ for all $t \geq 0$. Then $t \to \mathbf{x}(t, \mathbf{x}_0)$ must be one to one on every interval $[0, \hat{t}]$ since otherwise, $\mathbf{x}(t, \mathbf{x}_0) = \mathbf{x}(t + T, \mathbf{x}_0)$ for some T and Lemma H.0.56 would imply that $t \to \mathbf{x}(t, \mathbf{x}_0)$ is a periodic orbit so as above, $\Lambda^+ = \Lambda_+$ and this is assumed not to happen.

Then $\Lambda^+ \neq \emptyset$ because $\mathbf{x}(t, \mathbf{x}_0)$ remains in a bounded set so has a subsequence which converges to something in Λ^+. Let $\mathbf{z} \in \Lambda^+$. Then there exists a sequence $\hat{t}_n \to \infty$ such that $\mathbf{x}(\hat{t}_n, \mathbf{x}_0) \to \mathbf{z}$. Let l be as above. That is, for $\mathbf{x} \in B(\mathbf{z}, \varepsilon)$,

$$\mathbf{f}(\mathbf{x}) \cdot (a, b) > 0 \text{ and } \mathbf{z} \cdot (a, b) = c$$

Then as before, let

$$g(t, \mathbf{u}) \equiv (a, b) \cdot \mathbf{x}(t, \mathbf{u}) - c$$

Then

$$g(0, \mathbf{z}) = (a, b) \cdot \mathbf{x}(0, \mathbf{z}) - c = (a, b) \cdot \mathbf{z} - c = 0$$

and

$$g_t(0, \mathbf{z}) = (a, b) \cdot \mathbf{x}'(0, \mathbf{z}) = (a, b) \cdot \mathbf{f}(\mathbf{z}) > 0$$

so restricting ε if necessary, the implicit function theorem implies there is $(-\delta, \delta)$ and $B(\mathbf{z}, \varepsilon)$ such that there is a unique $t(\mathbf{u}) \in (-\delta, \delta)$ for each $\mathbf{u} \in B(\mathbf{z}, \varepsilon)$ such that $g(t(\mathbf{u}), \mathbf{u}) = 0$. That is, $\mathbf{x}(t(\mathbf{u}), \mathbf{u})$ is on the line segment l inside $B(\mathbf{z}, \varepsilon)$. Thus there exists a sequence $t_n \to \infty$ such that $\mathbf{x}(t_n, \mathbf{x}_0)$ is on l inside $B(\mathbf{z}, \varepsilon)$. Note the difference between this and what was done earlier. Here t_n is not bounded and the point $\mathbf{z}$ is not on the orbit of $t \to \mathbf{x}(t, \mathbf{x}_0)$.

Now consider $\Lambda_{\mathbf{z}}^+$ which is the limit set starting with $\mathbf{z}$ and $\Lambda_{+\mathbf{z}} \equiv \cup_{t \geq 0} \mathbf{x}(t, \mathbf{z})$. Thus, from the above lemma, these are equal and are a periodic orbit if $\mathbf{f}(\mathbf{z}) \neq \mathbf{0}$. Is $\Lambda_{\mathbf{z}}^+ \subseteq \Lambda^+$? This appears to be the case because from the above lemma, $\Lambda_{\mathbf{z}}^+ = \cup_{t \geq 0} \mathbf{x}(t, \mathbf{z}) = \mathbf{x}([0, T], \mathbf{z})$ for some $T > 0$. Now

$$\mathbf{x}(t, \mathbf{z}) = \lim_{n \to \infty} \mathbf{x}(t, \mathbf{x}(t_n, \mathbf{x}_0)) = \lim_{n \to \infty} \mathbf{x}(t + t_n, \mathbf{x}_0) \in \Lambda^+.$$

Therefore, the above argument will apply to every $\mathbf{w} \in \Lambda_{\mathbf{z}}^+$ and shows that $\mathbf{x}(t, \mathbf{u}_0)$ spirals in towards the periodic orbit $\Lambda_{\mathbf{z}}^+$. As before, this is because the solution curve must always cross l in the same direction and must spiral in towards $\Lambda_{\mathbf{z}}^+$ and not away from it because ε is arbitrarily small.

The line segment l is called a transversal. It has the property that the solution curves cross it and are not tangent to it in a small ball.

The following is the Poincare Bendixon theorem, the main result which summarizes the above discussion.

Theorem H.0.58 *Let D be the closure of a bounded region of the plane such that $\mathbf{f}$ is a C^1 function which has no zeros in D, and suppose that $\mathbf{x}(t,\mathbf{x}_0)$ stays in D for all $t \geq 0$ if $\mathbf{x}_0 \in D$, where this is the solution to*

$$\mathbf{x}' = \mathbf{f}(\mathbf{x}), \ \mathbf{x}(0) = \mathbf{x}_0$$

Then letting $\Lambda_+ = \cup_{t \geq 0} \mathbf{x}(t,\mathbf{x}_0)$, it follows that Λ_+ is either a periodic orbit or $t \to \mathbf{x}(t,\mathbf{x}_0)$ spirals toward a periodic orbit.

There is a very interesting consequence of the above major theorem which I will try to make plausible. Suppose you have one of these two dimensional autonomous systems and a closed orbit Γ_0. Denote by G_0 this orbit along with the inside of the simple closed curve, that is the bounded component of Γ_0^C. Thus G_0 is a compact set. More generally, let $\mathcal{G}$ denote the set of all such $G \subseteq G_0$ where the boundary of G denoted as Γ is a closed orbit of the O.D.E. A chain of these sets of $\mathcal{G}$ is a subset of $\mathcal{G}$ where any two are related in the sense that one is a subset of the other. Methods from set theory will allow you to obtain a maximal chain $\mathcal{C}$. This is a chain with the property that there is no chain which properly contains it. If $G, \hat{G} \in \mathcal{C}$, then one is larger than the other. Say $G \subseteq \hat{G}$. By uniqueness, $G \subseteq \text{interior}\left(\hat{G}\right)$. Suppose there are no equilibrium points in G_0.

Now if $\mathbf{z} \in G$, the Poincare Bendixon theorem says that $t \to \mathbf{x}(t,\mathbf{z})$ either spirals toward a closed orbit $\hat{\Gamma}$ or is itself a closed orbit $\hat{\Gamma}$. By uniqueness, $\hat{\Gamma} \cap \Gamma = \emptyset$ and so $\hat{G} \subseteq \text{interior}(G)$ or possibly $\hat{G} = G$. Now take the intersection of the maximal chain above. It is nonempty because each set is nonempty and compact. See Proposition E.2.19. Consider $\mathbf{z} \in \cap \mathcal{C}$. Suppose none of these points of G_0 are equilibrium points. Let $\Gamma_{\mathbf{z}}$ be the attracting orbit of $t \to \mathbf{x}(t,\mathbf{z})$ which exists by the Poincare Bendixon theorem.

Suppose $\Gamma_{\mathbf{z}}$ is contained on the interior of each G. Then this is not possible because it would violate the maximality of the chain $\mathcal{C}$. You could add in $G_{\mathbf{z}}$ and get a strictly larger chain. Hence for some $G \in \mathcal{G}$, we have $\Gamma_{\mathbf{z}}$ is the boundary of G. Then this is also the smallest G in $\mathcal{G}$. If you had a smaller one, $\hat{G}$, then you could not have $\Gamma_{\mathbf{z}}$ be the boundary of G because $\mathbf{x}(t,\mathbf{z})$ would need to cross $\hat{\Gamma}$ which is impossible by uniqueness. Thus for every $\mathbf{z} \in \cap \mathcal{C}$, you must have $\Gamma_{\mathbf{z}} = \Gamma$, the boundary of G. Also, the interior of G is contained in $\cap \mathcal{C}$ since G is the smallest. But now, pick $\mathbf{z}$ in the interior of G and let $\overline{B(\mathbf{z},\delta)}$ be in this interior. If none of these points are equilibrium points, then by the Poincare Bendixon theorem, $t \to \mathbf{x}(t,\mathbf{w})$ spirals toward Γ the boundary of G. Let $\mathbf{f} : \overline{B(\mathbf{z},\delta)} \to \partial \overline{B(\mathbf{z},\delta)}$ be defined as follows. $\mathbf{f}(\mathbf{w}) = \mathbf{x}(t,\mathbf{w})$ where t is the first time $\mathbf{x}(t,\mathbf{w})$ hits $\partial B(\mathbf{z},\delta)$. Thus this gives a continuous retraction of the ball onto its boundary which is known to be impossible. This is a contradiction. Hence there must be an equilibrium point in the interior of G.

This explains the following interesting result.

Theorem H.0.59 *If you have a periodic orbit of a solution to an autonomous two dimensional differential equation, $\mathbf{x}' = \mathbf{f}(\mathbf{x})$, then it must go around some equilibrium point.*

The above argument could be modified to rule out the case of a single saddle point. You can't expect to have $G \in \mathcal{G}$ very small around the saddle point because of the presence of the stable manifold. Thus there would be an open ball consisting of other points and these would lead to stable orbits as in the above argument, and a similar contradiction. This is just a rough idea of why the following corollary is valid.

Corollary H.0.60 *If you have a periodic orbit of a solution to an autonomous two dimensional differential equation,* $\mathbf{x}' = \mathbf{f}(\mathbf{x})$, *and there is only one equilibrium point on the inside of the closed orbit, then this equilibrium point cannot be a saddle point.*

Appendix I

Selected Solutions

I.1   Prerequisite Topics

1. Let s be in the left. Then it is not in the union of the sets of $\mathcal{C}$ and so it fails to be in any of them. Thus s is in A^C for each $A \in \mathcal{C}$. Therefore, s is in the right. If s is in the right, then it fails to be in A for each $A \in \mathcal{C}$. Therefore, it fails to be in the union of these sets and so it is in the left. Since each side is a subset of the other, they must be equal.

3. This is true for $n = 3$ by inspection. $\sum_{k=1}^{3} \frac{1}{\sqrt{k}} = 2.2845$ while $\sqrt{3} = 1.7321$ and so this case is all right. Now suppose the inequality is true for $n > 3$. Then by induction,

$$\sum_{k=1}^{n+1} \frac{1}{\sqrt{k}} > \sqrt{n} + \frac{1}{\sqrt{n+1}} = \frac{\sqrt{n(n+1)} + 1}{\sqrt{n+1}} > \frac{\sqrt{n^2} + 1}{\sqrt{n+1}} = \sqrt{n+1}$$

5. The theorem is true if $n = 0$ or 1 by inspection. Suppose it is true for n. Then

$$\sum_{k=0}^{n+1} \binom{n+1}{k} x^{n+1-k} y^k = x^{n+1} + \sum_{k=1}^{n} \binom{n+1}{k} x^{n+1-k} y^k + y^{n+1}$$

$$= x^{n+1} + \sum_{k=1}^{n} \left(\begin{array}{c} \binom{n}{k} \\ +\binom{n}{k-1} \end{array} \right) x^{n+1-k} y^k + y^{n+1}$$

$$= x^{n+1} + x \sum_{k=1}^{n} \binom{n}{k} x^{n-k} y^k + \sum_{k=0}^{n-1} \binom{n}{k} x^{n-k} y^{k+1} + y^{n+1}$$

$$= x^{n+1} + x \sum_{k=1}^{n} \binom{n}{k} x^{n-k} y^k + y \sum_{k=0}^{n-1} \binom{n}{k} x^{n-k} y^k + y^{n+1}$$

$$= x^{n+1} + x \left((x+y)^n - x^n \right) + y \left((x+y)^n - y^n \right) + y^{n+1}$$

$$= x^{n+1} + x (x+y)^n - x^{n+1} + y (x+y)^n$$

$$= x^{n+1} + (x+y)^n (x+y) - x^{n+1} = (x+y)^{n+1}$$

so the theorem holds for all n.

As to the other claim, it is true by inspection if $n = 1$. Suppose true for $n \geq 1$. Then

$$\binom{n}{k} + \binom{n}{k-1} = \binom{n+1}{k}$$

On the left, you have

$$\frac{n!}{(n-k)!k!} + \frac{n!}{(n+1-k)!\,(k-1)!} = \frac{(n+1-k)\,n!}{(n+1-k)!k!} + \frac{kn!}{(n+1-k)!k!}$$

$$= \frac{(n+1)!}{(n+1-k)!k!}$$

and so this equals $\binom{n+1}{k}$. Thus the formula holds.

7. Let $z = 2 + i7$ and let $w = 3 - i8$. Find $zw, z + w, z^2$, and w/z.

$(2 + i7)(3 - 8i) = 62 + 5i$

$2 + i7 + (2 + 7i) = 4 + 14i$

$(2 + 7i)^2 = -45 + 28i$

$(3 - 8i)/(2 + i7) = -\frac{50}{53} - \frac{37}{53}i$

9. They are equally spaced around the circle of radius 2 with one of them being the number 2. The four fourth roots are equally spaced around the circle of radius 2 and they are $-2i, -2, 2i, 2$

11. Yes it does. It obviously holds if $n = 0$. In this case, you get 1 on both sides. As to $-n$ for $n > 0$,

$$[r(\cos t + i \sin t)]^{-n} \equiv \frac{1}{[r(\cos t + i \sin t)]^n} = \frac{1}{r^n(\cos(nt) + i \sin(nt))}$$

$$= r^{-n} \frac{\cos(nt) - i \sin(nt)}{1} = r^{-n}(\cos(-nt) + i \sin(-nt))$$

13.

$$|z|(\cos(\theta) + i \sin(\theta))\,|w|(\cos(\phi) + i \sin(\phi))$$
$$= |z|\,|w|(\cos(\theta)\cos\phi - \sin(\theta)\sin(\phi) + i(\sin(\theta)\cos(\phi) + \sin(\phi)\cos(\theta)))$$
$$= |z|\,|w|(\cos(\theta + \phi) + i \sin(\theta + \phi))$$

15. $x^3 + 27 = (x + 3)(-3x + x^2 + 9)$

17. $\left(x^2 - 2\sqrt{2}x + 4\right)\left(x^2 + 2\sqrt{2}x + 4\right) = x^4 + 16$

19.

$$\overline{p(x)} = \overline{a_n x^n + a_{n-1}x^{n-1} + \cdots + a_1 x + a_0}$$

and this equals

$$\overline{a_n}\,(\bar{x})^n + \overline{a_{n-1}}\bar{x}^{n-1} + \cdots + \overline{a_1}\bar{x} + \overline{a_0}$$

and since these coefficients are real, this is just $p(\bar{x})$.

21. There is no such thing as a square root of negative one. There are two square roots of -1. There are also two roots of 1. This above nonsense occurs because it is meaningless symbol pushing without ever considering whether there is meaning to the expressions pushed.

23.

$$(\cos\theta - i\sin\theta)^n = (\cos(-\theta) + i\sin(-\theta))^n = \cos(n\theta) + i\sin(-n\theta)$$
$$= \cos(n\theta) - i\sin(n\theta)$$

Thus it is true if you have n an integer. The verification for negative integers was presented earlier.

25. This is just a repeated application of the Euclidean algorithm for polynomials. First suppose $a_n = 1$. Then $p(x) = (x - z_1)k(x) + r(x)$ where the degree of $r(x)$ is less than 1. Thus it is a constant. The constant can only be zero because z_1 is a root. Hence $p(x) = (x - z_1)k(x)$. Now $k(x) = k_2(x)(x - z_2)$ by the same reasoning. Continuing this way gives the result. In case $a_n \neq 1$, you get $\frac{p(x)}{a_n} = (x - z_1)(x - z_2)\cdots(x - z_n)$ so you multiply by a_n to get what is claimed.

27. a. $x^2 + 2x + 1 + i = 0$, $\sqrt{4 - 4(1 + i)} = (1 - i)\sqrt{2}$, $x = \frac{-2 \pm (1-i)\sqrt{2}}{2}$

I.2 The Idea of a Differential Equation

1. a. $y' = -y$
   On this one, $\lim_{t\to\infty} y(t) = 0$.

   b. $y' = y - \frac{1}{4}y^3$
   Here $y(t) \to 2$ if initial condition is in $(0, 5)$ and converges to -2 if the initial condition is in $(-5, 0)$

3. It is just like the above only this time, when y is too large, it gets even larger and when it is too small, it gets even smaller. In fact, the absolute value of the slope gets increasingly large as y moves away from the point $-b/a$.

5. The area of a sphere is $4\pi r^2$ and so when you solve for r in the volume, you get $r = \left(\frac{3V}{4\pi}\right)^{1/3}$ and so the differential equation is $\frac{dV}{dt} = k4\pi \left(\frac{3V}{4\pi}\right)^{2/3}$.

7. $m\frac{dv}{dt} = mg - kv^2$. Terminal speed is $\sqrt{\frac{mg}{k}}$.

9. $\frac{dA}{dt} = .05A - r$
   If $A_0 > r/.05$ then the slope is positive and so the amount of money continues to grow. If $A_0 < r/.05$ the slope is negative and so A gets smaller and so the slope keeps on getting more and more negative so he will eventually run out. Thus, unless he is the US government, he will have to quit spending and so the equation will no longer hold.

11. From the initial condition, $y(0) = 0, y'(0) = 1$, and now from the equation, $y''(0) = -y(0) = 0, y''' = -y'$ so $y'''(0) = -y'(0) = -1$. Then as before, $y^{(4)}(0) = 0, y^{(5)}(0) = 1$ etc. Thus $y^{(2n)}(0) = 0, y^{(2n-1)}(0) = (-1)^{n-1}$. It follows that the power series obtained is

$$\sum_{n=1}^{\infty} (-1)^n \frac{x^{2n-1}}{(2n-1)!} = \sin x$$

Then you can verify right away that in fact this solves the equation and initial condition.

13. The differential equation is $\frac{dV}{dt} = -kA(y) + r$ where r is the small constant rate. Of course this will not do. We need to write in terms of a single unknown function. However, $V(y) = \int_0^y A(x)\, dx$ and so $\frac{dV}{dt} = \frac{dV}{dy}\frac{dy}{dt} = A(y)\frac{dy}{dt}$. Thus the differential equation is

$$A(y)\frac{dy}{dt} = -kA(y) + r, \quad \frac{dy}{dt} = -k + \frac{r}{A(y)}$$

You start adding water the level cannot continue increasing because eventually if y large enough, the right side of the equation is less than 0 and so y' will then be negative. On the other hand, if y is too small, then y' will be positive and so the depth should stabilize.

I.3 Finding Solutions to First Order Scalar Equations

1. a. $y' + 2ty = e^{-t^2}$, Exact solution is: $\left\{ Ce^{-t^2} + \frac{t}{e^{t^2}} \right\}$

    c. $y' + \cos(t)\, y = \cos(t)$, Exact solution is: $\left\{ Ce^{-\sin t} + 1 \right\}$

3. $y = \frac{7}{3}t - \frac{1}{3t^2}$

5. a. $\ln(t)\, y' + \frac{1}{t}y = \ln(t)$,
       $y(2) = 3, y' + \frac{1}{t\ln t}y = 1, (\ln(t)\, y)' = \ln(t)$,
       $\ln(t)\, y(t) = t\ln t - t + C$
       $y(t) = t - \frac{t}{\ln(t)} + \frac{C}{\ln(t)} = t - \frac{t}{\ln(t)} + \frac{\ln 2 + 2}{\ln(t)}$

    c. $y' + \tan(t)\, y = \cos^3(t), y(0) = 4$,
       $y = 4\cos t + \frac{1}{8}\sin t + \frac{1}{8}\sin 3t + \frac{1}{2}t\cos t$

8. a. $(x^3 + y)\, dx - x\, dy = 0, xy' - y = x^2, \left(\frac{1}{x}y\right)' = x, \frac{y}{x} = \frac{x^2}{2} + C, y = \frac{x^3}{2} + Cx$

    c. $y(y^2 - x)\, dy = dx, x = Ce^{-\frac{1}{2}y^2} + y^2 - 2$

9. It is enough to show that $rt = \lim_{n\to\infty} tn\ln\left(1 + \frac{r}{n}\right) = rt\lim_{n\to\infty}\frac{\ln(1 + r/n)}{r/n}$ but an application of LHospital's rule shows this.

11. $\frac{dT}{dt} = k(10 - T)$, Thus $T = 60e^{-kt} + 10$ It says $50 = 60e^{-5k} + 10$, The solution is: $\frac{1}{5}\ln 3 - \frac{1}{5}\ln 2 = k$

13. $y' + y = \sin t$, $y = \frac{1}{2}\sin t - \frac{1}{2}\cos t + Ce^{-t}$

Now C depends on the initial condition but this term does not matter as $t \to \infty$ and so the $u(t) = \frac{1}{2}\sin t - \frac{1}{2}\cos t$

15.   a. $y' + 2xy = xy^3$, $y(1) = 2$, $y = \dfrac{\sqrt{2}}{\sqrt{-\frac{1}{2}e^{-2}\left(e^{2x^2} - 2e^2\right)}}$

    c. $y' + 2y = x^2 y^3$, $y(1) = -1$, $y = -\dfrac{1}{\sqrt{\frac{1}{2}x^2 + \frac{1}{4}x + \frac{3}{16e^4}e^{4x} + \frac{1}{16}}}$

17. $y = \sqrt{3}\dfrac{e^{3t}}{\sqrt{e^{6t} + 2}}$, $\lim_{t\to\infty} \sqrt{3}\dfrac{e^{3t}}{\sqrt{e^{6t} + 2}} = \sqrt{3}$.

19. The equation implies it will have a terminal speed which is $\sqrt{\frac{g}{k}}$.

21. $\frac{1}{2}e^{x^2} + \ln|y| = C$

23. $\dfrac{1}{2\sqrt{g}\sqrt{k}}\left(\ln\left(\sqrt{k}v + \sqrt{g}\right) - \ln\left(\sqrt{k}v - \sqrt{g}\right)\right) = t + C$

25. $f_x(x,y) + f_y(x,y)\,y' = 0$, $y' = \frac{-f_x}{f_y}$ for the given level curves. Thus you need to have $y' = \frac{f_y(x,y)}{f_x(x,y)}$ or $f_x(x,y)\,dy = f_y(x,y)\,dx$ for the new curves perpendicular to the given ones.

26.   a. $y' = y^2(y-1)$, $0, 1$ are equilibrium points. 0 is semistable. 1 is unstable.

    c. $y' = \sin(y)$, The equilibrium points are $n\pi$ where n is an integer. The derivative at this point is $(-1)^n$. The points are stable when the derivative is negative so the function is decreasing through the equilibrium point. Thus stability happens exactly at points $(2n-1)\pi$.

29. $\left(\frac{1}{y} + \frac{1}{K-y}\right)dy = r\,dt$, $\ln(y) - \ln(K-y) = rt + C$. The initial condition implies $\ln(\alpha K) - \ln(K - \alpha K) = C$ so $C = \ln\left(\frac{\alpha}{1-\alpha}\right)$. Now consider the time T to double. You have

$$\ln(2\alpha K) - \ln(K - 2\alpha K) = rT + \ln\left(\frac{\alpha}{1-\alpha}\right)$$

Then

$$\ln\left(\frac{2\alpha}{1-2\alpha}\right) - \ln\left(\frac{\alpha}{1-\alpha}\right) = rT$$

Then

$$\ln\left(2\frac{\alpha-1}{2\alpha-1}\right) = rT, \; T = \frac{1}{r}\ln\left(2\frac{\alpha-1}{2\alpha-1}\right)$$

31.

$$y' = a - y^2$$

where a is a real number. Show that there are no equilibrium solutions if $a < 0$ but there are two of them if $a > 0$ and only one if $a = 0$. Discuss the stability of the two equilibrium points when $a > 0$. What about stability of equilibrium when $a = 0$?

It is clear there are no equilibriums if $a < 0$ and if $a = \delta^2$, then you have two, namely $\pm\delta$. Then $a = 0$ you clearly have a semistable equilibrium. If $a > 0$, the point $-\delta$ is not stable and the point δ is stable.

33. In this case, if $a > 0$ you have a stable point at a and an unstable one at 0. If $a = 0$ you have a semistable point at 0. If $a < 0$ you have 0 is stable and $-a$ is unstable.

35.   a. $y' = \frac{1}{x^2}\left(x^2 + y^2 + xy\right), (1,1), xv' = v^2 + 1,$
       $\arctan(v) - \ln|x| = \frac{\pi}{4}, \ \arctan\left(\frac{y}{x}\right) - \ln|x| = \frac{\pi}{4}$

      c. $y' = \frac{1}{x^2}\left(x^2 + 9y^2 + xy\right), (3,1), \ xv' = 9v^2 + 1$
       $\frac{1}{3}\arctan 3v - \ln|x| = \frac{1}{12}\pi - \ln 3, \ \frac{1}{3}\arctan\left(3\left(\frac{y}{x}\right)\right) - \ln|x| = \frac{1}{12}\pi - \ln 3$

37. $xv' = 1 - \sin(v) - v$ where $v = y/x$.
    $\int \frac{dv}{1 - \sin v - v} = \ln|x| + C$

39. $xv' = -4v^2 - 7, \frac{dv}{4v^2 + 7} + \frac{dx}{x} = 0,$
    $\int \frac{dv}{4v^2 + 7} = \frac{1}{28}\sqrt{7}\left(2\arctan\frac{2}{7}\sqrt{7}v\right),$
    $\frac{1}{28}\sqrt{7}\left(2\arctan\frac{2}{7}\sqrt{7}\left(\frac{y}{x}\right)\right) + \ln|x| = \ln 2 - \frac{1}{14}\sqrt{7}\arctan\frac{1}{7}\sqrt{7}$

41. $xv' = \frac{-\left(x^3 - 7x^2(xv) - 5(xv)^3\right)}{\left(7x^3 + 5x(xv)^2\right)} - v, xv' = -\frac{1}{5v^2 + 7}, \ \left(5v^2 + 7\right)dv + \frac{dx}{x} = 0$
    $5\frac{v^3}{3} + 7v + \ln|x| = \ln(3) - \frac{418}{81}, \ \frac{7}{x}y + \frac{5}{3x^3}y^3 + \ln|x| = \ln(3) - \frac{418}{81}$

43.   a. $\left(2y^3 + 2\right)dx + \left(3xy^2\right)dy = 0, (1,1), \mu = x$
       Scalar potential is $x^2y^3 + x^2.x^2y^3 + x^2 = 2$ is solution.
      c. $\left(2xy^2 + y + 2xy\cos x^2\right)dx + \left(2\sin x^2 + 3x^2y + 2x\right)dy = 0, (2,1)$
       $\mu = y$, Scalar potential is $y^2\left(x + \sin x^2 + x^2y\right)$
       $y^2\left(x + \sin x^2 + x^2y\right) = 4\sin 1 + 12$

45. We need $\frac{dy}{dx}$ times the slope of the line tangent to the level curve $f(x,y) = C$ equals -1. This last is obtained as follows. $f_x + f_y y_x = 0$ so $y'(x) = \frac{-f_x}{f_y}$ and so we need to have the following for the new curves.

$$y'(x)\frac{-f_x}{f_y} = -1, dy(-f_x) = -f_y dx, \ f_x dy - f_y dx = 0$$

which is equivalent to what was to be shown. Now this is an exact equation because

$$\left(-f_y\right)_y = \left(f_x\right)_x$$

due to the fact that $f_{xx} + f_{yy} = 0$.

49. $y = t\cos t - 2\cos t$

51. $y' = \frac{2x-(7/2)y}{x-\frac{9}{4}y}$, $xv' = \frac{2x-(7/2)(xv)}{x-\frac{9}{4}(xv)} - v = -\frac{1}{9v-4}\left(9v^2 - 18v + 8\right)$,

$\frac{9v-4}{9v^2-18v+8}dv + \frac{dx}{x} = 0$, $\frac{4}{3}\ln\left|v - \frac{4}{3}\right| - \frac{1}{3}\ln\left|v - \frac{2}{3}\right|$ is antiderivative of first term.
Thus inserting $v = y/x$,
$\frac{4}{3}\ln\left|\frac{y}{x} - \frac{4}{3}\right| - \frac{1}{3}\ln\left|\frac{y}{x} - \frac{2}{3}\right| + \ln|x| = \frac{2}{3}\ln 2 - \ln 3$

53. $y' = -\frac{x^3-6x^2y-y^3}{6x^3+xy^2}$, $xv' = -\frac{x^3-6x^2(xv)-(xv)^3}{6x^3+x(xv)^2} - v = -\frac{1}{v^2+6}$

$(v^2 + 6)\,dv + \frac{dx}{x} = 0$, $\frac{v^3}{3}+6v+\ln|x| = \ln 2 - \frac{81}{8}$, $\frac{1}{3x^3}y^3+\frac{6}{x}y+\ln|x| = \ln 2 - \frac{81}{8}$

55. $\int 5\cos(3t)\,dt = \frac{5}{3}\sin 3t$, $\exp\left(\frac{5}{3}\sin 3t\right)$ is integrating factor.
Then $\left(y\exp\left(\frac{5}{3}\sin 3t\right)\right)' = 2\cos(3t)$ and so $y(t)\exp\left(\frac{5}{3}\sin 3t\right) = \frac{2}{3}\sin 3t + C$
which shows that $y = \left(\frac{2}{3}\sin 3t + C\right)\exp\left(-\frac{5}{3}\sin 3t\right)$. From conditions, $C = 2$
so $y = \left(\frac{2}{3}\sin 3t + 2\right)\exp\left(-\frac{5}{3}\sin 3t\right)$.

57. $y' = \frac{3x+\frac{19}{4}y}{4x+\frac{9}{4}y}$, $xv' = \frac{3x+\frac{19}{4}(xv)}{4x+\frac{9}{4}(xv)} - v = \frac{3}{9v+16}\left(-3v^2 + v + 4\right)$. Then
$\frac{9v+16}{3v^2-v-4}dv + \frac{3dx}{x} = 0$,

$$4\ln\left|v - \frac{4}{3}\right| - \ln|v+1| + 3\ln|x| = 4\ln\frac{2}{3} - \ln 3,$$

$$4\ln\left|\frac{y}{x} - \frac{4}{3}\right| - \ln\left|\frac{y}{x} + 1\right| + 3\ln|x| = 4\ln\frac{2}{3} - \ln 3$$

59. $y' = -\frac{y}{x+4y}$
$xv' = -\frac{xv}{x+4(xv)} - v = -2v\frac{2v+1}{4v+1}$
$\frac{4v+1}{2v(2v+1)}dv + \frac{dx}{x} = 0$

61. $y' = -\frac{6x^2-xy+5y^2}{x^2}$, $xv' = -\frac{6x^2-x(xv)+5(xv)^2}{x^2} - v = -5v^2 - 6$, $\frac{dv}{5v^2+6} + \frac{dx}{x} = 0$
$\frac{1}{60}\sqrt{30}\left(2\arctan\frac{1}{6}\sqrt{30}v\right) + \ln|x| = \frac{1}{60}\sqrt{30}\left(2\arctan\frac{1}{6}\sqrt{30}\right) + \ln|3|$
$\frac{1}{60}\sqrt{30}\left(2\arctan\left(\frac{1}{6}\sqrt{30}\left(\frac{y}{x}\right)\right)\right) + \ln|x| = \frac{1}{60}\sqrt{30}\left(2\arctan\frac{1}{6}\sqrt{30}\right) + \ln|3|$

63. $y' = -\frac{y-2x}{\frac{9}{2}y-x}$
$xv' = -\frac{(xv)-2x}{\frac{9}{2}(xv)-x} - v = -\frac{1}{9v-2}\left(9v^2 - 4\right)$
$\frac{9v-2}{9v^2-4}dv + \frac{dx}{x} = 0$
$\int\frac{9v-2}{9v^2-4}dv = \frac{1}{3}\ln\left|v - \frac{2}{3}\right| + \frac{2}{3}\ln\left|v + \frac{2}{3}\right|$
$\frac{1}{3}\ln\left|v - \frac{2}{3}\right| + \frac{2}{3}\ln\left|v + \frac{2}{3}\right| + \ln|x| = \frac{8}{3}\ln 2 - \ln 3$
$\frac{1}{3}\ln\left|\left(\frac{y}{x}\right) - \frac{2}{3}\right| + \frac{2}{3}\ln\left|\left(\frac{y}{x}\right) + \frac{2}{3}\right| + \ln|x| = \frac{8}{3}\ln 2 - \ln 3$

65. $y' + 2ty = te^{t^2}$, $y = \frac{1}{4}e^{t^2} + Ce^{-t^2}$

67. $y' + \tan(2t) y = \cos 2t$, $y(0) = 2$,

$\int \tan(2t)\, dt = -\frac{1}{2} \ln(\cos 2t)$, $\left(\sqrt{\sec(2t)} y\right)' = \sqrt{\cos(2t)}$,

$\sqrt{\sec(2t)} y(t) - 2 = \int_0^t \sqrt{\cos(2s)}\, ds$, $y = \frac{2}{\sqrt{\sec(2t)}} + \frac{1}{\sqrt{\sec(2t)}} \int_0^t \sqrt{\cos(2s)}\, ds$

69. $y = 0$ is obviously a solution. Also $\frac{dy}{y^{1/(2n+1)}} = dt$ and so

$\frac{2n+1}{2n} y^{2n/(2n+1)} = t$ is another solution. Thus $y = \left(\frac{2n}{2n+1} t\right)^{(2n+1)/2n}$, $t > 0$ is another solution.

71. You would let $z = y'$ solve the first order equation for z getting one constant and then you would integrate $z = y'$ to get another constant and find y.

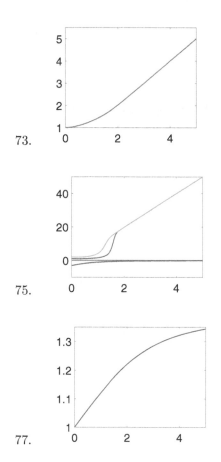

73.

75.

77.

81. Subtract and divide by t/n to get $\frac{y_{k+1} - y_k}{t/n} = y_k$. That $y_n = \left(1 + \frac{t}{n}\right)^n y_0$ follows from iterating the method. Now when $n \to \infty$, the limit is indeed e^t. This follows from a use of L'Hospitals theorem or other elementary calculus procedure.

I.4   Homogeneous Linear Equations

1. Find the solution to the given equation and initial condition.

   a. $14y + 9y' + y'' = 0$. $y(0) = -2, y'(0) = 1$.
      General solution: $C_2 e^{-2t} + C_1 e^{-7t}$.

      Solution to the initial value problem: Need to solve $\begin{aligned} C_1 + C_2 &= -2 \\ -7C_1 - 2C_2 &= 1 \end{aligned}$,

      solution is: $\frac{3}{5} e^{-7t} - \frac{13}{5} e^{-2t}$

   c. $23y' - 20y - 6y''$. $y(0) = 2, y'(0) = 2$.
      General solution: $C_1 e^{-\frac{5}{2}t} + C_2 e^{-\frac{4}{3}t}$.

      Solution to the initial value problem: Need to solve $\begin{aligned} C_1 + C_2 &= 2 \\ -\frac{5}{2}C_1 - \frac{4}{3}C_2 &= 2 \end{aligned}$,

      solution is: $6e^{-\frac{4}{3}t} - 4e^{-\frac{5}{2}t}$

2. Find the solution to the given equation and initial condition.

   a. $2y' - 8y + y'' = 0$. $y(2) = 1, y'(2) = 4$.
      General solution: $C_2 e^{2t} + C_1 e^{-4t}$.

      Solution to the initial value problem: Need to solve $\begin{aligned} C_2 e^4 + C_1 e^{-8} &= 1 \\ 2C_2 e^4 - 4C_1 e^{-8} &= 4 \end{aligned}$

      , solution is: $\frac{4}{3e^4} e^{2t} - \frac{1}{3e^{-8}} e^{-4t}$

   c. $4y - 4y' + y'' = 0$. $y(0) = -2, y'(0) = 2$.
      General solution:$C_1 e^{2t} + tC_2 e^{2t}$.

      Solution to the initial value problem: Need to solve $\begin{aligned} C_1 &= -2 \\ 2C_1 + C_2 &= 2 \end{aligned}$, so-

      lution is: $6te^{2t} - 2e^{2t}$

3. A solution to the equation is given. Find the general solution.

   a. The second solution is $e^{qt} \int e^{-qt} (b(t))^p \, dt$
   c. Another solution is $e^{-t} \int t^2 e^t dt = t^2 - 2t + 2$.

5. One solution is $x - 2$.

7. A solution is x^3.

9. General solution: $C_3 e^{-5t} + C_1 e^{\frac{1}{3}t} + C_2 e^{\frac{3}{2}t}$.

   Solution to the initial value problem: Must solve

   $$C_1 + C_2 + C_3 = 2$$
   $$\tfrac{1}{3}C_1 + \tfrac{3}{2}C_2 - 5C_3 = 7$$
   $$\tfrac{1}{9}C_1 + \tfrac{9}{4}C_2 + 25C_3 = 3$$

then this results in the solution:

$\frac{388}{91}e^{\frac{3}{2}t} - \frac{225}{112}e^{\frac{1}{3}t} - \frac{53}{208}e^{-5t}$

11. General solution: $C_1e^{3t} + C_3e^{-t} + C_2e^{\frac{1}{2}t}$.

Solution to the initial value problem: Must solve

$$C_2e^{-1} + C_3e^2 + C_1e^{-6} = -1$$

$$\tfrac{1}{2}C_2e^{-1} - C_3e^2 + 3C_1e^{-6} = 2$$

$$\tfrac{1}{4}C_2e^{-1} + C_3e^2 + 9C_1e^{-6} = 0$$

then this results in the solution:

$\frac{3}{20e^{-6}}e^{3t} - \frac{17}{12e^2}e^{-t} + \frac{4}{15e^{-1}}e^{\frac{1}{2}t}$

13. General solution: $C_1e^{3t} + C_3e^{2t} + C_4e^{-\frac{3}{2}t} + tC_2e^{3t}$.

Solution to the initial value problem: Must solve

$$C_1 + C_3 + C_4 = 0$$

$$3C_1 + C_2 + 2C_3 - \tfrac{3}{2}C_4 = -3$$

$$9C_1 + 6C_2 + 4C_3 + \tfrac{9}{4}C_4 = -3$$

$$27C_1 + 27C_2 + 8C_3 - \tfrac{27}{8}C_4 = 0$$

The solution to the initial value problem is then

$\frac{27}{7}e^{2t} - \frac{119}{27}e^{3t} + \frac{104}{189}e^{-\frac{3}{2}t} + \frac{10}{3}te^{3t}$

15. General solution: $C_1e^{-3t} + t^2C_3e^{-3t} + t^3C_4e^{-3t} + tC_2e^{-3t}$.

Solution to the initial value problem:

$$C_1 = 2$$

Must solve
$$C_2 - 3C_1 = 3$$

$$9C_1 - 6C_2 + 2C_3 = 3$$

$$27C_2 - 27C_1 - 18C_3 + 6C_4 = 1$$

The solution to the initial value problem is then

$\frac{175}{6}e^{-3t} + 9te^{-3t} + \frac{39}{2}t^2e^{-3t}$

17. General solution: $C_3e^t + C_1e^{-t} + C_2e^{-2t} + C_4e^{-3t}$.

The solution to the initial value problem is then

$y = 0$

19. Suppose that $\sum_{i=1}^{n} C_iy_i(t) = 0$ for all t. Then differentiating repeatedly, this says

$$\begin{pmatrix} y_1(t) & \cdots & y_n(t) \\ \vdots & & \vdots \\ y_1^{(n-1)}(t) & \cdots & y_n^{(n-1)}(t) \end{pmatrix} \begin{pmatrix} C_1 \\ \vdots \\ C_n \end{pmatrix} = \begin{pmatrix} 0 \\ \vdots \\ 0 \end{pmatrix}$$

Now the determinant of the matrix on the left is the same as the Wronskian which is not zero. Therefore, all of these scalars C_i are zero and so the functions are linearly independent.

21. Suppose that $\sum_i C_i y_i(t) = 0$ for all t. Then

$$
\begin{pmatrix}
y_1(\hat{t}) & \cdots & y_n(\hat{t}) \\
\vdots & & \\
y_1^{(n-1)}(\hat{t}) & \cdots & y_n^{(n-1)}(\hat{t})
\end{pmatrix}
\begin{pmatrix}
C_1 \\
\vdots \\
C_n
\end{pmatrix}
=
\begin{pmatrix}
0 \\
\vdots \\
0
\end{pmatrix}
$$

and the matrix on the left is invertible because it has nonzero determinant. Therefore, each $C_i = 0$. Thus these functions are linearly independent.

23. Recall that if $A(t)$ is a matrix with ij^{th} entry $a_{ij}(t)$,

$$
\det A(t) = \sum_{j_1 \cdots j_n} \operatorname{sgn}(j_1, \cdots, j_n) a_{1j_1}(t) \cdots a_{nj_n}(t)
$$

Then by the product rule, you get

$$
\begin{aligned}
\frac{d}{dt} \det(A(t)) &= \sum_{j_1 \cdots j_n} \sum_{k=1}^{n} \operatorname{sgn}(j_1, \cdots, j_n) a_{1j_1}(t) \cdots a'_{kj_k}(t) \cdots a_{nj_n}(t) \\
&= \sum_{k=1}^{n} \sum_{j_1 \cdots j_n} \operatorname{sgn}(j_1, \cdots, j_n) a_{1j_1}(t) \cdots a'_{kj_k}(t) \cdots a_{nj_n}(t)
\end{aligned}
$$

which is the sum of the determinants of the matrices in which the kth row is differentiated. You could have gotten the same thing with columns. Now in the case of the Wronskian, there is a repeated row except for the last one. Thus

$$
W(y_1, \cdots, y_n)'(t) =
\begin{vmatrix}
y_1(t) & y_2(t) & \cdots & y_n(t) \\
\vdots & \vdots & \cdots & \vdots \\
y_1^{(n-2)}(t) & y_2^{(n-1)}(t) & \cdots & y_n^{(n-2)}(t) \\
y_1^{(n)}(t) & y_2^{(n)}(t) & \cdots & y_n^{(n)}(t)
\end{vmatrix}
$$

24. In the above problem, let

$$
r_k(t) \equiv -a_{n-1}(t) y_k^{(n-1)}(t) - \cdots - a_1(t) y_k'(t) - a_0(t) y_k(t)
$$

Then if you have that y_k is a solution to $Ly = 0$, it follows that

$$
y_k^{(n)}(t) = r_k(t)
$$

It follows that

$$W(y_1, \cdots, y_n)'(t) = \begin{vmatrix} y_1(t) & y_2(t) & \cdots & y_n(t) \\ \vdots & \vdots & \cdots & \vdots \\ y_1^{(n-2)}(t) & y_2^{(n-1)}(t) & \cdots & y_n^{(n-2)}(t) \\ r_1(t) & r_2(t) & \cdots & r_n(t) \end{vmatrix}$$

Remember that determinants are linear in each row and equal zero if any row is a multiple of another. Thus the above reduces to $W(y_1, \cdots, y_n)'(t) =$

$$\begin{vmatrix} y_1(t) & y_2(t) & \cdots & y_n(t) \\ \vdots & \vdots & \cdots & \vdots \\ y_1^{(n-2)}(t) & y_2^{(n-1)}(t) & \cdots & y_n^{(n-2)}(t) \\ -a_{n-1}(t)\, y_1^{(n-1)}(t) & -a_{n-1}(t)\, y_2^{(n-1)}(t) & \cdots & -a_{n-1}(t)\, y_n^{(n-1)}(t) \end{vmatrix}$$

$$= -a_{n-1}(t) \begin{vmatrix} y_1(t) & y_2(t) & \cdots & y_n(t) \\ \vdots & \vdots & \cdots & \vdots \\ y_1^{(n-2)}(t) & y_2^{(n-1)}(t) & \cdots & y_n^{(n-2)}(t) \\ y_1^{(n-1)}(t) & y_2^{(n-1)}(t) & \cdots & y_n^{(n-1)}(t) \end{vmatrix}$$

$$= -a_{n-1}(t)\, W(y_1, \cdots, y_n)(t)$$

Letting $A'(t) = a_{n-1}(t)$, it follows that

$$(\exp(A(\cdot))\, W(y_1, \cdots, y_n))' = 0$$

and so

$$W(y_1, \cdots, y_n)(t) = C \exp(-A(t))$$

which shows that the Wronskian either vanishes for all t or for no t.

25. An Euler equation is one which is of the form

$$t^2 y'' + aty' + by = 0$$

where we are interested in the solution for $t > 0$. Let $t = e^s$. Then show that for $y(t) = y(s)$,

$$\frac{d^2 y}{dt^2} = -\frac{1}{t^2}\frac{dy}{ds} + \frac{1}{t^2}\frac{d^2 y}{ds^2}, \quad \frac{dy}{dt} = \frac{1}{t}\frac{dy}{ds}$$

Show that the Euler equation can be studied in the form

$$y''(s) + (a-1)\, y'(s) + by(s) = 0$$

Use this transformation to find the general solution to the following Euler equations.

a. $t^2 y'' - 2y = 0$, $y = \frac{1}{3}\frac{C_8}{t} + C_9 t^2$

c. $t^2 y'' + 3ty' + y = 0$, $y = \frac{C_{17}}{t} - \frac{C_{16}}{t}\ln t$

I.5 Scalar Linear Nonhomogeneous Equations

1.   a. $y'' - 2y' - 3y = -t - 3e^{-2t}$. $y(0) = -1, y'(0) = 4$.

General solution to homogeneous equation:
$C_2 e^{-t} + C_1 e^{3t}$. $y_{p_1}(t) = -\frac{3}{5}e^{-2t}$

$y_{p_2}(t) = \frac{1}{3}t - \frac{2}{9}$

$y_p(t) = \frac{1}{3}t - \frac{3}{5}e^{-2t} - \frac{2}{9}$

General solution to the differential equation: $\frac{1}{3}t - \frac{3}{5}e^{-2t} + C_2 e^{-t} + C_1 e^{3t} - \frac{2}{9}$

Now you have to find the constants. When you do this, you get $\frac{103}{180}e^{3t} - \frac{3}{4}e^{-t} + \left(\frac{1}{3}t - \frac{3}{5}e^{-2t} - \frac{2}{9}\right)$

   c. Find the general solution to the equation $y'' - 2y' - 3y = t + e^{-2t}$. Then determine the solution which also satisfies the initial condition $y(0) = 1, y'(0) = 3$.

General solution to homogeneous equation: $C_2 e^{-t} + C_1 e^{3t}$.

$y_{p_1}(t) = \frac{1}{5}e^{-2t}$

$y_{p_2}(t) = \frac{2}{9} - \frac{1}{3}t$

$y_p(t) = \frac{1}{5}e^{-2t} - \frac{1}{3}t + \frac{2}{9}$

General solution to the differential equation: $\frac{1}{5}e^{-2t} - \frac{1}{3}t + \frac{2}{9} + C_2 e^{-t} + C_1 e^{3t}$

Now you have to find the constants. When you do this, you get

$\frac{97}{90}e^{3t} - \frac{1}{2}e^{-t} + \left(\frac{1}{5}e^{-2t} - \frac{1}{3}t + \frac{2}{9}\right)$

3. The general solution is $C_1 e^{-t} - 2\cos t + C_2 e^{-3t}$. Then one needs to consider the initial conditions.

5. The general solution is $C_1 e^{-2t} - 3te^{-2t} + C_2 e^{3t}$.

7.   a. $y'' - y' - 6y = -15e^{3t}$. Then find the solution which satisfies the initial conditions $y(-1) = -3, y'(-1) = -3$.

The general solution is $C_1 e^{3t} - 3te^{3t} + C_2 e^{-2t}$. The solution with the given initial conditions is

$\left(-\frac{9}{5}e^3 - \frac{12}{5}e^{-3}e^3\right)e^{3t}$

$+ \left(-\frac{3}{5}e^{-2}\left(e^{-3} + 2\right)\right)e^{-2t} - 3te^{3t}$

   c. Find the general solution to the equation $5y - 4y' + y'' = -2e^{2t}$. Then find the solution which satisfies the initial conditions $y(0) = 5, y'(0) = 2$.

The general solution is

$e^{2t}\left(C_1 \cos t + C_2 \sin t - 2\right).$

The solution with the given initial conditions is obtained by solving equations.

Then the solution is

$-e^{2t}\left(8\sin t - 7\cos t + 2\right)$

9. Recall the easy way to do this is to look for a solution to $34y - 10y' + y'' = e^{-3it}$ and then consider real and imaginary parts. Thus you find a particular

solution of the form Ae^{-3it} and plug in to the equation $(25 + 30i) Ae^{-3it} = e^{-3it}$ and so you need to have $(25 + 30i) A = 1$. Thus

$$A = \frac{1}{61} - \frac{6}{305}i$$

This gives the solution

$$\left(\frac{1}{61} - \frac{6}{305}i \right) (\cos(-3t) + i \sin(-3t))$$

and so the real and imaginary parts are, respectively,

$$\frac{1}{61} \cos 3t - \frac{6}{305} \sin 3t, \ -\frac{6}{305} \cos 3t - \frac{1}{61} \sin 3t$$

Thus the desired particular solution is

$$y_p(t) = -\frac{3}{305} \cos 3t - \frac{33}{305} \sin 3t$$

11. Recall the easy way to do this is to look for a solution to

$$20y - 8y' + y'' = e^{2it}$$

and then consider real and imaginary parts. Thus you find a particular solution of the form Ae^{2it} and plug in to the equation $(16 - 16i) Ae^{2it} = e^{2it}$ and so you need to have

$$(16 - 16i) A = 1$$

Thus

$$A = \frac{1}{32} + \frac{1}{32}i$$

This gives the solution

$$\left(\frac{1}{32} + \frac{1}{32}i \right) (\cos(2t) + i \sin(2t))$$

and so the real and imaginary parts are, respectively,

$$\frac{1}{32} \cos 2t - \frac{1}{32} \sin 2t, \ \frac{1}{32} \cos 2t + \frac{1}{32} \sin 2t$$

Thus the desired particular solution is

$$y_p(t) = \frac{5}{32} \cos 2t - \frac{1}{32} \sin 2t$$

13. The solution to the homogeneous problem is $C_1 e^t + C_2 e^{3t} + C_3 e^{2t}$. A particular solution will be of the form $te^{3t} + te^t = y_p$. Then need to solve the following for the C_i

$$C_1 + C_2 + C_3 = 0$$

$C_1 + 3C_2 + 2C_3 + 2 = 1$, The solution is then

$$C_1 + 9C_2 + 4C_3 + 8 = 0$$

$$y = 4e^{2t} - \frac{3}{2}e^t - \frac{5}{2}e^{3t} + t\left(e^t + e^{3t}\right)$$

15. The solution to the homogeneous problem is $C_1e^{-2t} + C_2e^{-t} + C_3e^{2t}$. A particular solution will be of the form $e^{-t} + te^{-2t} = y_p$. Then need to solve the following for the C_i

$$C_1 + C_2 + C_3 + 1 = 1$$

$2C_3 - C_2 - 2C_1 = 0$, The solution is then

$$4C_1 + C_2 + 4C_3 - 3 = 0$$

$$y = \frac{3}{4}e^{-2t} + \frac{1}{4}e^{2t} + te^{-2t}$$

17. The general solution to the homogeneous problem is

$$e^{(3\sqrt{2}/2)t} \begin{pmatrix} C_1 \cos\left(\frac{3}{2}\sqrt{2}t\right) \\ +C_2 \sin\left(\frac{3}{2}\sqrt{2}t\right) \end{pmatrix} + e^{-\left(\frac{3}{2}\sqrt{2}t\right)} \begin{pmatrix} C_3 \cos\left(\frac{3}{2}\sqrt{2}t\right) \\ +C_4 \sin\left(\frac{3}{2}\sqrt{2}t\right) \end{pmatrix}$$

Then the particular solution is most easily found by looking for the imaginary part of

$$z'''' + 81z = e^{i3t}$$

So let $z_p = Ae^{3it}$. Inserting this in to the equation, we need to have $162A = 1$. Then the solution is $z = \frac{1}{162}e^{3it}$ and so $y_p = \frac{1}{162}\sin 3t$ and so the general solution is

$$e^{(3\sqrt{2}/2)t}\left(C_1\cos\left(\frac{3}{2}\sqrt{2}t\right) + C_2\sin\left(\frac{3}{2}\sqrt{2}t\right)\right)$$

$$+e^{-\left(\frac{3}{2}\sqrt{2}t\right)}\left(C_3\cos\left(\frac{3}{2}\sqrt{2}t\right) + C_4\sin\left(\frac{3}{2}\sqrt{2}t\right)\right) + \frac{1}{162}\sin 3t$$

19. The general solution to the homogeneous problem is

$$C_3\cos 4t + C_1e^{4t} + C_2e^{-4t} + C_4\sin 4t$$

Now consider

$$z^4 - 256z = e^{4it}$$

You have to multiply by t because e^{3it} solves the homogeneous problem. Then, inserting into the equation yields

$$-256iA = 1$$

$A = \frac{1}{256}i$. Thus

$$z = \frac{1}{256}ite^{4it} = -\frac{1}{256}t\left(\sin 4t - i\cos 4t\right)$$

and so

$$y_p = -\frac{3}{256}t\cos 4t - \frac{1}{128}t\sin 4t.$$

The general solution is

$$C_3\cos 4t + C_1 e^{4t} + C_2 e^{-4t} + C_4\sin 4t - \frac{3}{256}t\cos 4t - \frac{1}{128}t\sin 4t$$

21. $y_p = A\cos 5t + B\sin 5t$

$$A' = \frac{\begin{vmatrix} 0 & \sin 5t \\ -\frac{1}{\cos 5t} & 5\cos 5t \end{vmatrix}}{5} = \tfrac{1}{5}\tan 5t,\ B' = -\tfrac{1}{5},$$
$$A = \int \tfrac{1}{5}\tan 5t = -\tfrac{1}{25}\ln\left(\cos 5t\right)$$
$$B = \int -\tfrac{1}{5}dt,\ y_p = -\tfrac{1}{5}t\sin 5t - \tfrac{1}{25}\cos 5t\ln\left(\cos 5t\right)$$

23. Then the particular solution is $2t^2 e^{-3t}\left(\ln 3t - 1\right) - \tfrac{1}{2}t^2 e^{-3t}\left(2\ln 3t - 1\right).$

25. Then the particular solution is $t^2 e^{-5t}\left(\ln 3t - 1\right) - \tfrac{1}{4}t^2 e^{-5t}\left(2\ln 3t - 1\right).$

27. $y = \left(\int_0^t \cosh t^2\left(-\tfrac{1}{4}e^s\right)ds\right)\left(e^{-t}\right) + \left(\int_0^t\left(-\cosh t^2\right)\left(-\tfrac{1}{4}e^{-3s}\right)ds\right)\left(e^{3t}\right)$

29. Another solution is $\tfrac{1}{t^3}$, $y = tC_1 + \tfrac{1}{t^3}C_2 + \tfrac{1}{7}t^4\ln\tfrac{1}{t}$

31. Another solution is $\tfrac{1}{t}$

33. Another solution is $\tfrac{1}{t^2}\ln t$, $y = \tfrac{1}{8}t^6\ln t - \tfrac{1}{64}t^6\left(8\ln t - 1\right) + \tfrac{1}{t^2}C_1 + \tfrac{1}{t^2}C_2\ln t$

37. You can use $\sin^2\left(2t\right) = \frac{1-\cos(4t)}{2}$

$$y = \frac{1}{24}\cos 4t - \frac{1}{6}\cos 2t + \frac{1}{2}\sin 2t + \frac{1}{8}$$

39. a. $y'' + 3y' + 2y = \ln\left(1+t^2\right), y\left(0\right) = 1, y'\left(0\right) = 2.$

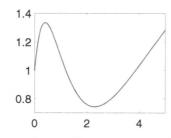

c. $y'' + \left(1+.1t^2\right)y' - y = \sin\left(\ln\left(1+y^2\right)\right), y\left(0\right) = 1, y'\left(0\right) = 0.$ On this one, graph on $[0, 2]$.

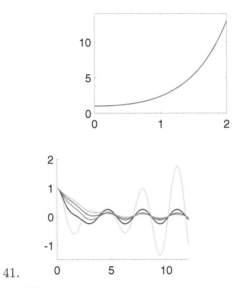

41.

When $a = 0$, you see resonance occurring. However, when a is larger, this time, the time interval is long enough relative to the damping coefficient a that you see the oscillations being squashed.

I.6   Laplace Transform Methods

1. a. $-\frac{2}{(s-3)^2-3}$, Is Laplace transform of $-\frac{2}{3}\sqrt{3}\left(\sinh\sqrt{3}t\right)e^{3t}$

    c. $3\frac{s}{s^2+9}$, Is Laplace transform of $3\cos 3t$

3.

$$Y'(s) = \int_0^\infty y(t)\lim_{h\to 0}\frac{e^{-(s+h)t}-e^{-st}}{h}dt = \int_0^\infty y(t)\left(-te^{-st}\right)dt$$

$$Y''(t) = \int_0^\infty y(t)(-t)\cdot\lim_{h\to 0}\frac{e^{-(s+h)t}-e^{-st}}{h}dt$$
$$= \int_0^\infty y(t)(-t)(-t)e^{-st}dt = \int_0^\infty y(t)(-1)^2 t^2 e^{-st}dt$$

Continuing this way or using math induction, this establishes the desired conclusion.

5. a.

$$4y - 4y' + y'' = 5t^3 e^{2t},\ y(0) = 1, y'(0) = 5$$

$$y = e^{2t} + 3te^{2t} + \frac{1}{4}t^5 e^{2t}$$

c. Using Laplace transforms, give a solution to the following initial value problem.

$$9y - 6y' + y'' = 2t^2 e^{3t}, y(0) = 1, y'(0) = 2$$

You take Laplace transforms using the table and solve for the Laplace transform.

$$\mathcal{L}(f) = \left(\frac{s-4}{s^2 - 6s + 9}\right) + \left(\frac{1}{s^2 - 6s + 9}\right)\left(\frac{4}{(s-3)^3}\right)$$

$$\frac{1}{s^2 - 6s + 9} = \frac{1}{(s-3)^2}, \quad \frac{s-4}{s^2 - 6s + 9} = \frac{1}{s-3} - \frac{1}{(s-3)^2}$$

and so the solution is

$$y = e^{3t} - te^{3t} + \frac{1}{6}t^4 e^{3t}$$

7. Take Laplace transforms of both sides.

$$s^2 \mathcal{L}(y) - (s+2) - 4(s\mathcal{L}(y) - 1) + 3\mathcal{L}(y) = \frac{1}{s} - \frac{e^{-2s}}{s} + e^{-2s}\frac{1}{s^2 + 1}$$

Then

$$\left(s^2 - 4s + 3\right)\mathcal{L}(y) = s - 2 + \frac{1}{s} - \frac{e^{-2s}}{s} + e^{-2s}\frac{1}{s^2 + 1}$$

so solve for $\mathcal{L}(y)$.

$$\mathcal{L}(y) = \frac{1}{s}\frac{s-1}{s-3} - e^{-2s}\frac{1}{s(s^2 - 4s + 3)} + e^{-2s}\frac{1}{(s^2 + 1)(s^2 - 4s + 3)}$$

$\frac{1}{s}\frac{s-1}{s-3} = \frac{2}{3(s-3)} + \frac{1}{3s}$, $\frac{1}{s(s^2-4s+3)} = \frac{1}{6(s-3)} - \frac{1}{2(s-1)} + \frac{1}{3s}$, $\frac{1}{(s^2+1)(s^2-4s+3)} = \frac{\frac{1}{5}s + \frac{1}{10}}{s^2+1} - \frac{1}{4(s-1)} + \frac{1}{20(s-3)}$

It follows that

$$y = \frac{2}{3}e^{3t} + \frac{1}{3} - u_2(t)\left(\frac{1}{6}e^{3(t-2)} - \frac{1}{2}e^{t-2} + \frac{1}{3}\right) +$$

$$u_2(t)\left(\frac{1}{5}\cos(t-2) + \frac{1}{10}\sin(t-2) - \frac{1}{4}e^{t-2} + \frac{1}{20}e^{3(t-2)}\right)$$

Does it work? The I.C. clearly holds. If $t < 2$,

$\left(\frac{2}{3}e^{3t} + \frac{1}{3}\right)'' - 4\left(\frac{2}{3}e^{3t} + \frac{1}{3}\right)' + 3\left(\frac{2}{3}e^{3t} + \frac{1}{3}\right) = 1$. Sure worked on this interval. For $t > 2$, the function is

$$\frac{2}{3}e^{3t} - \frac{7}{60}e^{3t-6} + \frac{1}{5}\cos(t-2) + \frac{1}{4}e^{t-2} + \frac{1}{10}\sin(t-2)$$

$$\left(\begin{array}{c}\frac{2}{3}e^{3t} - \frac{7}{60}e^{3t-6} \\ +\frac{1}{5}\cos(t-2) + \frac{1}{4}e^{t-2} + \frac{1}{10}\sin(t-2)\end{array}\right)''$$

$$-4\left(\begin{array}{c}\frac{2}{3}e^{3t} - \frac{7}{60}e^{3t-6} \\ +\frac{1}{5}\cos{(t-2)} + \frac{1}{4}e^{t-2} + \frac{1}{10}\sin{(t-2)}\end{array}\right)'$$

$$+3\left(\frac{2}{3}e^{3t} - \frac{7}{60}e^{3t-6} + \frac{1}{5}\cos{(t-2)} + \frac{1}{4}e^{t-2} + \frac{1}{10}\sin{(t-2)}\right)$$

does equal $\sin{(t-2)}$.

9. Take Laplace transforms of both sides.

We have

$$\mathcal{L}(y) = \frac{4s+1}{s^2-1} - \frac{3}{s(s^2-1)} +$$

$$\frac{5}{s}\frac{e^{-2s}}{s^2-1} + \frac{1}{s^2}\frac{e^{-2s}}{s^2-1}$$

Therefore, from the table,

$$y = e^t + 3$$

$$+ (5)\,u_2\,(t)\left(\frac{1}{2}e^{2-t} + \frac{1}{2}e^{t-2} - 1\right) + u_2\,(t)\left(\frac{1}{2}e^{t-2} - \frac{1}{2}e^{2-t} - t + 2\right)$$

13. a. $y^{(4)} - 3y - 4y' + 2y'' + 4y''' = 3e^{-t}$, $\quad y_p = \frac{3}{32}e^t - \frac{3}{16}e^{-t} + \frac{3}{32}e^{-3t} - \frac{3}{8}t^2e^{-t}$

 c. Using Laplace transforms, give a particular solution to the following equation.

$$y^{(4)} + 8y - 4y' - 6y'' + y''' = e^{-2t}$$

$$\left(s^4 + 8 - 4s - 6s^2 + s^3\right)\mathcal{L}(y) = \frac{1}{s+2}, \quad \mathcal{L}(y) = \frac{1}{(s+2)(s^4+8-4s-6s^2+s^3)}$$

$$y = \frac{37}{1728}e^{-2t} - \frac{1}{27}e^t + \frac{1}{64}e^{2t} + \frac{7}{144}te^{-2t} + \frac{1}{24}t^2e^{-2t}$$

15. a.

$$y(t) = -4 + 4\int_0^t \sin{(u-t)}\,y(u)\,du$$

Find $y(t)$.

$$\mathcal{L}(y) = -\frac{4}{s}\frac{s^2+1}{s^2+5}, \quad -\frac{4}{s}\frac{s^2+1}{s^2+5} = -\frac{16}{5}\frac{s}{s^2+5} - \frac{4}{5s}, \quad y(t) = -\frac{16}{5}\cos{\sqrt{5}t} - \frac{4}{5}$$

 c. The function $y(t)$ is a solution to the integral equation

$$y(t) = -3 + 2\int_0^t \sin{(4u-4t)}\,y(u)\,du$$

Find $y(t)$.

$$\mathcal{L}(y) = -\frac{3}{s}\frac{s^2+16}{s^2+24} - \frac{3}{s}\frac{s^2+16}{s^2+24} = -\frac{s}{s^2+24} - \frac{2}{s}, \quad y(t) = -\cos{2\sqrt{6}t} - 2$$

16. It obviously exists because

$$\int_0^R e^{-t^2}\,dt \le \int_0^1 dt + \int_1^\infty e^{-t}\,dt < \infty$$

Now

$$I^2 = \int_0^\infty \int_0^{\pi/2} e^{-r^2}\,r\,d\theta\,dr = \frac{1}{4}\pi$$

and so $I = \frac{\sqrt{\pi}}{2}$.

17. Change variable. Then $\Gamma(1/2) = \int_0^\infty e^{-u^2}\left(u^2\right)^{-1/2} 2u\,du = 2\int_0^\infty e^{-u^2}\,du = \sqrt{\pi}$.

19.

$$du = -\frac{1}{2\sqrt{\alpha^2 - y}}dy, \; -\frac{1}{2}dy = u\,du$$

$$\int\sqrt{\frac{1}{y}(\alpha^2 - y)}\,dy = -2\int\frac{u^2}{\sqrt{\alpha^2 - u^2}}du$$

Now you use a trig substitution. Thus the above is

$$u\sqrt{\alpha^2 - u^2} - \alpha^2\arcsin\left(\frac{u}{\alpha}\right) + C$$

In terms of y,

$$\sqrt{\alpha^2 - y}\sqrt{y} - \alpha^2\arcsin\left(\frac{\sqrt{\alpha^2 - y}}{\alpha}\right) + C$$

Then you get $x =$

$$\sqrt{\alpha^2 - y}\sqrt{y} - \alpha^2\arcsin\left(\frac{\sqrt{\alpha^2 - y}}{\alpha}\right) + C$$

You want $(0,0)$ to be on the graph and so

$$0 = -\alpha^2\arcsin(1) + C$$
$$C = \alpha^2\frac{\pi}{2}$$

Thus

$$x = \sqrt{\alpha^2 - y}\sqrt{y} - \alpha^2\arcsin\left(\frac{\sqrt{\alpha^2 - y}}{\alpha}\right) + \alpha^2\frac{\pi}{2}$$

Now recall $\alpha = \sqrt{\frac{2}{\pi}gT_0^2}$. Plug this in to get

$$x = \frac{1}{\sqrt{\pi}}\sqrt{y}\sqrt{2gT_0^2 - \pi y} - \frac{2}{\pi}gT_0^2\arcsin\left(\frac{1}{2}\sqrt{2}\frac{\sqrt{2gT_0^2 - \pi y}}{\sqrt{gT_0^2}}\right) + gT_0^2$$

21. $\mathcal{L}(g) \equiv \int_0^\infty e^{-st}\int_0^t f(u)\,du\,dt = \int_0^\infty\int_u^\infty f(u)e^{-st}dt\,du = \int_0^\infty f(u)\frac{1}{s}du = \frac{1}{s}\mathcal{L}(f)$

22. Let g be as just mentioned. Then from the above problem, for all s large enough,

$$\mathcal{L}(g)(s) = \frac{1}{s}\mathcal{L}(f)(s) = 0$$

and so $g(t) = \int_0^t f(u)\,du = 0$ for all t. Now at every point of continuity of f, the fundamental theorem of calculus implies $g'(t) = f(t) = 0$.

23. This is real easy. Let $t \to l(t)$ be a linear map, $l(t) = a + bt$ which is one to one and such that $l([0,1]) = [a,b]$. Now if g is a continuous function on $[a,b]$, then $g \circ l$ is a continuous map on $[0,1]$ and so for any $\varepsilon > 0$, there exists a polynomial $p(t)$ such that

$$\max_{t \in [0,1]} |g(l(t)) - p(t)| < \varepsilon$$

Now just let $x = l(t), t = l^{-1}(x)$. Then

$$\max_{x \in [a,b]} |g(x) - p(l^{-1}(x))| = \max_{t \in [0,1]} |g(l(t)) - p(t)| < \varepsilon$$

But $p(l^{-1}(x))$ is clearly a polynomial.

25. The left side of * equals

$$\frac{1}{s} \int_0^\infty \frac{\sin\left(\frac{u}{s}\right)}{u} e^{-u} du$$

and this clearly converges to 0 as $s \to \infty$. Now the derivative of the expression on the left is of the form

$$\int_0^\infty \frac{\sin(t)(-t)}{t} e^{-st} dt = -\int_0^\infty \sin(t) e^{-st} dt$$

$$= -\frac{1}{1+s^2} = \frac{d}{ds}\left(\int_s^\infty \frac{1}{1+u^2} du\right)$$

Now the two sides of * have the same derivative and they both converge to 0 as $s \to \infty$. Therefore, they are equal. You can then take the limit as $s \to 0$ on both sides. On the left, you get $\int_0^\infty \frac{\sin(t)}{t} dt$ which exists by considering

$$\int_0^1 \frac{\sin t}{t} dt + \int_1^\infty \frac{\sin(t)}{t} dt$$

the first obviously exists because the function is continuous if it is defined to equal 1 at 0 and it is bounded. The second improper integral exists by a simple integration by parts. Thus

$$\int_0^\infty \frac{\sin(t)}{t} dt = \int_0^\infty \frac{1}{1+u^2} du = \frac{\pi}{2}$$

27.   a. $\sin(t^2)$

$$\sin(t^2) = \sum_{k=0}^\infty \frac{(-1)^k (t^2)^{2k+1}}{(2k+1)!} \text{ so } \mathcal{L}(\sin(t^2)) = \sum_{k=0}^\infty \frac{(-1)^k}{s^{4k+3}} \frac{\Gamma(4k+3)}{(2k+1)!}$$

c. $t^p \sin t$

$$\mathcal{L}(t^p \sin t) = \sum_{k=0}^\infty \frac{(-1)^k}{s^{2k+p+2}} \frac{\Gamma(2k+p+2)}{(2k+1)!}$$

I.7 Power Series Theory

1. a. $\sum_{k=1}^{\infty} \left(\frac{x}{2}\right)^n$, $|x| < 2$

    c. $\sum_{k=0}^{\infty} k! x^k$, Need $\lim_{k \to \infty} \frac{(k+1)! |x|^{k+1}}{k! |x^k|} < 1$ and this happens only if

    $$\lim_{k \to \infty} (k+1) |x| < 1$$

    which happens only if $|x| = 0$ so radius is 0.

9. From the equation, $y^{(k)}(0) = 1$ and so the power series is $\sum_{k=0}^{\infty} \frac{1}{k!} x^k$. Of course e^x is the solution to the equation and so this gives its power series thanks to uniqueness of the solutions to the initial value problem.

11. Series solution is:

    $x - \frac{1}{12} x^4 + \frac{1}{504} x^7 - \frac{1}{45\,360} x^{10} + O\left(x^{14}\right)$

13. Consider the first derivative.

    $$\lim_{x \to 0} \frac{e^{-1/x^2}}{x} = \lim_{y \to \infty} y e^{-y^2} = 0$$

    which follows from an easy application of L'Hospital's rule. The derivative for $x \neq 0$ is

    $$\frac{2}{x^3} e^{-\frac{1}{x^2}}$$

    and so similar reasoning will show the second derivative equals 0 at 0. In general, you can see that the nth derivative is of the form $\frac{q(x)}{p(x)} e^{-1/x^2}$ where $p(x), q(x)$ are polynomials. Hence the derivative $n + 1$ is of the form

    $$\lim_{x \to 0} \frac{\frac{q(x)}{p(x)} e^{-1/x^2}}{x} = \lim_{y \to \infty} \frac{P(y)}{Q(y)} e^{-y^2} = 0$$

    where $P(y), Q(y)$ are just polynomials obtained by replacing $1/x$ with y. The above limit is obvious because for y large enough, $Q(y)$ is bounded away from 0.

15. To show this, consider $|f(x) - f(y)|$. Let $\varepsilon > 0$ be given. Let N correspond to $\varepsilon/3$ in the above definition. Then for $y \in S$, and fixed $n > N$,

    $$|f(x) - f(y)| \leq |f(x) - f_n(x)| + |f_n(x) - f_n(y)| + |f_n(y) - f(y)|$$

    The first and last terms on the right sum to no more than $2\varepsilon/3$. Then by continuity of f_n, the middle term is less than $\varepsilon/3$ provided $|y - x|$ is sufficiently small. Hence f is indeed continuous at x.

17. It follows right away from the definition of the integral that if $f \geq 0$, then $\int_a^b f(x)\,dx \geq 0$. Therefore, you have

    $$\int_a^b (|f(x)| - f(x))\,dx \geq 0, \quad \int_a^b (|f(x)| + f(x))\,dx \geq 0$$

and so

$$\int_a^b |f(x)|\, dx \geq \int_a^b f(x)\, dx, \quad \int_a^b |f(x)|\, dx \geq -\int_a^b f(x)\, dx$$

Now exactly one of the two on the right is $\left| \int_a^b f(x)\, dx \right|$ and so

$$\int_a^b |f(x)|\, dx \geq \left| \int_a^b f(x)\, dx \right|$$

If you like, just assume f is continuous. This is the case of interest in differential equations, but it is also true for arbitrary Riemann integrable functions as claimed.

21. $\frac{1}{t+1} = \sum_{k=0}^\infty (-1)^k t^k, |t| < 1$ and so, integrating the series termwise,

$$\sum_{k=0}^\infty (-1)^k \frac{x^{k+1}}{k+1}$$

whenever $|x| < 1$. So what about $x = -1$? Then the series obviously diverges because it is the harmonic series and in fact $\ln(0)$ is also undefined. So what about $x = 1$? In this case, the series converges by alternating series test. Does it converge to $\ln 2$?

$$\ln 2 = \int_0^1 \frac{1}{1+t}\, dt$$

Now $\frac{1}{1+t} = \sum_{k=0}^n (-1)^k t^k + \frac{(-1)^{n+1} t^{n+1}}{1+t}$. Therefore,

$$\ln 2 = \sum_{k=0}^n (-1)^k \frac{1}{k+1} + \int_0^1 \frac{(-1)^{n+1} t^{n+1}}{1+t}\, dt$$

This last term converges to 0 as $n \to \infty$ because

$$\left| \int_0^1 \frac{(-1)^{n+1} t^{n+1}}{1+t}\, dt \right| \leq \int_0^1 \frac{t^{n+1}}{1+t}\, dt \leq \int_0^1 t^{n+1}\, dt = \frac{1}{n+1}$$

Therefore, taking the limit as $n \to \infty$, we get

$$\ln 2 = \sum_{k=0}^\infty (-1)^k \frac{1}{k+1}$$

and so this series converges to $\ln(1+x)$ for $x \in (-1,1.]$.

23. $\sin(x^2) = \sum_{k=0}^\infty (-1)^k \frac{\left(x^2\right)^{2k+1}}{(2k+1)!}$

You know it is the right thing because the power series is unique.

25. It is a binomial series. $f(x) = \sum_{k=0}^\infty \begin{pmatrix} -1/2 \\ k \end{pmatrix} \left(-x^2\right)^k$ if $|x| < 1$.

27. $\int_0^1 \sum_{k=0}^7 (-1)^k \frac{\left(x^2\right)^{2k+1}}{(2k+1)!} dx = 0.310\,27$

$\int_0^1 \sin\left(x^2\right) dx = 0.310\,27$

29. $y'' + y = 0, y\left(0\right) = 0, y'\left(0\right) = 1. S'\left(x\right) = \sum_{n=1}^\infty \left(-1\right)^{n+1} \frac{x^{2n-2}}{(2n-2)!}$

$= 1 + \sum_{n=2}^\infty \left(-1\right)^{n+1} \frac{x^{2n-2}}{(2n-2)!}$

$S''\left(x\right) = \sum_{n=2}^\infty \left(-1\right)^{n+1} \frac{x^{2n-3}}{(2n-3)!} = \sum_{n=1}^\infty \left(-1\right)^n \frac{x^{2(n+1)-3}}{(2(n+1)-3)!} =$
$\sum_{n=1}^\infty \left(-1\right)^n \frac{x^{2n-1}}{(2n-1)!}.$

$S''\left(x\right) + S\left(x\right) = 0.$ Also $S\left(0\right) = 0, S'\left(0\right) = 1$ from the above.

31. It follows from the above problems that $C' = -S, S' = C$. Suppose y_i is a solution to the above initial value problem for $i = 1, 2$. Then $w \equiv y_1 - y_2$ solves $w'' + w = 0, w\left(0\right) = 0, w'\left(0\right) = 0$. Then multiply both sides of the differential equation with w'. This yields

$$\frac{d}{dx}\left(\frac{(w')^2}{2}\right) + \frac{d}{dx}\left(\frac{w^2}{2}\right) = 0$$

and so

$$\frac{(w')^2}{2} + \frac{(w)^2}{2} = C$$

From the initial conditions, this constant can only be 0. Therefore, $w = 0$ and so $y_1 = y_2$. The solution to $y'' + y = 0$, $y\left(0\right) = a$, $y'\left(0\right) = b$ is unique. What works?

$aC\left(x\right) + bS\left(x\right)$ works. Since any solution has some sort of initial condition involving y, y', this shows that all solutions are of this form.

32. Fix y. Then differentiate $C\left(x + y\right) - \left(C\left(x\right) C\left(y\right) - S\left(x\right) S\left(y\right)\right)$ twice with respect to x using $C' = -S, S' = C$.

$$C''\left(x + y\right) - \left(C''\left(x\right) C\left(y\right) - S''\left(x\right) S\left(y\right)\right)$$

$$+C\left(x + y\right) - \left(C\left(x\right) C\left(y\right) - S\left(x\right) S\left(y\right)\right) = 0$$

Also $C\left(y\right) - \left(C\left(0\right) C\left(y\right) - S\left(0\right) S\left(y\right)\right) = C\left(y\right) - C\left(y\right) = 0$ while

$$C'\left(x + y\right) - \left(C'\left(x\right) C\left(y\right) - S'\left(x\right) S\left(y\right)\right)$$
$$= -S\left(x + y\right) - \left(-S\left(x\right) C\left(y\right) - C\left(x\right) S\left(y\right)\right)$$

and evaluating at $x = 0$, this gives $-S\left(y\right) + S\left(y\right) = 0$. Now by uniqueness, it follows that for each y

$$C\left(x + y\right) - \left(C\left(x\right) C\left(y\right) - S\left(x\right) S\left(y\right)\right) = 0$$

The other identity can be established similarly. You could also differentiate the identity just obtained. This yields

$$-S\left(x + y\right) - \left(-S\left(x\right) C\left(y\right) - C\left(x\right) S\left(y\right)\right) = 0$$

Hence

$$S(x+y) = S(x)C(y) + S(y)C(x)$$

When $x = 0$, $S^2 + C^2 = 1$. Now differentiate.

$$2SS' + 2CC' = 2SC - 2CS = 0$$

and so this is a constant. Since it equals 1 when $x = 0$, it equals 1 for all x.

I.8   Series Methods for Scalar O.D.E.

1. You put in the infinite series. Thus

$$\left(\sum_{k=2}^{\infty} k(k-1)a_k x^{k-2} - \sum_{k=0}^{\infty} 2a_k k x^k + \sum_{k=0}^{\infty} 2na_k x^k \right) = 0$$

$$\left(\sum_{k=0}^{\infty} (k+2)(k+1)a_{k+2} x^k - \sum_{k=0}^{\infty} 2a_k k x^k + \sum_{k=0}^{\infty} 2na_k x^k \right) = 0$$

Thus you get

$$(k+2)(k+1)a_{k+2} = (2k - 2n)a_k$$

This yields

$$a_{k+2} = \frac{2(k-n)}{(k+2)(k+1)} a_k$$

Say n is odd. Say $n = 2l - 1$ Then let $a_0 = 0, a_1 = 1$. Then all even terms are 0. When $k = 2l - 1$, all further odd terms will be 0. Similar considerations will apply if n is even. You would then let $a_0 = 1$ and $a_1 = 0$.

3. See the chapter for the first part. For the second, you need consider

$$\lim_{x \to 1} \frac{(x-1)(-2x)}{1-x^2}, \lim_{x \to 1} \frac{(x-1)^2(n(n+1))}{(1-x^2)}$$

and determine whether these limits exist. However, they both exist.

$\lim_{x \to 1} \frac{(x-1)(-2x)}{1-x^2} = 1$, $\lim_{x \to 1} \frac{(x-1)^2(n(n+1))}{(1-x^2)} = (n(n+1))$,

$\lim_{x \to 1} \frac{(x-1)^2}{(1-x^2)} = 0$ The case at -1 is similar.

7. You put in the series and you get

$$\left(\begin{array}{c} \sum_{k=0}^{\infty} a_k k(k-1)x^{k-2} - \sum_{k=0}^{\infty} a_k k(k-1)x^k \\ - \sum_{k=0}^{\infty} 3a_k k x^k + \sum_{k=0}^{\infty} a_k n(n+2)x^k \end{array} \right) = 0$$

$$\left(\begin{array}{c} \sum_{k=0}^{\infty} a_{k+2}(k+2)(k+1)x^k - \sum_{k=0}^{\infty} a_k k(k-1)x^k \\ - \sum_{k=0}^{\infty} 3a_k k x^k + \sum_{k=0}^{\infty} a_k n(n+2)x^k \end{array} \right) = 0$$

Then a recurrence relation is

$$a_{k+2}\left(k+2\right)\left(k+1\right) = a_k\left(k\left(k+2\right) - n\left(n+2\right)\right)$$

$$a_{k+2} = \frac{k\left(k+2\right) - n\left(n+2\right)}{\left(k+2\right)\left(k+1\right)}$$

To get one solution, let $a_0 = 1$ and $a_1 = 0$ and to get the other, let $a_0 = 0$ and $a_1 = 1$. Then use the recurrence relation to specify the two series. If n is odd, the second of these two yields a polynomial and if n is even, it is the first of the two solutions which yields a polynomial.

9. $\frac{1}{2}x^4 + \frac{1}{3}x^3 - \frac{1}{2}x^2 - x + 1 + O\left(x^5\right)$

11. $-\frac{1}{6}x^4 - 3x^3 + x^2 + 3x - 2 + O\left(x^5\right)$

13. Multiply by x and divide by $\left(1 - x\right)$. Then

$$\left(x^2 y'' + \frac{x\left(\gamma - \left(1 + \alpha + \beta\right)x\right)}{1 - x} - \frac{\alpha\beta xy}{1 - x}\right) = 0$$

Then the associated Euler equation is

$$x^2 y'' + x\gamma + 0y = 0$$

and so the indicial equation is $r\left(r - 1\right) + r\gamma = 0$. Thus $r\left(r - 1 + \gamma\right) = 0$ so the two roots are $r = 0, r = 1 - \gamma$.

19. It is true if you have the product of two functions. Suppose then that

$$\frac{d^n}{dx^n}\left(fg\right) = \sum_{k=0}^{n} \binom{n}{k} f^{(n-k)} g^{(k)}$$

Then

$$\frac{d^{n+1}}{dx^{n+1}}\left(fg\right) = \left(\sum_{k=0}^{n} \binom{n}{k} f^{(n-k)} g^{(k)}\right)'$$

then by the product rule,

$$= \sum_{k=0}^{n} \binom{n}{k} f^{(n+1-k)} g^{(k)} + \sum_{k=0}^{n} \binom{n}{k} f^{(n-k)} g^{(k+1)}$$

$$= f^{(n+1)}g + \sum_{k=1}^{n} \binom{n}{k} f^{(n+1-k)} g^{(k)} + \sum_{k=0}^{n-1} \binom{n}{k} f^{(n-k)} g^{(k+1)} + fg^{(n+1)}$$

$$= f^{(n+1)}g + \sum_{k=1}^{n} \binom{n}{k} f^{(n+1-k)} g^{(k)}$$

$$+ \sum_{k=1}^{n} \binom{n}{k-1} f^{(n+1-k)} g^{(k)} + fg^{(n+1)}$$

$$= f^{(n+1)}g + \sum_{k=1}^{n} \left(\binom{n}{k} + \binom{n}{k-1} \right) f^{(n+1-k)}g^{(k)} + fg^{(n+1)}$$

$$= f^{(n+1)}g + \sum_{k=1}^{n} \binom{n+1}{k} f^{(n+1-k)}g^{(k)} + fg^{(n+1)}$$

$$= \sum_{k=0}^{n+1} \binom{n+1}{k} f^{(n+1-k)}g^{(k)}$$

21. You have

$$y_n'' - xy_n' + ny_n = 0$$
$$y_m'' - xy_m' + my_n = 0$$

Multiply the top by y_m and the bottom by y_n and subtract. Then you get

$$W'(x) - xW(x) = (m - n) y_n(x) y_m(x)$$

Then do the usual thing and integrate. Then

$$\int_{-\infty}^{\infty} \frac{d}{dx} \left(W(x) e^{-x^2/2} \right) dx = (m - n) \int_{-\infty}^{\infty} e^{-x^2/2} y_n(x) y_m(x) dx$$

The polynomial growth assumption implies that the integral on the left yields 0.

23. If $y \neq 0$ is a solution to the differential equation, then there exists z such that $C_1 y + C_2 z$ describes the general solution. To see this, there is a general solution coming from y_1, y_2 and $y = cy_1 + dy_2$. If $d = 0$ or $c = 0$, then let $z = y_2$ or y_1. Thus you can assume both $c, d \neq 0$. Then let $z = cy_1 - dy_2$. Then one can obtain both y_i as a linear combination of y, z and so the general solution is indeed $C_1 y + C_2 z$. It follows that y is one of a pair which gives the general solution, y, z the Wronskian $W(y, z)(t) \neq 0$. But this can't happen if both $y(t), y'(t) = 0$. Thus, since $y'(t) \neq 0$ whenever $y(t) = 0$, you have

$$y(s) = 0 + y'(t)(s - t) + o(t - s)$$

and so there is an interval around t such that there is only one zero on it. Indeed, for δ small enough and $s \in (t - \delta, t + \delta)$,

$$|y(s)| = |y'(t)(s - t) + o(t - s)| \geq |y'(t)| |t - s| - \frac{|y'(t)|}{2} |t - s| > 0$$

Given such a nonzero solution w to a second order linear differential equation, the next zero is $\hat{t} \equiv \inf \{s : s > t, w(s) = 0\}$. By continuity, $w(\hat{t}) = 0$ and so it is the next zero. It has to be larger than t because of the first part.

25. The equation is of the form $x^2 y'' + 2y' + \frac{1}{x} y = 0$ which is not of the right form for 0 to be a regular singular point. Indeed, it is spoiled nicely by $1/x$. Now suppose you have $\sum_{n=0}^{\infty} a_n x^{n+r}$ is a solution. Then plugging this in, you get

$$0 = \begin{pmatrix} \sum_{n=0}^{\infty} a_n (n+r)(n+r-1) x^{n+r+1} \\ + \sum_{n=0}^{\infty} 2a_n (n+r) x^{n+r} + \sum_{n=0}^{\infty} a_n x^{n+r} \end{pmatrix}$$

$$\begin{pmatrix} \sum_{n=1}^{\infty} a_{n-1}(n+r-1)(n+r-2) x^{n+r} \\ + \sum_{n=0}^{\infty} 2a_n (n+r) x^{n+r} + \sum_{n=0}^{\infty} a_n x^{n+r} \end{pmatrix}$$

thus there is a recurrence relation for $n \geq 1$

$$-a_{n-1}(n+r-1)(n+r-2)\frac{1}{3} = a_n$$

and also you need to have $2a_0 r + a_0 = 0$ so if $a_0 \neq 0$, you must have $r = -1/2$. Then in the recurrence relation,

$$a_n = -a_{n-1}\left(n - \frac{3}{2}\right)\left(n - \frac{5}{2}\right)$$

this cannot possibly result in only finitely many nonzero terms and

$$\left|\frac{a_n}{a_{n-1}}\right| = \left|\left(n - \frac{3}{2}\right)\left(n - \frac{5}{2}\right)\right|$$

so you have $\lim_{n \to \infty} \left|\frac{a_{n+1}}{a_n}\right| = \infty$ which requires that the radius of convergence must equal 0. Indeed, if $x \neq 0$,

$$\lim_{n \to \infty} \left|\frac{a_{n+1} x^{n+1/2}}{a_n x^{n-1/2}}\right| = \lim_{n \to \infty} \left|\frac{a_{n+1}}{a_n}\right| x = \infty$$

Thus the series fails to converge at any positive value of x. Does it even converge at 0? No, because it is not even defined there. Thus the only possible series solution is totally worthless.

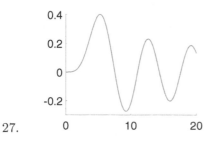

27.

I.9 First Order Linear Systems, Theory

1.

$$\Theta'(t) = \begin{pmatrix} \mathbf{x}_1'(t) & \cdots & \mathbf{x}_n'(t) \end{pmatrix} = \begin{pmatrix} A\mathbf{x}_1(t) & \cdots & A\mathbf{x}_n(t) \end{pmatrix}$$
$$= A\begin{pmatrix} \mathbf{x}_1(t) & \cdots & \mathbf{x}_n(t) \end{pmatrix} = A\Theta(t)$$

7. Say $\Theta(0)$ is invertible. Then $\Phi(t) = \Theta(t)\Theta(0)^{-1}$ and so determinant of both sides is nonzero. Hence $\det \Theta(t)$ is never zero. Also, the general solution is of the form $\Phi(t)\mathbf{v}$ and so there are vectors $\mathbf{v}_i$ such that the ith column of $\Theta(t)$ is $\Phi(t)\mathbf{v}_i$. Therefore, $\Theta(t) = \Phi(t)\begin{pmatrix} \mathbf{v}_1 & \cdots & \mathbf{v}_n \end{pmatrix}$. Thus $\Theta(t)$ is invertible for all t if and only if $\Theta(0) = \begin{pmatrix} \mathbf{v}_1 & \cdots & \mathbf{v}_n \end{pmatrix}$ is invertible.

9. See the earlier treatment.

11. It is pretty obvious
$$\|\Psi(s) - \Psi(t)\| = \max_{i,j} |\Psi_{ij}(s) - \Psi_{ij}(t)|$$
which is small for $|s - t|$ small if each of the component functions is continuous.

13. You just multiply the matrices and the result must be real. In fact, the product of these equals

$$\begin{pmatrix} 2 & 2 & 0 \\ 3-i & 3+i & -1 \\ -2 & -2 & 1 \end{pmatrix} \begin{pmatrix} \cos(t) + i\sin(t) & 0 & 0 \\ 0 & \cos(t) - i\sin(t) & 0 \\ 0 & 0 & e^t \end{pmatrix}$$
$$\cdot \begin{pmatrix} \frac{1}{4} - \frac{1}{4}i & \frac{1}{2}i & \frac{1}{2}i \\ \frac{1}{4} + \frac{1}{4}i & -\frac{1}{2}i & -\frac{1}{2}i \\ 1 & 0 & 1 \end{pmatrix}$$

which equals the matrix whose columns are

$$\begin{pmatrix} \cos t + \sin t \\ \cos t - e^t + 2\sin t \\ e^t - \cos t - \sin t \end{pmatrix}, \begin{pmatrix} -2\sin t \\ \cos t - 3\sin t \\ 2\sin t \end{pmatrix}, \begin{pmatrix} -2\sin t \\ \cos t - e^t - 3\sin t \\ e^t + 2\sin t \end{pmatrix}$$

which is indeed real.

I.10 Methods for First Order Linear Systems

1.
$$\left(\begin{pmatrix} s & 0 \\ 0 & s \end{pmatrix} - \begin{pmatrix} -11 & -4 \\ 42 & 15 \end{pmatrix} \right)^{-1} = \begin{pmatrix} \frac{s-15}{s^2-4s+3} & -\frac{4}{s^2-4s+3} \\ \frac{42}{s^2-4s+3} & \frac{s+11}{s^2-4s+3} \end{pmatrix}$$

$\begin{pmatrix} \frac{s-15}{s^2-4s+3} & -\frac{4}{s^2-4s+3} \\ \frac{42}{s^2-4s+3} & \frac{s+11}{s^2-4s+3} \end{pmatrix}$, is the Laplace transform of the matrix with columns

$$\begin{pmatrix} e^{2t}(\cosh t - 13\sinh t) \\ 42(\sinh t)e^{2t} \end{pmatrix}, \begin{pmatrix} -4(\sinh t)e^{2t} \\ e^{2t}(\cosh t + 13\sinh t) \end{pmatrix}$$

3. $\left(\begin{pmatrix} s & 0 \\ 0 & s \end{pmatrix} - \begin{pmatrix} -38 & -12 \\ 117 & 37 \end{pmatrix}\right)^{-1} = \begin{pmatrix} \frac{s-37}{s^2+s-2} & -\frac{12}{s^2+s-2} \\ \frac{117}{s^2+s-2} & \frac{s+38}{s^2+s-2} \end{pmatrix}$

$\begin{pmatrix} \frac{s-37}{s^2+s-2} & -\frac{12}{s^2+s-2} \\ \frac{117}{s^2+s-2} & \frac{s+38}{s^2+s-2} \end{pmatrix}$, is the Laplace transform of the matrix with columns

$\begin{pmatrix} e^{-\frac{1}{2}t}\left(\cosh\frac{3}{2}t - 25\sinh\frac{3}{2}t\right) \\ 78\left(\sinh\frac{3}{2}t\right)e^{-\frac{1}{2}t} \end{pmatrix}, \begin{pmatrix} -8\left(\sinh\frac{3}{2}t\right)e^{-\frac{1}{2}t} \\ e^{-\frac{1}{2}t}\left(\cosh\frac{3}{2}t + 25\sinh\frac{3}{2}t\right) \end{pmatrix}$

This is the fundamental matrix.

7. $\mathcal{L}(\mathbf{x}) = \left(s\begin{pmatrix} 1 & 0 \\ 0 & 1 \end{pmatrix} - \begin{pmatrix} 0 & 4 \\ -\frac{1}{2} & 3 \end{pmatrix}\right)^{-1}\left(\begin{pmatrix} -2 \\ -2 \end{pmatrix} + \begin{pmatrix} \frac{4}{s^2} \\ \frac{1}{s} \end{pmatrix}\right)$

$= \begin{pmatrix} 4\frac{\frac{1}{s}-2}{s^2-3s+2} + \left(\frac{4}{s^2}-2\right)\frac{s-3}{s^2-3s+2} \\ 2s\frac{\frac{1}{s}-2}{2s^2-6s+4} - \frac{\frac{4}{s^2}-2}{2s^2-6s+4} \end{pmatrix}$ so $\mathbf{x} = \begin{pmatrix} 8e^t - 6t - 5e^{2t} - 5 \\ 2e^t - t - \frac{5}{2}e^{2t} - \frac{3}{2} \end{pmatrix}$

9. $\mathcal{L}(\mathbf{x}) = \left(s\begin{pmatrix} 1 & 0 \\ 0 & 1 \end{pmatrix} - \begin{pmatrix} -\frac{16}{7} & \frac{15}{7} \\ \frac{10}{7} & \frac{9}{7} \end{pmatrix}\right)^{-1}\cdot\left(\begin{pmatrix} 3 \\ -2 \end{pmatrix} + \begin{pmatrix} \frac{1}{s} \\ \frac{2}{s^2} \end{pmatrix}\right)$

$= \begin{pmatrix} 15\frac{\frac{2}{s^2}-2}{7s^2+7s-42} + (7s-9)\frac{\frac{1}{s}+3}{7s^2+7s-42} \\ \frac{\frac{1}{s}+3}{\frac{7}{10}s^2+\frac{7}{10}s-\frac{21}{5}} + \left(\frac{7}{10}s+\frac{8}{5}\right)\frac{\frac{2}{s^2}-2}{\frac{7}{10}s^2+\frac{7}{10}s-\frac{21}{5}} \end{pmatrix},$

$\mathbf{x} = \begin{pmatrix} \frac{64}{21}e^{-3t} - \frac{1}{7}e^{2t} - \frac{5}{7}t + \frac{2}{21} \\ -\frac{16}{21}t - \frac{2}{7}e^{2t} - \frac{64}{63}e^{-3t} - \frac{44}{63} \end{pmatrix}$

11. $\mathcal{L}(\mathbf{x}) = \left(s\begin{pmatrix} 1 & 0 \\ 0 & 1 \end{pmatrix} - \begin{pmatrix} \frac{13}{7} & \frac{2}{7} \\ -\frac{4}{7} & \frac{22}{7} \end{pmatrix}\right)^{-1}\cdot\left(\begin{pmatrix} 3 \\ -2 \end{pmatrix} + \begin{pmatrix} \frac{1}{s-4} \\ \frac{4}{s^2} \end{pmatrix}\right)$

$= \begin{pmatrix} 2\frac{\frac{4}{s^2}-2}{7s^2-35s+42} + (7s-22)\frac{\frac{1}{s-4}+3}{7s^2-35s+42} \\ \left(\frac{7}{4}s-\frac{13}{4}\right)\frac{\frac{4}{s^2}-2}{\frac{7}{4}s^2-\frac{35}{4}s+\frac{21}{2}} - \frac{\frac{1}{s-4}+3}{\frac{7}{4}s^2-\frac{35}{4}s+\frac{21}{2}} \end{pmatrix}$

$\mathbf{x} = \begin{pmatrix} \frac{4}{21}t + \frac{22}{7}e^{2t} - \frac{46}{63}e^{3t} + \frac{3}{7}e^{4t} + \frac{10}{63} \\ \frac{11}{7}e^{2t} - \frac{26}{21}t - \frac{184}{63}e^{3t} - \frac{2}{7}e^{4t} - \frac{23}{63} \end{pmatrix}$

13. $\mathcal{L}(\mathbf{x}) = \left(s\begin{pmatrix} 1 & 0 \\ 0 & 1 \end{pmatrix} - \begin{pmatrix} 11 & -9 \\ 12 & -10 \end{pmatrix}\right)^{-1}\left(\begin{pmatrix} 3 \\ 2 \end{pmatrix} + \begin{pmatrix} \frac{3}{s-4} \\ \frac{4}{s^2} \end{pmatrix}\right)$

$$= \begin{pmatrix} \frac{3}{s^2(s^3-5s^2+2s+8)} \cdot \\ (s^4 + s^3 - 6s^2 - 12s + 48) \\ \frac{2}{s^2(s^3-5s^2+2s+8)} \cdot \\ (s^4 + 3s^3 - 8s^2 - 30s + 88) \end{pmatrix},$$

$$\mathbf{x} = \begin{pmatrix} 18t + \frac{54}{5}e^{-t} - 3e^{2t} + \frac{21}{5}e^{4t} - 9 \\ 22t + \frac{72}{5}e^{-t} - 3e^{2t} + \frac{18}{5}e^{4t} - 13 \end{pmatrix}$$

21. First the general notion. $t\alpha \mathbf{z} t^{\alpha-1} = A\mathbf{z}t^{\alpha}$ so this happens if and only if $A\mathbf{z} = \alpha\mathbf{z}$ and so $(\alpha, \mathbf{z})$ is an eigen pair.

a. $\begin{pmatrix} 3 & 4 \\ -2 & -3 \end{pmatrix}$, eigenvectors: $\left\{ \begin{pmatrix} -1 \\ 1 \end{pmatrix} \right\} \leftrightarrow -1, \left\{ \begin{pmatrix} -2 \\ 1 \end{pmatrix} \right\} \leftrightarrow 1$

General solution: $C_1 \begin{pmatrix} -1 \\ 1 \end{pmatrix} t^{-1} + C_2 \begin{pmatrix} -2 \\ 1 \end{pmatrix} t$

c. $\begin{pmatrix} 3 & 2 \\ -1 & 0 \end{pmatrix}$, eigenvectors: $\left\{ \begin{pmatrix} -1 \\ 1 \end{pmatrix} \right\} \leftrightarrow 1, \left\{ \begin{pmatrix} -2 \\ 1 \end{pmatrix} \right\} \leftrightarrow 2$

23. The differential equations are

$$x_1' = -\frac{5}{10}x_1 + \frac{4}{10}x_3 + 1$$
$$x_2' = \frac{5}{10}x_1 - \frac{5}{10}x_2$$
$$x_3' = -\frac{x_3}{10} + \frac{5}{10}x_2 - \frac{4}{10}x_3$$

This is

$$\begin{pmatrix} x_1 \\ x_2 \\ x_3 \end{pmatrix}' = \begin{pmatrix} -\frac{1}{2} & 0 & \frac{2}{5} \\ \frac{1}{2} & -\frac{1}{2} & 0 \\ 0 & \frac{1}{2} & -\frac{1}{2} \end{pmatrix} \begin{pmatrix} x_1 \\ x_2 \\ x_3 \end{pmatrix} + \begin{pmatrix} 1 \\ 0 \\ 0 \end{pmatrix}$$

The eigenvalues all have negative real parts so the limit is equal to a particular solution which is a constant. This works out to $x_1 = 10, x_2 = 10, x_3 = 10$ so the above limit is 10. Here is the graph. The case where the initial condition is 12 is similar.

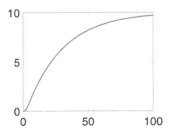

25. The above problem shows that $\Psi(t) = \Phi(t)M$ where $\Psi(0) = M$.

27. Say $\Psi(t) = \begin{pmatrix} \mathbf{v}_1(t) & \cdots & \mathbf{v}_n(t) \end{pmatrix}$. Consider $\Psi(t)\Psi(0)^{-1}$. It equals I when $t = 0$. Also $\left(\Psi(t)\Psi(0)^{-1}\right)' = \Psi'(t)\Psi(0)^{-1} = A\Psi(t)\Psi(0)^{-1}$ and so by uniqueness, $\Phi(t) = \Psi(t)\Psi(0)^{-1}$.

I.11 Theory of Ordinary Differential Equations

1. Potential energy needs to be computed first and then kinetic energy. Let x be the distance from the point at which the rod mass system is suspended. The potential energy of the chunk of the rod located at x corresponding to the angle θ measured from the vertical would be $(x - x\cos\theta)\rho dx$ and so the total potential energy is

$$\int_0^l (x - x\cos\theta)\rho dx = \rho(1 - \cos\theta)\frac{l^2}{2}$$

This is for the rod. Then you need to add in the potential energy for the mass which is $m(l - l\cos\theta)$. Thus the potential energy is

$$\rho(1 - \cos\theta)\frac{l^2}{2} + m(l - l\cos\theta)$$

Other than forces of constraint, the force would be

$$-\frac{\partial}{\partial\theta}\left(\rho(1 - \cos\theta)\frac{l^2}{2} + m(l - l\cos\theta)\right)$$

Now consider the kinetic energy. The moment of inertia is $\int_0^l \rho x^2 dx = \rho\frac{l^3}{3}$. This is for the rod. Therefore, the kinetic energy for the rod is

$$\frac{1}{2}\rho\frac{l^3}{3}(\theta')^2$$

and then there is the kinetic energy of the mass which is $\frac{1}{2}m(l\theta')^2$. Thus the Lagrangian is

$$\frac{1}{2}m(l\theta')^2 + \frac{1}{2}\rho\frac{l^3}{3}(\theta')^2 + \left(\rho(1 - \cos\theta)\frac{l^2}{2} + m(l - l\cos\theta)\right)$$

Then the equations of motion are

$$\left(\left(ml^2 + \rho\frac{l^3}{3}\right)\theta'\right)' + \left(\rho\frac{l^2}{2} + ml\right)\sin\theta = 0$$

3. Graph:

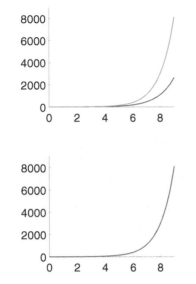

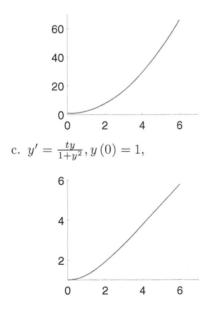

5.

In this case, you can't even see the difference between the two graphs, $y = e^t$ and $y(t)$ the solution from the Runge Kutta method. $y(9) = 8.0991 \times 10^3$ and the real answer is 8.103×10^3

7.   a. $y' = 5t - \sqrt{y}, y(0) = 1$,

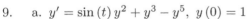

   c. $y' = \frac{ty}{1+y^2}, y(0) = 1$,

9.   a. $y' = \sin(t) y^2 + y^3 - y^5, \ y(0) = 1$

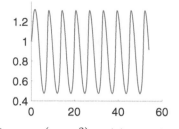

c. $y' = \tan\left(1 + y^2\right), y\left(0\right) = 0$ On this one, be sure to adapt the step size to compensate for large slopes. What happens if you don't do this?

If you don't adapt the step size, the algorithm will take you past the point where $1 + y^2 = \pi/2$ and you will get total nonsense involving wild jumping around. If you do adapt the step size, it will give you a good solution but will not go all the way to the point at which the solution blows up. Here is the type of things you get.

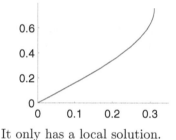

It only has a local solution.

11. Pick T. There is a local solution. Take dot product with $\mathbf{x}$

$$\frac{d}{dt}\left(\frac{\left|\mathbf{x}\left(t\right)\right|^2}{2}\right) + \left(\lambda\mathbf{x}\left(t\right) - \mathbf{f}\left(t, \mathbf{x}\left(t\right)\right)\right)\mathbf{x}\left(s\right) = \lambda\left|\mathbf{x}\left(t\right)\right|^2$$

$$\frac{\left|\mathbf{x}\left(t\right)\right|^2}{2} - \frac{\left|\mathbf{x}_0\right|^2}{2} \leq \lambda\int_0^t \left|\mathbf{x}\left(s\right)\right|^2 ds$$

and so

$$\left|\mathbf{x}\left(t\right)\right|^2 \leq \left|\mathbf{x}_0\right|^2 e^{2\lambda T}$$

Thus the solution is bounded so the solution exists on all of $[0, T]$. You just project onto a large ball of radius $\sqrt{2\left|\mathbf{x}_0\right|^2 e^{2\lambda T}}$, and consider the initial value problem

$$\mathbf{x}' = \mathbf{f}\left(t, P\mathbf{x}\right), \mathbf{x}\left(0\right) = \mathbf{x}_0$$

which has a global solution on $[0, T]$. However, P does not change anything thanks to the above estimate.

13. a. $y' = \left(1 + t\right)y^2, y\left(0\right) = 1$

$\begin{array}{l} y' = \left(1 + t\right)y^2 \\ y\left(0\right) = 1 \end{array}$, Exact solution is: $\left\{-\frac{2}{2t+t^2-2}\right\}$

The interval of existence is approximately $[0, .73]$

c. $y' = 1 + t\left(y + y^2 + y^4\right), y\left(0\right) = 1$

Here is a graph.

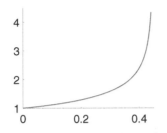

Here is the method used to obtain this graph.

```
f=@(t,x)[1+t*(x+x^2+x^4).];

n=3000; h1=.05;y(1)=1; t(1)=0;

hold on

for r=1:n

y(1)=1; t(1)=0;

hold on

for r=1:n

h=h1/(f(t(r),y(r))^2+1);

k1=f(t(r),y(r));

k2=f(t(r)+h/2,y(r)+k1*(h/2));

k3=f(t(r)+h/2,y(r)+k2*(h/2));

k4=f(t(r)+h,y(r)+k3*h);

y(r+1)=y(r)+(h/6)*(k1+2*k2

+2*k3+k4); t(r+1)=t(r)+h;

end

plot(t,y,'linewidth',2)
```

Thus it appears that the interval of existence is about $[0, .43]$.

I.12   Equilibrium Points and Limit Cycles

1.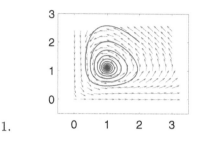

3. We are looking for an equilibrium point which is not the origin. Thus

$$a - by - \varepsilon x = 0$$

$$-c + dx - \delta y = 0$$

A solution is

$$\left(\frac{a\delta + bc}{\delta\varepsilon + bd}, \frac{ad - c\varepsilon}{\delta\varepsilon + bd} \right)$$

This is positive because ε, δ are small.

Now consider the stability of this equilibrium point.

$$a \left(\tfrac{a\delta+bc}{\delta\varepsilon+bd} + u \right) - b \left(\tfrac{a\delta+bc}{\delta\varepsilon+bd} + u \right) \left(\tfrac{ad-c\varepsilon}{\delta\varepsilon+bd} + v \right) - \varepsilon \left(\tfrac{a\delta+bc}{\delta\varepsilon+bd} + u \right)^2$$

$$-c \left(\tfrac{ad-c\varepsilon}{\delta\varepsilon+bd} + v \right) + d \left(\tfrac{a\delta+bc}{\delta\varepsilon+bd} + u \right) \left(\tfrac{ad-c\varepsilon}{\delta\varepsilon+bd} + v \right) - \delta \left(\tfrac{ad-c\varepsilon}{\delta\varepsilon+bd} + v \right)^2$$

Now expand this to consider an equilibrium point at $(0, 0)$.

$$-\tfrac{u\varepsilon+bv}{\delta\varepsilon+bd} \left(a\delta + bc + u\delta\varepsilon + bdu \right)$$

$$-\tfrac{v\delta-du}{\delta\varepsilon+bd} \left(ad - c\varepsilon + v\delta\varepsilon + bdv \right)$$

It is of the form $A\mathbf{z} + \mathbf{g}(\mathbf{z})$ where $\mathbf{g}$ is $\mathbf{o}(\mathbf{z})$. The matrix A is

$$\begin{pmatrix} -\varepsilon\frac{a\delta+bc}{\delta\varepsilon+bd} & -b\frac{a\delta+bc}{\delta\varepsilon+bd} \\ -d\frac{c\varepsilon-ad}{\delta\varepsilon+bd} & \delta\frac{c\varepsilon-ad}{\delta\varepsilon+bd} \end{pmatrix}$$

$\det(A) = (a\delta + bc)\frac{ad-c\varepsilon}{\delta\varepsilon+bd} > 0$ because ε is assumed very small. Thus the eigenvalues have both real parts positive or both real parts negative. So consider the trace. It equals

$$-\frac{1}{\delta\varepsilon + bd} \left(ad\delta + bc\varepsilon + a\delta\varepsilon - c\delta\varepsilon \right)$$

which is negative due to the assumption that δ, ε are both very small relative to a, b, c, d. Thus this equilibrium point is stable.

5. You have

$$x' = (xy)\left(\tfrac{a}{y} - b\right)$$
$$y' = (xy)\left(c - \tfrac{d}{x}\right)$$

Now

$$[(a\ln y - by) - (cx - d\ln x)]' = \left(\frac{a}{y} - b\right)y' - \left(c - \frac{d}{x}\right)x'$$
$$= \left(\frac{a}{y} - b\right)(xy)\left(c - \frac{d}{x}\right)$$
$$- \left(c - \frac{d}{x}\right)(xy)\left(\frac{a}{y} - b\right) = 0$$

It follows that the solution curves are of the form

$$(a\ln y - by) - (cx - d\ln x) \;=\; C$$
$$\text{in other words, } \ln\left(y^a x^d\right) - (cx + by) \;=\; C$$

So where does this have a local maximum?

$$D_x\left((a\ln y - by) - (cx - d\ln x)\right) \;:\; \tfrac{1}{x}\left(d - cx\right)$$
$$D_y\left((a\ln y - by) - (cx - d\ln x)\right) = \tfrac{1}{y}\left(a - by\right)$$

Thus the local maximum point is in the first quadrant and is given by $\left(\tfrac{d}{c}, \tfrac{a}{b}\right)$. You can see what it looks like by graphing the function of two variables given above. You could also consider the second derivative test to see it is a local maximum.

$$\begin{pmatrix} -\frac{c^2}{d} & 0 \\ 0 & -\frac{1}{a}b^2 \end{pmatrix}$$

This is the Hessian matrix and the eigenvalues are clearly negative.

8. If $x + y = L$, then since $L < K$, it follows that x' is positive, $y' = 0$ and so the solution curve moves in to the region. If $x + y = K$, then $x' = 0$ but $y' < 0$ so again the solution curve moves in to the region. Of course if $x = 0$, then $x' = 0$, $y' < 0$ so it can't cross the y axis. Similarly, it won't cross the x axis. Thus, if it starts in this region, it must stay there.

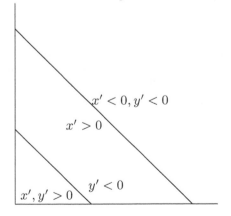

This also shows how y will become extinct. You start in the desired region and y' is always negative while x' increases and so $(x(t), y(t))$ must converge to the lower right corner. Even if the solution does not start in the region described, it will be led to it and then eventually y will become extinct and $x(t) \to K$.

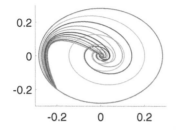

9.

You are varying a damping constant and graphing the solutions which result. The big circle corresponds to the damping constant equal to 0.

11. To show stability, note that $\sin x = x + o(x)$ and so the system is of the form

$$x' = y$$
$$y' = -\alpha y - x + g(x)$$

where $g(x)/x \to 0$. Then the matrix of interest is

$$\begin{pmatrix} 0 & 1 \\ -1 & -\alpha \end{pmatrix}$$

The determinant is positive and the trace is negative so the eigenvalues have negative real parts. Thus $(0,0)$ is stable.

13. You can't determine stability by looking at eigenvalues because the matrix is

$$\begin{pmatrix} -1 & 0 \\ 2 & 0 \end{pmatrix}$$

and it has a zero eigenvalue. A Lyapunov function is $x^2 + y^4 = W$ and $\nabla W \cdot f = -2x^2$. The limit set is then contained in the line $x = 0$. It is invariant also. Consider an initial condition $(0, y_0)$. Then the solution starting at this point does not stay on the y axis. This follows from the first equation. Therefore, the limit set is just $(0,0)$. Now if you have any initial data close to $(0,0)$, and look at $\frac{d}{dt} W(x(t), y(t)) = \nabla W \cdot f \leq 0$, it follows that the solution is bounded and so there is a limit set for any initial condition. Since there can only be one, this must be the limit. Thus $(0,0)$ is stable.

15. This is the same as the above except now we have $\nabla W \cdot f =$

$$-2x^2 y^2 - 2x^2$$

Thus the limit set is contained in $x = 0$. But if you consider an initial condition $(0, y_0)$, then the top equation will yield the solution leaving this set. Hence the only point in the limit set is $(0, 0)$. Thus if you begin close to the origin, the solution will converge to $(0, 0)$.

17. It follows from the Bendixon Dulac theorem with $R = 1$.

$$\left(x^3 - \sin y + x\right)_x + \left(y + \ln\left(1 + x^2\right) - \sin\left(\cos\left(x^2\right)\right)\right)_y = 2 + 3x^2$$

19.   a. $\begin{aligned} x' &= yx^2 + 2y \\ y' &= y^3 + 3y - x \end{aligned}$, $(0, 0)$.

It is already an almost linear system. The matrix with the decisive eigenvalues is

$$\begin{pmatrix} 0 & 2 \\ -1 & 3 \end{pmatrix}$$

and the eigenvalues are $\begin{pmatrix} 0 & 2 \\ -1 & 3 \end{pmatrix}$, eigenvalues: $2, 1$ so the origin is not stable.

c. $\begin{aligned} x' &= x^2 + 5x - 6y + 6 \\ y' &= 4x - 4y + xy^2 - 2xy + 4 \end{aligned}$, $(0, 1), (1, 2)$

At $(0, 1)$,

$\begin{pmatrix} 5 & -6 \\ 3 & -4 \end{pmatrix}$. The determinant is negative so one eigenvalue has negative real part. Thus $(0, 1)$ is not stable. It is a saddle point.

At $(1, 2)$,

$\begin{pmatrix} 7 & -6 \\ 4 & -2 \end{pmatrix}$ The determinant is positive and the trace is also so $(1, 2)$ is unstable.

23.   a.   Write as a first order system and use MATLAB to graph a vector field for $\mu = 1$.

$$\begin{pmatrix} x(1) \\ x(2) \end{pmatrix}' = \begin{pmatrix} x(2) \\ \mu\left(1 - \frac{1}{3}x(2)^2\right)x(2) - x(1) \end{pmatrix}$$

c. Determine whether there exists a limit cycle which attracts solutions beginning in some region.

You can do this by plotting some trajectories and noting that they can't cross. Thus there is a region for which initial data in this region lead to solutions which stay in it. These must spiral toward a limit cycle.

25. The top equation set equal to zero yields $y = \frac{1}{3}x^3 - x + k$. How many points of intersection can exist between this equation and $x + .8y = .7$? You have from the second, $y = \frac{.7-x}{.8}$. Then place this in the first. $0.875 = \frac{1}{3}x^3 + \frac{1}{4}x + k$. The function on the right is strictly increasing and so there is only one value of x which works.

Let (a, b) be the equilibrium point for k. Then let $x = a + u, y = b + v$. Find the equation in terms of u, v. Next determine stability according to the values of a.

$$u' = 3\left(-a^2u - au^2 - \frac{1}{3}u^3 + u + v\right), v' = -\frac{1}{3}(u + .8v)$$

The matrix is

$$\begin{pmatrix} -3a^2 + 3 & 1 \\ -1/3 & -.8/3 \end{pmatrix}$$

The characteristic equation is

$$\lambda^2 + \left(3.0a^2 - 2.733\,3\right)\lambda + \left(0.8a^2 - 0.466\,67\right) = 0$$

The eigenvalues are complex when

$$9.0a^4 - 19.600a^2 + 9.337\,6 < 0$$

This happens on

$$(-1.214, -0.839\,03) \cup (0.839\,03, 1.214)$$

Then for a on this interval, the real part of the eigenvalues is

$$\frac{-\left(3.0a^2 - 2.733\,3\right)}{2}$$

Thus the equilibrium point is stable on

$$(-1.214, -0.954\,52) \cup (0.954\,52, 1.214)$$

and unstable on $(-0.954\,52, 0.954\,52)$. When $|a| \geq 1.214$ the eigenvalues are real.

$$\det\begin{pmatrix} -3a^2 + 3 & 1 \\ -1/3 & -.8/3 \end{pmatrix} = 0.8a^2 - 0.466\,67$$

For these values of a, this determinant is positive and so the eigenvalues have the same signs. What of the trace? $-3a^2 + 3 + (-.8/3) = 2.733\,3 - 3.0a^2$. For these values of a, this is negative and so the equilibrium point is stable since both eigenvalues are negative. In summary,

$$\text{stable} \; < \; -.954 < \text{unstable}$$
$$< \; .954 < \text{stable}$$

What values of k correspond to these values of a?

We have $0.875 = \frac{1}{3}x^3 + \frac{1}{4}x + k$ and a was the solution to this equation. Thus the k which corresponds to this is $0.875 = \frac{1}{3}(-.954)^3 + \frac{1}{4}(-.954) + k$, solution is:1. 402 9. $0.875 = \frac{1}{3}(.954)^3 + \frac{1}{4}(.954) + k$. The solution is 0.347 08. thus we have unstable for k on $(.347, 1.402)$ and stable off this interval.

27. The matrix of interest is

$$\begin{pmatrix} -10 & 10 & 0 \\ r & -1 & 0 \\ 0 & 0 & -\frac{8}{3} \end{pmatrix}$$

Then the characteristic polynomial is

$$q(\lambda, r) = \lambda^3 + \frac{41}{3}\lambda^2 + \left(\frac{118}{3} - 10r\right)\lambda - \frac{80}{3}r + \frac{80}{3}$$

solve for r in $q(\lambda, r) = 0$. This gives

$$r = \frac{11}{10}\lambda + \frac{1}{10}\lambda^2 + 1$$

and then you also have the line $\lambda = -8/3$. That is, the level curve where $q(\lambda, r) = 0$ consists of the graph of the above function along with the line $\lambda = -8/3$. There is a minimum of this parabola at $\left(-\frac{11}{2}, -\frac{81}{40}\right)$. It intersects the r axis at 1. Thus for $r \in \left(-\frac{81}{40}, 1\right)$ there are three negative eigenvalues and this means the origin is stable. For $r > 1$, the origin is not stable because there is a positive eigenvalue for such values of r. For $r < -\frac{81}{40}$, there are two complex eigenvalues and a negative real eigenvalue. What about the real part of these two complex eigenvalues? They are $a + ib$ and $a - ib$. The matrix is

$$\begin{pmatrix} -10 & 10 & 0 \\ r & -1 & 0 \\ 0 & 0 & -\frac{8}{3} \end{pmatrix}$$

The trace is negative and the eigenvalues are $-8/3, a+ib, a-ib$. The trace is the sum of these eigenvalues. Thus $2a - 8/3 = \frac{80}{3}r - \frac{80}{3}$ and so

$$2a = \frac{8}{3} + \frac{80}{3}r$$
$$-\frac{80}{3} = \frac{80}{3}r - 24$$

which is clearly negative for these values of r. Thus a is negative and so the equilibrium point $(0,0,0)$ must be stable for these values of r. Thus $-\infty <$ stable$< 1 <$ unstable.

29. It is because there is only one equilibrium point and this point is a saddle. In fact the eigenvalues of the matrix are $-1, 1$. Recall that if the only equilibrium is a saddle point, there can't be any orbit containing it.

31. If $\mathbf{x} = \mathbf{0}$, there is nothing to show. Otherwise, $|\mathbf{x}/|\mathbf{x}|| = 1$ and so

$$\frac{1}{|\mathbf{x}|}|A\mathbf{x}| = \left|A\frac{\mathbf{x}}{|\mathbf{x}|}\right| \leq \|A\|$$

so multiplying on both sides by $|\mathbf{x}|$ gives the result. As to the other assertion,

$$\begin{aligned}\|AB\| &= \sup_{|\mathbf{x}|=1}\|AB\mathbf{x}\| \leq \sup_{|\mathbf{x}|=1}\|A\|\|B\mathbf{x}\| \\ &= \|A\|\|B\|\sup_{|\mathbf{x}|=1}|\mathbf{x}| = \|A\|\|B\|\end{aligned}$$

32. If $\mathbf{x}_0$ is given, explain why the solution is

$$\mathbf{x}(t) = \sum_{k=0}^{m-1} r_{k+1}(t)\,P_k(A)\,\mathbf{x}_0+$$

$$\sum_{k=0}^{m-1}\int_0^t r_{k+1}(t-s)\,P_k(A)\,\mathbf{f}(\mathbf{x}(s))\,ds \qquad (*)$$

where $r_k(t) = e^{\lambda_k t}(p_k(t) + \sum_{i=1}^{m_k} e^{\alpha_i t} q_i(t))$ where $\operatorname{Re}\alpha_i < 0$.

This follows from the chapter in which an explicit description is given of the fundamental matrix and it was shown that $r_k(t)$ is of the form claimed above. Here the minimal polynomial is

$$p(t) \equiv \prod_{i=1}^{m}(t - \lambda_i),\, m \leq n$$

Explain why $\mathbf{x}_0$ can be chosen such that $P_{m-1}(A)\,\mathbf{x}_0 \neq \mathbf{0}$.

Recall $P_{m-1}(A) = \prod_{i=1}^{m-1}(A - \lambda_i I)$. If no such $\mathbf{x}_0$ exists, then

$$\prod_{i=1}^{m-1}(A - \lambda_i I)\,\mathbf{x} = \mathbf{0}$$

for all $\mathbf{x}$ which implies that $p(t)$ was not the minimal polynomial after all.

If $|\mathbf{x}(s)| \leq \varepsilon$ for all $s > 0$, where ε is sufficiently small in comparison to $P_k(A)\,\mathbf{x}_0$, argue that the highest order terms from the integral are dominated by the highest order terms of $r_m(t)\,P_{m-1}(A)\,\mathbf{x}_0$. Thus it can't happen that $|\mathbf{x}(s)| < \varepsilon$ for all s. Hence $\mathbf{0}$ cannot be stable.

Let $0 < \delta < \frac{\operatorname{Re}\lambda_m}{5}$ and let ε be small enough that if $|\mathbf{x}| < \varepsilon$, then $|\mathbf{f}(\mathbf{x})| \leq \delta|\mathbf{x}|$.

Use the part a. and reduce attention to those terms which have exponential $e^{\lambda_m t}$ times t^p where p is the largest of all exponents in any of the polynomials occuring in the definition of $r_k(t)$. Recall these are of the form $p_k(t)\,e^{\lambda_k t}$ and the degrees of the p_k are increasing in k.

Let $C = \sup \{|P_{m-1}(A)\mathbf{x}| : |\mathbf{x}| = 1\}$ and choose $\mathbf{x}_0, |\mathbf{x}_0| \in \left(\frac{\varepsilon}{2}, \varepsilon\right)$ with

$$|P_{m-1}(A)(\mathbf{x}_0/|\mathbf{x}_0|)| > C/2$$

so

$$|P_{m-1}(A)(\mathbf{x}_0)| > C|\mathbf{x}_0|/2 > \frac{C\varepsilon}{4}$$

Argue that $\int_0^t (t-s)^p e^{\lambda_m(t-s)} ds$ is a polynomial of degree p times $e^{\lambda_m t}$ and the coefficient of the t^p term is $(1/\lambda_m)$.

Now obtain a contradiction from an assumption that $|\mathbf{x}(s)| < \varepsilon$ for all $s > 0$ by comparing the highest order terms.

In the above let ε be small enough that if $|\mathbf{x}| < \varepsilon$, it follows that $|\mathbf{f}(\mathbf{x})| \leq \delta|\mathbf{x}|$ where δ is very small. In particular, we assume that it is at least small enough that

$$\delta < \frac{\operatorname{Re}\lambda_m}{5}$$

Then the terms in the integral and the first sum in $*$ which dominate all others are

$$r_m(t) P_{m-1}(A)\mathbf{x}_0 \text{ and}$$

$$\int_0^t r_m(t-s) P_{m-1}(A)\mathbf{f}(\mathbf{x}(s)) ds$$

Recall that $r_m(t) =$

$$(p_m(t) + \text{terms which converge to } 0) e^{\lambda_m t}.$$

The terms which dominate all others correspond to the highest power of $p_m(t)$. Thus the terms which matter are of the form

$$t^p e^{\lambda_m t} P_{m-1}(A)\mathbf{x}_0 \text{ and}$$

$$\int_0^t (t-s)^p e^{\lambda_m(t-s)} P_{m-1}(A)\mathbf{f}(\mathbf{x}(s)) ds, \tag{*}$$

where p is the highest power. Let $C = \sup \{|P_{m-1}(A)\mathbf{x}| : |\mathbf{x}| = 1\}$. Choose $\mathbf{x}_0$ such that $|\mathbf{x}_0| \in \left(\frac{\varepsilon}{2}, \varepsilon\right)$ and

$$|P_{m-1}(A)(\mathbf{x}_0/|\mathbf{x}_0|)| > C/2$$

so

$$|P_{m-1}(A)\mathbf{x}_0| > \frac{C}{2}|\mathbf{x}_0| > \frac{C\varepsilon}{4}$$

The size of the first term is

$$t^p e^{\operatorname{Re}\lambda_m t}|P_{m-1}(A)\mathbf{x}_0| \geq t^p e^{\operatorname{Re}\lambda_m t}\frac{C\varepsilon}{4}$$

Assuming $|\mathbf{x}(s)| < \varepsilon$ for all s, the size of the second is dominated by

$$\int_0^t (t-s)^p e^{\operatorname{Re}\lambda_m(t-s)} C\varepsilon\delta ds$$

Now integrating by parts, one obtains

$$\int_0^t (t-s)^p e^{\operatorname{Re}\lambda_m(t-s)} ds = l(t) e^{\lambda_m t}$$

where $l(t)$ is a polynomial of degree p. In fact, its term having degree p will be of the form

$$\frac{1}{\lambda_m} t^p$$

and so the highest degree term of the integral is dominated by

$$\frac{C\varepsilon\delta}{\operatorname{Re}\lambda_m} t^p e^{\operatorname{Re}\lambda_m t}$$

If $|\mathbf{x}(s)| < \varepsilon$ for all s then it follows that it must be the case that

$$|P_{m-1}(A)\mathbf{x}_0| \le \frac{C\varepsilon\delta}{\operatorname{Re}\lambda_m}$$

since otherwise, the first term in $*$ would cause $|\mathbf{x}(s)|$ to blow up as $s \to \infty$ because you would have for some $a > 0$ a term of the form $at^p e^{\operatorname{Re}\lambda_m t}$ where p is larger than all other exponents of terms of polynomials and $\operatorname{Re}\lambda_m > 0$ is the largest real part of all eigenvalues. Therefore,

$$\frac{C\varepsilon}{4} \; < \; |P_{m-1}(A)\mathbf{x}_0|$$
$$\le \; \frac{C\varepsilon\delta}{\operatorname{Re}\lambda_m} < \frac{C\varepsilon}{5}$$

since $\delta < \operatorname{Re}\lambda_m/5$ and so this is a contradiction. Hence $|\mathbf{x}(s)|$ cannot stay less than ε and so $\mathbf{0}$ cannot be stable.

I.13   Boundary Value Problems and Fourier Series

1. The cosine series is being requested. The coefficients are

$$a_0 = \frac{1}{\pi}\int_0^\pi \sin(x)\, dx = \frac{2}{\pi}$$

$$a_n = \frac{2}{\pi}\int_0^\pi \sin(x)\cos(nx)\, dx = \begin{cases} 0 \text{ if } n \text{ is odd} \\ -\frac{4}{\pi(4k^2-1)} \text{ if } n = 2k \end{cases}$$

$\frac{2}{\pi}\int_0^\pi \sin(x)\cos(2kx)\, dx$

$$= \frac{2}{\pi}\left[\sin(x)\frac{\sin(2kx)}{2k}\Big|_0^\pi - \int_0^\pi \frac{\sin(2kx)}{2k}\cos(x)\right] = -\frac{2}{\pi}\left[\begin{array}{l} \frac{-\cos(2kx)}{(2k)^2}\cos(x)\,|_0^\pi \\ -\int_0^\pi \frac{\cos(2kx)}{(2k)^2}\sin x\, dx \end{array}\right]$$

$$= \frac{2}{\pi}\left[\begin{array}{l} \frac{\cos(2kx)}{(2k)^2}\cos(x)\,|_0^\pi \\ +\int_0^\pi \frac{\cos(2kx)}{(2k)^2}\sin x\, dx \end{array}\right] = \frac{2}{\pi}\frac{-2}{4k^2} + \frac{2}{\pi}\frac{1}{4k^2}\int_0^\pi \cos(2kx)\sin x\, dx$$

$\frac{2}{\pi}\left(1 - \frac{1}{4k^2}\right)\int_0^\pi \cos\left(2kx\right)\sin\left(x\right)dx = \frac{-4}{\pi 4k^2}$

$= \frac{2}{\pi}\int_0^\pi \cos\left(2kx\right)\sin\left(x\right)dx = -\frac{4}{4\pi k^2 - \pi}$

The series one obtains by letting $x = \pi/2$ is

$$1 = \frac{2}{\pi} - \sum_{k=1}^\infty \frac{4}{\pi\left(4k^2 - 1\right)}\cos\left(2k\frac{\pi}{2}\right) = \frac{2}{\pi} - \sum_{k=1}^\infty \frac{4}{\pi\left(4k^2 - 1\right)}(-1)^k$$

and so

$$\frac{2}{\pi} - 1 = \sum_{k=1}^\infty \frac{4}{\pi\left(4k^2 - 1\right)}(-1)^k$$

3. This time, you want the cosine series.

$a_0 = \frac{1}{\pi}\int_0^\pi x^2 dx = \frac{1}{3}\pi^2, a_k = \frac{2}{\pi}\int_0^\pi x^2 \cos\left(kx\right)dx$

$\int_0^\pi x^2 \cos\left(kx\right)dx = \frac{\sin(kx)}{k}x^2\big|_0^\pi - \int_0^\pi \frac{\sin(kx)}{k}2x$

$= \left[\frac{\cos(kx)}{k^2}2x\big|_0^\pi - 2\int_0^\pi \frac{\cos(kx)}{k^2}\right] = \frac{(-1)^k}{k^2}2\pi$

Then $a_k = \frac{2}{\pi}\frac{(-1)^k}{k^2}2\pi = 4\frac{(-1)^k}{k^2}$ and so the series is

$f\left(x\right) = \frac{\pi^2}{3} + \sum_{k=1}^\infty 4\frac{(-1)^k}{k^2}\cos\left(kx\right)$

at $x = \pi$ we get

$$\pi^2 = \frac{\pi^2}{3} + \sum_{k=1}^\infty 4\frac{1}{k^2}, \quad \frac{\pi^2}{6} = \sum_{k=1}^\infty \frac{1}{k^2}$$

5. From the theorem in the book, you can get it by differentiating formally. Thus it is $\sum_{k=1}^\infty 4\frac{(-1)^{k+1}}{k^2}k\sin\left(kx\right) = \sum_{k=1}^\infty 4\frac{(-1)^{k+1}}{k}\sin\left(kx\right)$

7. Yes, it does. $f\left(x + 2L\right) = f\left(x\right)$. Now differentiate both sides to find that $f'\left(x + 2L\right) = f'\left(x\right)$.

9. a. $y'' + \left(\frac{11\pi}{2}\right)^2 y = 0,\ y\left(0\right) = 0, y'\left(\frac{2}{2}\right) = 0$

    The solution is $C_1 \cos\left(\frac{11\pi}{2}x\right) + C_2 \sin\left(\frac{11\pi}{2}x\right) = y$ and the first condition requires that $C_1 = 0$. The second condition requires that $C_2 \cos\left(\frac{11\pi}{2}\right) = 0$. There are infinitely many nonzero solutions.

    c. $y'' + \left(\frac{2\pi}{7}\right)^2 y = 0,\ y\left(0\right) = 0, y'\left(\frac{7}{2}\right) = 0$. The solution is $y =$

    $$C_1 \cos\left(\frac{2\pi}{7}x\right) + C_2 \sin\left(\frac{2\pi}{7}x\right)$$

    and the first condition requires that $C_1 = 0$. The second condition requires that $C_2 \cos\left(\frac{2\pi}{2}\right) = 0$.

    There is no nonzero solution.

APPENDIX I. SELECTED SOLUTIONS

11. You just do the integration by parts using the boundary conditions and you get

$$\lambda \int_0^L (y')^2 \, dx = \int_0^L (y'')^2 \, dx$$

from which the result follows.

13. The hint is good. $(p(x) y')' z - (p(x) z')' y = ((p(x) y') z - (p(x) z') y)'$ follows from the product rule. Now you can just integrate. This gives

$$\int_a^b ((p(x) y') z - (p(x) z') y)' \, dx + \int_a^b (\lambda_1 - \lambda_2) q(x) y(x) z(x) \, dx = 0$$

Then you have

$$(p(b) y'(b)) z(b) - (p(b) z'(b)) y(b)$$
$$- [(p(a) y'(a)) z(a) - (p(a) z'(a)) y(a)]$$
$$+ \int_a^b (\lambda_1 - \lambda_2) q(x) y(x) z(x) \, dx = 0$$

From the boundary conditions, you know that $C_1 y(a) + C_2 y'(a) = 0, C_1 z(a) + C_2 z'(a) = 0$ and not both $C_i = 0$. Therefore, you must have $y(a) z'(a) - y'(a) z(a) = 0$. This shows the term

$$(p(a) y'(a)) z(a) - (p(a) z'(a)) y(a) = 0$$

Similar reasoning implies the other term in the boundary terms equals 0. Thus we are left with

$$\int_a^b (\lambda_1 - \lambda_2) q(x) y(x) z(x) \, dx = 0$$

and shows the orthogonality condition which is desired.

15. The equation is

$$(z(x) y(x))' + q(x) y(x) = z'y + zy' + q(x) y$$
$$= z'y + z^2 \frac{1}{p(x)} y + q(x) y = 0$$

Now cancel the y to get the equation for z

$$z' + \frac{1}{p(x)} z^2 + q(x) = 0$$

17. It is obvious that $\sin(nx)$ is a solution to the boundary value problem. Showing this directly was done in the chapter with use of a trig. identity. Now we do it by parts.

$$\int_{-\pi}^{\pi} \sin(nx) \sin(mx) \, dx = -\frac{\cos(nx)}{n} \sin(mx) \Big|_{-\pi}^{\pi}$$
$$+ \int_{-\pi}^{\pi} \frac{\cos(nx)}{n} m \cos(mx) \, dx$$

$$= \frac{\sin(nx)}{n^2} m \cos(mx) \,|_{-\pi}^{\pi} + \int_{-\pi}^{\pi} \frac{\sin(nx)}{n^2} m^2 \sin(mx)\, dx$$

and so

$$\left(1 - \frac{m^2}{n^2}\right) \int_{-\pi}^{\pi} \sin(nx) \sin(mx)\, dx = 0$$

and since $m \neq n$, the desired result follows.

19. $z'(x) = \alpha J_n'(\alpha x)$, $z''(x) = \alpha^2 J_n''(\alpha x)$. Thus

$$x^2 \frac{z''}{\alpha^2} + x \frac{z'}{\alpha^2} + \left(x^2 - \frac{n^2}{\alpha^2}\right) z = x^2 J_n''(\alpha x) + \frac{1}{\alpha} x J_n'(\alpha x)$$
$$+ \left(x^2 - \frac{n^2}{\alpha^2}\right) J_n(\alpha x)$$

Now multiply both sides by α^2 and you get

$$\left((\alpha x)^2 J_n''(\alpha x) + (\alpha x) J_n'(\alpha x) + (\alpha^2 x^2 - n^2) J_n(\alpha x)\right) = 0$$

because J_n is a solution of the Bessel equation.

21. Let $y(x) = z(\delta x)$. Thus $z(u) = y\left(\frac{u}{\delta}\right)$. Then from the chain rule,

$$x^2 \delta^2 z''(\delta x) + \delta x z'(\delta x) + \left(\delta^2 x^2 - n^2\right) z(\delta x) = 0$$

Letting $\delta x = u$, this shows that

$$\left(u^2 z''(u) + u z'(u) + \left(u^2 - n^2\right) z(u)\right) = 0$$

and so z is a solution of the Bessel equation. By assumption $z(\delta L) = 0$ and so $\delta = \alpha/L$ where α is some zero of z.

23. We know they must be positive. Also we know that any solution to this boundary value problem has $C_2 = 0$ because the solution is bounded near 0. It follows from the earlier problem above that $y = C J_n(\alpha x/L)$ where α is a zero of J_n. Note this is a specific function. To do it directly, let $z\left(\sqrt{\lambda}x\right) = y(x)$. Then

$$\left(\lambda x^2 z''\left(\sqrt{\lambda}x\right) + \sqrt{\lambda} x z'\left(\sqrt{\lambda}x\right) + \left(\lambda x^2 - n^2\right) z\left(\sqrt{\lambda}x\right)\right) = 0$$

Hence, letting $u = \sqrt{\lambda}x$, you have z is a solution to Bessel's equation. Then the boundary conditions show that z is a multiple of J_n and also $\sqrt{\lambda}L = \alpha$ for some α a zero of J_n. Thus $\lambda = \alpha^2/L^2$.

25. In this case, $p(x) = x$, $q(x) = x$ and so the form of the equation is, from the above,

$$z'' + \lambda z+$$

$$\left(\frac{\frac{d}{dx}\left(-\frac{1}{4}p^{-1/4}q^{-5/4}\frac{d}{dx}\left(x^2\right)\right)}{\left(x^{1/2}\right)} + \frac{r\left(x^2\right)^{-1/4}}{\left(x^{1/2}\right)} \right) z = 0$$

$$z'' + \lambda z + \left(\frac{\frac{d}{dx}\left(-\frac{1}{2\sqrt{x}}\right)}{\left(x^{1/2}\right)} + \frac{\left(-n^2/x\right)\left(x^2\right)^{-1/4}}{\left(x^{1/2}\right)} \right) z = 0$$

$$z'' + \lambda z + \left(\frac{1}{4x^2} + \left(-4n^2\right)\frac{1}{4x^2} \right) z = 0$$

$$z'' + \lambda z + \left(\frac{1 - 4n^2}{4x^2} \right) z = 0$$

Of course, you can take $t = x$ in this case and so you get simply

$$z'' + \lambda z + \left(\frac{1 - 4n^2}{4t^2} \right) z = 0$$

27. Showing this is just a repeat of what was done in the chapter concerning Fourier series.

29. By this problem, there is a polynomial $q(x) = \sum_{k=0}^{n} a_k p_k(x)$ such that

$$\max_{x \in [-1,1]} |f(x) - q(x)| < \sqrt{\frac{\varepsilon}{2}}$$

Then by the orthonormal property of the q_k, it follows that the Fourier series does an even better job of approximation. Hence

$$\int_{-1}^{1} |f(x) - S_n f(x)|^2 \, dx \leq \int_{-1}^{1} |f(x) - q(x)|^2 \, dx < 2 \left(\sqrt{\frac{\varepsilon}{2}} \right)^2 = \varepsilon$$

31. This function is odd so $b_n = 2 \int_0^1 \sin(n\pi t)\, dt = -\frac{2}{\pi n}\left((-1)^n - 1\right)$. Thus

$$f(t) = \sum_{n=1}^{\infty} \left(-\frac{2}{\pi n}\left((-1)^n - 1\right) \right) \sin(n\pi t)$$

$\sum_{n=1}^{7} \left(-\frac{2}{\pi n}\left((-1)^n - 1\right)\right) \sin(n\pi t)$. Now we solve $y'' + 3y = \sin(n\pi t)$. A particular solution is $-\pi n \frac{\sin(t\pi n)}{(\pi^2 n^2 - 3)\pi n}$. Therefore, a particular solution is

$$y_p(t) = \sum_{n=1}^{\infty} \left(-\frac{2}{\pi n}\left((-1)^n - 1\right) \right) \left(-\pi n \frac{\sin(t\pi n)}{(\pi^2 n^2 - 3)\pi n} \right)$$

33. It is just a matter of plugging in $A \exp\left(i\frac{n\pi t}{L}\right)$ and solving for A. As to the main observation, if y is the solution, then $y = u + iv$ and you have

$$u'' + 2au' + bu + i\left(v'' + 2av' + bv\right) = \left(\cos\left(\frac{n\pi t}{L}\right) + i \sin\left(\frac{n\pi t}{L}\right) \right)$$

From equating real and imaginary parts, you see that the claim is true.

35. It will likely break because you will have resonance with respect to this term and its amplitude will grow increasingly large till the equation is no longer satisfied and when this happens, it will likely involve the physical properties of the bridge changing enough that it will break or at least be seriously damaged.

37. For the first part, if $x > 0$ then eventually $1/n < x$ and so $f_n(x) = 0$ for all n sufficiently large. Thus $f_n(x) \to 0$. Of course, if $x = 0$, then $f_n(0) \equiv 0$ and so convergence takes place in this case also. However, $\|f_n\| = \int_0^{1/n} (\sqrt{n})^2 \, dx = 1$ and so it sure does not converge to 0 in the mean square sense. As to the second part, each of these functions f_n has the property that $\int_0^1 f_n(x)^2 \, dx = 0$ and so they obviously converge to 0 in the mean square sense, but they fail to converge pointwise because the equal 1 on every point of the form 2^{-n}.

I.14 Some Partial Differential Equations

1. a.
$$u_t = u_{xx}, u(0,t) = 0, u(4,t) = 0, \ u(x,0) = 1$$

The eigenvalue problem is $y'' + \delta^2 y = 0$ so with the boundary conditions, we need $y_n(x) = \sin\left(\frac{1}{4}\pi n x\right)$. Then the solution desired is $\sum_{n=1}^{\infty} b_n(t) y_n(x)$ and so we need to have $b_n'(t) y_n(x) = -b_n(t) \left(\frac{1}{16}\pi^2 n^2\right) y_n(x)$ and so we need to have $b_n(t) = b_n \exp\left(-\frac{1}{16}\pi^2 n^2 t\right)$ and so the solution is $u(x,t) =$

$$\sum_{n=1}^{\infty} b_n \exp\left(-\frac{1}{16}\pi^2 n^2 t\right) \sin\left(\frac{1}{4}\pi n x\right)$$

where b_n needs to be determined. It needs to equal $\frac{1}{2}\int_0^4 \left(\sin\left(\frac{1}{4}\pi n x\right)\right) dx$.

$$u(x,t) = \sum_{n=1}^{\infty} \left(-\frac{1}{2\pi n}\left(4(-1)^n - 4\right)\right) \exp\left(-\frac{1}{16}\pi^2 n^2 t\right) \sin\left(\frac{1}{4}\pi n x\right)$$

c.
$$u_t = 3u_{xx}, u(0,t) = 0, u(2,t) = 0, \ u(x,0) = 1 - x$$

$$u(x,t) = \sum_{n=1}^{\infty} \left(\frac{2}{\pi n}\left((-1)^n + 1\right)\right) \exp\left(-\frac{3}{4}\pi^2 n^2 t\right) \sin\left(\frac{1}{2}\pi n x\right)$$

3.
$$\sum_{n=1}^{\infty} \left(\begin{array}{c} \frac{2}{5\pi^3 n^3}(750(-1)^n - 750)\cos\left(\frac{2}{5}\pi n t\right) \\ + \left(-\frac{1}{\pi^2 n}(30(-1)^n - 5)\right)\sin\left(\frac{2}{5}\pi n t\right) \end{array} \right) \sin\left(\frac{1}{5}\pi n x\right)$$

5. The hint is good. Using the product rule and chain rule, w solves

$$
\begin{aligned}
w_{tt} &= 2w_{xx}, \\
w(0,t) &= 0, w(5,t) = 0, \\
w(x,0) &= -x(x-5), \\
w_t(x,0) &= x^2
\end{aligned}
$$

Now it is the same process as before and then you find $u = e^{-2t}w$.

7. The eigenvalue problem is $y'' + \delta^2 y = 0$ so with the boundary conditions, we need $y_n(x) = \sin\left(\frac{1}{2}\pi n x\right)$. Then the solution desired is $\sum_{n=1}^{\infty} b_n(t) y_n(x)$. Now you need to find $\cos x$ as a Fourier series.

$$
\sum_{n=1}^{\infty} b_n'(t) y_n(x) = -1 \sum_{n=1}^{\infty} \left(\frac{1}{2}\pi n\right)^2 b_n(t) y_n(x) + \sum_{n=1}^{\infty} g_n y_n(x)
$$

$$
g_n = 1 \int_0^2 (\cos x) \sin\left(\frac{1}{2}\pi n x\right) dx
$$

You need to solve

$$
b_n'(t) + \left(\frac{1}{4}\pi^2 n^2\right) b_n(t) = g_n
$$

Thus

$$
b_n(t) = b_n(0) \exp\left(\left(-\frac{1}{4}\pi^2 n^2\right) t\right) + \left(\frac{4}{\pi^2 n^2}\right) g_n
$$

Then the solution is

$$
\sum_{n=1}^{\infty} B(n,t) \sin\left(\frac{1}{2}\pi n x\right),
$$

$B(n,t) =$

$$
\left(
\begin{array}{c}
b_n(0) \exp\left(\left(-\frac{1}{4}\pi^2 n^2\right) t\right) \\
+ \left(\frac{4}{\pi^2 n^2}\right) \left(\frac{2\pi n - 2(-1)^n \pi n \cos 2}{\pi^2 n^2 - 4}\right)
\end{array}
\right)
$$

Now we need to find $b_n(0)$ to satisfy the initial condition. Plugging in $t = 0$,

$$
\sum_{n=1}^{\infty} \left(b_n(0) + \left(\frac{4}{\pi^2 n^2} \frac{2\pi n - 2(-1)^n \pi n \cos 2}{\pi^2 n^2 - 4}\right)\right).
$$

$$
\sin\left(\frac{1}{2}\pi n x\right) = (1-x)
$$

It follows that

$$
b_n(0) + \left(\frac{4}{\pi^2 n^2} \frac{2\pi n - 2(-1)^n \pi n \cos 2}{\pi^2 n^2 - 4}\right) = \frac{2}{\pi n}\left((-1)^n + 1\right)
$$

$$
b_n(0) = \left(\left(\frac{2}{\pi n}\left((-1)^n + 1\right) - \frac{4}{\pi^2 n^2}\right)\left(\frac{2\pi n - 2(-1)^n \pi n \cos 2}{\pi^2 n^2 - 4}\right)\right)
$$

Thus the final solution is

$$\sum_{n=1}^{\infty} \begin{pmatrix} \left(\left(\frac{2}{\pi n} \left((-1)^n + 1 \right) - \frac{4}{\pi^2 n^2} \frac{2\pi n - 2(-1)^n \pi n \cos 2}{\pi^2 n^2 - 4} \right) \right) \cdot \\ \exp \left(\left(-\frac{1}{4} \pi^2 n^2 \right) t \right) \\ + \left(\frac{4}{\pi^2 n^2} \frac{2\pi n - 2(-1)^n \pi n \cos 2}{\pi^2 n^2 - 4} \right) \end{pmatrix} \sin \left(\frac{1}{2} \pi n x \right)$$

9.

$$\sum_{n=1}^{\infty} \begin{pmatrix} \left(\frac{2}{5\pi^3 n^3} - \frac{2}{\pi n} \left(2 \left(-1 \right)^n - 1 \right) \right) \exp \left(\left(-5\pi^2 n^2 \right) t \right) \\ + \left(-\frac{2}{5\pi^3 n^3} \right) \end{pmatrix} \sin \left(\pi n x \right)$$

13.

$$\sum_{n=1}^{\infty} \left(-\frac{1}{\pi^2 n^2} \left(2 \sin \pi n - 4 \sin \frac{1}{2} \pi n \right) \right) \exp \left(-5\pi^2 n^2 t \right) \sin \left(\pi n x \right)$$

19. The eigenfunctions should satisfy

$$y'' + \lambda y = 0, \; y'(0) = 0, y'(2) + y(2) = 0$$

First of all, what is the sign of λ? Multiply by y and integrate. Thus

$$\int_0^2 y'' y dx + \lambda \int_0^2 y^2 dx = y' y |_0^2 - \int_0^2 (y')^2 \, dx + \lambda \int_0^2 y^2 dx$$
$$= y'(2) y(2) - \int_0^2 (y')^2 \, dx + \lambda \int_0^2 y^2 dx$$
$$= -y(2)^2 - \int_0^2 (y')^2 \, dx + \lambda \int_0^2 y^2 dx$$

Thus you must have $\lambda > 0$ since otherwise, $y(2)^2 + \int_0^2 (y')^2 \, dx \leq 0$ and so y' is a constant which also implies that the constant is 0 thanks to the $y(2)^2$. Thus $\lambda = \delta^2$ and the eigenfunctions are of the form

$$y = C_1 \cos (\delta x) + C_2 \sin (\delta x)$$
$$y' = -\delta C_1 \sin (\delta x) + \delta C_2 \cos (\delta x)$$

Then the left boundary condition says that $C_2 = 0$ and so you need to have $y = C_2 \cos (\delta x)$. Thus you have $-\delta \sin (\delta 2) + \cos (\delta 2) = 0$ and so, you need to have $\delta = \cot (2\delta)$. The equation $x = \cot (2x)$ clearly has infinitely many positive solutions. Finding them is another matter, but you can see from drawing a graph of the π periodic function $x \to \cot x$ that for large n, δ_n will be close a multiple of π. Thus, assuming the eigenfunctions are suitable to expand a function, the solution is of the form

$$u(x, t) = \sum_{n=1}^{\infty} b_n e^{-\delta_n^2 t} \cos (\delta_n x),$$

b_n chosen appropriately, Thus the limit is 0 as $t \to \infty$. Indeed, $|u(x, t)| \leq e^{-\delta_1^2 t} |\sum_{n=1}^{\infty} b_n \cos (\delta_n x)|$.

21. Suppose you have two solutions, u, v. Then letting $w = u - v$, $\Delta w = 0$ and $w = 0$ on the circle of radius R bounding B_R denoted as C_R. Then

$$
\begin{aligned}
0 &= \int_{B_R} (\Delta w)\, w\, dV = \int_{B_R} \nabla \cdot (\nabla w)\, w\, dV \\
&= \int_{B_R} (\nabla \cdot (w \nabla w) - \nabla w \cdot \nabla w)\, dV \\
&= \int_{C_R} w \nabla w \cdot \mathbf{n}\, dS - \int_{B_R} |\nabla w|^2\, dV = - \int_{B_R} |\nabla w|^2\, dV
\end{aligned}
$$

It follows that, since $\nabla w = 0$ that w must be a constant. However, it equals 0 on C_R and so $w = 0$ which means $u = v$. Now if you let $f(\alpha) = 1$ it follows that $u(r, \theta) = 1$ must be the solution. It works and there is only one. Thus $1 =$

$$
\int_0^{2\pi} \frac{1}{\pi} \left(\frac{1}{2} + \sum_{n=1}^{\infty} \frac{r^n}{R^n} \cos(n(\theta - \alpha)) \right) d\alpha
$$

23. From the above problems,

$$
\begin{aligned}
&|u(r, \theta) - f(\theta)| \\
&= \left| \int_0^{2\pi} \frac{1}{2\pi} \frac{R^2 - r^2}{R^2 - 2(\cos(\theta - \alpha))Rr + r^2} (f(\alpha) - f(\theta))\, d\alpha \right| \\
&\leq \int_{|\alpha - \theta| \geq \delta} \frac{1}{2\pi} \frac{R^2 - r^2}{R^2 - 2(\cos(\theta - \alpha))Rr + r^2} |f(\alpha) - f(\theta)|\, d\alpha \\
&+ \frac{1}{2\pi} \int_{|\alpha - \theta| < \delta} \frac{R^2 - r^2}{R^2 - 2(\cos(\theta - \alpha))Rr + r^2} |f(\alpha) - f(\theta)|\, d\alpha
\end{aligned}
$$

There, by uniform continuity of f, there exists $\delta > 0$ such that if $|\alpha - \theta| < \delta$, then $|f(\alpha) - f(\theta)| < \varepsilon/2$. Also, let $M \geq |f(\alpha)|$ for all α. Then the above is no larger than

$$
d(\delta, r)(2M)(2\pi) + \int_0^{2\pi} \frac{1}{2\pi} \frac{R^2 - r^2}{R^2 - 2(\cos(\theta - \alpha))Rr + r^2} \frac{\varepsilon}{2} d\alpha
$$

$$
= d(\delta, r)(2M)(2\pi) + \frac{\varepsilon}{2} < \varepsilon
$$

whenever r is close enough to R. That is, $\lim_{r \to R-} |u(r, \theta) - f(\theta)| = 0$. Of course, there are interesting results related to more information about the way this convergence takes place but this is the most familiar result.

Bibliography

[1] **Apostol, T.M.** *Calculus Volume II Second edition,* Wiley 1969.

[2] **Apostol, T.M.** *Mathematical Analysis,* Addison Wesley Publishing Co., 1974.

[3] **Baker, R.** *Linear Algebra,* Rinton Press 2001.

[4] **Boas, M.** *Mathematical Methods in Physical Science,* John Wiley and Sons, 1966.

[5] **Boyce, W. and DiPrima, R.** *Elementary Differential Equations and Boundary Value Problems,* John Wiley and Sons, 2005.

[6] **Braun, M.** *Differential Equations and Their Applications,* Springer Verlag, 1975.

[7] **Churchill, R. and Brown, J.** *Fourier Series and Boundary Value Problems,* McGraw Hill, 1978.

[8] **Carr, J.** *Applications of Center Manifold Theory,* Springer Lecture notes in Math. 1980.

[9] **Coddington, E. and Levinson, N.** *Theory of Ordinary Differential Equations* McGraw Hill 1955.

[10] **Davis, H. and Snider, D.** *Introduction to Vector Analysis,* William C. Brown 1997.

[11] **Eves, H.** *An Introduction To The History of Mathematics,* Holt Rinehart and Winston 1976.

[12] **Gray, A. and Mathews, G. B.** *A Treatise on Bessel functions and Their Applications to Physics,* Macmillan, N.Y. 1952.

[13] **Hardy, G.** *A Course Of Pure Mathematics, Tenth edition,* Cambridge University Press 1992.

[14] **Hochstadt, H.** *Differential Equations a Modern Approach,* Dover 1963.

[15] **Ince, E.L.** *Ordinary Differential Equations,* Dover 1956.

[16] **Kuttler, K.** *Calculus Theory and Applications Vol. 2,* World Scientific 2011.

[17] **Leighton, W.** *An Introduction to the Theory of Differential Equations,* Mc-Graw Hill, 1952.

[18] **Putzer, E.J.** Avoiding the Jordan canonical form in the discussion of linear systems with constant coefficients, *American Mathematical Monthly,* Vol. 73 1966, 2-7.

[19] **Redheffer, R.** *Differential Equations Theory and Applications,* Jones and Bartlett, 1991.

[20] **Rudin, W.** *Principles of Mathematical Analysis,* McGraw Hill, 1976.

[21] **Widder, D.** *Advanced Calculus,* Prentice Hall, 1961.

[22] **Yosida, K.** *Lectures on Differential and Integral Equations,* Dover 1991.

Index

$\cap$, 1
$\cup$, 1
n^{th} term test, 421

Abel's formula, 103, 176
absolute convergence, 418
absolute value
    complex number, 6
adjugate, 451
affine linear equations, 54
affine linear procedure, 57
algebraic multiplicity, 456
almost linear, 299
alternating series, 422
alternating series test, 422
alternative norm, 269
analytic, 177
annuities, 28
arcwise connected, 487
    connected, 487
asymptotically stable, 43, 298
autonomous, 45, 278, 298
    non-intersection of trajectories,
        278

Banach fixed point theorem, 270
Banach space, 268
basis, 438
basis of eigenvectors
    diagonalizable, 460
beams
    buckling, 358
beat, 119
Bernoulli equation, 37
Bessel equation
    ν an integer, 210
    eigenvalues, 360

    parameter not integer, 208
Bessel equations, 207
Bessel function
    integral identity, 214
Bessel functions
    addition formula, 213
    generating function, 213
    zeros, 360
Bessel's equation, 173
binomial series, 163
binomial theorem, 14, 163
boundary operator, 494
boundary problem, 333
boundary value problem
    Sturm-Liouville , 358
buckling beams
    eigenvalues, 358

capacitance, 121
catenary, 42
Cauchy condensation test, 419
Cauchy product, 427
    series, 166
Cauchy sequence, 481
Cauchy sequence, 405
Cauchy-Schwarz inequality, 13, 340
Cavendish, 69
Cayley-Hamilton theorem, 452
Ceasaro means
    convergence, 349
    uniform convergence, 350
center, 305
characteristic equation, 85
characteristic polynomial, 292, 452
characteristic value, 455
cofactor, 448
compact sets

intersection, 482
companion matrix, 131
comparison test, 418
complete, 269
completeness, 407
complex conjugate, 6
complex eigenvalues
    procedure, 95
complex numbers, 5
complex numbers
    arithmetic, 5
    roots, 8
    triangle inequality, 6
compound interest, 34
computer algebra system
    maple, 60
computer algebra systems
    mathematica, 61
    MATLAB, 62
    scientific notebook, 65
conjugate
    of a product, 15
connected, 485
    open balls, 487
connected component, 486
connected components, 486
    equivalence class, 486
    equivalence relation, 486
    open sets, 487
connected set
    continuous function, 488
connected sets
    intersection, 485
    intervals, 486
    real line, 486
continuity
    limit of a sequence, 482
continuous function, 477
contraction mapping
    fixed points, 270
    new norm, 268
converge, 473
convergence
    Fourier series, 343
    midpoint of jump, 343
    uniform, 270
converges
    uniformly, 270

convolution, 142
Cramer's rule, 451
critically damped, 119
curvilinear coordinates
    orthogonal, 365

De Moivre's theorem, 8
defective, 456
defective eigenvalue, 456
DeMorgan's laws, 2
derivative
    product of matrices, 225
determinant, 443
    alternating property, 445
    cofactor expansion, 448
    expansion along row (column),
        448
    matrix inverse formula, 450
    product, 447
    transpose, 445
diagonal matrix, 459
diagonalizable, 459
diameter, 475
difference equations, 28
differentiable dependence, 277
differential equation
    nonlinear and linear, 24
    different notation, 33
    first order, 24
    linear uniqueness, 32
differential equations, 24
    homogeneous, 47
    linear
        nonlinear, 24
    separable, 39
differential equations of motion
    derivation, 265
differential operator, 99
direct sum, 465
Dirichlet kernel, 345
Dirichlet test, 422
divergence
    orthogonal curvilinear coordi-
        nates, 367
divergence of gradient
    curvilinear coordinates, 368
divergence theorem, 368
Duffing's equation, 132

eigenfunctions
    Bessel functions, 360
eigenvalue, 455
eigenvalues, 292, 452
eigenvector, 455
electrical circuit, 121
empty set, 2
equations of motion
    spherical coordinates, 266
equilibrium point, 43, 298
    stable, 43
Euler equation, 191
Euler method
    improved, 281
Euler's formula, 11
    uniqueness, 259
Euler's method, 280
even extension, 355
exact equations, 48
    procedure, 49
exchange theorem, 437
existence and uniqueness, 88
    linear systems, 273
expansion
    Legendre polynomials, 362
exponential growth, 133

factoring polynomials
    synthetic division, 81
falling body
    air resistance, 19
Fejer kernel, 349
field axioms, 5
finite intersection property
    compact sets, 482
first order linear system
    Laplace transform, 239
first order system
    Laplace transform, 239
    MATLAB, 240
first order systems
    MATLAB, 239
    numerical methods, 243
Fitzhugh-Nagumo equations, 328
fixed points contraction mapping, 270
focus, 292
Fourier series, 339
    integrating term by term, 351

    pointwise convergence, 343
    term by term differentiation, 355
    term by term integration, 353
frequency
    amplitude, 124
frontier, 500
fundamental matrix, 76, 225, 251
    alternative definition, 228
    defective matrix, 233
    existence, 255
    group identity, 258
    independent columns, 228
    Laplace transform, 242
    no exact eigenvalues, 241
    uniqueness, 256
fundamental set of solutions, 90
fundamental theorem of algebra, 11
fundamental theorem of algebra
    plausibility argument, 12
    rigorous proof, 13

gamma function
    existence and convergence, 413
    properties, 413
general solution
    nonhomogeneous, 105
    particular solution, 237
geometric multiplicity, 456
geometric series
    sum, 416
geometric series, 416
global solutions
    estimates, 276
gradient
    orthogonal curvilinear coordinates, 367
    spherical coordinates, 367
grating, 493
Gronwall's inequality, 249, 276, 287

hanging chain, 41
heat equation
    derivation, 369, 370
    solutions, 372
Hermite polynomials, 215
    coefficients of power series, 218
homogeneous, 84
homogeneous equation, 99

procedure, 48
homogeneous equations, 47
Hooke's law, 116
Hopf bifurcation, 327
hyperbolic, 320
hypergeometric equation, 220

implicit function theorem, 489
impulse function, 141
independence
    distinct eigenvalues, 229
inductance, 121
infinite series, 415
    alternating series test, 422
    Cauchy condensation test, 419
    Cauchy product, 427
    combinations, 417
    comparison test, 418
    Dirichlet's test, 422
    double sums, 424
    limit comparison test, 419
    Merten's theorem, 428
    ratio test, 423
    root test, 423
infinite series of matrices, 254
initial condition, 30
initial conditions
    continuous dependence, 276
initial data
    continuous dependence, 277
initial value problem, 30
    local solutions, 274
integral
    uniform convergence, 171
integral curves, 39
integrating factor, 30, 37, 50
    Euler, 51
    procedure, 54
interest and principal, 28
interior point, 472
intermediate value theorem, 486
intersection, 1
intervals
    notation, 2
inverse
    left inverse, 451
    right inverse, 451
inverses and determinants, 450

invertible, 435

Jordan arc, 499
Jordan curve, 500

kinetic energy, 263

Lagrange
    mechanics, 265
Lagrange remainder, 430
Lagrangian mechanics, 263
lake pollution, 35
Laplace equation
    circular disk, 386
    rectangles, 383
Laplace expansion, 448
Laplace transform, 133
    first order linear system, 239
    fundamental matrix, 242
    linear, 133
    no problem for defective matrix,
        240
    step function, 134
    table, 134
Laplace transforms
    derivation of table, 136
    justification of method, 145
Laplacian, 365
    cylindrical coordinates, 369
    polar coordinates, 369
    spherical coordinates, 368
Lebesgue number, 476
Legendre equation, 181
    eigenvalues, 362
    existence of bounded solutions,
        362
    polynomial solutions, 183
Legendre polynomials
    expansion, 362
    generating function, 185
    orthogonality, 189
limit point, 472
linear and nonlinear equations
    differences, 59
linear combination, 436, 446
linear differential equations
    zeros of solutions, 222
linear equation

procedure, 37
linear equations
    MATLAB, 124
    scientific notebook, 127
    solution, 57
linear operator, 86
    superposition, 88
linear systems
    general solution, 227
linear transformations
    commuting, 466
linearly dependent, 436
linearly independent, 436
Liouville
    transformation, 361
Lipschitz condition, 267
local existence and uniqueness, 275
logistic equation, 42
Lorenz equations, 328
Lotka-Volterra, 304
Lotka-Volterra equations, 322

math induction, 3
mathematical induction, 3
matrices
    similar, 459
    transpose, 434
matrix
    injective, 440
    left inverse, 451
    lower triangular, 451
    main diagonal, 459
    right and left inverse, 440
    right inverse, 451
    surjective, 440
    upper triangular, 451
matrix multiplication
    properties, 433
matrix norm, 251
mean square
    poinwise, 364
mean square norm, 340
Merten's theorem, 428
minor, 448
multistep method
    Adams-Bashforth, 286

nested interval lemma, 407, 474

Newton
    second law, 115
Newton's laws, 116
node
    stable, 291
nondefective eigenvalue, 456
nonhomogeneous
    particular solution, 237
nonhomogeneous equation, 99
nonhomogeneous problems
    solutions, 380
nonlinear equation
    no uniqueness, 59
nonlinear equations
    local solutions, 58
norm
    mean square, 340
    uniform, 340
normal matrix, 260
numerical methods
    MATLAB, 243
numerical solutions
    scientific notebook, 243

odd extension, 355
Ohm's law, 121
open set, 472
operator norm, 298
ordinary point, 177
    finding solutions, 181
orthogonal matrix, 462
orthogonality conditions, 337
orthonormal, 462
oscillation
    critically damped, 119
    over damped, 119
    underdamped, 119
over damped, 119

p series, 420
partial differential equations
    nonhomogeneous, 378
partial summation formula, 422
partial sums
    Cauchy sequence, 417
particular solution, 105, 237
partition, 409
periodic, 337

permutation, 442
Picard iteration, 267, 268
piecewise continuous, 351
point at infinity, 220
pointwise
    mean square, 364
Poisson's equation
    representation, 389
polar form complex number, 7
polynomial
    division, 79
polynomial
    factoring, 80
polynomials
    factoring, 9
power series, 157
    binomial series, 163
    binomial theorem, 163
    Cauchy product, 194
    coefficients, 161
    cosine and sine, 162
    differentiating, 159
    integration, 164
    interval of convergence, 158
    limitations, 164
    multiplication, 165
    of inverse, 194
    operations on power series, 159
    polynomials, 166
    radius of convergence, 158
    sine and cosine, 162
    uniqueness, 161
power series methods
    limitations, 178
pumpkin
    flying, 36
Putzer's method, 294

quadratic formula, 9

ratio test, 423
rational root theorem, 79
reduction of order, 98
regular singular point, 195
    Euler equation, 195
    exponents of singularity, 198
    Frobenius method procedure,
        206

solution procedure, 203
regular singular points
    Euler equations, 195
    finding them, 196
    indicial equation, 198
resistance, 121
Riccati equation, 68, 359
Riemann-Lebesgue lemma, 343
Rodrigues formula, 184
Rodriques formula, 184
root test, 423
row operations
    inverse, 432
Runge-Kutta
    error estimate, 287
Runge-Kutta method
    syntax, 283

saddle point, 291
scalar linear equations
    real roots, 85
scaling factor, 366
Schur's theorem, 462
semi-stable
    equilibrium point, 70
separable differential equations, 39
separable equations
    procedure, 46
separated sets, 485
separation of variables, 372
sequence, 473
sequence of partial sums, 415
set notation, 1
sgn, 441
    uniqueness, 442
sign of a permutation, 442
similarity relation, 459
simply connected, 313
skew symmetric, 434
slope lines, 20
solution by power series, 177
solving heat equation, 372
span, 436, 446
spectrum, 455
spherical coordinates
    volume, 366
squeezing theorem, 403
stability from graphs, 45

stable, 43, 299
stable equilibrium point, 43
stable manifold, 319
Sturm
    separation theorem, 359
Sturm-Liouville problem, 358
subsequence, 473
subspace, 436
    basis, 438
    dimension, 438
superposition, 88
    linear systems, 226
Sylvester's theorem, 464
symmetric, 434
synthetic division, 81, 82

Taylor polynomial
    sine, 430
Taylor series, 157, 158
    coefficients, 161
Taylor's formula, 429
    many variables, 174
term by term differentiation
    Fourier series, 355
term by term integration
    Fourier series, 353
transient terms, 123
transpose, 434
    properties, 434
triangle inequality
    complex numbers, 6
trig. identities, 337
trivial, 436
two point boundary value problem,
    333

undamped oscillation, 116, 117

under damped, 119
undetermined coefficients
    procedure, 110
uniform continuity, 483
uniform convergence, 171
    integral, 171
uniform norm, 270, 340
uniformly close, 340
union, 1
unitary matrix, 462

variation of constants formula, 257
variation of parameters
    procedure, 113
vector space
    axioms, 433
    basis, 438
    dimension, 438

wave equation
    derivation, 375
Weierstrass
    approximation theorem, 143
Weierstrass approximation
    estimate, 143
Weierstrass approximation theorem,
    144
well ordered, 3
well ordering, 3
Wronskian, 89
    alternative, 89, 90
    another solution, 176
    general solution, 90
Wronskian alternative
    linear systems, 227
Wronskian and ratios, 92